FUNDAMENTALS OF
PLANT BIOTECHNOLOGY

FUNDAMENTALS OF

PLANT BIOTECHNOLOGY

Dr. B.S. Singh, Ph.D.
Cambridge Institute of Technology
Cambridge Village, Tatisilwai
Ranchi, (Jharkhand)

Dr. M.P. Singh, Ph.D.
University Professor & Chairman
Department of Forest Sciences
Birsa Agricultural University,
Ranchi (Jharkhand)

SATISH SERIAL PUBLISHING HOUSE
403, Express Tower, Commercial Complex, Azadpur, Delhi - 110033 (INDIA)
Phone : 011-27672852 Fax : 91-11-27672046
E-mail : info@satishserial.com, hkjain1975@yahoo.com
Website : www.satishserial.com

Published by :

SATISH SERIAL PUBLISHING HOUSE
403, Express Tower, Commercial Complex, Azadpur, Delhi - 110033 (INDIA)
Phone : 011-27672852 Fax : 91-11-27672046
E-mail : info@satishserial.com, hkjain1975@yahoo.com

ISBN 81-89304-29-1

Printed at
SALASAR IMAGING SYSTEMS, DELHI-35

Preface

Biotechnology includes all the industrial processes, mediated by living organisms, involved in the production of substances useful to humanity, which were previously obtainable only by more difficult and expensive processes. This is an area of unparalleled expansion of applied biology which involves genetic engineering and its industrial application to produce substance such as antibiotics, enzymes, hormones and a host of other substances. The environmental crisis involves the erosion of valuable genetic material. The economic opportunity spells increased breading capability (Mooney, 1983). The third world countries people become the experimental users to study the effects of new varieties on their health.Man, with all his powers and prowess, depends totally on other biological entities for all his nutritional needs; he is even unable to synthesize certain biochemical's essential for his own advantage; this began with plants and animals, and was soon extended to microbes and eventually to cultured animal and plant cells and tissues.

Clearly, man had always benefited, either directly or indirectly, from and suffered due to microbes without ever realizing this feet. It was only very recently in the history of mankind that Leeuwenhoek first described the microbes in 1877, and Pasteur reported the involvement of microbes in lactic acid fermentation in 1857. Since then micro-organism strains have been consciously developed and used in various industrial production processes. Subsequent developments in biology, especially molecular biology, have made it possible to create novel capabilities in micro-organisms for generating highly valuable compounds often of rare occurrence in nature. In addition, cells and tissues of plants and animals are being utilized for generating products and/or processes that enhance the quality of human life. Enzymes are used to produce various foods and beverages, and more effective detergents. These and various other technologies based on living agents or their components constitute biotechnology.

Commercial potential of biotechnology is immense since the scope of its activities covers the entire spectrum of human life and since it draws on the capabilities and resourcefulness of the entire biological world and even beyond (for examples, modification of existing genes like, crystal protein gene of *Bacillus thuringiensis* to enhance the level of its expression and to combat the problem of gene silencing in transgenic plants) . Biotechnology is visualized, with lot of apparent justification, by many scientists an the technology of the twenty - first century.

In view of the nature and scope of the subject, a comprehensive textbook is an ideal never to be achieved. Therefore, most foreign publishers have attempted to launch a series of textbooks covering different aspects of the subject. The students, however, often may not have access to all the volumes of the series and even if accessible, usually may not have the time to consult them all. These considerations promoted this effort to fulfill the needs of students as well as teachers of the subject.

The book is divided into 27 chapters and 9 appendixes. Chapter 1 introduces the subject matter and the scope of the biotechnology. Plant tissue culture: Principles and methodology, is dealt with in Chapter 2. Chapters 3,4, 5 and 6 deal Micropropagation in Plants, Protoplast culture, Somatic embryogenesis Principles, Concepts and application and Somaclonal and Gametoclonal variant selection respectively. Cell Culture, Plant tissue culture, Crop improvement and Transgenic Plants are described in Chapters 7, 8, 9,10 and 11 respectively. Chapters 12, 13, and 14 related to Biotechnology and its relation with crop improvement in India, in forestry and nitrogen fixation and plant productivity, respectively. Chapters 15, 16, 17, and 18 discuss the applications of biotechnology to human health, while Chapter 19 and 20 deals with utilization of Biotechnology in saving biodiversity, and utilization of enzymes as bio-accelerators, respectively. Although, chapter 21 is devoted to economic utilization of micro-organisms. Applications of enzymes for the generation of products/services is described in Chapters 22, 23, 24, 25, 26 and 27 deal with the contributions of biotechnology to foods and beverages, fuels, and environmental management respectively. Nine appendixes of glossary related to practical approaches of tissue culture procedures.

Obviously, it is unrealistic to claim any degree of competence in all the diverse areas covered in the book. The differences in the extent and the level of presentation of the different chapters reflect mainly the professional bias of the author and his understanding of the various subjects covered therein. This, however, is conductive to all kinds of errors, which may be minimized by and active interactive cooperation suggestions for improvement as these alone can serve as the guidepost for an author of a book of this nature.

We are grateful to a number of individuals, too many to name, for their varied contributions to the development of this book .We wish to thank Professor V. L. Chopra, Member of Planning Commission, Govt. of India, New Delhi for encouraging us to take up this task, Dr. H. Y. Mohan Ram (Retd.) Professor of Botany Delhi University, Delhi for meticulous correction of manuscripts . We are also thankful to a number of teachers / scientists for their assistance namely R. N. Sahgal, Dr. P. K. Khosla, Dr. D. N. Tiwary, Dr. M. P. Nayer, Dr. C. R. Basu have kindly spared valuable suggestions. Special appreciation to Mr. H.K Jain of Satish Serial Publishing House, New Delhi for taking pain for publishing this book in a very short period.

Ranchi
Date: 20/5/2006

B. S. Singh
M. P. Singh

Content

CHAPTER - 1

Introduction

Biotechnology includes all industrial processes, mediated by living organisms, involved in the production of substances useful to humanity, which were previously obtainable only by more difficult and expensive processes. This is an area of unparalleled expansion of applied biology which involves genetic engineering and its industrial application to produce substances such as antibiotics, enzymes, hormones and a host of other substances. The term biotechnology was coined during the late 1970s when the advances in molecular biology, biochemistry, and genetics, especially in the development of certain techniques, catalyzed new ventures to exploit these advances for the benefit of man kind. In India, the importance of biotechnology was emphasized at the 69th session of the Indian Science Congress in 1982 and consequently a 'National Biotechnology Board' was constituted under the Department of Science and Technology to coordinate and encourage research in this direction. Now there are many centers of biotechnological studies in India engaged in research on various aspects of biotechnology. An International Center for Genetic Engineering and Biotechnology (ICGEB) is also established for developing countries under the auspices of United Nations. This center has two locations; one in New Delhi and another in Triesta, Italy. An Interdepartmental Committee on Biotechnology in UK has defined biotechnology as "the application of scientific and engineering principles to the processing of materials by biological agents to produce goods and services" (Coleman, 1986).

The application of biotech methods has been in practice since ancient period in agriculture and the brewing industry, but the developments in genetic engineering and great advances in bioreactor design and computer-aided process control have given it a new dimension which greatly extends the present range of technical possibilities and has a strong potential to revolutionize several facets of medical, agricultural, and industrial practices. Many developments in biotechnology, such as those in recombinant DNA, monoclonal antibodies (MAbs), and immobilized enzymes, are directed towards producing a better product through application of sofirticated process. Thus, hitherto limited biological sources of hormones and growth regulators are being increasingly replaced by the use of genetically-transformed microbes, thereby providing a greatly increased scale of production. With greatly increased production and complete safety the modern vaccines, resulting from the absence of ineffectively inactivated virus, is a strong merit of the genetically-engineered antigen.

The word 'Biotechnology' first appeared in an article entitled "Biotechnology" published in 1933 (Anonymous, 1933) which have the title, the word biotechnology but nothing is mentioned about.In 1947, biotechnology was described as "the branch of technology concerned with the development and exploitation of machines in relation to the various needs of human beings" (Bud, 1989; Taylor and Boelter, 1947). In 1962 Dr. Elmer Gaden, Editor of the *Journal of Microbiological Technology and Engineering* changed the name of the journal to *Biotechnology and Bioengineering,* with the thinking that "biotechnology implied all aspects of the exploitation and control of biological systems". This very wide definition greatly contributed to the gradual popularity of the word biotechnology. On the other hand, the concept of biotechnology as a broad area suffered a setback in 1979 when E.F. Hutton obtained a trademark on the word biotechnology to describe a magazine dealing with genetic engineering. During the 1980s, the word came to be associated more and more with genetic engineering, thereby overshadowing the earlier wider concept adopted by Gaden.

There are two contrasting views on the concept of biotechnology. As per industrialists the term has been overused and is not specific enough to be of much value, while others complain that due to frequent usage the term has become more associated with genetic engineering (Kennedy, 1991). Attempts to deal with the former criticism include prefixing biotechnology with such adjectives as algal, fungal, plant, animal, industrial, microbial, marine, etc., with a view to overcoming the problem of too broad a definition. Another replacement for the term biotechnology has involved the creation of such 'bio' prefixed words as biocomputing, biocontrol, biocatalysis and biobusiness, biosensors, etc. (Kennedy, 1991). (Dia. 1.1).

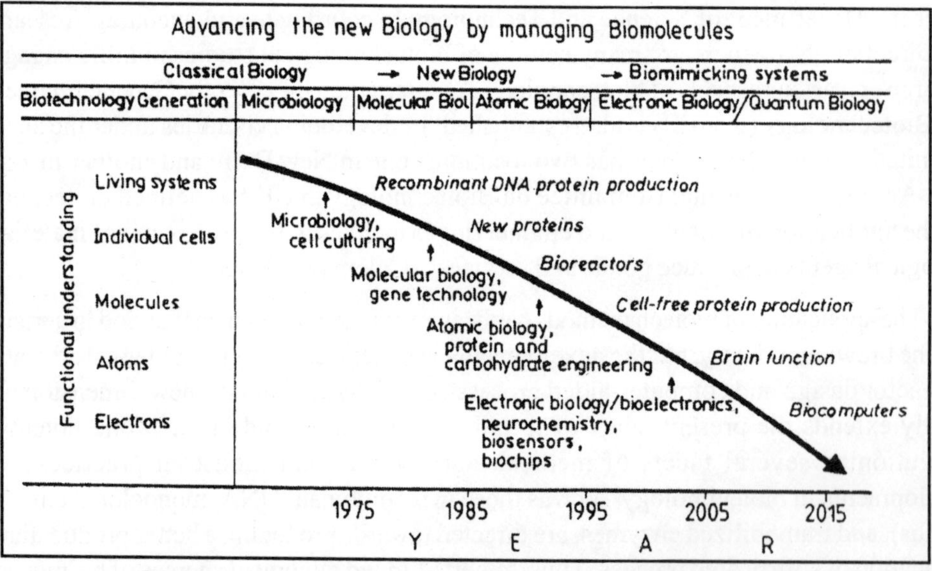

Diagram 1.1. Advances in biotechnology and modern biology trend in recent years with the projections, for future.

The history of biotechnology may be divided into four periods. The pre-Pasteur era witnessed the empirical practice of selection in animal and plant breeders and fermentation for food preservation. This era lasted until the second half of the 19th century and its foundation stone was just experience. The selection and breeding of plants was a slow, trial and error process carried out without any knowledge of the underlying genetic processes or the laws of inheritance. In its turn, fermentation was used for millennia without understanding that it was a biological phenomenon produced by the activity of living organisms. Serendipity was sometimes responsible for biotechnology application, as in the case of penicillin. But it was practical understanding without theory or scientific base, the techniques being transmitted from father to son. It was a technology without science.

The second phase started with the identification of microorganisms as the cause of fermentation by Pasteur which was followed by Buchner's discovery that the extracted enzymes from yeast have the capacity to convert sugar into alcohol. This development provided some impulse to fermentation technique in the food industry through the production, *inter alia,* of baker's yeast and citric and lactic acids, and in chemical industry, for the production of acetone, butanol, and glycerol through the use of bacteria.

The third era started with technological developments which reduced the impetus of biotechnology development in certain areas while giving a new push in other areas. The expansion of the petrochemical industry displaced the chemical processes based on fermentation. At the same time, A. Fleming's discovery of penicillin in 1928 made possible the large-scale production of antibiotics in the 1940s, and there were spectacular increases in yield of corn in the corn belt of the United States, which in due course heralded the Green Revolution.

The last (modern) era started with the exciting discoveries, such as, discovery by F.H.C. Crick and J. Watson of the double helical structure of DNA in 1953, which was followed by the process of replication, transcription, translation, enzyme immobilization, the first experiments on genetic engineering in 1973 by Cohen and Boyer, and the discovery of the hybridoma technology for producing monoclonal antibodies in the 1970s by Milstein and Kohler.

The greatest sorrow the present-day world is that whereas most of the world's germplasm for important crop plants is found in the underdeveloped countries i.e. third world countries, the technical and scientific expertise and facilities for exploiting materials are available in the developed countries. As a rule, the developing countries are users, rather than producers, of new technologies. Although the third world countries provide the germplasm for the development of new varieties, in return they have to pay a heavy price in importing these varieties from the advanced countries which evolved these new varieties. Germplasm now poses for the less developed countries a political problem, an environmental crisis, and an economic opportunity. The political problem relates to control and exchange of the germplasm. The environmental crisis involves the erosion of valuable genetic material. The economic opportunity spells increased breeding capability (Mooney, 1983). The third world countries people become the experimental users, to study the effects of new varieties on their health. Diagram 1.2 shows the history and future projection *of ex situ* genetic conservation.

The quintessence of most bio-technological processes is the conversion of relatively cheap raw materials into fairly or highly valuable products or services. The development of efficient processes, a prerequisite for the commercialization of new products or services, which requires a coordinated coupling of unit operations.

Although much of the scope in biotechnology has been generated by discoveries and advances in molecular genetics, the role of the process engineer to translate these discoveries into usable processes, on a large scale, is by no means insignificant. This is the area where scientists cooperate with technologists or engineers. This meaningful synthesis can be diagrammatically shown as in diagram 1.3 which shows an overview of any typical biotechnological process, with particular reference to the living strains or materials used. In this process, the central place is occupied by the bioreactor, whose successful operation requires adequate upstream processing. Similarly, the recovery of the final product requires a series of steps collectively called *downstream* processing.

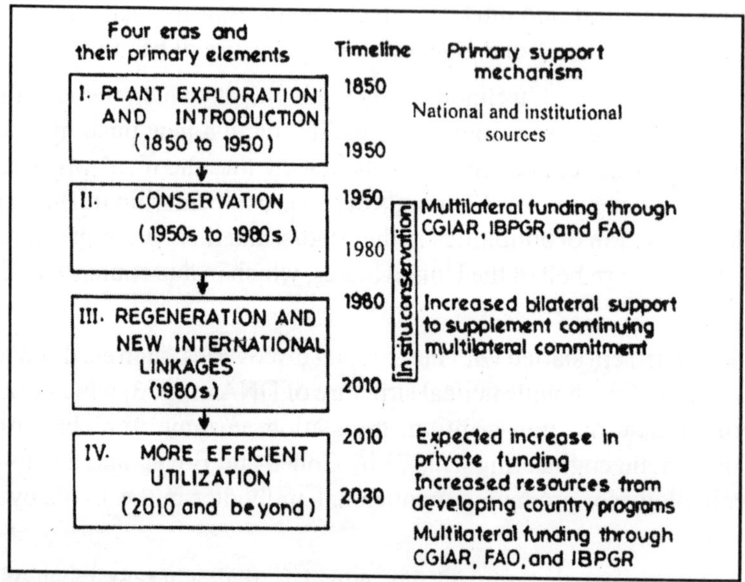

Diagram 1.2. Four eras of *ex situ* genetic resource conservation and use, with timeline of conservation events (modified from Cohen *et al.*, 1991).

The problems of gas, liquid, and solid handling prior to the use of these materials for a bioconversion, form important aspects of upstream processing. While over 75% of the world's total fermentation capacity is anaerobic and hence not requiring gas compression, the remaining capacity is aerobic and generates highly valuable products. For these aerobic processes, gas compression assumes great importance as a critical upstream operation. Other important aspects include air and media sterilization and/or filtration, and removal of heat from the bioreactor.

Many of the products produced through modern biotechnology are intracellular proteins. Cells must be broken up to release these proteins. High-pressure homogenizers and high-speed ball mills are commonly used to disrupt the cells.

Following the completion of fermentation, a solid-liquid separation constitutes part of the downstream processing. Centrifugation *and* filtration are the methods of choice for separating cells from broth. Membrane filtration technology is important not only for separation of cells from broth but also for concentration of protein solutions (Cooney, 1985).

Liquid-liquid extraction is used for the recovery of..antibiotics and other low molecular weight organic materials produced in fermenters that need high speed centrifugation. Chromatography constitutes another versatile unit operation of downstream processing. It is based on separation by charge, hydrophobicity, size, or molecular recognition. Ion-exchange chromatography is used for recovering antibiotics as well as proteins. Molecular sieve chromatography is another modern technique used for recovering proteins.

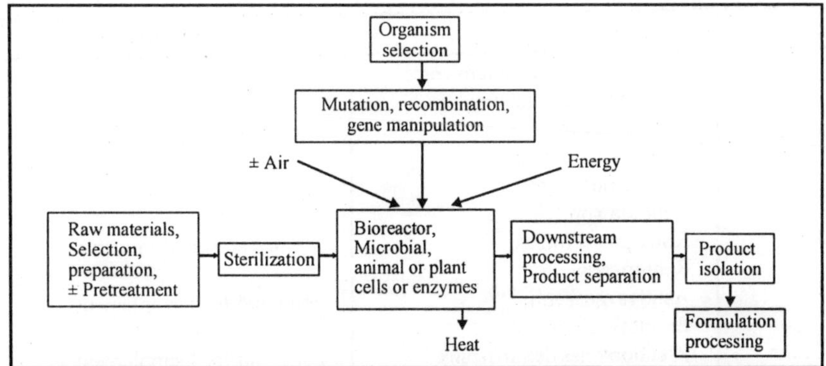

Diagram 1.3. A typical flowchart of any biotechnological process, with reference to the microbe or living material employed (after Smith, 1984).

Man has exploited certain microorganisms in the food and beverage industries for many centuries. In the present century, microbial activities and products have already benefited the pharmaceutical and effluent treatment industries. During the last two decades of the twentieth century, yet another exciting possibility for exploiting microbes has achieved by the development of recombinant DNA technology. The potential of diverse microbes to produce several valuable products has been appreciated and, in conjunction with DNA manipulations, can usher in the new biological revolution in which biotechnologists may tailor microbes that might utilize some cheap substrate or waste material to synthesize a useful and costly end-product. The simplest bacterium *Escherichia coli* acts as a hospitable host for genetic material derived from other organisms, which is then expressed to produce either, valuable proteins which are excreted, or useful enzymes which in turn can catalyze metabolic reactions to yield new products.

Microbes are a valuable source of primary metabolites, involved in their anabolism and catabolism. More important examples of these metabolites that have numerous uses in the food and chemical industries (Table 1.1) include certain amino acids, nucleotides, vitamins, solvents, and organic acids (Britz and Demain,1985).

Different aerobic microbes oxidize many of the hydrocarbons present in crude oil, by using as the source of carbon and energy. These microbes produce several metabolites of utility in oil service industry; specific examples of these useful metabolites include surfactants and polysaccharide biopolymers. Genetic manipulation techniques are now being in practice to create strains that may aid dewaxing, desulphurization, and oil recovery operations, or may prove conducive to enhanced synthesis of surfactants and biopolymers. Recent researches have underscored the potential of microbial technology for gainful application in the oil service industry.

Modern biotechnology has played a significant role in the development of the health care chemical industries (Table 1.2). It has made possible the availability of several diagnostic, prophylactic, and therapeutic products. Most of the products in the pharmaceutical industry are typically of high potency, low volume (Table 1.3), very costly materials. These products are commonly made by *aerobic submerged* cultivation of certain microorganisms.

Table 1.1. Some industrially-important primary metabolites and their uses (after Britz & Demain, 1985)

Metabolite(s)	Organism(s)	Present or potential use(s)
Lysine, threonine	Corynebacterium glutamicum, Brevibacterium flavum	Food supplements
Glutamate	C.glutamicum, Brevibacterium spp.	Flavouring agent in food
Citric acid	Aspergillus niger, Candida spp.	Food and pharmacy
Fumaric acid	Rhizopus arrhizus, R. nigricans	Plastics and food industry
Riboflavin	Ashbya gossypii, Eremothecium ashbyii	Food and feed supplement
Cyanocobalamin (Vitamin B_{12})	Pseudomonas denitrificans, Propionobacterium shermanii	Food and feed supplement
Xanthan gum, dextran	Xanthomonas campestris, Leuconostoc mesenteroides	Thickening, stiffening, and setting agent in food, pharmaceutical, and textile industries
Acetic acid	Saccharomyces spp., Acetobacter spp., Clostridium spp.	Vinegar, chemical feedstocks, polymer and food industries
Butanol, acetone	Clostridium acetobutylicum	Solvent and thinners, synthetic polymers
Ethanol	Saccharomyces cerevisiae, Zymomonas mobilis, Clostridium spp.	Beverage industry, solvent, fuel extender

Table 1.2. Some important drugs for human and animal health care, in whose development biotechnology has played essential role.

Drugs for	Use / Target
Diabetics (e.g., insulin)	Human health care
Antiinflammatories	Human health care
Renal system	Human health care
Cardiovascular system	Human health care
Nervous system / mental disorder	Human health care
Vaccines and biologicals	Human health care
Antiparasitics	Animal health care
Gastrointestinal disorder	Animal health care
Growth promotion (feed efficiency)	Animal health care

Table 1.3. Relationship between volume and value of some biotechnological products or activities (modified from Bull *et al.*, 1982)

Products / Activities	Volume	Value
Methane, ethanol, biomass, animal feed	High	Low
Amino acids, organic acids, baker's yeast, acetone, butanol, polymers, foods	High	Moderate
Antibiotics, enzymes, vitamins, pharmaceuticals	Low	High

ANTICANCER AGENTS

Being one of the greatest sorrow of the present-day civilized world, cancer has naturally attracted the attention of biotechnologists and medical microbiologists. Attempts have been made to discover any cytotoxic agents that might inhibit mammalian cell proliferation. To this end, thousands of naturally-occurring compounds produced by living organisms have been screened. By the early 1980s, a number of anticancer fermentation products were being produced commercially (Table-1.4).

Table 1.4. Some anticancer products produced commercially by biotechnology/fermentation (after Flickinger, 1985).

Products(s) (Trade Name)	Manufacturer	Country
Adriamycin	Farmitalia Carlo	Italy
Cerubidine	Ives Labs	USA
Cosmegen	Merck, Sharp and Dohme	USA
Mutamycin, blenoxane	Bristol	USA
Bestatin, bleo, pepleo injection	Nippon Kayaku	Japan
Toyomycin	Takeda	Japan
Mitomycin-C-Kyowa	Kyowa	Japan
Crasnitin	Farbenfabriken Bayer AG	Germany
Rubidazome injuection	Rhone-Poulenc	France

SCOPE, POTENTIAL AND ACHIEVEMENTS

The new biotechnologies may be classified into the following four broad categories:

A. Techniques for cell and tissue culture likely to produce substantial impact on agriculture.

B. Technological development associated with the fermentation processes, particularly those in the chemical sector which includes the enzyme immobilization technique. These techniques are already creating some impacts in several industrial branches, e.g., production of enzymes and amino acids.

C. Techniques that apply microbiology for the screening, selection and cultivation of cells and microorganisms.

D. Techniques for the manipulation and transfer of genetic material.

The above four groups are mutually supporting. However, there is one basic difference between the first three categories on the one side and the fourth on the other. The first three groups as well as all old or traditional technologies were based on the empirical or scientific understanding of the characteristics and behaviour of microorganisms and the intentional use of their characteristics for the fulfilment of economic objectives. The enormous potentialities of the fourth category come from the capacity of scientists to manipulate the structural and functional characteristics of organisms and the application of this capacity to overcome their natural limits in performing specific tasks of some economic and social importance (Bifani, 1989).

Any technological revolution usually has the following five characteristics:

1. A drastic reduction in cost of several products and services.
2. A dramatic improvement in the technical properties of processes and products.
3. Social and political acceptability in the sense that innovation is socially accepted but it involves modification in the legislative and regulatory patterns of society and some changes in management and labour attitudes.
4. Environmental acceptability.
5. Pervasive effects through the economic system.

These five conditions are likely to be met by future biotechnological developments. The productive activities from mining and agriculture to manufacturing and services may be radically affected by developments in biotechnology which are in fact blurring the traditional boundaries between productive sectors and between these and services. Biotechnology overcomes the traditional sectoral classification systems of macroeconomic accounting. This implies greater flexibility of the economic activities which will be reflected in scale operation and in a new approach to integrated management of resources (Bifani, 1989).

Biotechnology depends on the interrelated efforts of different scientific disciplines. Technologies should therefore be operated by multidisciplinary effort supported by a wide range of information networks and services provided by different disciplines. Its development, application and optimal use can be accelerated by the growing capacity to gather, store, retrieve, manage, and interpret scientific, technological, and economic information (Bifani, 1989).

The ambitious achievement of biotechnology mostly rests on the capabilities of the living cell which acts as a protein factory, as a chemical plant, and as a source of extracellular proteins. When acting as a protein factory, the cell's potentiality may be scaled up on a large scale for single cell protein, and on a smaller scale as a source of enzymes and hormones. Likewise, the cell as a chemical factory can give us ethanol on a large scale and fine chemicals and antibiotics in a small-scale production unit. Various enzymes and antibodies can be obtained in a small-scale process, utilizing the potentiality of the cell to yield extracellular proteins (Fairtlough, 1986). Recent discoveries in microbial genetics have opend a new era of possible applications. Some of the most impressive discoveries in genetic engineering have involved the capacity to manipulate or recombine genetic material in the cells of microorganisms. Although the first applications of the new technology have been in medicine, their potential extends over a wide range. The development of the genetic engineering technology has, in turn have transformed life into a productive, vital force.

Biotechnology is expected not only to exert significant effects on human and animal food, energy and chemicals, waste and pollution treatment, medical care, and crops and minerals, but even to create entirely new industries in the not too distant future. Many more microbially-synthesized mammalian proteins should become available in the next few years. Some of the medically-important proteins that can be produced by the application of biotechnology include interferon, hormones, vaccines, and antibodies. The bacterial cell can be used as a factory to make these and other proteins. Table 1.5 lists some valuable proteins having pharmaceutical applications and being developed by recombinant DNA technology.

Table 1.5. Some pharmaeeutically-important substances being developed by recombinant DNA technology (after OTA, 1984).

Substance(s)	Functions(s)
Somatostatin	Inhibits growth hormone secretion
Somatomedins	Mediate action of growth hormone
Growth hormone release factor	Stimulates pituitary hormone release
Calmodulin	Mediates calcium's effects
Calcitonin	Inhibits bone resorption
Parathyroid hormone	Prevents excretion of calcitonin, mobilizes calcium
Luteinizing hormone	In females, induces ovulation; in males, stimulates androgen secretion
Follicle-stimulating hormone	Induces ovarian growth
Relaxin	Uterine relaxation
b-Endorphin	Analgesic
Encephalins	Analgesic
Interleukin-2	Promotes growth and activity of T-cells
Thymic factor	Restores delayed-type hypersensitivity
Thymopoietins	Inhibit B-cell differentiation

Viral hepatitis is one of the most common problem in several developing countries is caused due to polluted drinking water. In many cases, the hepatitis virus (hepatitis B type) is persistently retained in the liver. Much work has been done on prophylactic immunization of human against the hepatitis B virus. Till recently, the plasma-derived hepatitis B vaccine has been used in these immunizations, but the supply of such a vaccine is limited by the amount of available plasma from hepatitis B carriers. Another constraint is the cumbersome and tedious process of antigen purification. The gene of hepatitis B surface antigen can be cloned into yeast cells, yielding the recombinant vaccine (Hilleman. 1987). This recombinant vaccine has the potential to eradicate hepatitis B worldwide.

Encapsulated bacteria such as *Haemophilus, Pneumococcus, Streptococcus,* and *Salmonella* have caused dreadful diseases since time immemorial. Biotechnology has given us the means to prevent some of the diseases caused by these pathogens by designing and administering polysaccharide vaccines. These vaccines act against the encapsulated bacteria present in the bloodstream of the person (Robbins and Schneerson, 1987). One advantage of these vaccines is that immunization of adults with the polysaccharides frequently confers a virtual lifelong immunity.

Recent advances in molecular genetics have opend the door, for developing vaccines or other means for tackling four of the leading diseases, namely, malaria, trypanosomiasis leprosy and cancer. The circumsporozoite gene of the malaria parasite has already been cloned. We have a better understanding of the surface glycoproteins of the trypanosome parasite. *Mycobacterium leprae* bacteria have been purified from tissues of the armadillo and several antigens are now being examined with a view to determining their relation to pathogenesis to counter-attack leprosy.

The cancer-producing virus (Rous sarcoma virus) was first described over 70 years ago. This and other mammalian cancer viruses have since been studied. Many viral oncogenes are now known to have counterparts in the genome of the host cell, often being present in the normal genetic complement of man. The *one* gene products have been biochemically identified and found to be related to some known growth-promoting substances. The product of the Rous sarcoma virus *src* gene, designated pp60, is a tyrosine-specific protein phosphokinase. Another related finding was that the product of the *v-erbB* oncogene of avian erythroblastosis virus resembles the receptor for epidermal growth factor, a factor already known to be a glycoprotein with an intrinsic tyrosine-specific protein kinase that is stimulated upon binding to the epidermal growth factor. This kind of emerging relationship between growth factors and proteins can go a long way in unravelling some of the black boxes in our understanding of cancer by precisely defining some of the components of the complex system embracing the cell surface, the cell membrane, the cytoplasm, and the nucleus. The tools of biotechnology are now available to fill the gaps in our knowledge, and may soon help find a way to prevent/cure this dreadful disease.

AIDS (Acquired Immunodeficiency Syndrome) has expanded its hand greatly and widely and kills about 100,000 people a year worldwide, compared with 1 million malaria deaths, 4 million from diarrhoeal disease, and 12 million from cardiovascular diseases (WHO statistics). But it can spread fast and it is expected that by the end of century the annual death toll may reach 400,000. There is a potential for further rapid spread in the more than 200 million new cases of other sexualiy transmitted diseases that appear in the world each year. Worldwide, perhaps 10 times as many people are HIV-positive as have AIDS (that is, they are infected with the virus which invariably seems to lead to the ultimately fatal illnesses grouped together under the name AIDS). During its long incubation of up to 10 years, few if any symptoms may be apparent, but people carrying the virus can infect others. The incubation period means that even if all HIV transmissions were to halt immediately, the number of AIDS cases would continue to grow during the next decade at an average rate of 10% per annum. Up to 12 million adults are estimated to be infected with HIV, or one in 250 of the world's adult population. One million children had contracted HIV by early 1992. More than 80% of all these cases are in developing countries, because HIV has been the high correlation that exists between poverty and vulnerability to the virus. As yet no vaccine or cure, one of the most effective way to check the spread is to couple care with prevention.

Heterosexual intercourse accounted for 70-75% of all infections by 1992. Other common means of transmission include blood transfusion, injecting drug use and mother-to-child transmission. The proportion of cases resulting from heterosexual transmission is currently showing a rising trend while the proportion due to transmission through contaminated blood or blood products is falling. As it is predominantly transmitted through sex, it kills many people in the 20-40 age group, the most economically productive section of society. According to the Asian Development Bank, by the year 2000, most of the projected 40 million HIV infections and 10 million adult AIDS cases worldwide will be in *Asia*.

Chimeric toxins and immunotoxins appear to have great potential as chemotherapeutic agents for the treatment of cancer, AIDS, and some other diseases. Bacteria and fungi produce powerful toxins that kill many animal cells. Toxins have been attached to growth factors and antibodies to create specific cytotoxic agents active on cells bearing appropriate receptors or antigens. Conventionally, these agents are made by chemically coupling the two protein molecules. Recently receptor specific chimeric toxins have been produced in *E. coli* by gene fusion. Bacterial toxins, a Pseudomonas exotoxin (PE), and diphtheria toxin (DT) have been used to make single-chain immunotoxins directed at the human transferrin receptor. In single-chain immunotoxin made with PE, the antigen binding portion (Fv) of a monoclonal antibody directed at the human transferrin receptor is placed on the amino terminus of the toxin, while in that made with DT it is placed on the carboxyl end of the toxin. Both single-chain immunotoxins specifically kill cells bearing human transferrin receptors (Batra, 1994).

Restrictocin is a fungal toxin produced by *Aspergillus restrictusis.* It inhibits protein synthesis in eucaryotic cells. However, unlike PE and DT, restrictocin is a single-chain polypeptide lacking any cell binding activity. It acts on eucaryotic cells by hydrolyzing a single phosphodiester bond in the 28SrRNA. A very limited immunogenicity has been shown to be associated with restrictocin which makes it a very attractive molecule for construction of immunotoxins and chimeric toxins.

Biotechnology have a great impact on food and drink industries. This involves the use of industrial enzymes to manufacture high fructose corn syrups which are widely employed as sweeteners in diverse soft drinks. Recombinant DNA techniques have been employed to make rennin which is used to clot milk for making cheese. Till recently, most of this rennin has been extracted from the stomachs of young calves (after slaughtering the calves) but now recombinant DNA technology, using suitable bacterial hosts for making this protein.

Recent researches have opened up exciting possibilities of combining genetic manipulation with fermentation technology. One instance is the production of specially-designed varieties of corn for biogas. Until 1981, there was no commercial application of a genetically-engineered process, but by 1983, several commercial companies came out, using and exploiting recombinant DNA technology industrially, mostly in the medical and pharmaceutical fields.

Recent advances in the culture of plant cells, tissues, and organs has led to two thrusts of biotechnological applications. The first relates to the micropropagation of plants, and the second is concerned with the production of special chemicals. Plant micropropagation through tissue culture has made it possible to produce large numbers of virus-free plants. It is also possible to introduce new, desirable traits into chosen plants through the techniques of selection, protoplast fusion, and somatic cell hybridization. These traits may, *inter alia,* include disease resistance, salt tolerance, enhanced yield, and composition or yield of some natural product. Tissue culture techniques have found commercial applications in several horticultural plants such as orchids and, in speciality crops, remarkable success has been achieved with oil palm, jojoba, and citrus. Not much success so far has been achieved in the case of cereals.

Advances in cell culture technology have catalyzed the development of processes using fermenters for the production of high-value secondary metabolites usually derived from the whole plants. Two such processes, relating to commercial production of shikonin and berberine, are already operational in Japan.

Fowler (1986) has proposed a general strategy for developing a suitable plant cell culture process, highlighting those areas where substantial progress should be made if the production of speciality chemicals is to gain wider commercial notice. Diagram 1.3 outlines this strategy.

Biotechnology has provided a great stimulus for agricultural improvement. Its application to crop improvement depends on our ability to grow plant cells or tissues, and to induce their organized growth and development into whole plants. This technology makes it possible to propagate elite genotypes especially if clonal propagation can be achieved in large numbers. Recent progress in the tissue culture and genetic engineering of crop plants has made it possible to (1) achieve large-scale, rapid multiplication (Diagram 1.4) of genetically uniform plants from elite specimens; (2) select novel and improved varieties using somaclonal variation technology; (3) develop new hybrids between different cultivars and species by means of protoplast fusion; and (4) use recombinant DNA techniques to introduce new and desirable genetic traits into plant cells (Ammirato *et at.,* 1984). These achievements have generated a considerable optimism for the future growth and potential of agriculture to meet the increasing needs of the world in the coming decades.

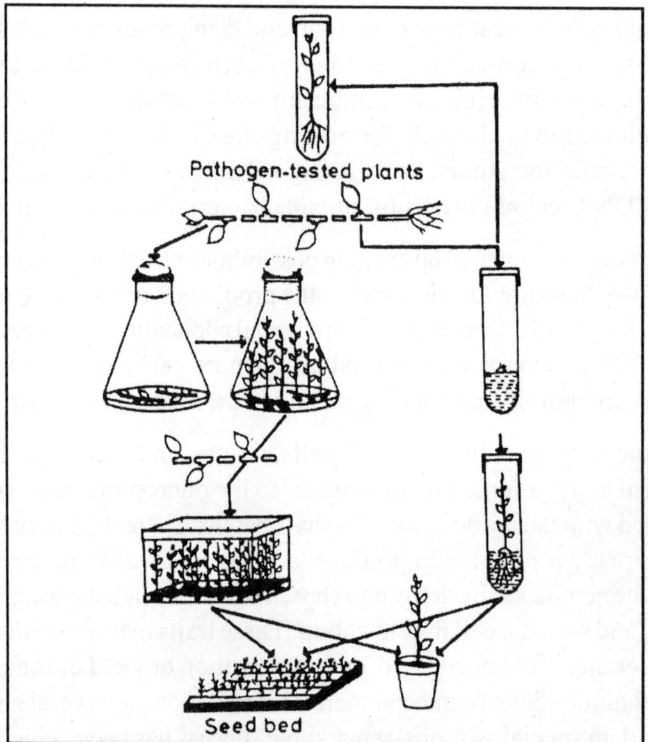

Diagram 1.4. Procedure for *in vitro* propagation. The leaf axillary buds are induced to sprout by cuting the shoot into pieces, each containing one or more axillary buds. Each bud produces a new shoot which in turn can again be cut into pieces. This process can be repeated *ad infinitum*. (After Schilde-Rentschler, 1986).

The various basic approaches in tropical agriculture and some of their characteristics are listed in Table 1.6. Traditional agriculture and Green Revolution agriculture are often not sustainable, the former has a low level of productivity and can only be sustained at certain (maximum) levels of population and with a low demand for external consumer goods. The Green Revolution approach, as originally conceived, *is* not sustainable. Packages of chemical-based technologies have led to an accelerated use of non-renewable resources, to pollution of the air, water and soil and to a loss of the biological diversity. High production levels may not be maintained for long.

In Green Revolution agriculture, initiatives are now being taken to use external inputs'more efficiently, leading to several forms of integrated agriculture, such as Integrated Pest Management (IPM), or Integrated Plant Nutrient Systems (IPNS).

Green Revolution agriculture and integrated agriculture are based on optimizing the production conditions for genetically uniform plants and animals, with genetic modifications playing an important role. Much experience has been gained by the application of a wide range of biotechnologies. Current emphasis in Green Revolution agriculture is on developing pesticide (herbicide) tolerant crop varieties. Within integrated agriculture, the emphasis is on reduction of use of chemical inputs. Biotechnology can contribute by increasing pest resistance of crops and by promoting the use of biofertilizers and biopesticides (Haverkort and Hiemstra, 1993). For both LEISA (low external input sustainable agriculture) and organic agriculture, the farm system is more diverse and complex: intensification occurs by well-designed diversification; the genetic resource base is broad; multiple cropping systems prevail where the interaction between a range of different organisms and process play a major role; pest management is guided by the principle of prevention and prefers the use of natural processes to counter the effects of pests. Here the ultimate goal is not so much maximization of production for the market, but rather sustainable and stable production for local consumption.

It seems that for LEISA and organic agriculture, genetic modification may not be the most relevant biotechnology. The development of technologies for the production and use of biofertilizers, biopesticides, local food processing, and low-cost tissue culture techniques for disease free production of planting material may be more important.

Haverkort and Hiemstra (1993) have proposed the following priorities for research and development in the domain of biotechnology under low-input conditions:

1. Improving the understanding of the biological and physical processes involved in local practices in the domain of microorganism management.
2. Refining and improving the processes involved (e.g., support farmers' practices in genetic improvement by selection and breeding, improve the use of biofertilizers and biopesticides; reducing toxic or antinutritional components, increase vitamin or protein content, improve hygiene, reduce fuel needs, substitute scarce elements, reduce labour needs).
3. Improving the utilization of locally produced biotechnological products.

Table 1.6. Some characteristics of development approaches to tropical agriculture (source: Haverkort and Hiemstra, 1993)

	Green Revolution agriculture	Sustainable agriculture			Traditional agriculture
		Integrated agriculture	Organic agriculture	Low external input and sustainable agriculture (LEISA) (improved traditional agriculture)	
Basic characteristics	Science and technology based for favourable (irrigated) conditions and monocultures	Integrated Pest Management (IPM) and Integrated Plant Nutrient Systems (IPNS)	Locally adapted farming systems	Complex and integrated systems based on maximum synergy, minimum losses, farmer's local knowledge and agroecological sciences	Complex systems with complementarity between crops, animals, and people, based on local farmer's knowledge
Goal	Economic: maximum production for the market	Economic and ecological: reduce use of damaging chemicals	Multiple economic, ecological, and social goals: optimize productivity for self-sufficiency and the market and conserve resource base		Self-sufficiency
Level of external inputs	High	Medium	Medium low (only organic)	Medium low	Very low
Pest management	Chemical: eliminate or reduce pests	Chemical and natural: reduce pesticides; plant breeding and natural enemies	Agroecosystem diversity and stability to minimize pest outbreaks		
			Natural	Natural and limited chemical	Natural
Fertilization	Mineral fertilizers	Balance production and conservation: mineral and organic	Create favourable soil conditions by managing organic matter, enhancing soil life, and balancing nutrient flows		Fallowing, flooding, other sources of natural fertilization
			Organic	Organic and limited mineral	Organic
Diversity of farm system	Low (specialized farms)	Rather low	HIGH		
Present focus of biotechnology	Genetic modification (herbicide tolerance) in addition to range of other technologies	Genetic modification (pest resistance), biofertilizers and biopesticides	Building on indigenous biotechnologies: optimize use of symbiotic microorganisms, biofertilizers, biopesticides, ethno-veterinary practices, and improve process technologies		Indigenous management of microorganisms

One of the most cherished objectives of scientists engaged in agriculture is to produce new, better varieties of crops and other plants. Diagram 1.5 outlines the procedure for achieving this goal.

Crop plants are susceptible to several viral, bacterial, or fungal diseases which extract a heavy toll in terms of growth and yields. The epidemiology of many of the plant pathogens has yet to be investigated. The failure to diagnose these pathogens properly and subsequent counter checking contributes to substantial decreases in agricultural productivity of important food crops. Many of these diseases occur worldwide and result in substantial crop losses. These problems can be tackled by developing high quality diagnostic mechanism capable of identifying viral, bacterial, or fungal pathogens. A serious thought is now being given to utilize monoclonal antibody technology to produce antibodies for use in diagnostic and epidemiological studies of specific plant diseases. Priorities deserve to be given to developing antibodies to the following agents:

Rice	Dwarf, grassy stunt, tungro
Corn	Chlorotic leaf spot, rough dwarf, streak
Citrus	*Xanthomonas citri, Spiroplasma citri*
Potato	Leafroll virus; potato viruses M, S, X, Y
Prunus	Necrotic ring spot, prune dwarf
Tobacco	Streak, ring spot viruses
Tomato	Ring spot

The use of biotechnology in incorporating abiotic stress tolerance in crop species depends on a better understanding of the physiology of stress tolerance, and the identification and introduction of specific genes determining tolerance to a specific stress. However, molecular markers may well serve in manipulating quantitatively inherited traits of stress tolerance (Kuo, 1992).

Physiological processes such as source-sink relationships, translocation, interrelations between plant parts, water status, hormonal levels and balance are crucial in determining a plant's response to stress. It is therefore necessary to study whole seeds, seedlings or plants, rather than excised parts, and to characterize individual genotypes when assessing stress response (Kuo, 1992).

Biotechnology rescent advances have prompted plant breeders to look at novel approaches to clone genes for stress tolerance. For example cloning of proline biosynthetic genes and their introduction into plants has conferred drought tolerance in certain crop plants.

It is important to know the extent to which stress tolerance genes enhance the breadth of adaptation to stress, and whether they increase or decrease tolerance to stress. Abiotic stresses elevate levels of heat-shock, cold or anaerobic-response proteins for long periods but the question remains: will expression of response proteins ahead of stress really protect plants?

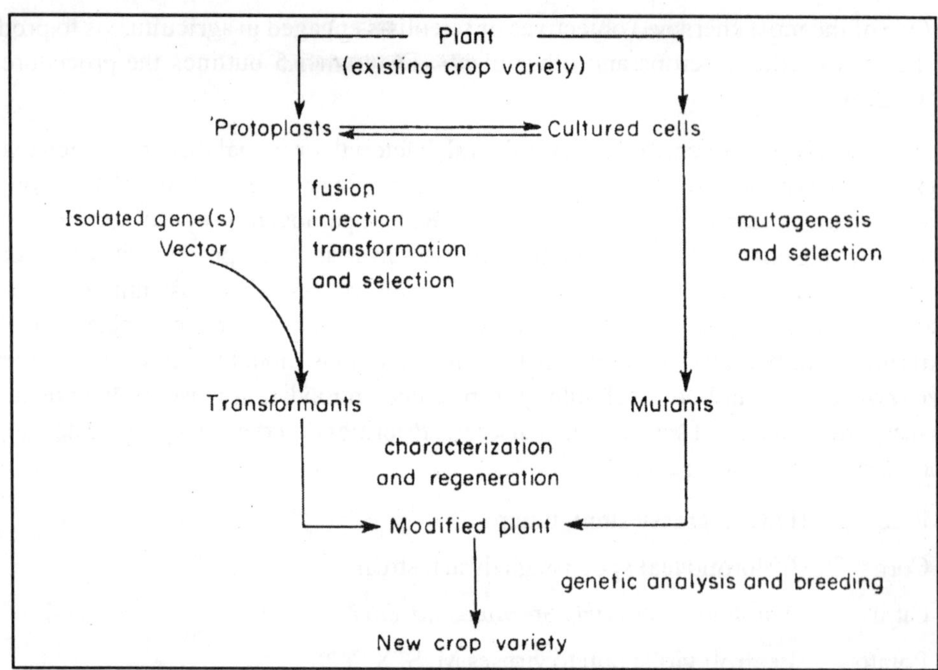

Diagram 1.5. Procedure for producing a new plant variety by using modern biotechnology.

Breeding for stress tolerance is hampered by the breeder's capacity in selecting for stress-tolerant genotypes which is governed by the person's ability to secure the desirable tract in large populations with minimum cost and time. Restriction fragment length polymorphism (RFLP), random amplified polymorphic (RAP) DNA analysis or indirect marker selection (IMS) may meet this requirement. Molecular markers can serve as a powerful tool to monitor gene introgression from wild and related species (Kuo, 1992).

Biotechnology is largely concerned with a gainful exploitation of biocatalysts which come from microbial, plant, and animal cells. Only a limited number of species have so far been developed as biocatalysts and there is a vast scope for screening many more organisms to select newer, wider range of microbial and cultured cell types for pest control, mineral processing, health care, and food production.

Significant advances in recombinant DNA research, molecular genetics, and in blastomere manipulation have brought within reach the technology to insert genetic material in plant cells as well as animal cells. As between plants and animals, it seems likely that greater potential benefits may be realized in the cultivation of the domesticated plants rather than in the production of domestic animals.

Several methods are underway to inject functional proteins, or antibodies raised against such proteins, into living cells. The term microinjection denotes direct pressure injection of macromolecules into cells through glass microcapillaries or needles. For many applications, needle microinjection is fairly suitable and its popularity is likely to increase with continuing development of microscopic techniques. Apart from this simple technique, called needle microinjection, there are certain other approaches which fall into two broad categories, namely, (1) membrane-vesicle methods, in which preloaded membrane vesicles such as liposomes and protoplasts are made to fuse with cultured cells and release their contents into the cytoplasm; and (2) physical methods, which involve physical diffusion of macromolecules into cells through holes transiently introduced in their plasma membranes (Richardson, 1988). Lipid vesicle-mediated injection is preferred for incorporating membrane proteins into cells and protoplast fusion is advantageous for protein engineering *(vide infra)*.

OTA (1984) have identified a number of areas where bioprocesses and modern biotechnology may be exploited. These include the production of complex substances such as antibiotics (Diagram 1.6) and proteins where there is no practical alternative, where microbes can execute a number of sequential reactions, and where microbial processes can give fairly high yields. Another promising area relates to the exclusive production of some specific kind of isomeric compound. The chief advantages of a bioprocess over conventional chemical processes are the milder reaction conditions, use of renewable resources such as biomass as raw materials for producing high-value chemicals, and much less hazardous operations, involving reduced environmental hazards. At the same time, there are some disadvantages associated with bioprocesses. These relate to the frequent generation of complex product mixtures necessitating tedious separations and purifications, problems arising from the relatively dilute aqueous environments in which bioprocesses operate, susceptibility of most bioprocesses to contamination by foreign organisms, and inbuilt variability of bioprocesses arising from genetic heterogeneity and raw material variability (OTA, 1984). Another drawback is the stringent safety and containment requirements of any work involving recombinant DNA technology (OTA, 1984).

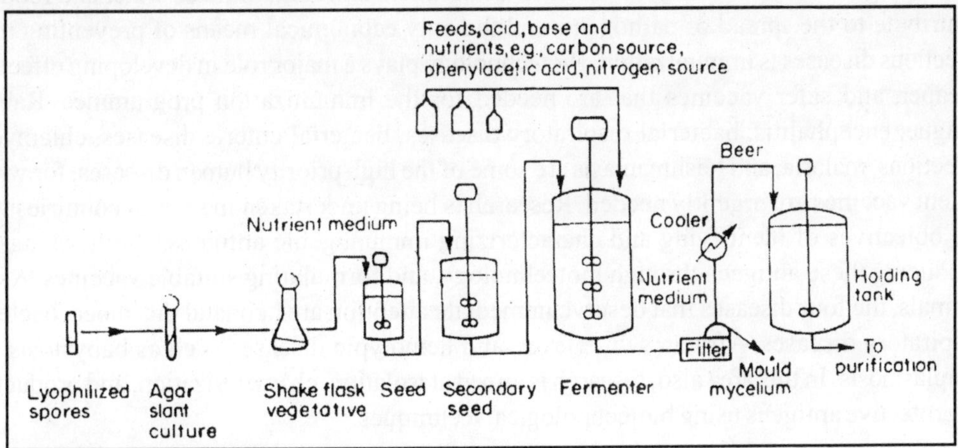

Diagram 1.6. Flow diagram to illustrate the stages in industrial antibiotic production.

Biotechnology has already produced a significant impact on clinical medicine. The cellular and molecular cloning techniques used to develop monoclonal antibodies and DNA probes have found applications not only in basic research but also clinical medicine. There have been remarkable developments which influence the fight against infectious diseases. These developments can be broadly considered under the following four categories (Heden, 1985):

(a) *Improved diagnostic tools:* Monoclonal antibody kits; kits for nucleic acid hybridization (e.g., Chlamydia trachomatis); Fluorescent and luminescent labelling of reagents.

(b) *Improved laboratory techniques:* Isolation and purification of antigens by monoclonal antibody methods; large-scale production of protein antigens via hybridomas or by cloning (e.g., herpes simplex and hepatitis B viruses); incorporation of several foreign antigens in vaccinia virus; & protein engineering for antigen design (e.g., rabies & polio).

(c) *Immunological therapeutic* methods. Monoclonal antibodies for passive immunization or for drug-targeting.

(d) *Novel vaccines:* Polysaccharide-based vaccines; pneumonia infections caused by Streptococcus B and Haemophilus influenzae; genetically-attenuated or self-destructing living vaccines (typhoid, hepatitis A, and diarrhoea).

Monoclonal antibodies have greatly aided tumour diagnostics, and have also been used in diagnosis of infectious diseases. DNA probes have likewise proved useful in the study and diagnosis of genetic diseases.

In fact, it will not be far wrong to state that the monoclonal antibody constitutes a refined tool *par excellence* in the field of clinical diagnosis. The ability to study individual antigenic sites on a virus, bacterium, or parasite is sure to find application in diagnosis and in basic research. The characterization of tumour cell lines using monoclonal antibodies facilitates diagnosis. The usage of appropriate labelling methods permits easier detection and localization of tumours.

Various kinds of infectious diseases are a major cause of human mortality in several developing countries. Poverty, malnutrition, starvation, and contaminated water and food, all contribute to the spread of pathogens, and the only economical means of preventing most infectious diseases is immunization. Biotechnology plays a major role in developing effective, cheaper, and safer vaccines that are needed for the immunization programmes. Rabies, dengue, encephalitis, bacterial respiratory diseases, bacterial enteric diseases, chiamydial infections, malaria, and leishmaniasis are some of the high priority human diseases, for which potent vaccines are urgently needed. Research is being undertaken in several countries with the objectives of identifying and characterizing immunogenic antigens, synthesizing and producing these antigens through biotechnology, and formulating suitable vaccines. As for animals, the four diseases that deserve immediate attention are neonatal diarrhoea, bacterial respiratory diseases, African swine fever, and hemotypic diseases such as babysiosis and anaplasmosis. In this area also, research is aimed at isolation, characterization, and production of protective antigens using biotechnological techniques.

In addition to the foregoing diseases of humans and animals, mycobacterial tuberculosis afflicts both humans and animals all over the world. It would be desirable to direct research toward (1) the development of improved diagnostic tools to distinguish in humans the BCG (Bacille Calmette Guerin) vaccine reactions from those resulting from infection; (2) improved production of TB-specific antigens; (3) proper evaluation of the potency of the BCG vaccine currently in use; and (4) production of a more effective vaccine, including bio- or organic synthesis of the immunogen, that can be used in areas of high incidence.

Human Health and Medicinal Plants

The plants which in one or more of its organs contains substances that can be used for therapeutic purposes or which are precursors of chemopharmaceutical semisynthesis is known as medicinal plants. About 20,000 plants fall within this category.

Biologically active compounds from plant sources have had a dramatic impact in medicine, such as quinine for treatment of malaria; reserpine for controlling hypertension; cocaine as a muscle relaxant; and vincristine for treating children with leukaemia. Tropical plants have an enormous, untapped potential to yield novel drugs and medicines. Only a small proportion of the world flora has been examined for pharmacologically active compounds, but with the ever increasing danger of plants becoming extinct, there is a real risk that many important plant sources with its gene pool may be lost for ever.

According to Farnsworth *et al.* (1985), over 75 different chemical compounds of known structure derived from higher plants are represented in medicinal prescriptions. Of these, only the following seven are commercially produced by synthesis: emetine, caffeine, theobromine, theophylline, pseudoephedrine, ephedrine, and papaverine. Other naturally occurring pharmaceuticals have been synthesized, but commercial production of such important drugs as morphine, codeine, atropine, digoxine, digitoxine, and reserpine is not yet feasible. During the past 20 years, at least one novel compound from higher plants has been marketed every 2.5 years (Deans and Svoboda, 1990).

The antileukaemic alkaloid vincaleukoblastine has been isolated from the Madagascar periwinkle *Catharanthus roseus.* One million kilograms of steroids were processed industrially, and about 75% of the world total supply was derived from plant sources, notably *Dioscorea* species.

Success has been achieved in culturing several medicinal plants in laboratory bioreactors (Wilson *et al.,* 1987; Yeoman, 1986), and in some cases patents have been obtained to protect the processes.

Since long, plants have served as the primary source of useful natural products. While fermentation products such as penicillins and cephalosporins now account for 12% of pharmaceutical sales, some 25% of therapeutic drugs are still derived from plants (Table 1.7) see Mann, 1989; UNIDO, 1987.

There is a current resurgence of industrial interest in naturally occurring substances as sources of novel pharmaceuticals, crop protectants and the development of 'mode of action' bioassays, including immunoassays, capable of detecting picogram quantities of potentially useful compounds.

The relative ease of isolating novel strains of microorganisms and of optimizing fermentation conditions has attracted attention as a source of new products. However, plants produce a highly individual range of natural products, which vary widely from species to species and are mostly structurally distinct from microbial metabolites. Although only a small percentage of plants has been screened, thousands of phytochemicals have already been isolated, and many of these have been shown to possess useful biological activity. A vast storehouse of valuable new phytochemical products still remains to be discovered.

Developing countries are extremely rich in plant genetic resources. Much of the world's genetic diversity is found in twelve scattered sites, known as the Vavilov Centres, nine of which are located in the Third World. Seven 'megadiversity' countries which contain a high percentage of the world's plant species are Brazil, Colombia, Indonesia, Australia, Mexico, Zaire, and Madagascar. Many commercially important products are derived from the flora of developing countries. Traditional plant remedies have yielded a number of widely used pharmaceutical products, e.g,, the antihypertensive reserpine, and the anticancer alkaloids, vincristine, and vinblastine.

Table 1.7. Some phermaceuticals products produced from tropical plant sources (after Joffe and Thomas, 1989).

Plant Source	Clinical Activity	Pharmaceutical
Catharanthus roseus (Vinca rosea)	Anticancer	Vinblastine and vincristine
Cinchona spp.	Antimalarial	Quinine
Dioscorea spp.	Contraceptive precursor	Diosgenin
Erythroxylum coca	Anaesthetic	Cocaine
Rauvolfia serpentina	Antihypertensive	Reserpine
Strychnos spp.	Muscle relaxant	Tubocurarine
Artemisia annua	Antimalarial	Qinghaosu
Aspilia spp.	Antibiotic	Thiarubrine-A
Camptotheca acuminata	Anticancer	Hydroxycamptothecin
Cannabis sativa	Antiemetic	Nabilone

Pyrethrum are cultivated in several countries, e.g., Kenya and Tanzania, its flower contain active constituents, including the pyrethrins, which are insecticidal.

The greatest losses in the world's plant resources are occurring in Third World countries. Approximately half of all plant species are found in the tropical forests but these areas are under increasing pressure, primarily from the spread of agriculture and from unsustainable logging. In tropical forests some 100,000 square kilometres and 10,000 plant species are lost each year. In this way there is a loss of many undiscovered phytochemicals and undescribed forest plants. At the present rate, the remaining forests may completely disappear within the next seven or eight decades.

Plant Tissue Culture and its Impact on Agriculture, Pharmaceutical Industries and Food

Techniques of plant tissue culture (e.g., hairy root culture, Aird *et al.*, 1988) are now powerful tools for studying plants. These methods have found wide commercial application in the propagation of plants, in the preservation of biological material, and in the elimination of pathogens.

The term 'plant tissue culture' broadly refers to the *in vitro* cultivation of plant parts. such as meristems, apices, axillary buds, young inflorescences, leaves, stems, and roots under a controlled aseptic environment and a suitable nutrient medium. These essential nutrients include inorganic salts, a organic or carbon compound as energy source, vitamins and growth regulators. The basic technology can be divided into five classes, depending on the material being used: callus, organ, meristem, protoplast, and cell culture (Deans and Svoboda, 1990).

The potential of *in vitro* methods in agriculture lies in intensification of clonal propagation, in variety development, genetic modifications, and in the establishment of specific pathogen free plants. Techniques of embryo, ovule, ovary, anther, and microspore culture are used and can yield genotypes that cannot easily be produced by conventional methodology. Protoplasts can be manipulated to form somatic hybrids while somaclonal variation (originating in cell and tissue cultures) may be of some value in crop improvement.

This variation can affect the morphological yield, quality, or biochemical characteristics. Increased resistance to phytotoxins, herbicides and antibiotics, along with salt and metal tolerances are some examples of the latter type of variation.

Tissue culture techniques have already been applied to such agricultural crops as rice, sugarcane, coffee, and potato.

TECHNOLOGY FOR ENZYME

Technology for enzyms involves the synthesis,purification, and immobilization of enzymes and their application in industry, health care, cosmetics, diagnostics and therapeutics (Table 1.8).

Table 1.8. Utilization of enzymes in four areas (after Towalski, 1983).

Class code*	Enzyme class	Classified	Diagnostics	Industry	Therapy	Cosmetics
EC1	Oxidoreductases	575	26	11	6	1
EC2	Transferases	572	8	4	2	0
EC3	Hydrolases	577	15	36	35	3
EC4	Lyases	231	8	8	6	0
EC5	Isomerases	96	0	32	0	0
EC6	Ligases	87	0	2	0	0

* As per IUPAC and IUB nomenclature.

The history of enzyme technology extends back to over half a century now. Out of more than 2000 enzymes that have been identified, some 150 are being used commercially. Denmark and Netherlands are the two leading countries in the worldwide production of industrial enzymes. A significant fraction of the enzyme market involves the enzymes used for producing high fructose corn syrup (i.e., isoglucose) and the use of alkaline proteases in detergents. The history of fermentation technology is even longer and richer than that of enzyme technology.

Enzymes are useful as industrial biocatalysts in view of their non-polluting biodegradable nature and in view of efficacy at physiologically-mild conditions (such as pH, temperature, and pressure).

At present about 150 of the 2000 known enzymes find commercial applications, and another 200 are available for use in genetic engineering, which include restriction endonucleases, ligases, and editing enzymes ("editases").

Table 1.9 gives some current applications of enzymes.

Several factors influence the commercial production of enzymes. Animals, plants, and microbes are the three important biological sources of enzymes. These organisms are themselves influenced by climatic, edaphic, hydrological, and other factors. Some of these influences are lessened in the case of microbes which are usually grown in sterile cultures under controlled conditions. Diagram 1.7 shows some methods of choice for the immobilization of enzymes.

Table 1.9. Current applications of enzymes (after Towalski and Rothman, 1986).

Enzyme(s)	Region of Application	Use(s)
Dextranase	Cosmetic/Health care	Dental hygiene
Proteases	Cosmetic/Health care	Skin preparations
Glucose oxidase	Diagnostics	Blood glucose
Urease	Diagnostics	Urea
Cholesterol oxidase	Diagnostics	Cholesterol
Streptodornase	Therapeutics	Antithrombosis
Pepsin	Therapeutics	Digestion
Catechol oxygenase	Therapeutics	Poison ivy treatment
Trypsin	Therapeutics	Wound cleaning
Superoxide dismutase	Therapeutics	Antiinflammatory
Lysozyme	Therapeutics	Antibacterial activity
Amylases	Food and food processing	Baking
Rennet	Food and food processing	Dairying
Pectinase	Food and drink industry	Fruit juices
Glucose oxidase	Food and drink industry	Antioxidant, glucose removal
Proteases	Leather industry	Leather
Invertase	Food and drink industry	Confectionery
Lipases	Food and drink industry	Fat synthesis
Glucose isomerase	Food and drink industry	Fructose production
Subtilisin	Chemicals industry	Detergents
Amylases	Chemicals industry	Paper making, fuel alcohol
Proteases	Textile industry	Desizing cotton
Amylases	Textile industry	Degumming silk

Until a few years ago it was only practical to use immobilized systems containing whole cells or specific enzymes but recently, more complex systems that regenerate cofactors outside living cells have been developed; coimmobilization of enzymes (Diagram 1.8) cells, and subcellular organelles from different organisms has brought within our reach a notable improvements in the industrial utility of immobilized biocatalysts. Four selected examples of industrial applications of immobilized biocatalysts are shown in Table 1.10.

Already, genetic engineers are attempting to design organisms with improved enzyme profiles and specifically-tailored individual enzymes. The objective is to use these genetically manipulated strains both as sources of single enzymes and as more complex, whole-cell biocatalysts. Recent advances in the technology of downstream processing are likely to catalyze the availability of more and more enzymes at reasonable, affordable cost.

Proteolytic Enzymes

Over one-half of the industrial enzyme market is accounted for by proteolytic enzymes. Commercially-significant proteases are produced from microbial, animal, and plant sources. The oldest known examples of proteolytic enzymes are the milk-clotting enzymes used for transforming milk into cheese. More modern examples are detergent proteases, animal and microbial rennets, and proteases of *Aspergillus oryzae* used in baking.

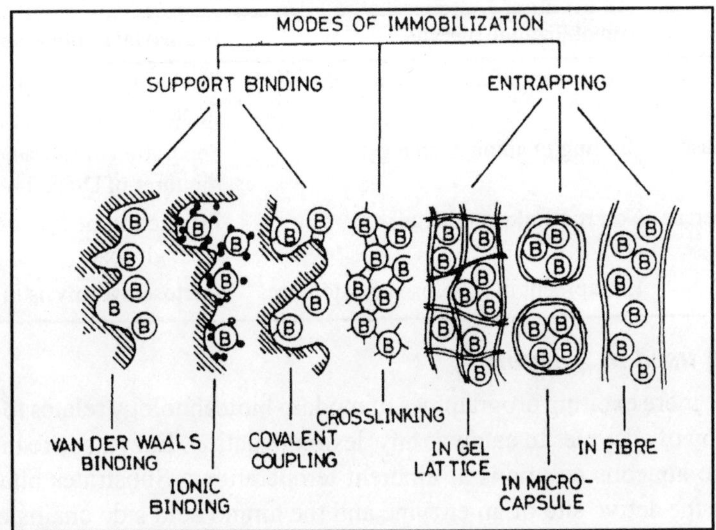

Diagram 1.7. Schematic sketches to illustrate how enzymes can be immobilized in various ways.

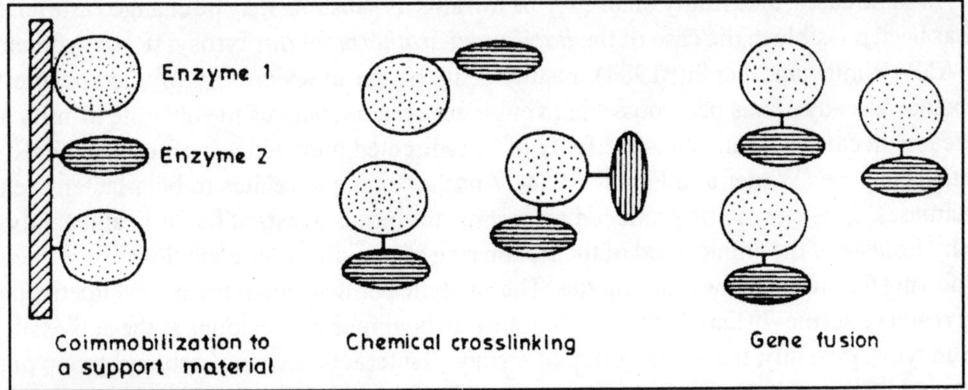

Diagram 1.8. Three basic methods of generating proximity between two enzymes: (1) coimmobilization of the enzymes to a support material (e.g., agarose); (2) chemical conjugation utilizing crosslin king reagents; and (3) gene fusion of the corresponding structural genes. (After Bulow & Mosback, 1991).

Proteinases hydrolyze large polypeptides into smaller molecules that can be assimilated by the organisms. Proteolytic enzymes also regulate various metabolic processes such as blood coagulation, fibrinolysis, complement activation, phagocytosis, and blood pressure control. Proteolytic activity is quite essential during cellular differentiation. Papain, bromelain, and microbial proteases are often incorporated into animal feeds to improve their nutritional value. Urokinase is produced from kidney cells in tissue culture and is used for treatment of clotting disorders. Proteases are believed to be involved in the modulation of gene expression, and in the modification or secretion of enzymes.

High-yielding microbial strains are used in surface or submerged fermentation systems for the production of proteases (Diagram 1.9). The enzymes are formed extracellularly. Their recovery involves separation of the spent medium by filtration or centrifugation.

Table 1.10. Industrial applications of some immobilized biocatalysts

Catalyst	Immobilization method	Industrial applications
L-Aspartase	Entrapment of *Escherichia coli* cell	Production of L-aspartic acid from fumaric acid
L-Amino acylase	Binding to anion exchanger	Separation of L-amino acids from mixtures of D-and L-amino acids.
Glucose isomerase	Co-crosslinking with gelatin	Production of fructose-rich syrups from glucose
Lactase	Entrapment in cellulose acetate fibres	Lactose hydrolysis in milk

Engineering used in Enzyme

One of the more exciting programmes of modern biotechnology relates to the designing and construction of enzymes to catalyze any desired reaction. Enzymes are highly specific, acting in dilute aqueous solutions at ambient temperatures. Substrates attach in precise orientations in the active site of an enzyme and the amino acid side chains of the enzyme assist catalysis by attacking or destabilizing the substrate molecules.

In some cases, the affinity of an enzyme towards its substrate may be changed artificially, as has been possible in the case of the *Bacillus stearothermophilus* tyrosyl tRNA synthetase for ATP (Winter and Fersht, 1984), resulting in change in specificity of^the enzyme for tyrosine. Already, it has been possible to engineer certain changes in substrate affinity and increases in catalysis rate and specificity by site-directed mutagenesis of the tyrosyl tRNA synthetase gene (Winter and Fersht, 1984). Another example relates to beta-lactamase (β-lactamase). This enzyme is produced by certain antibiotic-resistant bacteria and catalyzes the hydrolysis of the amide bond of the lactam ring of penicillins or cephalosporins, thereby conferring resistance to these antibiotics. The catalytic pathway includes an acyl intermediate and residues serine-70 and threonine-71 appear to be important residues at the active site of the enzyme. Possibly, the —OH group of serine-70 interacts with the carbonyl group of the beta-lactam ring. Threonine-71 is also essential but its mechanism is not understood. When serine-70 is replaced by threonine (by site-directed mutagenesis), the product shows no beta-lactamase activity. The serine-70 residue in the enzyme from wild-type cells can also be replaced by cysteine, producing a thiol-β-lactamase. These mutations change the specificity of the enzyme, the mutant enzyme being much more resistant to trypsin digestion than the wild-type enzyme. This difference appears to be due to increased thermal stability of the mutant enzyme—a major goal of enzyme engineering (Walker, 1985).

One of the most promising targets for enzyme engineering is subtilisin, a protease that cuts polypeptides after small aliphatic groups. This enzyme is secreted abundantly by *Bacillus subtilis* and is an ingredient of bioactive detergents. Its crystallographic structure is known and the gene has been cloned. Attempts are underway to produce a mutant showing high activity at fairly low (about 40°C) temperature for low-temperature washes of delicate clothes. An alternative approach is to change its proteolytic specificity to a defined sequence of side chains; this kind of tailoring of subtilisin could pave the way toward using it for cleavage of certain genetically-engineered fusion proteins. Some scientists have already succeeded in isolating subtilisin gene mutants having altered affinities for synthetic peptides.

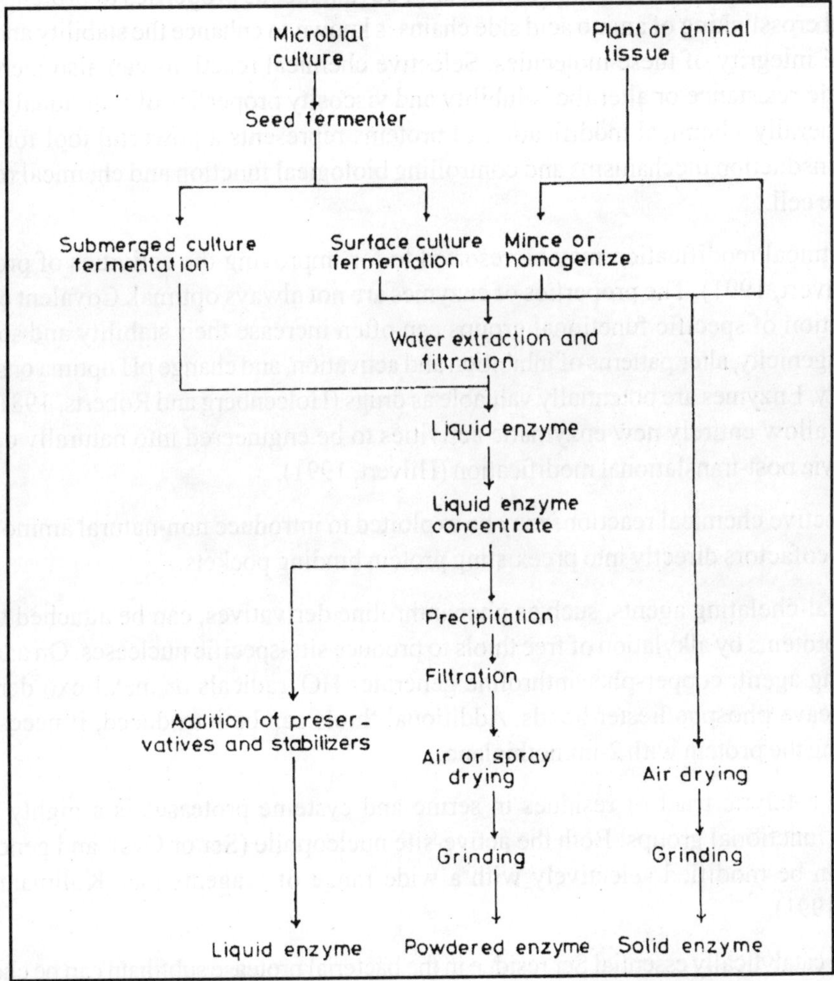

Diagram 1.9. General flow diagram for production and extraction of industrial enzymes from living organisms (after Ward, 1985).

Some other suitable candidates for enzyme engineering may be glucose isomerase, alpha-amylase, and para-hydroxybenzoate hydroxylase.

Protein engineering may usher in the next major boom in biotechnology, offering the promise of tailor-made industrial enzymes and therapeutic proteins. Already, some improved proteins for specific industrial and therapeutic uses have been produced (Bryan, 1987). It has been shown that tailoring enzymatic properties for the non-physiological substrate conditions, altering pH optima, changing substrate specificity, and improving stability are feasible.

Selective chemical modification is now being used to design novel proteins, particularly enzymes and antibodies, with altered specificities and catalytic activities *in vitro*. Modification strategies now being developed are expected to yield a wide spectrum of novel biomolecules with optimum activities for specific industrial processes or therapeutic application. Post-translational modification confers a number of advantageous properties to proteins *in vivo*. Chemical crosslinking of amino acid side chains is known to enhance the stability and overall structural integrity of these molecules. Selective chemical reactions can also increase the proteolytic resistance or alter the solubility and viscosity properties of individual proteins. More generally, chemical modification of proteins represents a powerful tool for altering signal transduction mechanisms and controlling biological function and chemical reactivity within the cell.

Chemical modification may be resorted to for improving the activities of proteins *in vitro* (Hilvert, 1991). The properties of enzymes are not always optimal. Covalent chemical modification of specific functional groups can often increase their stability and solubility, mask antigenicity, alter patterns of inhibition and activation, and change pH optima or substrate specificity. Enzymes are potentially valuable as drugs (Holcenberg and Roberts, 1981). These methods allow entirely new enzymatic activities to be engineered into naturally occurring proteins via post-translational modification (Hilvert, 1991).

Selective chemical reactions may be exploited to introduce non-natural amino acids or catalytic cofactors directly into preexisting protein binding pockets.

Metal-chelating agents, such as phenanthroline derivatives, can be attached to DNA-binding proteins by alkylation of free thiols to produce site-specific nucleases. On addition of a reducing agent, copper-phenanthroline generates HO^- radicals or metal-oxo derivatives which cleave phosphodiester bonds. Additional thiols can be introduced, if necessary, by pretreating the protein with 2-iminothiolane.

The catalytic triad of residues in serine and cysteine proteases is a highly reactive group of functional groups. Both the active-site nucleophile (Ser or Cys), and general base (His), can be modified selectively with a wide range of reagents (see Kullmann, 1987; Hilvert, 1991).

The catalytically essential Ser residue in the bacterial protease subtilisin can be chemically converted into a Cys residue, yielding large amounts of pure thiolsubtilisin. Also non-natural amino acids may be chemically introduced into the protein binding site.

Restriction endonucleases and DNA ligases have made it possible to make manageable-sized fragments of genetic material accessible for manipulation and study. These restriction endonucleases are wonderful molecular scissors occur naturally in several bacteria where they safeguard the DNA of the host by degrading and hence making ineffective any invading foreign DNA molecules. These enzymes were discovered by two American scientists Hamilton Smith and Daniel Nathans in 1978. Over 200 restriction enzymes have been purified from different species of bacteria. Design of new agents able to bind and cleave large DNAs site-specifically will further facilitate cloning and mapping of genomic DNA. Rationally designed catalysts able to cleave large RNAs site-specifically should also aid in studies of RNA structure and function (Hilvert, 1991).

Modern technology has led to the formation of multiprotein structures which mimic virus particles. These particles lack genetic material, are highly immunogenic, and elicit an immune response which protects against infectious virus challenge. The formation of virus-like particles (VLPs) using this new technology offers a novel approach in vaccinology. Most of the current viral vaccines are prepared using attenuated live or inactivated (killed) virus. However, insufficient attenuation or incomplete inactivation of the virus is always a threat to animal and human health. Recent recombinant DNA technology has provided novel approaches to designing safe vaccines. This involves synthesis of relevant viral proteins carrying antigenic determinants which elicit protective immune responses. Systems capable of synthesizing such proteins are known as expression vectors. These can be derived from bacteria, yeast, animal, plant, viral or other sources. Successful vaccine development requires systems where the products of expression vectors resemble the authentic proteins. Ideally, expressed proteins should be produced in large quantities using fairly easy techniques.

Baculoviruses have received attention as vectors for high-level expression of various genes. These expression vectors have been used to synthesize blue tongue viral proteins and viral-like particles, and their ability to protect sheep against blue tongue disease has been tested (Roy, 1991).

Blue tongue virus (BTV) (genus: *Orbivirus)* not only causes disease in sheep, but also infects cattle, goats, buffaloes, and camels, as well as wild ruminants. In sheep, the disease is acute and mortality may be high whereas in cattle and goats the disease is usually milder. In a typical case in sheep, the onset of the disease is marked by high fever lasting about a week. By 7-10 days, distinctive lesions appear in the mouth, and the tongue can be severely affected, turning blue. In contrast to sheep, infected cattle experience prolonged viraemia, and infection during pregnancy can often cause teratogenic defects in calves and abortion of the foetus (Erasmus, 1990). Blue tongue disease occurs in South Africa, South East Asia, and some other countries.

Baculoviruses can only replicate in particular arthropods, e.g., in moths and butterflies. They do not infect vertebrates, other invertebrates, microorganisms or plants. In view of this restricted host range, baculoviruses cannot be used directly as vectors of immunogens for vaccination (either in man or in other vertebrates). Recombinant baculoviruses can, however, be used for the production of subunit vaccines, either *in vitro* (cell culture) or in their host species (e.g., caterpillars). Therefore, not only are they safe to use (due to their restricted host specificity) but are also cost-effective as large quantities of immunogens can be made.

Redox catalysts: Enzymes can use metal ions, vitamins, and various cofactors to catalyze certain reactions that cannot be catalyzed by protein side chains alone. This is especially true for oxidative functional group transformations. Likewise, artificial oxidoreductases can be prepared by covalently attaching redox-active prosthetic groups to existing active sites.

Alkylation of protein binding sites with a reactive 10-methylisoalloxazine derivative yields semisynthetic flavoenzymes that combine the reactivity of the catalytic cofactor (electron transfer, thiol, and dihydronicotinamide oxidation) with the specificity of the template protein.

Affinity labelling is a powerful strategy for incorporating catalytic groups into antibody combining sites. Use of a cleavable affinity reagent places a free thiol proximal to the binding pocket after treatment with dithiothreitol (DTT). The thiol is a convenient handle for attaching chemical functionality (e.g., imidazoles) (Hilvert, 1991).

Semisynthetic antibodies: It has now become possible to incorporate catalytic groups selectively into antibody combining sites via chemical modification (Hilvert, 1991). Catalytic antibody technology allows the creation of catalysts for virtually any chemical transformation, even reactions that have no physiological counterpart.

Cofactor Engineering

All enzyme-catalyzed reactions involve the interaction of the enzyme, its substrate and the immediate environment (e.g., solvent). Changing the properties of the enzymetic reaction involves manipulating one or more of these three components. Some examples of engineering enzyme reactions include site-directed mutagenesis or selective chemical modification of the enzyme; derivatization of the substrate to better suit the enzyme or environment; or use of organic solvents or additives to modify catalytic activity.

For more than 50% of known enzymes, either a cofactor or coenzyme is also required in the reactions they catalyze: this provides yet another way of manipulating reactions. Cofactor engineering is a good approach for improving bioconversion for specific applications (Duine, 1991).

Some potential areas of cofactor engineering are listed below:

1. Regenerating the required redox form of a coenzyme such that it is not a rate-limiting factor has long been a problem in optimizing enzyme catalysis. One approach has involved attempts to attach the coenzyme NAD to dehydrogenases so as to let NAD function as a cofactor (prosthetic group). In this way, escape of the valuable NAD is prevented, although the problem of regeneration is shifted now from the coenzyme to the enzyme. Elegant solutions exist, however, for the latter problem; for example, NAD-dependent glucose dehydrogenase was engineered to a variant containing a cysteine residue in a position where the NAD analogue covalently coupled to it could not only participate in catalysis by the glucose dehydrogenase, but could also serve in

catalysis by NAD-dependent lactate dehydrogenase present in the same solution (see Duine, 1991).

2. Modification of enzymes by incorporating, a cofactor which is normally associated with a quite different type of enzymes. An example is the alkylation of a cysteine residue in the active site of papain with a flavin derivative transforming the hydrolase into an oxidoreductase (Duine, 1991).

Recombinant DNA Technology

Recombinant DNA Technology provides molecular genetics to involve directed manipulation of genetic material and the transfer of genetic information between species which cannot interbreed. This includes certain techniques that allow fragments of DNA from an animal, plant, or microbe to be transferred to a host bacterium (or some other microbe) which in turn incorporates the fragments into its own genome, thereby gaining new capabilities for synthesis through biochemical reactions. The host of choice in most experiments has been the bacterium *Escherichia coli,* but other microorganisms or even cultured cells of higher plants and animals can now be used as hosts.

How exactly is the genetic information moved from the donor to the host? This is done by means of certain vectors whose good examples include bacteriophages and restriction enzymes. The former are viruses which infect bacteria; the latter are synthesized naturally by bacterial cells and are capable of nicking DNA molecules at specific sites where there is a complementary or specific sequence of nucleotide bases. Two characteristic properties of most vectors are the ability to move from organism to organism, and reproducing themselves as the cells divide. One can cut out a fragment from the donor DNA molecule by means of a suitable restriction enzyme and insert the fragment into a vector which carries the donor DNA into the host cell (Diagram 1.10).

The host cells that have received the alien DNA through vectors by the foregoing technique can sometimes be coaxed to synthesize fairly large quantities of a novel protein which the unmodified host does not synthesize naturally. The first noteworthy example of a tangible achievement of this technology was the production of human insulin by *E. coli.*

Recombinant DNA technology and genetic engineering techniques find several useful applications in the areas of vaccines, foods, antibiotics, alcohols, hormones, and monoclonal antibodies.

It has become possible in some cases to diagnose genetic defects by use of the restriction mapping technique. This is based on the fact that the base sequence in defective genes differs from that in normal genes, leading to the production of different-sized DNA fragments when a gene is cut up with a restriction endonuclease.

One interesting application is the creation of gene libraries or gene banks, which store genes of rare organisms inside bacterial hosts until needed.

Genetic engineering techniques have perhaps found one of the most important uses for the production of insulin (Diagram 1.11) and somatostatin.

Table 1.11 lists some applications of cloned genes.

It has now become possible to insert foreign genes into cells, not just anywhere in the host genome but exactly where desired. This new technique of targeting a transferred gene to some specific site on a chromosome is bound to improve the chances of achieving effective (repair gene defects) gene therapy for such hereditary diseases as sickle-cell anaemia.

Targeted gene transfer can also be used to introduce specific mutations into mice to generate mice of any desired genotype; such mice could serve as models for human genetic diseases (Marx, 1988a).

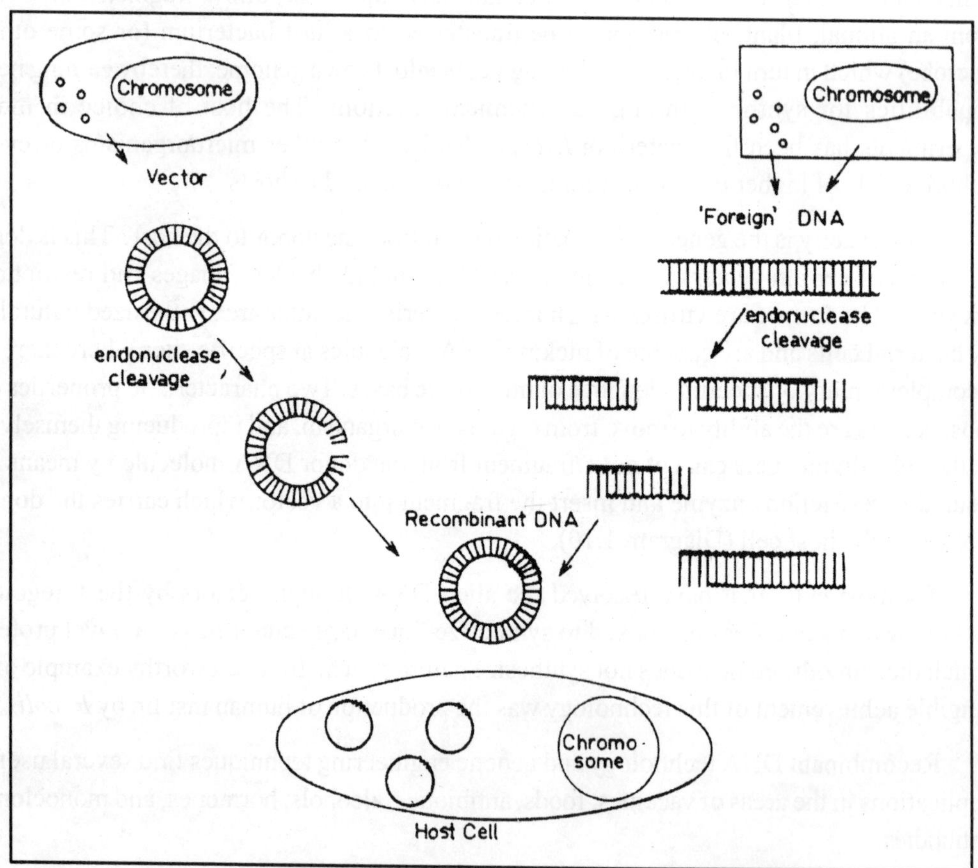

Diagram 1.10. Transfer of vectors from organism to organism.

Human beta-globin gene sequences have already been successfully inserted into the beta-globin gene of the recipient cells by homologous recombination, which is the basis for all targeted gene transfer; the vector used to introduce the new gene into cells carries nucleotide sequences identical to those of the DNA at the chromosomal site where one wants the gene to integrate.

GENE TRANSFER

The ability to produce recombinant DNA, which is ligated DNA from different organisms, started from the discovery of restriction endonucleases in 1970. These enzymes cleave DNA at specific sites. In 1972 the method of joining DNA fragments was discovered. In 1973, the first plasmid vector was constructed.

Gene transfer in animals involves four steps: (1) a method of cutting and joining DNA, (2) a vector (gene carrier) that can replicate itself and a foreign DNA segment that has been inserted in it,'(3) a method of producing enough DNA for insertion into the germline of animals (cloning), and (4) a method of introducing the cloned DNA into germ cells (Turton, 1989). The first three of these steps became possible in the early 1970s; the fourth step could be achieved several years later. In 1977, rapid and accurate methods of identifying the precise nucleotide sequences of genes were designed by Sanger, Nicklen, and Coulson at Cambridge, and Maxam and Gilbert at Harvard. This opened the way for the molecular dissection of genes, and elucidation of the way in which they function.

In the mid-1970s, it had been shown that viral DNA could be introduced into mouse embryos, for example simian virus 40 DNA can be placed into the blastocyst cavity.

Cline *et al.* (1980) were the first to insert a gene into a living animal albeit into bone marrow cells rather than germ cells. Also in the same year, Gordon *et al.* (1980) demonstrated insertion of cloned DNA into the mouse genome by microinjection of the male pronucleus of the one-celled embryo. The way was now clear for germ line transfer of DNA to produce transgenic animals. In the procedure, embryos which survive are implanted into recipient females through microinjection and some offspring developing from injected eggs carry the foreign gene in all cells, integrated into a host chromosome (Turton, 1989).

The integrated DNA usually occurs in multiple copies. In livestock, microinjection is technically more demanding than in mice, because their eggs are almost opaque, making the pronucleus difficult to see.

Retroviruses can be used as gene vectors both with cell systems and also with multicellular embryos (Anderson, 1986). Retroviruses have a gene coding for the enzyme reverse transcriptase, which catalyzes the production of double-stranded DNA complementary to the RNA core of the retrovirus. This DNA can integrate into the DNA of host cells, as a single copy called provirus. The virus multiplies in host cells by transcribing retroviral RNA from the proviral DNA, using the host cells own RNA polymerase. Some of this RNA is used to produce the proteins required for the retrovirus envelope using the cells system for translating RNA to protein. Provirus genes coding for protein may be deleted by standard genetic engineering techniques, and replaced by foreign DNA. When viral genes are deleted, the provirus can no longer replicate on its own, and is now known as "defective virus". However, a non-defective helper virus can be used to infect the cells, and supply the missing gene functions to the defective virus.

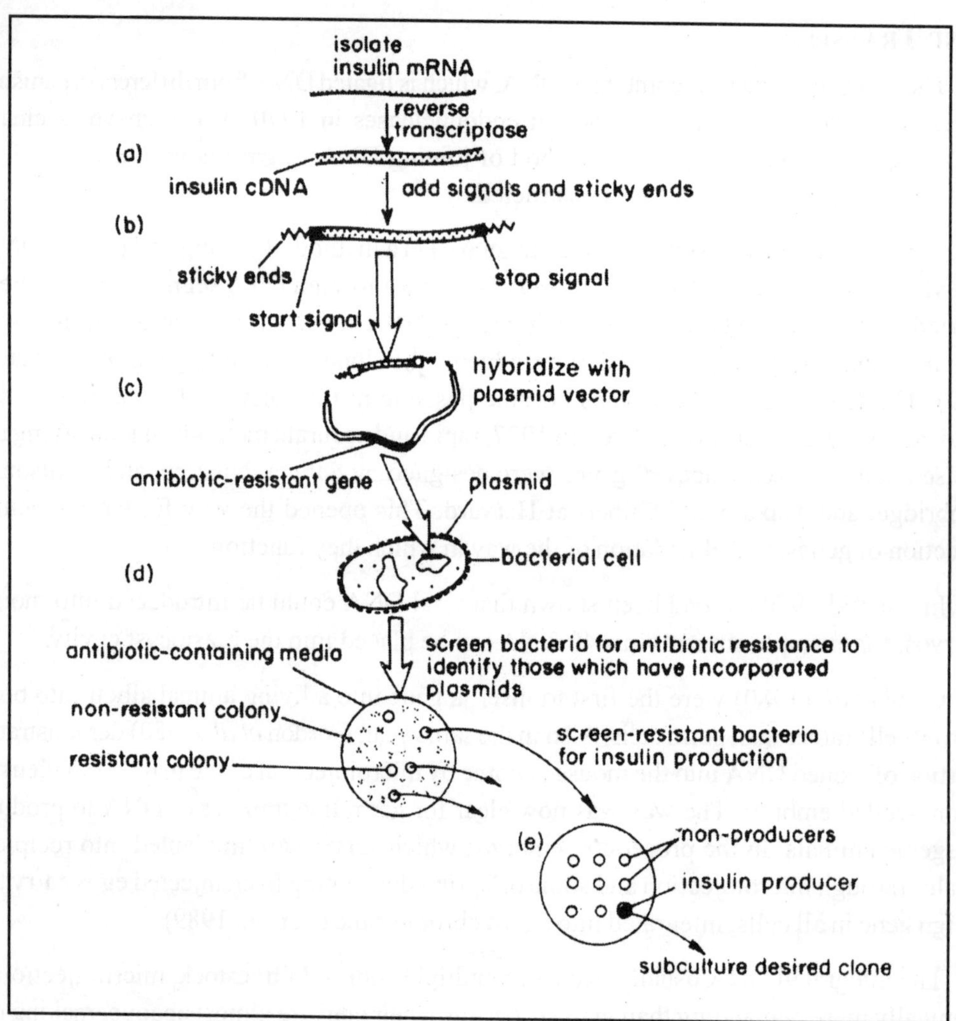

Diagram 1.11. Production of insulin by recombinant DNA technology (after Ingle, 1986).

Table 1.11. Applications of cloned genes.

Gene(s)	Special application(s)	Medical implication(s)
Globins	Haemoglobinopathies	Prenatal diagnosis, gene therapy (?)
Phenylhydroxylase	Phenylketonuria	Prenatal and heterozygote diagnosis
Ferritin	Iron overload and iron metabolism	New chelators (?)
Ig	Antibodies, somatic mutations (diseases)	Interactions with oncogenes
Apo-LP	Cardiovascular disease	Polygenic inheritance

Retroviruses have some limitations as vectors. The size of the foreign DNA sequence that can be inserted in provirus is limited, and the nucleotide sequences at each end of the provirus, the so-called long terminal repeats, which are necessary for viral transcription, can interfere with or override signals determining expression of foreign DNA (Clark *et al*., 1987; Turton, 1989).

For successful commercial use, transgenes must function in the right place (the chosen target tissue), at the right time in the animal's life history, and deliver the right amount of product.

When the DNA of a gene transcribes messenger RNA, (mRNA) it does so under the action of RNA polymerase. The site on the DNA where the enzyme attaches is known as the promoter. In higher animals, the promoter alone allows only very inefficient transcription of structural genes. Other elements in the genome must regulate or 'open' the transcription initiation site(s). Specific regulatory elements, called enhancers, appear to affect genes even though they are not located next to them.

Antisense DNA and RNA

Only one of the two strands of DNA of a gene codes for the gene's product. The other is known as antisense, as its nucleotides are those which pair with the nucleotides of the coding strand, e.g., adenine with thymine and guanine with cytosine. It is thus possible to block gene action by creating antisense RNA, as the heteroduplex RNA formed from sense and antisense RNA is not translated to form the protein product of the gene. There are two potential applications of antisense RNA. The first is in the study of gene function, by observing cells or animals with functioning and inactivated genes. The second is to use antisense inhibition for therapy in genetic diseases (Turton, 1989).

Any antisense genes must be active in the same tissues as the corresponding normal gene in order for the expression of the latter to be inhibited. Also, antisense genes must produce a large excess of RNA relative to the amounts produced by the normal, or sense, gene. For such an excess to be obtained *in vivo*. the promoter and enhancer(s) attached to the antisense gene must be more active than those of the sense gene, and have the same tissue specificity (Turton, 1989).

Restriction Fragment Length Polymorphism (RFLP)

Non-protein coding makes up over 90% of all DNA. Within this non-coding DNA, variation occurs in the nucleotides that individual animals have at the same base positions. This variation appears to have no phenotypic effect. If some of these differences occur within the cleavage sites of restriction endonucleases, then variation will occur between individuals in the length of restriction fragments produced by a particular restriction enzyme. This constitutes RFLP, and it is detected by gel electrophoresis of DNA. Variation in the number of tandem repeats of sequences between restriction sites (DNA fingerprinting) is another source of RFLP (Turton, 1989).

Genetic techniques such as restriction fragment length polymorphism (RFLP), variable number of tandem repeats (VNTRs) and random amplification of polymorphic DNA (RAPD) are expected to aid the large-scale gene mapping programmes in progress. RFLP-based maps have already been constructed for *Hordeum vulgare, Arabidopsis thaliana,* and some other species.

Reverse genetics is another important approach to genome analysis. Instead of relating observed phenotypes to biochemical changes, it is now possible to identify a protein of interest and work backwards towards the gene. Reverse genetics had previously involved identifying probes around the gene of interest, cloning the whole area and then working along the cloned area to the chosen gene. However, this technique, called chromosome walking, is quite laborious, and not more than 10 human genes have so far been mapped using chromosome walks. But, recent advances in yeast artificial chromosome (YAC) techniques may revolutionize the procedure. A gene requiring 3 years to map via a chromosome walk while it may be mapped by YAC techniques in just one day!

Transgenic plants and animals may also turn out to be particularly useful in gene mapping since they eliminate the need for complex breeding strategies to manipulate genes of interest.

When DNA of higher animals is cut by restriction enzymes, so many fragments of different size are produced that a continuous smear is produced on electrophoresed gel. It becomes necessary to reduce the number of fragments to get gels with discrete bands. This is accomplished by using DNA probes on gels that have been transferred to nitrocellulose filters (Southern blotting). These probes used in RFLP detection can be with any DNA sequence that hybridizes with only some of the fragments on a restriction enzyme-digested gel. The probe which is radioactively labelled, hybridizes with certain fragments; the unhybridized fragments can then be washed off. The filter is dried, and placed against X-ray film; the radiolabelled bands on the gel produce an image on the film (autoradiography). Variations in fragment length due to nucleotide differences within the restriction site recognized by a specific restriction enzyme are inherited in a Mendelian manner, and if particular fragments can be associated with specific traits in animals, then the RFLPs can be used as a means of gene mapping. This is being done very extensively not only in humans, but also in animals and plants (Turton, 1989).

DNA FINGERPRINTING

The sophisticated technology that facilitates the identification of individuals at the genetic level is popularly known as 'DNA fingerprinting' or more appropriately call it 'DNA profiling'.

In the early 1980s, DNA regions (hypervariable minisatellite DNA) were discovered which varied in nucleotide sequences between individuals, so that no two individuals, except identical twins, had identical, hypervariable regions. Ten to 15 kb long core sequences were found to be common to hypervariable regions in all individuals, which can not be altered in his/her life time and therefore could be used as genetic markers of such regions, by means of DNA hybridization probes (Jeffreys *et al.*, 1985). It is known that 99 per cent of base sequence is same in the DNA of all human beings. Only the sequence of very short stretches of DNA, sprinkled over the total DNA of a cell which is about three million base pairs, differs from person to person. Of the total DNA, about one in 1000 base pairs is a site of variation in the population. For matching the DNA of two persons one cannot scan the entire length of their respective DNA molecules.

The hypervariable regions consist of short nucleotide sequences that are repeated many times, and it is the number of repeats that is the key to DNA fingerprinting. Use is made of restriction endonucleases to cut the DNA into pieces. These enzymes recognize specific sequences of 4-6 nucleotides, and create breaks in the double-stranded DNA.

In case of variation in the number of short-sequence repeat's between two restriction sites, DNA pieces of different length will be produced. When the cut DNA is electrophoresed, DNA fragments of different length will move through the gel to different distances from the start point, and occur as transverse bands in the gel. The gel is then transferred by blotting to a nitrocellulose filter, hybridized to a radioactive DNA probe for a core region, and the hybridizing bands may be identified by autoradiography. The technique is used in forensic science to identify the offending male in rape cases by examination of DNA extracted from semen, and also in parenthood testing. Tommie Lee Andrews was the first man to be convicted in USA on basis of DNA profiling technology, who committed sexual assault of two women in Orland USA.Use of the technique in animals has followed in the wake of the human applications (Turton, 1989).

Georges *el al.* (1988) have carried out DNA fingerprinting of cattle, horses, pigs, dogs, fowls, and a fish, using four probes, namely, wild type M13 bacteriophage DNA, a plasmid containing a human alpha globin hypervariable region sequence, a plasmid containing a mouse DNA fragment related to a *Drosophila* gene, and a plasmid containing 25 tandem copies of a core sequence. Individual specific-fingerprints could be detected in all the species for one or more probes.

In farm livestock, inaccuracy in parentage registration, particularly in artificially inseminated cattle, is often higher than is generally realized. Currently, most checking involves typing for blood groups and sometimes for polymorphic biochemical markers. In the future, DNA fingerprinting is likely to be the method of choice, since it has brought about a revolution in the diagnosis of infectious diseases besides its plethora of uses in food industry and agriculture. Predictive testing of genetic disorders by detecting faulty genes is also possible.

Gene Amplification

A new gene amplification method has greatly facilitated DNA analysis and has found several applications. The method and its uses are here briefly described. The technique is called *polymerase chain reaction* (PCR). It works with intact or broken DNA pieces, even as small as about 50 base pairs. It is essentially an *in vitro* method for copying simultaneously the two complementary DNA strands which make up a gene sequence. By this technique it is possible to synthesize millions of copies of a single sequence in only a few hours. The specific DNA segment to be amplified is selected by using primers (short segments of DNA that have been synthesized to have sequences complementary to the DNA flanking the target region). These primers help define the ends of the DNA to be duplicated. Upon heating of the DNA sample, the two strands separate, allowing the primers to bind to the flanking sequences, one on each strand. Thereafter, the primers initiate the synthesis of two daughter strands, complementary to the parental strands, in the presence of DNA polymerase. This kind of cycle involving heating and DNA synthesis can be repeated several times. The enzyme DNA polymerase used in this PCR technology is that isolated from the thermophilic bacterium *Thermus aquaticus* and is quite heat-stable. Twenty cycles can amplify the DNA by a factor of about a million.

The PCR technology is stimulating efforts to track down the cellular changes involved in cancer. It makes it easier to detect even a single base-pair mismatch in the human genome, by amplifying it. The DNA sequences of the human papilloma virus can be detected in samples of cervical cancer tissue by means of the PCR technique, even in old samples in which DNA may have degraded. The PCR has made it possible to perform immediate tests on hypotheses linking the presence or absence of specific DNA sequences with a disease or its prognosis (Marx,1988b).

Biomass Technology

Biomass constitutes an inexhaustible and renewable source of important ingredients of fodder, feed, and foodstuffs, as well as of substances that are used in medicine, chemistry, energetics, agriculture, and environment. Much attention is now paid to the production and utilization of microbial, plant, and animal biomass.

The lignocellulosic materials of plants now provide a self-renewable source of raw materials and energy for those countries which have limited sources of fossil fuels such as oil (Diagram 1.12). Man has, of course, utilized phytomass since time immemorial, especially as firewood for cooking in developing countries. What is new is the development of forest energetics as a branch of science that extends the supply of phytomass by purposeful planting of such fast-growing trees as elder, poplar, willow, and birch. This approach is coupled to the technology for microbial conversion of cellulose into glucose, catalyzed by cellulases which are found in several fungi and bacteria.

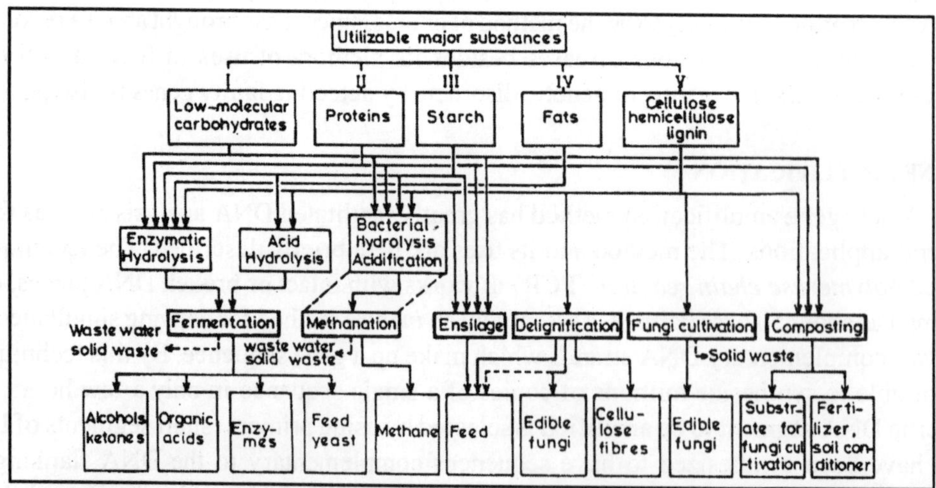

Diagram 1.12. Biotechnical processes for extracting useful materials from biogenic residues and byproducts (after Baader and Weiland, 1991).

Biomass production technologies are instrumental in slowing down the processes of deforestation and soil erosion that pose serious threats to traditional subsistence agriculture in developing countries. Of course, the poor people already depend on such biomass resources as firewood, charcoal, and agricultural residues and they will continue to rely heavily on these sources. The newer biomass technologies will simply increase the availability of these materials and ensure that they are used more effectively than hitherto.

Biogas generation has perhaps been the most enthusiastically promoted technology in developing countries. This technology requires some capital investment, abundant source of animal dung or other substrate, and also a certain level of technical skill to obtain sufficient quantities of the gas. With some exceptions, successful operation of biogas generators is intimately associated with sophisticated and integrated systems of waste management based on cattle, poultry or swine production, in which the gas produced constitutes a byproduct rather than the main product. Biogas is no panacea for the poor; it is unlikely to provide much fuel for the poor until they acquire other resources, e.g., cattle, and have risen above the poverty line.

Biomass and waste materials such as pulpmill sludge have been used for producing alcohol (Diagram 1.13) in many countries. In an interesting exercise, the World Bank (1980) recorded the ratios between production of alcohol from biomass energy, and agricultural self-sufficiency for several countries. This relationship is illustrated in Diagram 1.14 the context of this (alcohol) matter, examples of agricultural-surplus but energy-deficient countries include Brazil and USA, those of agricultural as well as energy surplus countries are Australia and Canada; agricultural as well as energy deficient countries include the UK and Italy, whereas agriculturally deficient but energy-surplus are the USSR and Egypt. India occupies a rather distinctive position more or less on the crossroads, being just self-sufficient in agriculture but deficient in energy.

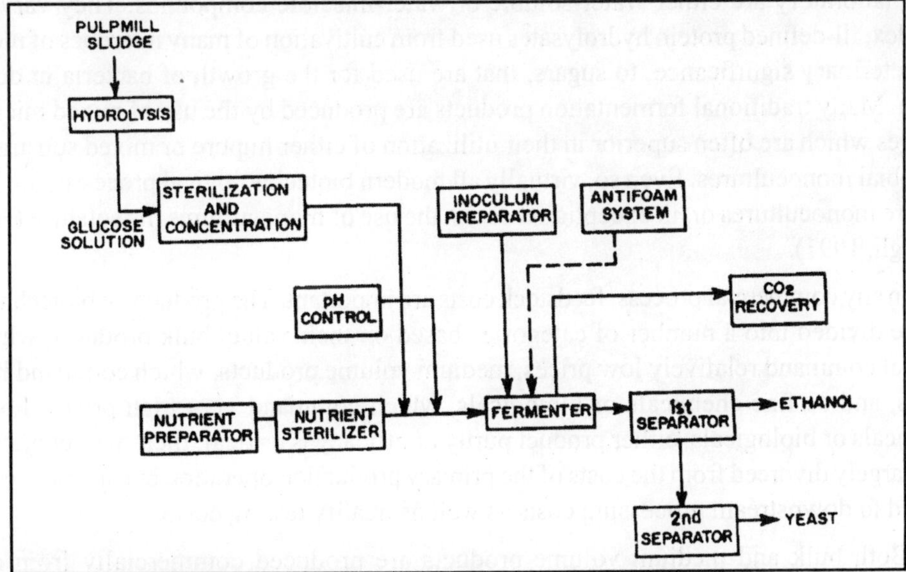

Diagram 1.13. Flowchart for ethanol production from pulpmill sludge (after Moo-Young *et* al., 1986).

Alcohol has, of course, occupied a prominent place in the life of man since time immemorial, and people have been brewing it since long. In fact, brewing and wine-making along with production of fermented foods constitute the core of classical or old biotechnology. Diagram 1.15 is an outline of the brewing process, leading to the production of beer.

FEEDSTOCKS

Regardless of the scale of operation, the technology employed and the product produced, all commercial manufacturing processes require either a single feedstock or a range of feedstocks. The fundamental concept of commercial processing ventures is the generation of profit by risking capital investment in a processing plant. That plant, with an input of both know-how and energy, can transform or convert feedstocks by chemical, biological, mechanical or physical means into products of enhanced economic value relative to that of the feedstock utilized, such that after allowing for costs, charges, and taxes incurred, some net profit accrues (Hamer and Egli, 1991).

In the case of microbially mediated processes, feedstocks include only the major substrates and nutrients required by the microbes. Heterotrophic microbes have an obligate requirement of organic compounds for their growth. Autotrophs on the other hand can use carbon dioxide as their carbon substrate for growth and either light (in the case of the photoautotrophs) or energy derived from the oxidation of reduced inorganic chemicals (in the case of the chemoautotrophs) as separate energy sources, respectively. Facultative autotrophs can additionally utilize organic compounds as combined carbon/energy substrates. These combined carbon/energy substrates can be water-soluble solids and water-miscible liquid organic compounds, or water-immiscible liquid and insoluble solid organic compounds (frequently encountered in industrial fermentation processes).

Most of the carbon/energy substrates commonly used for the growth of microbial cultures in the laboratory are either water-soluble or water-miscible compounds. They vary from complex, ill-defined protein hydrolysates used from cultivation of many microbes of medical and veterinary significance, to sugars, that are used for the growth of bacteria in defined media. Many traditional fermentation products are produced by the use of mixed microbial cultures which are often superior in their utilization of either impure or mixed substrates to microbial monocultures. Even so, virtually all modern biotechnological processes are based on pure monocultures or, in exceptional cases, the use of monocultures in sequence (Hamer and Egli. 1991).

In any commercial process, feedstock costs are important. The products of biotechnology can be divided into a number of categories based on their value: bulk products, which in general command relatively low prices, medium volume products, which command higher prices, and as fine chemicals or biologicals, which command very high prices. For fine chemicals or biologicals, either product purity or efficacy is critical; their market prices are very largely divorced from the costs of the primary production operation but are more closely related to downstream processing costs as well as quality testing costs.

Both bulk and medium volume products are produced commercially from cheap feedstocks. For products, medium-cost feedstocks are widely utilized for production processes.

Diagram 1.14. Ratios between production of alcohol from biomass energy and agricultural self-sufficiency, for some countries, as of 1975. (Based on the data from World Ban k, 1980).

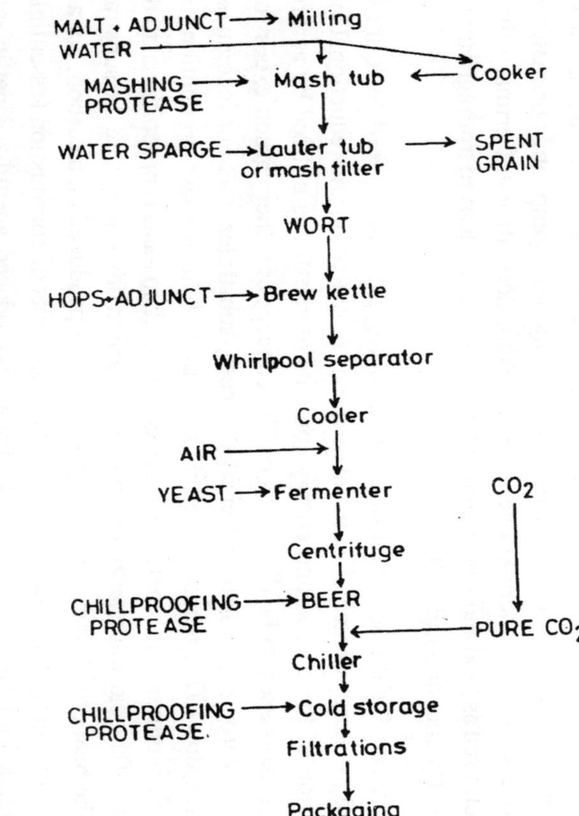

Diagram 1.15. Summary of the brewing process (after Ward, 1985).

In the case of production processes mediated by heterotrophic microbes, carbon feedstocks also act as energy sources. These feedstocks include compounds or materials that have an alternative use for thermal energy production, derivatives of primary fuels, agricultural products used as food and feed ingredients and byproducts derived from agricultural product processing (Hamer and Egli, 1991).

Two major criteria for the selection of a biotechnological process feedstock are its carbon content and its energy content, i.e., its relative state of oxidation and reduction. Until recently the question of the carbon and energy contents of substrates utilized by aerobic heterotrophic organisms used to be disregarded. The concept of either carbon excess but energy deficient, or carbon deficient and energy excess, substrates was not recognized. Linton and Stephenson (1978) demonstrated that, for microbial biomass production, the maximum attainable biomass yield coefficient on a carbon-carbon basis is proportional to the heat of combustion of the carbon/energy substrate on a carbon basis, provided the substrate does not exceed the heat of combustion of the dry microbes produced on a carbon, but ash free, basis. All carbon/energy substrates that show higher heats of combustion (essentially higher degrees of reduction) than does the microbial biomass produced, are utilized inefficiently by microbes.

Virtually any carbonaceous compound that is biodegraded by heterotrophic microbes can be considered as a potential process feedstock. However, this does not automatically make the compound an attractive process feedstock. As far as commercial processes are concerned, it is the rate at which the feedstock is utilized and the completeness of the overall process that are critical.

Feedstocks for Water Soluble Carbon-energy

Any monosaccharide (e.g., glucose, fructose) that finds application as the carbon/energy feedstock for either a medium or bulk volume microbially-mediated production process will be derived from either an oligosaccharide (sucrose, maltose or lactose) or a polysaccharide (starch, glycogen, or cellulose); with the exception of D-glucose, D-fructose, and possibly D-xylose, they are rarely available in either sufficient quantities or of appropriate purity to be considered as essentially single compound process feedstocks that can stand alone for evaluation. The most widely utilized fermentation process feedstocks are either an oligosaccharide such as sucrose, in an essentially pure form, or fluid byproduct streams from oligosaccharide production, specifically molasses (sucrose) and whey (lactose).

In many countries with extensive dairy industries, whey is an important oligosaccharide containing feedstock. It is a byproduct of cheese manufacture and, in the form in which it is produced, is relatively dilute for use as a feedstock. Production is either widely distributed or centralized.

Certain alcohols are both fermentation process products and fermentation process feedstocks. Primary substrates as process feedstocks are methanol, ethanol, and glycerol.

All three alcohols occur naturally. Traditionally, methanol is produced by the destructive distillation of wood, ethanol by the fermentation of carbohydrates, and glycerol by the hydrolysis of fats. Today methanol is usually produced on a large scale from either natural gas or a liquid hydrocarbon fraction such as naphtha, although production from coal is equally feasible. Ethanol and glycerol are produced in significant quantities by the traditional technologies mentioned above. As far as its use as a fermentation process feedstock is concerned, fermentation ethanol is clearly a non-starter because the carbohydrate feedstock used for its production would inevitably substitute for it. However, where ethanol is manufactured via a petrochemical route from ethylene, its potential as a fermentation process feedstock depends on the economics of its production. For many years petrochemical ethanol was cheaper to produce than the fermentation alcohol, but today respective prices are clearly controlled by the complex politics and strategic arguments surrounding agricultural commodities and oil (energy) (Hamer and Egli, 1991). Glycerol is also produced on large scale by petrochemical process routes from propylene. An alternative biological route for glycerol production involves production by osmophilic species of unicellular green microalgae of the genus *Dunaliella* which are able to synthesize very high intracellular glycerol concentrations when growing photoautotrophically under conditions of high salinity. If glycerol is to become a major fermentation process feedstock in the future, its most likely source will nevertheless be via a petrochemical, rather than a biological, route unless strategic policies invalidate conventional economic principles. From the technical viewpoint, glycerol represents an ideal substrate for microbial biomass production because its carbon and energy contents are essentially balanced.

The primary application of methanol as a feedstock has so, far been for the production of SCP.

Provided that crude oil prices remain low and more closely associated with production costs, ethanol produced by petrochemical routes is also a potential major feedstock for fermentation processes in the future.

The shorter chain length carboxylic acids are also both microbial substrates and microbial products. Carboxylic acids can be regarded as either carboxylic derivatives of hydrocarbons or derivatives of water in which a hydrogen atom has been replaced by the $C_nH,_{w+|}CO$- radical. With increasing chain length, the odour of short chain carboxylic acids becomes increasingly obnoxious, but in spite of this, acetic, propionic, butyric, and valeric acids can all be considered to be either actual or potential fermentation process feedstocks. Formic acid, the first member of the carboxylic acid series, is not a potential process feedstock in view of its being one of the most energy-deficient carbon substrates.

The carboxylic acids are, of course, natural products with acetic acid being the most widely available for potential utilization as a feedstock. It was traditionally produced either by the microbial-mediated oxidation of ethanol or by recovery from pyroligneous acid produced by the distillation of wood.

Gaseous Carbon as Energy Feedstocks

Gaseous alkanes are hydrocarbons with the general formula C_nH_{2n+2} and differ from each other by CH_2 or multiple thereof. As potential biotechnological process feedstocks, the group comprises only 5 compounds, methane, ethane, propane, n-butane, and iso-butane; of which methane has been widely studied and iso-butane has been very largely neglected with respect to their potentials as carbon energy substrates for microbes.

These feedstocks, in their commercially available forms, are likely to contain significant concentrations of impurities. Methane and other gaseous alkanes are much less preferred feedstocks than is the methanol which is manufactured directly from them by chemical means.

When hydrogen and carbon dioxide mixtures are used as substrates for microbial growth, the energy for growth is derived from oxidation of the hydrogen while the carbon for cell and product formation is provided by the carbon dioxide; if the ratio of hydrogen : carbon dioxide supplied can be varied at will, an optimum balance with respect to carbon and energy requirements is possible.

The use of microbial processes based on the utilization of hydrogen/carbon dioxide mixtures by "Knall gas bacteria" was proposed by Schlegel (1969) when he introduced the concept of food production from electricity via the electrolysis of water. During the decade that followed the autotrophic route for single cell protein production from hydrogen and carbon dioxide was further researched. However, it was not single cell protein but rather polyhydroxybutyrate that was to prove to be the commercial product of greatest potential that could be produced from hydrogen by Knall gas bacteria. Polyhydroxybutyrate can be produced from a range of substrates, including lower alcohols, lower carboxylic acids, sucrose, dextrose, and hydrogen, and hydrogen appears to be the least convenient. But judged on the basis of feedstock cost per tonne of product, hydrogen might turn out to be the cheapest feedstock if it were available as a pipeline distributed fuel; if not, methanol would be the most cost-realistic.

Immiscible Liquid Carbon as Energy Feedstocks

During the second half of the 20th century, mineral oil has been used as the major feedstock for chemicals production. For the present and near future, few economically-realistic alternatives for its replacement exist. Mineral oil is, of course, a highly complex mixture of a broad spectrum of hydrocarbons; before refining it is an unrealistic microbial process feedstock.

According to Miall (1980), provided they are available at competitive prices, n-alkanes can be used as substitute feedstocks for a wide range of conventional fermentation products; the only difficulty lies in the need for process cultures to produce appropriate intermediates that can be incorporated into the tricarboxylic acid cycle.

In view of their immiscibility with water, n-alkanes have to be emulsified before their effective uptake by microbes. Most hydrocarbon-utilizing microbes in fact produce such emulsifying agents thereby facilitating utilization. Nevertheless, with n-alkanes there is a risk of feedstock residues being associated with products, particularly if the product is either biomass or a product derived from it.

In most proposed fermentation processes involving *n*-alkanes the substrate is present as the dispersed phase, with water as the continuous phase. The phases might conceivably be reversed, allowing markedly higher saturation concentrations of dissolved oxygen in the continuous phase and hence enhanced driving forces for oxygen transfer. In addition, heat transfer for effective cooling might also be facilitated but product separation and purification could become markedly more difficult and expensive (Hamer and Egli, 1991).

The best known fats and oils of animal origin are tallow from cattle and sheep, the traditional raw material for soap and candle manufacture, lard from swine, butter, whale oil and a range of fish oils. Their counterpart vegetable oils are derived from soybeans, palm, rapeseed, sunflower, olive, cottonseed, corn, and more recently, sun flower. Both animal and vegetable oils are widely used in foodstuffs and are frequently referred to as "edible oils".

Vegetable oil production has tended to increase in recent decades. Soybean oil is dominant among edible oils and like all commodity products is subject to major price fluctuations depending on the supply and demand. However, soybean oil is cushioned by the interlinked supply and demand position for soybean meal which is the major protein ingredient in many compounded animal feeds.

The acceptability of any particular oil for inclusion in human foods has, in the past decade, been very strongly influenced by its polyunsaturated nature, which are believed to reduce the risk of heart attacks. Such a hypothesis does favour the use of soybean, rapeseed, and sunflower oils in human food. It discriminates strongly against palm oil, decreasing its price and thereby making it an attractive fermentation process feedstock. Some other plant oils are also in a similar position.

Biotechnology and Third World Concerns

The farm sector is a major export-earning enterprise for developing countries. There is genuine concern about the potential adverse impact of genetic engineering research directed at finding substitutes for natural products. Some examples are: high fructose corn sweetener as a substitute for sugarcane sugar and an extract from *Dioscorea* species substitutes for vanilla, cocoa, and diosgenin. Another concern is the safety aspects of genetic engineering research as tests be done in the Third World which are not permitted in the industrialized and developed countries?

Will the biotechnology revolution help resource-poor farmers to increase the productivity with the help of farm grown inputs? How can we design mutually reinforcing packages of technology, services, and public policies which can ensure that rural people can derive economic and social benefit from new biotechnologies? Is it possible to design a pro-poor biotechnology programme?

What will be the impact of the extension of intellectual property rights to individual genes and genotypes on the availability of improved material to resource-poor farmers? Will intellectual property rights be exclusively reserved for rewarding formal innovation, even though the informal innovation system has played and is playing a key role in the identification and conservation of plant and animal genetic resources? What are the rights of the farm families who have conserved and selected genetic diversity in contrast to the rights of the breeders who have used them to produce novel genetic combinations?

Will priorities in biotechnology research be solely market-driven or will they also take into consideration the larger interests and the long-term well-being of mankind, whether rich or poor?

In agriculture, while the "green revolution" technologies arising from research funded by 'Rockefeller' and 'Ford Foundations' and by governments of developing and industrialized countries were available to all farmers who could derive benefit from them. The "gene revolution" technologies associated with biotechnological research may not likewise be available, since they owe their origin mostly to investments made by private companies and may be protected by patent rights. Where should the line be drawn between private profit and public good, in a world characterized by glaring economic inequities?

The potential risks of unintentional releases of genetically-modified organisms, and the lack of predictable behaviour of these in the sown environment are the cause of some concern. This concern is much greater with deliberate releases.

In the sector of pharmaceuticals, well-known bacteria or lower eucaryotic organisms produce useful proteins after proper genetic manipulation. The same technology can sometimes be used for organisms designed for environmental purposes as for example agricultural biopesticides or detoxifying waste "cleaners". There is, however, a clear difference between the industrial and the environmental application of gene technology. The former is carried out with weak, non-competitive organisms under physically contained conditions, whereas the latter uses strong, competitive organisms growing under non-contained and uncontrolled conditions.

Extrapolations from laboratory experiments to natural conditions have to be made with caution, because the regulatory diversity of the organisms and the number of possible behavioural patterns are so great that the organisms react unpredictably when introduced into the complex, natural environments. This leaves some lacunae in regard to predictability of microbial activities. Molin and Kjelleberg (1993) have discussed design of microorganisms for release to the environment on the basis of two major arguments:

1. One hurdle in designing efficient strains for specific tasks outside the laboratory is the lack of knowledge and experience concerning the bacteria to be used. This lack of specific knowledge poses the same problems for the strain designers as it does for the risk assessors. The build-up of general knowledge concerning bacterial life and activity

in the environment is equally important for the commercial organism constructors and the regulatory authorities and for risk assessment.

2. The use of many different unknown bacterial strains in the extremely complex natural environment makes it impossible to deal with risk problems in the same way as we do with respect to the contained use in industry. In industry, the combination of personnel training, physical containment, and the use of weak strains has removed much of the public fear of this type of biotechnology. None of these factors is relevant when organisms are deliberately placed in the environment and left to themselves (Molin and Kjelleberg, 1993). According to these authors, "as long as we know so little about microbial life and activities in the environment, and about the interaction between these and the other ecological participants, we should be concerned about the environmental concentrations of the released microorganisms". The environment may be expected to cope up with fairly low concentrations of our engineered microorganism, and therefore these concentrations over time should be reduced to the lowest possible levels.

The aim of biological containment is to increase predictability about the behaviour or fate of a released microbe. The concept of biological containment in this context comprises either the use of crippled, non-competitive strains (a passive containment approach) or the introduction of specific functions (suicide) (Contreras *et al.*, 1991), which under particular conditions eliminate the engineered organisms (active containment). Further there is much concern about the transfer of genetic material from the engineered organisms to the natural populations; consequently, biological containment also deals with reduction of the transfer capacity of the manipulated strains. The overall purpose of biological containment is to limit the capacity of the organism to invade the environment and establish permanently itself (Molin and Kjelleberg, 1993).

Involving suicide functions in the design of biological containment systems covers the following aspects: (1) the killing efficiency must be very high, (2) the lethal genes should be of bacterial origin, (3) the killing proteins should be active in a broad spectrum of host bacteria, and (4) resistant mutants should be rare or absent. The most important problem can be leakiness, i.e., elimination should be 100% effective.

In most cases, the released organisms will be non-pathogenic, and probably without immediate ecological effects. The main concern, therefore, is the large numbers of cells which are simultaneously introduced locally; they may have unpredictable long-term effects, possibly involving horizontal transfer of genetic material to the indigenous microflora.

Socio-Economic Impact

It is possible to impart pro-poor biomass technology development and dissemination. For example, agricultural biomass is the most important feedstock available to poor countries. In rural areas, biomass refineries can help to get value-added product from such biomass.

Biotechnology in Relation to Poverty and Employment

The goals of social and economic development are not conflicting goals; they must not be counterpoised to one another but must be integrated into a single process of sustainable development. The process of development is more than merely the growth of incomes. The goals of development are unattainable if the environment in which human beings live is polluted or degraded. Anthropogenic pressures on natural resources have increased considerably, particularly in developing countries. The poor are the worst victims of the environmental crisis. For the poor environmental degradation often entails the loss of occupations, of sources of fuel, and fodder and of access to hitherto common property resources. Working people, street dwellers, and other sections of the poor are the worst victims of industrial pollution and disaster. They are worst affected by the pollution of water sources and by unhealthy living and work environments. Hence there is a need to reorient development strategies in a manner that the quality of human life improve while living within the carrying capacity of supporting ecosystems. To improve the life of the poor people, it is necessary to:

1. arrest and reverse the damage to land and water resources and to forests,
2. conserve and develop biodiversity and balanced ecosystems,
3. maintain watersheds and hydraulic cycles,
4. prevent soil erosion and degradation,
5. reduce the silt loads of tanks and rivers,
6. recycle garbage and sewage for energy generation and composting,
7. promote environmental hygiene, and
8. develop safe drinking water resources.

It may perhaps be appropriate to shift food security considerations solely from the global and national angles to the level of individual households and to link the livelihood security of rural and urban communities with the ecological security of nations. Biotechnology can play an important role in the poverty-elimination programmes, through tissue culture and micropropagation techniques, animal health care, propagation of elite forest tree species, aquaculture, and establishment of biomass refineries.

Biodiversity and Biofuture

There is considerable controversy on methods of saving and sharing the global biological wealth. The Keystone International Dialogue Series on Plant Genetic Resources has tried to throw light on methods of resolving opposing viewpoints. Terms such as "Farmers' Rights" and "Breeders' Rights" are freely used to indicate the importance of according recognition to the informal innovation system in conjunction with the rights already accorded to plant breeders in the 20 developed nations which have so far adhered to the rules of the International

Union for the Protection of New Varieties of Plants (UPOV). The ongoing discussions at the General Agreement on Tariffs and Trade (GATT) on Trade-Related Intellectual Property Rights (TRIPs) are also important in the context of North-South relationships in germplasm conservation and exchange. Fourteen developing nations have proposed to the Negotiating Group on TRIPs at the Uruguay Round of Multilateral Trade Negotiations that plant or animal varieties or essentially biological processes for the production of plants or animals should not be subjected to patent protection.

The polyunsaturated/saturated argument appears to constitute yet another false trade barrier against the South. Palm oil is a product from developing countries, particularly Malaysia and Indonesia, whereas soybean, rapeseed, and sunflower oils are products of the North. In the specific case of soybean, it should be remembered that its cultivation was introduced on a significant scale into the U.S. only in 1930. Since then, the Soybean Producers Association has made soybean a major crop.

Stowell (1987) has reviewed the potential of animal and vegetable oils as carbon-energy feedstocks for antibiotic fermentations. The potential of edible oils as fermentation process feedstocks generally exceeds that of mineral oil.

Insoluble Solid Carbon as Energy Feedstocks

Lignocellulosic waste material is an interesting carbon-energy feedstock for biotechnological processes. Both lignin and cellulose have proved challenging carbon/energy substrates in laboratory studies.

Lignocellulose is an abundant renewable resource. Although its composition varies with source, lignin, cellulose, and hemicellulose are always present as major components.

The main lignocellulosic products are sawn timber, wood chips for pulping, and wood for fuel. For biotechnological processing the most widely available feedstocks in this category are either the byproducts and waste products of sawn timber and wood chips or those of agricultural origin such as cereal straw and bagasse. At the present time, very large quantities of lignocellulosic wastes are subjected either to natural decay or to wasteful combustion. However, certain processes can use up lignocellulosic wastes: these include sensible combustion pulping and reconstitution. Biotechnologies for the utilization of lignocellulosic wastes include composting, methane generation in anaerobic bioreactors or landfills, and occasional utilization as ruminant feeds or as feedstocks for SCP production. Saccharification involves the production of fermentable sugars from lignocellulosic materials to yield either glucose syrup for direct consumption as a food-grade product or as a fermentable feedstock stream for bulk chemicals or fuel ethanol production (Hamer and Egli, 1991).

The exploitation of lignocellulosic wastes as an economically attractive feedstock for the biotechnological process industries depends upon the effective hydrolysis of the original waste material. Until this is achieved, lignocellulosic materials will remain potential rather than actual biotechnological process feedstocks.

Microbial biomass is commonly regarded as a product rather than a feedstock. However, very large quantities of waste microbial biomass are produced by the fermentation industry, and from municipal sewage and industrial wastewater treatment.

Waste microbial biomass can be considered as a potential process feed both for processes where other microbes grow on the waste biomass and for those involving the separation of components present in the waste biomass. In biotechnological processes involving whole microbial cells as the process feedstock it is important to differentiate the process microbes which grow, but also suffer death or lysis, and the non-growing feedstock (substrate) microbes which are subject only to degradative processes. Unlike lignocellulosic materials, which are either continuous or porous solids, microbes as process feedstocks comprise a complex essentially fluid cytoplasm surrounded by a rigid cell wall and membrane which maintains the physical and chemical integrity of the cell. When microbial cells are biodegraded the cell wall/membrane is punctured or burst and both readily soluble matter and cell wall fragments are released.

Starch is the most important carbohydrate reserve material present in plants. As a traditional fermentation process feedstock, starch in the form of barley is used for malting and brewing beer.

Some primary sources of starch are barley, wheat, maize, potatoes, rice, sorghum, and cassava. Cassava is one of the crops considered to be of primary interest in the production of "food crops" specifically for industrial purposes and after hydrolysis it clearly has potential as a feedstock for fuel ethanol production.

Although most aerobic processes do not incur raw material costs by using air as their oxygen source, substantial capital and operating costs do accrue to processes because of the need to both compress and sterilize the air prior to use. In commercial processes it is most important to minimize such costs by achieving maximum conversion of the oxygen supplied. According to Hamer and Egli (1991): "The fractional conversions achieved for either oxygen or gaseous carbon and/or energy substrates only slightly soluble in water (methane, carbon monoxide, hydrogen, etc.) rarely exceed 0.3 in high intensity laboratory bioreactors while in conventional large industrial-scale bioreactors fractional conversions for oxygen are typically about 0.15 when air is used as the source of oxygen. Even with such low fractional conversions, gaseous substrates or nutrients frequently become limiting with respect to process rate and whereas this might be technologically desirable in some cases for carbon and/or energy substrates, oxygen limitation in aerobic processes can seriously affect overall process performance. A further unfortunate feature is that high fractional conversions of gaseous substrates and nutrients are frequently inconsistent with the maximization of respective transfer rates and clearly a techno-economic optimum must be sought to ameliorate the conflict between transfer rate and conversion. The operating mode of the bioreactor will also markedly affect gaseous substrate or nutrient conversion and the superiority of systems where the gas phase can be maintained in essentially plug flow should be noted in this respect. One of the most frequently proposed means of enhancing fractional conversions in bioreactors is the introduction of recycle, an approach of equal applicability to either the gaseous or the liquid (aqueous) phases in biotechnological processes, but one that is not without numerous ramifications."

Biotechnology, in Proper Management of Environment

Developments in biotechnology can make significant contributions to world health, food supplies, and environmental protection.

World population is projected to double from 5-10 billion by 2050, with 9 out of 10 people living in developing countries. Advances in biotechnology can help to meet present and future needs for:

1. Food, by increasing yields and nutritional value of plants, microbes, animals, and fish. Agricultural output can be improved and made healthier by biological pest controls.
2. Fuel, by increasing yields of biomass.
3. Health, by improving nutrition and developing vaccines for various diseases. Safe, reversible and long-lasting fertility-regulating mechanisms are needed.
4. Safe water, by detecting hazards such as parasites and pathogens.
5. Environmental protection and recovery, by improving yields and providing materials for reforestation and recovery of degraded land. It can also help preserve biological diversity.

Focus is required on increasing food productivity, improving health, enhancing environmental protection, enhancing safety and international information exchange, and establishing methods for applying environmentally-sound biotechnology.

Biotechnology, A Threat to Third World

There is some apprehension that current biotechnology developments may threaten rather than support sustainable rural development in the Third World. Market forces direct the agricultural research programmes of private companies or public research institutes with emphasis on industrial applications and capital intensive, well-controlled large-scale farming systems and the substitution of agricultural raw products. The growing influence of biotechnology in plant breeding is resulting in the introduction of procedures often requiring high levels of investment and which are directed at developing new, patentable varieties of crops. These patents would enable plant breeders in developed countries to obtain property rights to genetic resources which have become available as a result of the efforts of local communities in developing countries who have maintained and developed local varieties of crops for centuries.

A recent international debate organized by the Vrije University, Amsterdam, and the Netherlands Organization for International Development and Cooperation at the University of Amsterdam took the assumption that current biotechnological research does have the potential to yield interesting and beneficial results as a starting point for the debate. The debate focussed on (1) how to protect small-scale farmers in developing countries by strengthening their legal position with regard to the genetic resources they have maintained and conserved; and (2) how to develop biotechnological innovations, appropriate for small-scale farming systems. These are two areas where the science planners in India should also be putting some thought.

Biotechnology and Biodiversity

Modern biotechnologies offer great potential for improving the quality and increasing the productivity of agriculture, forestry, and fisheries. Genes from organisms that flourish in the forests, fields, and seas of the developing world are the strategic raw materials for the commercial development of new pharmaceutical, agricultural, and industrial products. Whereas genetic wealth, especially in tropical areas such as rainforests, was once a relatively inaccessible trust fund, it is now becoming a highly valuable currency. One may compare the introduction of modern biotechnologies in the developing world to the Green Revolution. While the latter Revolution involved introducing new varieties of primarily wheat and rice in selected areas, biotechnology can potentially affect all crops and tree species, as well as fish and livestock, in any corner of the globe. The Green Revolution was introduced to the Third World largely by international institutions, but the "Gene Revolution" is primarily in the hands of the private sector, with transnational corporations being the leading players. As scientific and technical capacity in the biosciences is mostly centred in the industrialized world, biotechnology research does not focus on the needs or interests of poor farmers in marginal areas of the world. Emerging biotechnologies have considerable potential to enhance food and agricultural production in the developing world, but they could also add to existing inequities by displacing traditional agricultural products, accelerating genetic erosion and introducing new environmental hazards (FAO, 1993).

Molecular biology is one of the most powerful tools of biotechnology. In the area of genetic engineering, scientists can transfer genes between unrelated species endowing such "transgenic" organisms with properties that they could probably never have acquired in nature. As yet, only a handful of genetically-engineered products are available commercially, but hundreds are in the pipeline. Genetic engineers can design crop varieties containing natural insecticidal genes, fish with human growth hormones, and faster growing trees. However, genetic engineering consists essentially of mixing and matching genes from different species. It cannot create new genetic material, replace lost material, or eliminate the need to conserve living resources.

Molecular biology is important in characterizing and conserving biodiversity. For example, molecular markers can help establish the extent of diversity within a species and to identify genes of interest to breeders. Such techniques can also help establish priorities for conservation.

Biotechnology already assists the conservation of plant and animal genetic resources through: new methods for collecting and storing genes (as seed and tissue culture); detection and elimination of diseases in gene bank collections; identification of useful genes; improved techniques for long-term storage; and safer and more efficient distribution of germplasm to users.

Tissue culture technique, which involves growing small pieces of plant tissue or individual cells in culture, is a quick, good way of taking many cuttings from a single plant. Entire plant may sometimes be regenerated from a single totipotent cell. After selecting a disease-free cutting, for example, one can mass-produce genetically identical copies. This is plant cloning, or micropropagation of plant.

In gene banks, tissue culture is now used routinely to preserve the genetic information of plants which have seeds that do not store well, are sterile or have poor germination rates. Plant cells maintained on a growth medium in a test-tube replace seeds or plants. Plants stored in this way include sweet potatoes, bananas, plantains, apples, cocoa, and many tropical fruits.

In certain areas, modern biotechnology may hinder development or create serious hardship for rural communities. The economies of developing countries are threatened by biotechnology research that promises to eliminate or displace traditional export commodities, often a primary source of foreign exchange earning. Biosynthesis in the laboratory of high-value ingredients such as vanilla, pyrethrum, and rubber could ultimately transfer production out of farmers fields and into industrial bioreactors, wreaking havoc on already weak economies. Further, biotechnology may threaten the genetic diversity on which it depends. In the absence of conservation, commercial biotechnology can unleash a new era of genetic erosion. Commercial semen and embryo transfer services for domestic animals have generated concern about the displacement of traditional livestock breeds. Cloning could accelerate replacement or dilution of indigenous stock by imported breeds, leading to a loss of genetic diversity.

A related concern involves the ecological risks of introducing genetically-engineered plants into centres of diversity. Transgenic varieties, some of them resistant to herbicides, have been produced in more than 40 crop plants. Gene flow to weeds from resistant plants could have far-reaching consequences. The resulting herbicide-tolerant weeds might prove difficult to control, harming the surrounding ecosystem.

Biotechnologists could develop new varieties and breeds adapted to low-input agriculture or harsh conditions, or improve processing. Biotechnology may help to create markets by developing new industrial, medicinal, and aromatic crops. Given their richness in biodiversity, several developing countries that have the capabilities, such as China, India, and Brazil, could produce new high-value products based on local flora. The congenial agroecological settings and availability of relatively cheap labour should be conducive to large-scale production of new high-value crops, enabling such countries to maintain their comparative advantage in these commodities.

Biodiversity and Agriculture

Biodiversity provides the raw materials and gene combinations that produce the plant varieties and animal breeds upon which agriculture depends. Thousands of different and genetically unique varieties of crops and animal breeds owe their existence to 3000 million years of natural biological evolution and to careful selection and nurturing by our farming and herding ancestors over centuries of agricultural practice..The genetic resources of plants and animals are a valuable global asset to humankind. As genetic diversity erodes, our capacity to maintain and enhance crop, forest, and livestock productivity decreases along with the ability to respond to changing conditions. Genetic resources hold the key to increasing food security and improving the human condition.

Crop Plants

The plant genetic diversity used in agriculture is being lost at an alarming rate. Just nine crops (wheat, rice, maize, barley, sorghum/millet, potato, sweet potato/yam, sugarcane, and soybean) make up over 75% of the plant kingdom's contribution to human dietary energy. Although none of the staple crops is likely to disappear, they, too, are threatened—not by the loss of any single crop species but by the loss of diversity within species.

All major food crops, the staple crops grown and consumed by the vast majority of the world's population, have had their origins in the tropics and subtropics of Asia, Africa, and Latin America. Wheat and barley originated in the Near East, for example. Soybeans and rice came from China. Sorghum, yams, and coffee came from Africa. Potatoes and tomatoes originated in the Andes of South America, and maize in South and Central America.

Crop genetic diversity is still concentrated mainly in regions (Diagram 1.16) known as "centres of diversity", and located in the developing world. Farmers in these areas, who still practise traditional agriculture, cultivate local varieties known as "land races" that have been selected over many generations. Closely related species that survive in the wild are known as "wild relatives" of crops. Together, land races and their wild relatives are the richest repositories of crop genetic diversity (FAO, 1993).

Thousands of genetically distinct varieties of major food crops owe their existence to organic evolution and to careful selection and nurturing by our farmer ancestors over the centuries. This diversity protects the crop and helps it to meet the demands of different environments and human needs. Potatoes, for instance, originated in the Andes (Diagram 1.16) but nowadays they can be found growing below sea level behind Dutch dykes or high in the Himalayan mountains. One variety of rice survives on just 60 centimetres of annual rainfall while another grows floating in 7.5 metres of water.

Since the beginning of this century, about 75% of the genetic diversity of agricultural crops has been lost. We increasingly depend on fewer and fewer crop varieties and a rapidly diminishing gene pool. The primary reason is that commercial, uniform varieties are replacing

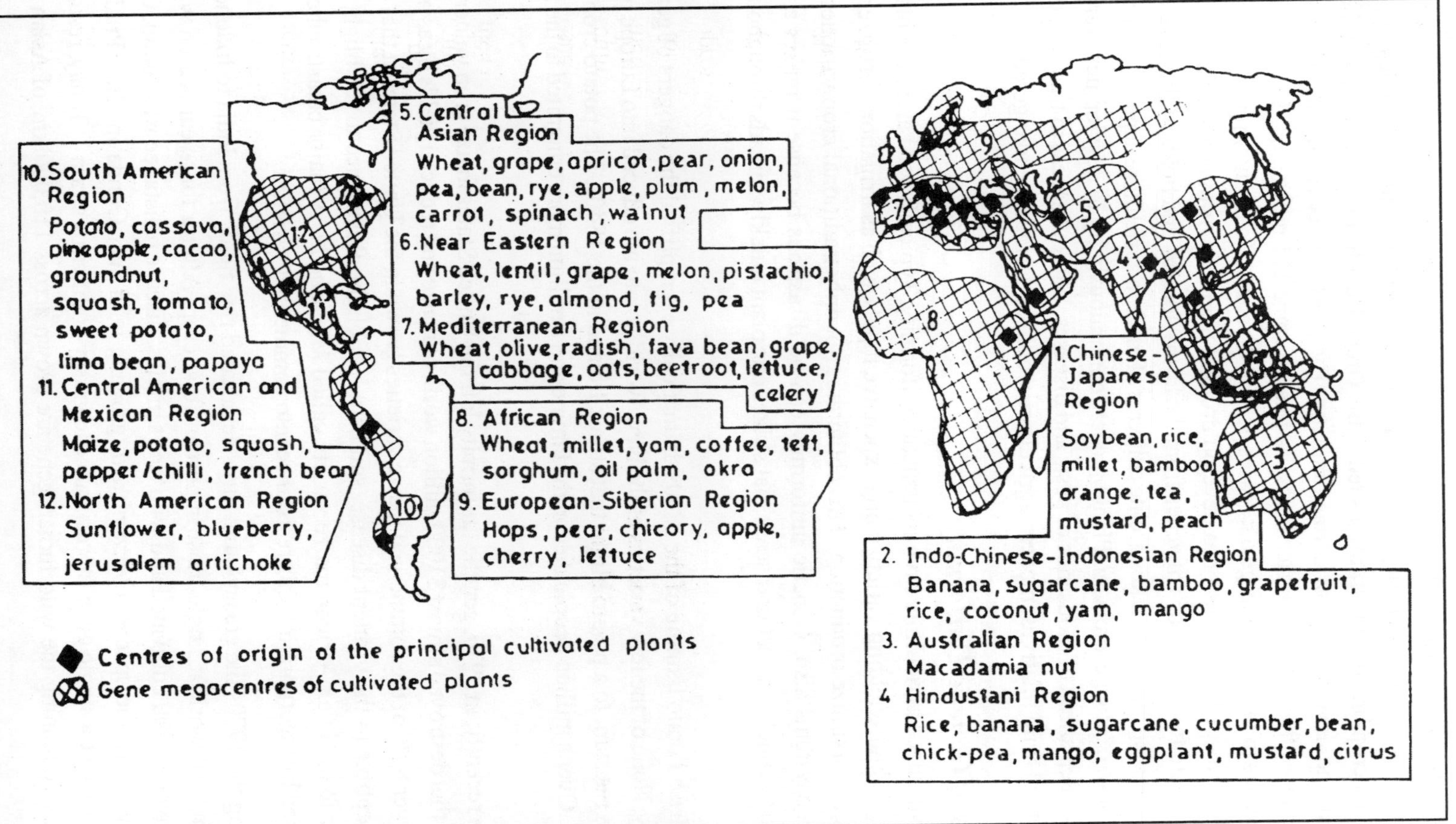

Diagram 1.16. The 12 megacentres of cultivated plants. Panels show selected food crops (source: FAO, 1993).

traditional ones. When farmers abandon native land races to plant new varieties, the traditional ones die out. The introduction, beginning in the 1950s, of high-yielding grains developed by international crop breeding institutions led to the Green Revolution. The spread of the new varieties in the developing world was dramatic. By 1990 they covered half of all wheat lands, and more than half of all rice lands—a total of some 115 million ha. This resulted in large increases in yields, but large decreases in crop diversity. To maintain pest and disease resistance in major food crops or to develop drought tolerance or improved flavour, plant breeders require fresh infusions of genes from the farms, forests, and fields of the developing world. Developing the high-yielding, elite cultivars of modern agriculture depends on a steady stream of new, exotic germplasm. Plant breeders continuously try to develop new varieties to keep one step ahead of thousands of pests and diseases. Without access to traditional land races and their wild relatives, modern agriculture would be seriously endangered.

Dangers of Genetic Uniformity

Industrialized agriculture favours genetic uniformity. Typically, vast areas are planted with a single, high-yielding variety using expensive inputs such as irrigation, fertilizer, and pesticides to maximize production. This obliterates not only traditional crop varieties, but farming ecosystems also. Genetic uniformity can be disastrous because it makes a crop vulnerable to attack—a pest or disease that strikes one plant quickly spreads throughout the crop.

The Irish Potato Famine of the 1840s dramatically exemplified the dangers of genetic uniformity. None of the few varieties of the New World potato introduced into Europe in the 1500s was resistant to a potato blight that struck Ireland in the 1840s. The potato crop was wiped out. Over a million people died in the famine and a million more emigrated to the New World.

More recently, in 1970, genetic uniformity left the United States maize crop vulnerable to a blight that destroyed almost $1000 million worth of maize and reduced yields by as much as 50%. Over 80% of the commercial maize varieties grown in the United States at that time were susceptible to the virulent disease, southern leaf blight. Resistance to the blight was eventually found in an African maize variety called *Mayorbella.* A major catastrophe has been averted by incorporating this resistance into commercial varieties.

During the 1970s the grassy-stunt virus devastated rice fields from India to Indonesia, endangering the world's single most important food crop. After a four-year search which screened over 17,000 cultivated and wild rice samples, disease resistance was found. Only one population of the species *Oryza nivara,* growing wild near Gonda in Uttar Pradesh, was found to have a single gene for resistance to grassy-stunt virus strain 1. Today, resistant rice hybrids containing the wild Indian gene are grown across 110,000 km^2 of Asian rice fields.

For feeding an increasing world population, the genetic resources of wild relatives have to be tapped. Modern plant breeding as well as new biotechnologies offer the potential to exploit little-known plant species as sources of food, and to enhance the qualities of those plants that are underutilized—especially traditional plants of special significance to poor people such as local grains, legumes, oilseeds, fruits, and vegetables.

Traditional food crops are usually drought resistant, can be grown without expensive inputs and have good storage qualities. For developing countries, self-reliance in food production will depend on low-input agriculture in poor production environments. The capacity to grow varieties, particularly those resistant to pests and diseases and adapted to marginal lands, is therefore essential for sustainable agriculture and food security.

Who Owns Biodiversity?

International cooperation with respect to biodiversity has been complicated by the efforts of some industrialized countries to extend intellectual property rights to genes, plants, animals, and other living organisms, which inevitably leads to restrictions on access to genetic resources. With the advent of genetic engineering, for example, the biotechnology industry has promoted the extension of industrial patenting regimes to living organisms—an approach popularly known as "life patenting".

Proponents of patenting argue that it stimulates innovation by rewarding patent holders and enables companies to recoup their research investment. In the 1980s, precedents have been established for extending the concept through "life patenting". As a result, genes, plants, animals, and microorganisms—whether discovered in nature or manipulated by genetic engineers—could be made the intellectual property of private interests.

The patenting of useful genes found in nature is particularly controversial. For farmers and consumers in the developing world it could mean paying royalties on products that are based on their own biological resources and knowledge. Under patent law, a farmer breeding a patented animal and selling its offspring without payment of royalties would be contravening the law. Similarly, it would be illegal for farmers to save seed from a patented variety for replanting. The danger is that claiming intellectual property rights, without reciprocal benefits and compensation for developing nations, could set up formidable barriers preventing access to genetic resources. In the wake of new intellectual property proposals, developing nations are questioning the concepts of free access and heritage of humankind. They may react by restricting access to germplasm on their territories. Clearly, present proposals could have grave implications for future economic development and world food security.

Domesticated Animals

Animal genetic resources include all species, breeds, and strains that are of economic, scientific, and cultural interest to mankind for agriculture. Common relevant species include sheep, goats, cattle, horses, pigs, buffaloes, chickens, camels, donkeys, elephants,s reindeer, rabbits, and rodents are also important to different cultures and regions of the world.

Animal domestication began some 10 centuries ago when people began selecting animals for food, fibre, draught, and other agricultural uses. Livestock provide valuable products, such as hides, wool, and manure, that are important both for subsistence and as sources of income for rural communities. Livestock process forage and crop waste, inedible to humans, into nutritionally important food products.

Approximately 40% of the total land available in developing countries can only be used for some form of forage production. An estimated 12% of the world's population lives in areas where people depend almost entirely on products obtained from ruminant livestock—cattle, sheep, and goats. There are now thousands of genetically diverse breeds of domestic animals adapted to a wide variety of environmental conditions and human needs such as resistance to parasites or disease and adaptation to humidity, drought or extremes of heat and cold.

Animals account for about 20% of the world's food basket directly, and they also provide draught power and fertilizer for crop production. Livestock also serves as an important form of cash reserves in many of the mixed farming systems.

In Europe, half of the breeds that existed at the beginning of the century have become extinct; a third of the remaining 770 breeds are in danger of disappearing over the next 2 decades. Less is known about breeds in the developing world. Domestic animal diversity is greatest in the developing world. Asia is home to more than 140 breeds of pig, while North America can claim only 19.

Worldwide, the greatest threat to domestic animal diversity is the highly specialized nature of modern livestock production. In the developed world, commercial livestock farming is based on very few breeds that have been selected for the intensive production of meat, milk or eggs in highly controlled and regulated conditions. The spread of intensive production systems to the developing world places thousands of native breeds at risk.

After thousands of generations of controlled interbreeding, most domesticated animals no longer have wild relatives from whom germplasm can be obtained. When a variety becomes extinct, an already narrow genetic base shrinks irreversibly.

The genetic diversity now found in domestic animal breeds allows farmers to select stocks or develop new breeds in response to unpredictable changes in the environment, threats of disease, market conditions, and societal needs. Indigenous livestock breeds often possess valuable traits such as disease resistance, high fertility, good maternal qualities, longevity, and adaptation to harsh conditions and poor-quality feed, all desirable qualities for low-input, sustainable agriculture.

There is already less genetic diversity in farm animals than in crop plant species and over a third of the remaining animal genetic resources are now at risk (Diagram 1.17).

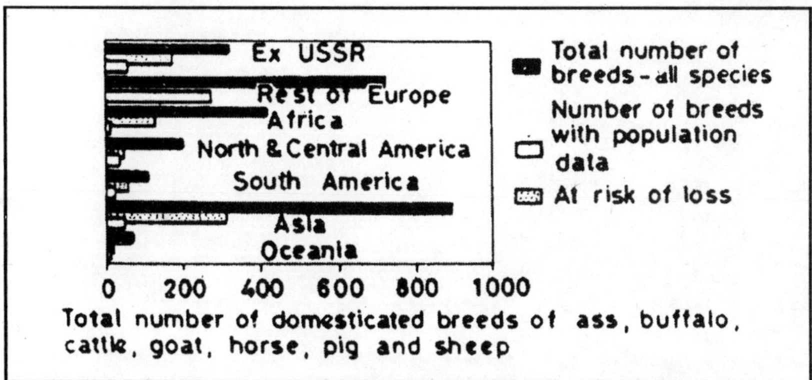

Diagram 1.17. Domestic animals at risk (source: FAO, 1993).

□□□

CHAPTER - 2

Essentials Concept of Biotecnology

Biotechnology is the fastest growing industry in the world. It could transform major industries, agriculture, livestock breeding and forestry. The science of biotechnology is as old as civilization. Production of liquors and vinegar, processing of tea , coffee, tobacco etc., can be traced to time immemorial. Vaccine production to control small pox, rabbies, and antibiotics, antidotes against snake venom and tetanus etc. are few examples which relieved humanity from untold sufferings.

ORIGIN

The story of the use of biological systems for the benefit of human beings perhaps started in 6000 B.C. when Sumerians and Babylonians made a liquor, called beer through fermentation. Greatest revolution till now commenced with 1970's and 1980's, when a product of interaction between the science of biology and technology called biotechnology came into wider existence. Fermentation, antibiotic production, baking and brewing are included under old biotechnology, whereas, techniques related with cell culture, fusion, bioprocessing, genetic engineering etc. were named *as* new biotechnology. Besides several educational institutions, many commercial companies also engaged themselves in biotechnology research for potential gains.

Biotechnology has a wide spectrum of application like tissue culture, microbial culture, determination of bio-chemical pathways and mechanisms, growth and enzyme kinetics and biomedical applications. Biotechnology can provide specific products like antibiotics and other health products, food, feeds, bio-fuels, beverages, industrial chemicals and enzymes to treat industrial wastes and pollutants. Production of bio-fertilizers, bio-pesticides and plant growth nutrients can increase forestry yield manifolds. In India, National Biotechnology board has chosen genetic engineering , photosynthesis, tissue culture, enzyme engineering, alcohol fermentation, immuno-technology as major areas of research and application. There are nearly sixty laboratories and Research Institutes, where plant tissue culture work is presumed with vigour.

Biotechnology is the application of biological organisms, prokaryotes, eukaryotic algae, glycophytes and halophytes systems or processes to manufacture and service industries, comprises a number of technologies based upon increasing understanding of biology at the cellular and molecular level. The technique includes recombinant DNA manipulations, monoclonal antibody preparation, tissue culture, protoplast fusion, protein engineering, immobilized enzyme, cell catalysis, sensing with the aid of biological molecules, etc. The term *biotechnology* gained several definitions from different group as follows:

Old & New Definitions of Biotechnology

There are many definitions of biotechnology but are *more confusing,* as its meaning differs in University and Polytechnic Laboratories, Government and private research institutes and large and small industries, each of which involved to some extent in research, development, testing, evaluation and production exercises. Clearly what goes on in industry, is better defined as Process Biotechnology as it enables exponents to convert raw materials to final products when either the raw material and/or a stage in the production process involves biological entities. Biotechnology in the broad sense, however, involves the cloners, the hybridisers and the molecular and cell biologists. It involves the conjoint interaction of two identifiable sub-components; BioScience and Bio Technology.

a. Biotechnology is also defined as *the applications of scientific and engineering principles to the processing of material by biological agents to provide goods and services.*

b. The Spinks Report (1980) defined biotechnology as *the application of biological organisms. systems or processes to the manufacturing and service industries.*

c. United States Congress's Office of Technology Assessment defined biotechnology as "any technique that used living organisms to make or modify a product, to improve plants or animals or to develop microorganisms for specific uses". The document focuses on the development and application of modern biotechnology based on new enabling techniques of recombinant-DNA technology, often referred to as genetic engineering.

d. British Biotechnologist: Application of biological organisms, system or processes to manufacturing and service industries'

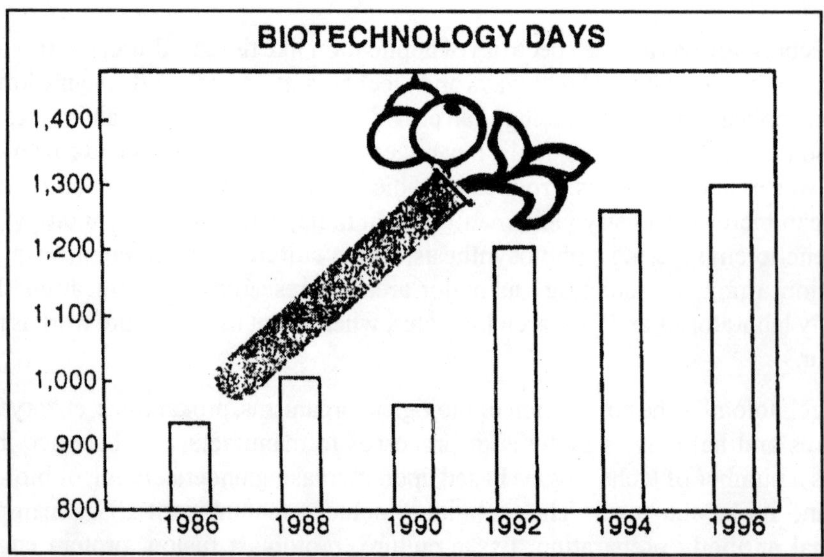

Diagram 2.1. Biotechnology Days. Despite the potential gains, the number of companies engaged worldwide in biotechnology research remains stagnant.

e. European Federation of Biotechnology: The integrated use of biochemistry, microbiology and engineering sciences in order to achieve technological (industrial) application of the capabilities of micro-organisms, cultured tissue cells and parts thereof

f. Japanese Biotechnologists: A technology using biological phenomena for copying and manufacturing various kinds of useful substances

g. US National Science Foundation: The controlled use of biological agents, such as microorganisms or cellular components for beneficial use.

Unfortunately, biotechnologies, are still in the process of early development, do not possess a sharp and easily defined form. Only in last fifteen years progress have been made by microbiologists and genetic engineers, and we are hopeful to solve many fold problems of the present day, specially energy and food crisis to cater the need of growing population of the world. Mineral ore deposits are also becoming more scarce and expensive to recover from earth's crust. Microorganisms can be used to enhance the recovery of metals from low-grade ores and from effluents containing undesirable quantities of heavy metals or other toxins.

When these technologies are applied at industrial level, they constitute bio-industry *(Table 2.1}* which include, on the one hand, industrial activities where biotechnologies can replace technologies normally or currently in use and, on the other hand, industrial activities where biotechnologies play an important driving role. There are several areas in which these technologies are being used, the important are chemical industry, food industry (mass production of yeast, algae and bacteria with a view of providing proteins, amino acids, vitamins, and use of enzymes), agricultural productivity, pharmaceutical industry, environmental protection and abatement of pollution, etc.

Table 2.1. Schematic distribution of the principal products of bio-industry

Fermentation Technology	
a. Health	Antibiotics, Vitamins, Enzymes, Amino acids, Polysaccharides (dextran), Nucleotides, Steroids, Alkaloids, Diagnostic reagents, etc.
b. Food & Agricultural	Citric acid, Lactic acid, Gluconic acid, Malic acid, Amino acids, Nucleotides, Enzymes, Biopolymers, etc.
c. Agriculture	Biopesticides, improvement of varieties, etc.
d. Energy	Ethanol, Acetone, Butanol, Biogas, etc.
e. Chemical industries	Chemistry of ethanol, Ethylene, Acetaldehyde, Acetone, Butanol, etc.
Enzymatic Engineering	
a. Food & Agricultural	Isoglucose, Glucose syrup, etc.
b. Energy	Ethanol
Genetic Engineering and Cell Cultures	
a. Food and agriculture	Single cell protein, clones, etc.
b. Health	Interferons, Hormones, Vaccines, Blood products, Monoclonal antibiotics, etc.

Historical Background

The oldest biotechnological processes are found in microbial fermentations, as born out by a Babylonian tablet dated circa 6,000 B.C., unearthed in 1881 and explaining the preparation of beer. The Sumenians were able to brew as many as twenty types of beer in the third millennium BC. In about 4000 BC. leavened bread was produced with the aid of yeast. *Table, 2.2* presents chronological history of biotechnology. The term biotechnology was described in a Bulletin of the Bureau of Biotechnology published in July, 1920 from the office of the same name in Leeds in Yorkshire. The articles in this bulletin described the varied roles of microbes in leather industry to pest control.

Table 2.2. Biotechnology - a review of the past

Year		Work
Before 6000 BC.	:	Yeast employed to make wine and beer
Approx. 4000 BC.	:	Leavened bread produced with the aid of yeast.
Before AD. 1521	:	Aztecs harvested algae from lakes as a source of food.
Before 1670-1680	:	Copper mined with aid of microbs, Rio Tinto, Spain.
	:	Antoine van Leeuwenhoek first observed microbes with newly designed microscope.
1876	:	Louis Pasteur identifies extraneous microbes as a cause of failed beer fermentations.
Approx. 1890	:	Alcohol first used to fuel motors.
1897	:	Edurad Buchner discovered that enzymes extracted from yeast can convert sugar into alcohol.
Approx. 1910	:	Large scale sewage purification systems employing microbes, are established.
1912-1914	:	Three important industrial chemicals (acetone, butanol and glycerol) were obtained from bacteria.
1928	:	Alexander Flaming discovered penicillin.
1944	:	Large scale production of penicillin.
1950s	:	Introduction of many new antibiotics.
1953	:	Double helix structure of DNA revealed.
1962	:	Mining of uranium with the aid of microbes begins in Canada.
1973	:	Brazilian government initiates major fuel programme to replace oil with alcohol.
	:	First successful genetic engineering experiments.
1975	:	Hybridomas which make monoclonal antibodies were first created.
	:	US outline guidelines for genetic engineering.
1976	:	US National Institute of Health introduces guidelines on genetic engineering.
1980	:	Rank Hovis McDougall receive permission to market fungal food for human consumption in UK.
	:	Court decides that genetically engineered microbes can be patented.
1981	:	Monoclonal antibodies receive US approval for use in diagnosis.
	:	Biotechnology firm Cetus. Sets Wall Street record for first public offer of stock ($ 115 million).

1982	: Genetically engineered insulin approved for use in diabetics in UK and USA.
1984	: Animal interferons approved for protection against cattle diseases.
Mid Eighties	: Genetically engineered growth hormone approved for treatment of dwarfism.
	: Interferon used to treat some viral diseases. Monoclonal antibodies widely employed in diagnosis.
	: New antibiotics produced by cell fusion.
	: Commercial production of dyes and industrial chemicals from algae.
	: Genetically engineered proteins used to treat heart attacks and strokes.
	: Monoclonal antibodies employed to boost the body's defence against cancer and other diseases.
	: New vaccines against foot and mouth disease.
	: Growth hormones used to increase yields of meat and milk from cattle.
Late Eighties	: Raw materials for plastic industry obtained from microbes.
	: Interferon employed to treat certain type of cancer.
	: More industrial chemicals produced by microbes.
Nineties	: An yeast strain capable of producing 12-14 per cent ethanol from Institute of Microbial Technology, Chandigarh (India)
	: Animal birth control product (Talsur)
	: Aquaculture technology developed for prawn and major crops, economically viable technology
	: Biofertilizer, improved Rhizobium particularly for legume crops and nitrogen fixation.
	: Biopesticides (viral, bacterial and fungal sources)
	: Diagnostic kit for filaria
	: DNA fingerprinting and Bkm probe
	: Embryo transfer and split embryo technology
	: Genetically engineered microbes help in extracting oil from the ground.
	: Microbes widely employed to extract metals from factory wastes.
	: Monoclonal antibiotics used to guide anti-cancer drugs to cancerous tissues.
	: Number of vaccines
	: Oral polio vaccines (Bharat Immunological and Biological Corporation, Ltd.)
	: Production of elite trees from micropropagation (Delhi University, Delhi)
	: Sericulture involving improvement in production and quality of Indian silk
	: Small scale production of hydrogen from bacteria.
	: Tissue culture propogation of cardamom and oil palm for commercial applications

Scope and Importance of Biotechnology

Biotechnology is the applied science and has made advances in two major areas, viz., molecular biology and production of industrially important biochemicals (including enzymes). Contribution of above cited scientists including and may Nobel Laureates is land mark in this field. The scientists are now diverting themselves toward biotechnological companies, this has caused the development of many biotechnological industries. In USA alone about 225 companies have been established and successfully working, like Biogen, Cetus, Geneatech, Hybritech, etc. In world, USA, Japan, and many countries of Europe are leaders in biotechnological researchers encouraged by industrialists. These companies are working for human welfare and opted following areas for research and development.:

a. Automated bioscreening.
b. Bioprocessing alkenes to valuable oxides and glycoles.
c. Developing immobilized cell and enzyme systems for chemical process industries.
d. Engineering of a series of organisms for specific industrial use.
e. Genctical improvement of microorganisms for production of pharmaceutical products.
f. Human gene therapy.
g. Improved production of Vitamin B_{12}.
h. Large scale production of fructose from inexpensive forms of glucose.
i. Manufacturing ethanol by continuous fermentation.
j. Microbiological based production of human insulin and interferons.
k. Microbiologically upgradation of hydrocarbons.
l. Production and development of vaccine to prevent calibacillosis (a disease develops in newborn calves and piglets)
m. Production of biopesticide and biofertilizers.
n. Production of diagnostic kits for toxoplasmosis identification.
o. Production of monoclonal antibodies for organ transplant tissue typing.
p. Production of photosynthetically efficient plants.
q. Production of transgenic plants and animals.
r. Production of xanthan gum in oil fields for recovery of crude mineral oils.

Genetic Engineering

Genetic engineering is the most fundamental mechanics of biotechnologies and is a recent offshoot of biotechnological research. It involves gene splicing, recombinant DNA cloning and tissue culture technology. The technique overall involves into two steps:

1. The *in vitro* incorporation of the gene or segment of DNA of interest into a small, self-replicating chromosomes and,
2. The introduction of the recombinant *minichromosome* into a host cell where it will replicate. Step one involves synthesis of recombinant DNA and step two is the Gene cloning.

These two, recombinant DNA and gene cloning technologies are the most powerful tools developed in field of biology. Genetic Engineering involves manipulation of the genetic material of an organism to give an altered expression of our choice. It deals with identification and isolation of desired gene and then joining this gene of interest into another organism. *The* desired gene expresses itself in that organism by the gene product. Various steps of genetic engineering are:

a. *Gene isolation:* Desired gene is identified, isolated and purified. This DNA of interest is also called donor DNA or target DNA.

b. *Selection of vector:* A vector is a self-replicating molecule of DNA or replicon to which desired gene is linked. Vector molecule with foreign DNA inserted is known as chimeric DNA. Vector acts as carrier and transports the gene into the host cell. Thus, it is also known as a cloning vehicle or a carrier molecule. Suitable vector is identified for a system; commonly used vectors are plasmids and viral DNA molecule. However, recent techniques involve physical delivery of DNA by various methods.

c. *Cloning of desired gene:* Multiple copies of desired gene can be obtained by placing them in host cell with the help of vectors. Here, the desired gene along with the vector is amplified. Large number of identical copies of gene of interest are produced for subsequent gene transfer into target cell. Now, gene cloning is a fast and mechanized process, using polymerase chain reaction (PCR) machines.

d. *Specific gene transfer:* The gene of interest is finally transferred to host cells. Transformed cells are selected, multiplied and produce transgenic plants-by tissue culture technique.

e. *Expression of desired gene:* The desired gene produce, the product in new environment of host, thus the desired traits.

Genetic Engineering includes the propagation of chimeric DNA in different host organisms. The ability to cross natural species barriers and place genes from any organism in an unrelated host organism is one important feature of gene manipulation. A second important feature is the fact that a defined and relatively small piece of DNA is propagated in host organism. Thus genetic engineering opened the door to a range of molecular biological opportunities including nucleotide sequence determination, site-directed mutagenesis, and manipulation of gene sequences to ensure very high-level expression of an encoded polypeptide in a host organism. If used wisely genetic engineering promises to enhance the quality of human life. However, only future will determine the scope and final outcome of this technology.

Tissue Culture facts

The term *tissue culture* is actually a misnomer borrowed from the field of animal tissue culture. It is a misnomer because plant micropropagation is concerned with the whole plantlet and not just isolated tissues, though the explant may be a particular tissue. The *terms plantlet culture* or *micropropagation,* therefore, are more accurate. Plant tissue culture is the technique of growing plant cells, tissues and organs in an artificial prepared nutrient medium static or liquid, under aseptic conditions. Better control of environmental conditions like light, temperature, gas mixtures and nutrients can be achieved for plant tissues growing *in vitro.* But whether we call it cloning, tissue culture, micropropagation, or growing *in vitro,* the process remains the same; it is a vegetative method for multiplying plants.

Tissue culture is the ever-ready tool for specialists who hybridize plants by either sexual or asexual means. It is a clean and rapid way for genetic engineers to grow material for identifying and manipulating genes or to transfer individual characteristics from one plant to another. It plays a role in a wide array of fields, such as botany, chemistry, physics, genetic engineering, molecular biology, hybrid development, pesticide testing, and food science.

A piece of a plant, which can be anything from a piece of stem, root, leaf, or bud to a single cell, is placed in that tiniest of greenhouses, a test tube. In an environment free from microorganisms and in the presence of a balanced diet of chemicals, that bit of plant, called an *explant,* can produce plantlets that, in turn, will multiply indefinitely, if given proper care. The *medium* (plural, *media)* is the substrate for plant growth, and in the context of plant tissue culture it refers to the mixture of certain chemical compounds to form a nutrient-rich gel or liquid for growing cultures, whether cells, organs, or plantlets. The process of tissue culturing plants from the explant stage to the final stage of transferring a mature plant to field or greenhous conditions involves 4 basic stages. The 4 stages of culture growth are: Stage I, explant establishment or initiation; Stage II, multiplication; Stage III, rooting; and Stage IV, acclimatization or hardening off. These stages can overlap in certain cases, and the requirements of each stage vary widely from plant to plant.

Tissue culture can serve a number of purposes, and growers have started their own commercial production laboratories for a variety of reasons. Growing plants from seeds or cuttings can be unacceptable or impractical due to some of the following (and other) factors:

— Seed-grown products lack uniformity	— Seed-grown products are not ture to type
— Seeds take too long to grow to mature plants	— Seeds are difficult to handle
— Seeds are not available	— Cuttings grow too slowly
— Cuttings have poor survival rate	— Cuttings require too much care
— Tissue culture is often the only practical way to produce the large numbers required.	— There is a shortage of stock plants from which to take cuttings because there is: (a) only one hybrid, (b) only one virus-free plant, (c) only one desirable mutant
— There is insufficient room for stock plants	— It is not cost effective to maintain the stock plants.

If an ample supply of seeds is available, or if plants from cuttings are acceptable and cutting stock is available and not a problem to maintain, then tissue culture may not be the most practical option because it can be an expensive, labor-intensive process, especially if less than a thousand plants are needed. If plants from seeds are acceptable but there are not enough seeds, then seeds or excised embryos can be used as starting material for tissue culture.

Every year excessive amounts of growers time, labor, and space are spent on unproductive seeds, cuttings, and grafts. Significant numbers of young plants are lost to pests, diseases, or other environmental factors. Tissue cultured plantlets are less subject to such attacks and disasters because in the sterile environment of the laboratory they are not exposed to the pathogens or extreme conditions that afflict many plants grown in the field or greenhouse. Material usually comes out of culture as well-started plantlets or microcuttings with a stockpile of nutrients and vigor often superior to that of conventional cuttings. It is no secret that healthy plants are the first line of defense against diseases.

Tissue culture avoids and enormous amount of the daily care that is required with cuttings and seedlings. Cultures usually need to be divided and transferred to a fresh medium every 2 to 6 weeks, but between transfers there is no need to water or tend to the cultures, other than casual surveillance. How different this is from the daily watering and weeding requirements accompanying greenhouse growing!

The simplest tissue culture hobby is the multiplication of easy, fast-growing plant material, such as *Kalanchoe,* Boston ferns *(Nephrolepis),* African violets *(Saintpaulia),* or *Begonia;* next in order of complexity are carnations *(Dianthus),* strawberries *(Fragaria),* or *Syngonium.* The chemist looking for a tissue culture hobby may be challenged to explore the field of plant by products; dyes, flavorings, medicinals, and oils are just some of the by-products of certain plants.

The earlier researchers explored problems of academic interest, then the emphasis shifted to applied aspect such as haploid from pollen, triploids from endosperms, somatic hybridization, tissue culture of cereals, legumes and oil crops, and clonal multiplication of elite species of plants. Our knowledge of cell and tissue culture is developing with a fast speed specially in many areas, like totipotency, differentiation, cell division, cell nutrition, metabolism, radiobiology, cell preservation, etc. We are now in position to cultivate cells in quantity, or as clones from single cell, to grow whole plant from isolated meristems, to induce callus or even single cell to develop into complete plant either by organogenesis or directly by embryogenesis *in vitro.* The production of haploid through tissue culture from anthers or isolated microspores and of protoplasts from higher plant cells has served as the basic tools for genetic engineering and somatic hybridization. Protoplasts can also be used as genetic material present in nuclei and chloroplasts as well as isolated DNA molecules. This technique provides the opportunity to combine by fusion the genotype of species which are sexually incompatible and to introduce foreign genetic material such as organelles or DNA into the genome.

Micropropagation and Protoplast Culture

The hybridization programme is hampered in some cases because of sexual incompatibility. Synthetic production of hybrids has, however becomes possible through the novel technique of protoplast culture and their fusion. The isolation, culture and fusion of protoplast are one of the most fascinating fields of research, though still in developing stage. The protoplast culture technique can be suitably used for microinjection and other genetic engineering experiments. The techniques are important, specially, because of their far-reaching effects on crop improvement by somatic hybridization and cell modification. The protoplast culture can be regenerated into an entire plant. It is imperative that the technique in future will be one of the most frequently used research tools for tissue culturists, physiologists, pathologists, molecular biologists, cytogeneticists, and biotechnologists. The discovery of enzymes which could separate the cells and remove the cell wall revolutionized the concept of plant tissue culture. It has provided a tool for isolating protoplasts and exploring the possibilities of genetic engineering.

Biosensors, Biosurfactents, Biochips and Biofilms

The natural biosensors are the *sense organs* primarily the chemical sensors of smell and taste. Biosensors technically in their various forms share a reliance of biological materials as sensing elements. Technical biosensors have been under intense development since the middle of 1960 with the prospects of commercial potential offered by biotechnology. Biosensors are the combinations of biologically active material displaying characteristic specifically with chemical or electronic sensor to convert the response into electrical signals. Biosensors have their roots in military research, as means of detecting nerve gases and other chemical warfare toxins. Their applications have branched out to include simple-t-use *alternate site* diagnostic devices for home, doctor's office, or drug use screening; medical and surgical monitors (small enough to fit inside a blood vessel) and environmental quality monitors.

Biochips are made from different biological materials and can control the computer by replacing silicon chips. Biomolecular computers thus made, promise to be ten to thousand times smaller than the best super computers with much faster switching times and extremely low power dissipation. They are made up **of semi-conducing molecules** inserted into the protein framework and fix the whole on to a protein support. They can be used in implanting of several sorts in human body, like regulation of heart beats, responses to nerve impulses by artificial limbs (bearing such computer device), overcoming of blindness and deafness, etc. Biochips may be infected by microbes since they are made up of proteins.

A biofilm is an accumulation of microbial cells and inorganic components held together in a polymeric matrix and firmly attached to a substratum. Accumulation of biofilms is encountered in many natural and modulated environments. It may be fundamental to process performance i.e., for fixed-film biological waste water treatment, sudden deterioration of water quality and deterioration of substrata.

Biosurfactents are surface active compounds produced by microorganisms. Biological surfactants possess a number of potential advantages over their chemically manufactured counterparts, including lower toxicity, biodegradability, a wide variety of possible structures, and ease of synthesis from inexpensive, renewable feed stocks. Consequently, biosurfactents may have applications in numerous areas, particularly for enhance oil recovery and in foods, beverages, cosmetics, pharmaceutical preparations, etc.

Uses of Biotechnology to Agriculture

Improvements in agricultural production and in the food and nutrition situation depend on land, water and energy resources which are generally considered as limited, in spite of the fact that it is possible to increase their availability. Increasing the biological productivity is an area of active research in the field of life sciences. In this wide field, research aims to obtained increased production, higher nutritional value, greater plant resistance to adverse weather, pathogenic agents and pests. The contribution of plant genetics to this research is essential as much as its concepts and methods developed rapidly because of recent discoveries in molecular biology and the exploitation of certain characters that are peculiar to plants.

With reference to the agriculture's of the Third World countries, the ever-increasing trend in world population cannot be ignored. Recent estimates of the world population growth rate of 2.0% per annum implies a 50% increase in population by the turn of the century. Biotechnology must have a very significant role to play in rendering more remunerative agricultural industries of Third World countries. It is also certain that the agricultural industries in many parts of the world will increasingly make a major contribution to the production of crops destined for, and in certain instances specifically designed for a **vegetable plant based chemical industry.** Some of the objectives currently being pursued in the application of biotechnology to agriculture are presented in Table 2.3. Up to the present it is true that successful commercial application has been limited - the development of certain vaccines and, also of novel protein sources for inclusion in human and farm livestock diets are examples.

Table 2.3. Some applications of biotechnology to agriculture.

Crops	Improvement in yields, ratio of primary to secondary products, nutritive value of products, physical properties - i.e. bread making, resistance to insecticides and pesticides, capacity to resist stress.
Livestock	Improvement in disease control, efficiency of reproduction, yields of livestock products, i.e., meat, milk, wool, eggs, composition of livestock products - i.e., leaner meat, feed value of low quality feeds - i.e. straw, Introduction of novel protein sources
Man	Introduction of novel protein sources for man, i.e. meat analogues

The main development in crop improvement with the use of biotechnological researches are (1) plant cell, tissue and organ culture, (2) genetic engineering leading to transformation followed by regeneration of plants to give transgenic plants carrying desirable traits like disease resistance, insect resistance and herbicide resistance, eventually this may also be used of increasing photosyn-thetic efficiency, nitrogen fixing ability, improved storage protein, hybrid crops for food processing, (3) somatic hybrids between sexually incompatible species permitting transfer of desirable traits from wild or unrelated crop species to our crop plants, (4) transgenic animals produced in mice, pig, goats, chicken, cows, etc., it is proposed that some of these will eventually be used as bioreactors to produce drugs through their milk, blood or urine. This field is some times described as **molecular farming.**

Modem biotechnologies can add greater precision and speed to plant breeding. Trarisgenics have already been reported in more than 40 crop plants, including maize, -rice, soybean, cotton, rape-seed, potato, sugar beet, tomato, potato and alfalfa, but the new varieties are yet to be used commercially. Near future opportunities for commercial exploitation include vegetables and fruits (potato, tomato, cucumber and squash), followed by legumes (alfalfa), oilseed crops (rapeseed) and a few resistant plants whose widespread use is somewhat controversial.

Tissue culture techniques are currently in wide use for micro propagation of elite clones and for freeing planting materials from pathogens. Monoclonal antibodies are also in use as diagnostic aids in the detection and identification of viruses and viroids. Anther culture and microspore culture giving rise to haploids are being used in variety improvement to facilitate and accelerate breeding. Molecular maps and markers are being widely used to identify genes of interest to accelerate conventional breeding programmes. Efficient biological nitrogen fixation systems and strains for efficient utilization of soil nutrients are being genetically engineered. Other long-term objectives are the genetic manipulation of photosynthesis patterns and the production of hybrid seed through apomixis. A very distant possibility is that of providing nitrogen fixation capacity to cereals.

Plant Breeding

An enhanced production of cereal grains during the last forty five years due to improved varieties of seed, developed by the application of classical genetics and plant breeding. The application of genetic engineering technology will circumvent such restrictions and allow plant breeders to access a much more diverse range of genes. The new technologies will only a new dimension to plant breeding not replace it and, that currently, successful application of the new technologies is being hampered by the lack of a sound understanding of the genetic processes concerned with crop productivity. Genetic improvement in the major agricultural crops up to the end of the present century will be mostly the result of conventional plant breeding practice.

The establishment of a plant breeding industry based on a deep understanding of cellular and molecular biology offers the possibility to overcome such limits to production. For example the pathway of photosynthesis in temperate plants involves fixation of carbon dioxide under the action of the enzyme ribulose-1, 5-biphosphate carboxylase. It has been suggested that small alterations in the characteristics of the enzyme via changes in the coding sequence of appropriate gene may result in enhanced efficiency. Another possibility could be the replacement of the three carbon pathway of fixation characteristics of temperate region plants by the four carbon pathway of carbon dioxide fixation to phosphoenol pyruvate which exists in certain tropical fodder plant.

Table 2.4. In vitro responses of important fruit crops.

Common name	Botanical name	Responses elicited	Type of explant
Apple	*Malus pumila*	Multiple shoots, callus and adventitious shoot	Shoot tip, lateral bud, apical meristem, seedling explants
Banana	*Musa* spp.	Axillary shoot proliferation, Hybrid plantlets and multiple shoot proliferation	Rhizome tip, floral tissue (excised embryo)
Ber	*Zizyphus mauritiana*		
Cape gooseberry	*Physalis peruviana*	Triploid plant regeneration	Anther
Grapefruit	*Citrus paradisi*	Callus, embryoids and adventitious shoots	Stem and leaf segments
Lemon	*C. limon*	Direct adventive embryo formation	Nucellus
Lime	*C. aurantifolia*	Callus, deformed shoots and roots plantlets	Nucellus, anther
Mandarin orange	*C. reticulata*	Callus and shoot differentiation	Shoot segment
Pumello	*C. grandis*	Callus and shoot differentiation	Shoot segment
Rough lemon	*C. limon* var. *Jambhiri*	Callus and embryogenesis	Nucellus

Common name	Botanical name	Responses elicited	Type of explant
Sweet orange	*C. sinensis*	Differentiation	Unpollinated ovary, shoot segment
Custard apple	*Anona squamosa*	Organogenesis, haploid plants	Leaf, anther
Date palm	*Phoenix dactylifera*	Multiple shoots, somatic embryogenesis	Shoot tip, seedling
Fig	*Ficus carica*	Elongation and rooting of shoot tips	Shoot tip
Grape	*Vitis* spp.	Multiple shoots, callus formation and somatic embryogenesis	Shoot tip
Guava	*Psidium guajava*	Multiple shoots	Shoot tip
Jackfruit	*Atrocarpus heterophyllus*	Multiple shoots	Stem tip
Jamun	*Syzygium cumunii*	Regeneration of plantlets	Leaf axillary bud
Mango	*Mangifera indica*	Somatic embryogenesis, hybrid plantlets and somatic embryogenesis	Nucellus, embryo
Papaya	*Carica papaya*	Multiple shoots, callus and somatic embryogenesis, hybrid plantlet production, haploid plantlets	Shoot tip, ovules embryo, anther
Passion-fruit	*Passiflora edulis*	Adventitious buds	Stem and leaf
Peach	*Prunus persica*	Axillary shoots	Shoot tip
Pear	*Pyrus communis*	Callus and adventitious shoots and roots	Seedling tissues
Pineapple	*Ananas comosus*	Multiple shoots, callus	Lateral meristem, shoot tip
Plum	*Prunus domestica*	Adventitious shoot proliferation	Shoot tip
Pomegranate	*Punica granatum*	Multiple shoots	Mature stem tip
Sapota	*Achrus sapota*	Callus formation	Endosperm
Swet cherry	*Prunus* spp.	Axillary shoot proliferation	Meristem tip
Strawberry	*Fragaria* spp.	Axillary and adventitious shoots	Shoot tip

Quality and Productivity Improvement in Horticulture

High plant biotechnology has been witnessing major advances in recent years. Some technologist have found extensive use while others are on the anvil or in the developmental stages. The alliance of basic science with the applied sciences in the latter half of the 70s led to the foundation of biotechnology, which has since emerged as a versatile tool packing in powerful manipulation of life-forms.

If the Green Revolution of the 60s ushered in an era of high yielding varieties, biotechnology has put us on the threshold of the second green revolution or biorevolution with tremendous implications in agriculture and related fields like industry, health care, environmental control etc. India's export of horticultural produce in the world market is only of the magnitude of 0.45 per cent. The potential for growth in fruits, vegetables and floriculture items seems unlimited considering our tropical and agro-climatic conditions. All these years we have been concentrating only on perfecting production technologies and our attempts-most of the time through government agencies have been lop-sided.

Private sector initiatives and efforts are making a small impact. Such as, the Bangalore based Indo-American Hybrid Seeds and the Cochin-based AVT and others have proved that a highly motivated private sector initiative could create records in crops like banana, cardamom and vegetable seeds and ornamental flowers etc. Here is a write up on high tech banana as developed by IAHS.

Private sector is introducing latest technologies including developing the glasshouse technology for operating climate-controlled glass-house to produce large volumes of tissue culture plants and flowers that would ensure high quality in the export market.

Out of the 29 species of plants accounting for 90% of the world food production, horticulture constitutes one fourth. India produces about 24 million tones of fruit and 46 million tones of vegetables netting in Rs. 12,184 crore to the exchequer annually.

However, this yield falls far short of the requirements of both domestic and export to other developing countries. It is here that biotechnology steps in to envisage enhanced horticultural production.

Pathogen-free Plant Production

Virus, as we know, is a piece of bad news wrapped in protein. Worse still, viruses decide to harbour in many horticulturally important crops with disastrous consequences. Fortunately, these and other pathogens, in most cases, are unable to invade the meristem. Hence, plant meristems offer themselves as ideal explants for initiating chief virtue of this technique is the production of virus-free propagules in commercial quantities which has been amply demonstrated in the case of potato. This is of tremendous significance for small farmers. Some private companies in the country have already gone into brisk business with cardamom clones, since the *Katte Virus* and other devastating diseases have attracted much attention of the planters. Meristem culture has been successfully used in carnation, chrysanthemum, papaya, banana etc. and cured plants have been obtained. The technique when coupled with thermotherapy and chemotherapy, as has been done with potato, demonstrates greater efficiency.

An Efficient Method for Virus Elimination in Meristem Cultures and Regeneration of High Quality Plants

Since garlic is propagated vegetatively, it is universally infected with viruses and other pathogens which causes a decline in the harvest. Selection of vigorous clones based on the absence or a reduction in leaf symptoms resulted on average in a 25% higher yield. These results indicated that elimination of viruses by tissue culture might be beneficial. Therefore, meristem tip cultures, sometimes in combination with thermotherapy were used several times for the regeneration of virus-free garlic. Efficient methods for micropropagation of garlic, resulting in *in vitro* bulb formation were also published. But there have been few reports of the field performance of virus-free plants compared with infected stocks, such as one was an extensive agronomic evaluation of virus-free and virus-infected garlic published by Walkey and Antill (1989). They reported 30-90% higher yields in OYDV (onion yellow dwarf virus) free garlic.

Generally, commercially sized bulbs resulting from plantlets from meristem culture are obtained after three vegetation periods but occasionally two cycles are sufficient.

The identification of garlic viruses is complex and all viruses found are not yet well characterised. Most authors have described mixed infections of garlic with poty and carlaviruses. OYDV was found all over the world. Preliminary studies of Slovenian garlic *(Allium sativum* L. cv. *Ptujski jesenski)* by electron microscopy showed that this cultivar is totally infected with different viruses, among them OYDV and CLV (Carnation Latent Virus) related viruses. Therefore, a set of experiments was initiated to produce virus-free garlic. This reports the optimal conditions for the meristem tip-cultures, thermotherapy, and transfer to soil. Data are also given on yield comparisons of virus-free and infected plants grown *in vivo* in the first and second vegetative seasons.

Nitrogen-Fixation & Biotecnology

Another, potentially significant approach is that of introducing symbiotic, nitrogen-fixation into non leguminous plants, Novel symbiotic vectors like gall forming *Agrobacterium tumefaciens* (Ti-plasmids) and root nodule bacteria *(Rhizobia)* have delivered *osm* genes to leguminous plants. *Agrobacterium* transfers foreign genes to host cells; where as the *Rhizobium* does not transfer DNA to the host itself but functions as a chemical factory converting raw material into useful products that are passed to the host plants. In addition, a segment of DNA has also been identified that permit high level synthesis of foreign genes in root nodule bacterioids. These foreign genes are expected to enhance energy efficiency of nitrogen fixation, delivery of plant growth regulators and perhaps even pest resistance.

There is much speculation whether such plants would be less productive than their nitrogen-dependent relative, due to the high-energy requirement for nitrogen-fixation. But this may not prove to be such a very significant factor in food production in the Third World countries where the lack of dependence on nitrogenous fertilizer could be very important.

Disease Resistance Increased

Additional improvements are likely to be associated with increased, and long-lasting resistance to insects, pests in plants and also increased resistance by plants to cost-effective herbicides and pesticides. The technique of *in vitro* tissue culture and selection has already allowed Molecular Genetics, Inc. to develop maize plants that are resistant to a family of herbicides which act by inhibiting an enzyme in the pathway of synthesis of the three branched-chain amino acids. Tissue culture selection of material in which the enzyme's specific binding properties are modified have rendered the herbicides non-toxic to the plants.

Quaity Improved of Plant Products

Another important area in which plant biotechnology is likely to have a considerable impact on the agricultural industry is in the development of plant varieties whose seed proteins have been modified to the needs of man. One such example is the improvement in bread making quality of high yielding, home grown wheat which is largely, though not entirely dependent upon the characteristics of the glutein proteins present in the seed endosperm (Miflin *et al.,* 1983). A second such objective is to improve the nutritive value of the cereal proteins for man and also pigs and poultry, by alteration in overall amino acid composition of the proteins present; cereal proteins are limiting in lysine and to a lesser extent threonine. It is often overlooked that cereal proteins provide the major dietary source of protein for man.

Tree Improvement thorough Biotecnology

Biotechnology promises to have great beneficial impact on tree improvement and forestry. Plant breeding has already resulted in higher yielding plants and also resistance to some diseases and insect pests. However, in forest trees, conventional methods of selection, breeding and progeny testing are very slow and difficult. The problem is due to the long life cycle of trees (20 to 50 years). Conventional methods also requires large spaces. Tissue culture technique could reduce the time for selection and propagation of desirable traits. Selection of genetic variants through cell and tissue culture (somaclonal variation) requires very little space. Advances in biotechnology will rapidly accelerate tree breeding. They offer new ways to increase forest productivity and to make trees harder and resistant to pests, diseases and stress.

Improvement for Resistance of Plants to Stress

The potential for improved crop plant productivity arising from the presence of mycorrhizal fungi is being investigated. Such fungi benefit their symbiotic host plants by facilitating water and nutrient uptake, particularly of phosphorous and some of the trace elements, and increased resistance to stress.

Salt Tolerance and Biosaline Concepts Improving

Salinity is a problem in a very large area of the world. In order to extend intensive cultivation to marginal environments, the problem is to be faced and it has been reiterated time and again that solution shall come from an integrated or interdisciplinary approach. In that pursuit information has been generated through working with prokaryotic unicells, eukaryotic algae and higher plants. Osmoregulation with organic and inorganic solutes, cellular compartmentation, vacuolar sequestering of ions, patch clamp studies, acidic proteins and genetic engineering of plants are some of the dimensions which represent an interface of basic and applied knowledge. The genes for production of osmoprotective molecules *(osm* genes) are present in plants but these have been studied in detail in bacteria (Rains et al., 1980). Some of the possibilities are as follows:

a. It is possible to convert osmo-sensitive bacteria into osmotolerant one.

b. *osm* genes conferring stress tolerance are carried as a segment of DNA comprising about 10,000 base pairs.

c. The segment codes for enzymes catalysing first two steps in proline pathway.

d. Enzyme synthesized through *osm* gene losses its sensitivity to feed back inhibition by proline, which is the requisite property important for overproduction of metabolites and imparting salt tolerance.

The biosaline concept has proved its relevance in the modern context of biotechnology. The concept was promulgated as poor soils, high solar insulation and saline water, which prevail in lands should be viewed as useful resources rather than as disadvantages, and that these can be used for non-traditional production of food, fuels and chemicals. Since then much information has been generated on **Bio-saline** concept (Pasternak and San Pietro, 1985). In addition to the application of genetics and molecular biology in developing salt tolerant glycophytes and halophytes (Gallagher, 1985; Gorham et al., 1985). Reed et al.(1985) have described that filamentous blue green alga *Spirulina* can be cultured on saline water as a major source of single cell protein for animal nutrition. At NFTRI, Mysore it has been demonstrated that the algae is good for human consumption also.

The presence of the organic storage compound poly p-hydroxy-butyrate in blue green algae grown in saline media may find further commercial applications in plastic industry. Along other potentially useful chemicals which can be obtained through algal culture on saline water are sulfated polysaccharides from unicellular red alga- *Porphyridium alginates* and **carrageenan** from seaweeds (McLachalan, 1985) and glycerol from *Dunaliella* which inhibit saline water upto 5M NaCl concentration (Ben-Amotz and Avron, 1980).

❑❑❑

CHAPTER - 3

Plant Tissue Culture : Principles and Methodology

Introduction & Definition

Plant tissue culture broadly refers to the growth and development of plant segments, tissues or cells nurtured in a contaminant free environment. The plant materials are grown on gel or liquid media, which typically contain salts, sugars, vitamins and hormones that regulate plant growth i.e. induce sequences of cellular differentiation leading to the formation of root and shoot which ultimately develops as a test tube plant.

The ability of many plant cell to regenerate entire plants through cell-culture make it possible to exploit this property for introducing large-scale cloning in horticulture. Cell culture also provides a good way to extend studies in plant pathology by establishing causal relationships in tumour formation, host-parasite interactions, and in sanitation of pathogen-infected stock of crop plants. Crop improvement through somatic cell hybridization is entirely based on cell culture methods.

Plant tissue culture is the technique of growing plant cells, tissues and organs in an artificial prepared nutrient medium static or liquid, under aseptic conditions. It has advanced the knowledge of fundamental botany, specially in the field of agriculture, horticulture, plant breeding, forestry, somatic cell hybridization, phytopathology and industrial production of plant metabolities, etc.

Importance of Tissue Culture Technique

Although different cells of different organs of an organism passess same genenitic material but they function differently, as per the need of the organ. Our knowledge of cell and tissue cultures has been developing with increasing knowledge, specially in biotransformation, forestry, genetic engineering, morphogenesis, somatic hybridization, maintaining pathogen free plants and rapid clonal propagation, totipotency, differentiation, cell division, cell nutrition, metabolism, radio biology, cell preservation, etc.

It is now possible to cultivate cells in quantity, or as clones from single cells; to grow whole plant from isolated meristems and to induce callus or even single cell to develop into complete plant either by organogenesis or directly by embroygenesis in vitro.

The production of pure haploid plants through tissue culture from anthers or isolated microspores and of protoplasts from higher plant cells has served as the basic tools for genetic engineering and somatic hybridization. Tissue culture technique helps to propagate plants of economic importance such as orchids and other ornamental plants in large numbers by their meristem culture or by other in vitro methods. This provides them virus-free plantlets. Propagation of valuable economic plants through tissue culture based on the principle of **totipotency** *(every cell within the plant has the potential to give rise a whole plant).*

In plant breeding, embryo, ovary and ovule culture as well as in vitro pollination have been employed to overcome morphological and physiological sterility and incompatibility. In recent years, plant tissue culture technique is in increasing use for producing haploids from anthers or isolated microspores, and of protoplasts from higher plant cells and the recognition of the potential of these materials in genetics and plant breeding. One of the most significant developments in the field of plant tissues culture during recent years are the isolation, culture and fusion techniques which have their special importance in studies of plant improvement by cell modification and somatic hybridization.

Plant tissue culture technique is a boom in the studies of the biosynthesis of secondary metabolites and provides an efficient means of producing economically important plant products (fine chemicals).

History of Plant Tissue Culture Researches

The first works in the field of excised plant tissue cultivation date back to the end of the last and the beginning of this century and are linked with the names of the three outstanding scientists: Vochting, Rechinger, and Haberlandt. Vochting (1878) studied plant polarity, and attempted to grow small lumps of tissue in vitro. He demonstrated that polarity was present even in the plant cell itself. Rechinger (1893) placed segments of poplar stem and pieces of beet and candeloin roots on a wet filter paper, and observed callus formation. Since the start of 19th century people started thinking about the culture of plant tissues. Haberlandt (1902), a German botanist tried but invein to obtained plant tissues from the isolated mesophyll cells of *Lanium* on artificial Knop's medium in vitro. Though he did not succeed in his attempts, but he pointed out towards a new technique- *Plant tissue culture.* In 1904, Harming succeeded in culturing nearly mature embryos of *Raphanus* and *Cochlearia* to maturity. Realizing the complexity of nutritional requirements of younger embryos, van Overbeek and co-workers, for the first time, used **coconut milk** for *Datura* embryos and demonstrated its stimulatory effect. Subsequently, coconut milk and other substances were added to the nutrient medium (containing mineral salts, vitamins, amino acids and sugars) but it failed to promote the growth into cultured tissues. Lampercht (1918), Kundson (1919) and Nemec (1924) have made numerous attempts to find suitable media and optimal conditions for growing organs, tissues, and cells excised from whole plants.

Many zoologists have also tried to culture mammalian tissues. The workers like Skvortsov (1886), Garrison (1907), Carrel and Burrows (1911), and Krontovskii (1917), developed the nutrient media for growing mammalian tissue. Zoologists like Czech (1927), Prat (1927) and other workers attempted to grow excised plant tissues on plant extracts.

Knop and Prdifer developed synthetic media. The approach of Mlliard (1921) is considered to be an original who used segments of root and hypocotyl of young radish shoots. These tissues, possessing embryonic activity, grew in the culture, but the cell did not divide, and new tissues were not formed. The utility of embryo culture technique soon became apparent when Laibach (1925, 1929) reared hybrid embryos *(Linum perenne* x *L. austriacum)* from non-viable seeds to maturity. Impetus was thus provided for further work on other tissues as well, and prompted Robbins (1922) and Kotte (1922) to try root culture. By using a synthetic medium containing inorganic salts, pyridoxine, thiamin, nicotinic acid, an iron source, yeast extract and sucrose, White (1937) could maintain tomato root culture for almost 30 years (1934-1968).

Philip White (U.S.A.) and R. Gautheret (France) devoted many years to conduct tissue cultivation experiments. They could get success because of their fortunate choice of material for investigation, careful selection of nutrient media suitable for growing plant tissues. They are regarded as the founders of methods for cultivating excised tomato roots, root tips, root meristem and callus of cambial origin. The workers from France, U.S.A., Czechoslovakia, Hungry, Germany, Italy, Switzerland, China, India and Russia, etc. have done sincere efforts in developing tissue culture techniques in various fields of normal and pathological plant physiology, biochemistry, cytology, and genetics which have now been used in biotechnological studies. R. Gautheret (1937) cultivated undifferentiated carrot tissues.

Isolation and culture of single cell, remain elusive until Muir (1953) devised the technique of agitating callus tissue in liquid medium (on a shaker). Isolated plant cells were grown in culture in 1954 by Muir and coworkers. The technique developed by them was perfected in France by Lutz. Bergmann who used bacterial culture techniques to grow plant cell suspensions. H.E.Street (UK.) developed bacterial culture techniques for tissue culture. Skoog and Miller (1957) obtained roots and stems from callus treated with auxin and kinetin. Morel used gibberellin for induction of proliferation of meristems and their differentiation into whole plants. The regeneration process has also been developed by use of tissue culture techniques.

The single cell thus obtained could be mechanically picked up and grown by using the nurse culture method. Later, a micro culture method was designed by Jones et al. (1960). F. C. Steward of Cornell University has shown that it is possible to produce a complete plant from a single cell. Steward isolated cells of carrot root, which grew in a medium containing coconut milk, the fluid that nourishes the coconut embryo. He found that these isolated cells, which under ordinary circumstances would not divide again, began to grow. Some of them showed highly abnormal growth patterns, but others, as they divide, organized themselves into perfect duplicates of normal carrot embryos. Such embryos can develop into normal mature carrot plants, with normal roots, stalks, flowers, and seeds. Vasil and Hilderbrandt (1975) used this technique to raise complete plant from single cell culture of *Nicotiana.*

Besides, the academic interest, the significance of tissue culture technique in rapid and clonal propagation of plants was soon realized.

Observations over the past several years have led to a general belief that cells grown in culture for an extensive period of time are genetically unstable due to increased polyploidy and spontaneous mutation. This often results into decrease or loss of morphogenic potential, decline in biosynthetic capability of secondary metabolites and/or reversion of valuable mutant in wild types.

Tissue Culture Researches in India

In India, the significance of in vitro culture technique was realized quite early by Professor P. Maheshwari. He initiated investigation with complex tissues such as ovary, ovule, and control of fertilization. The initial exploratory work on these organs laid the foundation of several achievements in plant cell, tissue and organ culture in India. During 1956 to 1966, attempts were initiated to culture the pollinated ovaries of *Aerva tomentosa, Allium cepa, Althaea rosea, Anethum graveolens, Foeniculum vulgare, Hyoscyamus niger, Iberis amara, Linaria maroccana* and *Tropaeolum majus*. Success with the culture of pollinated ovaries made it possible to culture ovaries and ovules at even earlier stages. Cultured ovaries from pollinated flowers of *Zephyranthes* developed seeds with fully differentiated embryos. Excised ovules containing zygote and primary endosperm nucleus also matured into normal-looking seeds but were without fully-differentiated embryos. The seeds were however always smaller than the natural seeds (M. Kapoor, 1959). The in vitro formed fruits of *Anethum* were larger than the in vivo fruits. Another important feature was the occurrence of polyembryony due to cleavage and/or budding of zygotic embryos, on a culture medium. Polyembryony is useful in the multiplication of plants producing a limited number of seeds. This is of a special significance in umbellifers because of low seed-set and germination.

The method of ovary and ovule culture is also appropriate for rearing hybrid seeds. In *Gossypium hirsutum,* Joshi (1962) was able to grow ovules with 12-celled proembryo and 500 endosperm nuclei to maturity. The hybrid ovules of a cross between *G. arboreum* and *G. hirsutum* could be grown to maturity and viable seeds were obtained.

Of considerable importance are the adventive embryos from nucellus which genetically uniform, and reproduce the characters of the maternal parent. Excised nucellus from micropylar-half of pollinated ovules of *Citrus microcarpa,* a polyembryonic species, proliferated on a medium and the callus differentiated *pseudobulbils* which, eventually, developed into plantlets (Rangaswamy, 1961). Embryogenesis from nucelli of unfertilized ovules of *Citrus sinensis* and *C. aurantifolia* is reported by Mitra and Chaturvedi (1972). The embryos arose from nucellar cells or the callus. It was demonstrated that the pollination/fertilization stimulus is not essential.

The organogenic potentiality of endosperm has been demonstrated by Johri and Bhojwani (1965) in *Exocarpus cupressiformis*. In *Scurrula* and *Taxillus,* shoot buds differentiated on a cytokinin medium, whereas auxin is also essential for *Dendrophthoe* and *Tolypanthus*. Continuously-growing tissue and organogenesis have also been observed in the mature endosperm of autotrophic taxa such as *Croton honplandianum, Jatropha panduraefolia, Putranjiva roxburghii, Codiaeum variegatum, Santalum album, Oryza sativa, Nigella damascena, Euphorbia geniculata* and *Emblica officinalis.*

Khuspe et. al. (1980) described methods for in vitro growth of immature hybrid embryos of *Carica papaya* x C. *cauliflora* and development of F_1 plants. The resultant seedlings could be transferred to pots and field. The culture of young floral primordia has provided a potential tool for understanding the morphogenic sequences leading to flowering. Studies in floral morphogenesis have been conducted in *Ranunculus sceleratus, Phlox drumondii, Kalanchoe pinnata* and *Browallia demissa.* In *Ranunculus* spp., cultured floral buds proliferated rapidly, and differentiated roots, shoot buds and embryoids. The buds developed into shoots terminating in a flower bud. Flowers were also born directly on the callus. Rarely, mature seeds were also formed. The embryoids passed through the stages comparable to in vivo embryogenesis. These plants also flowered in vitro.

The work on haploids provided a novel experimental system for speeding up of homozygosity, analyses of genetic combinations, and detection of mutations. In cultured anthers of *D. innoxia* (with uni- or bi-nucleate pollen), numerous embryoids emerged on Nitsch-medium. The embryoids germinated to form plantlets (Guha and Maheshwari, 1967). George and Rao (1979) induced triploids through anther cultures of *Physalis* spp. Direct microspore-derived embryos in anther culture of two cultivars of *Brassica juncea* have been demonstrated by Sharma and Bhojwani (1985). On B5-medmm with 2% sucrose the embryos developed into plantlets. Anthers of *B. nigra* cultured on B5-medium developed callus and/or embryos. The most significant result is the in vitro flowering of these plants in 2-months-old cultures. Other examples of anther culture leading to the formation of pollen embryoids/plantlets are *Petunia axillaris* by Doreswamy and Chacko (1973), *Solatium melongena* by Raina and lyer (1973), *Petunia obconica* by Bajaj (1981), and *Physalis ixocarpa, Datura innoxia* and *Petunia hybrida* by Chandra and coworkers (Chandra and Kothari, 1987). By anther culture technique, homozygous recombinants of *Hyoscyamus niger* with higher alkaloid content could be obtained . Such a technique has proved very useful in the improvement of crop plants. Haploids of cereals are used for one-step transfer of genotypes of inbred lines. For the induction and characterization of mutants also, haploids are used. The pollen grains are extensively used for the isolation of protoplasts. Shrivastava and Johri (1988) reviewed Indian contribution to morphogenesis and plant tissue culture studies (Indian Review of Life Sciences, 8:249). A. F. Mascarenhas has compiled a Hand Book of Plant Tissue Culture published by ICAR. New Delhi (Feb,1993)

Table 3.1 : Land marks in plant tissue culture research.

Worker	Year	Their Contribution
G. Haberlandt	1902	Was first to culture isolated plant cells in vitro on artificial medium.
W.J. Robbins, W. Kotte	1922	Culture of isolated roots-organ.
P.R. White	1934	Demonstrated capability of indefinite culture of tomato roots (long periods).
R.J. Gauhtheret, P. Nobecourt	1939	First culture cambial tissues isolated from carrot.

P.R. White	1939	Callus culture of tobacco tumour tissue from interspecific hybrid of Nicotiana glauca x N. langsdorfii.
J. Van Overbeck	1941	Discovered nutritional value of liquid endosperm of coconut (coconut milk) for culture.
P.R. White, A.C. Braum	1942	Experiments on crown-gall and tumour formation in plant growth of bacteria-free crown-gall tissue
A. Caplan, E.C. Steward	1948	Use of coconut milk and (2, 4 Dichlorophenolacetic acid) for proliferation of cultured carrot and potato tissues.
G. Morel	1950	Culture of monocot tissues using coconut milk.
W.H. Muir	1953	Developed technique for culture of single isolated cells (nurse culture method).
W. Tubecke	1953	Haploid cultures from pollen of gymnosperm (Ginkgo sps.)
C.O. Miller, F. Skoog	1955	Discovery of cytokinins, e.g. kinetin, as potent cell division factor
E. Ball	1955	Culture of gymnosperm tissue (Sequioa)
F. Skoog, C.O. Miller	1957	Predicted hypothesis that shoot and root initiation in cultured callus is regulated by the proportion of auxin and cytokinin in the culture medium
E.C. Cocking	1960	Enzymatic isolation and culture of protoplasts
G. Morel	1960	Development of shoot apex culture technique
G. Morel	1964	Use of modified shoot apex technique for orchid propagation
S.G. Guha, S.C. Maheshwari	1966	Cultured pollen and anthers can produce haploid embryos
J-P. Nitsch	1974	Culture of microspores of Datura and Nicotiana, to double the chromosome number, and to harvest seeds, from the homozygous diploid plants just within 5 months
R.R. Hendre & coworkers	1975	Established technique for obtaining virus free sugarcane, citrus, potato and cassava.
G. Melchers	1978	Production of somatic hybrid from fusion of protoplasts isolated from potato and tomato.
S.K. Mukherjee, S. Bhaskaran	1980	Isolated auxotrophic mutants for vitamin B components in Nicotiana.
A. Mascarenhas, V. Jagannathan	1982	Isolated spontaneous varients from wheat callus cultures.
K.A. Barton, W.J. Brill,	1983	Insertion of foreign genes attached to a plasmid vector into naked plant protoplasts
M.D. Chilton	1983	Production of transformed tobacco plants following single cell transformation or gene isertion
H. Kohn and coworkers	1985	Somatic hybrids in tobacco mediated by electrofusion.
E. Sundberg and K. Glemelius	1986	Somatic hybrids in Brassicaceae
R. Nadgauda, A. Mascarenhas	1986	Isolated and planted variants of sugarcane resistance to mosaic virus and turmeric plants of the variety 'Tekurpeta' in the field.

Plant Tissue Culture: Principles

The technique has developed around the concept that a cell is totipotent that it has the capacity and ability to develop into whole organism. The principles involved in plant tissue culture are very simple and primarily an attempt, whereby an explant can be to some extent freed from inter-organ, inter-tissue and inter-cellular interactions and subjected to direct experimental control.

The most common culture in plant tissue is callus, which is wound tissue composed of undifferentiated, highly vacuolated and unorganized cells.

Callus Culture

For raising the callus tissues, a tissue culturist must have clear understanding of some basic principles. A cell from any part of the plant like shoot apex, bud, leaf, mesophyll cells, epidermis, cambium, anthers, pollen, fruit etc., when inoculated in a suitable medium under aseptic laboratory conditions can able to differentiate and multiply. This results into the formation of an amorphous mass of cells known as callus, which can be induced to re-differentiate on appropriate medium to develop embryoids which directly develop into the plantlets, eventually giving rise to a whole viable plant

The term clone (from the Greek *klon,* meaning: a slip or twig suitable for plant propagation) was suggested by Webber (U.S.A.) in 1903 to explain those plants which were obtained by a sexual reproduction, it is even applied to DNA multiplication (cloning of genes in bacteria). In strict scientific sense, cloning means *an organism obtained from a single cell through mitotic divisions.*

Meristem Culture

When a meristem is cultured in vitro, then it produces a small plant bearing 5 or 6 leaves. This could be obtained within a few weeks. Then the stem is cut into 5-6 small micro cuttings, which under favourable conditions, become fully grown plants.

Organ Culture

A body of higher plants has complex inter-relationships between different organs like root, shoot, apical meristem, leaf primordia, floral buds, ovary, ovule, anther lobs, pollen grains, fruit, seed, etc. In this method a particular organ is isolated and cultured under laboratory conditions in a chemically defined medium where they retain their characteristic structures and other features and continue to grow as usual. In organ culture, organs are not induced to form callus, therefore, it differs from the callus culture where the organization of the intact tissues is lost.

This technique provides an experimental system to define the nutrients and growth factors that are usually received by the organ from other organs of the plant body and from surrounding environment. It also helps us in understanding the inter-dependence of organs with respect to various physical and chemical growth factors including growth hormones. Organ culture technique also provides the knowledge about the various problems of morphogenesis and the sites of biosynthesis of specific metabolites and growth compounds. It may be used as a tool for improvement of various economically important crops.

Organ culture may be grouped into two major categories: vegetative organs (root culture, leaf culture, and shoot tip culture) and reproductive organs (complete flower culture, isolated ovary culture, isolated ovule and embryo culture, pollen mother cell culture, seed and fruit culture).

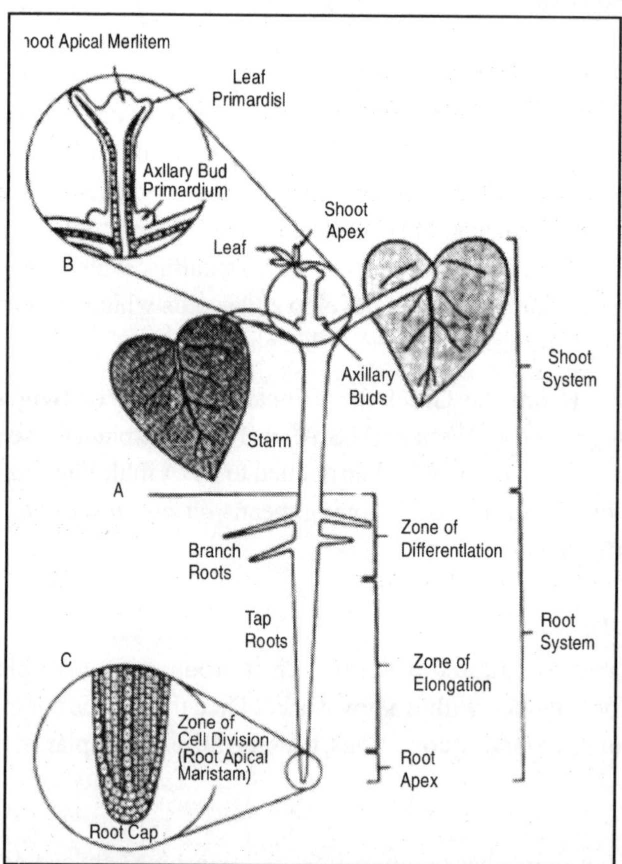

Diagram 3.1 Primary organisation and growth of dicot plant. (A) Overall morphology of a small bean seedling. (B) Longitudinal section of the shhot apex. (C) Longitudinal section of the root apex.

THE CONCEPT OF TOTIPOTENCY OF CELLS

Some concepts in science become inherently accepted long before they practicaly demonstrated. This was so in the concept of the totipotency of cells of higher plants. Even in the mid-twenties one encountered the tacit view that, apart from inherent practical difficulties, there was no theoretical reason why one should not rear a begonia plant from a single leaf hair cell. This view was traceable first to the then well recognized principle that as cells divide mitotically, they do so equationally to produce daughter cells in facsimile.

Secondly, the concept was due to G. Haberlandt's historic but then unsupported claim, expressed in *1902,* that one day it should be possible to rear plants from isolated surviving cells of flowering plants. Haberlandt's insight was, in fact, the more remarkable because he even stated that out of surviving somatic cells artificial embryos would be reared asexually.

In higher plants, embryos do develop in situ from appropriately stimulated somatic cells but without any prior sexual act. At the outset, the special significance of the fertilized egg in plants should be seen as restricted to its role as the genetically unique product of a fusion of male and female gametes. Having established the genetic constitution of a given individual, the zygote (or fertilized egg) really behaves develop-mentally as a very general kind of living cell, a cell with a built-in capacity to grow in an organ, the ovule, that fosters that growth. Moreover, as will be shown, and in a satisfying number of cases, isolated somatic cells may return to a simulated zygotic state and grow into embryos and to plants under the appropriately applied conditions which furnish the requisite nutrients and stimuli.

Contrasts between Plant and Animal Cells

Some inherent differences between plants and animals are relevant here, for these so far to explain why the clonal development of individuals from somatic cells is intrinsically more difficult in animals than it is in plants. The much heralded new biology rightly emphasizes that the molecular basis of metabolism, whether of plants, animals, or bacteria, is essentially the same. The classical events are essentially the same from microorganisms to man. However, the differences in plants and animals are organized and provide their cells and organs in situ with the requirements for what they do?, where they are?; moreover, they profoundly affect the ability of animal cells to develop in isolation.

The green flowering plants are autotrophic and can live in an inorganic world. Even so, the non-green cells of plants, which are necessarily heterotrophic for carbohydrate, have such simple other requirements that they may often be grown (even though slowlý) in tissue explants or organ cultures (for example, of roots) in a relatively simple, largely inorganic, medium. Plant cells remain capable of receiving their requirements from the outside world (CO_2 and H_2O, NO_3, SO_4, and PO_4 and mineral elements), where they are very dilute. They can survive in varied composition of their ambient media. Plants are bounded by a cellulose wall and so may utilize their turgor to maintain a thin protoplasmic lining around large enclosed vacuoles; this thin, living layer is, however, in easy contact with the outer air or medium with which the cells of plants are in constant interchange.

In contrast, higher animals are heterotrophic for a wide range of metabolites, and they are adapted to receive, digest, and assimilate their nutrients from complex sources and to distribute these via the blood in a complex way, without direct parallel in plants, to the cells and organs of the body. Moreover, the animal body creates and maintains within narrow limits an internal environment to which its living cells are exposed and which furnishes their requirements in relatively concentrated form. For example, a mammalian embryo develops from a fertilized egg, it rapidly establishes a complex nutrient supply with the parent via the placenta; or, in other forms, the embryo develops first a free living larval form at the expense of the nutrients prestored in the egg.

In higher plants the provision of nutrients to the embryos of higher plants in the embryo sac and ovule, via the endosperm for their attachment in the ovule via a suspensor serves a mere mechanical purpose, while the pro-embryo and its later formed cotyledons absorb nutrients and stimuli directly over their outer surfaces. Thus, if somatic plant cells can be induced to form immature embryos, it is very much simpler to provide for their nourishment externally than for embryos of higher animals, which need a blood supply.

This may well account for the fact that testicular teratomas of the mouse, lacking a blood supply, fail to develop. However, these structures do embark upon a simulated embryogeny and indeed similar asexual forms have long been known to occur in other animals. By contrast, and as will be shown below, proembryos of several plants which originate from free somatic cells can now be caused to form embryos, and plantlets by the thousands; this occurs even in isolation and in a limited volume of an appropriate aqueous solution which behaves, in this respect, like an artificial ovule or embryo sac.

There is, however, another obvious reason why totipotent development from somatic cells is more feasible in plants than in higher animals. Such animals develop their organs early, they assign to them highly specialized functions and then integrate them by means of the circulating tissue of the blood, by a nervous system spinal cord, and brain, and by highly specialized hormones which are the product of discrete glands. Moreover, these developmental events in animals essentially occur but once. By contrast, higher plants create their organs repetitively and, often, virtually indefinitely. Plants do all this by arranging, in their growing regions, living cells which retain the essential characteristics of the plant and which, during development, depart from, or return to, the situation to be seen in the primary growing apices of shoot and root by relatively small and often easily reversible steps. Diagram 3.1 shows the organization of a shoot and root tip, to emphasize both the unrestricted growth and repetitive organ formation and the many sites in the plant body in which living cells occur, and from which they may be explained and cultured. Thus, the organization, nutrition, and cell physiology of higher plants and animals are very different in ways which render the clonal development of higher plants in large numbers from free cells much more feasible than it is, at present, for higher animals.

The Autonomous Organelles
Their Behaviour During Growth Induction and Morphogenesis

The stimuli that cause quiescent cells of carrot to proliferate and grow also affect all their organelles (mitochondria, plastids, dictyosomes, reticulum and polysomal aggregates of ribosomes in the ground cytoplasm) in various ways. The same stimuli affect the metabolism of the tissue and its ability to form residual or secondary biochemical products.

For example, deep orange-red and carotene-rich plastids occur in the carrot root, but, even so, they become bright green in explants cultured on coconut milk in the light. Combinations of light intensity and of temperature and of diurnally fluctuating light and temperature all have their effects on the pigmentation, but none have yet produced in free cells or proliferating explants the bright orange-red, carotene-rich cells of the normal carrot root. Thus, it is not only what the potentially totipotent cells are that determines what they will do, nor in fact how readily they may be caused to grow, but where, morphologically, they develop.

Sites and Modes of Action of Growth Regulating Substances

How the totipotent cells respond to the exogenous messages that they received would be gratifying if, in response to the appropriate stimuli, the nuclear genes were to pour out messenger RNA's into cytoplasm and there stimulate the appropriate proteins and enzymes to cause the cells to do what they should, when and where they are. This mechanism, essentially bacterial in nature, may in due course be observed to operate in flowering plants, but, so far, attempts to demonstrate it have had meagre success. In fact, it seems as if carrot cells on a basal medium may possess all the idling mechanisms of protein synthesis and require only certain external chemical triggers, which have been described, to set it into directed motion. All this raises the question *whether plant cells really need to ask the permission of their nuclei,* in terms of activated or de-repressed genes, for every task they need to perform.

The Behaviour of Carrot Cells in vitro

The genetic inheritance of the zygote, transmitted at cell division to all the daughter cells, conveys to every organelle the inherent capacity to carry out its various functions. However, thereafter the autonomous organelles, each with their own genetic information, may be called into appropriate activity by the great array of simple substances which, singly or in combination, act as growth regulators; these release, wholly or in part, the inherent **totipotency** of angiosperm cells. Thus, there is a vast area open to exogenous chemical control over the responses of otherwise totipotent cells.

Clonal Development from Animal and Plant cells

There is now good evidence that the nuclei of certain animal cells (e.g. of the intestinal wall of a tadpole) retain intact all the genetic information, but they only release this, as yet, in contact with the cytoplasm of an enucleated egg. Perhaps eventually the equivalent of the plant techniques here described may induce somatic cells of animals to embark directly upon an embryological development & perhaps, if this is done, they may even be furnished in some way with the equivalent of a parental blood supply.

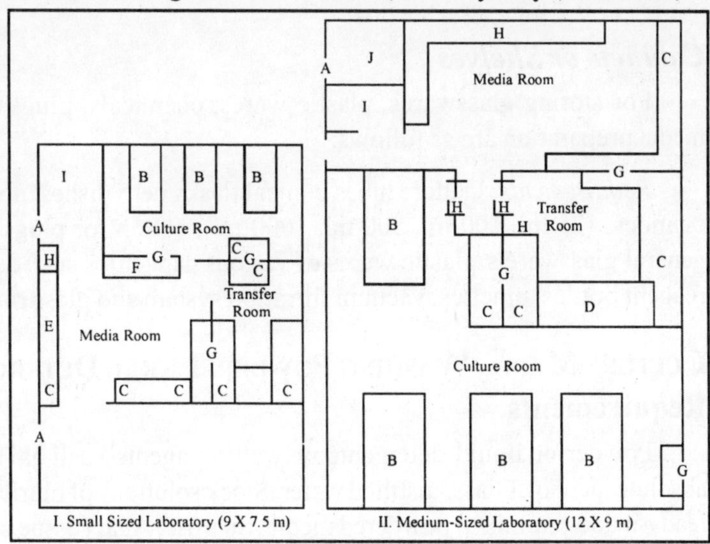

Diagram 3.2 Various designs of tissue culture laboratory (A) entrance (B) shelves (C) counter (D) laminar air-flow cabinet (E) sink (F) gas outlet/burner (G) window (H) refrigerator and deep freezer (I) store (J) Conference room (K) library

Such steps present, however, obviously greater orders of difficulty, even if they do not constitute inseperable barriers to the free clonal development of higher animals from somatic cells. Thus, until these problems show signs of being resolved, it seems best to regard the free clonal growth of higher animals from somatic cells as a distant, even if it is a desirable, goal. Meanwhile, however, the techniques of clonal development of plants from free cells open up various technically rewarding possibilities, which are being pursued in this and many other laboratories.

PLANT TISSUE CULTURE: METHODS

The setting up of a tissue culture laboratory needs proper planning. It depends upon space availability, volume of work to be carried out and funds. Usually, the available laboratory space, is be divided into five distinct laboratory areas. These are: *Media preparation area/ room, Aseptic transfer chamber area, Environmentally controlled culture room, Analytical room and Acclimatization room.*

MEDIA PREPARATION: AREA/ROOM

Requirements

Laboratory tables or benches and revolving stools with adjustable height, Hot plate and magnetic stirrers, analytical loading single pan balance with precision of ± 0.001 g, (weighing range 0.1 mg-180g, digital read out), top pan loading balance for quick weighing, (range 100 mg-500g capacity, sensitivity O.lg), refrigerator and freezer, water purification and storage system, glassware washing facility with proper drainage, gas outlet, electric hot-air oven (range upto 250 ± 2°C), microwave oven, digital pH meter, (range 0-14 pH, accuracy 0.1 pH) temperature compensation 0-100°C, autoclave preferably horizontal, and continuous supply of single and double distilled water.

Cabinets or Shelves

For storing glasswares, plastic wares, chemicals, plugs and appliances required for media preparation are as follows:

Requirements: Culture tubes/conical flasks/petri dishes of various capacities, measuring cylinders (25 ml, 100 ml, 500 ml, 1000 ml), cotton for plugs/plastic caps (autoclavable), general glassware's/plastic wares of various capacities such as volumetric flasks, beakers, reagent bottles, pipettes, vacuum filtration system and glass rods.

CULTURE MEDIA, WASHING POWDER/ LIQUID DETERGENT, DISINFECTANTS

Requirements

Powder or liquid detergents or wetting agents such as Tween-20, 70% alcohol and absolute alcohol. Glass, distilled water, Stock solutions of nutrients of tissue culture media or ready-made/pre-mixed powdered media, Sucrose, Agar (tissue culture grade), Sterile culture, vessels with distilled water, Chemicals of analytical grades (Inorganic, organic salts, vitamins, ammo acids, growth regulators/hormones and activated charcoal), Coconut milk, yeast extract malt extract, casein hydrolysate and extracts of potato, carrot and tomato.

Other Requirements

Spatulas, weighing butter paper/boats, stirring bars (magnetic) for magnetic stirrers and stirring retrievers, Brush (flask and test tubes), gloves (disposable 23 cm), mop (household), scoop, towels (household), wastebasket (large and medium), Service lines such as gas, water, electricity, vacuum pump and generator, A low bench/table/desk for culture evaluation and data recording.

Aseptic Transfer Chamber Area

Requirements

Laminar air-flow cabinets to provide constant flow of air across working area, Dissecting and stereo zoom microscopes and lens magnifier, Micro-dissecting scissors, scalpel handles with blades, forceps (various sizes), catheter trays, needles and inoculating loops, Gas outlet, vacuum facility, Tissue paper/filter paper, Sterilizer (dry heat with glass beads) and pipette dispenser (automatic).

Environmentally Controlled Culture Room

Requirements

Racks with light arrangements on timers and controlled temperature (25 ± 2°C) maintained with air cooler or by window air conditioners fitted with temperature indicators, Incubators having dark, and light/dark photoperiod with controlled temperature. Rotary shakers of variable speed from 80-220 rpm to take 100 ml or 250 ml. Erlenmeyr flasks with arrangement of lighting to provide an intensity of 2000 to 4000 lux. Lux meter to measure intensity of light in culture room, incubator and shaker.

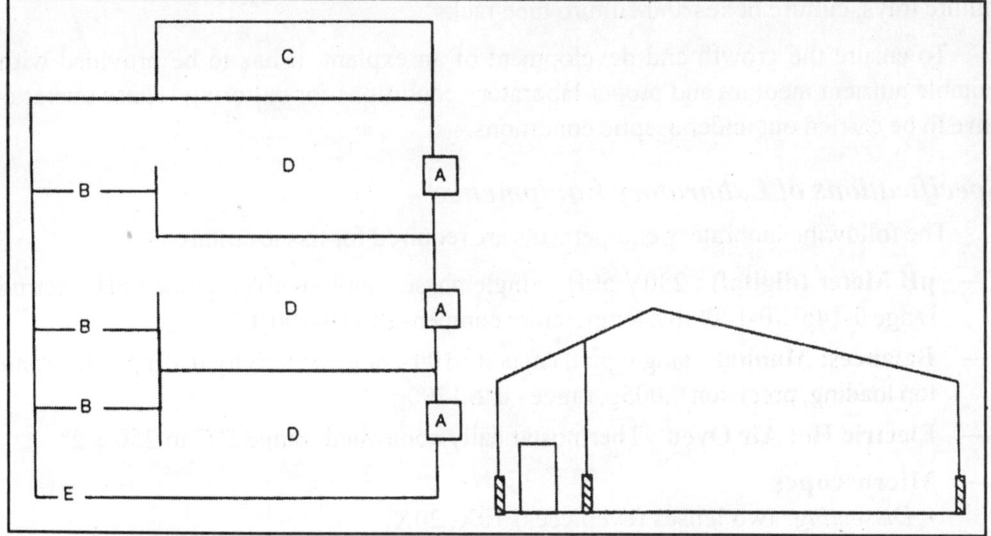

Diagram 3.3. Design and elevation of greenhouse with three compartments having desert coolers. (A) Desert-cooler (B) Air-vents (C) Water tank (D) Greenhouse chamber (E) Entrance.

The cultures are usually incubated at 25 ± 2°C under 16:8 ligrrt : dark photoperiod. The source of light for the cultures in racks should be make available with cool-day-light (fluorescent tube lights of 40 watts, 2000 lux). It is advisable to connect the controlled culture room with power generator for emergency power supply. Some space should also be kept for incubating cultures in continuous darkness.

Analytical Room

Requirements

Inverted microscope for bright field, dark field with compensating wide angle eye pieces, preferably with photo-micro graphic attachment, Colorimeter for chemical estimation such as chlorophyll, starch, nucleic acid, phenols, oxidising enzymes etc. Low speed centrifuge with continuous variable electronic speed control, Chemical reagent racks for qualitative and quantitative chemical analysis. Viscosity meter, Gas outlet.

Acclimatization Room

The hardening chamber needs high illumination (4,000-10,000 lux) and high humidity (90-100%. through mist and fog systems). Humidity is required for conditioning tissue culture plants after taken out from rooting media and transfer to pots under greenhouse.

Miscellaneous Items

Requirements

Air conditioners, uninterrupted power supply (UPS) and emergency light, Bunsen burners, Permanent markers, tapes (autoclave indicator), tape label (self-adhesive), aluminium foil and parafilm, Fluorescent lamps/tubes, Trays and baskets for cultures, Plastic carboys to store water and other solvents, UV germicidal lamp, Metal racks to keep test-tubes in culture room, Gas lighter/match box, Fire saving, equipment and first-aid box etc., Filter paper, culture trays, culture boxes and culture tube racks.

To ensure the growth and development of an explant, it has to be provided with a suitable nutrient medium and proper laboratory conditions for culturing. These operations have to be carried out under aseptic conditions.

Specifications of Laboratory Equipments

The following laboratory equipements are required for tissue culture.

— **pH Meter (digital) :** 230V 50H_z, single phase supply with combined pH electrode, range 0-14pH/0-1400mv, temperature compensation 0-100°C.

— **Balances: Manual -** single pan, capacity 100-200g, sensitivity 0.1 mg. Electronic - top loading, precision 0.005g, range - 0 to 1200g.

— **Electric Hot Air Oven** : Thermostatically controlled, range 5°C to 250 ± 2°

— **Microscopes**

- *Dissecting:* two lenses (eyepieces) 10X, 20X.
- *Laboratory:* with facilities for bright field, dark field, phase contrast, wide angle eyepiece with photomicrographic attachment.

- *Inverted:* Wide angle eyepieces, quintuple objective turret adjustable specimen 55 mm petridish holder and photomicrographic attachment.

— **Centrifuge** (low speed): Variable electronic speed control, speed indicator, Amp. meter, timer, dynamic break, 0 starting switch, 230V $50H_z$.

— **Electric Autoclaves**

- *Vertical:* with safety valves, pressure gauge, steam-release cock.
- *Horizontal:* mounted to tubular stand, heavy hings
- *Steamers:* With immersion heaters of ejection safety device, size: 31cm x 13cm x 10cm to 61 cm x 20cm x 15cm.

— *Filter Sterilization Equipment:* Syringe filter holder: 2.5cm

— *Pressure Filter Holder:* 4.7cm

— *Manestry Stills:* Electrical water stills, 2 lit capacity/hr of 1.5 kw.

— *Double Distilled Water Equipment:* Built-in energy regulator all glass water stills, heat resistant boiling flask with heater.

— *Environmental Growth Cabinets:* Cabinet with controlled temperature, light and humidity, temperature range 4° to 45°C with timer to regulate photoperiods.

— *Gyratory Shakers:* Capacity: 50 to 1000 ml flasks, speed: 80 to 200 rpm.

— *Laminar Air-flow Cabinets:* Constant flow of purified air, sizes: 0.6m, 1.2 m and 1.8m.

The other miscellaneous equipments which are required for tissue culture are: air conditioners, arrow heads, bunsen burners, deep-freeze, dissecting needles, glasswares, forceps, florescent lamps/tubes, heaters, hot plates with magnetic stirrer, inoculation cabinets, metal trays and bowls (for transport of cultures), tubes, refrigerators, UV germicidal lamps, wooden or metal racks, etc.

Aseptic Technique

In in vitro condition plant cells and microbes have basically same requirements. When the culture medium contains sugar (as carbon source) it attracts a variety of microorganisms which grow faster than that of the cultured tissue in medium and they ultimately kill the plant cells. It is, therefore, necessary to have complete aseptic condition around the culture equipments which prevents contamination of the culture medium. Following are the three main,sources of contamination of the medium and the subsequent methods to check them:

1. The microorganisms may be present in the nutrient medium at the time of its preparation. These microorganisms can be destroyed by proper plugging and autoclaving the culture tubes/flask. The medium can be completely sterilized by maintaining it at 120° C for about 20 minutes at 15 Ib pressure in the autoclave.
2. The explant (plant part to be cultured) may carry microorganisms with it, therefore, the plant part should be surface-sterilized by mercuric chloride (1 to 2%) or by sodium hypochloride solution for 30 minutes.

3. Precautions must be observed to prevent the entry of microorganisms when the plug of a culture is removed during transfer of the plant material to the medium or from one medium to another. The inoculation chamber may be sterilized by UV-radiations.
4. Correct pH of the medium is important. Highly alkaline or acidic pH affects the nutrient uptake in culture tissues. Therefore, the tissue culture medium is adjusted to a pH of 5.6 to 6.0 before autoclaving.
5. Semi-solid and liquid media are most commonly used for growing plant cells. A high concentration of gelling agent (agar-agar, gelatin, silica gel) makes the medium very hard and decreases the nutrient uptake by the tissues. Agar at 0.8% to 1.0% concentration is widely used.

Sterilization of Plant Tissues

It is essential to remove dirt and debris from the plant tissue and should be washed in a weak detergent solution and rinsed several times with distilled water prior to sterilization. Some woody tissues, such as buds and twigs, are cleaned by immersing them briefly in a 70% ethanol solution, which wets and spreads over the tissue surfaces more effectively than a higher concentration alcohol solution.

— *Sodium Hypochlorite (NaOCl):* It is the most common chemical agent used to sterilize plant tissues (0.025%-0.25% NaOCl). Diluted household bleach can also be used for this purpose, which normally contains 5.25% NaOCl. It is equally effective and considerably less expensive.

— *Calcium hypochlorite (CaOCl):* It can be used as a substitute for NaOCl. CaOCl causes slightly less damage to plant tissues but tends to precipitate out of solution. To avoid the accumulation of CaOCl on the plant tissue surfaces, sterilization solutions should be filtered or decanted prior to use.

— *Hydrogen Peroxide* (H_2O_2) *Solution:* Plant tissues can also be surface sterilized, by using H_2O_2 (3%-10%). It is much easier to remove from tissues than NaOCl and CaOCl.

— *Other Substances:* Plant tissue can also be surface-sterilized by bromine water (1%-2%), silver nitrate ($AgNO_3$ 1%) and mercuric chloride ($MgCl_2$ 0.1%-1%).

Cleaning (Preparation of Glasswares/Plastic wares (Autoclavable)

1. Clean the glasswares/plastic wares in 10% commercial detergent liquid or powder for 1 hr and then in HCl for 2 hrs.
2. Remove the traces of detergent and acid by thorough washing with tap water
3. Rinse vessels with double distilled water and allow them to dry over night at room temperature.

Sterilization

1. Plug glasswares such as conical flasks or test tubes with non-absorbent cotton or cover by the plastic caps. Wrap the petridishes with aluminium foil. Place forceps and

scalpels in test tubes, plug the tubes with cotton or cover with aluminium foil. Plug the mouth end of the pipettes with cotton. Wrap them individually in aluminium foil.

2. Autoclave glasswares and instruments at 121 °C for 1 hr.
3. For dry-heat sterilization, metal instruments should be sterilized in an oven at 140-160°C for 2hr.
4. Filter sterilization for heat labile amino acids, vitamins, growth regulators, antibiotics, natural complexes should be through millipore filtration assembly using filter membranes of 0.45 or 0.22 μm porosity. Plug the receiver flask with cotton. Assemble the filtration assembly and wrap the filtration unit with paper or foil. Autoclave the receiver and filtration unit at 121°C for 1 hr. Attach the filtration unit with receiver flask with vacuum pump in a laminar flow bench pour solution to be sterilized into the filtration unit. Apply slight air pressure to start filtration. Transfer the desired volume to sterile flasks under laminar air-flow bench. Use a sterile pipette for drawing filter sterilized solution to autoclaved medium.

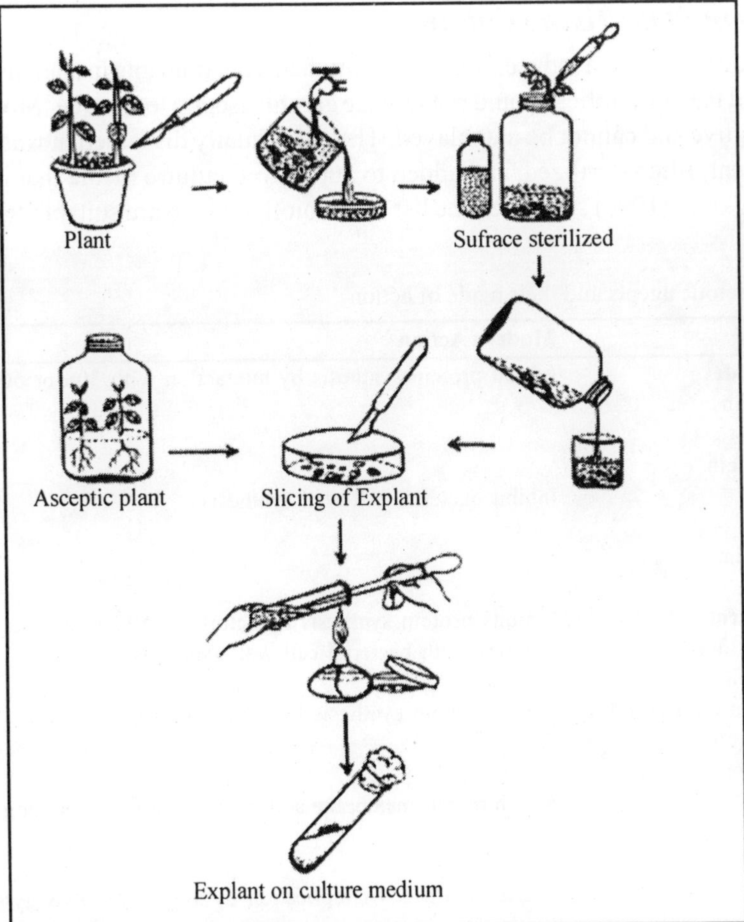

Diagram 3.4. Diagrammatic presentation of the procedure for surface sterilization of plant material and inoculation of explant for culture.

Surfactants

- *Tween 20 or Triton X-100:* These are scientific reagent-grade surfactants and are often added in low concentrations (0.05%) to chemical sterilization solutions. Their use ensure that the sterilizing agent come in contact with the entire plant tissue surface.
- *Stirring of the Tissues:* Good surface contact is also facilitated by stirring the tissues during sterilization.
- *Ultrasonic Bath:* It is an effective method to ensure good surface contact during sterilization treatment in an ultrasonic bath like those used to clean dentures.

This technique is particularly useful for sterilizing buds and woody tissues that have many small surface crevices and cracks.

After surface sterilization, a minimum of three sequential rinses with sterile distilled water are recommended to remove any remaining chemical sterilizing agent.

Control of Bacterial and Fungal Contaminants by Antibiotics in Plant Tissue Culture

Antibiotics are used to reduce, control, or eliminate contaminants in plant tissue cultures. It is suggested that antibiotics should not replace careful aseptic technique. Most antibiotics are heat-sensitive and cannot be autoclaved. They are usually dissolved in water or another suitable solvent, filter-sterilized, and added to autoclaved culture media that has cooled to 45°-50°C. Falkiner (1990) suggested the list of antibiotics which are suitable for plant tissue culture.

Table 3.2 Antibiotic agents and their mode of action

Antibiotic	Mode of Action
Amino glycosides	Inhibit protein synthesis by interaction with 30S or 50S ribosomes.
Manamycin	
Neomycin	
Streptomycin	
β-Lactams	Inhibit bacterial cell wall synthesis.
Ampicillin	
Carbenicillin	
Penicillin	
Chloramphenicol	Inhibits protein synthesis by acting on 50S ribosome.
Glycopeptides	Interfere with bacterial cell wall synthesis.
Vancomycin	
Macrolides and lincosamides	Inhibit protein synthesis by acting on 50S ribosome.
Erythromycin	
Lincomycin	
Polymixins	Attach to cell membrane and modify ion flux, resulting in cell lysis.
Polymixin B	
Polymixin E	
Quinolones	Interfere with DNA replication by inhibition DNA gyrase.
Nalidixic acid	
Norfloxacin	
Ofloxacin	

Antibiotic	Mode of Action
Rifampicin	Interferes with mRNA formation by binding to RNA polymerase.
Tetracyclines	Inhibit protein synthesis by acting on 30S ribosome
Trimethoprim and Sulphonamides	Inhibit synthesis of tetrahydrofolate.

In Vitro Environment

A piece of plant tissue taken out from original site of plant and transferred to an artificial tissue culture media for the growth or maintenance, is called as explant material. The choice of tissue depends upon on ultimate goal of the tissue culture project. Any piece of the plant tissue can be used as an explant material.

Various factors of an explant tissue source influence the culture on tissue culture media. These are:

1. Physiological and ontogenetic age of organ or tissue
2. Quality of source plant
3. Season in which explant tissue is obtained
4. Size of the explant
5. Aim of the culture

Goal	Explant Tissue
1. Bud culture	Apical and axillary bud
2. Meristem culture	Apical meristem
3. Micropropagation	Shoot apex or lateral bud or embryo
4. Root culture	Lateral roots from adult or seedling
5. Callus culture/ Somatic embryogenesis	Cotyledons, hypocotyle, stem, leaf, root, any part of seedling
6. Haploid culture	Anther and pollen

Pretreatment to Explant Tissues Prior to Culture

Many explant tissues specially woody tissues produce phenolics into tissues culture medium, which causes browning of explant material. Antioxidants such as L-ascorbic acid, free acid 0.1 g/1; citric acid, free acid anhydrous 0.15 g/1; L-cysteine HCl 0.05 g/1 are commonly used for checking browning problem. The browning pigments are toxic to the plant growth in cultures. Best results are obtained when explants are treated with freshly prepared antioxidant solutions.

Table 3.3 Commonly used disinfectants for plant tissue culture

Disinfectant	Concentration (%)	Exposure (min)
Calcium hypochlorite	9-10	5-30
Sodium hypochlorite	0.5-5	5-30
Hydrogen peroxide	3-12	5-15
Ethyl alcohol	70-95	0.1-5.0
Silver nitrate	1	5-30
Mercuric chloride	0.1-1.0	2-10
Benzalkonium chloride	0.01-0.1	5-20

Procedure

1. Prepare the mixture of L-ascorbic acid (100 mg/1) and citric acid (150 mg/1) in double distilled water
2. Filter sterilize the solution through a 0.22 μm filter unit
3. Store the explant material in a cold antioxidant mixture and incubate explants in refrigerator at 0°C for 5-30 min. to allow the tissue to soak in antioxidant solution
4. Commercial bleach contains about 5% sodium hypochlorite and thus may be used at a concentration at 10-20% which is equivalent to 0.5-1.0% sodium hypochlorite

Sterilization procedures may be enhanced by

1. Place the material in a 70% ethyl alcohol solution before treating with another disinfectant solution. The use of a two step (two source) sterilization procedure is beneficial.
2. Use of a wetting agent (Tween 20 or 80 respectively) is beneficial. It reduces surface tension and allow better surface contact.
3. Conduct the sterilization process under vacuum. This helps in the removing air bubbles and provides efficient sterilization process.

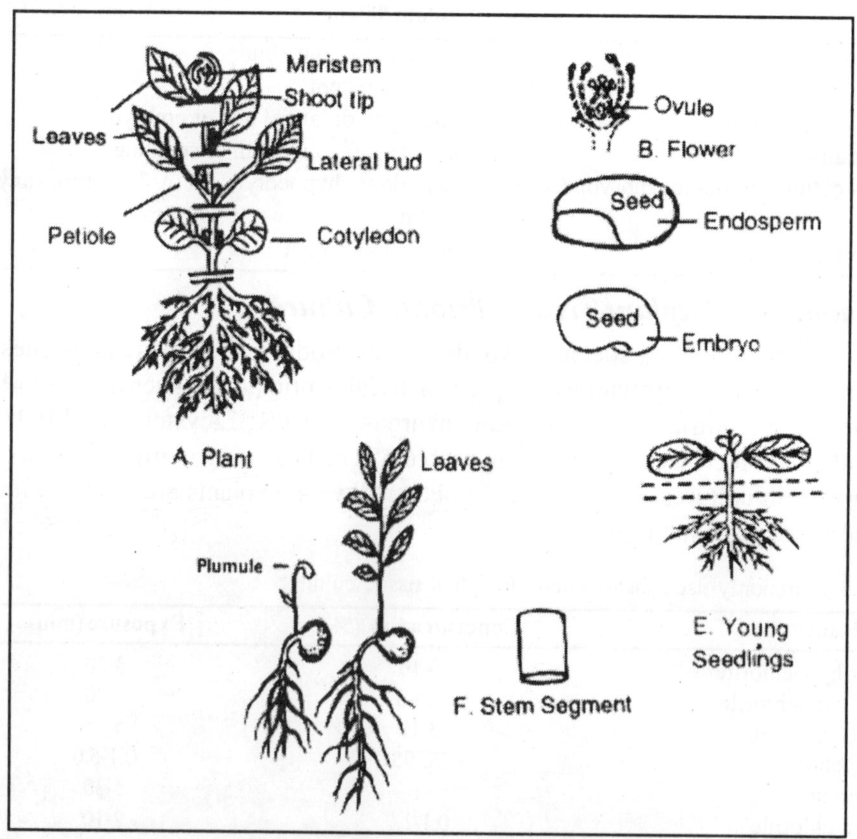

Diagram 3.5. Presentation of various plant parts.

Procedure

1. Wash explants in a mild detergent before treatment with disinfecting solution (exclude herbaceous material which may not need this treatment).
2. Rinse explants thoroughly under running tap water for 10-30 min
3. Submerge explants into the disinfectant solution and gently agitate.
4. Under sterile conditions, decant the solution and rinse explants for several times with sterile distilled water.
5. The explants drawn from adult woody species are often contaminated heavily with microorganisms. The sterilization procedure can be improved by a two step (two different disinfectants) sterilization procedure. The ethanol solution (70%) may enhance the sterilization procedure. A brief alcohol rinse or swabbing with alcohol wetted cheese cloth may be used.
6. Use of wetting agents such as Tween-20 or 80 to the disinfecting solution to reduce surface tension and allow better surface contact
7. Sterilization under vacum also improves the sterilization process.

Controlled *in vitro* Environment	
Aerial Physical Environment	**Root Zone Environment**
Temperature	**Physical Environment**
	Temperature
	Water potential
	Osmotic pressure
	Gas and liquid diffusivity in the medium
	Hardness or compactness of the medium
Light	**Chemical Environment**
Photosynthetic	Inorganic substance composition
Photomorphogenic	Organic substance composition
	pH
	Dissolved oxygen
Gaseous Composition	**Biological Environment**
Carbon dioxide	Symbiotic Microorganism
Oxygen	Competitive association
Ethylene	
Other gases	
Gas diffisuivity (air movement)	
Pressure	

Procedure

Aeration of Cultures: Cultured plant should be given proper aeration (supply of oxygen). The plant cell/callus/organs when grown on the semi-solid medium, it does not require any special device for aeration. However, in the liquid medium the plant cells get submerged and require some special device for proper aeration. In such case aeration is provided by shaking the flask or tubes on an gyratory shaker which have 80 to 220 rpm. The supply of oxygen to the plant part and shaking of the medium also check clumping of cells.

Why Aeration is Essential?: Effective aeration is essential for optimal growth of plant cells in suspension culture. One of the most commonly used fermenters for growing plant cell cultures is a V-shaped design (Kurz and Constable, 1979) in which the culture is agitated by a Teflon-coated magnetic stirrer supported on a small glass rod at the bottom. The air is introduced through a hypodermic needle.

Isolation of Plant Material

The tissue or plant part (2 to 4 mm^3 sterile segment) removed from the plant body for culture is called **explant** and the plant from which it is removed is known as **stock plant.** The age of stock piant and location on the stem from which the explants are removed can greatly effect the establishment of callus tissue in vitro.

Age of the Plant Tissue

Only young and healthy plant parts are used to establish callus on nutrient medium. Any meristematic tissue and rarely mature cells are used in raising callus tissue. Stem apex and seeds are the plant materials, commonly used to raise callus.

Tissues injured during explant excision from the stock plant often causes the release of various compounds that are air oxidized (oxidized by peroxidases or polyphenoloxidases and turn brown or black resulting in darkening of both tissue and culture medium). Often, if left unattended, a lethal browning of the explant will occur. It is, therefore, necessary to use a sharp scalpel to excise explant from stock plant.

Callus Tissue and Organogenesis

Callus (pl. calli) *on a wounded plant parts or on a culture medium is made of an amorphous aggregate of loose parenchyma cells which proliferate from the mother cells.* Callus is either homogenous parenchymatous mass or treachery elements or sieve elements or submerized cells or secretory cells or the trichomes. Callus formation has been found in angiosperms, gymnosperms, pteridophytes, and bryophytes. Callus contains no organised meristems. Callus is somewhat an abnormal tissue which has the potentiality to produce normal roots and embryoids and in turn it develops into plantlets. Callus may be hard (due to lignification of cell walls) or brittle and sometimes soft.

Suspension Cell Culture from Callus in vitro

After few subcultures, small bits from a soft callus can be cut and inoculated into liquid medium where they give rise to a suspension culture. Cell clones can be raised in the same manner, by plating a suspension of cells on agar plates. Colonies are formed, each representing a clone. They can be picked up individually and inoculated into liquid medium. Most plant cells can now be grown in suspension culture. Within a few days, cells proliferate and a callus culture is obtained. In this way, dividing cells form a layer of meristem and build a globular mass of non-dividing parenchyma. Alternatively, they may form small meristematic and build a globular mass of non-dividing parenchyma. Alternatively, they may form small meristematic zones interspersed in non-meristematic regions, yielding a sort of nodulated callus.

Organogenesis is the development of adventitious organs or primordia (embryoid) from undifferentiated cell mass (callus) in tissue culture. It is controlled mostly by a balance between cytokinin and auxin. A relatively high ratio of auxin: cytokinin induces root formation in callus tissues whereas, a low ratio induces shoot formation. **Caulogenesis** is a type of organogenesis by which only adventitious shoot bud initiation takes place in the callus tissue. When it is applicable for root, it is known as **rhizogenesis.** Anomalous structures when develop during organogenesis is called **organoids.** The localized meristematic cells on a callus which give rise to shoots and/or roots is termed as **meristemoids.**

MICROPROPAGATION THROUGH ORGANOGENESIS

The principle of micropropagarion is based on the phenomenon that shoot tip when cultured on tissue culture medium, can develop large number of shoots identical to the parent plant.

The technology provides many advantages over conventional propagation methods such as product development in highly speedy manner, uniforming in product development, large population can be produced in small growing space, quick multiplication of elite and genetically engineered products, and in vitro storage of germplasm. The technique also helps in enhancing tissue culture products via specific product format, rooted or unrooted micro cutting, bulblets, and embryos, certified plant material free of pathogens and production can be obtained throughout the year, as most of the multiplication phase is completed under artificially controlled conditions.

MAJOR STAGES OF MICROPROGRATION

⇓

Selection of an elite mother plant

⇓

Explant pieces

⇓

Trimming

⇓

Surface sterilization

⇓

Washing

⇓

Establishment of cultures on appropriate growth medium

⇓

Transfer of multiplication medium

⇓

Rapid shoot or embryo formation

⇓

Embryo shoots or plantlets transferred to sterilized soil or artificial medium by different gradual weaning processes (gradual reduction in humidity and nutrient levels)

Diagram 3.6. Procedure of removing plant parts.

Through the use of auxin and cytokinin, hormonal control of organogenesis became feasible. Skoog and Miller (1957) demonstrated that the differentiation of root, shoot, and both root and shoot in tobacco pith tissue is the result of auxin-cytokinin ratio (i.e., the differential regulatory role is concentration-dependent). A high level of cytokinin promoted shoot, while auxin promoted root formation. An equal concentration of cytokinin: auxin resulted in callusing. However, the requirement of exogenous supply of growth hormones for differentiation is dependent upon the endogenous level of these substances. To prove the efficacy of somatic cells to behave as zygote, Reinert (1958) and Steward (1959) reported the first somatic embryogenesis from the tissues of carrot.

By culturing shoot tips of *Lupinus* and *T.opaeolum,* Ball (1946) raised complete plants which could be transplanted. Shoot tip culture helped Morel and Martin (1952) to recover virus-free plants *of Dahlia* and, subsequently, in rapid multiplication of orchids.

Culture Media and Préparation

The success in cell, tissue and organ culture technology is related to the selection or development of the culture medium. As no single medium will support the growth of all tissue. Various media compositions which are frequently used for tissue culture. A literature search is useful for selecting the appropriate culture medium as a starting points in developing

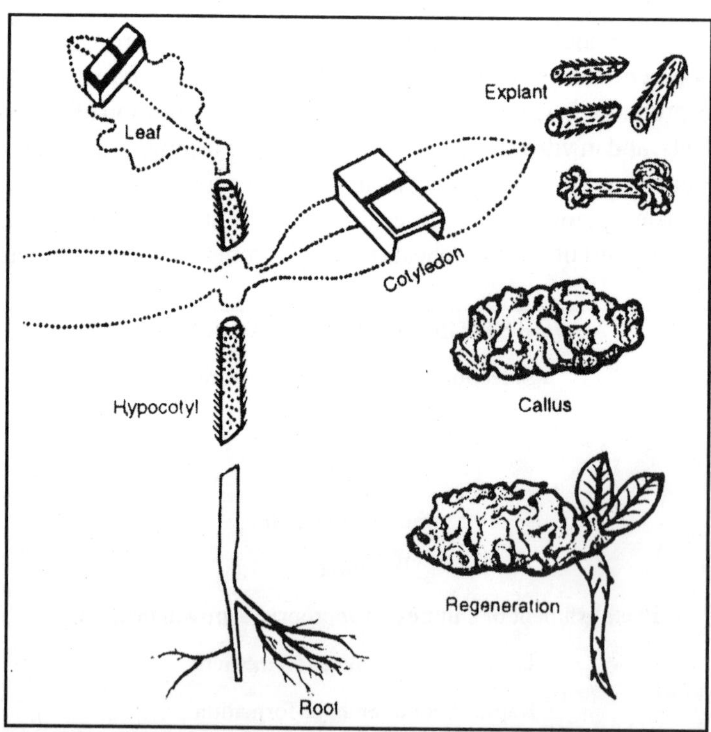

Diagram 3.7. Schematic representation of static callus culture of plant tissue.cultures therefore modifications in the nutritional component including growth regulators are often necessary for different types of growth responses in a single explant material.

a medium for specific purpose such as callus induction, axillary bud proliferation, organogenesis, somatic embryogenesis, anther culture etc. A nutrient medium generally contains inorganic salts, vitamins, growth regulators, a carbon source and gelling agent. Other components added for specific purposes include organic nitrogen compounds, hexitols, amino acids, antibiotics and plant extracts. The Murashige-Skoog medium (MS) (1962), Revised Murashige-Skoog medium (Raj Bhansali and Arya, 1978), White's medium (1963), Linsmaier and Skoog (LS) (1965), B5 (Gamborg et. al., 1968), Nitsch and Nitsch (1969), Woody plant medium (Llyod and McCown, (1981), Somatic embryogenesis medium (Raj Bhansali, 1988, 1990) and derivatives of these media have wide applications for different plant species and for different culture objectives. The decision on using type of media for the metabolic needs of the cultured cells and tissues, is a major factor of success in plant regeneration process.

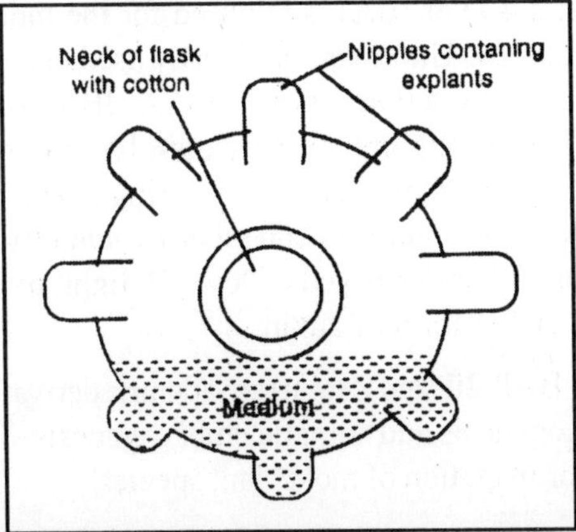

Diagram 3.8. Tissue culture techniques are now practiced on a grand scale. Special nippled flasks have proved very useful for the liquid culture of single cells and cell clumps.

Media Components

Inorganic Salts

A relatively small number of mineral salts are used as component of media for plant tissue culture. The inorganic salt formulations can vary in various reported media, however MS formulation is most widely used with or without modifications. The distinguishing feature of MS inorganic salts is their high content of nitrate, potassium and ammonium in comparison to other salt formulations. The stocks are prepared at 100 X (times) the final medium concentration. Each stock is added at the rate of 10 ml per 1000 ml of medium prepared.

The Na-FeEDTA stock should be protected from light stored in bottle that is amber coloured or wrapped in aluminium foil. Concentrated salt, stocks enhance the accuracy and speed of media preparation. Guidelines for maintaining stock solutions,

1. All salt stocks should be stored in the refrigerator and are stable for several months
2. Always prepare stocks with glass distilled or demineralized water
3. Label the stock solutions clearly with date

4. Reagent grade chemicals should be used to ensure maximum purity
5. The nitrate stock usually precipitate out and must be heated until crystals are completely dissolved before using
6. Any stock showing cloudy or has bacterial or fungal growth should be discarded
7. Do not combine the stock to other stocks unless they are stable and compatible.

Plant Growth Regulators

The four classes of growth regulators are commonly used in tissue culture media i.e., auxins, cytokinins, gibberellins and abscisic acid. The type of growth regulators and concentration used will vary according to the cell culture purpose.

Aauxin (IAA, NAA, 2,4-D or IBA) is required for the induction of cell division and root initiation in cultured tissues. The auxins are mostly used in combination with cytokinins. The 2,4-D is used for callus induction where as IAA, IBA, and NAA are used for root induction. Auxin stocks are usually prepared by dissolving in ethanol, I N NaOH or 1 N KOH until crystals dissolved (not more than 0.3 ml/10 mg of auxin), rapidly adding 90 ml of distilled water and increasing the volume to 100 ml in a volumetric flasks. Though the auxins are thermostable, however IAA is destroyed by low pH, light, oxygen and peroxidases. The NAA and 2,4-D are most stable form of auxin.

The **cytokinins** (KN, BAP, 2iP and Zeatin) are adenine derivatives, promote cell division, shoot proliferation, organogenesis and somatic embryogenesis. They have essential role in differentiation and micropropagation of most plant species.

The cytokinin stocks are prepared in a few drops of In HCl and water to dissolve crystals. Gentle heating is usually required for complete dissolving of crystals. Bring the stock up to the desired volume by adding double distilled water in a volumetric flasks. Cytokinin stocks can be stored for several months in the refrigerator. Cytokinins are thermostable during autoclaving in media.

The **gibberellins** are infrequently used in plant tissue cultures as it can inhibit callus growth but for meristem culture, after shoot primordia formation, are used in plant regeneration and elongation.

The stock solution of GA_3 can be prepared by dissolving in water and adjusting the pH 5.7. They are not thermostable therefore should be filter sterilized.

The **abscisic acid** is useful in embryo culture and somatic embryogenesis. Abscisic acid is heat stable but light sensitive. Stock solutions can be prepared in double distilled water and stored in coloured bottle in refrigerator. Dilution of stock solutions may be as per the requirement.

Vitamins: Vitamins have catalytic functions in enzyme reactions. The most commonly used vitamins in tissue culture media are nicotinic acid (B_3), thiamine (B_1) and pyridoxine (B_6). They are added in medium before autoclaving. The stocks are usually prepared in water at 100 X or 1000 X (10 ml per 1000 ml medium or 1 ml per 1000 ml medium) and stored in a freezer.

Carbon Source

The carbohydrates in form of sucrose or glucose (2-5% W/V), as a carbon source are essentially required in tissue culture as cells or tissues are generally not photosynthetically active. Lower levels of a carbohydrate may be used in protoplast culture but higher levels are required for embryo or anther culture.

Sugars undergo caramelization prolonged on autoclaving (too long period) and will react with amino acid compounds. Sugars are degraded and form melanoidin, which are brown, high molecular weight compounds that can inhibit cell growth.

Hexitols: Among hexitols, myo-inositol has been found very important ingredient in tissue cultures, it is considered as growth promoter in tissue cultures. This has an action like carbon source as well as vitamin. Mannitol or sorbitol are good osmotica for protoplast isolation. It is water soluble and stock can be made up at the strength of 100 X (10 ml aliquots are used for 1000 ml medium).

Gelling Agent

In tissue cultures, washed or purified agar of TC grade or Difco-bacto agar grade is used. The agar must be kept in motion while dissolving, otherwise it will burn on the bottom of the flask. The agar must be completely dissolved before it is dispensed into the culture vessels.

The agar can also be melted in a autoclave or in a foil capped Erlenmeyer flask for 15 min. at 121° C and dispensed aseptically into sterile containers by using laminar air and low bench before solidification of agar.

Amino Acids and Amides

The amino acids and amides are very important in tissue cultures specially for the morphogenesis. All L-forms of amino acids are commonly used, as L-tyrosine can contribute to shoot initiation, L-arginine can facilitate rooting, and L-serine can be used in haploid embryos induction in microspore cultures. L-cysteine is used for controlling phenol leaching from explant tissues. Amides such as L-glutamine and L-asparagine can induce somatic embryogenesis.

Antibiotics

The various fungicides and bactericides are used in case plant explants on cultures excessively contaminated. These chemicals are toxic not only to contaminants but also to cultures or explant materials so restricted use should be made for additions into the culture medium. The antibiotics are soluble in water should be made fresh and be added to the medium after autoclaving by filter sterilization.

Natural Complexes

The natural complexes such as coconut (endosperm) milk (CM), yeast extract (YE), malt extract (ME), tomato juice, potato extract, casein hydrolysate (use enzyme digest) and fish emulsion are used in tissue cultures for various purposes. Addition of these complexes in the medium make the medium undefined, since variation in growth promoting or inhibiting compounds in these complexes, exist.

Antioxidants: The antioxidants such as citric acid, ascorbic acid, pyrogallol, phloroglucinol and L-cysteine are used in tissue culture to reduce excessive browning of the explants. Adabsorbents like PVP and activated charcoal are also used for checking excessive browning.

ADDITIONAL REQUIREMENTS

Quality of water, chemicals and natural complexes: Demineralized or double distilled water of high purity are used in making stocks and medium. Glass distilled water is most desirable and stored in clean containers.

Callus-induction medium

Murashige and Skoog medium	1.01
2,4-D	1.0 mg
Agar	8.0 g

Prepare 1 litre of standard MS medium. Add the 2,4-D and adjust the medium pH to 5.5 using 0.1 M NaOH. Dissolve the agar in the medium in a steam bath. Dispense the medium into culture tubes or vessels and autoclave for 20 min at 121° C.

Chlorate Selection Medium

$MClO_3$	600.0 mg
$Ca(NO3)_2$ $4H_2O$	118.0 mg
$MgSO_4$ $7H_2O$	19.5 mg
K_2HPO_4	19.7 mg
P-IV trace metal solution	3.0 ml

Dissolve the salts in 900 ml of distilled water. Add the P-IV trace metal solution and adjust the medium pH to 7.0. Add additional distilled water to make 1 litre of medium.

Chlorate Selection Medium Overlay

Chlorate selection medium	500.0 ml
Gel-rite™ gelling agent	4.0 g

Slowly add the Gel-rite a little at a time to the chlorate selection medium while stirring the mixture with a magnetic stirrer. Set the mixture in a steam bath to dissolve the Gel-rite. Dispense 4 ml of the overlay medium into each culture tube; 4 ml should spread out as a very thin layer over the surface of the media in plates.

Embryo Culture Medium

Murashige and Skoog medium	1.01
Agar	8.0 g

Prepare 1 litre of standard MS medium. Adjust the medium pH to 5.5 using 0.1 M NaOH. Dissolve the agar in the medium in a steam bath. Autoclave for 25 min at 121°C. Allow the medium to cool to 50°C in a temperature controlled water bath. Pour the medium into sterile 100 mm petri plates.

L/10 Green Algae Medium

$Ca(NO_3)_2$ $4H_2O$	0.118/g
$MgSO_3$ $7H_2O$	0.0195 g
K_2HPO_4	0.0197 g
P-IV trace metal stock	3.0 ml

Dissolve all of the salts in 1 litre of distilled water. Adjust the medium to pH 7.0 by adding 1 M HCl or 1 M NaOH.

Micropropagation Medium

Murashige and Skoog medium	1.01
Indolebutyric acid (IBA)	1.0 mg
Benzylaminopurine (BAP)	3.0 mg

Prepare 1 litre of standard MS medium. Add the IBA and BAP and adjust the medium pH to 5.5 using 0.1 M NaOH. Dissolve the agar in the medium in a steam bath. Dispense the medium into culture tubes or vessels and autoclave for 20 min at 121°C.

Murashige and Skoog (MS) Medium

Macro salts		**Micro salts**	
NH_4NO_3	1.65 g	$FeSO_4$ $7H_2O$	27.80 mg
KNO_3	1.90 g	Na_2EDTA $2H_2O$	33.60 mg
$CaCl_2$ $2H_2O$	0.44 g	Kl	0.83 mg
$MgSO_4$ $7H_2O$	0.37 g	H_3BO_4	6.20 mg
KH_2PO_4	0.17 g	$MnSO_4$ $4H_2O$	22.30 mg
		$ZnSO_4$ $7H_2O$	8.60 mg
		Na_2MoO_4 $2H_2O$	0.25 mg
		$CuSO_4$ $5H_2O$	0.025 mg
		$CoCl_2$ $6H_2O$	0.025 mg
		Organic Supplements	
		Myoinositol	100.00 mg
		Nicotinic acid	0.05 mg
		Pyrodoxine HCl	0.05 mg
		Thiamine HCl	0.05 mg
		Glycine	0.20 mg
		Sucrose	20.00 g

Dissolve the salts and organics in 800 ml of distilled water. Adjust the medium pH to 5.7 by adding 1 M NaOH. Add additional distilled water to adjust the final volume to 1 litre.

MS/C Medium

MS salts and organic supplements	As described above
IAA solution	8.0 ml
Kinetin solution	2.5 ml
Agar	8.0 g

Dissolve the salts and organics in 800 ml of distilled water. Add the IAA and kinetin solutions. Adjust the medium pH to 5.7 by adding 1 M NaOH or 1M HC1. Add additional distilled water to adjust the final volume to 1 litre. Add the agar and heat the medium on a hot plate or in a steam bath until the agar melts. Stir the medium occasionally until all the agar is dissolved and the solution is clear. Do not let the medium boil. Dispense 8 ml aliquots in 20 X 150 mm culture tubes (approximately 120 tubes of medium).

Myriophyllum aquaticum Shoot-Induction Medium

Murashige and Skoog medium	1.0 litre
[2-isopentenyl] adenine (2iP)	2.0 mg
Agar	8.0 g

Prepare 1 litre of standard MS medium. Add the 2iP and adjust the medium pH to 5.7 using 0.1 M NaOH. Dissolve the agar in the medium in a steam bath. Dispense the medium into culture tubes or vessels and autoclave for 20 min at 121°C.

M. aquaticum Stock Plant Medium

Murashige and Skoog medium	1.01
Agar	8.0 g

Prepare 1 litre of standard MS medium. Adjust the medium pH to 5.7 using 0.1 M NaOH. Dissolve the agar in the medium in a steam bath. Dispense the medium into culture tubes or vessels and autoclave for 20 min at 121°C

P-IV Trace Metal Solution

Na_2 EDTA	0.750 g
$FeCl_3$ $6H_2O$	97.0 mg
$MnCl_2$ $4H_2O$	41.0 mg
$ZnCl_2$	5.0 mg
$CoCl_2$ $6H_2O$	2.0 mg
Na_2 MoO_4	4.0 mg

First dissolve the Na_2EDTA in 500 ml of distilled water, then dissolve the remaining metal salts. For greater accuracy, it may be easier to prepare 10 X concentration stock solutions of the Zn, Co, and Mo salts and add 1/10 of these stocks to the P-IV stock solution.

Pandorina Ammonium Medium

NH_4Cl	27.0 mg
$CaCl_2$ $2H_2O$	100.0 mg
$MgSO_4$ $7H_2O$	19.5 mg
K_2HPO_4	19.7 mg
P-IV trace metal solution	3.0 ml

Dissolve the salts in 900 ml of distilled water. Add the P-IV trace metal solution and adjust the medium pH to 7.0. Add additional distilled water to make 1 litre of medium.

Pandorina Nitrate Medium

$NaNO_3$	35.0 mg
$CaCl_2\,2H_2O$	100.0 mg
$MgSO_4\,7H_2O$	19.5 mg
K_2HPO_4	19.7 mg
P-IV trace metal solution	3.0 ml

Dissolve the salts in 900 ml of distilled water. Add the P-IV trace metal solution and adjust the medium pH to 7.0. Add additional distilled water to make 1 litre of medium.

Pandorina Hypoxanthine Medium

Hypoxanthine	68.0 mg
$CaCl_2\,2H_2O$	100.0 mg
$MgSO_4\,7H_2O$	19.5 mg
K_2HPO_4	19.7 mg
P-IV trace metal solution	3.0 ml

Dissolve the salts in 900 ml of distilled water. Add the P-IV trace metal solution and adjust the media pH to 7.0. Add additional distilled water to make 1 litre of medium.

Pandorina Uric Acid Medium

Uric acid	84.0 mg
$CaCl_2\,2H_2O$	100.0 mg
$MgSO_4\,7H_2O$	19.5 mg
K_2HPO_4	19.7 mg
P-IV trace metal solution	3.0 ml

Dissolve the salts in 900 ml of distilled water. Add the P-IV trace metal solution and adjust the medium pH to 7.0. Add additional distilled water to make 1 litre of medium.

Potato Dextrose Agar

Wite potatoes, sliced	250 g
Dextrose	20 g
Agar	15 g

Boil the potatoes in 500 ml of distilled water for 15 min until it get soften. Filter this mixture through cotton to remove most of the particulate matter. Dissolve the dextrose in 200 ml of the potato infusion. Add 800 ml of distilled water. Adjust the final pH to 3.5-4.0. Dissolve the agar in a steam bath or on a hot plate. Autoclave at 121°C for 25 min.

Trypticase-Soy Broth Medium

Trypticase	17.0 g
Phytone	3.0 g
NaCl	5.0 g
K_2PO_4	2.5 g
Glucose	2.5 g

Dissolve the ingredients in 1 litre of distilled water. Adjust the medium pH to 7.3 by adding 1 M NaOH. Dispense 10 ml aliquots in 20 X 150 mm culture tubes, (approximately 100 tubes of medium).

Yeast Extract Broth (YEB)

Yeast extract	1.0 g
Beef extract	5.0 g
Peptone	5.0 g
Sucrose	5.0 g
$MgSO_4$ $7H_2O$	0.5 g

Dissolve all the ingredients in 1 litre of distilled water. Adjust the medium pH to 7.0 by adding 1 M NaOH. Dispense and autoclave for 25 min. at 121° C.

Yeast Extract Indicator Medium (YI)

Yeast extract	1.0 g
Lactose	10.0 g
Agar	20.0 g

Dissolve the yeast extract and lactose in 1 litre of distilled water. Adjust the medium pH to 7.0 by adding 1 M NaOH. Add the agar and dissolve it by heating the mixture in a steam bath or on a hot plate. Autoclave for 25 min at 121°C.

STERILIZATION OF MEDIA

Tissue culture media are generally sterilized by autoclaving at 121°C and 1.05 kg/ cm^2 (15-20 psi). The time required for sterilization depends upon the volume of medium in the vessel. Dispense medium in small aliquots whenever possible as many media components are broken down on prolonged exposure to heat. Medium exposed to temperatures in excess of 121°C may not properly get or may result in poor cell growth.

Minimum Autoclaving Time

Volume of Medium per Vessel (ml)	Minimum Autoclaving (min)	Volume of Medium per Vessel (ml)	Minimum Autoclaving (min)
25	20	500	35
50	25	1000	40
100	28	2000	48
250	31	4000	63

Minimum autoclaving time includes the time required for the liquid volume to reach the sterilizing temperature (121° C) and 16 min at 121° C. Times may vary due to differences in autoclaves. Validation with system is recommended. Several medium components are considered thermo-labile and should not be autoclaved. Stock solutions of the heat labile components are prepared and filter sterilized through a 0.22 mm filter to sterile container. The filtered solution is aseptically added to the culture medium which has been autoclaved and allowed to cool to approximately 35-45°C. The medium is then dispensed under sterile conditions.

Use and Storage of Coconut Water

Coconut water has been shown to stimulate shoot proliferation in many species of plants. Coconut water is prepared from selected coconuts by filter sterilized and frozen prior to use. This precipitation should not effect the growth of the plant tissue. Coconut water can be divided into smaller aliquots, corresponding to standard medium batch size, and refrozen until needed. Coconut water should be used at a concentration of 5-20% (v/v).

Related Procedures

Ultraviolet Light

UV light may be divided into three wave length groupings near UV (315-400 nm), mid range UV (280-315 nm) and far UV (200-280 nm). Maximal sensitivity in humans is at about 280 nm. Exposure to direct or indirect mid-range or for UV can cause acute eye irritation after a latent period of 2-24 hrs. Because retina is not sensitive to UV eye damage may result without the subject being aware of the exposure. Skin is also sensitive to UV which may cause for skin cancer. Hence protect your eyes and skin from the effects of UV irradiation by wearing goggles with side shields by clothing, and by limiting exposure.

Preparation of Phenol

All crystalline phenol must be redistilled at 160° C to remove contaminants that cause or cross linking of DNA or RNA. Soon after distillation add 0.1% hydroxyquinoline. The melted phenol is extracted several times with an equal volume of 1.0 M Tris pH 8.0 followed by 0.1 M Tris pH 8.0 and 0.2% β-mercaptoethanol, until pH of the aqueous phase is 7.6. Phenol is stored in aliquots at 4° C under equilibration buffer for periods upto 1 month.

Phenol is widely used as a disinfectant and germicide. It is a dangerously toxic materrial that can produce poisoning when ingested, inhaled or absorbed through the skin. The toxic effect include headache, dizzines, nausea, weakness, difficulty in breathing, unconciousness and death. Phenol is corrosive to skin, initially producing a softened area followed by severe burns.

1. If phenol is spilled on the skin, flush immediately with large amounts of water. Do not use ethanol.
2. If eyes are contaminated, wash them with running water for about 15 min, call for medical help.

Working with 32P Labelled Compounds

β-particles with an energy of 1.71 MeV (6.1 meter range in air) is emitted by32p Hence - 32P labelled compounds must be handled carefully with much caution using shields. When β particles hit targets, electromagnetic radiation known as Bremsstrahlung is produced, the yield of which is directly proportional to the density of material used for shielding. Therefore, a low density material may be added to absorb the Bremsstrahlung emitted. Always ear gloves (two pairs if necessary), protect eyes and use dosimeters. Always cover the work area with absorbant papers and use survey meter to check spillage. Eating, smoking or drinking while handling radioactive compounds should be banned. Use special tape to label containers and tubes in which radioactive materials are kept. The maximum permissible burden of 32P is 30 uCi but the maximum permissible burden for bone is only 6 uCi.

Silanization of Plastic and Glassware

Place the plastic and glasswares to be silanized in a desiccator. Add about 1 ml of dichlorodimethyl silane in a small container in the desiccator. Pull vacuum on the system until dichlorodimethyl silane boils. Close the system and allow to sit for about 2 hrs. Open the desiccator in a fume hood and allow to air out several hours. Rinse the plastic and glassware with water and autoclave.

10 X Restriction Endonuclease Buffers (Refrigerate)

Low	100 mM Tris-HCl (pH 7.5) 100 mM $MgCl_2$ 10 mM Dithiothreitol (DTT)
Medium	500 mM NaCl 100 mM Tris-Cl (7.5) 10 mM DTT
High	1 M NaCl 500 mM Tris-Cl pH 7.5 100 mM $MgCl_2$ 10 mM DTT

Preparation of Dialysis Tubing

Cut dialysis tubing into convenient length. Boil them for 10 min in a large volume of 2% sodium bicarbonate and 1 mM EDTA, cool and rinse thoroughly in distilled water and again boil for 10 min. in distilled water. Cool and store inrefrigerator submerged in water. Just before use rinse with water. Wear gloves while handling the dialysis tubing.

Lengths and Molecular Weights of Common Nucleic Acids

Nucleic Acid	Number of Nucleotides	Molecular weight
LAMBDA DNA	48,502 (Circular, dsDNA)	3.0×10^7
pBR322 DNA	4,363 (dsDNA)	2.8×10^6
28SrRNA	4,800	1.6×10^6
23S rRNA	3,700	1.2×10^6
18S rRNA	1,900	6.1×10^5
16S rRNA	1,700	5.5×10^5
5S rRNA	120	3.6×10^4
tRNA (E.coli)	75	2.5×10^4

Standards

1 kb of ds DNA (sodium salt)	6.6×10^5 Daltons
1 kb of ss DNA (sodium salt)	3.3×10^5 Daltons
1 kb of ss RNA (sodium salt)	3.4×10^5 Daltons
The average MW of a deoxynucleotide base	324.5 Daltons

Common conversions of Nucleic acids and Proteins

I. Spectrophotometric Conversions

IA $_{260}$ unit of ds DNA	50 mg / ml
IA $_{260}$ unit of ss DNA	33 mg / ml
IA $_{260}$ unit of ss DNA	40 mg / ml

II. Protein Molar Conversions

100 pmoles of 100,000 MW protein	10 mg
100 pmoles of 50,000 MW protein	5 mg
100 pmoles of 10,000 MW protein	1 mg

III. Protein / DNA Conversions

1 kb of DNA = 333 amino acids of coding capacity	3.7 x 10^4 MW
10,000 MW protein	270 bp DNA
30,000 MW protein	810 bp DNA
50,000 MW protein	1.35 kb DNA
100,000 MW protein	2.7 kb DNA

Half life of Important Radioisotopes Used

Radionucleotide	Half life
Tritium	12.43 years
Carbon - 14	5.730 years
Sulphur-35	87.4 years
Phosphorous-32	14.3 years

Problems and Possible Solutions in Plant Tissue Culture Work

Symptoms	Possible Causes	Possible Solution
Culture contamination	Source heavily infested/ infected with micro organisms Poor sterilization	Improve sterilization method (70% Ethanol, drop of detergent, antimicrobial agents) clean plant material/select unexposed tissue.
Explant dies	Strong disinfectants	Use weaker disinfectant/change sterilizing agents
	Media too strong	Use 1/2 or 1/4 strength
	Wrong stage of growth	Obtain explants at different stage of growth
Culture blackens and dies	Contaminated by micro-organisms	Discard with care. Review sterile technique and sanitation
	Bleeding	Subculture immediately. Transfer more frequently Try different agar concentrations
	Agar problem	Check water purity Try different formula
	Water problem	
	Wrong formula media constituents	

Symptoms	Possible Causes	Possible Solution
Explant live but no growth	Dormant	Chill for a month
	Media too harsh	Use explant at different stage of growth
	Wrong formula	Lower salts and hormones try different formula
Culture live but no growth	Too cold	Increase temperature
	Wrong medium	Try different medium
Shoots tool long (leggy), and poor multiplication leaves are yellow, watery	Too little cytokinin	Increase cytokinin Change cytokinin/auxin ratio
Leafy shoots too short	Hormones too strong	Decrease or omit hormones
No multiplication	Too little cytokinin	Increase cytokinin
	Needs chilling	Cold store 4-8 weeks
	Too cold	Run cytokinin / auxin grid
	Requires dormacy period	Increase temperature Cold treat 3-8 weeks
Fat stems, small and pale leaves	Too much cytokinin	Decrease cytokinin/increase auxin
Unwanted callus	Wrong hormones	Decrease or omit hormones Run cytokinin/auxin grid
Leaves chlorotic	Contaminant	Index for contaminants
	Too hot	Decrease temperature
	Wrong formula	Try different medium
Leaves succulent (watery), virtrification, abnormal stem, embryos	Osmotic potential upset	Decrease temperature
	Too high cytokinin	Increase agar strength
	Wrong agar	Decrease hormones
	Culture too old	Try different agar
		Transfer more often
Premature rooting	Wrong hormone balance	Transfer more often
		Run cytokinin/auxin grid
		Increase cytokinin/decrease auxin
Red stems/embryos/cells	Stress	Change light and/or temperature
		Lower sucrose content
		Increase nitrate (NO_3)
		Transfer more frequently
Non-friability of callus	Too much sugar Not enough NO_3 Culture too old Media composition Plant source	Using number of media having range of hormones. Select most rapidly growing callus. Change to liquid culture and after friable to agar medium.

Source : Bhansali, R.R. (1995)

INITIATION OF EMBRYOGENESIS IN SUSPENSION CULTURE

Steward and coworkers (1958) observed the initiation and development of embryos from differentiated tissue (somatic tissue). However, earlier researchers had already observed these formations in culture, without recognizing them as embryos.

These embryos are named as **somatic embryos,** and their origin is different from zygotic embryos, which result from the fusion of gametes. This phenomenon is also called adventive or **asexual embryogenesis.**

Somatic embryos closely resemble zygotic embryos in their structure, as well as in their physiological properties, especially with wild carrot. Nevertheless, the organization of these somatic embryos is generally more variable than that of zygotic embryos. Abnormalities in shape and size are frequent, and because of this variation they were, at first, called **embryoids.**

In contrast to buds, somatic embryos have a bipolar axis with an apical meristem and a root, and as they have no vascular connection with the mother callus, they are easily detached by the swirling action of the agitated medium.

Three types of medium are required to induce somatic embryogenesis in suspension.

1. The first medium is solidified by agar and contains auxin; this dedifferentiates the somatic tissue and provokes initiation of embryogenic cells (primary culture).
2. The second medium, which is liquid and also contains auxin, ensures the multiplication of these cells.
3. The third medium, which is also liquid but without auxin, allows the cells to express their embryogenic potential.

It is not very easy to carry out culture in a liquid medium. When the tissue is inoculated, it sinks and rapidly dies because of lack of oxygen. To prevent asphyxia, the liquid must be agitated to dissolve air in the medium. The culture of somatic embryos in a liquid medium, however, has numerous advantages. The swirling medium naturally separates the embryos, which are then easily observed. Thus, the embryos can be fractionated according to their stages. They can be obtained in great quantity and used as a basis for a large-scale micropropagation.

Materials

1. Growth chamber for plant culture. Temperature (25° C) and lighting are regulated. Lighting must be bright (5 W/m^2), discontinuous (16 h of lighting/d), and provided by Grolux fluorescent tubes (F36W/GRO). This growth chamber is optional, since plant material can also be collected from outdoors.
2. Culture room for tissue culture. Temperature has to be regulated at approximately 25° C. Continuous light (1 W/m^2) can be supplied by ordinary fluorescent tubes placed above the rotary shaker.
3. Actively growing callus in tubes.

4. Culture medium: The mineral solution of Murashige and Skoog can be used for culture of somatic embryos. The solution described in chapter was evolved for the culture of zygotic embryose because of its low toxicity. It must have a pH of 5.5. In addition, it contains $NH4^+$ and K^+ ions, which are considered to be promoters of somatic embryo growth.
5. For mature embryo-growth, use half-strength MS medium containing sucrose at 5 g/1.
6. Sieves (1-mm grid) made of stainless screens (tea sieves from hardware store) to fraction suspension culture and transfer only the smaller embryos into the new flask. Sieves can also be made with nylon mesh filtration cloth (industrial nylon mesh filter).
7. Subsequent plantlet development requires filter-paper bridge apparatus and small pots containing soil mixture.

Method

Generally, three steps are needed to obtain embryogenic suspension culture. First, embryogenic tissue is initiated by culture on an agar medium with auxin; it is then transferred to liquid medium with auxin, and eventually transferred into a liquid medium without auxin.

Initiation of Embryogenic Tissue

Growth initiation is better on solidified medium containing 2,4-D (1 mg/1). When the callus has developed the potentiality for embryogenesis, that is to say 5 wk at least after explant isolation, it is transferred to a liquid medium.

Establishment of Embryogenic Suspension Culture

1. The callus should be removed from the culture tube with sterile forceps and transferred to a Petri dish containing filter paper.
2. Discard brownish parts of the callus, and inoculate a fragment of about 2 g in a 250-ml, Erlenmeyer flask containing 50 ml of medium.
3. Smaller Erlenmeyer flasks can be used, but the volume of liquid in relation to the size of the flask must be, for adequate aeration, about 20% of the volume of the flask. These flasks are placed on a gyrotory or orbital shaker and are agitated at 100-150 rpm.
4. When the plant material is first placed in the medium, there is an initial lag period prior to cell division. This is followed by an exponential rise in cell number. Finally, the cells enter a stationary stage.
5. In order to maintain the viability of the culture, the cells should be subcultured at the beginning of this stationary phase. It is reached in 2-3 wk, and the suspension has to be transferred to fresh medium at regular intervals within this period. In addition, a minimum density must be achieved when cells are transferred to fresh medium, to maintain embryogenic potential.

6. To subculture embryogenic suspension, wait several minutes until the cells are settled at the bottom of the flask and decant almost all the medium. Resuspend the suspension by gently rotating the flask, and transfer one-fourth of the entire population to fresh medium.

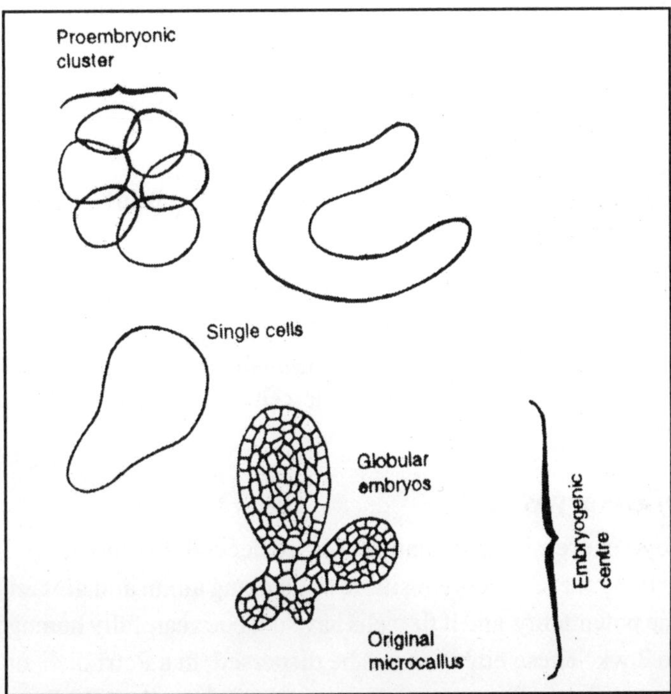

Diagram 3.9. The embryogenic cells divide in the liquid medium and constitute proembryonic clusters. A certain number of new embryogenic centers will give rise to globular embryose of different sizes (wild carrot).

1. All these operations can also be carried out, under safer conditions, using a micropipet. In that case, when the cells have settled, aspirate almost all the supernatant above the cells then, transfer one-fourth of the cell population into a new medium.

2. A population of cells typically shows a wide range of proembryonic stages visible under the dissecting microscope. It is a mixture of single cells, proembryonic cell clusters, and new embryogenic centers, and older embryos.

3. To obtain a certain degree of uniformity, it is necessary to sieve the inoculum when transferring the suspension.

4. The proembryo suspension is passed through a 200 mm sieve and then a 100 mm sieve. In the second sieve, differentiated embryos from the late globular stage are retained and may be examined, while proembryonic cells pass through.

5. When the suspension, which has passed the sieves, is settled decant most of the medium to obtain a suspension with a high density of cells.

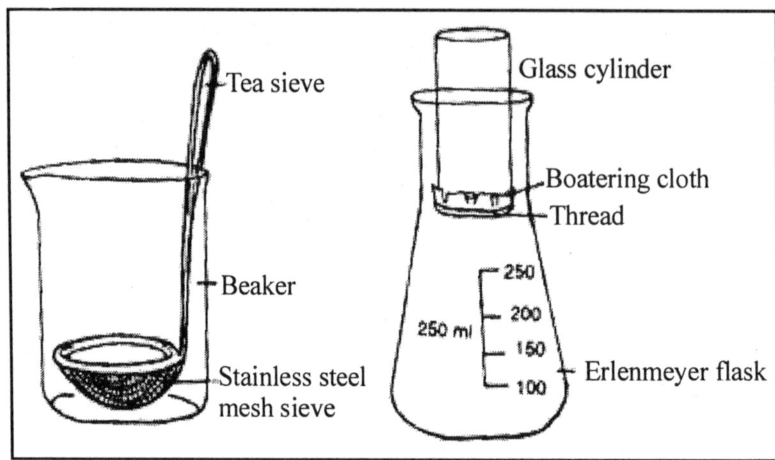

Diagram 3.10. The metal sieve, on the left, is used principally to remove the larger embryos before transferring the embryonic population into a fresh medium. The nylon sieve, on the right, is used to obtain a homogeneous population of embryogenic cells and synchronize the different stages of embryo development.

Maturation of Embryos

When embryo expression and maturation are needed, the inoculum is transferred to a liquid medium having the same composition, but lacking auxin and glutamine. Cells express their embryogenic potentiality, and if the cells have not been carefully homogenized by sieving, a mixture size in 2 wk. These embryos can be dispersed, in a Petri dish, on an agar medium without auxin, but with small amounts of cytokinins (0.1 mg/1 of BAP or zeatin).

When the embryos are mature, they are transferred to tubes on filter paper bridges. The medium is composed of Murashige and Skoog mineral solution diluted by half and containing a low concentration of sucrose (5 g/1). To stimulate the formation of leaves, the tubes must be well illuminated. When the plantlets are tall enough, with a large number of leaves and roots, they are transferred into pots with a mixture of peat, soil, and vermiculite for subsequent development. After their transfer, plantlets have to be protected from descication by covering the pots with plastic film for a few days.

Important Parameters for Consideration

Simplification of the Technique

1. Suspension culture can be started by inoculating the liquid medium containing 2,4-D (1 mg/1) with an explant of differentiated tissue isolated directly from the plant.
2. The initiation of multiplication of cells requires the inoculation of a large amount of tissue-about 3 g for 100 ml of medium.
3. The dividing cells will gradually free themselves from the inoculum because of the swirling action of the liquid.

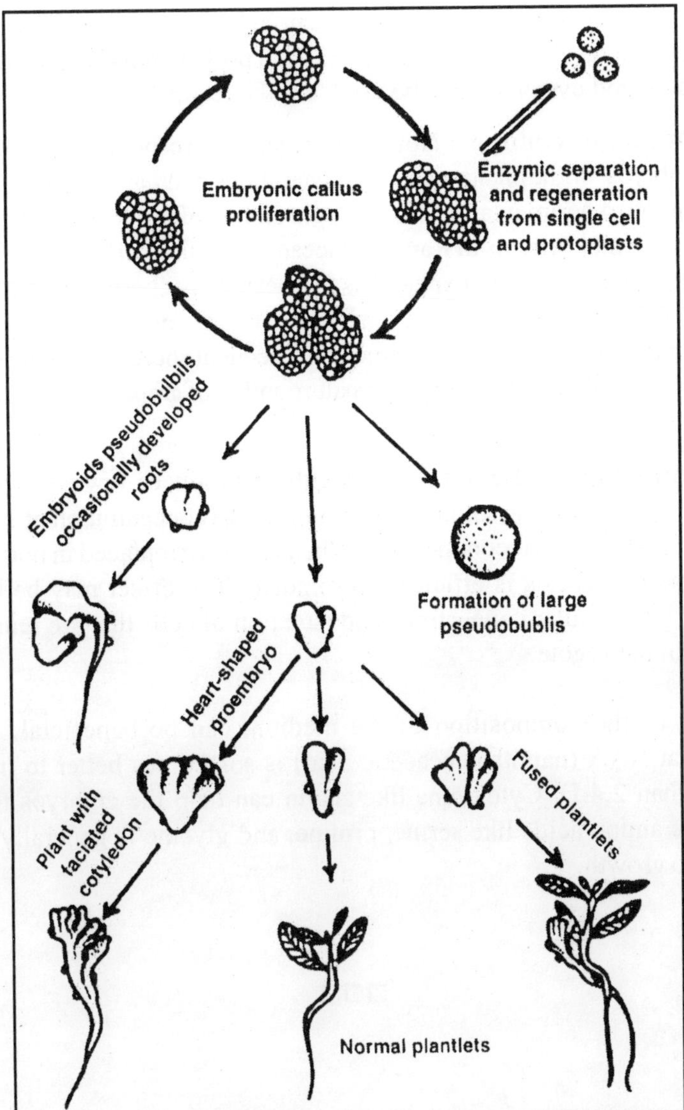

Diagram 3.11. llustrative presentation of embryonic callus proliferation, enzymic separation and regeneration from single cell and protoplast.

Different Causes of Failures (and Remedies)

1. **Darkening of the medium:** In a number of cases, especially with explants isolated from woody plants, the liquid medium becomes dark after a few days. The radicle ends of the embryos turn brown, and the embryos rapidly stop growing and show necrosis.

   Several procedures can help to reduce this process. More frequent transfer into fresh medium can prevent browning. After transfer to liquid, for the first days, when the proembryos are particularly sensitive, the culture must be performed in complete darkness. After this period, the culture can again be exposed to light. This lethal browning is the result of formation of oxidized polyphenols, so antioxidants, such as ascorbic acid and cysteine, may be used, but the results are variable.

2. **No appearance of embryos:** Not all plant species respond to culture and produce somatic embryos. Some plants give only adventitious buds, and others, no organogenesis at all. If a good experimental model is sought , it is preferable to select some species in families like Umbelliferae an Ranunculaceae, since they offer an easy way to obtain somatic embryogenesis. The explant has to be carefully chosen, since different regions of the plant body do not have the same embryogenic potential. Young material like zygotic embryos and reproductive tissue is an excellent source to initiate embryogenesis. For example, the diploid tissue of an immature anther of grapevine develops into somatic embryos.

3. **Insufficient number of embryos:** If a culture produces few embryos, suspending the cells in a plasmolyzing solution containing a high concentration of sucrose (1M) for a short time (45 min) is recommended. When these are replaced in normal conditions, somatic embryogenesis is efficiently promoted. The effect may be because of the rupture of intercellular connections and isolation of cells that are reinstated into the condition of the zygote.

Alterations of the composition of the medium can be beneficial. Some authors demonstrated that NAA (naphthaleneacetic acid) is sometimes better to initiate somatic embryogenesis than 2,4-D. Cytokinins like zeatin can help the embryos to mature. The addition of some amino acids like serine, proline, and glycine is generally considered to stimulate embryo growth.

❑❑❑

CHAPTER-4

Micropropagation in Plants

Men as per its natural habit to do something new, started to develop superior plant material through vegetative propagation, because of its beauty, growth form or yield for centuries. It is only during the 20 years that micropropagation has been started to increase the levels of efficiency and control over these systems. Micropropagation provides method to maintain the original genetic stock of the organism.

It is important, if one has to produce many plants true to its mother, from a single individual that the starting material be of the right clonal selection. A clone is a population of genetically identical cells or organisms. Micropropagules or clones of plants are easily produced by vegetative propagation or by tissue culture methods. Cloning is a method of preservation of superior genotypes in organisms which otherwise might get lost along with the death of the individual. For effective micropropagation plant must be in an active, vigorous, and healthy state. Micropropagation cannot be regarded as a rehabiliation phase for poor, sickly material; if it is poor and sick on its own roots, it is going to be equally substandard in culture. The technique has been successfully employed in various ways for the benefit of man , such as cloning of nitrogen fixation (nif) region, cloning of penicillin G acylase gene, cloning of genes of human peptide hormones, cloning of human leucocyte interferon gene etc.

Material for Micropropagation

The starting material for a micropropagation programme is a shoot tip, not necessarily as small as the meristematic apical dome, but something a little larger with leaf primordia already developing around it. This material must initially be surface sterilized. Any tissue affected by exposure to sterlize material is trimmed and the remaining material placed on one of a variety of different media, since it is initially unknown which medium will be suitable for a specific variety or species involved. Concentration of growth regulators should be adjusted in order to allow the plant tissue to initiate into culture. This is the *establishment phase* of the tissue in vitro conditions and can take, in some species, as long as 3 or 4 months, usually it can be completed within a few weeks. Under the optimal conditions the small shoot tip will initiate to grow in normal way, giving rise to a small, unrooted shoot in culture.

The initiated shoot, after a suitable period in the initiation medium can be subcultured such that nodal sections of the shoot are placed onto a multiplication medium. This medium differs from the initiation medium in that, it contains higher levels of cytokinins which are likely to give rise to precocious shooting of the axillary buds. The normal period between subcultures is three weeks and during this time the number of shoot tips available for subculture multiply as a result of breaking of nodal axillary buds anywhere from 2 to 12 times. Thus, after three weeks on multiplication medium, somewhere on an average of about 5 shoot tips can be placed onto fresh medium, and these in turn will multiply 5 times in the subsequent 3 weeks, and thus micropropagation has led to a 25-fold increase in the plant material available over a total of 6 weeks.

For rooting in micro-shoots, transfer the shoot to media which is rich in auxins and will allow the induction of roots in vitro. The roots formed in vitro are usually different morphologically than that of normal. They are be slightly swollen and brittle, and upon transfer to the soil new roots grow out from the plant; before any extensive growth of the shoot can occur and the initial in vitro roots die back. This lag in the establishment of the plants can be critical and certainly costly to a grower. It can be avoided by the direct rooting of the micropropagated shoots into a natural or artificial potting mix. If the micropropagated shoots are treated as micro-cuttings, and placed in artificial potting mix maintained at high humidity, roots will develop within 2 to 3 weeks. These are normal physiological roots and not like the in vitro roots mentioned above. There is no lag phase when the material is potted on. It represents important part in the overall strategy of plant establishment.

Once the micropropagated plants are satisfactorily rooted in soil or soil substitutes, they must be weaned to a normal humidity level. During the rooting process they are maintained at close to 100% humidity, and this does not allow the normal development of cuticle and stomatal activity associated with drier conditions. The rooting normally takes place under reduced light. After rooting plants are exposed to full sunlight. This is done slowly for a period of 7 to 10 days. Such small plants are now autonomous, photosynthesi/ing and can fixed carbon. During the rooting phase under the reduced lighting conditions, they are living on carbohydrate reserves built up during micropropagation period in vitro. Once this weaning process is completed, the hardened can be incorporated into normal growing systems for their further development.

Advantages of Micropropagation

This technique is an alternative method of vegetative propagation, and is applied with the objective of enhancing the rate of multiplication. The number of plant propagules that can be developed in a short time are extremely large. Currently, worldwide use of micropropagated plants is about 50 million per annum. It is likely that the market itself could absorb more than 250 million per annum, if sufficient facilities are available. The significance of the system lies in the fact that if one sprout or hybrid plant becomes available from a grower or breeder, it is possible to use hundreds of thousands of plants within a relatively short time - possibly 18 months to 24 months through micropropagation. This allows the rapid introduction of new lines and varieties into the marketplace. It is estimated that this system shortens the average length of time taken to introduce new lily varieties by about 50% from 15 to 16 years down to 7 to 8 years. The same is true of many other plant species.

The constant demand for new varieties and novelty within the ornamental industry means that the micropropagation process can answer an industry requirement by allowing for the introduction of new varieties on a rapid basis.

A major advantage of micropropagation happens to be the minimum growing space is required in commercial nurseries. Several thousand million plants can be maintained inside vials on a shelf space built into a room of about 3 m x 3 m x 5m. This makes possible the propagation of clones on a commercial scale for large number of horticultural species (African violet, banana, eucalyptus, ferns, orchids, gloxinia, gerbera and rhododendrons etc.).

The investment in commercial tissue culture business will depend to a large extent on cost of the laboratory set-up, type of plant to be propagated and the skill involved. It is advisable that a commercial nursery man start with those crop species for which published methods are available.

Table 4.1. Comparison of conventional and in vitro breeding system

Factors	Conventional breeding	*in vitro* breeding
Growth cycle	Several decades	25-26 hr
Size		One to several dozen metres, 50-100 mm in diameter
Space (106 offspring per year)	Several hectares	101 of suspended cells
Quantities required for mutation rate investigation	107 trees	107 cells
Quantities required for character investigation	103 trees	103 cells
Time required for seed production	5-15 years	One month produces 106 somatic embryos for use as artificial seeds
Predictability of seed production	Seed-setting has on-and off-years and is affected by natural factors	Seed production is completely under artificial control
Genetic uniformity	Much variation occurs during sexual propagation and pollination, and selection should be controlled	Less variation in somatic embryos
Multiplication	Cutting or grafting at slow speed	Propagation coefficient of in vitro culture can increase by millions
Ploidy	Difficult to obtain haploid and homozygous lines; diploid materials	Easy to get haploid and pure lines
Breeding methods	Selection is most common; hybridisation and mutagenisation are available in some species; self-cross and back-cross are difficult or impossible	Combination of hybridisation and anther culture can accelerate breeding process and increase selection efficiency; mutation and mutant screening are conducted at cellular level with high efficiency; pure lines can be achieved through haploid culture and chromosome doubling; genetic engi neering methods can be used

Table 4.2. List of Orchids which have been clonally propagated *in vitro*

Plant	Explant
Anacamptis pyramidalis	Shoot tip
Aranda	Shoot tip, axillary bud
(Arachnis hookeriana x Vanda lamellata)	Shoot tip
Aranthera	Shoot tip
Arundina bambusifolia	Shoot tip (from young seedling)
Ascofinetia	Inflorescence segment (with flower primordia)
Brassocattleya	Axillary bud
Calanthe	Shoot tip
Cattleya	Shoot tip, axillary bud, lateral bud, leaf base, leaf tip
Cymbidium	Shoot tip
Dendrobium	Shoot tip, Nodal segment, Flower stalk segment (with vegetative buds)
Epidendrum	Leaf tip
Laelia	Axillary bud
Laeliocattleya	Axillary bud
Lycaste	Shoot tip
Miltonia	Shoot tip
Neostylis	Inflorescence segment (with flower primordia)
Neottia nidus-avis	Root
Odontioda	Shoot tip
Odontoglossum	Shoot tip
Odontonia	Shoot tip
Oncidium	Shoot tip
Oncidium papilio	Flower-stalk segment (with dormant apical buds)
Phajus	Shoot tip
Phalaenopsis	Flower-stalk segment, Shoot tip, Leaf segment, stem segment, root segment
Pleione	Shoot tip
Rhynchostylis gigantea	Shoot tip, lateral buds
Schomburgkia superbiens	Lateral bud
Vanda (Terete-leaf)	Shoot tip, Stem section
Vanda (Strap-leaf)	Shoot tip
Vanda hybrid	Shoot tip
(V. teres x V. hookeriana)	Axillary bud, root segment
Vascostylis	Inflorescence segment (with flower primordia)
Vuylstekeara	Shoot tip

Plant Health

Good plant health is the major advantage in the development of elite mother stock material for both the herbaceous ornamental producer and the woody plant grower. Before micropropagation, plant material can be put through stringent tests to bacterial and virus index. The presence of fungal contaminants or pathogens can also be detected and eliminated prior to micropropagation. Thus, the plants produced by the system can be certified as being disease indexed, and they represent clean, elite mother stock material for future conventional propagation methods.

Germplasm Storage

The concept of a tissue bank (a collection of valuable breeding or growth material) in a relatively confined space under controlled environmental conditions is particularly attractive for plant breeders. Many plant breeders wish to maintain specific breeding or hetrogenous individuals in the field or greenhouse for subsequent years for testing or seed production. Micropropagation techniques applied to these materials allow them to over winter the material in the safety of a controlled environment and also give rise to the potential for multiplying up individuals for specific crossing or testing.

This elimination of the seasonal nature of much breeding and growing work has been another advantage often suggesting with respect to micropropagation. However, a word of caution is necessary here, although it is possible to micropropagate plant species throughout the year, in the case of biennial species requiring vernalization, it is necessary to do the micropropagation at a suitable time of the year to allow the tissue to become vernalized for flowering for next year.

METHODS OF MICROPROPAGATION

Shoot tip is the starting material for micropropagation which is surface sterilized then trimmed and the remaining material placed on one of a variety of different media. The selection of specific medium is necessary with special reference to concentrations of particular growth regulator. This is the phase for acclimatization of the tissue to the in vitro conditions which requires about 3 to 4 months (varies from species to species). Under the suitable laboratory conditions the small shoot tip grows well in a normal fashion, giving rise to a small, unrooted shoot in culture.

The initiated shoot can, after a suitable period on the initiation medium, be subcultured so that the nodal sections of the shoot are placed onto a multiplication medium. This medium differs from the initiation medium in that it contains higher levels of cytokinins which are likely to give rise to precocious shooting of the axillary buds. The normal period between subcultures is 3 weeks and during this time the number of shoot tips available for subculture will multiply as a result of breaking of nodal axillary buds anywhere from 2 to 12 times. After 3 weeks on multiplication medium, somewhere on an average of about 5 shoot tips can be placed onto fresh medium, and these in turn will multiply 5 times in the subsequent 3 weeks, and thus micropropagation has led to a 25-fold increase in the plant material available over a total of 6 weeks. Rooting of above cultured micro-shoots can occur in two ways. One can transfer the shoot to media rich in auxins which will allow for the induction of in vitro roots. In some species it is necessary since they need to be actively induced to root.

Which Material Should be Used?

1. The starting material for a micropropagation programme is a shoot tip, not necessarily as small as the meristematic apical dome, but something a little larger with leaf primordia already developing around it.
2. This material must initially be surface sterilized.

3. Any tissue affected during to sterlization of material is trimmed and the remaining material will be placed on one of a variety of different media.

Stages of Micropropagation

1. Selection and sterilization of elite plants
2. Establishment of axillary buds in culture
3. Multiplication in culture
4. Rooting of in vitro plants and transfer to compost

METERIALS AND METHODS:

Stage I: Selection and Sterilization of Elite Plants

1. Use a clean sharp blade, carefully excise axillary buds of the desired variety and store in distilled water until enough buds have been obtained.
   - Commercial preparations of M&S and B5 salt (minus sucrose glutamine and hormones) are available in packets that make up 1 1 of media.
   - pH adjusted using, IM HC1 and 0.1M KOH before agar added and prior to autoclaving. The amount of agar needed varies depending on the brand used.
   - Varying the hormones allows the above media to be used for axillary bud culture of a wide range of species.
2. The practical should be conducted in a laminar flow bench.
3. Put a maximum of 20 buds into a sterile test tube.
4. Fill to the brim with ethanol solution and leave for 1-1.5 min.
5. Decant the ethanol.
6. Fill to the brim with chloros solution, replace the top, and agitate at 120 strokes/ mm for 12 min, either manually or with a shaker.
7. Decant the chloros solution and refill with sterile distilled water.
8. Rinse in sterile distilled water.
9. Store in sterile distilled water until reedy to continue (not longer than 2h).

Table. 4.3. Growth medium for potato *(Solanum tuberosum)*

Osmocote	57 g
Sulphated potash	21 g
Epsom salts	28 g
Frit 252 A	14 g
Lime	170 g
Phosphate per 300 1 peat sand mix	85 g

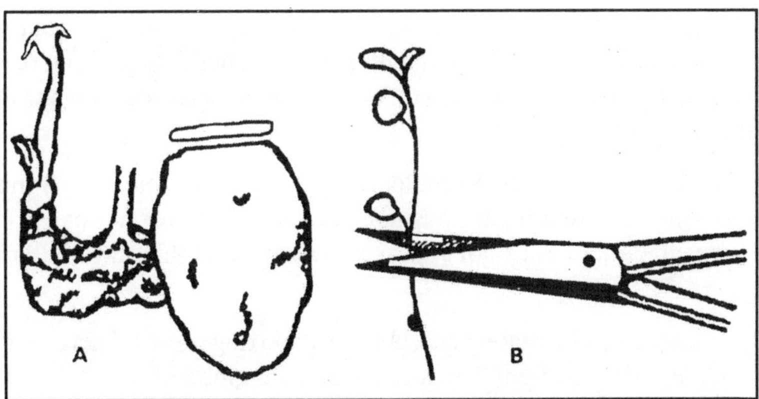

Diagram 4.1. (A) Axillary bud being excised from a sprout of *Solarium tuberosum.* (B) Subculturing of in vitro grown shoot of *S. tuberosum* by internodal cutting.

Stage II: Establishment of Axillary Buds in Culture

1. Empty the water plus axillary buds into an empty sterile petridish, for ease of handling.
2. Place up to 4 buds in a 50 mm Petri dish or 10 buds in an 85 mm Petri dish containing medium 1.
3. Make sure that the base of the bud is stuck firmly in the medium.
4. Take care that the bud is not buried.
5. Seal the Petri dish with laboratory sealing film, ensuring adequate gaseous exchange by puncturing the film 2-4x with a fine sterile needle.
6. Transfer to growth room or incubator, and leave it for 1 week to 2 months depending on the variety.

Stage III: Multiplication in Culture

1 ml sterile jars containing approximately 30 ml of medium 2 or medium 3. Environmental conditions as described for Stage II.

Stage IV: Rooting of in vitro Plants and Transfer to Compost

1 ml sterile jars containing approximately 50 ml of medium 4, Propagation trays, plastic bags, or seed trays covered with glass sheets or a misting or fogging system, 50 mm diameter pots containing compost (3:2, peat:sand) supplemented with nutrients.

Precautions for Sterile Culture

Use a laminar flow bench, and swab down the bench, roof, and sides with disinfectant. Always wear gloves (thin latex gloves are test), and swab gloves and bench surface frequently during any experimental work. Sterilize and flame all instruments before use. Finally, avoid passing anything over the top of opened containers.

Stage I: Choose as juvenile a tissue as possible to initiate cultures. Do not try to sterilize too many different genotypes at once. Start with 1 or 2, and build up to about 6. Problems of contamination and toxic effects of the sterilant increase with increased numbers of genotypes being treated simultaneously.

Stage II: Condensation can be a problem in Petri dishes. This can be minimized by stacking the Petri dishes and rotating the position of each dish in the stack daily. The intensity of light reaching each culture does not appear to be critical at this stage, although light is necessary.

Gibberellins tend to produce elongated, abnormal looking shoots. These revert to normal growth when transferred to medium that does not contain gibberellins.

Callusing will occur around the base of the bud; only shoots that can be clearly seen to have derived from extension of the preexisting axillary bud should be transferred. Adventitious shoot formation in callus tissue generally produces high levels of variation and is the basis for producing somatic variants.

Stage III: Murashige and Skoog medium and Gambourg B5 medium are those which most frequently used in the culture of axillary buds. They often need to be modified, mainly by varying the vitamins and growth regulators, depending on the species or variety being used.

1. If the intent is to use the axillary bud cultures as a source of protoplasts, then all media need to be made up using distilled deionized water or water of equal purity other additions may also have to be made to the culture medium.
2. Vitrification can be a serious problem in in vitro cultures. In this condition, the shoots tend to elongate and look glassy, the leaves become reduced or malformed, and if not treated the culture will die. The causes of vitrification are not fully understood. However, the risk of vitrification can greatly reduced if there is adequate gaseous exchange between the culture and the external environment which is achieved by piercing laboratory film seals or by loosening jar lids slightly.

Stage IV: Rooting and transfer of plants to compost can be difficult. If the auxin level is too high, subsequent root development after transfer to compost is reduced. Some species can be transferred to compost without a rooting step. Some fruit tree species need a dark treatment as well as additions of auxin for successful rooting to take place (5). The transition from culture to compost is critical. High humidity must be maintained, but care must be taken not to waterlog the compost. If plants are transferred to the field, they need protection from wind damage, drought, and birds.

Methods of Micropropagation (In Nutsel):

1. Shoot tip is the starting material for micropropagation.
2. It is surface sterilized then trimmed and the remaining material placed on one of a variety of different media.

3. The selection of specific medium is necessary with special reference to concentrations of particular growth regulator.
4. This is the phase for acclimatization of the tissue to the in vitro conditions which requires about 3 or 4 months and varies from species to species.
5. Under the suitable laboratory conditions the small shoot tip grows well in a normal fashion, giving rise to a small, unrooted shoot in culture.
6. The initiated shoot can, after a suitable period on the initiation medium, be subcultured so that the nodal sections of the shoot are placed onto a multiplication medium.
7. This medium differs from the initiation medium in that it contains higher levels of cytokinins which are likely to give rise to precocious shooting of the axillary buds.
8. The normal period between subcultures is 3 weeks and during this time the number of shoot tips available for subculture will multiply as a result of breaking of nodal axillary buds anywhere from 2 to 12 times.
9. After 3 weeks on multiplication medium, somewhere on an average of about 5 shoot tips can be placed onto fresh medium, and these in turn will multiply 5 times in the subsequent 3 weeks, and thus micropropagation has led to a 25-fold increase in the plant material available over a total of 6 weeks.
10. Rooting of above cultured micro-shoots can occur in two ways. One can transfer the shoot to media rich in auxins which will allow for the induction of in vitro roots. In some species it is necessary since they need to be actively induced to root.
11. Once the micropropagated plants are satisfactorily rooted in soil or soil substitutes, they must be weaned to a normal humidity level.
12. During the rooting process they require about 100% humidity, which inhibits the normal development of cuticle and stomatal activity associated with drier conditions.
13. The rooting normally takes place under conditions of reduced lighting, therefore, the plants must not be exposed for the first time to full sunlight. This is done slowly for a period of couple of weeks.
14. This procedure will develop small, autonomous and photosynthesizing plants. During the rooting phase under the reduced lighting conditions, they live on carbohydrate reserves built up during micropropagation period in vitro.
15. Once this weaning process is completed, the hardened can *be* incorporated into normal growing systems for their further development.

Once the micropropagated plants are satisfactorily rooted in soil or soil substitutes, they must be weaned to a normal humidity level. During the rooting process they require about 100% humidity, which inhibits the normal development of cuticle and stomatal activity associated with drier conditions. The rooting normally takes place under conditions of reduced lighting, therefore, the plants must not be exposed for the first time to full sunlight. This is done slowly for a period of couple of weeks. This procedure will develop small, autonomous and photosynthesizing plants. During the rooting phase under the reduced lighting conditions, they live on carbohydrate reserves built up during micropropagation period in vitro. Once this weaning process is completed, the hardened can be incorporated into normal growing systems for their further development.

Table 4.4 Relative cost components associated with micropropagation of a typical crop

Cost component	Per cent
Laboratory, direct labour	15
Utilities	9
Depreciation	7
Supervision (laboratory and greenhouse)	23
Planting, direct labour	9
Other production costs	37
Total	**100**

APPLICATIONS OF MICROPROPAGATION

Let us consider the end product of a micropropagation programme. Provided the systems have been properly developed and controlled, one is left with a very large number of clonal individuals. These individuals are genetically similar and under ideally controlled conditions they behave similarly, showing the same disease resistance, tolerance, environmental stresses, fruiting times and yields. In dioecious species, if fruit is to be the final product, clones must be mixed in such a way that there is a composite population of male and female individuals in order to bring about fertilization and crop development. There may be other reasons for mixing clones. Because the plants may contain similar disease tolerance and resistance, the spread of epidemic disease through a micropropagated monoculture is an ever present threat. In these circumstances, multiline planting is necessary in order to circumvent some of the problems of monoculture. This requires that several similar clones are selected prior to the micropropagation phase. Multi-line planting has been shown not only to be extremely effective in epidemic disease control in cereals, but it is also important in terms of harvest Uniform harvest date is of paramount importance when one utilizes machine harvesting methods. However, uneven harvest dates can have their advantages as well. A spread harvest, lasting several weeks, can be handled manually and provides for an even supply to the limited market size, rather than one-time flooding of the market.

Its main influence has been in the area of ornamental horticulture where enormous advances have been made both in the plant health status and in variety selection of a number of species.

Sometimes it is presumed that micropropagated material will never be able to compete with seed in terms of its price. Certainly in the extensively planted crops such as the cereals, this is probably true. However, there are examples in the vegetable and ornamental crop areas where micropropagation methods may be applicable. The technique of so-called artificial seed production, using somatic embryos rather than conventionally micropropagated propagules, may have sufficient advantages to warrant its commercial use.

Micropropagation has already a proven record of increased yield and vigor through high health programmes in floriculture, and a big role to play in the rapid introduction of new varieties of both ornamental and vegetable crops into the market. Micropropagation also holds centre stage in many of the novel genetic manipulation and genetic engineering techniques that are being developed, for it alone holds the key to the rapid bulking up of elite germplasm. The technology also has important roles in the potential hastening of plant breeding, both in ornamental and crop plants. It is clear that if the correct strategies for the use of micropropagation are developed and used in the appropriate circumstances, the technology will have a major effect in the next decades on both the nature of the crops that are grown internationally and also on the whole structure of horticulture and agronomy worldwide.

CHAPTER - 5

Protoplast Culture

In 1965 H. Haris and J. F. Watkins of Oxford reported for the first time that cells from different animal species (mouse and man) can be made to fuse to form hybrid cells. Before this, it was not in thought even, that unrelated species and genera may fuse. Conventional breeding experiments show that only related species can be made to mate. Besides several attempts, no animal hybrids have been produced by this technique; although a hybrid between two different species of tobacco and another one between two different species of petunia has been successfully synthesised. However, recently the success has been met in fusion between protoplasts of plant and animal cells.

Protoplasts are plant cells contents with a plasma membrane, without the cell wall. The absence of cellulosic cell wall permits advantages to fuse protoplast of similar or of different species and the fused product can generate into whole plants. This is called somatic fertilization or hybridization or fusion of two vegetative cells to form hybrids. Each nuclear material of each species behave function independently. Thus, it provides different characteristics of different species within a single species. Protoplast fusion also facilitate to over come from natural sexual barriers and genetic elements of sexually isolated organisms may be made to mix. This short of hybridization is known as parasexual hybridization.

Importance of Protoplast Isolation and Culture

The isolation, culture and fusion of protoplast are one of the most fascinating fields of research. The techniques are important for the following reasons:

1. To develop novel hybrid plant through protoplast fusion, genetic engineering would continued to be an exciting area of research in modern plant biotechnology. This technology holds great promises to synthesize a plant of desired characteristics.

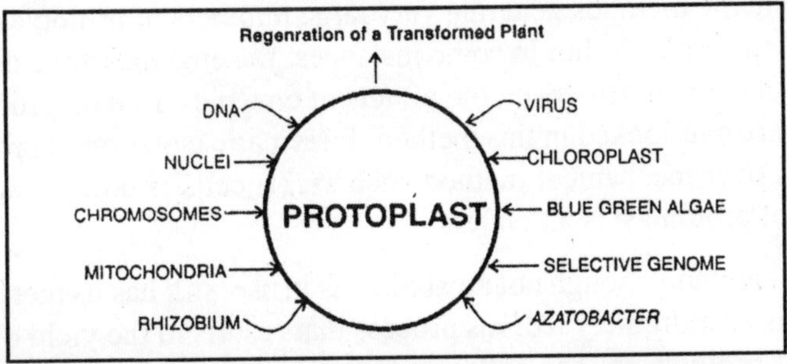

Diagram 5.1 Hypothetical illustration for the uptake of different cell organelles and introduction of genetic material into plant protoplasts. This technique results into the regeneration of new plant bearing desired characters.

2. This helps in crop improvement by incorporating more protoplasts in a single cell.
3. The protoplast in culture can be regenerated into an entire plant.
4. The technique in future will be one of the most frequently used research tools for tissue culturists, physiologists, pathologists, molecular biologists, cytogeneticists, and biotechnologists.
5. It provides a tool for isolating protoplasts and exploring the possibilities of genetic engineering.

Isolation of Protoplast from Various Plant Parts

Plant tissues	Young leaves, roots, stems
Flower parts	Petal, reproductive organs
Callus/cells	Friable callus tissues/cells
Suspension cultures	Fast growing cells
Aseptically grown plantlets	Leaves, stems, roots

Protoplast is the living material of the cell where as an isolated protoplast is the cell from which the cell wall is removed. Protoplasts can be isolated from almost all plant parts viz., roots, leaves, fruits, tubers, root nodules, endosperms, crown gall tissues, pollen mother cells, pollens, pollen tetrads and cells of callus tissues grown in vitro. Protoplasts are isolated from cells by two methods:

1. Enzymatic isolation
2. Mechanical isolation, and

Table 5.1 Application of plant protoplast in fundamental research of plant sciences

1. **Structure**	Cell wall, Plasma membrane, Organelle-structures and biosynthesis
2. **Function**	Biochemistry, Photosynthesis, Absorption
3. **Genetics**	Fusion, Organelle and DNA uptake
4. **Plant Pathology**	Virus, RNA uptake, Pathotoxin
5. **Differentiation**	Morphogenesis, Embryogenesis

1. Enzymatic method is most often used in isolation of protoplasts from various plant parts. Enzymatic methods generate very large numbers of protoplasts in comparison to mechanical methods, but in some instances, the enzymes have deleterious effects on plant metabolism. However, the effects of enzymes used on protoplasts (which is unknown) are overlooked in this method. Enzymatic isolation of protoplast has some advantages over mechanical method such as: (i) cells is not damaged, (ii) Osmotic shrinkage of protoplast is much less.

2. Mechanical method, though not most often is in use, still has its merit. It is rarely used because it is an extremely tedious process that results in the yield of only very small numbers of protoplasts. The cells are plasmolysed, causing the protoplast to shrink away from the cell wall. The protoplasts obtained from this method are then cultured on suitable culture medium.

ENZYMIC ISOLATION OF PROTOPLASTS: METHOD

The starting material is sterile in vitro shoot cultures or any plant part. In these cases protoplasts are enzymetically isolated in a two-step or a one-step process.

1. *First:* Isolated cells are prepared (maceration) by treating the leaf or other tissue segment with macroenzyme (a pectinase) in 13% mannitol. The isolated cells obtained from this incubation are purified by filtration through a nylon mesh.
2. *Second:* The protoplasts are prepared by incubating these isolated cells in 2 per cent cellulase for about 90 minutes. In the *one-step* method both enzymes (cellulase + pectinase) are used simultaneously. Here 2 to 3 hours at 20 to 22° C is adequate to release significant numbers of protoplasts.

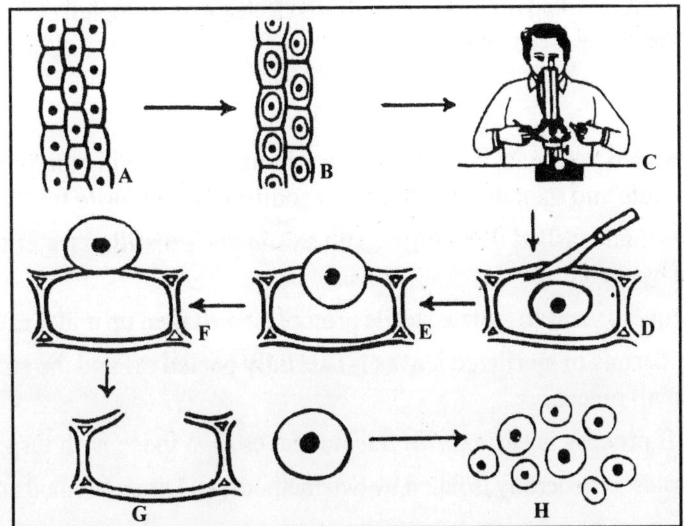

Diagram 5.2 Method of mechanical isolation of protoplasts. (A) A small piece of peeled epidermis, (B) Plasmolysis of the same cells, (C-F) Steps of removal of protoplasts from the cell after incesion, (G & H) Free isolated protoplasts.

ISOLATION OF PROTOPLAST FROM LEAVES

The process of isolating protoplasts from leaves involves a number of steps; these include surface sterilization, removal of epidermis, enzymatic digestion and cleaning of the protoplasts.

For isolation, the physiological state of the leaves is extremely important. The leaves normally chosen for experimental purposes should be fully expanded and from young plants (vegetative state, e.g., 10-14 weeks in tobacco). The lower epidermis is removed and is used to isolate epidermal and guard cell protoplasts. The remaining peeled leaf is used for isolation of mesophyll protoplasts. The isolation of protoplast involves four main steps which are as follows:

1. Surface sterilization of leaves.
2. Peeling off the epidermis.
3. Enzymatic treatment.

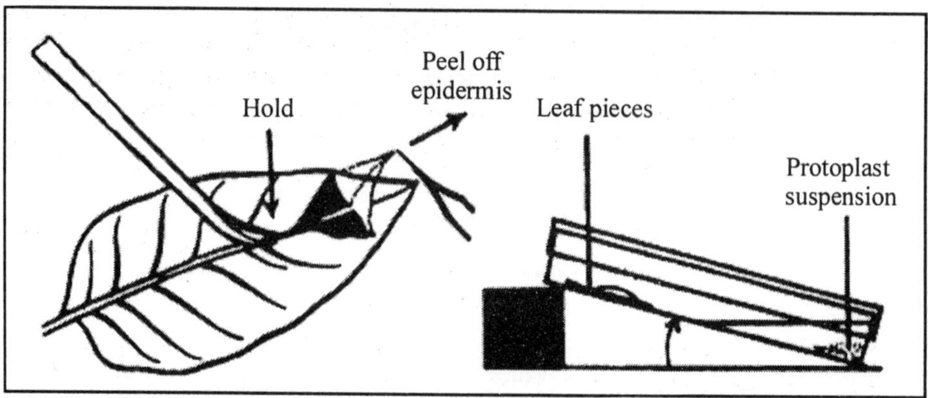

Diagram 5.3. (A) Peeling off leaf epidermis. (B) Release of protoplasts from peeled leaf pieces after enzymatic treatment.

Method

1. Fully expanded young leaves are surface sterilized by dipping them into 70% ethanol for one minute and then in 2% solution of sodium hypochlorite for 20-30 minutes.
2. The leaves than washed three times with sterile glass distilled water to remove traces of sodium hypochlorite.
3. The subsequent steps to isolate sterile protoplast are taken up under aseptic conditions.
4. Lower epidermis of sterilized leaves is carefully peeled off and the stripped leaves are cut into small pieces.
5. Peeling off process is easier with flaccid leaves than those with turgid leaves.
6. The protoplast is generally isolated by two methods: (i) Direct Method and (ii) Sequential Method.

(i) Direct Method

1. Peeled leaf pieces are placed with lower surface down, into 20 ml solution of 13% mannitol and inorganic salts contained-in petridishes (90 mm) for one hour.
2. The mannitol-inorganic salt solution is pipette out with the help of Pasteur pipette and an enzyme mixture (20 ml) in mannitol (0.5% macerozyme + 2% Onozuka cellulase in 13% sorbitol or mannitol at pH 5.4) is added and incubated overnight (15 to 18 hours) at 20 to 22° C.
3. The petri plates are then gently agitate to facilitate the release of the protoplasts, pushing the larger pieces of leaf-material to one side with a sterile forceps and keeping the petridish at an angle of 15°.
4. After sometimes protoplasts will settle down in the petridish and then protoplast-enzyme mixture is transferred to 15 ml screw-capped centrifuge tubes using Pasteur pipette.
5. Centrifuge the material at 100g for one minute. This process is repeated for 2 to 3 times and protoplasts are washed with 13% sorbitol solution.

6. For the final washing, 20% sucrose solution is used in place of sorbitol and centrifuged at 200g for one minute.

7. The cleaned protoplasts float and debris settles down.

8. The floating protoplasts are carefully pipette out with a Pasteur pipette leaving behind the remains of mesophyll cells.

9. This method is useful for the isolation of protoplast both from spongy and pallisade parenchyma.

(ii) Sequential Method

In this method of protoplast isolation cells are first separated and then transformed into protoplasts. This method is completed into two steps:

Step I. Isolation of Cells

1. The peeled leaf segments are placed into a petridish (140 mm) containing an enzyme mixture (mecerozyme 0.5%, potassium dextran sulphate 0.3% in 13% mannitol at pH 5.8).

2. The perridishes are put into a desiccator for vaccum infiltration of enzyme mixture for about 4 to 5 minutes, then transferred to a water bath at 25° C.

3. Agitate them slowly.

4. After 15 to 20 minutes the enzyme mixture is replaced by fresh one and is again incubated for one hour. By this process cells get completely separated from one another.

5. The isolated cells are filtered through a nylon mesh, then centrifuge at lOOg for one minute.

Step II. Isolation of Protoplasts

1. Isolated cells are now treated with an another enzyme mixture containing 2% cellulase in a 13% solution of mannitol at pH 5.4 for 90 minutes at 30°C and then centrifuged at 100g for one minute.

2. The supernatant is decant off and the protoplasts are washed three times with mannitol.

3. The protoplasts are finally cleaned with 20 to 25% sucrose solution. This method can be used for isolation of protoplasts from pallisad cells.

Isolation of Protoplast from Shoot Apex

The use of sterile shoot cultures as a starting material for leaf and stem protoplast isolation is popular. One key advantage is that the material is already sterile and does not have to be surface sterilized prior to incubation in the digestive enzyme mixture. Sterilc shoot cultures are used as starting material in a wide range of plant species, from potato and tobacco, and a wide range of tree species.

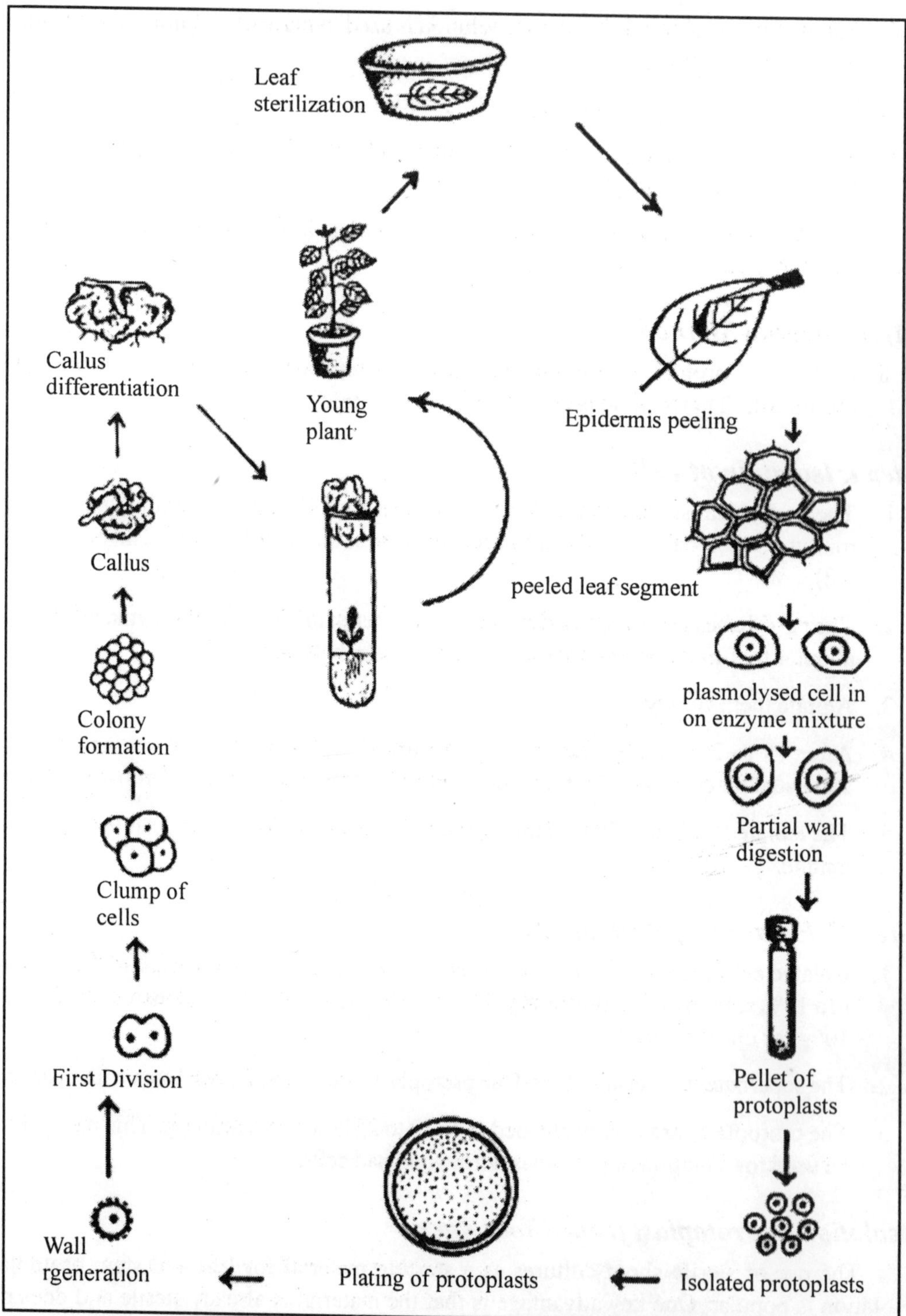

Diagram 5.4. Systematic illustration of the technique used for isolation, culture and regeneration of plants from leaf protoplasts.

Isolation of Protoplast from Root and storage organs

Protoplasts have been isolated directly from a wide selection of roots and storage organs. These include potato tubers, artichoke tubers, onion roots and root nodules. The enzyme treatments are sometimes slower and often a more complex enzyme mixture is required.

Isolation of Protoplast from Pollen Grains

Pollen protoplasts have some advantages over the somatic cells being found in uniform in chromosome number (ploidy). Protoplast obtained from mature pollen could be used in the study of pollen ontogeny, mutation, somatic hybridization, cell modification and regeneration of haploids. Fusion causes the formation of a normal diploid instead of doubling the chromosome number (to tetraploid) of the new hybrid. It is also possible to regenerate haploid plantlets from isolated pollen protoplasts.

Pollen grain has highly sculptured and chemically complex cell wall. It is covered in a highly resistant coat of sporopollenin, as polymer of carotenoid esters and can be removed by a pretreatment with strong oxidizing solutions.

Pollen grains taken out of the anther lobes are sterilized with 2% sodium hypochlorite solution for 10 to 12 minute and subsequently washed 2 to 3 times with sterile distilled water. The concentration and combinations of the enzyme incubation to pollen grains depend upon their stage of development. Protoplasts are released in many ways described as follows:

Protoplast ooze out through germ pores due to dissolution of pecto-cellulosic membrane by enzyme action. The enzyme mixture used normally contains high concentrations of p-l,3-gluconase, which is capable of digesting both cellulose and callose.

1. Due to mechanical rupture of the exine.
2. By partial dissolution of the exine.

Isolation of Protoplast from Cultured Cells

Actively dividing cells in suspension cultures are proved as-most suitable material for isolation of non-germ protoplasts in large amount. Morever, protoplasts isolated from various somaclones can also be used by plant breeders as raw material for somatic hybridization and cell manipulation. Protoplast isolation through enzymatic activity from older plant parts is very difficult. Protoplasts have been isolated from suspension culture of many plant species like, carrot, rose, *Atropa* sps., sugarcane, etc. Protoplasts can be easily isolated from such cultures by treating the filtered suspension with 2 to 4% Onozuka cellulase in 0.6 M mannitol, for 4 to 6 hours at 32° C in a gently shaking water bath.

Other Sources: Protoplats have also been isolated from coleoptiles, flower petals, aleurone layers and plant cell rumours or galls. The basic isolation technology is the same in all cases.

Methods of Protoplasts Culture

Laboratory Facilities

Protoplasts can be cultured in liquid as well as solid medium. Concentration of enzyme, intensity of light (300-3000 lux) and temperature (28-45°C) play vital role in protoplasts culture.

Table 5.2 Plant species in which protoplast have been successfully cultured.

Source	Enzyme mixture used	Culture medium	Growth response
Mesophyll Cells			
Brassica napus L.	Cellulase Onozuka P 1500 (0.5%) + Rhozyme HP 1500 (0.5%) + Driselase (0.5%) + Hemicellulase (0.5%) in sorbitol (4.5%) + mannitol (4.5%)	B + NAA (10-6M) + BA (10-6M)	Plants
Cucumis sativus	Preplasmolysis in a salt medium and cellulase (0.3%) + pectinase (0.4%) + PDS (0.5%) in 11% mannitol	Harad's medium (1973)	Cell clusters
Datura innoxia	Macerozyme (1%) + cellulase Onozuka (3%) in 0.5M mannitol BAP (0.4 mg/1) + NAA (0.4 mg/1)	Durand et al. (1973)	Plants
Hyscyamus niger	Pectinol (0.02%) + cellulase Onozuka R 10 in KCl (2.5%) + $MgSO_4$ (1%) + 0.6 M mannitol	Durand et al. (1973)	Callus
Nicotiana tabacum cv. xanthi (2x)	Macerozyme (0.5%) + cellulase Onozuka (2%) NAA (3 mg/1) + 6-BAP (1 mg/1) + 0.7 M mannitol	Nagata & Takebe (1971)	Plants
Callus			
Saccharum officinerum	Cellulase Onozuka 4S (5%) + glusulase (1.5%) + sorbitol + 0.4M + glucose 5.5 mM	Mod. White's medium + yeast extract	Callus
Stem Callus			
Atropa belladonna	Cellulase R 10 (1.5%) + in 0.5 M sorbitol + NAA (2 mg/1) + kinetin (0.5 mg/1)	Mod. MS	Embryoids & plants
Ovular Callus			
Citrus sinensis	Cellulase (1%) + pectinase (1%) + PDS (0.3%) in 0.14 M sucrose + 0.28 M mannitol + 0.28 M sorbitol	Murashige & Tucker's (1969) basal medium	Embryoids
Pericarp Callus			
Vitis vinifera	Cellulase Onozuka SS (2%) + meserozyme (1%) in 0.1 M $CaCl_2$ + 0.14 M KCl + (2000 mg/1) kinetin + 0.2 mg/1) + NAA (0.1 mg/1)	B5 + Casein hydrolysate	Callus
Crown Gall			
Parthenocissus tricuspidata	Mecerozyme R 10 (0.01%) + cellulase + NAA (0.1 mg/1) + Onozuka R 10 (2%) in 13% mannitol	Nagata & Tabeke (1971) medium	Callus

Enzyme Mixture and Osmotic Stabilizer (Osmotica)

The composition of the cell wall indictates that the digestion mixture should be able to degrade cellulose, hemicellulose and pectin, and in some cases callose. The number of commercial enzyme preparations available is somewhat limited and very often they are contaminated with a wide range of biologically active impurities of other enzymes- i.e., proteases, nucleases and lipases. They have a degrading effect on the plasma membrane of the isolated protoplasts.

Following are the commercial enzyme preparations used for isolation of protoplasts.

1. Onozuka R_{10} cellulase (Kinki Yakult Biochemical, Nishinomiya, Japan)
2. Cellulase R_{10} (Kinki Yakult Biochemical, Nishinomiya, Japan)
3. Macerozyme R_{10}, pectinase (Calbiochem-Behring, San Diego, California, USA)
4. Hemicellulase (Rohm and Hass Co. Philadelphia, PA., USA)

5. Driselase, cellulase, (Kyowa Hakko Kogyo Co., Tokyo, Japan)
6. Cellulysin, cellulase (Calbiochem-Behring, San Diego, California, USA)
7. Pectinase (Sigma, St. Louis, MO, USA)
8. Macerase (Calbiochem-Behring, San Diego, California, USA)
9. Rhozyme HP 150, Hemicellulase (Rohm and Mass Co. Philadelphia, PA., USA)

The enzyme mixture is normally dissolved in culture media together with an osmotic stabilizer. Calcium (2 to 10mM) is a necessary component and phosphate (0.5 to 2.0 mM) appears to stabilize the isolated protoplasts.

After the digestion of cell wall the isolated protoplast is subject to osmotic stress. If an osmotic stabilizing agent is not included in the medium the isolated protoplasts would take in water by the process of osmosis and would eventually burst as there is no cell wall to constrain the cells. The osmotic agents may be sugar, alcohols, sorbitol or mannitol (13% w/v). Sucrose can be used for this purpose. The concentration of osmoticum should not be high because it causes protoplast shrinkage.

Protoplast is cultured in liquid medium in Erlenmeyer flasks or in microchambers. Microchamber is prepared by placing two 22 mm^2 coverslips about 18-22 mm apart on a drop of heavy mineral oil on a slide. A drop of medium containing protoplasts is placed in between the two coverslips. A third coverslip is then gently placed on the medium and finally sealed with mineral oil or parafilm. Through this method protoplast can be cultured in a small amount and for a limited time.

Purification of Isolated Protoplasts

Protoplasts contain cell debris and broken cell organelles in medium. A number of methods have been used to separate the intact protoplasts from this surrounding cellular debris.

Sedimentation and Washing

1. The crude protoplast suspension in osmotically adjusted medium is decanted into a conical tip centrifuge tube.
2. It is centrifuged at low speed (50-100 x g/5 min). Under these conditions the intact protoplasts accumulate in the form a soft pellets in the tip of the tube.
3. The supernatant, containing the broken cells debris, cell wall, and cell organells etc., are then carefully pipetted off.
4. The pellets are then gently resuspended in fresh culture medium containing mannitol, and rewashed. This process is repeated two or three times.

Flotation

As intact protoplasts have a relatively low density as compared to other cell organelles, many types of gradients have been used that allow the protoplasts to float and for the sedimentation of cell debris.

A concentrated solution of mannitol, sorbitol or sucrose may be combined with the enzyme-protoplast mixture in a Babcock bottle and then centrifuged at low speed.

Protoplasts can be pipetted off from the top of the tube. The concentration of sucrose used for flotation varies from 0.3 to 0.6 M. The flotation method causes less loss due to damage than the pelleting and washing method. However, the high osmotic concentrations of solution may cause osmotic stress to the protoplasts which may delay the wall formation or leads to eventual loss of the viability.

Other Purification Methods

Other more complex pelleting and flotation systems are also used for protoplast purification. Some workers have layered crude protoplast preparation of *Hyoscyamus muticus* onto 20% Percol (polymer of solicos). After centrifugation debris was in the lower half of the gradient and protoplasts floated to the top.

After obtaining a reasonably pure isolated protoplast preparation, it is important to determine what proportion of the protoplasts are alive and to be able to culture.

Protoplast Viability Testing

In early days the viability of protoplast is determined by observing its cytoplasmic streaming, exclusion of Evans Blue dye, changes in protoplast size induced by changes in the level of osmoticum and measure of photosynthetic and respiratory activity. Most of these methods have proved to be rather unreliable for determining viability.

The most frequently used methods for estimation of viability is the use of fluorescein diacetate (FDA). As the FDA accumulates within the plasma membrane, viable protoplasts fluoresce green or white. Protoplast preparation treated with FDA (normally 0.01%) should be observed within 5 to 1 5 minutes, because after this period FDA dissociates from the membrane.

Another stain commonly used is phenosafranine (0.01%). This stain detects dead protoplasts, which turn red in the presence of the dye. Viable protoplasts remain unstained even after 2 hours of staining.

Isolated protoplast can be cultured in several ways of which agar embedding technique in small petridish is commonly followed. In petridishes, embedding of protoplasts in solid agar medium is known as plating of protoplasts or protoplast planting.

General Steps of Protoplast Culture

The medium having agar and store at 30-35° C until required. Higher temperature is injurious to protoplasts. Mix gently but quickly by rotating petri dish. Allow the medium to set and seal petri dishes with parafilm. Incubate petri dishes down in an incubator at 25 ± 2° C with continuous illumination at 2000 lux. Protoplasts can also be cultured by suspension culture (without agar) and droplet culture method by placing of protoplast suspension containing protoplasts plate (1 x 10^4 or 1 x 10^5 ml). Incubation procedure is same as described above.

The plated protoplast can be handled very easily and the agar medium gives good support to the protoplast. Though various techniques have been developed for the culture of plant protoplasts however, the style of the culture often depends upon the objective of the experiment.

Culture Media for Protoplast Culture

Protoplast Culture Media for PC I Group

Table. 5.4. F.5 Medium (Frearson et al., 1973)

Constituents	Concentrations (mg/1)	Constituents	Concentrations (mg/1)
$CaCl_2$ $2H_2O$	850.0	$ZnSO_4$ $7H_2O$	4.3
$CuSO_4$ $5H_2O$	0.025	BAP	1.0
$CoSO_4$ $7H_2O$	0.15	Biotin	0.05
$FeSO_4$ $7H_2O$	13.9	Folic acid	0.5
H_3BO_3	3.1	Glycine	1.0
KH_2PO_4	353.6	Mannitol	130,000.0
KI	0.498	Meso-inositol	100.0
KNO_3	525.0	NAA	2.0
$MgSO_4$ $7H_2O$	739.0	Nicotinic acid	5.0
$MnSO_4$ $4H_2O$	11.15	Pyridoxine HCl	0.5
Na_2EDTA	18.75	Sucrose	10,000.0
Na_2MoO_4 $2H_2O$	0.125	Thiamine HCl	1.0
NH_4 NO_3	412.25	pH	5.8

Table 5.5. Nagata and Takeba Medium (1971)

Constituents	Concentrations (mg/1)	Constituents	Concentrations (mg/1)
$CaCl_2.2H_2O$	220.0	Na_2EDTA	37.3
$CuSO_3.5H_2O$	0.025	$Na_2MoO_4.2H_2O$	0.25
$CoSO_4.7H_2O$	0.030	NH_4NO_3	825.0
$FeSO_4.7H_2O$	27.8	$ZnSO_4.4H_2O$	8.6
H_3BO_3	6.2	BAP	1.0
KH_2PO_4	680.0	Mannitol	130,000.0
KI	0.83	Meso-inositol	100.0
KNO_3	950.0	NAA	3.0
$MgSO_4.7H_2O$	1233.0	Sucrose	10,000.0
$MnSO_4.4H_2O$	22.3	Thiamine HCl	1.0
		pH	5.8

Table 5.6. V 47 (Binding, 1974)

Constituents	Concentrations (mg/1)	Constituents	Concentrations (mg/1)
$CaCl_2.2H_2O$	735.0	BAP	0.4
$CuSO_4.5H_2O$	0.015	Biotin	0.04
$CoCl_2.6H_2O$	0.015	Folic acid	0.4
H_3BO_3	2.0	Glycine	1.4
KH_2PO_4	68.0	Glucose	99,000.0
KI	0.25	Mannitol	8,19,000.0
KNO_3	1480.0	Meso-inositol	10.0
$MgSO_4.7H_2O$	984.0	NAA	1.52
$MnSO_4.H_2O$	5.0	Nicotinic acid	2.0
Na_2EDTA	37.0	Pyridoxine HCl	0.7
$Na_2MoO_4.2H_2O$	0.1	Sucrose	1,70,000.0
NH_4NO_3	1444.0	Thiamine HCl	4.0
$ZnSO_4.7H_2O$	1.5	pH	5.8

Murashige and Skoog Medium

For protoplast culture, use MS culture medium as described for tissue culture supplemented with 13% mannitol (w/v).

Table 5.7. Modified B5 Medium (Kartha et al., 1974).

Constituents	Concentrations	Constituents	Concentrations
+ Sorbitol	4.55% w/v	$CaCl_2.2H_2O$	0.0875%
Mannitol	4.55% w/v	2,4-D	2.3×10^{-6}M
d-glucose	0.25% w/v	BAP	10^{-6}M
D-ribose	0.125% w/v	NAA	10^{-6}M
N-Z anime	0.015% w/v	Calculated molarity	0.52

Protoplast Culture Media PC II Group

Media calculated under PC I Group with minor changes. The changes are: the concentration of mannitol/sorbitol present in media individually or in combination reduced by half. Concentration of agar in the medium is 0.8%.

Proroplast Culture Media PC III Group

Media described PC I Group with following minor changes:

1. Mannitol/sorbitol/D-glucose/D-ribose individually or in combination are omitted in the media.
2. The sucrose concentration in all media is adjusted to 2% w/v.
3. PC III Group media may be prepared with 1.6% or 0.8% agar depending on the mode of protoplast culture.

Composition of the digestion mixture, wash medium and regeneration medium used for isolation and culturing of protoplasts.

Table 5.8. Composition of the digestion mixture, wash medium and regeneration medium used for isolation and culturing of protoplasts.

Digestion Mixture	
Salts and Vitamins of Murashige & Skoog (MS)	
Cellulysin	1.0%
Macerase	0.5%
Rhozyme	1.0%
$CaCl_2$	4.5 mM (0.5 g/1)
Mannitol	0.3 M (54.7 g/1)
Sorbitol	0.3 M (54.7 g/1)
MES	0.3 mM (640 mg/1)
pH	5.7
Filter Sterilize Washing Medium (MS)	
$CaCl_2$	4.5 mM
Sorbitol	0.3 M
Mannitol	0.3 M
pH	5.7

Autoclave Regeneration Medium (MS)		**Vitamins, Sugar and sugar alcohols (pH 5.7)**		**Sugar and Sugar Alcohols (g/1)**	
Sucrose	20 g/1	Nicotinamide	1.0	Fructose	0.125
2,4-D	0.5 mg/1	Pyridoxine HCl	1.0	Fibose	0.125
Kinetin	1.0 mg/1	Thiamine HCl	1.0	Xylose	0.125
Coconut water	2.0% (v/v)	D-calcium pantothenate	0.5	Mannose	0.125
Inositol	100 mg/1	Folic acid	0.2	Cellobiose	0.125
Casein hydrolysate	200 mg/1	p-aminobenzoic acid	0.01	Sorbitol	54.7
Sodium pyruvate	5.0 mg/1	Choline chloride	0.5	Manitol	54.7
Citric acid	10.0 mg/1	Riboflavin	0.1		
Malid acid	10.0 mg/1	Ascorbic acid	1.0		
Fumaric acid	10.0 mg/1	Vitamin B12	0.01		

Agar Embeded Culture

In this method protoplasts may be allowed to regenerate a new cell wall in liquid culture before embedding or may be embedded directly after isolation. The original method of Nagata and Takebe for the isolated protoplasts is mixed with 1.0% agar-culture medium and maintained at 40 to 45° C. Small amount of the agar (liquid) protoplasts mixture is then poured into sterile petriplates.

Microchambers

This methods for the observation of plant cells is quite common. The number of possible designs of chamber are enormous. A chamber is constructed of a small plastic ring with a coverslip on top. The joints of the system are sealed with mineral oil. The simplest form is to hold a droplet between two coverslips with a ring of mineral oil around to form a seal.

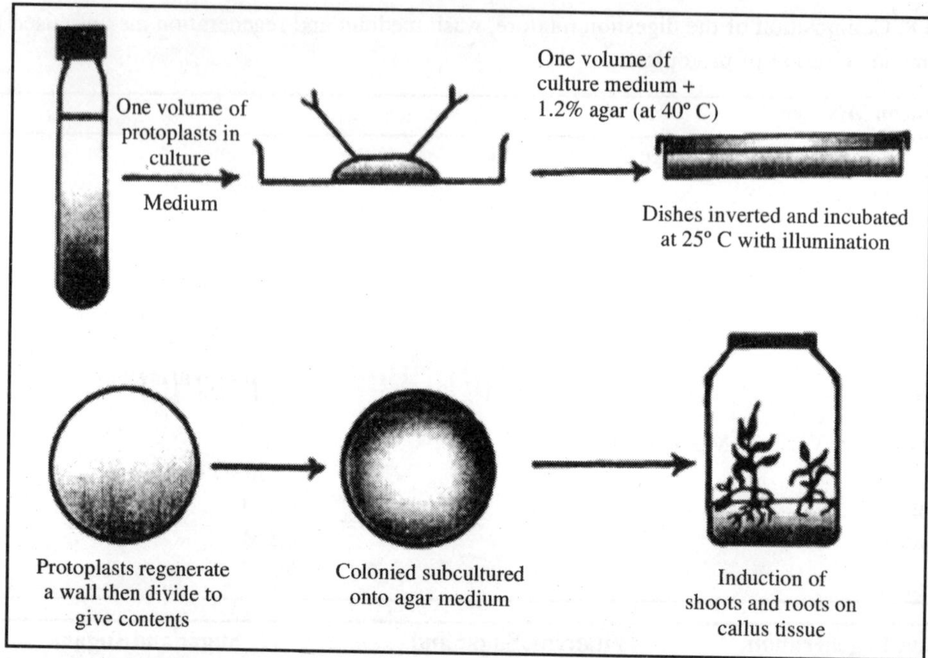

Diagram 5.5. Diagrammatic presentation of culture of protoplast by agar embedding technique in small petridishes.

Hanging Drop Cultures (HDC) Techniques

To avoid evaporation of the media while observing protoplast, this method is used. Drop can be placed in the depression of a specially prepared hanging drop-slide which is then undersealed with a coverslip and oil, or the drops are placed on the lid of a petridish. The petridish contains mannitol solution to maintain the humidity. The dish is sealed with parafilm.

Multidrop Array (MDA) Techniques

This is a refinement and amplification of the hanging drop technique. Through this technique it becomes easier to screen a wide range of nutritional and hormonal factors rapidly (using a small amount of plant material, less than 1g of plant tissue). The system can be described as follows:

1. Prepare a dilution series of two variables A and B, so as to have tubes A_1 to A_{10}, B_1 to B_{10}, likewise 100 possible combinations can be prepared, i.e., A_1B_1, A_2B_2 ..... $A_{10}B_{10}$, etc.
2. To these 100 different media (1.0 cm³ each) add 1.0 cm³ of protoplast suspension prepared in basal media at 20 times the minimum planting density.
3. For each combination, i.e. A_1B_1, A_2B_2, etc. make, 49 drops + 4 control drops in order on the lid of a petridish and label the dish, i.e. A_1B_1.
4. To each drop add 20 µl of another variable media component previously prepared. Thus each dish contains 49 different media combinations.

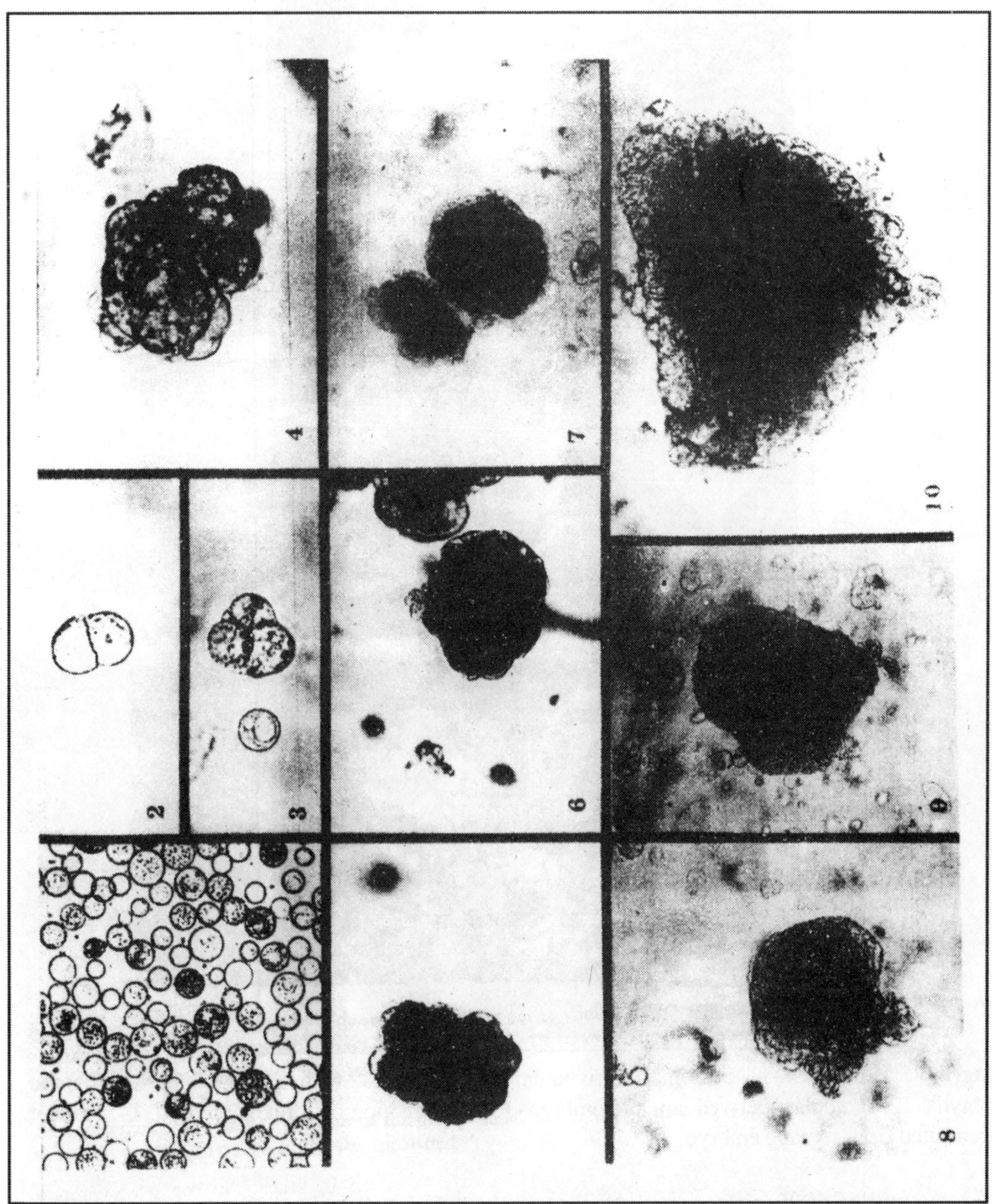

Diagram 5.6. Diagram 5.6, (1) Formation of wall around the protoplasts, (2) Reconstituted cells increases in size and first division taken place (within 7 days), (3) to (10) Subsequent divisions give rise to small cell colonies (within 2 – 3 weeks)

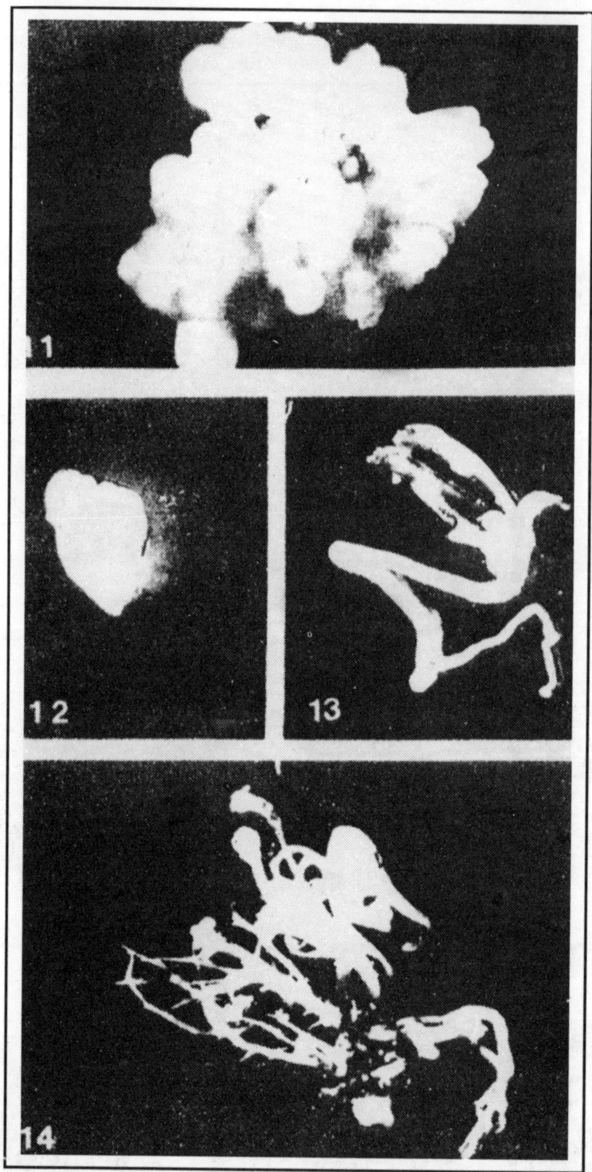

Diagram 5.6 (contu...). 11. Protoplast-derived embryogenic callus with different stages of somatic embryos. 12. Protoplast-derived somatic embryo. 13. Germination of somatic embryo. 14. Plantlet regenerated from somatic embryo.

5. A total of 4900 different media compositions each inoculated with cells have thus been prepared on only 100 petriplates.
6. 20 cm³ of mannitol is added to the base of the dish. The lid with attached droplets is inverted into position and the plates are sealed and incubated.
7. After a suitable incubation period the plates can be easily evaluated microscopically to determine which media combination, has given the most successful growth.

8. The best method for the culture of protoplast is by plating them in a medium containing 1.2% agar. In this condition, protoplasts grow best by good number of divisions and can be handled conveniently. Moreover, it could.be observed under the microscope.

Protoplast Fusion

Plant prptoplasts represent the finest single cell system that could offer exciting possibilities in the fields of somatic cell genetics and crop improvement. The isolated protoplasts are surrounded by plasma membrane only. The lack of the cell wall allows the plasma membrane of two or more protoplasts to come into intimate contact. The important aspect has been that incompatibility barriers do not exist during the cell fusion process at interspecific, intergeneric, or even interkingdom levels. Under certain conditions they will stick together rather like two soap bubbles. Later, if they are given an appropriate stimulus they will fuse together forming a single mass surrounded by a common membrane. Kuster was able to see occasional spontaneous fusion of protoplasts. However, with the development of enzymic methods for producing large numbers of isolated protoplasts, great interest has now been developed in the use of protoplast fusion (somatic hybridization) as a possible tool of plant breeding. Many people believe that protoplast fusion offers a useful tool to plant breeders to make crosses between sexually incompatible species for transfer of nuclear or cytoplamic characters. Protoplast fusion can be used to make crosses within species (intraspecific), between species (interspecific), within genera (intrageneric) and between genera (intergeneric).

Growth and Division of Protoplast

For the growth and nuclear division, the regeneration of cell wall is not seems to be a prerequisite in protoplast culture. In *Convolvulus* sp. protoplasts undergo one or two nuclear division prior to cytokinesis. The same was observed in *Haplopappus* spp.

It has been observed that for the rapid growth of protoplasts frequent subculturing in mannitol free medium is necessary. Protoplasts donot grow in large colonies on high osmotic medium. Its growth gradually shows a downward trend which ultimatly inhibited altogether. The colonies, therefore, should be picked up along with small pieces of agar and then transferred on the top of another medium to avoid any damage to protoplasts by rough handling with the forceps.

Protoplast Culture: Regeneration of Cell Wall

In culture, protoplasts start developing a wall around itself within few hours and it takes only few days to complete the process. Wall materials are progressively deposited at the surface of the plasmalemma. The cellulose is deposited either between the plasmalemma and the multilamellar wall material or directly on the plasmalemma. The nature of biosynthesis of the cell wall depends on the plant material and the system of protoplast culture.

The newly built cell wall can be observed either by plasmolyzing the protoplast by transferring it in a hypertonic solution, or by staining the cell wall with **calcofluor white** fluorescent stain (0.1%). However, electron microscopic studies and freeze etching studies

Table 5.9. Name of some of the families and species in which shoot differentiation of plant regeneration has been achieved from cultured protoplasts.

Family	Species
Compositae	*Cichorium intybus, Lactuca sativa cultivars (L. serriola, L. saligna), Petasites japonicus, Senecio vulgaris*
Cruciferae	*Arabidopsis thaliana, Brassica campestris (B. carinata, B. juncea, B. napus, B. nigra, B. oleraceae var. capitata), Sinapis alba*
Cucurbitaceae	*Cucumis sativus*
Euphorbiaceae	*Manihot esculenta*
Gramineae	*Bromus inermis, Oryza sativa, Pennisetum americanum, Saccharum spp., Triticum aestivum*
Leguminosae	*Glycine argyrea, G. canescens, G. clandestina, G. max, Medicago arborea, M. coerulea, M. difalcata, M. falcata, M. glutiniana, M. hemicyla, M. sativa cultivars (M. varia), Psophocarpus tetragonolobus, Trifolium hybridum, T. repens, T. rubens*
Liliaceae	*Asparagus officinalis, Hemerocallis sp.*
Linaceae	*Linum usitatissimum, L. strictum, L. lewissii*
Magnoliaceae	*Liriodendron tulipfera*
Ranunculaceae	*Ranunculus sceleratus*
Rutaceae	*Citrus aurantifolia, C. grandis, C. limon, Citrus medica, C. paradisi, C. reticulata, C. sinensis*
Salicaceae	*Populus tremula, P. alba, P. grandidentata, P. nigra var. Butulifolia, P. trichocarpa*
Santalaceae	*Santalum album*
Solanaceae	*Atropa belladonna, Capsicum annum, Datura metel, D. meteloides, D. innoxia, Hyoscyamus muticus, Lycopersicon esculentum, Nicotiana acuminata, N. alata, N. debneyi, N. glauca, N. langsdorffii, N. longiflora, N. otophora, N. paniculata, N. plumbaginifolia, N. suaveolens, N. sylvestris, N. tabacum, Petunia hybrida, P. inflata, P. parodii, P. parviflora, P. violacea, Salpiglossis sinuata, Solanum aculeatissimum, S. aviculare, S. brevidens, S. chacoense, S. dulcamara, S. etuberosum, S. fernandezianum, S. gilo, S. khasianum, S. luteum, S. lycopersicoides, S. melongena, S. nigrum, S. pennellii, S. phureja, S. phureja, S. chacoense, S. pinnatisectum, S. torvum, S. tuberosum cultivars (S. tuberosum tetraploid clones, diploid clones), S. uporo, S. viarum, S. xanthocarpum*
Ulmaceae	*Ulmus species*
Umbelliferae	*Daucus carota, Foeniculum vulgare*

have revealed much about the structure and progressive development of cell wall around the protoplast in culture medium.

1. Observe regularly the regeneration of cell wall, cell division and small callus formation under inverted microscope
2. Examine cell wall formation in protoplasts with a droplet of 0.1% calcofluor white R, American Cyanamid, Bound Brook, N.J. USA, in 0.4 M sorbitol solution on a slide. The cell wall regenerated protoplasts fluoresce.
3. Small cluster of calli are observed after 2-3 weeks of culturing protoplasts

4. Subculture the cell clusters on a freshly prepared protoplast culture medium with or without 1/2 the intial mannitol and 0.8-1.6% agar.

Determination of Protoplast Plating Efficiency

Calculate the average number of cell colonies per cm^2 of the petri dish by using following formula :

$$\text{Plating efficiency} = \frac{\text{X x Area of petri dish x 100}}{\text{no. of protoplast plated/petri dish}}$$

$$X = \frac{S_1 + S_2 + S_3 + S_4 + S_5}{5}$$

Development of Callus/Regeneration of Whole Plant

Soon after the formation of wall around the protoplasts, the reconstituted cells show considerable increase in size and first divisions usually occur within 7 days. Subsequent divisions give rise to small cell colonies. After 2-3 weeks macroscopic colonies are formed which can be transferred to an osmotic free medium to develop a callus. The callus may be induced to undergo organogenic differentiation, or whole plant regeneration through embryogenesis.

Organogenesis

Organogenesis takes place from a callus and not directly from a single cell. When isolated protoplasts are put into culture under appropriate conditions of plant inoculum, medium and environmental factors they go through a set series of events.

1. Wall regeneration
2. Early mitotic division and callus formation
3. Organogenesis

Embryogenesis

Plant tissue in vitro can induce to form somatic haploid embryos. Steward and coworkers first observed the phenomenon of somatic embryogenesis in carrot. Somatic embryos can be induced in cultural conditions from three different sources:

1. vegetative cells of mature plants;
2. reproductive tissues other than the zygote; and
3. hypocotyls and cotyledons of embryos

Embryoids are initiated in callus from small superficial clumps of cells associated with dense cytoplasm and large starch grains. The developing embryoids in vitro then pass through a sequence of growth stages that is exactly the same as seen in the development of a seed embryo.

Like cells, protoplasts are genetically unstable in vitro. Somaclonal variations have also been detected in the plants regenerated from protoplasts. Chromosomal variation has also been observed in plants regenerated from protoplasts. Shepard and his associates conducted detailed studies of plant regeneration from protoplasts of Russet Burbank potatoes. A significant variations were detected for growth habit, tuber colour, fruit production and tuber uniformity. Several of these traits are considered to be improved over the widely grown commercial variety. They further reported somaclones with resistance to early blight *(Alternaria solani)* and late blight *(Phytophthora infestans).* Our knowledge is insufficient to understand the mechanism of such instability in culture.

In *Nicotiana tabacum,* protoplast cultures were regenerated into cells and callus tissues. These calli were then transferred to fresh medium supplimented with IAA (4 mg/1) and cytokinin (2 mg/1). Calli differentiated into shoots and occasionally in roots within 3 to 4 weeks. The plantlets are then subsequenctly transplanted to pots.

General Steps of Protoplast Fusion

An efficient method employing polyethylene glycol (PEG) has found wide application in experiments on protoplast fusion. Many fusion inducing agents such as sodium nitrate, high Ca^{++}, high pH 10.5, high temperature (37°C) and polyvinyl alcohol have been used for fusion of protoplasts.

Procedure

1. Place a drop (1 drop = 50 μ) of each protoplast suspension in sterile petridish, gently shake the plate to ensure proper mixing of suspensions from callus or suspension culture with leaf protoplast. The concentration of each species should be about 1 x 10^5/ml before mixing.
2. Wait for 5 min. to allow the protoplasts at settle to the bottom of petridish
3. Slowly add 300-450 ml PEG solution to the edge of the protoplast suspension, then in the centre of the protoplast, and wait for 15 min
4. Add drop by drop 1 ml of 0.7 M mannitol solution to dilute the PEG solution
5. Raise one side of the plate and wash the protoplast clumps adhering to the plastic surface of plate upto with 9 ml of 0.7 M mannitol
6. Remove the PEG and mannitol residues from petri dish. Add a few drop of mannitol solution to the fused cells
7. After 5 min observe the fused product on the inverted microscope. The process of adhesion or agglutination of protoplasts in various combinations can be seen
8. Study the fused and unfused protoplasts for the synthesis of cell wall, first cell division, formation of homokaryons/heterokaryons, frequency of multinucleate homokaryons from heterokaryons and mitotic index, cytologically examination under microscope.

The technique of protoplasts fusion has developed new vistas for the plant breeders where conventional methods failed or the plant species are incompatible. Protoplasts fusion

Protoplasts (Leaf)	Protoplast Fusion	Protoplasts (Cell suspension culture)
1	⇓ ⇓ Combine ⇓ ⇓ ⇓	PEG High Ca^{++} and pH (10.5) PEG Dilution
2	Fusion ⇓ Mitosis ⇓ Hybrids	Culture Medium

Diagram 5.7. Presentation of stages in protoplast fusion and production of hybrids. 1 = Parental protoplasts, 2 = Aggulation adhesion.

results in somatic hybrids, the phenomenon is known as somatic hybridization. Protoplast fusion may be spontaneous or induced.

The protoplasts fusion may be of three kinds:

1. Spontaneous fusion
2. Mechanical fusion
3. Induced fusion

Spontaneous Fusion

Only interspecific protoplasts have been found very often fusing through their plasmodesmata during enzyme treatment. Nearby protoplasts in a Perti dish have been observed to form large multinucleate (coenocytic) mass. Protoplasts of young leaves are prone to fuse spontaneously with one another.

Mechanical Fusion

The giant protoplasts of *Acetabularia* have been fused mechanically. This kind of fusion is not dependent uopn the presence of fusion-inducing agent. However, in this procedure protoplasts are likely to get injury.

Induced Fusion

This kind of fusion can be done with the help of (a) $NaN0_3$ treatment, (b) High pH/Ca^{++} treatment, (c) Polyethylene glycol (PEG) treatment, (d) Electrofusion. This method can bring together both intra- and interspecific protoplasts. For this, some inducing agent is necessary. The inducing agent is called **fusogens.** In animals **sendai virus** plays this role but in plants both physical as well as chemical methods are used.

Several chemicals like sodium nitrate, PEG, poly-L-ornithine, poly-D-lysine, cytocholasin B, protomine sulphate, lysozyme, glycerol, dimethyl sulphoxide, proteins and calcium ions at high pH have been used as fusogens. Kuster (1909) for the first time used sodium nitrate to plasmolyse onion protoplasts which underwent fusion on deplasmolysis. Thereafter, Power et al. (1970) employed the same technique and successfully fused root tip protoplasts of maize and oat.

Keller and Melchers (1973) employed calcium ions at high pH (0.05 M $CaCl_2$ $2H_20$ in 0.4 M mannitol at pH 10.5) for fusion of tobacco protoplasts. They spinned the isolated protoplast in the fusion inducing solutions for 3 minutes at 50g and then kept the tubes in water bath at 37° C for 40-50 minutes nearly 20-50% protoplasts undergone fusion process.

The one most widely used fusogen is polyethylene glycol. The effects of PEG are not specific and it promotes the aggregation and fusion of protoplasts from the same or different species or inter-generic protoplasts.

Zimmermann has shown that if protoplasts are placed into a small culture cell containing electrodes and a potential difference is applied, then the protoplasts will lie between the electrodes. Protoplasts fusion can be induced after applying extremely short, square wave electrical stocks. This method has become very popular for its highly controllable fusions property. It is possible to fuse two single protoplasts.

Handling of Regenerated Plantlets

Plantlets regenerated by the process of embryogenesis or organogenesis are then used to transfer to non-sterile conditions. This can be very critical period when the plantlet is removed from sterile controlled environment and transplanted to the real world of agriculture. As the plantlet is moved to non-sterile conditions care must be taken to allows gradual release of these careful controls.

Anatomical analysis of these plantlets have shown that under these conditions often the cuticle (covering the leaf surfaces), is extremely thin and the root hairs are poorly developed. If these plantlets are suddently move to a situation where they are subjected to water stress they would be unable to control their water balance adequately and will die. Both the plantlets should first allow to develop a good root system before transferring them in vivo conditions and care should be taken for least possible damage of the roots. If the plants have been grown on agar solidified medium, this can be removed by gentle washing in a basin of warm water.

These plantlets are then carefully taken out from the test tubes or flasks and adhered agar medium is removed from them. The in vitro plants with roots are then carefully planted in small pots. The care is to be taken that roots should make quick and good contact with the potting mix because the plant may suffer water stress. The planting mix should be well balanced for water retention and sterile. Small, sterile peat pots (Jiffy pots) are excellent for this purpose.

PROTOPLAST CULTURE: UPTAKE OF FOREIGN MATERIALS

Isolated protoplast of plant cells have the remarkable property of taking up micro- as well as macromolecules like ferritin, polystrene, latex, protein, DNA, chromosomes, whole nuclei, chloroplasts and even virus and bacteria. This unique property of protoplast can be utilized for the incorporation of desirable traits into plants to get what we call *transferred* or *noval varieties* from wild ones.

Incorporation of Foreign DNA

Unlike transformation of bacteria by exogenous DNA, transgenosis in cells or protoplasts of higher plants are now possible and have been tried successfully in some plant species. Doy and his associates transformed haploid tomato callus culture by lambda phage. The transformed higher plant cells can grow on a medium containing galactose and lactose as the only carbon source, that is considered to be due to phage DNA incorporation. This kind of transformation phenomenon is termed as **trasgenosis.** These studies have given a major break through in agricultural biotechnology.

Incorporation of Nuclei

Intra- and interspecific nuclear incorporation into protoplasts of *Nicotiana tabacum, Petunia hybrida* and *Zea mays* have been observed. These exogenous nuclei enter the protoplasts without any damage to plasmalemma. However, the role of these incorporated nuclei in the metabolism of protoplasts in culture is yet to be confirmed.

Incorporation of Chloroplasts

In addition to nuclear material and nuclei as a whole protoplast can take up certain cell organelles like chloroplasts, mitochondria etc. Transplantation of chloroplasts into albino plant species could solve a great physiological deficiency. Potrykus (1973), Bonner and Eriksson (1974) have reported the uptake of chloroplasts by albino *Petunia* and carrot protoplasts. Nass (1969) and Gills and Sarafis (1971) have incorporated spinach chloroplast into animal cell cultures. They have also reported that these chloroplast survive and remain matabolically active in animal cells. Transplantation of chloroplast in plant having insufficient or deficient photosynthetic system will have farreaching implications in plant improvement programme.

Several methods for the induction of chloroplast into plant protoplasts have been reported, all of them depending either on some form of osmotic effect or on hypothesized changes in the charge pattern on the plasmalemma.

Bonnett and Eriksson used a method involving plasmolysis by polyethylene glycol (PEG) to induce uptake of *Vaucheria* chloroplasts by carrot protoplasts. A dense mixture of protoplasts and chloroplasts was made up to a volume of 0.5 ml with protoplast culture medium. A volume of 1.5 ml of the solution containing 3 ml of protoplast culture medium and 7 ml of a 56% aqueous solution of PEG was added to the protoplast-chloroplast mixture. The final concentration of PEG was about 30%, a concentration known to promote cell aggregation and cell fusion without reducing cell viability. After 10 min the mixture was diluted to 10 ml with a solution of 0.1 M $CaCl_2$ in 0.3 M sorbitol, and centrifuged for 3 min at 150x g. The pellet was resuspended in 10 ml of the same solution, recentrifuged and resuspended in 2 ml of the protoplast culture medium.

Incorporation of Cyanobacteria

Experiments have been performed to incorporate certain cyanobacteria or blue-green algae *(Gleocapsa, Anacystis,* etc.) in protoplast of higher plants. The method is used to co-incubate the algal preparation with the isolated protoplast preparation in the presence of 25% polyethylene glycol, when high Ca concentration is added the protoplasts begin to engulf the algal cells. The method of uptake has been shown to be by envagination of the plasma membrane.

Incorporation of Bacteria

Incorporation of nitrogen fixing bacteria *(Rhizobium, Azotobactor)* blue green algae and m/-genes into non-leguminous plants specially cereals is a important task in the hands of geneticists and agronomists. Davey and Cocking (1972) introduced a symbiotic nitrogen fixing bacterium *(Rhizobium)* in legume chloroplasts. These chloroplasts containing nitrogen fixing bacteria could be fused with non-legume chloroplast to impart the capability of nitrogen fixation in non-leguminous cash crops. *Agrobacterium tumefaciens* have been adapted to become efficient and reliable gene vectors for genetic engineering of dicotyledonous plants. Chimeric-genes have been constructed which upon transfer into differentiated plant cells allow the identification of DNA sequences involved in the regulation of gene expression. Recently, direct DNA transfer and expression of a bacterial gene in protoplasts of *Triticum monococcum* and *Nicotiana tabacum* have been reported; and in the case of tobacco transformed protoplasts were regenerated into plants which obeyed Mendelian law of inheritance.

Incorporation of Virus

Virus, an obligate parasite in nature enters the plant cells, leaving behind its protein coat only in those plant cells whose cells wall is damaged. The virus genome (RNA) enters the protoplasts of the cell and then multiply in cytoplasm. Tatebe and Otsukic (1974) reported that protoplasts can be infected by one or more than one virus. Cocking (1966) observed the uptake of TMV by tomato fruit protoplast.

The inclusion of these micro- and megamolecules in plant protoplasts have opened a new path to solve the genetic deficiency in economically important plants through biotechnology.

Incorporation of Non-biological Materials

Incorporation of non-biological materials have been tried such as polystylene in to plant protoplasts. These type of studies have proved useful as a method of determining the maximum size of an object that can be engulfed by endocytosis. However, no direct biological value can be seen to these studies at the present time.

Selection of Fusion Hybrids

Visual Selection

This method can be used to select fusion products of protoplasts that have distinct physical characters. For example, fusion between leaf mesophyll protoplasts of green colour and a cultured cell that has an anthocyanin- containing red protoplast. A single fusion product between the two will have both chloroplasts and an anthocyanin vacuole together with two nuclei. The same method can be applied to fusions between leaf mesophyll and hyaline cultured cell protoplasts.

Obviously, the drawback with this method is that it is very laborious and only a very limited number of fusion products can be selected.

Fluorescent Labels

1. In this method fluorescent labelled dyes are used to detect fusion products.
2. If the two original protoplast cultures are pre-incubated for 12-15 hours, one in octadeconyl aminofluorescein and the other in octadecyl palamine B, each group of protoplasts takes on a specific fluorescence colour. The dyes are non-toxic and do not affect viability, wall regeneration or growth.
3. After fusion of the protoplasts fusion products may be identified by their fluorescence characteristics under a fluorescence microscope.

Fluorescence Activated Cells Sorting

Protoplasts are fluorescently labelled as above described. After fusion three types of fluorescence can be found; two original fluorescence of the two parental populations and the fluorescence of the fused hybrids. A machine now exists to separate these populations. A droplet containing a single cell is held between two fluorescence detectors. If it is one of the two parent populations the machine charges two electrical plates, as the droplet falls it is deflected by the electrical charge. If the fluorescence is of the hybrid, no electrical charge is applied to the plates and the droplet fall vertically. Thus the protoplasts are automatically separated into the parent types and the fusion products. The fusion products can then be plated into an appropriate culture system.

Nutritional Selection

In some cases, Carlson et al. were able to separate fusion products by their ability to grow on a media that would not permit growth of the parent protoplast lines. For example hybrid of *Nicotinia glauca* and *N. langsdoifu* could be cultured in auxin free culture medium in contrast to its parents protoplast.

Drug Sensitivity and Resistance

Power and co-workers used sensitivity to actinomycin-D to select fusion products between *Petunia hybrida* and *Petunia parodii.* The two parental species are sensitive to actinomycin-D; however, the hybrid is resistant. Again it would be of great benefit to the field of somatic cell hybridization if a wider range of resistance mutants could be isolated. Some research groups are now attempting to produce large numbers of resistant mutants for use in protoplast fusion studies.

Somatic Cell Hybridization or Cybrid or Cytoplast

Success in protoplast fusion between diversified plant species gave great impetus to geneticists and plant breeders though the technique has some shortcomings. Nevertheless, it has stimulated research activities in many laboratories around the world.

When two protoplasts fuse along with their nuclei true hybrid are formed. However, often the two nuclei will survive independently in the mixed cytoplasms, forming a heterokaryon. Chromosome loss can occur from one or both of the nuclei; if all the nucleus of one parent is lost, what is left is one nucleus in a mixture of both cytoplasms. This product is called a **cytoplast or cybrid.** Cybrids can also be prepared by fusing a normal protoplast with an enucleated protoplast. Presently, researchers are studying the use of cybrids to transfer resistance to some herbicides. This resistance character is carried on the chloroplast genome. Transfer of cytoplasmic male sterility can be controlled by transfer of the mitochondrial genome. The study of cybrids is also important for investigation of possible recombination within these extra-chromosomal genomes.

Protoplasts for Isolation of Cell Components

The danger of erroneous interpretation of protoplast metabolism as truly representing intact cell processes is hardly encountered when the aim is merely the isolation of cell components. The objective of isolating such components as plasmalemma, nuclei, and chloroplasts from protoplasts, rather than from cells or intact tissues, is in most cases to obtain them by convenient methods which will cause the least destruction. Therefore, if the functionality of these components is of primary importance, the evaluation of the rate and quality of damage (e.g. in nuclei) as well as their unquestionable identification and purity (e.g. of plasmalemma) are important for considerations. Most of the problems involved in the isolation of cell components from protoplasts are specific to plant cells because of the latter's rigid cell wall and their specific organelles, e.g. chloroplasts and vacuoles. These problems are very rare in animal cells but are common to plant cells.

Plasmalemma

Isolation of plasmalemma from cells of higher plants is a difficult task. It is hampered by two major factors: (a) high shear force used for cell homogenization will damage the plasmalemma (b) good markers are essential to identify the plasmalemma fraction. Therefore, protoplasts obtained from leaf mesophyll or from cell suspensions were isolated by enzyme treatment. Plant protoplasts bind concanavalin A (con A) is used as a plasmalemma marker.

Chloroplasts

Chloroplasts were first isolated from protoplasts by Wagner and Siegelman as a by-product of their isolated vacuoles. The protoplasts were ruptured osmotically by suspending them into a phosphate buffer containing Mg^{++} and dithiothreitol. Chloroplasts were then separated on discontinuous sucrose gradients and were found to be photochemically active. Isolation of cereal chloroplasts is especially problematic, Chloroplasts of these plants are obtained by suspending the protoplasts in an appropriately buffered osmoticum (containing bovine serum albumin and dithiothreitol) and passed through a 20μ nylon net. The Chloroplasts are then collected by centrifugation without fractionation. Nishimura developed a new technique for spinach leaf protoplasts, using a syrings needle to rupture the protoplasts.

Mitochondria

Large number of mitochondria can be obtained from plant tissues and cultured cells. Some of the methods mentioned above for chloroplast isolation from protoplasts may also be useful for mitochondria and the two types of organelles can be separated from the same sample of plant materials. When 0.05-0.1% bovine serum albumin is included and a sucrose gradient is employed for fractionation of the disrupted protoplasts, mitochondria will band quite sharply at a density of 1.18g. cm^3, and they can be identified by their fumerase activity, while Chloroplasts will peak at an appreciably higher sucrose density (1.22 g.cnf).

Vacuoles

Vacuoles are most fragile cell organells. Wagner and Siegelman reported that when protoplasts from various plant tissues are ruptured gently by transfering it into phosphate buffer, vacuoles could then be separated by a low-speed centrifugation and be further purified by layering over 5% Ficol containing 0.55 M sorbitol and 1 mM tris-MES buffer. The homogeneity of the vacuole preparation can be nicely demonstrated by the use of anthocyanin-containing tissues, in Tulipa petals.

Conclusion

The scenario presented in this chapter gives an overall view of protoplast culture. The technique has unravelled the complexities of protoplast culture and its application to agricultural biotechnology. The protoplast technique is used in studies of plant cell genetics specially in somatic embryogenesis and transfer of genetic information by DNA uptake and organelle implantation. Though scientists have developed many plantlets and new varients through this technique, care must be taken to achieve cost-effective results so that the in vitro technique of protoplast culture can be adapted for commercial exploitation of agricultural crops in general and plantation crops in particular.

CHAPTER-6

Somatic Embryogenesis Principles, Concepts and Applications

In zygotic embryogenesis, embryo i.e. zygote derive from fertilization of egg cell and male gamete, which on germination gives rise to plantlets. While somatic embryogenesis is the process by which the somatic cells or tissues directly develop into differentiated embryos, and each fully developed embryo is capable of developing into a plantlet (young or miniature plant). Embryos can be obtained either directly from cultured explants (the organised structures, for example, leaf, hypocotyl, stem and other plant parts) and anthers (or pollen) or indirectly from callus (an unorganized mass of parachymatous tissue derived from explant culture as a result of wound response) and isolated single cells in culture. The process of embryogenesis involves various stages of differentiation and development such as proembryo, globular, heart-shape and torpedo embryo.

More recently, the embryos derived from somatic cells (mostly) or pollens are being utilized in the preparation of artificial seeds which appear to be of great value in plant propagation programme.

There are many differences exhist between embryogensis and organogensis, such as a bipolar somatic embryo arises from a single cell that shows no vascular connection with the maternal tissue or explant; in fact it develops a closed radicular end. In organogenesis, on the other hand, the origin is multicellular and the monopolar structure develops procambial strands which establish a connection with the pre-existing vascular tissue dispersed within the callus or the cultured explant. However the main difference between embryogenesis and organogenesis is that organogenesis is a property of the somatic tissues, whereas embryogenesis pertains to the reproductive tissues.

The properties of a reproductive tissue which are acquired, *in vivo,* during the phase transition that occurs when a vegetative meristem becomes a floral bud (i.e. reproductive part), are acquired by the somatic cells *in vitro* through a process which is normally called **acquisition of totipotency,** but whose molecular mechanism is not yet understood.

Somatic embryogenesis has been described in more than a hundred species and it is likely that with the passage of time more, species will be included will be made to exhibit in the list. Of the several factors that affect the induction of somatic embryogenesis, foremost are the genotype of the starting material, origin of the explant and media composition, particularly with respect to hormones.

Carrot as a Embryogenic System

There are now a number of embryogenic systems that work satisfactorily but carrot, the first to be described, is still the best. A number of embryogenic lines of carrot have been maintained in suspension cultures, in some cases for more than ten years. When hormones are removed from the medium and the cell concentration is lowered, these lines start the embryogenic programme that leads to a plantlet through the morphologically defined stages of *globular, heart* and *torpedo-shaped* embryos.

Somatic and Zygotic Embryos

Somatic embryos show a few differences from a zygotic one. In the initial stages, the suspensor is less developed in somatic embryos (perhaps a reflection of its uselessness under in vitro conditions).

However, an important difference is that whereas the zygotic embryos after forming embryonal axis, with the apical meristems, prepare for dormancy, the somatic embryos do not undergo desiccation and dormancy and embryogenesis goes on from initial cell to plantlet without interruption. In vivo, development is similar to somatic embryogenesis up to the torpedo stage. Later, development of somatic embryos, particularly the developmental programme of the meristems, follows a path which, in a sense, shows more analogies with organogenesis (Ammirato, 1983; Sung et al., 1984; Nomura and Komamine, 1986).

Beside above difference, somatic embryogenesis shows similarity in structure, function and biochemistry to that of zygotic embryogenesis. Somatic embryos share many characteristic features with zygote embryos; for instance, both are bipolar structures and originate from a single cell and do not show vascular differentiation. Nevertheless, somatic embryos are large in size and sometimes show more than two cotyledons **(pluricotyledony)** as compared to zygote embryos which are smaller in size and exhibit two cotyledons (in case of dicots). Significantly, a large number of embryos potential use in volume and rapid micrppropagation of high value crops, including crops propagated by seeds. Despite its potential, this method has not yet been commercialized, except in case of oil palm *(Elaeis guineensis)*.

Totipotency and Role of Auxin

A tissue explanted in vitro consists of several cell types. The explant is treated with growth regulators so that its cells (or, rather, a fraction of them capable of responding to hormonal stimuli) de-differentiate and start to proliferate. In the presence of auxin, after several days (sometimes weeks) a population of small, compact cells emerges in this tissue which can be hand-picked or concentrated by a process of differential centrifugation through a Ficoll gradient.

Proembryogenic Masses and Role of Auxin

Those small, compact cells divide in an asymmetrical way and their daughter cells, sticking together, give rise to typical cell clumps which have been called proembryogenic

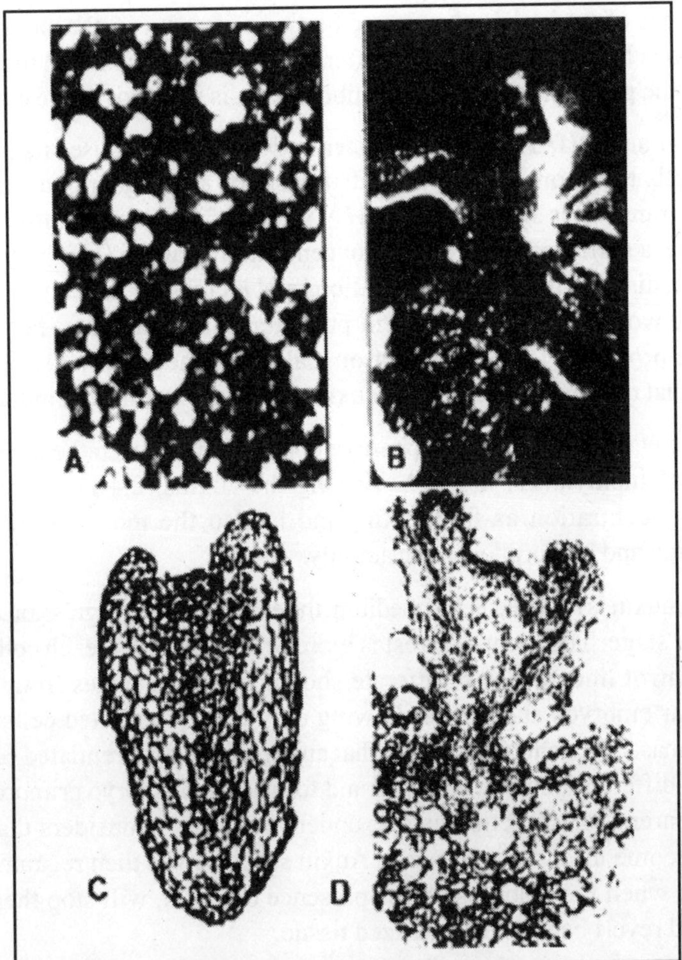

Diagram 6.1 (A) Suspension culture of carrot somatic embryos. (B) An organogenic callus of tomato. (C) Section of a carrot somatic embryo (torpedo stage). (D) Section of a primordium arising from an organogenic tobacco callus.

masses (Halperin, 1966) or **embryogenic clusters** (McWilliam et al., 1974). From these proembryogenic masses embryos develop upon dilution of cells and removal of hormones. Totipotency, once acquired, is a long-lasting capacity in the sense that a cell population that has acquired this property can be subcultivated in the presence of auxin for months and years and yet retain the capacity to generate embryos upon removal of the hormone.

Acquistion of Totipotency

It is clearly one of the most critical steps in somatic embryogenesis. However, the details of the process are not known. Treatment with auxin at high concentration is needed to cause de-differentiation and to elicit totipotency; carrot plantlets regenerated via somatic embryogenesis give rise to adventive embryos directly from the epidermal cells of the stem, but this has never been observed in the plants originated from zygotic embryos.

Various types of auxins have been used for this purpose, viz., the natural auxin IAA and the synthetic ones NAA and 2,4-D. The latter is the most efficient and the most commonly used auxin for the promotion of somatic embryogenesis in carrot tissue cultures.

The natural auxin (IAA) at a low concentration (10^{-6} M) causes its polar and induces the adjacent cells to become specific for its transport; at a higher concentration (10^{-5} M), however, polar transport is switched off and IAA slowly diffuses in all directions (Goldsmith, 1982). In nature, a shift from low to high concentration of auxin (as it may occur on cutting a root and, thus, disrupting the communications and increasing the concentration of IAA in the cells of the wound) causes a block in polar transport and, thus, favours the onset of morphogenetic processes (for example, from callus to a new root). Other examples of re-programming that require the presence of auxin are gall formation and induction of tumours.

LoSchiavo et al. (1989) found a positive correlation between the level of added auxin and the level of methylation in DNA. Among the auxins, 2,4-D can reach the highest intracellular concentration as free auxin, and is also the most efficient in promoting hypermethylation and in morphogenetic activity.

If, instead, auxin is retained in the medium, the embryogenic progression stops somewhere before globular stage. The point of arrest is more or less typical of each cell line. Therefore, there will be carrot lines which proliferate showing various stages from undifferentiated cells to globular embryos and others showing only undifferentiated cells and PEM (**pro-embryogenic** masses). It should be noted that auxin affects differentiated cells of an explant by inducing de-differentiation on one hand and formation of embryo primordia (PEM) on the other. This apparent contradiction can be understood if one considers that once PEM are formed they become insensitive to auxin. Auxin sensitivity is then regained at a later stage (post-globular) when the embryos, in the presence of auxin, will stop their differentiative programme and revert back to unorganized tissue.

Experiments on embryogenesis in the presence of TIBA (2,3,5 tri-iodobenzoic acid), a drug that blocks the polar transport of auxin from the cells with consequent increase in the intracellular hormonal levels revealed that during the PEM-globular stages the embryos do not produce auxin, because embryogenesis proceeds normally up to globular stage embryos. Thereafter, the embryos start callusing, which can be explained by assuming the production of IAA, block of its transport, internal accumulation, and callus formation. Hence, after the globular stage the embryos start producing auxin and become sensitive to it and, thus, behave like the somatic tissue.

Effects of Auxin Absence:

The onset of embryogenesis leads to a profound rearrangement of the cell metabolism. Such studies are usually made by comparing unorganized callus tissue with embryos the rate of protein synthesis as well as the rate of poly (A)$^+$RNA synthesis increases in embryos. DNA synthesis and polyamine levels also increases. Tubulin genes were also found to be expressed at higher levels during embryogenesis. Lipids also shows a certain amount of developmental variation.

Methylated Cytosine Level and Development

At the DNA level, changes in the level of methylated cytosine were recorded as a function of the developmental stage. As soon as the hormones are removed (the act that operationally marks the onset of embryogenesis) there is a sudden demethylation that may interest up to 30% of the sites. Then, with progress in the embryogenic process, the level of methylated cytosine goes up again and reconstitutes the level typical of the starting cell population. This methylation is, in part at least, site-specific so that DNA fragements from a genomic DNA library can be used as probes to analyze development-associated variations in cytosine methylation. The development-associated variations were found and, moreover, the pattern of methylation may, in turn, be associated with RNA transcription as measured in Northern blots.

The Genetic Approach of Somatic Embryogenesis

The Requirements

1. Somatic embryogenesis is studied because it is an interesting process, liable to produce results of economic interest.
2. The field of plant embryogenesis is not well developed and somatic embryogenesis looks a promising model system.
3. Somatic embryos provide an interesting possibility as it is possible to mutagenize a cell population, make it to embryogenize and look for mutants of conditional type, e.g. temperature-sensitive mutants.
4. The seed embryos, being wrapped in various coats, are not amenable to microsurgery as in case of animal embryos. Moreover, for biochemistry a large number of synchronized embryos is needed and this is not so easy to obtain, particularly if one is interested in early embryonic stages.

This has been done in three laboratories, with or without mutagenic treatment, starting from haploid or diploid cell lines, using 24-25°C as permissive temperature and 31-32°C as non-permissive one.

Now that genetic manipulation of cells has reached such a high level of refinement plant regeneration from cultured cells is extremely important. In this respect, somatic embryogenesis has many advantages over organogenesis because embryos, unlike shoots, originate from single cells, and the embryogenic cultures can be synchronized and purified so that one can deal with practically pure cultures of homogeneous material. Moreover, once embryogenesis has started, it goes by itself; no further intervention is needed to adjust the auxin/cytokinin ratio, to remove the embryos from undifferentiated callus and so on. Haploid embryos can be obtained by cultivating anthers and the possibility of raising triploids from endosperm has been suggested and, to a very limited extent, exploited.

The formation of transgenic plants and the use of artiflcal seeds, which are the main applications of this technique are so well established.

The Process of Somatic Embryogenesis

Induction of the Embryogenic State

Induction of the embryogenic state in differentiated explants often requires extensive proliferation through unorganized callus cycles, death or disruption of surrounding explant cells, and/or high levels or a synthetic auxin such as 2,4-dichlorophenoxyacetic acid (2,4-D) or picloram. Plasmolysis of explant cells was also shown by Wetherell (1984) to enhance somatic embryogenesis in wild carrot *(Daucus carota)*. These factors are presumed to alter the epigenetic state of cells, and may be related in their ability to disrupt the cell-cell interactions required to maintain coordinated patterns of development. The role of cell isolation in induction of somatic embryogenesis was reviewed by Williams and Maheswaran (1986). Smith and Krikorian (1989) showed that breakage or wounding of zygotic embryos at explanting led to formation of somatic embryos on a hormone-free medium. Constituent cells of the damaged tissues were presumably released from positional or chemical restraints and were thus able to express their innate embryogenic potential on a medium permissive for growth. A similar observation was made for *Dysosma pleiantha* by Chuang and Chang (1987). Embryogenic callus was induced from wounded zygotic embryos, while intact embryos failed to respond. Cell isolation is manifest early in somatic embryogenesis by the formation of a cuticle.

In embryogenic cultures, each proembryonic group of cells becomes separated from surrounding cells by thickened, cutinized walls on the outer surface (Williams and Maheswaran, 1986).

Direct somatic embryogenesis, especially in the absence of exogenous auxin, is normally associated with a relatively brief developmental period between the time of cotyledon initiation and the beginning of seed maturation (Maheswaran and William, 1986b). During this time, embryonic cell division appears to directly clone the existing early embryonic epigenetic state. Even within this developmental window, some differentiation away from the embryogenic state may be inferred from the nature of structures produced by cellular-proliferation. In flax *(Linum usitatissimum),* late cotyledonary stage embryos produced numerous accessory cotyledons in addition to well formed somatic embryos (Pretova and Williams, 1986a).

A similar observation was made by Young et al. (1987) for *Lycopersicon.* Older immature zygotic embryos gave rise to accessory cotyledons and shoot apices rather than recognizable somatic embryos.

Developmental gradients within a zygotic embryo are also evident by their response to different auxins. For example, in soybean *(Glycine max)* the tissue that preferentially produces somatic embryos in response to the auxin. When the stronger 2,4-D is used as the inducing auxin, somatic embryos are initiated from a more extensive subset of cotyledon cells.

Starting with a culture of embryogenic cells in the form of proliferating proembryonic masses, the process of regeneration requires, the initiation of bipolar differentiation to produce cotyledons, shoot apex and root pole, and then, in sequence, maturation of embryos, germination or conversion to plantlets, and transfer of plants out of culture into soil.

Recurrent Embryogenesis

It is also known as repetitive, accessory, proliferative, or secondary embryogenesis. The power of embryo cloning techniques and their exploitation for mass propagation, metabolite production, or genetic transformation have recurrent embryogenesis as their basis. It also occurs when primary somatic embryos fail to mature normally into plantlets and instead give rise to successive cycles of embryos, most commonly from superficial cells of the cotyledons or hypocotyl. The process is probably homologous with the proliferation of globular proembryos in standard embryogenic cultures, differing only with respect to the stage at which integrated control of development is lost.

Expressions of recurrent embryogenesis are best viewed as a continuum, with proliferation of globular PEMs or early globular stages at one exterme, and the development of early embryogenic stages on bipolar embryos or germinating plantlets at the other exterme.

Problems

Recurrent embryogenesis may become a problem if it cannot be controlled when germination and normal growth are required. Where it can be stimulated or prevented, it offers the advantage of greatly facilitating mass propagation.

Maintenance

The **maintenance** of recurrent cycles of somatic embryogenesis can be spontaneous as is the case with alfalfa *(Medicago saliva* L, Lupotto, 1983, 1986). The cycles are maintained in the absence of growth regulators.

More frequently, however, the initiation of recurrent cultures requires that the developing embryos be locked into a developmental stage beyond which they cannot proceed, thereby repeating a cycle. This can be achieved by initial exposure to a very high auxin concentration such as 40 mg/1 of 2,4-D, followed by maintenance of the recurrent system using a lower level of auxin, such as 5 mg/1 of 2,4-D (Finer and Nagasawa, 1988), which prevents the transition from proembryonic to embryonic development.

Auxin vs Cytokinin for Induction of Embryogenesis

Although a majority of studies have employed auxins, particularly 2,4-D and NAA, for the induction of somatic embryogenesis from immature embryonic explants, cytokinins have also been used as inducing agents in some instances. When young coryledonary embryos of white clover *(Trifolium repens)* are explanted and exposed to the cytokinin BAP (6-benzylamino purine), the hypocotyl responds with the formation of somatic embryos (Maheswaran and Williams, 1984). When 2,4-D is used instead of BA, it is the cotyledons that form somatic embryos. Likewise, pea *(Pisum sativum)* shows a positive response from hypocotyl cells in the presence of cytokinin, but it is the cotyledon cells that respond in the presence of auxin (Kysely et al, 1987). Genotype, tissue type, and developmental stage may all be determining factors in the comparative ability to respond to auxin or cytokinin.

How to Obtain Embryos from Embryogenic Cultures?

As stated earlier, a removal or decrease of the auxin concentration in the growth medium can break the cycle of **continuous proliferation** of PEMs, and permit embryos to develop to maturity.

Brawley et al. (1984) showed that in globular somatic proembryos of carrot *(Daucus carota)*, ionic currents flowed inward at the site of the future shoot and out at the site of the future root pole. These currents were identified as being largely a K^+ influx and H^+ efflux (Rathore et al, 1988). Similar ionic currents have been detected around haploid embryos forming from immature pollen grains of tobacco.

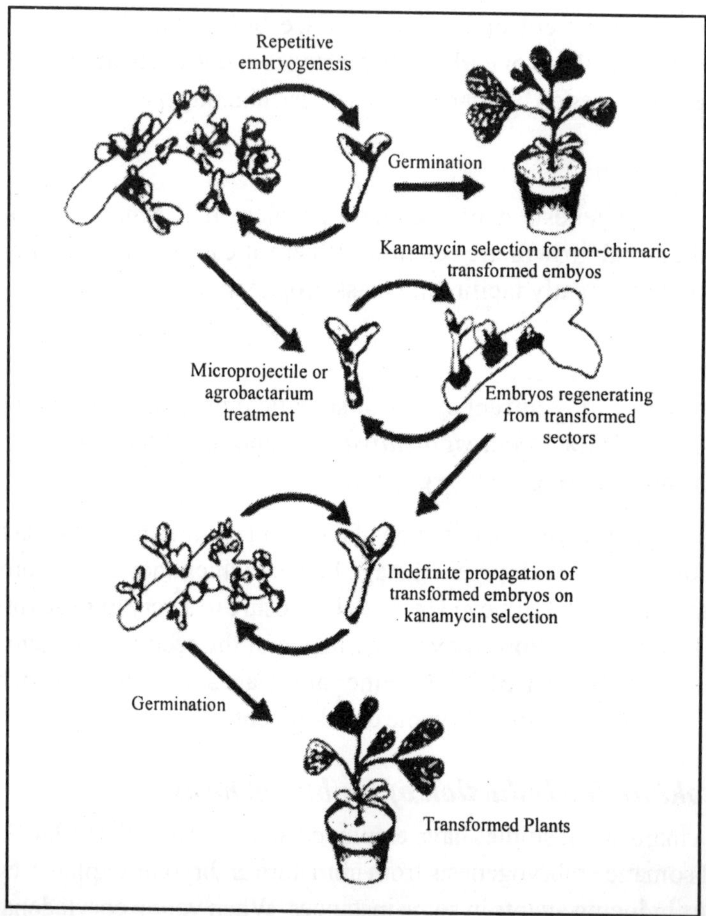

Diagram 6.2 Recurrent embryogenesis and its use in genetic transformation and mass propagation. Although new embryos can form from older embryos at any stage of development, this example depicts recurrent embryogenesis occurring from . cotyledonary stage embryos. Developing embryos can be exposed to *Agrobacterium* or bombarded with microprojectiles, transforming individual epidermal cells. As the embryo continues to grow and develop, transformed cells give rise to patches of transformed tissue from which transformed embryos develop on selection medium. In this example, kanamycin is used as the selection agent, but several other agents could be used, depending on the vector used in transformation. As long as the cycles of recurrent embryogenesis are maintained, transformed or nontransformed embryos can be propagated indefinitely.

In carrot somatic embryogenesis, electrical polarity is accompanied by asymmetry in the distribution of activated **calmodulin.** Timmers et al. (1989), after using fluphenazine fluorescence, showed that polarity in the distribution of activated calmodulin already exists in the globular proembryogenic masses before morphological polarity is visible. Activated calmodulin concentration remains higher in the region of the root pole. During the later stages of bipolar development, fluphenazine fluorescence is also strong in the region, forming shoot apex. Polarity in development of somatic embryos has also been detected as asymmetry in DNA synthesis by Nomura and Komamine (1986).

In direct somatic embryogenesis from immature zygotic embryos, suppression of normal polarity may play a role in embryo induction. Zygotic embryos in which growth of the main axis is weak or suppressed are more likely to give rise to somatic embryos from superficial cells.

Embryo Maturation and the Development of Germinability

Auxin Levels

It is essential that for germination a embryos must have functional shoot and root apices with the activity of meristematic growth. **High auxin** levels can inhibit development and growth of the shoot meristem if young proembryos are not transferred to a low-auxin or zero-auxin medium after induction. Activated charcoal may be added to the medium to remove excess auxin from the somatic embryos (Buchheim et al., 1989). At **low auxin** levels, shoot meristem formation is usually achieved early after the initiation of cotyledons, so that under inappropriate culture conditions, germination can occur prematurely to give weak or inviable plantlets.

Sucrose Level

For production of vigorous plantlets, a period of embryonic growth and maturation is needed before germination. This can be achieved by culturing at sucrose levels of 3 to 6%, although progressively increasing levels up to 40% have been used for some species. High concentration of sucrose in the medium may cause osmotic desiccation in the embryos, however, for some species, efficient conversion to plantlets also requires the imposition of temporary desiccation before germination. This procedure, which mimics seed maturation in vivo, may be necessary to trigger metabolic processes needed for germination and seedling growth. The gradual reduction in osmotic potential through desiccation of mature somatic embryos of wheat *(Triticum aestivum)* improved germination percentages has also been reported.

Heat Shock Treatment

The vigor of the plantlets from dried somatic embryos remains greater than that of plantlets derived from embryos. Heat shock treatment induced a degree of desiccation tolerance in alfalfa somatic embryos that was equivalent to that conferred by ABA treatment but did not have detrimental effects on subsequent growth of plantlets.

ABA Levels

Ammirato (1974) showed that ABA at 10^{-7} M prevented precocious germination of somatic embryos of caraway *(Carum carvi)* in suspension. Ammirato (1983) later reported that the same level of ABA had a similar effect on suspension-cultured carrot somatic embryos, producing embryos more similar to their zygotic counterparts than those grown without ABA. Based on these results, he proposed that regulation of embryo maturation by ABA might be used to facilitate large-scale batch cultures, mechanized planting, artificial induction of dormancy and incorporation into artificial seeds.

The role of ABA in initiating the accumulation of storage reserves has not been ruled out, especially as the initiation of reserve accumulation coincides with the highest levels of endogenous ABA. The role of ABA in the initiation of reserve accumulation has also been observed (Roberts et al., 1990). They found the presence of ABA essential for the stimulation of **protein accumulation** in somatic embryos of interior spruce (mixtures of *Picea glacua,* and *P. engelmannii* and their hybrids).

Water Saturation

Plantlets grown in vitro in a water saturated atmosphere show reduced development of cuticular waxes and abnormal stomatal function. On removal from culture, losses of such plantlets can be high if they are not protected from transpirational water loss while roots and normal leaves are developing. Acclimatization of culture-grown plantlets remains a problem in commercial micropropagation, since plantlets must usually be subjected to progressively reduced humidity over a period of week. Somatic embryogenesis offers some hope of avoiding or minimizing acclimatization problems if embryos can be removed from culture at physiological maturity and germinated under normal growing condition.

Genetic Control of Embryo Cloning

The capacity to undergo somatic embryogenesis from immature zygotic embryos is under some degree of genetic control. This control appears to be the result of a low number of genes, and is, therefore, highly heritable and amenable to selection. Consistent with the observation, that low-numbers of genes are responsible for regeneration capacity. Regeneration capacity appears to consist of two major parameters which are commonly measured in the literature.

1. The first is the frequency of explants which regenerate.
2. The second is the average number of somatic embryos formed per regenerating explant.

The mass propagation and genetic transformation of crops are most likely dependent on the breeding and development of germplasms with a high capacity to undergo recurrent somatic embryogenesis. Once capacity for regeneration has been backcrossed into elite lines and agronomically superior cultivars, embryo cloning techniques will finally be sufficiently efficient to play an important role.

Genotype is one of the factors that influences regeneration from cell culture, however, very little is known about the genetic components of somatic embryogenesis from immature zygotic embryos. The genetics of regeneration of alfalfa is especially well documented and individual genes have been identified and named.

Genetic Variability

Genetic variability for regeneration via somatic embryogenesis has been documented for a wide variety of species, including soybean, maize, rice, barley, wheat, etc.' Genetic control of regeneration capacity is largely additive and highly heritable in maize, rice, and wheat.

Cytoplasmic Effects

Cytoplasmic effects have also been important in maize, rice, wheat. In these crops cytoplasmic effects are sufficient to necessitate careful selection of maternal parents to ensure regeneration success. Non-embryogenic callus can be derived from embryogenic callus initiated from immature zygotic embryos of the cultivar Chinese Spring. The use of defined cytogenetic stocks has made it possible to further elucidate the nature of the genetic control of regeneration in wheat.

APPLICATIONS OF SOMATIC EMBRYOGENESIS

The high-volume multiplication of embryonic propagules is the most commercially attractive application of in vitro somatic embryogenesis. As commercially conceived, the system involves harvesting of maturing embryos from a continuously proliferating embryogenic culture of elite genotype, and converting the harvested **clonal embryos to seedling** transplants or synthetic seeds for delivery to the grower. Although the induction of normal embryo physiology, scale-up of culture volume, and design of field delivery systems have so far prevented industrial applications. Recent intensive work on synthetic seed systems shows commercial promise.

Mass Propagation

Somatic embryos have powerful advantages for mass propagation in comparison to both conventional clonal propagation methods (e.g., root cuttings, grafting) and other in vitro regeneration systems (e.g., micropropagation).

One advantage of propagation via somatic embryogenesis is the very high multiplication rates possible with many embryogenic systems. Unlimited numbers of embryos can be generated from a single explant. It depends on type of plant species.

In comparison, multiplication by root cuttings is limited to the amount of material available from the mother plant, and for most species, micropropagation also is characterized by relatively low multiplication rates. A second advantage of somatic embryogenesis is that, for many species, both growth of the embryogenic tissue and development of the somatic embryos can be carried out in liquid medium, making possible the handling of enormous numbers of embryos at one time.

Drew (1980) estimated that one liter of a carrot suspension culture contained 1.35 million somatic embryos. Thus, in comparison to root cuttings and micropropagation, somatic embryos offer the potential for high volume, large-scale propagation systems that can be translated into significant labor savings. Even greater economics of scale may be possible if bioreactor and continuous culture technologies can be applied to embryogenic systems.

Plants derived from somatic embryos are less variable than those derived via organogenesis. This may reflect an intolerance of somatic embryos to mutations in any of the numerous genes that must be necessary for ontogeny to be successfully completed. In contrast, vegetative meristems may be more tolerant to mutations and epigenetic changes.

Probably the most obvious advantage of somatic embryogenesis in comparison to other clonal propagation methods is the fact that the product is an embryo. The morphological and physiological similarity of somatic embryos to zygotic embryos means that they are almost complete propagules in themselves, with embryoinc roots, shoots and leaves (or at least cotyledons) and, most importantly, the programme to make a complete plant. Thus, unlike other clonal propagation systems, no separate shoot growth or rooting steps are required for plantlet production, again providing savings in labor.

Unlike organogenic or axillary branching systems, many embryogenic systems produce discrete embryos, and thus require no physical separation from mother tissue or other embryos in order to be handled, which once again means savings in labor.

Over the past few years, some of these special characteristics of somatic embryogenesis have been examined for possible commercialization purposes. The potential for somatic embryos to be grown in large volumes in continuous culture and employed as direct-delivered propagules has received much attention.

Scale-up Potential

The fact that both the growth of embryogenic cells and subsequent development of somatic embryos can be carried out in **liquid medium** gives somatic embryogenesis the potential to be combined with engineering technology to create large-scale mechanized or automated culture systems. Such systems are capable of producing huge numbers or propagules with low labor inputs. With the application of this technology, costs per propagule have the potential to be reduced to the point where they may be competitive with seed-derived plants, depending on the crop.

Use of Bioreactors

The first report of large-scale embryogenic cultures described an attempt to grow carrot cells in 20-liter carboys, which resulted in the formation of few embryos (Backs-Husemann and Reinert, 1970). The biological/mechanical system most often described for application to embryogenic systems is the stirred-tank bioreactor, a mass culture system originally developed for microbial fermentations, but more recently adapted for growing plant cells on a large scale (Wilson et al., 1971; Martin, 1980; Kurz and Constabel, 1981).

A major problem with adapting these bioreactor designs for use with plant cells is the high shear that stirring generates in these systems. **Air driven bioreactors,** with lower shear levels, have been tested as possible alternatives to the stirred-tank design, and have supported successful growth of a number of plant cell types. Air-lift bioreactors gave slightly higher yields of alfalfa somatic embryos compared to **propeller-stirred bioreactors** or cultures grown in flasks on a shaker.

Plant Biotech Industries, Ltd. has developed an automated system for large scale commercial propagation of plants, which makes use of somatic embryos as well as other propagules such as microtubers and bulblets. The system integrates a bioreactor with a bioprocessor in a closed system for separation, sizing and distribution of propagules into a culture vessel, and even employs an automated transplanting machine which transfers plantlets to soil mix in greenhouse trays at the rate of 8000 per hour. The so developed bioreactor-based system could cut production costs of plantlets by as much as 60% compared with conventional tissue culture propagation methods. Other benefits of bioreactor technology are lower contamination rates, savings in space, time and labor, accurate monitoring and control of temperature, pH, and gasses.

To date, the application of bioreactor technology has apparently not met its potential to produce hundreds or thousands of clonal embryos capable of growing into plants. To improve the capability of bioreactors to produce competent embryos, a group of researchers has recently developed a kinetic model of carrot somatic embryo development in suspension culture by monitoring substrate utilization, culture growth and embryo development over the time course of an embryogenic culture.

Protoplast Culture

Embryogenic callus and suspension cultures, as well as somatic embryos themselves have been employed as a source of protoplasts for a range of species. The logic of this approach is that isolation of protoplasts from cells or tissues that are themselves regenerable will likely yield protoplast cultures capable of forming whole plants (Shillito et al., 1989). Earliest application of the regenerative potential of protoplasts isolated from embryogenic material was made with embryogenic carrot suspension cultures.

Since then in three groups of plant species, viz. graminaceous species, citrus species, and forest trees (especially conifers), embryogenic cultures have proven to be especially valuable in providing a source of regenerable protoplasts.

In the Gramineae, regeneration of callus or even sustained cell divisions in mesophyll derived protoplasts could not be achieved following methods that had previously proven successful with mesophyll protoplasts of solanaceous species. Although there were many reports of sustained cell divisions in protoplasts isolated from nonmorphogenic cell suspension cultures of the Gramineae, the protoplast-derived calli failed to undergo morphogenesis. Therefore, Vasil and Vasil (1980) turned to embryogenic cultures derived from immature embryos of pearl millet *(Pennisetum glaucum)* as a source of protoplasts. These protoplasts could be cultured to give rise to cell masses, from which embryoids and eventually plantlets could be regenerated. Similar success was subsequently reported using embryogenic suspensions of several other graminaceous species.

Embryogenic citrus suspension cultures have not only provided a source of regenerable protoplasts, but also made possible the production of interspecific and even intergeneric somatic hybrid plants. Interspecific somatic hybridization in citrus was first achieved by Kobayashi et al. (1988). They fused protoplasts isolated from an embryogenic suspension culture of navel orange (C. *sinensis)* cultivar Washington with leaf protoplasts of satsuma mandarin *(C. unshiu)* cultivar Hayashi. Interspecific somatic hybrids have since been produced between a number of citrus species using similar techniques.

The ability to isolate protoplasts from embryogenic cultures of forest trees has had a large impact on regeneration studies for this group of plants, in particular coniferous species. Although a few researchers reported the growth of protoplasts isolated from conifer cotyledons, leaves, or suspension cultures to the colony or even callus/suspension stage.

The development of embryogenic callus and suspension cultures proved to be the key to the production of morphogenic protoplasts in conifers.

Embryogenic cultures have also been shown to be a valuable source of regenerable protoplasts in some hardwood forest tree species. Rao and Ozias-Akins (1985) isolated Protoplasts from embryogenic cell suspension cultures derived from proliferating shoot segments of a 20-year-old sandalwood tree *(Santalum album)*. The protoplasts could be cultured to form embryogenic cell aggregates, somatic embryos and eventually plantlets.

Similarly, embryogenic suspension cultures of yellow poplar provided protoplasts capable of regenerating whole plants via embryogenesis (Merkle and Sommer, 1987).

Embryo Cloning and Gene Transfer

Regeneration in several species, especially tress and large-seeded legumes, is limited via direct somatic embryogenesis from immature zygotic embryos. The embryos form directly on the original explant tissue. If callus is present, it grows concomitantly with the somatic embryos. The embryos do not originate from it, thereby bypassing any opportunity that a callus phase provides to sort transformed cells from non-transformed cells. Confounding the problem is the fact that these somatic embryos may have originated from groups of cells rather than from single cells within the explant.

As current **gene transfer techniques** transform single cells, not clumps of cells, the recovery of **chimeric embryos** consisting of transformed and non transformed tissues is virtually assured. Whereas the absence of a callus phase is the main factor that has limited genetic transformation in these species, this barrier has been overcome by the use of recurrent embryogenesis.

In recurrent embryogenesis, a cycle is initiated whereby somatic embryos continuously proliferate from previously existing somatic embryos. The cycles of embryo proliferation

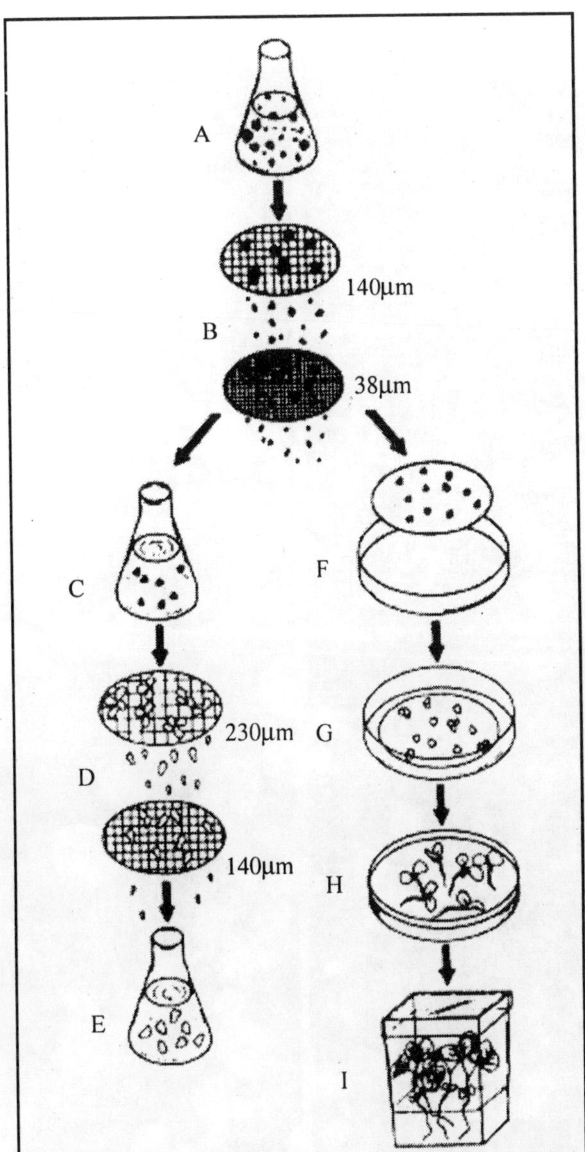

Diagram 6.3. Fractionation/synchronization of embryogenic yellow-poplar suspension cultures for mass propagation. A. Embryogenic suspension are grown in shaken flasks of liquid induction medium. B. PEMs are fractionated on stainless steel sieves, saving the fraction that passes through 140 μm, but not 38 μm mesh. C. Sived fraction is cultured for one week in liquid basal medium. D. Globular stage embryos are fractionated again to eliminate clusters and free cells, saving the fraction that passes through 230 μm, but not 140 μm mesh. E. Globular stage embryos are cultured an additional 7-10 days in basal medium to obtain synchronous heart-torpedo stage embryos. Alternatively, following the first fractionation, F. PEMs are immediately placed on filter paper, which is then plated on semisolid basal medium. G. PEMs are cultured on filter paper/basal medium for 12-14 days to obtain synchronous, mature embryos. H. mature embryos are transfered to basal medium without filter paper to promote germination. I. Germinants are transferred to plantlet development medium, from which plantlets are ready for transfer to soil mix in 6-8 weeks.

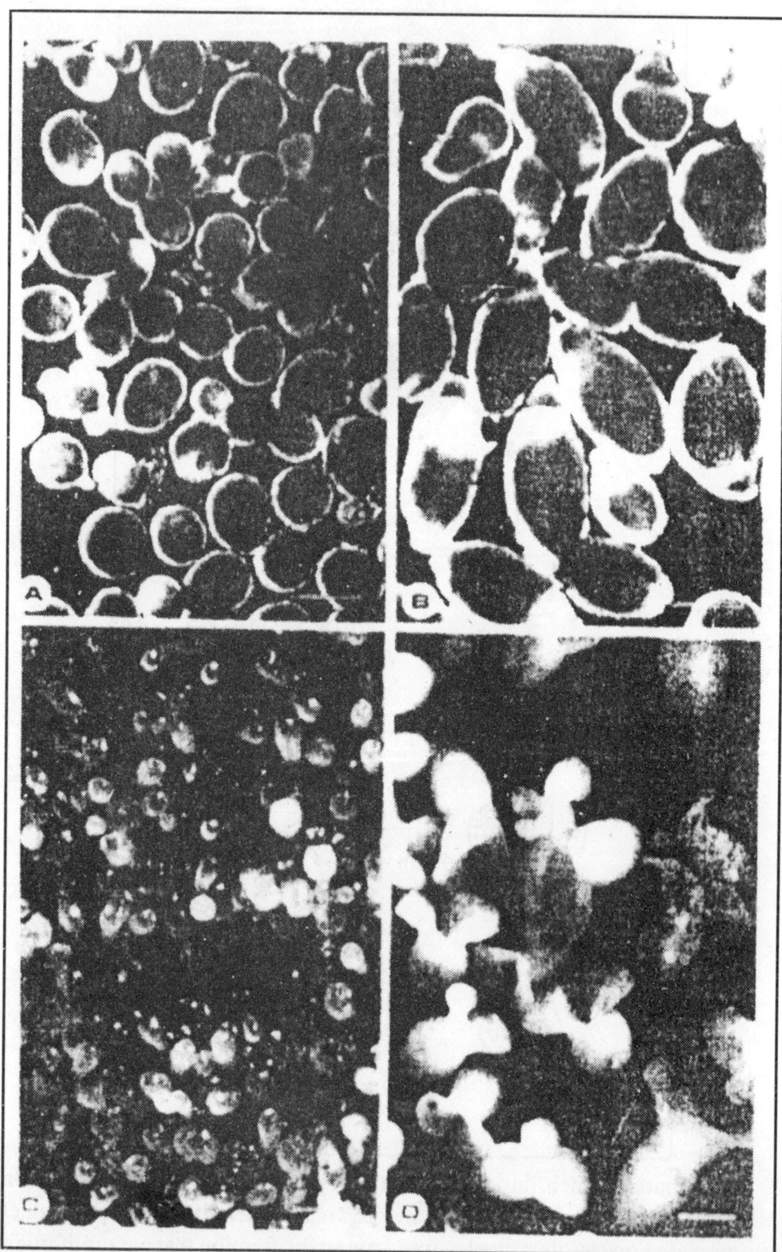

Diagram 6.4. Synchronous yellow-poplar somatic embryo populations obtained by fractionation of PEMs on stainless steel sieves. A. Globular-stage yellow poplar somatic embryos derived from fractionated PEMs 3 days after second sieving (bar = 500 μm). B. Early torpedo-stage yellow-poplar somatic embryos derived from fractionated PEMs 10 days after second sieving (bar = 500 μm). C. Roughly synchronous population of yellow-poplar somatic embryos 6 days after fractionation of PEMs and plating on filter paper placed on semisolid medium (bar = 500 μm). D. Mature yellow-poplar somatic embryos 14 days after fractionation and plating on filter paper placed on semisolid medium (bar = 500 μm).

effectively substitute for a callus phase. Even if a chimeric embryo is obtained in the first cycle of regeneration, it becomes possible to obtain a non-chimeric embryo from the patch of transformed tissue on the original embryo. The recurrent embryos appear to have an epidermal or subepidermal origin. The recurrent embryos have single cell origins. Consequently, if a transformation technique is applied to a primary somatic embryo instead of a zygotic embryo, it should become possible to obtain totally transgenic somatic embryos, and this has, in fact, been observed for walnut.

The nature of recurrent embryogenesis also makes it ideally suited to particle gun mediated transformation (Klein et al, 1987). Instead of relying on *Agrobacterium* to mediate the transfer of genes into plant cells, the particle gun literally shoots into plant cells DNA that has been precipitated onto particles of a heavy metal.

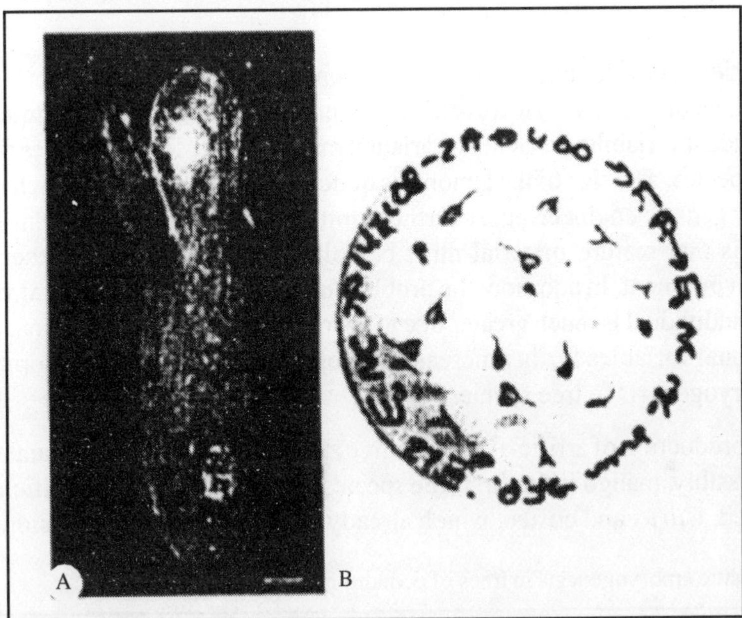

Diagram 6.5. Yellow-poplar and magnolia somatic embryo maturation, conversion, and plantlet acclimatization. A. Mature yellow-poplar somatic embryo, at the onset of germination, obtained from fractionation and plating (bar = 200 μm). B. Germinating yellow-poplar somatic embryos obtained from fractionation and plating of PEMs, following transfer from filter paper to fresh medium.

Somatic Embryogenesis in Trees

The technique and pattern of the development of somatic embryogenesis in tree species differs, depending on wheather the species is angiospermous or coniferous (Gymnsperms). Conifer somatic embryogenesis is analogous to zygotic cleavage polyembryony and for most examples, depends on an in vitro proliferation of non-callus, suspensor-like cells derived from mature or immature zygotic embryos. In *Picea abies,* the cells, known as the embryonal suspensor mass (EMS), proliferate on BM-1 medium containing high concentrations of kinetin. Differentiation of ESM to cotyledonary-stage, somatic embryos occurred on BM-1 medium containing 10 per cent of the original kinetin. However, few somatic embryos have been converted to whole plants.

Somatic embryogenesis in angiosperm tree species is either analogous to adventive in crassinucellate angiosperms, nuclear polyembryony as in *Citrus,* or adventive zygotic polyembryony as in immature zygotic embryos *of Juglans.* The key tissue culture manipulation in a number of species was the use of immature zygotic embryos or nucellar tissue as the original explant source. The cell proliferation stage in tissue culture is in the form of an undifferentiated callus.

Somatic embryogenesis has, also been obtained from non-zygotic explant tissue (coffee leaf), suggesting that somatic embryogenesis may be obtained from sexually immature tree, zygotic embryo or ovule tissue is not available, thus it accelerate the process of tree genetic engineering.

Difficulties in developing somatic embryogenesis systems for tree species are similar to those for herbaceous species. However, free species may also exhibit a unique set of tissue culture-dependent variabilities. Genetic variation may be higher for tree species than cultivated herbaceous species. The size of the genome is quite large for many trees, which is particularly important for genetic engineering. A further limitation for tree improvement using genetic engineering is that mature material must be collected from the field rather than from a controlled environment. In addition, the problem of ecotypic or physiological variation from individual to individual is much greater due to the overall heterozygosity of most tree species. These additional variables further increase the potential difficulty of developing repeatable somatic embryogenesis in tree species.

For the production of artificial seeds in free species, the recovery of plants (conversion) is crucial. Possibly, mango is the first free species for which the viable artificial seeds have been produced. *Citrus* and coffee, which already have reasonable conversion frequencies.

Table 6.1 Somatic embryogenesis in trees of economic importance.

Name	Common name	Germination	Conversion
PULP, PAPER AND TIMBER			
Gymnosperms			
Abies balsamea	Balsam fir	no	no
Larix decidua	European larch	yes	1% of calli
Picea abies	Norway spruce	5-10%	no
		24% of calli	no
		15% of calli	no
P. glauca	White spruce	yes	no
Pinus labertiana	Sugar pine	1-2%	no
P. taeda	Loblolly pine	yes	1-5/g callus
Pseudoesuga menziesii	Douglas-fir	no	no
Angiosperms			
Albizia lebbeck	Indian walnut	yes	yes
Eucalyptus citriodora	Gum	50%	yes

Name	Common name	Germination	Conversion
Fraxinus americana	White ash	yes	yes
Juglans regia	English walnut	yes	yes
J. hindsii	Walnut	yes	yes
Leucaona diversifolia	Popinac	yes	no
Liquidambar styraciflua	Sweet gum	yes	yes
Populus ciliata	Poplar	yes	yes
Pterocarya spp.	-	yes	yes
Quercus rubra	Red oak	no	no
FRUIT TREES			
Citrus aurantifolia	Lime	yes	no
C. aurantium	Sour orange	no	no
C. limonia	Lemon	yes	yes
C. medica	Citron	no	no
C. paradisi	Grapefruit	yes	32%
C. reticulata	Mandarin	yes	yes
C. sinensis	Sweet orange	yes	20%
Cocos nucifera	Coconut	no	no
Eriobotrya japonica	Loquat	no	no
Eugenia malaccensis	Malay apple	yes	no
Malus spp.	Apple	yes	yes
Mangifera indica	Mango	yes	yes
Musa ABB	Plantain	no	no
Myrciaria cauliflora	Jaboticaba	no	no
Phoenix dactylifera	Date palm	yes	no
Prunus cerasus	Sour cherry	no	no
Pyrus spp.	Pear	no	no
Syzygium jambos	Rose apple	yes	no
OILS, PHARMACEUTICALS, AND ORNAMENTALS			
Chamaedorea costaricana	Palm	no	no
Coffea arabica	Arabian coffee	yes	yes
C. arabica var. catimor	Coffee	yes	20-27%
C. canephora	Robusta coffee	yes	no
Elaeis guineensis	Oil palm	yes	no
Hevea brasiliensis	Para rubber	yes	no
Howea forsteriana	Paradise palm	no	no
Liriodendron tulipifera	Yellow poplar	yes	yes
Paulownia tomentosa	Paulownia	50%	no
Santalum album	Sandalwood	yes	yes
Theobroma cacoa	Cacao	yes	no
Thuja orientalis	Oriental arborvitae	no	no
Veitchia merrilli	Christmas palm	yes	no

For the production of artificial seeds in tree species, the recovery of plants (conversion) is crucial. Possibly, mango is the first tree species for which the viable artificial seeds have been produced. *Citrus* and coffee, which already have reasonable conversion frequencies.

❑❑❑

CHAPTER-7

Somaclonal and Gametoclonal Variant Selection

The term *somaclonal variation* was introduced by Larkin and Scowcroft (1981), which states, variant obtained from tissue culture is called **somaclones.** Evans et. al. (1984) coined the term **gametoclonal** variation for variant clones specifically raised from gametic or gametophytic cells. Occurrence of somaclonal variation in regenerated plants is considered a rule rather than exception since genetic variability decelops spontaneously during tissue culture.

Studies concerning different aspects of somaclonal variation are important for several reasons.

1. First, it is hailed as a novel source of genetic variation. Which, successful utilization depends upon its systematic evaluation and judicious utilization in breeding programmes, that needs appropriate experimentation.
2. Second, somaclongal variation is of interest as a basic genetic process, since it contradicts the concept of **clonal uniformity.** It is thought that the cell passes through the stress, as a result of which the genome, known for its plasticity, restructures itself to modulate the expression of genes as demanded by *in vitro* conditions.
3. Third, somaclonal variation is unwanted when the objective is micropropagation of elite genotypes or genetic transformation which partly involve tissue culture. It needs minimisation of variation which may be achieved through manipulation of media components, explant source, culture conditions etc.

Majority of studies undertaken on somaclonal variation are confined to early generations of somaclones. Therefore, information on the nature, inheritance pattern and stability of morphological and molecular changes expressed in the advanced stages of somaclones is lacking.

The different aspects of somaclonal variation investigated so far are as follows:

1. Generation of variation
2. Characterization of variants for morphological traits.
3. Analysis of biochemical and chromosomal basis of variation.
4. Relating the variations to alterations in DNA.

Genetic Variations and Crop Improvement

Genetic variation appears during or after culture in vitro at the stage of undifferentiated cells, isolated protoplasts, calli, tissues and morphological traits of regenerated plants. Variant selected in tissue cultures have been referred as **calliclones** (from callus culture), or **protectories** (from protoplast culture) The change occurs because of variation in chromosome number and structure. Cytological heterogeneity in cultures develops due to the following reasons:

1. The expression of chromosomal mosaicism or genetic disorders in cells of the initial explant. The chromosomal mosaicism may be pre-existing mosaicism (polysomaty - the phenomenon of polyploidisation of body cells) in explants used for culture initiation, or by nuclear fragmentation associated with first cell division of callus initiation or endoreduplication or endomitosis occurring during culture initiation and/or abnormalities of the mitotic process leading to aneuploidy.
2. New irregularities brought about by culture conditions.

Somaclonal variation causes problem for plant propagationists, since their objective is to maintain a specific plant genotype. While for plant geneticists somaclonal variation is a valuable source of new genetic information. It leads to crop plant improvement and a deeper understanding of the biochemical and molecular basis of inheritance (Chaleff 1983).

Somaclonal variation is one of the aspects of tissue culture technology and is widely recommended for crop improvement specially of desired traits for the salt, drought, temperature and disease tolerance. Even though relative data from deliberatly designed experiments is not available, indications of possible advantages of somaclonal *vis-a-vis* induced mutagenesis are:

1. The frequency of variation seems to be far greater than the yield of induced mutations.
2. The changes are very subtle and may not involve drastic altration in the genetic background.
3. Somaclonal variation occurs for trait of both nuclear and cytoplasmic origin. The variation of cytoplasmic genes obtained by this method is a distinct advantage.

In wide crosses somaclonal variations provide a mechanism of gene introgression. Immature embryos of the wide cross can be callused and plants with the introgressed desired gene (or gene complex) are selected among the regenerants of their progenies.

Cause For Variations

(i) Single Gene Mutation: Variants may also arise as a result of more suitable changes due to single gene mutations in cultures which have cells apparently showing no karyological changes. Recessive mutations are not detected in plants regenerated in vitro from any cell or tissue, but expressed in progeny. This shows that variants are the mutants. Single gene mutation responsible for somaclonal variation relates to transposable elements. (ii) Transposon-induced changes have been observed in maize, tobacco and wheat. (iii) Mitotic disorders: Somaclonal variation may also be due to changes caused by mitotic crossing over in regenerated plants. (iv) Cytoplasmic organelle DNA: Such changes may also occur due to changes in organelle DNA (cytoplasmic genom) isoenzyme and protein profiles e.g., in wheat, potato, maize, barley and flax, etc.

The Mechanism of Somaclonal Variation

According to Bhaskaran (1985) variations in somaclons occur due to the following reasons:

1. The pre-existing genetic variation in the explant tissue,
2. The spontaneous mutation which can accumulate during many division cycles that cells of the explant go through before differentiating into an in vitro plant. The recessive mutations will naturally require a method by which they can express even in diploid cells. Somatic crossing over followed by segregation is a likely mechanism, for the homozygosity and thus phenotypic expression of the recessive gene (Chopra and Sharma, 1988).
3. Numerical (polyploidy) and structural changes (chromosomal abberations or mutations) in chromosomes during in vitro growth.
4. Intracellular mutagenic agents produced during in vitro growth.
5. Activation of transposable elements or **jumping genes,** are genetic entities which have the locus at which they get integrated in matured.

Agriculturists are very hopeful about practical advantages of somaclonal variations and they are waiting when this technique is fully integrated with the conventional plant breeding procedures.

SOURCE MATERIAL AND CULTURE CONDITIONS

Plant cell, tissues and somatic embryos developed from various explant sources for generating somaclonal variation. Explants are generally taken from any tissue, namely leaves, internodes, ovaries, roots and inflorescence. The source of explant has often been considered a critical variable for somaclonal variation.

Determination of Cell Number

Take an aliquot of suspension and filter off the cultures through a wire mesh (300 mm). Note the volume of the filtrate (F) containing single cells and small clumps and place the drop of this suspension to haemocytometer to determine the number of cells by the equation

$$N = \frac{P \times 100 \times F}{0.1 \text{ mm}}$$

Where, N = total number of cells and clumps, P = number of cells in the squares of the haemocytometer, F = volume of the filtrate.

Detection and Isolation of Somaclonal Variants

There are several different approaches to detecting and isolating somaclonal variants from cultured plant cell populations.

1. Morphologically distinct cells such as nonphotosynthetic (nongreen) cells or cells that accumulate anthocyanins and other plant pigments are detected visually.
2. To isolate herbicide and antibiotic resistant variants, plant cells are simply grown on media containing the wild-type cells in a culture.
3. The surviving cells are then subcultured and retested for growth on herbicide or antibiotic supplemented medium.
4. Through this method one can eliminate any remaining wild-type cells that may have inadvertently survived the first round of selection.

Table 7.1. Somaclonal variation induced in morphological traits in some crop plants.

Crop	Characters
Sugarcane	Cane diameter, stalk length and weight, Cane yield, sugar yield, stalk number, length, diameter, volume, density and weight
Potato	Growth habit, maturity period, tuber unformity and skin color. Photoperiod requirement, fruit production.
Tobacco	Yield, Days to flowering, plant height, stem diameter, leaf number, leat length, leaf width and yield, Leaf Shape, leaf number, plant height, type of inflorescence and yield.
Rice	Number of tillers per plant, number of fertile tillers per plant, average panicle length, frequency of fertile seed, plant stature and flag leaf length. Flowering period, plant height, seed fertility and heading date.
Oats	Plant height, heading date, twin culms, yellow leaf stripe, awn morphology and fertility.
Maize	Twin stalks from a single node, reduced pollen fertility and male sterility.
Brassica	Altered leaf way, multiple branching of the stem, precocious flowering from the apex, stem or leaf, abnormal leaves, reduced lamina in leaves, spontaneously absorbing vegetative buds, slow growth, failure to flower, large pollen grains. Delayed flowering altered growth habit and gross morphology.
Pelargonium	Leaf shape, size and form, flower morphology, plant height, fasciation, pubescence and anthocyanin pigmentation.
Pineapple	Leaf colour, foliage density, leaf width and leaf spine formation
Tomato	Male sterility, jointless pedicel, tangerine virescent leaf, flower and fruit colour, lethal chlorophyll deficiency, mottled leaf appearance, fruit ripening and growth habit.

We can also use this direct selection technique for isolation of temperature-resistant variants because those cells which survive in an extended incubation periods at abnormally high or low temperatures are temperature-resistant.

Those somaclonal variants that cannot be detected visually or selected directly are isolated by indirect means.

Those auxotrophic plant cells which are unable to survive in absence of specific nutrient supplements not required by sensitive cells, which will not survive at temperatures above or below a certain threshold. This threshold will not affect normal wild-type cells. In absence of the necessary nutrient supplement or when grown at excessive temperatures, these cells become very weak and at this weakened state, these cells assimilate exogenously supplied cytotoxic compounds (arsenate, bromodeoxyuracil, or fluorodeoxyuridine) at lower rate than that of wild-type cells. Thus, treating a cell culture under restrictive growth conditions with one of these substances will tend to favour short term survival of the weaker auxotrophic or temperature-sensitive cells.

Mutagenesis and Somaclonal Variation

Mutagenic agents (chemical and physical) are used to produce particular heritable forms of somaclonal variation e.g., ultraviolet (UV) radiation, ethyl-methane sulfonate (EMS), nitrosuguanidine and sodium azide, etc.

UV radiation induces dimer formation between adjacent thymine residues in DNA. This produces lesions that cause frameshift mutations and base pair substitutions during DNA replication. EMS and nitrosoguanidine are alkylating agents that cross-link and sever DNA molecules, often resulting in gross DNA alterations. In contrast, sodium azide induces single base pair substitutions or deletions, resulting mostly in small point mutations.

Somatic Genetics of Nitrogen Metabolism

Plant cells in culture medium can metabolize ammonium, nitrate, and nitrite sources of inorganic nitrogen. The cells use ammonium directly while the nitrate is first reduced to nitrite (enzyme nitrate reductase, NR) and then nitrite is reduced to ammonium (enzyme nitrite reductase NiR). NR needs a molybdenum-containing cofactor (MoCo) for proper enzyme activity.

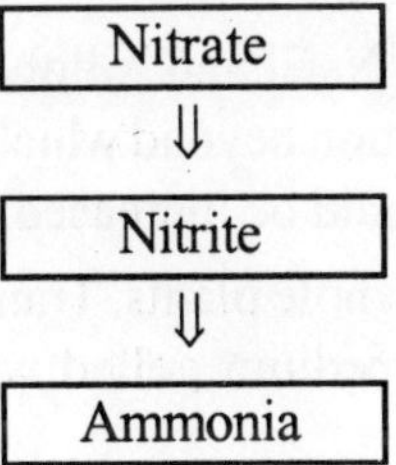

Diagram 7.1 The biochemical reduction of nitrate to ammonia

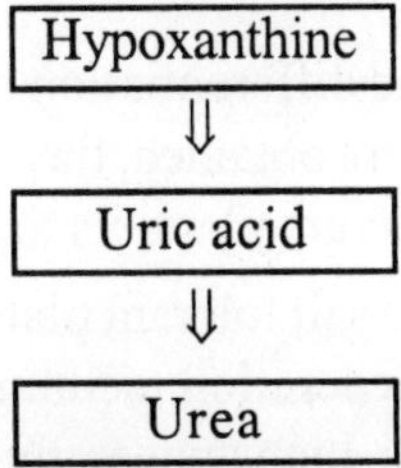

Diagram 7.2 The biochemical reduction of hypoxanthine to urea.

For isolation of variants which are unable to utilize nitrate, plant cells are grown on chlorate-supplemented media. Chlorate is a close analog of nitrate and will be reduced to chlorite, a patent cytotoxin that accumulates internally and eventually kills the cell. Certain metabolic mutants may survive from chlorate treatment. Some of these mutant cells carry mutations affecting NR expression or assimilation. Other mutations may affect MoCo expression or chlorate assimilation.

Method for Isolation of Desired Variant Cells for NaCl-Tolerant from Callus/ Suspension Cultures

1. Semi-solid liquid media is prepared and NaCl (500 mg/1) is added to the medium, pH is adjusted at 5.6. Distribute the medium in 30 ml aliquots into 150 ml flasks or 50 ml aliquots into 250 ml flasks and grow the callus on NaCl containing medium. Select the granular friable callus for further subculture.
2. 1 g of callus is inoculated to each 250 ml flask containing 50 ml of liquid medium then incubate the cultures on a gyratory shaker (100 rpm) for 21 days at 25 ± 2°C.
3. Pass the cell suspension through a stainless steel wire mesh (300 μm) and collect filtrate in 15 ml centrifuge tubes. Examine the filtrate for single cells and small aggregates (5-20 cells). Centrifuge the filtrate at 100 g for 5 min and discard the supernatant, add 5 ml of the sterilized medium to the pellet to obtain a cell suspension, and maintain the cell density of the suspension about 0.5-2.5 x 10^5 cell/ml. Add to the flasks containing with or without NaCl.
4. Incubate the cultures at 100 rpm at 25 ± 2°C and centrifuge the cultures grown in control and NaCl (500 mg/1) containing medium at 100 rpm for 5 min.
5. Transfer the pellets to fresh medium of the same composition. Incubate at 25 ± 2°C for 21 days and repeat these steps for 2-3 times to get a good growth rate of the NaCl tolerant cell line
6. Grow cultures on 1000 mg/1 NaCl and follow the same steps, as carried out at 500 mg/1 NaCl till a salt concentration beyond which the cells cease to grow. It is suggested that the salt concentration should be increased gradually by repeating steps
7. Regenerate variant cells into whole plants. Transfer 0.5 ml of cell suspension onto the surface of the regeneration medium gelled with agar by a wide mouth graduated sterile pipette.
8. In case of callus, inoculate 500 mg piece onto the surface of the organ/shoot forming medium having 500 mg/1 NaCl
9. In most of the plant species, the differentiation of shoots and roots occurs on different media. Hence once the shoots are obtained, they are transferred to the rooting medium. Transfer the selected NaCl tolerant plantlets to soil.
10. Use NaCl solution to water the salt tolerant plants. The concentration of NaCl should be the same as used in the regeneration medium. This procedure can also be used to develop tolerant plants of alkali, high temperature, drought, disease, herbicide by using in vitro selection system.

APPLICATIONS IN PLANT BREEDING

Somaclonal variation and gametoclonal variation are the important sources of introducing genetic variations that could be of value to plant breeders. Single gene mutation in the nuclear or organel genome usually provides the best available variety *in vitro* which has a specific improved character. Somaclonal variation are used to discover new variants retaining all the favourable characters along with an additional useful trait, e.g., resistance to diseases or an herbicide. These variants can then be field-tested to ascertain their genetic stability. Gametoclonal variation is induced by meiotic recombination during the sexual cycle of the F_1 hybrid, results in transgressive segregation to develop unique gene combinations.

Various cell lines selected *in vitro* and plant regenerated through it prove potentially applicable to agriculture and industry specially resistance to herbicide, pathotoxin, salt or aluminium, useful in the synthesis of secondary metabolites on a commercial scale, etc.

The techniques used for development of somaclonal and gametoclonal variation are relatively easier than recombinant DNA technology and is the appropriate technology for genetic manipulation of above cited crops.

Table 7.2. Examples of gametoclonal variation among some plants raised from gametophytic tissue cultures.

Species	**Altered characteristics observed**	**Genetic basis**	**Androgenetic (A) or Callus Mediated (CM)**
Brassica napus	Time to flowe, glucinolate content, leaf shape and colour, flower type, pod size and shape	Mutation	A
Hordeum vulgare	Plant height, yield, fertility, neck length, days to maturity	Not examined	CM
	Grain yield, culm length, heading date	Transgressive segregation, gametophytic selection	CM
Nicotiana sylvestris	Leaf colour, leaf shape, growth rate, DNA content, repeated sequences of DNA	Crumpled leaf is nuclear, germ amplification	A
N. sylvestris (several cycles)	Flower length, plant height, leaf shape, leaf colour, exerted stigma, capsule weight, foliar outgrowths	Nucleus genes	A
N. tabacum	Yield, grade, time to flower, plant height, number of leaves, leaf shape, total alkaloids, % reducing sugars	Not examined	A
	Yield, plant height, number of leaves, % nicotine	Not residual heterozygosity	A
	TMV resistance, root knot resistance	Variation in expected segregation ratios	A
	Days to flower, total alkaloids, yield per plant, leaves per plant	Residual heterozygosity; no cytoplasmic effects	A
	Yield reduction, plant height, days to flower, total alkaloids	Nuclear mutations except plant height is maternal; not residual heterozygosity	A

Species	Altered characteristics observed	Genetic basis	Androgenetic (A) or Callus Mediated (CM)
	Yield reduction, alkaloid content, leaf size, time of flowering, leaf quality	Induced changes, not residual heterozygosity	A
	Heterochromatin content, nuclear DNA content	Gene amplification	A
	Time of flowering, plant size, leaf shape, alkaloid content, number of leaves	Residual heterozygosity	A
Oryza sativa	Altered temperature sensitivity	Cytoplasmic mutant	A
	Chloroplast content, time to flower, plant height, fertility	Some single genes induced in culture	CM
	Plant height, seed and leaf size, seed weight, tillering, flowering, % open hulls	Transmitted to progeny	CM
	Seed size, seed protein level, plant height, level of tillering	Transmitted to progeny	CM
Saintpaulia	Waxy mutant	Single gene change	CM
ionantha	RuBCase activity	Not examined	A
Secale cereale x S. vavilotrii	Fertility	Not analyzed	CM

Table 7.3 Aspects of improvement desired in some crop plants through somaclonal variation

Crop plants	Improvement desired
Cereals	
Rice	Tungro virus and leaf hopper resistance
Wheat	Quality, high temperature tolerance
Sorghum	Borer and shoot fly resistance
Maize	Delinking of phytohormone effect from reproductive development
Legumes	
Pigeon-pea	Resistance to pod borer, fusarium wilt, high protein content
Bengal gram	
Oilseeds	
Groundnut	Resistance to *Cercospora*, *Fusarium*, *Aspergillus flavus*
Rapeseed	Male sterility, low pungency combined with insect resistance
Sunflower	Self compatibility, resistance to Alternaria leaf blight, *Rhizoctonia*, *Fusarium* complex
Crop plants	**Improvement desired**
Cotton	Insect resistance to gossypol free lines
Vegetables	
Potato	Virus and Phytophthora resistance
Tomato	Resistance to disease
Fruits	
Grapes	Resistance to disease, short duration
Citrus	Higher solid content, increased shelf life
Watermelons	
Other (s)	
Sugarcane	Resistance to disease, short duration
Castor	Mutants with high fatty acids, ricinoleic acid

Table 7.4. Somaclonal variation induced in morphological traits in some crop plants.

Crop	**Characters**
Sugarcane	Cane diameter, stalk number diameter volume, length and weight, cane yield and sugar yield.
Potato	Growth habit, maturity period, tuber uniformity and skin colour. Photoperiod requirement, fruit production.
Tobacco	Days to flowering, plant height, stem diameter, leaf number, leaf length, leaf width and yield leaf shape, leaf number, plant height, type of inflorescence and yield.
Rice	Number of tillers per plant, number of fertile tillers perplant, average panicle length, frequency of fertile seed, plant stature, flag leaf length, flowering period and heading date.
Oats	Plant height, heading date, twin culms, yellow leaf stripe, awn morphology and fertility.
Maize	Twin stalks from a single node, reduced pollen fertility and male sterility.
Brassica	Altered leaf way, multiple branching of the stem, precocious flowering from the apex, stem or leaf, abnormal leaves, reduced lamina in leaves, spontaneously absorbing vegetative buds, slow growth, failure to flower, large pollen grains. Delayed flowering altered growth habit and gross morphology.
Pelargonium	Leaf shape, size and form, flower morphology, plant height, fasciation, pubescence and anthocyanin pigmentation.
Pineapple	Leaf colour, foliage density, leaf width and leaf spine formation.
Tomato	Male sterility, jointless pedicel, tangerine virescent leaf, flower and fruit colour, lethal chlorophyll deficiency mottled leaf appearance, fruit ripening and growth habit.

❑❑❑

CHAPTER-8

Cell Culture and Biotechnology of Animals

INTRODUCTION

The practice of cell culture in case of plants is very popular but it is not so familiar in case of animals. However, many types of animal cells can now be grown in culture, such as tumour cells, steroid-producing adrenal cells, ACTH-secreting pituitary cells, growth hormone and prolactin-secreting cells from pituitary tumour, pigmented melanoma cells, teratoma cells capable of differentiation *in vitro,* and neuroblastoma cells. Functionally-differentiated cells from tumours can be readily cultured. Pigmented cells and cartilage cells *have* been cultured since the early 1920s. Cancerous mast cell lines producing serotonin and heparin *have* also been raised. Chick myoblasts are known to proliferate and fuse to form muscle straps. Rat myoblasts cannot only fuse in culture but also repair injury in crushed muscle (Sato, 1982).

Whereas cells of bacteria and yeasts can be grown freely suspended in deep suspension cultures, those of mammals can only be grown in industrial-scale cultures in multiple low-productivity roller bottles. This is because the mammalian cells require attachment to a suitable surface and consequently the maximum cell numbers are limited by the surface area available. This kind of limitation can now be overcome by growing mammalian cell-cultures on microcarriers such as beads of anion exchange resin. By employing conventional conditions of medium, serum, and oxygen, and using suitable beads as carriers, Thilly *et al.* (1982) have grown certain mammalian cells to densities as high as 5×10^6 cells/ml. By substituting fructose for glucose, they have been able to control the overproduction of lactate which, in many cell cultures, causes an abnormal lowering of the pH, thus limiting the growth of the cells.

During the last three decades, various types of additives have been used to protect freely suspended animal cells in culture from agitation and aeration damage. These include pluronic polyols, various derivatized celluloses and starches, protein mixtures, polyvinyl-pyrrolidones, dextrans. and, more recently, polyethylene glycol (PEG) and polyvinyl alcohol.

Damage of suspended cells in agitated and/or aerated bioreactors is usually due to the interactions of cells with bubbles and the rearrangement of gas-liquid interfaces. In bubble-column reactors, cell injury appears to be due to shear forces generated either by film drainage around bubbles (such as in unstable foams), or by bubble breakup. In agitated bioreactors, cultures of suspended cells appear to suffer damage by the following two fluid mechanical mechanisms:

1. Formation of a gas phase due to bubble breakup, either because of direct sparging or because of gas entrainment.
2. Cell damage occurs in the absence of a gas phase (and, therefore, the absence of bubbles) only at very high agitation rates by stresses in the bulk turbulent liquid. With this second mechanism, cell damage correlates with eddy sizes similar to or smaller than the cell size (9-15 urn).

Whether cells are damaged in viscometer is due to well-defined laminar flows, or due to bubble breakup and film drainage, or due to interactions with eddies, or by shear forces acting on the cells through the surrounding fluid layer (boundary layer), which is always in contanct with laminar flow. Cell damage occurring in such cases may therefore be referred to as either shear or fluid-mechanical damage (Papoutsakis, 1991a).

All additives that protect freely-suspended cells from fluid-mechanical injury must either decrease the fragility of the cells or affect the forces on the cells due to their interactions with gas-liquid interfaces.

Serum permits better cell growth in agitated and/or aerated cultures in a dosage-dependent fashion. Low serum or serum-free cultures are more susceptible to fluid-mechanical damage. Concentrations, up to 10%, of foetal bovine serum (FBS) tend to reduce cell death and allow growth of cells at substantially higher agitation rates in bioreactors with surface aeration where cell damage is due to air entrainment and bubble breakup. The protective effect of FBS appears to be largely physical. However, though the protective effect of FBS is primarily physical in nature, it can vary depending on the design and operational characteristics of the bioreactor, and the cells in question (Papoutsakis, 1991a).

The non-ionic surfactants Pluronic F68 and F88 block copolymer glycols of poly(oxyethylene) and poly(oxypropylene) protect cells from fluid-mechanical damage in agitated and aerated bioreactors. Several investigators have used these surfactants as medium additives in static, agitated and/ or aerated cell cultures (Papoutsakis, 1991a).

Since long, derivatized celluloses have been used for suspension cell culture technology; methylcelluloses (MCs) have not been unequivocally established as reliable shear protectants. MCs and other derivatized celluloses have been frequently included as media additives (in combination with serum, various protein mixtures and other defined additives) in the cultivation of a large variety of cells. It is not known whether these derivatized,celluloses are indeed needed as shear-protection additives in most formulations of modern serum-free media, since control experiments to demonstrate their shear-protection effect are lacking (Papoutsakis, 1991a).

In recent years, the use of MCs (in combination with other shear-protecting additives, such as serum, yeast extract or protein hydrolysates) has been somewhat restricted to the cultivation of insect cells probably because of the more widespread use of F68, which does not cause any of the cell-aggregation problems often associated with the use of MCs.

It seems that both the MCs and dextran increase the shear robustness of the insect cells through a biological mechanism. According to Papoutsakis (1991a), the effect of MCs as additives to protect animal cells from shear damage in agitated, bubble-column or otherwise mixed cultures may be cell-type, MC-grade, and MC-make dependent. In addition, MCs may elicit biological responses from cells, either positive or negative; these responses appear to be also cell-type and MC-grade and MC-make dependent.

Several protein mixtures (in addition to serum) and a protein have been used as shear protectants for the cultivation of various cells. Again, their protective effect is not definitely established and there is a lack of proper control experiments.

Bovine serum albumin (BSA), a major component of bovine serum, is a widely used additive in serum-free media, and has been frequently used as a shear protectant. It may be an effective additive against fluid-mechanical damage of hybridoma cells in an airlift bioreactor, although it has no effect on the cells in static or spinner cultures (Papoutsakis, 1991a).

While several options are available regarding shear-protecting additives, it is not clear if the use of these additives is suitable for all cell culturing and processing needs. For each cell type, the physiological and / or product expression effects of an additive must be assessed carefully under both static and bioreactor growth conditions. If there are no obvious detrimental effects, the effect of the additive on cell aggregation must be evaluated, in case cell aggregation is an undesirable processing property. The effect of the additive on a possible modification of the cell culture protein product as well as on the purification of this product also has to be assessed. Most additives may complicate membrane, adsorption, precipitation, and chromatographic processes. So, on the basis of a combination of chemical structure, molecular mass and protective-effect concentration, some additives may be more advantageous than others from the DSP point of view.

An autoclavable fluidized bed fermenter for culturing animal cells on carriers is shown in Diagram 8.1.

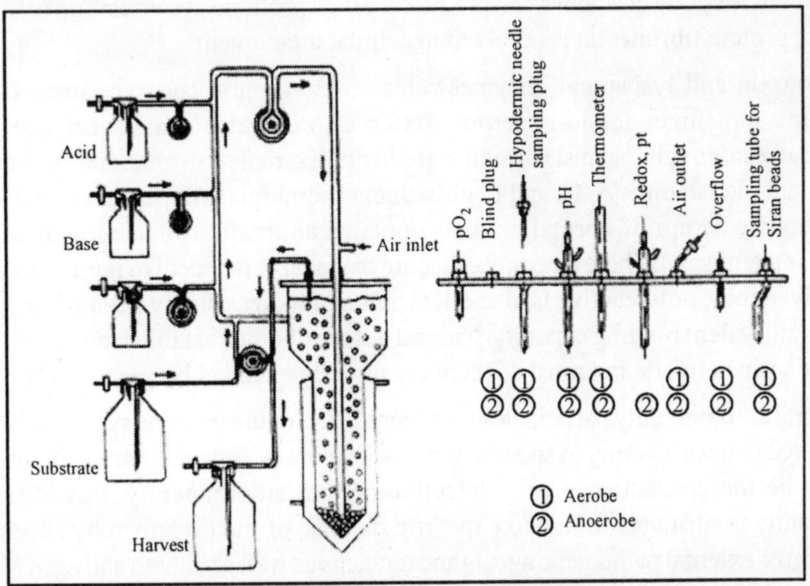

Diagram 8.1 The autoclavable fluidized bed fermenter for the cultivation of animal cells on carriers.

Two types of cell-bubble interactions that can damage animal cells in bioreactors are illustrated, in Diagram 8.2. Immobilized animal cells may be grown in monolayer culture in a multiplate vessel, shown in diagram 8.3.

SERUM

Ever since tissue cultures of animal cells were started, serum has been used as an essential component of the culture media. It appears that serum supplies hormones to the cells. This is borne out by the fact that serum can be replaced by complexes of hormones. For instance, the serum in media used to culture GH_3 cells (growth-secreting pituitary cell line started from a tumour) can be replaced with a mixture of insulin (pancreas), transferrin (liver), triiodothyroxine (thyroid), parathyroid hormone (parathyroid), TSH-releasing hormone (hypothalamus), fibroblast growth factor (pituitary), and somatomedin C (liver). Some of these substances are not hormones *sensu stricto.* In fact, it is now known that all animal cells which have the potential to grow can be made to grow in a serum-free defined medium containing a limited number of substances of the types we have mentioned.

Animal cells, tissues, and organs can be grown by using serum as the sole support of the culture. More recently, particular components in serum responsible for cell growth in culture have been identified and isolated. Now, instead of adding whole serum, we can add its desired components to the culture medium. However, most serum-free systems are highly cell-specific and lack the wide applicability of serum-supplemented media. Serum is a highly complex mixture of diverse molecules that together provide growth-promoting and growth-inhibiting factors to the cells. Some of the most useful constituents of serum include hormones, binding proteins (e.g., albumin), growth factors, transport proteins, attachment factors, and micronutrients. Glucocorticoids and other hormones present in serum stimulate the growth and differentiation of some cells and inhibit others. They can also change attachment sites on membranes. Peptide hormones can produce marked proliferative effects on cells in culture. The serum protein fibronectin is involved in cellular attachment.

Endotoxin and lysosomal enzymes released from burst cells are some of the toxic components sometimes found in serum. Serum also contains some substances that exert protective effects on cells against harmful ingredients. Examples of-some antitoxin components of serum include vitamins A, C, and E, glutathione, cemloplasmin, catalase, and superoxide dismutase. The serum of normal animals contains antibodies directed against their own antigens (autoantigens). These autoantibodies are the main component of normal autoimmunity, and mostly include polyreactive IgM and IgG autoantibodies usually having low monovalent but high multivalent binding capacity. Natural autoantibodies establish among themselves a dense idiotypic network and most of them are directly encoded by germline genes.

Natural autoimmunity participates in the regulation of the immune system and is probably also involved in the triggering of specific immune responses. Some immunopathological states appear to be the consequence of a defective natural autoimmunity. In addition, natural autoimmunity contributes in the non-specific defense of the organism by accelerating the elimination of external pathogenic agents and antigens as well as altered and aged autoantigens (Avrameas, 1994).

Immunopathology and autoimmune disease are driven by defects in immune regulation, rather than by mistakes in lymphocyte development or the accidental generation of autoimmune T-cells. This view is supported by the evidence that self-reactive T-cells occur frequently in the normal repertoire, where they are constrained by an unknown mechanism. According to Mitchison (1994), the most important form of immune disregulation is in the balance of T-cell cytokines. This is particularly evident in rheumatoid and reactive arthritis, which show distinct patterns of T-cell cytokine expression in the affected joints.

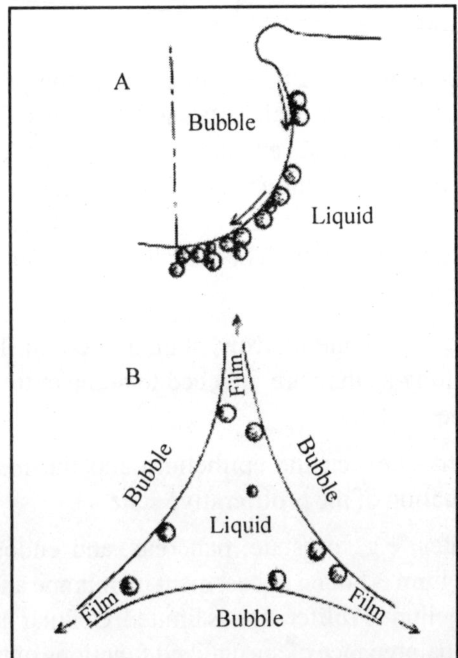

Diagram 8.2 Sketches showing two types of cell-bubble interactions which are most likely to damage cells. A. cells near the bubble interface experience large shear stresses during breakup when the bubble surface is collapsing at very high speed; B. cells trapped in the liquid, either between bubbles during foam formation or when a bubble reaches the free liquid surface, are sheared in the thinning liquid films either between bubbles or around bubbles. (After Papoutsakis, 1991b.)

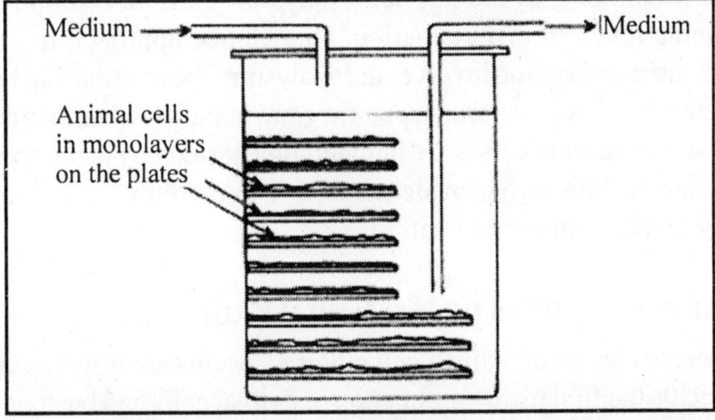

Diagram 8.3 Immobilized cell culture technique for animal cells.

Most workers have used foetal bovine serum as the supplement of choice for their culture media for growing mammalian cells. Horse serum can be used for the support of hybridomas and parent myeloma strains, and for mycoplasma. Bovine calf serum can sometimes substitute for foetal bovine serum. Porcine serum (from pigs) can be used as a growth supplement for culturing human cell lines.

Serum itself has to be diluted before use. Hormone-supplemented media often prove superior to serum-based media for raising cell cultures which show better behaviour in terms of growth and differentiation.

The assemblage of cells within a tissue interacts as the primary determinant of the regulation of growth and differentiation in all metazoan organisms. One such important cellular interaction is that between the epithelium and the mesenchymally-derived cells (Reid, 1982). This kind of interaction is in fact ubiquitous in all tissues of higher animals. It is known that substrates of tissue-specific extracellular matrix extracts (called biomatrix) are essential for long-term maintenance of epithelial cells. These systems represent some of the hypothetical classes of tissues now listed:

1. Proliferative tissues, e.g., bone marrow, skin, and colon. In these, committed stem cells proliferate so long as they are attached to some extracellular matrix known as basement membrane.

   The communications between the epithelium and the mesenchymal cells depend primarily on perpetuation of the proliferative state. .
2. Differentiative tissues, e.g., prostate, pancreas, and endocrine tissues. In these, a differentiated epithelium is bound to basement membrane and, in turn, associated with fibroblasts or endothelium. Proliferation is limited. Cellular interactions in these tissues mainly involve the maintenance of specialized functions of the epithelial cells.
3. Regenerative tissues, e.g., liver and kidney. In these, the epithelium is bound to basement membranes and is usually associated with endothelium.

Role of Plasma Membrane

The plasma membrane of animal cells plays an extremely important role in cell proliferation, interaction, and differentiation. One choice approach to understanding the mechanisms of membrane function involves the analysis of the relationship between structural and functional alteration resulting from specific gene mutation. Another powerful tool that has become available in recent years is the immunological study by using specific monoclonal antibodies. This new technique has made it possible to monitor the cell surface molecular changes that may result from gene mutation.

Cell Cultures as Sources of Valuable Products

Several processes based on animal cell culture systems are now in commercial use. In fact, as early as 1950s the first major application of animal cells had been developed, namely, the production of polio virus in cells cultured from primate neural and neural kidney tissues.

Animal cell products are typically proteinaceous, high molecular weight compounds. They include several enzymes, hormones, and animal vaccines. Immunobiologicals, prophylactic viral vaccines, monoclonal antibodies, immunobiological regulatory molecules such as interleukins and lymphokines, and insecticides are some other important products of animal cell cultures. Some important enzymes derived from animal cells are asparginase, collagenase, hyaluronidase, pepsin, rennin, trypsin, tyrosine hydroxylase, and urokinase. Four important hormones produced by animal cells are the luteinizing hormone, follicle-stimulating hormone, chorionic hormone, and erythropoietin.

Some important products made with animal cell lines are shown in Table 8.1.

Table 8.1 Animal cell lines as sources of some valuable products

Cell line	Product
Tumour, human	Angiogenic factor
Leucocytes, human	Interferon
Kidney, human	Urokinase
Kidney, dog	Canine distemper vaccine
Kidney, cow	Foot and mouth disease vaccine
Chick embryo fluid	Vaccines for influenza, measles, and mumps
Duck embryo fluid	Vaccines for rabies and rubella

Recent advances in molecular biology, somatic cell genetics, and cell biology have revealed the suitability of mammalian cells as extremely useful hosts for the expression of alien genes. Various approaches have become available to achieve a high-level expression of proteins in a variety of mammalian cells (Kaufman, 1987). The ability to engineer mammalian cells genetically to produce high levels of desired proteins is presently complemented by advances in biochemical engineering relating to the ability to grow these cells in fairly large volumes or at high densities with greatly reduced requirements for serum. Consequently, the cost of production of gram quantities from a mammalian host cell has fallen to compare favourably with that similarly derived from microbial systems (Kaufman, 1987). As compared to proteins from microbial systems, those from mammalian cells have certain advantages, viz., (1) the signals for synthesis, processing, and secretion of these proteins are better recognized; (2) the proteins can be readily synthesized and secreted into the growth medium; (3) the protein folding and disulphide bond formation are usually similar to those of the natural protein; and (4) the multimeric proteins can be correctly assembled.

Cultured mammalian cells constitute a good source of certain biochemicals of medical importance. Examples are different growth factors, monoclonal antibodies, and lymphoblastoid interferons. By culturing a permanent lymphoblastoid T-cell line, fairly sufficient quantities of the human lymphokine interleukin-2 or T-cell growth factor may be produced, if necessary, by mitogenic stimulation. The technology for large-scale cell culture for this purpose is already available (Tolbert *et al.*, 1982).

Genetic engineering methods make it possible to construct mouse fibroblasts that express glycosylated human β-interferon constitutively. Special bioreactors are used for the cultivation of the fibroblasts on suitable microcarriers, and yields as high as 5000 units β-interferon /ml/ day can be obtained.

The genetic engineering involved in the construction of the fibroblasts includes the following process. Induce the production of β-interferon by human primary cell cultures, with double-stranded RNA, or introduce the isolated; 3-interferon gene into heterologous cells showing immortalized growth. By either of these means, a mouse fibroblast cell capable of expressing glycosylated human β-interferon constitutively can be produced.

One of the more interesting applications of recombinant DNA technology has been the successful construction of interferon-producing mouse cell lines by genetic engineering. The humane-gene and a promoter substituent of the same gene has been transferred and expressed into various mammalian cell types. In fact, Hauser *et al.* (1984) have produced a high-yielding, continuously-producing cell line of mouse by using both an inducible and a constitutive promoter in mouse Ltk-cells. This cell line has then been used in a scaled-up fermentation process to obtain glycosylated human β-interferon.

Transgenic Animals

The first transgenic animals carrying foreign DNA in somatic and germ cells were produced in 1976 by exposing mouse embryos to infectious retrovirus (Jaenisch, 1976). This feat was soon followed by the technique of microinjecting recombinant DNA into a pronucleus of a zygote to produce transgenic animals. In these early experiments mice were the animals of choice. Attention was later given to farm animals. By now, gene transfer has been carried out successfully in many classes of animals such as mammals, birds, fish, insects, and worms. Success of genetic engineering in domestic animals depends not only on the identification of relevant genes but also on proper understanding of the regulation of the alien genes in transgenic animals. For instance, any attempts to use farm animals for molecular farming (i.e., for producing valuable human proteins) require a better knowledge of basic mechanisms of gene regulation. Also it is sometimes difficult or even risky to extrapolate the findings on mouse to larger animals.

Transgenic mice have been used for the analysis of the immune system (Iglesias, 1991; Bluethmann, 1991). Several transgenic mice expressing genes concerned with the immune system have been produced. Transgenic expression of immunoglobulin heavy and/or light chain genes of different specificities has facilitated better understanding of the processes involved in B-cell development such as allelic exclusion of immunoglobulins and B-cell tolerance (Iglesias, 1991). Transgenic mice expressing interleukin genes have been used to learn the modes of action of these important growth and differentiation factors in the context of the mouse immune system. Transgenic mice expressing Igs with specificities directed against mouse self components have been greatly helpful in unravelling the mechanisms involved in the tolerance of B-lymphocytes.

There exists an enormous diversity of T-cell receptor specificities. This enables the immune system to mount a specific immune response to virtually any given antigen encountered by the host (Bluethmann, 1991). The diversity is produced by somatic rearrangements of distinct germline gene segments during T-cell development and the addition of N regions (Davis and Bjorkman, 1988). Thymocyte precursors from the bone marrow colonize the thymus and soon proliferate and rearrange their T-cell receptor loci. Rearrangement and expression of these loci are determined in relation to time with lymphocytes expressing certain specific receptor loci appearing sequentially during thymic development. Transgenic mice carrying functionally rearranged T-cell receptor genes have advanced our knowledge of T-cell development and of thymic positive and thymic negative selection processes (Bluethmann, 1991).

Transfer of alien genes into the germline of cattle opens up revolutionary prospects for the modification of animal production traits, including the composition of milk. The mammary gland of a cow or buffalo is an efficient vat for the production of specific proteins, lipids, and sugars. Man is now contemplating to introduce changes in these constituents, especially in proteins, with a view to exploiting the farm animals for producing some human proteins needed for the treatment of disease. The gene transfer methodology has opened up new vistas for the production of novel proteins in milk. Milk protein genes may be selected and cloned and the sequences governing tissue specific hormonally-induced expression in the mammary gland may be identified. Studies with three genes, viz., bovine beta lactoglobulin, rat beta casein, and whey acidic protein of mouse and rat, suggest that beta casein genes can direct production of novel proteins in the milk of transgenic mice, sheep, rabbits, and pigs. These proteins were biologically active and usually comigrated with authentic proteins (Wilmut *et al.,* 1991).

Diagram 8.4. Shows some promising areas of possible investigation of domestic livestock using the transgenic technology.

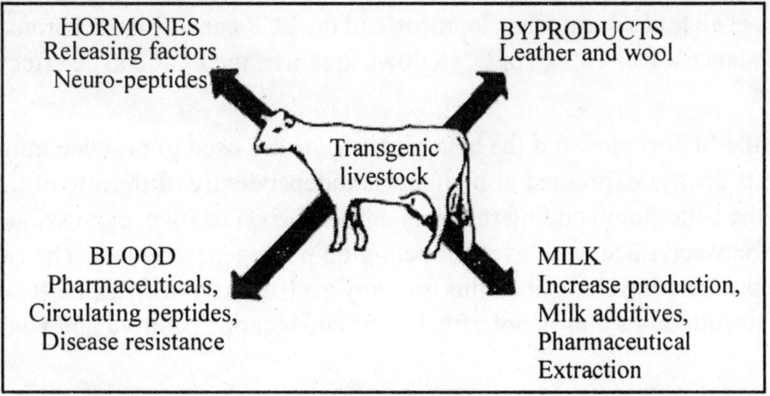

Diagram 8.4. Areas of possible investigation of domestic livestock, using the transgenic technology (after Ebert, 1989).

Mammals

The transfer of recombinant DNA by microinjection into embryonal pronuclei is a novel approach to manipulation of production traits in domestic animals. Historically, the genetic potential associated with such traits as wool growth, milk yield, and body weight has been improved by selective breeding whereby elite animals are used as the breeding stock. This classical approach has several limitations, especially the barrier to interspecific crossing which precludes the transfer of some desired gene from one species to another. The successful transfer of recombinant DNA into mouse embryos by microinjection into pronuclei of one-celled embryos has been achieved (Gordon, 1989; Palmiter and Brinster,1986). This technique can be used to alter the growth of mice (Palmiter *et al.,* 1983). It is now possible to alter the genetic properties of animals without resorting to conventional breeding. It is also possible to transfer small pieces of the genome instead of the entire chromosomes. The first transgenic pigs and sheep were reported in 1985 (Hammer *et al.,* 1985) and even cattle have now come in this category (Roschlau *et al.,* 1989). Indeed, most, if not all, major domestic animal species can be genetically modified by this technology. Some of the systems amenable to genetic manipulation include the endocrine system, the biochemical pathways, the structural proteins of textile fibres and milk, and the immune system (Ward and Nancarrow, 1991). By altering the concentration of the circulating growth hormone in transgenic mice, their growth rate and final body size can be significantly increased (Palmiter *et-al.,* 1983). Comparable success has, however, not been achieved in cattle, pigs or other large animals.

Globin gene switching: Some work has been done on gene switching mechanisms in animals, especially on the human beta-globin locus using transgenic mice. In human, a developmental switch occurs during embryonic development. At 8 weeks of development alpha- and gamma-globins are produced. The alpha-globins are also expressed in the adult, but the gamma-genes are turned off at birth and beta-globins persist only at a low level. This switch only occurs in humans and is of particular interest since diseases caused by mutations in the beta-globin gene can only be detected at birth.

The locus control region (LCR) has been identified as particularly important in controlling switching of globin genes. The LCR contains DNA hypersensitive sites which are present in erythrocytes at all times during development, and the LCR can influence chromatin structure over long distances. 100-150 kb of DNA downstream of the LCR can be affected as can 30 kb upstream.

When the LCR region and the beta-globin gene are used to produce transgenic mice, beta-globin is always expressed at high levels independently of the site of integration. In contrast, if the beta-globin coding region is introduced on its own, expression is erratic, its levels vary between mice, and levels of beta-globin are generally low. The LCR probably contains, some isolating element. In this way, any gene inserted with the LCR is independent of its new surroundings and is not affected by endogenous positive and negative control elements.

The relative distance between the genes and the LCR seems to be important for gene expression and a competitive interaction occurs, so that the balance between alpha- and beta-globins can be maintained. If the balance is disrupted, thalassaemia results.

Fish

Gene transfer is now being made into the embryos of several fishes such as trout, salmon, carp, catfish, and goldfish. As pronuclei are usually not visible, microinjection has to be done into the cytoplasm of early embryos (Houdebine and Chourrout, 1991). Millions of copies of the desired gene are injected to get some success in transgenesis. In medaka fish, transgenesis is achieved by injecting the alien gene into the nucleus of the oocyte. The injected DNA rapidly replicates during embryonic development and the survival of the injected embryos is reasonably good, with a large number reaching maturity.

In most fishes, fertilization is external and many embryos at various developmental states can be obtained readily. This facility makes fish a material of choice for gene transfer studies.

In those species where microinjection is not very successful, electroporation with whole early embryos has permitted gene transfer (Inoue *et al.,* 1990).

In early embryos of zebrafish, the microinjected DNA is amplified about ten-fold within a few hours after fertilization and only a small proportion of the replicated DNA is maintained after the gastrula stage (Stuart *et al.,* 1988). In trout also the microinjected DNA replicates rapidly and most of it disappears progressively.

All the transgenic fishes examined so far have been found lo be mosaics and a relatively large difference of alien gene copy numbers has been recorded among several tissues. In no case did any given tissue contain consistently more alien genes that the others, suggesting that integration is a random process (Houdebine and Chourrout, 1991). The observed mosaicism points to the fact that the integration step occurs later than the one-celled stage.

The spermatozoa of transgenic trout and zebrafish contain the alien DNA which can therefore be transmitted to the next generation at fertilization. Fish hatching from the fertilization of normal oocyte with sperm from transgenic fish is transgenic, though only around 10-50% of the F_1 offspring harbours the alien genes (Stuart *et al.,* 1988). The proportion of transgenic individuals tends to increase in the F_2 generation, the transgenes being transmitted in a Mendelian fashion.

Unambiguous expression of transgenes has been reported in several cases (Houdebine and Chourrout, 1991). Transgenic carps grow faster than the control fish, and the trait is transmitted to their progeny.

The overall rate of transgenesis in fish is higher than that of mammals in view of the ease with which foreign DNA can be microinjected into the nucleus of oocyte or, even more readily, into the cytoplasm of one-celled embryos. The foreign DNA is initially polymerized, then amplified and integrated, and finally stably transmitted in Mendelian fashion without undergoing rearrangement.

Birds

Gene transfer technology in birds has not developed as much as that in mammals. Progress in birds has been partly hampered by the avian reproductive and embryonic developmental system. Bird eggs are large, fragile, and contain much yolk. Embryonic development begins in the oviduct during egg formation. When the egg is laid, the blastoderm already contains over 50,000 cells organized into a *1-2* layer disc (Kochav *el al.,* 1980). This makes it difficult or ineffective to microinject DNA into isolated early embryos. In fact, no transgenic birds so far have been produced by the microinjection technique (Shuman, 1991).

The most successful method of gene transfer in birds is by retrovirus vector (Salter *et al.,* 1987; Shuman and Shoffner, 1986).

Retroviral vectors prove useful in situations where small genes are to be transferred and where regulated expression is not critical. For larger genes or when regulated expression of the transferred gene is desired, transfection or injection of DNA is necessary. The use of totipotent cells (such as embryonic stem cells) can be contemplated for targeted gene insertion using homologous recombination. This can allow gene replacement and/or targeting to a desired chromosome.

Retroviruses overcome the difficulty in accessing germline cells in multicelled embryos, the most convenient developmental stage for avian manipulation. A window is made in the shell of a freshly laid egg and the retrovirus is microinjected near the blastoderm (Salter *et al.,* 1987). The shell is then sealed and the egg incubated to hatch. The efficacy of retrovirus vectors has been demonstrated by germline insertion of replication-defective retrovirus vectors carrying bacterial marker genes. Retroviral vectors have also proved useful for the transfer and expression of genes in somatic cells. Further, germline transgenesis has been reported in both the chicken and the Japanese quail.

Gruenbaum *et al.* (1991) have reported that chicken sperm can be used to deliver alien genes to the ovum; spermatozoa appear to be feasible targets for gene transfer studies as they are easy to obtain and offer direct access to the germline. Chemically-mediated transfection is another promising method to introduce DNA into early embryos (Han *et al.,* 1991).

Attempts are currently underway to explore the feasibility of using transgenic chicken as a bioreactor for producing pharmaceutical and other proteins. One possibility is to express the pharmaceutical gene in the hen's oviduct so that the protein product is incorporated into the albumen of the egg (Shuman, 1990). Another is to strive for the expression of the gene in the liver followed by its manipulation so as to incorporate it into the egg yolk. The higher reproductive rate and relatively short generation time of chickens coupled with high protein ratio in their eggs makes them superior to mammals as prospective bioprotein production systems.

Animal Bioreactors

Transgenic animals designed to produce useful drugs or proteins have potential as a new industry. The American company Transgenic Sciences has produced transgenic mice which secrete human growth hormone in their milk at levels of up to 0.5 gram per litre. Unlike cattle and pigs carrying a foreign growth hormone construct, these mice have shown no adverse effects due to the expression of the transgene. This may be because the gene has been targeted so that it is expressed only in the mammary glands. Cattle and pigs expressing extra copies of growth hormone are often infertile.

The company has plans to scale up the production of growth hormone by introducing the gene into rabbits. Rabbits have a short gestation period and a high concentration of protein in their milk. Human growth hormone may conceivably be produced in rabbits at much less cost than in the bacterial cultures used at present.

In the UK, the Institute of Animal Physiology and Genetics Research, Edinburgh, has produced a transgenic lamb carrying a human alpha-antitrypsin (AAT) gene. Deficiency of AAT is a common and serious disorder in humans for which there is no treatment at present. AAT inhibits the enzyme elastase which clears the lungs by digesting foreign particles. If elastase is not inhibited by AAT, it digests lung tissue, causing fluid accumulation and death from emphysema. When the transgenic lamb lactates its milk will be tested for AAT. Even if the efficiency of AAT expression is only 10% of the one achieved in mice carrying the same construct, sheep will be a commercially viable means of producing AAT.

Mammalian Genome

Until 1976, our concept of the mammalian genome was that genes are encoded in continuous arrays of nucleotides, organized in simple loci containing only those genes corresponding to known alleles. It was also thought that genomic DNA was quite stable, changing slowly one base at a time so as to produce the kind of single amino acid substitutions that distinguish, for instance, normal from sickle-cell haemoglobin. This concept also supposed that genes do not easily move, especially during the lifetime of a somatic cell.

We now know that genes are not encoded in continuous sequences. They are not represented in simple loci reflective of their phenotype; rather, their loci are very complex, laced with extra copies of cryptic pseudogenes (Leder *et al.,* 1982). Genomic DNA does not change slowly but does so much more quickly, by inserting and deleting fairly large segments of DNA. Furthermore, DNA is not stable, and genes do move. In fact, genes not only move during evolution (as, for example, the globin and immunoglobulin genes) but they also move during somatic development (for example, in the immune system).

The existence of interrupted genes of mammals is an established fact now. Genetic loci consist of large arrays of related gene sequences, some of which encode active genes, whereas others encode inactive or pseudogene copies. The beta-globin locus of the mouse provides the best example of this situation; in this, there are at least seven β-like genes spread over approximately 50 kb of genomic DNA (Diag. 8.5). At one end (β-end) are found the β-globin major and minor genes that are expressed in the adult red cell. At the β-end, the most distal gene, Y2, is an embryonic gene, expressed only in the nucleated red cells that appear in yolk sac of the embryo. The remaining four β-like sequences are the pseudogenes. These latter closely resemble the β-globin genes but, having undergone alterations, they cannot encode a coherent globin polypeptide chain.

Pseudogenes are not translated and also appear not to *be* transcribed either in embryonic or adult erythrocytes.

Genetic Recombination in Mammalian Cells and Embryos

In the mid-1970s, it became possible to integrate foreign genetic material stably into the genome of mammalian cells in such a way that it could be expressed and transmitted to their offspring. This early work was done in mouse cells and involved the use of viruses, e.g., the mouse leukaemia virus, which act as gene vectors. These are retroviruses and their genome is made of two molecules of single-stranded RNA (Kelly, 1982). When a cell is infected, the reverse transcriptase synthesizes a complementary DNA molecule which becomes integrated into the cellular DNA as provirus. The latter behaves like an episome, and causes the disease.

Another virus that is sometimes used in the foregoing approach is SV40 which has a dsDNA molecule associated with histones. The SV40 DNA has been hybridized with non-viral DNA such as that of *E. coli* plasmids, and recombinant DNA molecules have been constructed which may be used as vectors for introducing genes into mammalian cells (Berg, 1981).

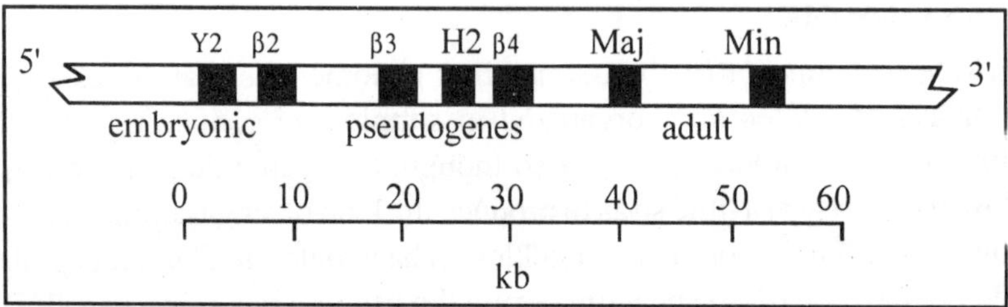

Diagram 8.5 Sketch of the mouse beta-globin gene locus. The filled regions represent the positions of beta-globin-like sequences. The two adult genes, beta-globin major and minor, are shown. The embryonic gene is shown on the left-hand side. The beta-like sequences between the genes are pseudogenes. (After *Leder et al.,* 1982.)

Yet another approach is to mix the DNA fragment containing the gene with a carrier DNA, followed by precipitating it with calcium phosphate. The precipitate tends to penetrate at least some cells, and the transferred gene is often expressed and passed on to the offspring. This technique also has been mostly applied to mouse cells.

Alien genes are also being introduced into embryonic cells with a view to following their fate in the transgenic (i.e., young transformed) animals. This kind of work has been done in eggs and embryos of the mouse. The desired gene can be microinjected into the egg by means of a fine glass micropipette, either before or after its fertilization. Before the injection, the envelope of the egg may be dissolved by treatment with proteolytic enzymes. The gene should be injected into the nucleus to increase the chances of its integration and expression. Using this approach, the gene for the herpes virus thymidine kinase has been successfully inserted into mouse cell genome, either alone or along with the gene for β-globin of human haemoglobin. The use of embryonic stem (ES) cells for the genetic manipulation of mice has several advantages over microinjection techniques. More precise genetic modifications are possible using ES cells, and cells carrying the required modification can be selected before being introduced into the host blastocyst. Some recent work has enabled transgenic rather than chimaeric mice to be produced using ES cells. Tetraploid mouse embryos can divide and implant but cannot survive to produce a viable foetus. Manipulated ES cells can be aggregated with tetraploid embryos and then implanted into a host animal.

The tetraploid cells die leaving the ES cells to take over and form an entirely ES-derived embryo. The only remaining contribution of the tetraploid embryos is in the extraembryonic membranes. Some of these embryos mature but no live young are produced.

Certain mutations can also be introduced into ES cells, using positive and negative enrichment strategies.

Ferrying Genes into Mammalian Cells

Certain genetic defects in cultured mammalian cells are amenable to alleviation by the introduction of normal genes. Some methods currently used to move genes or gene products into mammalian cells include chromosome-mediated transfer, liposomes, microinjection, calcium phosphate-mediated transfer, and viral vectors. A widely-used method is the calcium coprecipitation technique with dominant selectable markers. Eucaryotic viral vectors are also quite effective in ferrying DNA into cells, facilitating expression of alien genes. This method has several prospects in medicine, e.g., the development of recombinant pox viruses expressing the antigens of hepatitis B virus or polio virus. Another promising development relates to the potential of retroviral recombinants to move genes into hematopoietic cells. This method utilizes the patient's own tissue and eliminates allogenic rejection. Retroviral vectors have good potential as tools for somatic gene therapy. These vectors have a wide tissue and host range and are capable of transducing fairly long pieces (up to 7 kbp) of genetic material.

Much success has been achieved in the microinjection of genes into embryonic cells (Diag. 8.6). All the foregoing techniques can be used for various types of tissue culture experiments. However, only microinjection and retroviruses are effective in inserting functional genes into whole, intact animals (Andersen, 1986). An oft-observed drawback of the procedures for the introduction of alien genetic material into mammalian cells is that the newly introduced DNA fails to integrate into the host genome by homologous recombination, as happens in yeast. Instead, this DNA tends to integrate randomly diverse, unpredictable regions of the host chromosomes. This has potentially undesirable consequences as the introduced DNA could inactivate some essential gene, or it might produce some position effect aberrations.

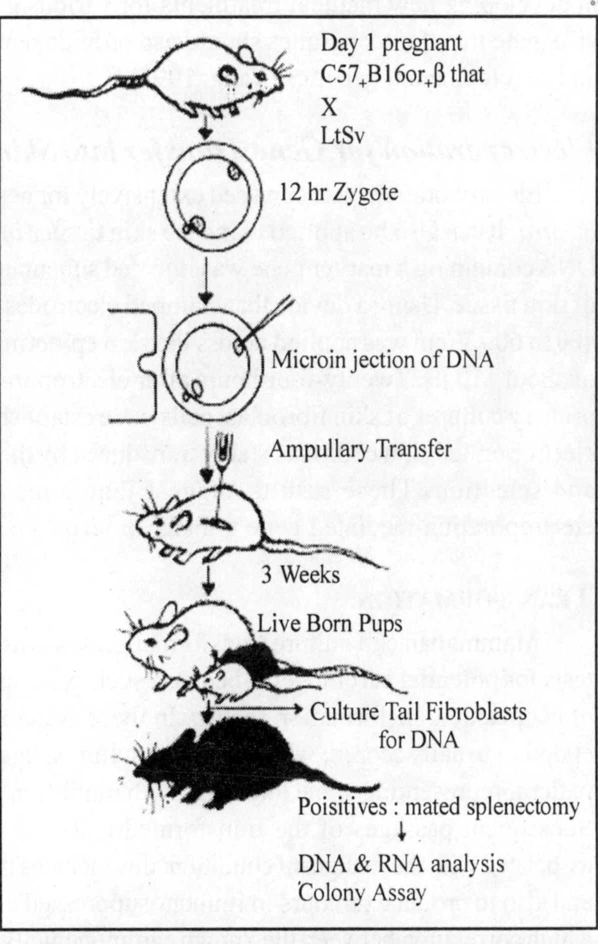

Diagram 8.6 Scheme for microinjection of murine zygotes (after Andersen, 1986).

Gene Transfer into Mammalian Somatic Cells

Direct gene transfer into mammalian somatic tissues *in vivo* is a developing technology with potential application for human gene therapy (Weatherall, 1991). The current strategy for this approach is to first identify the mutant gene(s) causing a genetic defect, then to supplement the defective somatic tissues with the correct functional gene(s). Direct transfer of functionally active foreign genes into mammalian somatic tissues or organs *in vivo* is a good strategy for gene therapy (Yang, 1992). Genetically-engineered retroviral vectors have been used successfully to infect live animals, effecting foreign gene expression in liver, blood vessels, and mammary tissues. Recombinant adenovirus and herpes simplex virus vectors have been utilized effectively for *in vivo* gene transfer into lung and brain tissues, respectively. Direct injection or particle bombardment of DNA has been demonstrated to provide a physical means for *in situ* gene transfer, while carrier-mediated DNA delivery techniques have been extended to target specific organs for gene expression. These technological developments in conjunction with the initiation of careful human gene therapy trials have marked a milestone in developing new medical treatments for various genetic diseases and cancer. Various *in vivo* gene transfer techniques should aso provide new tools for basic research in molecular and developmental genetics (Yang. 1992).

Electroporation for Gene Transfer into Skin

Electroporation has been used extensively for gene transfer into a variety of cell cultures *in vitro.* It can also be applied to mouse skin tissues *in vivo.* Prior to electroporation, plasmid DNA containing a marker gene was injected subcutaneously, thus exposing the dermal layer of skin tissue. Using a device that clamped electrodes onto taut skin tissue, a field strength of 400 to 600 V/cm was applied across the skin epidermal and dermal tissues, with a pulse time of about 150 μs. Twenty-four hours after electroporation the skin tissues were excised, and primary cultures of skin fibroblast cells were established. A significant population ($\sim 10^{-4}$) of electroporated skin cells was stably transduced by this method combining *in vivo* cultivation and selection. These results suggest that some tissues can be made susceptible to electroporation-mediated gene transfer *in vivo.*

TRANSFORMATION

Mammalian cell culture transformation systems make it possible to conduct short-term tests for potential carcinogens, because such systems could conceivably mimic the process of neoplastic transformation *in vivo.* In these systems, morphological transformation is the endpoint usually scored; with fibroblast cultures, this means piling up of cells in a crisscross pattern, representing some loss of growth inhibition and cell-cell orientation at confluency. Subsequent passages of the transformed cells can lead to the acquisition of other traits associated with the malignant condition; this includes the ability to grow in a semisolid medium, and also to produce tumours in itnmunosuppressed animals. Several workers have observed a high correlation between the known carcinogenicity *in vivo* of a given agent and its capacity to produce transformation in mammalian systems (Borek, 1979).

Several assays have been developed for the screening of suspected carcinogens. Some non-epithelial cell cultures that have proved useful in studying chemical transformation are primary and secondary embryo cultures (for example, of Syrian hamster, guinea pig, and rat or mouse infected with adenovirus or murine leukaemia virus) and fibroblast-like cell lines (including mouse prostate and hamster kidney).

It should, however, be noted that no single *in vitro* transformation system is really adequate to test different types of agents suspected to be carcinogenic.

Cell Fusion

It is possible now to induce the fusion of different types of animal cells to form hybrids which have diverse applications in biotechnology. Studies on the control of gene expression and differentiation, gene mapping, malignancy, viral replication, and antibody production have greatly benefited from experimental cell fusions.

Myoblasts fuse spontaneously, forming multinucleate muscle fibres. Macrophages furnish another example, as they are phagocytic, fusing around foreign bodies or bacterial cells in the tissues that are too big to be engulfed by single cells. Bone cells are also known to fuse. Viruses can induce cells growing in culture to fuse. Nucleated cells of different types sometimes fuse into a single cell, called heterokaryon. If the nuclei of a heterokaryon undergo synchronous mitotic divisions, uninucleate hybrid cells are formed. Hybrid cells from mixed cultures of two different mouse cell lines were successfully produced in the 1960s in France (Sidebottom and Ringertz, 1984). By now, cells from widely-different taxa can sometimes be fused. Sendai virus is the agent of choice to induce such fusions.

The essence of cell fusion is that it imposes a kind of artificial sexuality on otherwise somatic cells, thereby facilitating genetic analysis of somatic cells. The fact that cells coming from taxonomically remote animals can be fused suggests that there may not be any basic incompatibility between the membranes, nuclei, or other organelles of these different cells (Sidebottom and Ringertz, 1984). Thus, there appears to be no rejection mechanism operating at the intracellular level analogous to the immunological rejection mechanism working at the tissue level in whole animals.

Diag. 8.7 illustrates the process of cell fusion induced by Sendai virus. A heterokaryon is produced first and may then divide synchronously to give uninucleate hybrid cells.

Polyethylene glycol and certain other chemicals can also induce fusions of cells. Quite often, removal of surface carbohydrates is a necessary prerequisite for fusion.

Successful fusions have been achieved between cells in different phases of the cell cycle (e.g., in HeLa cells) and also between mitotic and interphase cells.

Multiple Ovulation in Farm Animals

Scientists have extracted, isolated, and purified the hormones that stimulate the ovaries, not only those of pituitary origin (e.g., luteinizing hormone, LH; follicle stimulating hormone, FSH), but also those of placental origin (pregnant mares' serum gonadotropin, PMSG).

These extracts have been used to induce either a small increase in ovulation rate, with the aim of inducing twin or triplet births, or a major increase in ovulation rate (superovulation), such that embryos can be collected from donor animals and then transferred to recipient animals (Webel and Day, 1982). Current superovulatory techniques show much variability. The newer reproductive technologies require continuous and reliable supply of a large number of eggs or embryos. Also, multiple and superovulatory procedures are required for an increasingly larger range of species. This industry has already spread from cattle and sheep to goats, horses and deer, and has the potential for the captive breeding of several endangered species (Price, 1991).

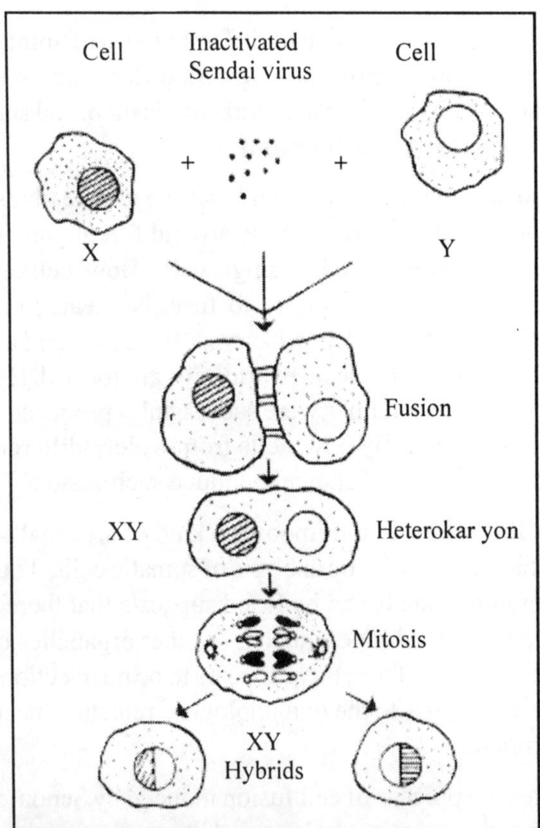

Diagram 8.7 Diagrammatic sketch of cell fusion induced by Sendai virus, resulting in the production of hybrid cells.

Price (1991) has reviewed the current practices of multiple ovulation and superovulation. He has shown that: (1) genetic selection for increased litter size has been of use only in sheep; (2) the administration of gonadotropic hormones is useful for the induction of superovulation in goats, sheep and cattle, but the response is highly variable and quite unpredictable; and (3) immunization against ovarian steroid hormones can increase litter size in sheep. He has suggested two major areas of research to be pursued to alleviate the problems associated with superovulation: first, the different molecular forms of FSH should be characterized in the farm species (some of these may be more potent stimulators of ovarian function than others); second, the actions of follicular proteins need to be examined, as these appear to influence the manner in which ovarian cell types respond to gonadotropic stimulation (Price, 1991).

The standard superovulatory technique in cattle involves the synchronization of the oestrous cycles and injections of gonadotropic hormones around the time of luteolysis. This hormone treatment usually takes the form of a total dose of 36-48 mg of FSH-P (a porcine pituitary gland extract), split into twice-daily injections of progressively declining doses, given over 4 days. Luteolysis is induced on the third day of treatment. The average results of superovulation are 10 or 11 embryos recovered, out of which a few (about 3-5) are found morphologically to be of sufficiently high quality, to be transferred to donor cows.

As regards the transfer of superovulatory technology to other species, tropical cattle commonly do not respond quite as well as the European breeds. Experiments with single doses of PMSG or multiple injections of FSH-P in *Bos indicus* have produced mean figures of 4 embryos recovered or less (Donaldson, 1984; Jordt and Lorenzini, 1988; Misra *et al.*, 1990). In sheep, the superovulatory responses to FSH are generally similar to those of European cattle.

Superovulatory techniques have been much less successful in the mare than in the goat, sheep or cow. Mares do not respond detectably to PMSG.

Production of Cattle Embryos *in Vitro*

In several countries now cattle embryos can be produced by *in vitro* procedures. In Ireland, the new technology is being used commercially and in research to expand the field of embryo transfer in cattle.

The raw material required for production of cattle embryos is the ovaries. These are brought to the laboratory within an hour or two of the animal's slaughter and are held at body temperature.

An ovary contains thousands of oocytes, but only about 20-30 can be recovered per animal. While collecting oocytes from follicles, due care should be taken to ensure that the cumulus cells surrounding the egg remain intact. Cumulus cells appear to play a vital role in oocyte maturation. The oocytes are released into a suitable tissue culture medium for maturation. This medium is supplemented with additional cumulus cells, some oestrous cow serum, gonadotropins (FSH and LH), and oestradiol.

The immature eggs are incubated for about 24 hours in the maturation medium at a temperatureof 39°C (core temperature of the cow) and a gas phase of 5% CO_2 in air. At the end of the maturation period, the oocytes emerge at the same stage of cytoplasmic and nuclear maturation (metaphase II) as they normally attain when released at the time of ovulation in the live cow (Gordon, 1989).

For *in vitro* fertilization (IVF), spermatozoa have to be properly capacitated; this is usually done by exposing them to glycosaminoglycan heparin.

The IVF of eggs is carried out in small microdroplets of culture medium, each containing 10 or more oocytes, and using sperm doses of 1 million per ml of medium. A sperm motility stimulating mixture, consisting of penicillamine, hypotaurine, and epinephrine is added as it improves sperm penetration.

The rabbit and sheep oviducts have been used as *in vivo* culture systems for the early bovine embryo. Cattle eggs are introduced into the ligated sheep oviduct a few hours after ovulation has occurred. Hundreds of cattle eggs can be introduced into each sheep oviduct, and many of these are recoverable a week later. At that time, a yield of 40% or higher of good quality embryos at the late morula/blastocyst stage of development has been recorded (Lu *et. al.,* 1987). Non-oviductal cells, such as cumulus cells, can also be used in a monolayer culture system to yield viable cattle embryos (Goto *et al.,* 1988; Gordon, 1989).

The birth of the first IVF calf was reported by Brackett *et al.* (1982) after they had succeeded in fertilizing eggs recovered after ovulation in the live cow. Since then, hundreds of IVF calves have been born in Ireland, United Kingdom, Japan, and other countries.

Sexed cattle embryos are now available commercially, but are very costly. The use of the polymerase chain reaction to amplify DNA sequences on the Y-chromosome so as to directly see the reaction product has made it possible to determine sex of human embryos on the basis of a single blastomere removed from the early embryo (Handyside *et al.,* 1989) and the same technology is already being applied commercially to cattle embryos in some countries (Gordon, 1989).

Diagram 8.8 shows the principle of using nuclear transplantation for cloning livestock embryos. Embryos can be microsurgically bisected by means of a microneedle or razor blade to form two semi-embryos, one of which is used for chromosome evaluation (karyotyping) and the other may be either transferred immediately or frozen for subsequent use (Diag. 8.9).

Gene targeting makes possible precise alterations to specific genes in animal cells. Use of these procedures in combination with embryonic stem cells enables the production of whole animals which carry the specific alteration in every cell of their body. This approach has several advantages over conventional methods of making transgenic animals, and promises to be a valuable aid in the study of human disease. These procedures have been developed in mice; their extension to farm livestock could also facilitate the development of commercial transgenic livestock (Melton, 1990). Many human genetic diseases and some cancers result from mutation of a single gene. By inactivating such genes in mice, it might prove possible to produce mouse models for human diseases and permit evaluation of novel forms of treatments, such as somatic gene replacement therapy, where the introduced functional genes would complement the genetic deficiency in an individual. Commercial benefits could also follow from the ability to make a precise germline alteration, leading to the overproduction of a particular natural gene product in an animal, or to the synthesis of a novel product under the transcriptional control of an animal's own genes (Melton, 1990).

Embryonic stem cells of mouse are valuable as a route into the mouse germline (Diag. 8.10). These ES cells are isolated from the inner cell mass of early embryos (blastocysts).

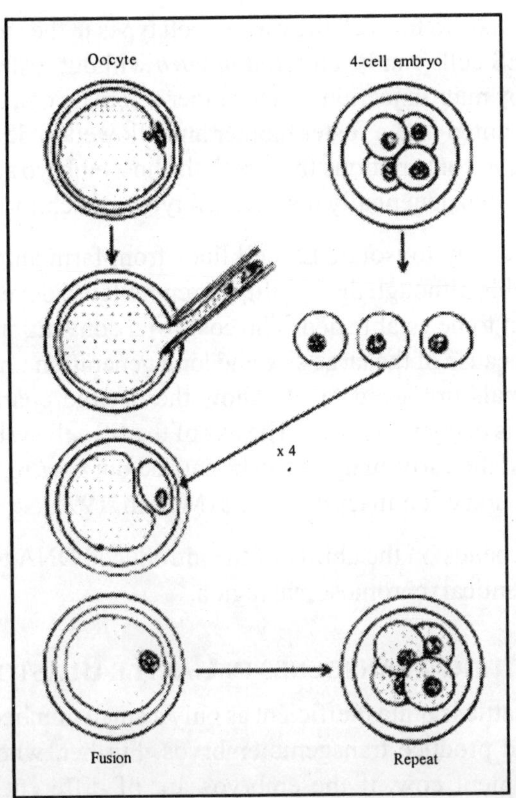

Diagram 8.8 Use of nuclear transplantation for the cloning of livestock embryos (after Picard and Betteridge, 1989).

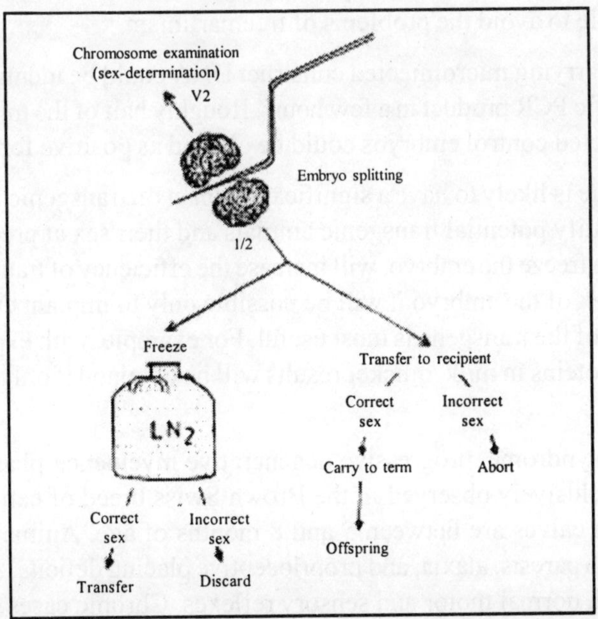

Diagram 8.9 Microsurgical bisection of embryo into two semi-embryos (after White, 1989).

They normally differentiate to form all the various cell types in the mouse. Under appropriate stringent conditions, ES cells can be cultured *in vitro* without losing this ability. Hence, if they are reintroduced by microinjection into the inner cell mass of a different embryo, which is subsequently reimplanted into a foster mother and allowed to develop, the tissues of the resultant mouse will have contributions from both the host embryo and the injected ES cells. Such an animal having two genetically distinct cell types is a chimaera.

Attempts are underway to isolate ES cell lines from farm animals. The pig cell lines seem particularly suitable although their ability to repopulate a developing embryo to give a chimaeric animal is yet to be established. The costs of a conventional transgenic livestock programme are high because of the large size and long generation time of the animals. Many of the transgenic animals produced fail to show the desired characteristics, but it takes several years before this becomes known. The use of the ES cell system is considerably less costly, because much of the early analysis can be carried out *in vitro,* with only the best lines selected for the production of chimaeric animals (Melton, 1990).

Gene targeting depends on the ability of the introduced DNA to locate and recombine with a homologous (identical) chromosomal region.

Sexing and Detecting Transgenes in Cattle Blastocysts by PCR

Gene transfer in cattle is quite inefficient as only a small number of injected blastocysts integrate the DNA and produce transgenic embryos. Further, when several embryos are implanted into a recipient cow, if the embryos are of different sexes the problem of freemartinism arises. By means of a new technique, blastocysts injected with a transgene can be tested for the presence of the construct and sexed. Only known transgenics can then be used for implanting. In addition, groups of embryos of the same sex can be implanted into the recipient female to avoid the problems of freemartinism.

Blastocysts carrying microinjected construct DNA could be identified by the presence of construct-specific PCR product in a few hours. Roughly half of the microinjected embryos and 2/16 non-injected control embryos could be classed as positive for the construct DNA.

This technique is likely to have a significant impact on transgenic technology in cattle. The ability to identify potential transgenic animals and their sex at pre-transfer stages, and without the need to freeze the embryo, will increase the efficiency of transgenesis procedures. By knowing the sex of the embryo it will be possible only to implant embryos of the sex in which expression of the transgene is most useful. For example, with experiments to express pharmaceutical proteins in milk, quicker results will be obtained if only female embryos are implanted.

The Weaver syndrome, progressive degenerative myeloence-phalopathy, is a genetic disease almost exclusively observed in the Brown Swiss breed of cattle. The syndrome is first noticed when calves are between 5 and 8 months of age. Animals show progressive signs of pelvic limb paresis, ataxia, and proprioceptive placing deficits. Affected animals are mentally alert with normal motor and sensory reflexes. Chronic cases show atrophy of the hip and stifle muscles. Afflicted animals have defective posture. The affected animals suffer several neurological changes including degeneration of the white matter of the spinal cord and loss of Purkinje cells in the cerebellum. Weaver carrier animals produce much more milk and fat per year than homozygous normal animals. The Weaver gene is tightly linked to a microsatellite marker, *TGLAI16,* with 3% recombination. The *TGLA 116 and* Weaver genes have been assigned to bovine synteny group U13 by analysis of a panel of bovine-rodent somatic cell hybrids. But this synteny group has not yet been conclusively mapped to a specific bovine chromosome. In man, all of the genes, except esterase D, are on chromosome 7, but in the mouse, the corresponding genes are mapped over several chromosomes (Georges, 1993). The identification of the *TGLA116* marker should allow breeders to select against the Weaver disorder without using lengthy and expensive progeny testing procedures. The marker identified for Weaver disease should also allow studies of the associated quantitative trait locus (QTL) for milk production.

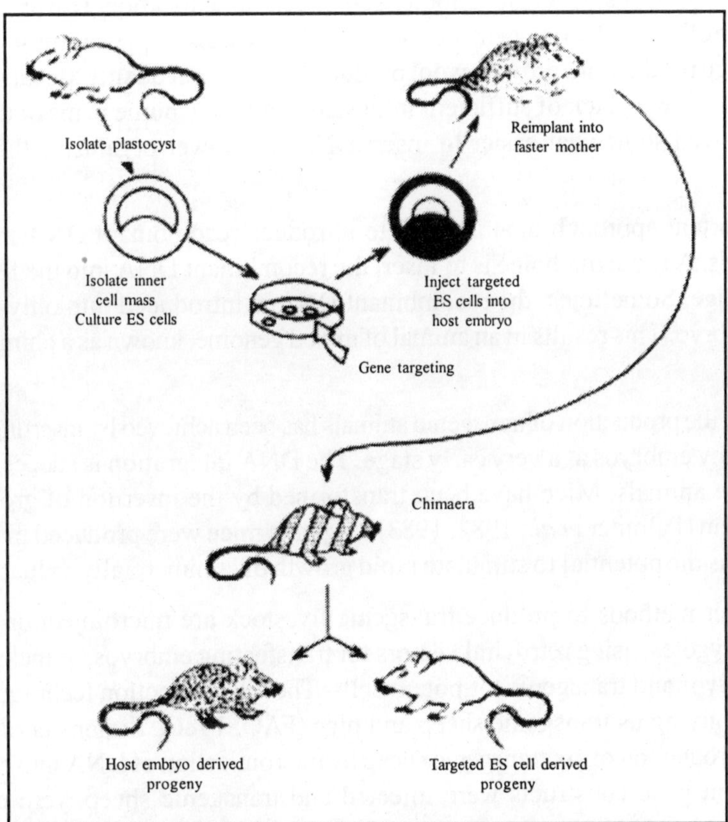

Diagram 8.10 Use of embryonic stem cells as a route into the mammalian germline (after Melton. 1990).

Early applications of agricultural biotechnology indicate that it can be an essential tool in increasing food production, under diverse ecological and socio-economic conditions. Rapid developments in agricultural biotechnology research over the past decade have induced several developing countries to devise national programmes aimed at realizing the potential benefits of biotechnology.

Biotechnology and Domestic Animals

Biotechnology has great potential for the improvement of our cattle and other domestic animals. Worldwide animal products contribute over 50 million tons of edible protein and more than 1 billion megacalories of energy annually. With its high biological value, animal protein is equivalent to more than 50 per cent of the protein produced from all cereals. Major advances have been made in hormone production, disease diagnostic kits, and vaccines; all these areas have a direct bearing on livestock production and betterment. Suitable changes in the genome of domestic animals can sometimes bring about marked improvements in their productivity.

The successful use of recombinant DNA technology for animal improvement depends on the identification of the appropriate part (segment) of the DNA (i.e., a gene), its isolation from the cell, and its insertion into another cell at a suitable location. However, some of the traits are controlled not by one gene but by several genes. Examples of these traits are milk production, egg production, growth, wool production, and meat quality. Because of these and other difficulties (e.g., lack of sufficient information about genome maps of the animals), it has so far proved somewhat easier to insert DNA into lower organisms than into higher animals.

An important approach in animals is to introduce recombinant DNA into their germ cells or gonads. A route of choice is to insert the recombinant DNA into the fertilized ovum at an early stage. Sometimes, the recombinant DNA is introduced into only some, not all, cells of an embryo. This results in an animal of mixed genome, known as a chimaera (Hodges, 1986).

Hitherto, the production of transgenic animals has been achieved by inserting recombinant DNA into many embryos at a very early stage. The DNA integration is successful in at least a few of these animals. Mice have been transformed by the insertion of growth hormone genes from man (Palmiter *et al.*, 1982, 1983). Gigantic mice were produced in this way. This technology has the potential to stimulate rapid growth of commercially-valuable animals.

Two other methods to produce transgenic livestock are microinjection of DNA into pronuclei of zygotes, using retroviral vectors for transfecting embryos, or making chimaeras between embryos and transgenic totipotent cells. The microinjection technique has already succeeded in giving us transgenic sheep and pigs (FAO, 1986). Simons *et al.* (1988) have reported the production of six transgenic sheep by microinjection of DNA into early embryos. Three different gene constructs were injected and transgenic sheep were obtained with each. The incorporation as well as transmission of transferred genes to the progeny of three of these sheep were demonstrated. Five founder transgenic sheep also carry certain genes that determine the production of human clotting factor IX or human al-antitrypsin in milk. The importance of such transgenic animals stems from their potential as a new source of these and other therapeutic proteins.

An animal's physiology can be potentially changed by using specific promoters or enhancers to direct expression of alien genes to some specific cell type. The transgenic technology has proved valuable for observing the consequences of oncogene development in the animal. Some problems that cannot be tackled satisfactorily in cell cultures may be solved by using transgenic mice, for instance, the spectrum of tissues that are susceptible to the transforming activity of an oncogene, and the effect of oncogenes on growth and differentiation.

Certain phenotypes induced in transgenic mice by expression of various viral gene products constitute model systems for pathological conditions, some of which resemble human diseases. Transgenic mice have served as important tools for studying the Ig gene expression. Functionally-rearranged Ig genes introduced into the germ line were found to be correctly activated and to alter the expression of the endogenous immunoglobulin repertoire, indicating that light and heavy chains may interfere by some feedback mechanism with further Ig gene rearrangement. Some recent work on transgenic mice suggests that expression of functional Ig genes can cause abnormalities in the immune system, and that somatic mutations and gene rearrangement are not concomitant processes. Ig genes from rabbit or chicken become rearranged in transgenic mice and produce hybrid Ig molecules, suggesting that the production of interspecies monoclonal antibodies may be achieved in genetically-engineered mice (Jaenisch, 1988).

Especially in the context of developing countries, Hodges (1986) has suggested the following promising approaches for the future:

1. Introduction of new functions, e.g., genes that confer upon the animals or increase their resistance to environmental stresses such as disease and water scarcity.
2. Enhancement of productivity by the substitution of alleles. The weaker or disabling alleles may be substituted by improved or better alleles. Some genes responsible for lower productivity may be removed and be substituted by alleles from the same or other species. Bindon (1984) mentions the specific case of the introduction of the newly-discovered Booroola gene for fecundity in sheep.
3. Enhancement of productivity by the duplication of sets of alleles. There is some possibility that multiplying flocks of certain segments of DNA may be conducive to increasing the genetic effect of a trait. This has the potential of producing novel genetic variations, some of which could be exploited gainfully.
4. Control of genome activity. This concerns the ability to insert recombinant DNA switches into the genome of a domestic animal with a view to regulating the timing of certain developmental functions (such as development of 'heat' or mating capability in cattle). This could be a potent way of triggering or hastening the onset of puberty (Hodges, 1986).

Priority Areas for Improvement of Domestic Animals

Four of the target areas for possible application in domestic animals are briefly explained here.

Reproduction and Genetics

Problems currently associated with livestock enterprises in many developing countries are related to late sexual maturity, low reproductive rate, long calving intervals, and rather low genetic potential for various production traits aggravated by high disease incidences and susceptibilities. Embryo transfer technique involves a series of steps starting with the stimulation of multiple egg production at the time of ovulation. Artificial insemination is followed by embryo recovery, usually providing up to 15 embryos from a single cow. These embryos can be stored for some time or may be immediately transferred to a surrogate mother. The chief use of embryo transfer is to increase the reproductive rate of valuable but slow reproductive rate cows. An established field technique is the use of superovulation in mammals, followed by artificial insemination, flushing out the embryos, and transplanting them into surrogate mothers. Superovulation means the production of significantly larger numbers of ova than the normal or average number. Superovulation can greatly increase the number of offspring. Gonadotrophins (LH, FSH, and PMSG) are used to induce superovulation in cattle. Likewise, equine pituitary extract can be used to induce superovulation in mares. Antisera to steroids and gonadotrophins have proved useful in sheep. The superovulation technique is also being applied to increase the populations of some endangered animals. Developments in recombinant DNA technology may be expected to increase the chances of inserting parts of genome from other or better (elite) animals or even alien species, to enhance the genetic capability and adaptation of the embryos. The possible benefits may be amenable to multiplication by cloning of the transgenic embryo, resulting in the production of many identical transgenic embryos of known sex for insertion in several surrogate mothers. In this way, elite embryos could be multiplied rapidly through a local domestic animal population of inferior quality. A genetically-superior donor, through its recipient cows, has the potential to produce as many as 50 superior calves in a single year; normally, one cow gets only one embryo per year. Furthermore, embryos can be frozen for international shipment. This makes it possible for any developing country to import a few superior embryos and achieve rapid progress in upgrading its cattle wealth (Bodde, 1982).

Health

Some serious diseases of animals are as follows:

1. *Neonatal diarrhoea* It affects cattle and pigs and is caused by bacterial and viral agents.
2. *Bacterial respiratory diseases* These affect many animals and are caused by such bacteria as *Pasteurella hemolytica, P. multicida,* and *Bordatella bronchiseptica.*
3. *African swine fever* Endemic in Spain, Portugal, and several African countries, it is caused by a virus.
4. *Tuberculosis* This is caused by mycobacteria, and is common in livestock. Table 8.2 lists some more infectious diseases with their present and prospective vaccine status.

Table 8.2 Some selected infectious diseases of animals

Disease	**Animals**	**Requires**		
		Vaccine or better vaccine	**Improved diagnosis**	**Disease control possible**
Brucellosis	Bovine		+	+
Leptospirosis	Bovine		+	+
Salmonellosis	Swine, poultry			+
Tuberculosis	Bovine	+	+	+
Ascariasis	Several	+		+
Cysticercosis	Several	+		+
Schistosomiasis	Several	+		+
Coccidiosis	Avian, ruminants	+		+
Trypanosomiasis	Several	+	+	+
Blue tongue	Ruminants	+	+	
Paralytic rabies	Bovine	+	+	+
Foot and mouth	Swine, bovine	+		+
Newcastle	Poultry			+
Rinderpest	Bovine	+	+	
Infectious anemia	Horses	+	+	

Biotechnology has contributed significantly to animal health improvement in three main aspects, viz., diagnostic methods, vaccine production, and treatment of infected animals. Recombinant DNA techniques can be used to produce attenuated strains of pathogens, resulting in the production of low-cost but highly potent, live attenuated vaccines. Monoclonal antibodies have been pressed into service for tackling some serious diseases of cattle, such as East Coast fever caused by single-celled parasite *(Theileria)* that is spread by ticks and which afflicts over 500,000 cattle annually in East Africa. *Theileria* tissue is first injected into mice causing antibody formation. Antibody-producing cells are removed from the mice and fused with bone marrow cells. The resulting hybrid cells are cloned as monoclonal antibodies. These are then used to control the fever. Efforts are also underway to remove embryos from cattle that are resistant to trypanosomiasis (sleeping sickness) disease caused by tsetse fly. These embryos may then be transplanted to sensitive cattle with a view to producing calves resistant to sleeping sickness. Monoclonal antibodies have been developed for specific organisms and antigens. Such MAbs are indispensable for disease diagnosis work.

Nutrition

The nutrition of ruminants has attracted much attention in recent years, especially in relation to the modification of rumen microflora. One approach is to develop suitable methods of identifying chemical inhibitors (secondary metabolites) in plants to permit more efficient use of indigenous fodder species. Another approach is the improvement of quality of animal feed by treating it with transgenic microbes, before ingestion by the animal, with a view to increasing its palatability and digestibility. Yet another way is the protein enrichment of locally-available starchy materials routinely fed, for instance, to poultry and pigs. This could be done by solid state fermentation of the materials (e.g., cassava), leading to their enrichment in microbial proteins. Some suitable microbes for this purpose are species of *Rhizopus, Aspergillus,* and *Candida.*

Wool production by sheep depends on the availability of sulphur-containing amino acids. There are several ways by which the amino acid availability to sheep may be increased to get more wool. These include (1) supplementation of the feed with a protein meal which escapes fermentation and is digested in the intestine, thereby augmenting amino acids from microbial protein; (2) stimulating microbial growth rates in the rumen; (3) preventing the growth of rumen protozoa, thus decreasing bacterial protein turnover in the rumen, thereby increasing the protein : energy ratio in the products of fermentive digestion; and (4) using recombinant DNA technology to produce animals capable of synthesizing S-amino acids.

By inserting pea genes which code for a high-sulphur protein into alfalfa, Australian scientists found (*Genetic Engineering News,* July/August, 1986) dramatic increases in wool production in sheep. For sheep, good amounts of cysteine and methionine (being constituents of wool) are essential. Forage plants do have high-quality proteins but most of the sulphur amino acids are quickly degraded to hydrogen sulphide and urea in the rumen, the first stomach. When sulphur amino acids are introduced into the sheep's second stomach (or intravenously), the rate of wool growth can often be substantially increased, even doubled.

The pea protein p-albumin contains over 10 per cent cysteine and one per cent methionine, and is quite resistant to breakdown in rumen. Australian scientists are attempting to use *E. coli* and *Agrobacterium tumefaciens* to transfer the *Pisum sativum* gene into alfalfa or clover using tissue culture methods.

Australian scientists have also succeeded in producing the world's first "transgenic sheep" by .inserting into an embryo a gene coding for the sheep growth hormone. This first transgenic sheep was born in April 1986. The objective of raising transgenic sheep is to produce larger, leaner, and faster-growing sheep *(Genetic Engineering News,* July/August, 1986).

The FAO (1986) has made the following recommendations on animal nutrition, growth, and lactation, with special reference to the needs of developing countries:

1. Identification of appropriate feed supplements and testing of their value in providing nutrients to rumen microbes and in providing nutrients that will bypass the rumen.
2. Development of suitable techniques for the processing of feed supplements so as to stimulate rumen fermentation or supply of bypass nutrients to improve animal feed conversion efficiency.
3. Further development of physico-chemical treatments for improving the rumen degradability of lignocellulosic materials for commercial application.
4. Enhancement of digestibility of plant materials via suitable microbes which solubilize lignin.
5. Genetic improvement of rumen microbes.

Over 2000 million metric tons of crop residues are produced annually in the world. These are sources of cellulosic feedstuffs and include straw, grain hull, sugarcane bagasse, wood and wood byproducts, grasses, and animal and other wastes which are nutritionally rather poor in quality, but can be improved through certain physical, chemical or microbial treatments or pretreatments. Sometimes, the nutritional value of these feedstuffs can be improved by as much as 25 per cent.

The major uses of lignocellulosic plant residues are as fuel, fodder, and paper manufacture. When used as fodder, the breakdown of lignocelluloses in the animal gut is limited by the crystallinity of the cellulose fibres and the barrier to gut microbes and enzymes created by the coating of the fibres with lignin and hemicellulose present in the plant cell walls. The white rot fungi are perhaps the most effective lignin degraders known; they depolymerize lignin, utilizing the lignin peroxidases which are secreted by the fungal mycelia. These enzymes act by single electron oxidation, generating cationic radicals in the lignin that stimuiaie polymer breakdown. The genes coding for lignin peroxidases have already been cloned and expressed in bacteria and several potential new applications for these enzyme preparations have been proposed.

Systemic Changes in Animal's Life Processes

From our viewpoint, the important life processes for animal production are growth, lactation, and work capacity.

Some of these processes (also reproduction) can be manipulated by exogenous applications of hormones. The importance of biotechnology in this context lies in the prospect of much cheaper production of the hormones by transgenic or genetically-engineered microbes. Richard *et al.* (1985) have reported that cows respond to bovine growth hormone by increased milk yields amounting to over 10% or increased fat content of milk by about 25%, without any concomitant increased intake of feed.

There is great interest now in the splitting and sexing of bovine embryos. Splitting is becoming a routine part of embryo transfer. Survival rates of split embryos vary from 40% to 60%. Some of the techniques for determining the sex of early embryos prior to transfer are sex chromosomal analysis, demonstration of H-Y antigen, metabolic activity of X-linked enzymes, and DNA probes. Of these, the best results are obtained with the H-Y antigen methods, and hence this technique is the most promising for use in commercial programmes provided that a simple kit becomes available.

Some work is underway on *in vitro* fertilization in farm animals. Two technological barriers in this effort are the inadequate knowledge about oocyte maturation and capacitation of the spermatozoa. The correct timing of oocyte recovery is critical to success, the best results being obtained when the oocyte is recovered immediately prior to, at the time of, or immediately after ovulation. At this time, the oocyte is in the second metaphase of meiosis and its maturation is complete. Capacitation of the spermatozoa is also essential for *in vitro* fertilization. The biotechnology programme involves the use of elite animals repeatedly as donors to supply many oocytes for improvement and multiplication of farm animals.

Controlled environments constitute a technology scarcely tapped for livestock production. They are extensively used now for poultry and pigs and are being tested for sheep and goats. Programmed lighting and temperature regulation deserve to be pursued with a view to determining their effects on dairy cattle productivity, feeding efficiency, rates of gain, hormonal relationships, and reproductive behaviour (Wittwer, 1986).

Insect Cell Culture

In the last decade, cultured insect cells have become an effective tool for abundant expression of a wide variety of heterologous gene products. These cells are good hosts for the efficient production of biologicals with applications in health care and in agriculture. Suitable baculovirus expression vectors have been developed (Agathos, 1991). These vectors are genetically-altered baculoviruses which exclusively infect invertebrates but neither vertebrates nor plants. Several alien proteins express in insect cells at levels much higher than in bacterial or eucaryotic hosts. Insect cell technology is now being widely used for producing recombinant proteins and improved viral insecticides. Cells of moths, butterflies, flies, mosquitoes, and certain other insects are being used for this purpose. The most widely used group of baculoviruses in biotechnological applications are the nuclear polyhedrosis viruses (NPV). These NPV have circular, double-stranded DNA; they are natural pathogens of lepidopteran cells, and are easily grown *in vitro* to infect many continuous insect, but not vertebrate, cell lines (Maramorosch, 1987). The most prominent NPV is the *Autographa californica* (alfalfa looper) NPV; it characteristically forms occluded viral particles or "occlusion bodies" that are embedded in polyhedron shaped proteinaceous particles in the nucleus of the infected insect cell. This NPV has a gene encoding polyhedrin protein. This gene is expressed quite late in the viral infection cycle and, accordingly, is not essential for viral replication. The construction of the vector entails the placement of the cloned gene of interest under the control of the strong viral polyhedrin promoter, which allows efficient expression of the alien gene. The desired protein can also be produced on a commercial scale *in vitro* by cultivation of a susceptible continuous insect cell line in a bioreactor, followed by infection of the cells with the genetically modified baculovirus, which leads to the release of the protein product into the medium following lysis of the insect cells. For crop management application, biopesticides can be produced using a strategy similar to the above, except that the insect cell line is infected with a wild-type virus for abundant production of occlusion bodies, the main product used in crop management (Agathos, 1991).

The alfalfa looper is used primarily in cell lines of the fall armyworm, *Spodopterafrugiperda.* Many types of protein, including membrane-bound, cytosolic, nuclear, and periplasmic, can be expressed at very high levels. In this system, lymphokines, biologically active enzymes, viral antigens, and oncogene proteins are some of the gene products that have been successfully overproduced. Indeed, this vector is now being used to produce HIV envelope proteins for an AIDS vaccine to be clinically tested on humans. An important merit of the alfalfa looper NPV vector *Spodoptera frugiperda* host cell system is that the baculovirus infected cells carry out many of the post-transcriptional modifications of the higher eucaryotes, such as phosphorylation, glycosylation, and proteolytic processing.

Cultured dipteran cells are being used for producing vaccines against arthropod-borne viruses, for human and veterinary applications. Cells of *Aedes aegypti* and *A. albopictus* are used for the *in vivo* .propagation of arthropod-viruses that cause infectious diseases mainly in developing countries. Table 8.3 lists some insect cell culture systems for applications in biotechnology.

For commercial production, insect cells are grown in suspension in suitable bioreactors, using spinner vessels or various fermenters. The bioprocess technology developed for microbial or mammalian cells may also be applied to insect cells, in some cases after some modifications. Some important factors that can affect foreign gene expression in cultured insect cells using baculovirus expression vectors include cell density, cell viability, origin of cell line (species and type of tissue), cell attachment characteristics, the nature of the substratum, dissolved oxygen level, and the composition of the culture medium.

Table 8.3 Selected insect cell culture systems for biotechnology applications

Common name	Technical name	Applications
Fall armyworm	Spodoptera frugiperda	Recombinant proteins (enzymes,
Cabbage looper	Trichoplusia ni	lymphokines, oncogene proteins, viral
Silk worm	Bombyx mori	antigens), bioinsecticides
Alfalfa looper	Autographa californica	Bioinsecticides
Cotton bollworm	Heliothis zea	Bioinsecticides
Tobacco bollworm	Heliothis virescens	Bioinsecticides
Gypsy moth	Lymantria dispar	Bioinsecticides
Yellow fever mosquito	Aedes aegypti	Arbovirus antigens, diagnostics, vaccine

While asepsis is required for work on insect cell cultures in the same way as for mammalian cells and in both the systems the growth of the cultured cells depends on the inoculum size, there are several differences in the cell propagation technology for the two systems. These differences are listed in Table 8.4.

Some insect cells have been grown in serum-free media. The development of new and low-cost serum-free media should contribute to a wider commercial applicability of insect cell cultures. Further, the biochemical engineering methodologies employed in studies of shear-induced cellular damage and the protection of cells from mechanical stresses by improved bioreactor design and medium supplements are sure to produce a favourable impact on the scaling-up of cultured insect cells (Agathos, 1991).

Table 8.4 Some differences between the cell propagation technologies for insect cells and mammalian cells (after Agathos, 1991)

Factor/Parameter	Insect cells	Mammalian cells
Maintenance of cell lines	Fairly easy	Difficult
Versatility of suspension/attachment	Yes	No
Immortality	Yes	Only for transformed cell lines
Contact inhibition	Absent or weak	Generally yes (exceptions : transformed and lymphoid cell lines)
Detachment from substratum surface	By gentle force	Trypsinization
Sensitivity to changes in pH, dissolved oxygen, temperature, osmotic swhock	Rather low	Fairly high

A gene which inactivates juvenile hormone (JH) can be introduced and expressed in insects. In normal insect development a drop in the level of JH is associated with the end of the larval feeding stage, and the start of metamorphosis into a pupa (which in time will produce a moth). The reduction in JH is apparently caused by the production of high levels of an enzyme, juvenile hormone esterase (JHE) which converts JH into inactive JH acid. Inhibition of JHE allows the JH level to remain high, leading to continued feeding and giant insects.

Sufficient amount of JHE has been purified for amino acid sequencing and antibody synthesis. Clones encoding JHE have then been obtained from a complementary DNA library made from fat bodies of the major insect pest *Heliothis virescens.* The JHE gene has also been expressed in an *in vitro* baculovirus system. The nuclear polyhedrosis virus (isolated from the moth *Autographa californica)* has proved a useful molecular tool and infects a broad range of insect hosts. Cells from the moth *Spodoptera frugiperda* have been co-transfected with the virus and the JHE gene. Cells containing both the virus and the gene expressed useful amounts of JHE. The larva injected with JHE showed marked blackening. This is used as an assay for the action of an anti-juvenile hormone. The effect of infection of larvae with the virus and JHE gene combination has been studied. The insect which was infected to prevent feeding and further development remained a tiny larva as compared to the control. The great reduction in feeding makes it possible to control insect pests by viruses. Thus, engineered JHE viral insecticides look quite promising.

□□□

CHAPTER-9

Plant Tissue Culture Some Related Aspects

The increase in population and decrease in plant diversity on earth and its related consequences has rang the alarm bell. Therefore, scientists are doing massive research work to increase the plants density with saving of natural germplasm. Due to deforestation a number of plant species, which are yet to be identified are lost. Besides a number of known species are also going in vein. In this context, plant tissue culture method may serve not only the human kind but also the earth in one or many ways.

Bio Village Concept

The tissue culture could become an essential component of the **bio village** concept being mooted by noted agricultural scientists Dr. M. S. Swaminathan. His foundation is already engaged in development of such villages in Pondicherry.

Dr. C. R. Raju of Kerala has created some sort of history by transforming the science of tissue culture into an art that is being practiced by school going children. Raju is turning matriculates into tissue culturists in a matter of 12 days. Thanks to Raju's ingenious the practice which was limited to botanists, is now within reach of the common man. The training can help villagers to start cottage industry of test tube grown flowers and herbs.

Tissue culture technology as practiced in the scientific laboratories is very expensive. Contrary to this the **Raju- technology** is of low-cost and appropriate one. Dr. Raju is associated with Centre of Science and Technology for Rural Development (COST-FORD) in Kochi, Kerala.

This technology can be applied in any place where the year round ambient temperature does not fluctuate beyond 18 and 34° C. This is specifically to avoid the need to have air-conditioning which involves high capital and recurring expenditure. Normal electricity supply is not required. A germicidal ultraviolet lamp can be run alternate lyon a 12 volt battery connected to an inverter. All the expensive equipments used usually in a regular laboratory are done away with.

Currently, Raju is trying to grow virus free healthy coconut seedlings from selected tissue in a test tube for the Central Plantation Crops Research Institute (CPCRI). He is translating his experience of working with coconut to other crops and is using coconut-water and coconut-milk to fortify the medium to grow his plants. Coconut-pith has been used to grow orchids.

Once perfected, the system is conveyed to the trainees in simple terms. The trainees practice the procedure leading to mass propagation step by step. Once trained, all that remains is the mechanical process of mixing chemicals, preparing culture media, collecting, disinfecting and introducing plant parts into the culture medium. From the test tube the healthy seedlings are taken out into pots with subsequent transfer to the fields.

He has used all kind of household glass-utensils including bottles to replace the conventional test tubes. These are properly disinfected with hot water. After evaluating different kinds of plants. Raju has found tropical orchids and medicinal plants as the best candidates to start with. He has valid reasons to choose the orchids and medicinal plants. Says he, growing orchids and medicinal herbs is profitable. It costs Rs. 5,000 to set up a tissue culture facility for 100,000 plants.

The technique has proved successful in multiplying several orchids which include 22 varieties of Dendrobium, 3 varieties of Phalaenopsis, 3 varieties of Vanda and one variety of Asconcenda. He has trained young boys and girls in at least four types of medicinal herbs.

Qualifications Required to Join the Training Course

Any matriculate who can pick up the names of chemicals is fit for the course. Besides, one must have a love for plants. Two basic aspects of the technique of tissue culture, however, must be learned thoroughly. These are the composition of the medium required at different stages of multiplication and the method of handling tissues and culture in sterile- germ-free conditions.

Employment for Rural Youth

Raju is confident that his short term course can open wide opportunities for unemployed rural youth to seek self employment. Following are advantages of popularizing tissue-culture in Kerala.

1. Climate is suitable to grow test tube plants.
2. There is plenty of coconut-water or coconut-milk used to nourish the growing plantlets in the test tube.

It was Kerala which provided an opportunity for the first Plant Tissue Culture Company of India. It was set up as a small scale laboratory at Manalaroo in Kerala. It has used tissue culture technology to grow cardamoms plants in test tubes. The technology was developed by National Chemical Laboratory (NCL), Pune.

The company at one time was making just about Rs, 2.5 crores in profit. It now hopes to net a profit of Rs. 12 crores. The phenomenal growth of the first biotech company has encouraged many others to join the fray. If companies can make profit, why can't poor villagers.

Dr. Raju's technique is worth emulating by research institutions and extension agencies in the country.

Efforts of Public Research Institutes

In India, research facilities in public research institutes on biotechnology are being improved with the assistance of UNIDO and other agencies. Private sector companies are also showing keen interest in the technology as is evident from their increased investments in research and development as well as the collaborative arrangements made with foreign companies. The research effort so far is concentrated more on plantation crops like tea, coffee, rubber, spices besides vegetables, flowers and fruits.

A great deal of biotechnology research in the West is looking into the replacement of reduced use of flavour and fragrance plants with laboratory processes. A produce called Nocordia is being developed by two companies which would eliminate the need for castor oil. Biotechnology offers the potential to displace sugar as an industrial sweetener through the development of new natural sweeteners from plants. Experiments in cocoa, rubber, vanilla are aiming to reduce the dependence on the developing countries. All this would have a profound impact on the livelihood of the Third World farmers. Such development should be constantly monitored so that timely and appropriate policy actions can be taken before they swamp the Indian farmers.

Production Criteria and Economics

Plant tissue culture is the only way of modern plant propagation where a large volume of plants needs to be produced regularly, as in the case of cut flowers. Plants produced by tissue culture technique are much healthy with desirable characteristics. In vitro culture systems have the potential for long distance shipment of propagules and allow long-term storage of clonal material.

In the conventional propagation system, about 75% of the greenhouse area is committed to mother plant maintenance. Micropropagation would obviate such necessity and the entire space of the greenhouse could be effectively planted with plants of high value received from tissue laboratories on a contractual basis. Micropropagation offers an environment-friendly industry to flourish and is a source of potential employment.

Selection of Crops for Micropropagation

Tissue culture production is market oriented. Therefore, it is essential that one should have a sound market-place reason for contemplating the use of plant tissue culture for a particular crop or clone. The use of plant tissue culture should offer economic advantages to the commercial laboratory as well to the grower. The tissue culture method of production can be employed only if it assures as increase in the net economic return, compared to that of the conventional propagation method, for a certain unit of land, water, time and capital.

For economic viability producer should know the costs of not only propagating a crop through tissue culture, but also the costs of propagation through conventional means. Under congenial conditions certain plants can be produced more economically, using conventional techniques. The micropropagation system should offer propagules at a cheaper rate as against conventional methods of propagation. Propagation through plant tissue culture is the only method wherein conventional propagation is not possible, as in the case of runnerless strawberries. The greater impact of tissue culture can be felt where the cost of field production of micropropagation progeny can be lowered, rather than in those crops whose productivity can be increased with yield increases.

Growing varieties resistant to pests, diseases, salinity or drought would save corresponding input costs or decrease operating costs associated with energy, irrigation and labour. Therefore, clones or varieties with such traits should be selected for micropropagation.

For example ten flowering plants in the floriculture trade can be arranged in the descending order of their turnover are: rose, *Chrysanthemum,* carnation, tulip, *Freesia, Gerbera,* lily, orchid, *Gypsophila* and *Iris.* Similarly, ten foliage plants are *Ficus, Dracaena, Begonia, Saintpaulia, Yucca, Azalea, Poinsettia, Kalanchoe, Dieffenbachia* and *Cyclamen.*

Selection of location

It is essential to consider following factors before selecting the location of a laboratory:

1. *Availability of water and electricity:* These are the major requirements of a tissue culture laboratory.
2. *Transport of sapling:* As cost incurred on transport is a limiting factor. The laboratory should be established near the place of a potential market. If the unit is export-oriented it should be near an international airport, which will reduce the time lag between packing and shipment. This will help in quick delivery of quality plants.
3. *Climate:* Maintenance of stock plants and the hardening and weaning of tissue-cultured plants require a conducive microclimate, either natural or controlled. The laboratory should be established in a place where conducive natural environment prevails. This will save considerable cost on cooling or heating, obviating the need for sophisticated greenhouses.
4. *Infrastructure:* Availability of inputs, skilled manpower and infrastructural facilities such as availability of engineering skill to repair the unit without losing time, should also form one of the major factors in selection of locality.

Optimising Size of Production

Maximum benefits from investments can be obtained by producing the maximum number of plants per unit capital and by proper management and labour availability. Whether, the facility is small or large (which are only relative terms), our aim should be to produce plants at competitive price levels.

The investments made in most of the capital equipment is not dependent on the volume of production. For example if we produce 1 million or 2 million plants, the investment made on a pure water system or autoclave remains constant. Large-scale production necessitates investment in certain facilities or equipment (such as change room and horizontal double-door autoclave) either to reduce the contamination or to increase the efficiency of operation. Such investments are beneficial to the industry in the long run. With increased volume of production the capital burden per plant will be minimal, reiterating that micropropagation is a production emphasised industry with the ability to produce plants well beyond the demand.

The advantage of large-scale production is that the range of crops micropropagated can be larger compared to the small-scale production unit with restricted range of crops. A wide range of products hold the price of produce stable, assuring higher profits for the large-scale units. Mass production of restricted crop range would result in heavy competition and slashing of prices well below the break-even point.

Available Facilities and Production Potential

Financial capability and market potential always decide the size of the production facility. Our next step is to arrive at a combination of various facilities or equipment that would go hand in-hand. The important items of investment that need careful study are growth room, laminar airflow cabinets and autoclave. These will decide the peak accumulated culture capacity to get the maximum growth-room space requirement. As demand for tissue-cultured plants is greater in certain months, therefore, one should take into account the peak growth-room space requirement of these months. The number of cultures handled to achieve such accumulated growth-room capacity will give the idea to install the number of laminar airflow cabinets or autoclaves in the unit; and these equipments should be subjected to use on shifts or 3-4 cycles in a day to reduce capital burden per plant. As this industry is fast growing and changing, enough thought should be given to further expansion or to adopt changes to produce a different kind of end product.

Economic Analysis of the Project

The economy of any micropropagation project is market driven and production emphasised and is affected by (Cumulative) volume of despatch, number of despatches and frequency of despatches. These factors are governed by market demand and beyond our control, however, the right combination of these factors will increase the profitability of the unit. An increase in the volume of despatch will increase the profitability throughout the year. The investment made in capital items can be minimised and manpower utilisation can also be uniform.

Though the tissue culture method of propagation is possible throughout the year, the market demand is not uniform and is characterised by peaks and falls. Many fruit and vegetable species which are being targeted by micropropagation businesses as crops for the future, have product delivery windows of only 2-3 weeks of the year. Even many flowering crops, for example *geraniums,* have production windows of less than two months for the deliver of 15-20 million units. Because of this erratic market trend it is necessary to have capacious growth rooms, extra autoclaves to prepare media in the peak work period and increased manpower during such peaks. In subsequent lean months, the facilities and manpower are underutilised. This calls for a constant search for a continuous market and efficient planning to produce the maximum number of plants, utilising the resources available.

Production Management

The implementation of any micropropagation project starts with the layout of physical and financial programmes in tune with the presumed production volume. The maximum possible production, per unit investment, can be possible only through efficient production management. The success of production management lies in understanding the art and science of propagation. Order processing, production planning and production monitoring are the main aspects of production management.

Order Processing

This needs macrolevel planning because all orders on hand are to be put into a master plan. This will to assess the requirement of growth-room space, cabinets, media and other consumables and manpower at a particular time, This also helps to know the potential production capacity and limitations of the laboratory. Orders should be accepted at least 6 months in advance (if the laboratory has enough starter cultures) and preferably with about 25% advance payment. The down payment indicates a commitment by the buyer to the laboratory in buying plants. It also enables the laboratory to meet a part of the requirement of recurring expenditure.

Planning for Production

Production planning forms the base for producing various crops, taking into consideration of the market demand, limited facilities of the laboratory and limitation of the nature of the plants themselves (that they multiply exponentially and ought to be subcultured at regular intervals). Tissue culture production forms the base for the entire plant production industry and hence all laboratory operations are time bound. Production planning tends to tap the maximum possible benefit from the existing market by the best use of available scarce resources of the laboratory. The production plan should be prepared for every individual crop, comprising all the crops of the laboratory. It is essential that producer should be aware of the nature of the end product which is to be delivered, e.g., microcurtings or ex-agar rooted plants. Therefore, it is essential to consider certain parameters like (i) multiplication rate, (ii) passage or transfer cycle, and (iii) operator efficiency. The plan should be finalised before considering these parameters.

Multirate

Currently, the *in-vitro* production methods are the extension of traditional vegetative multiplication methods. Therefore the expectation is to have a mutation frequency comparable to that of classical *in vivo* methods. Chances that variations will occur are much greater if the system is based on adventitious shoot formation and still greater with embryogenesis, callus and cell systems. Hence the multiplication ratio adopted should be the one that would give product plants having minimum variation and are repeatable in nature, i.e., the multirate is constant on each subculturing for a particular stage of plant. The system development of any crop should aim at arriving at an optimum multirate. Too low a multirate would not be economical and a high multirate, though it would help the laboratory to build up the stock in the initial period, would not be advisable as it would render the manpower insufficient to handle the huge volume of culture build up later.

It is important that on each culturing the cumulative multirate be adopted for further planning. Production planning prepared based on the multirate of the latest transfer alone would not be practical, as it might under or overestimate the subsequent volume of production.

The multiplication rate of four stages are to be borne in mind in working out the plan:

1. multiplication stage
2. multiplication to shooting stage
3. shooting to rooting or multiplication to rooting, and
4. Rooting to despatch.

Passage

The other important parameter in production planning is the transfer cycle or passage, which varies from crop to crop as well from stage to stage within a particular crop. The transfer interval should be repeatable and reliable. The optimum number of cycles required to produce plants with minimum variation is to be inferred from field experiments. For most of the crops the number of subcultures would be 15-20. Hence the initiation programme is to be meticulously planned to meet the periodic starter culture requirement of the production division.

Whenever the passage is shorter, more stress falls on manpower as frequent subculturing is required. On the other hand, whenever the passage is longer, more stress falls on growth-room, space.

During the peak season of production of shorter passage crops (say *gerbera)* we can manipulate the production by increasing the number of shifts or number of operating hours of the laminar airflow bench.

But if crops of longer passage are chosen (for instance lily), the requirement of growth-room space would be a limiting factor, which cannot be manipulated as in the previous case. We are left with no other option but to have capacious growth rooms. Hence a combination of crops having various transfer cycles is to be chosen for effective utilisation of cabinet space and growth-room space throughout the year.

The transfer cycle of the following stages is to be borne in mind in working out the plan.

1. During multiplication stage
2. During shooting stage
3. During rooting stage

Once we know the stage wise multiplication rate and transfer cycle of a particular crop, we have to work out a plan backwards, from the orders received, to check whether we have enough starter cultures or not. If we do not have enough starter cultures, we should plan forward and inform the customer about the quantity we can deliver in the periods of his requirement.

Operator Efficiency

After finalising the individual crop it is essential to calculate the requirements of manpower in handling, multiplying, shooting and rooting cultures.

Production Monitoring

Subsequent to the formulation of the annual plan, weekly work schedules should be prepared for implementation. After the completion of work, a statement of completed work is prepared and compared with the original plan. The deviation from the original plan in terms of cumulative culture position (growth-room stock), culture handled, efficiency of operators, multiplication rate, contamination etc., are constantly monitored by the date-processing section.

Frequent small experiments are to be conducted along with the mass production, to fine-tune the production system. If proper monitoring is not done or efforts are not made to smoothen the production process, deviations may occur, coming as a great shock. As micropropagation forms the base for the entire crop production, the laboratory should inform the buyer frequently of possible changes in the delivery volume or the time of delivery, according to the availability of stock.

Contamination: Economic Assessment

Contamination is a serious problem plaguing the micropropagation industry. Losses can be ascribed to carry-over of inoculum on the explant surface, failure of contamination control system in the laboratory, infestation with mites and thrips or poor sterile technique. More serious losses could occur in stage 2 as compared to stage 1 where the value of inputs is very high. Large-scale losses of this kind are most likely to be a consequence of poor monitoring of production combined with the clonal multiplication of sublimally infected stock. Production should only be based on multiplication of axenid stocks and should involve continuous monitoring. Whether an axenic status can be achieved for all the crops at an economic cost is still in question (Cassells, 1991).

Calculation of Cost Price

With a variety of plants on the production line, each exacting various production parameters, it if necessary that the producer should understand per plant production cost and various strategies to lower the (per plant) production cost. Wary eyes should be kept on the costs of inputs whose change will greatly reflect on the cost of the end product, such as the price of agar.

Quality Control in Commercial Plant Tissue Culture

The term quality is defined as *fitness for purpose.* In the commercial manufacturing environment it is important that the quality of the product match the expectations of the eventual customer. The object of quality control, therefore, is to produce a product of quality that will reflect to the following aspects:

1. satisfy the customer's need,
2. be deliverable according to scheduled requirements.

Thus it is clear that the object of quality control in the manufacturing process is not to produce an item of the highest possible quality, but, rather to satisfy the customers needs.

The concept of quality within the environment of a plant tissue culture laboratory has been largely overlooked for a number of years. This is probably due to the fact that commercial plant tissue culture has evolved from fundamental research into plant propagation systems. Examples of this are the salt mixtures which were originally developed to study the growth and development of tobacco (Murashige and Skoog, 1962; Linsmaier and Skoog, 1965). Such salts still form the basis of many media formulations and only a limited number of media have been formulated specifically for certain plants, e.g. rhododendron.

The research and development of plant propagation systems is often a complicated process wherein each stage presents a number of problems to be overcome:

1. efficient sterilisation of explants,
2. initiation of multiplying cultures,
3. formation of rooted plantlets,
4. establishment of weaned plantlets.

The results from micropropagation research projects are often published but their direct transference to a commercial plant propagation system is seldom possible without further development and refinement.

In a commercial environment a micropropagation system must be capable of producing a product of the desired quality. A number of factors come into defining the plant quality but basically the plant must satisfy the customers requirements. It is not sufficient for the laboratory to define its own quality standards; it must work closely with its customers to ascertain exactly what they require of the plant; otherwise their own standards may be too low or too high.

If the laboratory sets too low a standard of quality, the customer will go elsewhere to find a more suitable, better quality product.

If the laboratory sets too high a quality standard, the customer may be very pleased with the product but may not be willing to pay the increased price required to cover the increased production costs associated with very high-quality standards.

Obviously, there is a very wide difference between the aims of the research micropropagation laboratory and the commercial production laboratory. The production laboratory has to be able to produce a plantlet of sufficient quality within the constraints of the commercial operating environment.

❑❑❑

CHAPTER-10

Biotechnological Methods of Crop Improvement

Plantation of crops meet human requirements for food, timber, medicines, spices and beverages. It is necessary to increase the out put of crops several fold per unit area because of increasing human population associated with shrinking cultivated land area.

Crops plantation present certain unique problems for the plant breeder in terms of their improvement. Biotechnology is an important tool for rapid multiplication of several economically important crop plants which can be of immense use in the multiplication of true-to-type high yielding plants on a large scale for planning. There are several methods applied to achieve the target are dealt here.

THE GREEN REVOLUTION

Remarkable increase in productivity is also one of the achievements of what come to be known as green revolution and it could be possible by sincere efforts of agricultural scientists, specially plant breeders, agronomists, plant pathologists, and entomologists. After the Second World War, research work on the selection of new high-yielding varieties (specially of wheat) was started in Mexico. Many rice varieties were developed in Philippines. In sixties, new cultivars were disseminated all over the World which had increased remarkable agricultural production and productivity. Cross-breeding between many crop varieties and local hardy breeds made it possible to obtain cultivars that were even more adapted and which gave the better yield. The research work has also been done on millet and sorghum, triticale, maize and several species of legumes. This was the first phase of green **revolution.**

Mid 1970s was the second phase of green revolution. During this period the traditional methods of crop improvement like selection and cultivation of new plants, resistant to disease and drought, etc. were in practice but new research techniques were also emerged using cell, protoplast culture, and genetic recombination (gene transfer). These techniques helped to create cultural varieties by controlling the molecular and cellular mechanisms responsible for biological diversity. The definition of biotechnology with reference to crop improvement can be extremely broad and could be considered to include the general somatic and genetic manipulation by selective methods.

In India, the improvement of productivity of the major food, fibre and other crops has been impressive. The success is due to intensive research in plant breeding resulting in the development of new varieties with very high genetic yield potentials and suitable for wide range of agro-climatic conditions.

In broad sense, biotechnology has two impacts on the development of agriculture:

1. First with the improvement of genetic engineering techniques, the time for generating and evaluating new germplasm can be drastically reduced.
2. Secondly, chemical and biotechnological processes have opened new frontiers to products of improved value from agricultural raw materials.

Breeding Objectives

The problems come into focus when we examine the objectives like to increase yield, improve quality and reduce costs. The last item identify limiting factors in the many processes that contribute to yield. For example short straw has led to greatly increased harvest index in cereals by partitioning more fixed carbon into grains and less into stems and leaves. Increases in net rates of photosynthesis are a common objective, but in practice there have been few successful applications of selection for higher rates of photosynthesis that have resulted in improved varieties.

Yield and quality are interrelated in the sense that the breeder wishes to maximise both in the harvested product. When it is possible to control the expression of genes that direct the synthesis of particular products, such as the endosperm storage proteins in a grain crop, the amount of these products could be increased either by increasing the copy number of the genes or by altering their control. The excitement of molecular biology is that it provides new tools and ways of exploring the black boxes represented by our crop plants.

Basic Problems in Development of Useful Crop Varieties

Transfer of genes between plant species has played an important role in crop improvement for many decades. Useful traits such as resistance to disease, insects, and stress have been transferred to crop varieties from non cultivated plants. Recombinant DNA methods greatly extend (even outside the plant kingdom) the sources from which genetic information can be obtained for crop improvement. Genetic transfer system based on recombinant DNA are available for several crop species and are under development for others. The concerted use of traditional and more recent methods for plant genetic manipulation will contribute to crop improvement.

Improvement of Existing and New Cultivated Plants

Biotechnology plays two roles in the development of plant varieties. First, by the development of plant cell culture and gene analysis the period of conventional plant breeding is drastically-reduced. Secondly it is increasingly possible by genetic engineering to change the qualitative characteristics of plants in a purposeful way. Till date most work has been conducted with only about 20 plant species, most of them dicotyledons of less agricultural significance like tomatoes, rapeseeds, cotton and tobacco. Current efforts to support plant breeding by biotechnological methods, in general, pursue three following aims:

1. Higher yields are sought especially by introducing plants with resistance to diseases, pests and herbicides. Diagnostic methods based on DNA hybridization or monoclonal antibodies for proving plant diseases in the soil or the plant (viruses) belong to this category. Attempts to transfer genes from nodular bacteria into plants to develop nitrogen fixing in family Gramineae, are typical examples.
2. The improvement in quality of useful plants by increasing their nutritional value, taste or chemistry. This includes, e.g., the increase in the proportion of certain essential ammo acids like lysine in barley, the amount of fibre (as in tomatoes and flex), but also in the increase in amounts of chemically interesting compounds like oleic acid in sunflowers.
3. The development of new methods in plant protection. In this category falls, e.g., increasing frost resistance of useful plants by means of genetic engineered bacteria, or an increase in resistance to insects by building in defense mechanism from beneficial plants on the basis of insect viruses.

Plant Breeding and Concepts of Crop Improvement

Plant breeding as a science began in 19th century with discoveries how plant traits are inherited (Allard, 1960). In simple terms, plant breeding is the selection of plants with desired traits after the sexual exchange of genes by hybridization between two genotypically different parents. Here such parents may belong to the same variety, different varieties of the same species, different species of the same genus or species from different genera. Hybridization may be classified into following three classes:

1. Intervarietal or intraspecific hybridization
2. Interspecific hybridization
3. Intergeneric or distant hybridization

Breeders have expanded their search for new genetic variation to the entire crop species through new biotechnologies including their wild relatives. The intervarietal hybridization is common for transferring superior characters from one another in two varieties belongs to same species. In case of distant hybridization where crosses between different species of the same genus or of different genera are attempted in order to get specific character into the existing variety. Generally, the objective of such crosses is to transfer one or few simply inherited characters like disease resistance to a crop species. Sometimes, interspecific hybridization may be used for developing a new variety.

The definition of plant species rests on the concept of genetic isolation. Nevertheless, sexual exchange of genes between species can and does occur in nature without a plant breeder, e.g., such transfer has been documented between maize *(Zea mays)* and teosinte *(Zea mexicana)*.

Methods for Crop Improvement

Plant Tissue Culture Techniques

Plant tissue culture techniques are used to select somaclones that are resistant to the herbicide. Plant cells growing in culture have been shown to have varying genotypes due to growing under culture conditions. The problems associated with this technique are that the herbicide has to have the same affect on plant tissue growing in culture as it does on tissue in the intact plant, and the culture be able to regenerate whole plants so that the resistance can be transferred to desirable commercial varieties of the crop. The examples in which herbicide resistance has been selected for tissue culture, has been confined to a small number of species (i.e. tomato, tobacco, and carrot) that have limited commercial potential.

Cell Fusion method

Cell Fusion method was developed in 1960s. It includes the preparation of large numbers of single plant cell stripped of their cell wall (protoplasts). Protoplasts from diffeient species can be induced to fuse-by exposure to certain chemicals or electric current. This results into tissue from which a whole plant can be regenerated. The concept is based on the principle to combine the chromosomes of species that are sexually incompatible or, as a short cut, to combine the nuclear genome of one species with the cytoplasm (i.e., the organellar genomes) of another. Though much work has been done in this direction but commercial utilization is still in primary stage.

Cell fusion methods may be used in creating new crop varieties bearing the nuclear genome of one species in the cytoplasmic background of another (nuclear transfer) or in the mixed cytoplasm with organelles from both species (cybrids). The DNA of the cytoplasmic organelles (chloroplasts/mitochondria) encodes some of the proteins that make up the structure and metabolic machinery of the organelle few agriculturally important traits are the products of interaction between the nuclear and cytoplasmic genomes. For example, a form of male sterility that is useful in commercial production of hybrid seeds results from nuclear-mitochondrial interactions. Well studied examples include species of tobacco *(Nicotiana* spp.) and combinations of rapeseed *(Brassica napus)* nuclear genomes with cytoplasms from radish *(Raphanus sativus).*

Somatic Cell Hybridization

Somatic Cell Hybridization or Cybrid or Cytoplast is the process in which two protoplasts along with their nuclei fuse then a true hybrid can be formed. However, often the two nuclei will survive independently in the mixed cytoplasms, forming a heterokaryon. This product is called a **cytoplast or cybrid.** Cybrids can also be prepared by fusing a normal protoplast with an enucleated protoplast. The study of cybrids is also important for investigation of possible recombination within these extra-chromosomal genomes.

Somaclonal Variation

Somaclonal variation is the method refers to heritable changes which accumulate in the callus from a somatic explant and express in the progeny of in vitro regenerants obtained from callus. The important aspectsd are:

1. the frequency of variation seems to be far greater than the yield of induced mutations;
2. the changes are very subtle and may not involve drastic altration in the genetic background;
3. somaclonal variation occurs for trait of both nuclear and cytoplasmic origin. The variation of cytoplasmic genes obtained by this method is a distinct advantage; and
4. in wide crosses somaclonal variations provide a mechanism of gene introgression. Immature embryos of the wide cross can be callused and plants with the introgressed

desired gene (or gene complex) are selected among the regenerants of their progenies (Chopra and Sharma, 1988).

Organ Culture Techniques

Organ culture techniques have been used to culture isolated embryos. The conditions used are designed to supply the life support for the hybrid embryo, which the embryo obtain from endosperm. The younger immature embryos that can be cultured in vitro usually are those that show observable signs of differentiation.

Embryo developed in distinct crosses, suffers from post-fertilization developmental blocks (Chopra and Sharma,1988). This prevents the formation of normal seeds. The hybrid embryo can be rescued from damage/collapse by growing it in artificial medium (tissue culture technique). Following are the methods for **embryo rescue:**

Table 10.1 Example of agricultural important genes and traits transferred to crop plants by interspecific or intergeneric hybridization. Though selective, the examples given are representative of the plant families in which such transfers have been most successful. The two families dominating the list are the Gramineae (wheat, oat, rice, and maize), and the nightshade family, Solanaceae (tomato, potato, and tobacco).

Crop species	Donor species	Trait
A vena saliva (oat)	*A. sterilis*	Increase yield 25-30%
Beta vulgaris	*B. procumbens*	Sugarbeet nematode resistance
Brassica napus	*B. campestris*	Club root resistance
Cucubita pepo	*C. lundelliana*	Mildew resistance
Gossypium hirsutum	*G. tomentosum*	Nectorless (decreased incidence of boll rot)
Gossypium hirsutum	*G. raimondii*	Rust resistance
Lycopersicon esculentum	*L. birsutum*	Bacterial canker resistance
Lycopersicon esculentum	*L. peruvianum*	Nematode resistance
Lycopersicon esculèntum	*L. peruvianum*	Jointless (facilitates clean fruit harvest without stems)
Lycopersicon esculentum	*L. peruvianum*	TMV resistance
Lycopersicon esculentum	*L. pimpinellifolium*	Fusarium wilt race I resistance
Nicotiana tabacum	*N. glutinosa*	TMV resistance
Nicotiana tabacum	*N.longiflora*	Blackfire resistance
Oryza saliva (rice)	*O. nivora*	Grassy stunt virus resistance
Ribes nigrum	*R. sanguineum*	Mildew resistance
Ribes nigrum	*R. grossularium*	Gall midge resistance
Solanum tuberosum	*S. acaule*	potato virus resistance
Solanum tuberosum	*S. demissum*	Late blight resistance, leaf roll resistance, potato virus y resistance
Solanum tuberosum	*S. stoloniferum*	Late blight field resistance, potato virus A resistance, potato virus Y resistance
Triticum aestivurn	*Aegilops comosa*	Strip rust resistance
Triticum aestivum	*Aegilops ovata*	High kernel protein
Triticum aestivum	*Aegilops speltoides*	Stem rust resistance
Triticum aestivum	*Aegilops squarrosa*	Leaf rust resistance
Triticum aestivum	*Aegilops umbellulata*	Leaf rust resistance
Triticutn aestivum	*Aegilops elongatum*	Leaf rust resistance, drought tolerance
Triticum aestivum	*Secale cereal*	Yellow rust resistance, powdery mildew resistance, winter hardiness, leaf rust resisrance, stem rust resistance
Triticum aestivum	*T. monococum*	Stem rust resistance
Triticum aestivum	*T. timopheevi*	Stem rust resistance
Triticum aestivum	*T. monococcum*	Stem rust resistance
Zea mays (maize)	*Tripsacum* spp.	Northern corn leaf blight resistance

1. *Embryo culture:* This culture technique help in development of normal hybrid seedlings from enbryo and has been practiced in a number of interspecific crosses. For example in legumes, crosses involving *Phaseolus vulgaris* x *Phaseolus vitensis, Arachis hypogaea x A. montiola* or *A. glebrata* embryos have been successfully rescused and seedlings established from them.
2. *Embryo implantation:* The nutritional requirement of every plant-species in tissue culture is varied, and therefore, requires a standaridization in each case. This limitation can be overcome by transplanting hybrid embryo, before their collapse into a normal endosperm. This technique has successfully employed in *Hordeum* x *Triticale, Hordeum* x *Agropyron* and *Hordeum* x *Secale.*
3. *Ovule culture:* This is another method to overcome the difficulties of arriving at the right kind of complex nutrient medium for culturing hybrid embryos and for avoiding the difficulty of dissecting out uninjured embryos at the right stage for the purposes of transplantation. Cotton is a good example for a crop plant in which ovule culture has been used to rescue hybrid embryos.
4. *Ovary culture:* Culture of ovaries is also in practice. In crosses of *Brassica campestris* x *Brassica oleracea,* seeds with well developed embyros have been obtained when ovaries were cultured four days after pollination in White's medium containing caesine hydrolysate.

After a hybrid plant has been sucessfully recovered, differences in the number or compatibility of parental chromosomes may cause sterility. Manipulation by cytogenetic techniques have been found fruitful in obtaining stable gene transfer.

Through Genetic Engineering

In this technique gene expressing herbicide resistance is transferred to a previously susceptible species. If the gene product is expressed in the recipient species then it should become resistance to the herbicide. The problem with this approach is that a gene must be identified that will confer herbicide resistance to a plant. For most herbicides the actual site of action of the herbicide is unknown, which makes it extremely difficult to isolate a gene coding for a resistant enzyme. An alternative approach is to transfer a gene into a crop that codes for an enzyme that detoxifies the herbicide. Unfortunately most of these enzymes are not known and many detoxification systems involve several different enzymes working in concert, making it even more difficult to transfer this trait into a crop. Another problem with the genetic engineering approach, besides the expression of a new gene in a plant, is that the techniques have to be available to transfer a gene into a plant. Currently the only system available works exclusively in dicots and not in monocots.

1. *Intergeneric Gene Transfer:* There is evidence that some of our modern crop species, e.g., rapeseed *(Brassica napus),* tobacco *(Nicotiana tabacum),* and wheat *(Triticum aestivum),* originated in nature by hybridization between different species or genera, e.g., the ancestor of *B. campestris.* Similarly most important hybridization is between the species

from the genera *Secale* (rye) and *Triticum* (wheat) to create a new cereal crop, *Triticosecale* (triticale).

It is very difficult to use natural barriers for interspecific and intergeneric hybridization. Successful gene transfer is made by pollination of the flowers of one of the two species with pollen from other. As a result of fertilization, embryo is produced. Development of the embryo and the endosperm associated with it gives rise to a mature seed which after germination produces a hybrid plant.

Gene Transfer by Manipulating DNA Directly

The term transgenic is used to describe plants which have had DNA introduced into them by means of other than by the transfer of DNA from a sperm cell to an egg. The DNA transferred by these non-sexual processes may be DNA from same cultivator or crop, from another plant species, from an animal or from a prokaryot.

The methods for transferring DNA directly from one organism to another have established because DNA was confirmed as the chemical basis of genetic inheritance. The direct transfer of DNA into protoplast was found to be most successful.

Nonsexual DNA transfer techniques make possible manipulations that are outside the reach of breeding or cell fusion techniques. Genes can be accessed from exotic sources plant, animal, bacterial, even viral and introduced into crop. Because the DNA elements which control the expression of genes, can often must, be modified for proper function in the new host, it is possible to control timing, tissue specificity, and expression level of transferred genes. It is now possible to reprogrammed the endogenous plant genes by the process of reintroduction of the engineered gene. Therefore, through nonsexual DNA transfer methods we can increase the sources of variability available for crop improvement to include all living things, and can also manipulate to get quantitative control over gene expression.

Agrobacterium-mediated *Gene Transfer*

Fraley et al. (1983) and Am et al. (1985) exploited the natural ability of *Agrobacterium tumefaciens* to transfer DNA into plant chromosomes. *A. tumefaciens* is a plant pathogen which transfers a set of genes encoded in agerion celled T-DNA of the Ti-plasmid into cells at wound positions. The pathogen has a wide range of hosts among higher plants, including many dicotyledonous crop plants. The genes are usually transferred into a tumorous growth called crown gall in which T-DNA is stably integrated into host chromosomes. The tumor phenotype, which can be maintained indefinitely in tissue culture, results from the expression of genes on the T-DNA which changes the normal balance of growth substances (phytohormones) in transformed cells. The ability to cause gall disease can be eliminated by genes in the T-DNA without loss of DNA transfer and integration functions; an *Agrobacterium* strains that does not cause disease is said to be disarmed. In a disarmed strains, the DNA to be transferred is attached to broder sequences that define the end points of an integrated T-DNA.

Under laboratory conditions, disarmed *Agrobacterium* strains are used to transfer genes to prtoplasts with partially regenrated cell walls, suspension cell cultures, leaf pieces, and stem segments. Gene transfer by means of engineered *Agrobacterium* strains is very common now a days specially in the plants of family solanaceae such as tobacco,and petunia. Though such transfer in other families is very difficult, however, in soybean gene transfer has been demonstrated in cultured cells (Facciotti et al., 1985), but the ability to regenerate a complete plant from cultured cells containing T-DNA has not yet been reported. Data are available which reveal that *Agrobacterium* can transfer T-DNA to monocotyledons hosts (Hernalsteens et al., 1984).

Direct DNA Transfer Technology

It is now possible to transfer *purified* DNA into plant, either by direct DNA uptake or by microinjection. Direct DNA uptake involves physiochemical reactions which result in DNA transfer to protoplasts. The mechanical introduction of DNA into cellular compartments with microscopic pipettes known as **microinjection.** The direct gene transfer methods are not subject to host range restrictions, but practically are limited by the need to recover a whole plant from the target cells or tissue. The reports are now available which demonstrate that plant protoplasts can take up nucleic acids directly from the *culture* medium. This phenomenon was first demonstrated with viral RNAs, According to Krens and Schilperoot (1986), the treatments, such as polyethylene glycol (PEG) and the application of electric pulses (electroporation), which increase the permeability of membranes, can result in the transformation frequencies of one transformant per thousand protoplasts.

Microinjection

This is the most recent techique and is the useful addition to the repertoir of plant transformation methods, involves the introduction of DNA solutions under pressure into plant protoplasts by means of micropipettes. Crossway and coworkers (1986) presented a review on this technique. The successful transformation has been the development of methods for the immobilization of cells during injection and methods for their subsequent culture. In one study conducted by Crossway and his associates demonstrated that cell lines cultured from microinjected tobacco protoplasts were shown to have integrated the foreign DNA sequences into the nuclear DNA; the average transformation frequency depended on whether the injection was intranuclear (14%) or cytoplasmic (6%). 'Reich, Lyer and Miki also transformed cell lines cultured from intranuclear injection of alfalfa protoplasts and were identified by screening of enzyme activity encoded by the foreign DNA. This technique is in its infancy of practical knowledge and still we need more information in this regard. The main reason is that microinjection technique is the physical method of introducing DNA, and it should be capable of delevering genes into targets other than protoplasts. However, this method in principle be used with any crop species from which whole plants can be obtained from single transformed cells.

Virally Mediated Gene Transfer

This is based on the knowledge of viral-based gene expression systems for animals, both for experimental and therapeutic uses. Similar efforts have been made to develop vectors based on plant viruses for gene transfer into plants.

Gene-controlled Crop Yield

A number of characters, like yield, are controlled by many different genes of individually small effect. While it is possible that such genes may be manipulated in blocks, being transferred as chromosomes or parts of chromosomes, there is little or no practical application of such approaches because of the difficulty of analysing them and following their inheritance.

Restriction Fragment Length Polymorphisms

Restriction fragment length polymorphisms (RFLPs) may well provide a more precise tool for the breeder to detect such gene blocks in segregating progenies (Burr et al., 1983). In this way selection could be based in part on a characterisation of the genotype without the need to allow it to develop a phenorype.

Detection of Nucleic acid Sequence

The detection of nucleic acid sequence in plant material is also useful in identifying plants that carry virus or other infections. The cDNA probes are a cheaper and more easily produced means of detecting single-stranded RNA viruses than specific antisera used in the enzyme-linked immunosorbent assay system. Nucleic acid probes are also useful in discriminating cytoplasmic male sterile phenotypes in maize breeding programmes. In this instance the probe detects the presence of specific DNA sequences in the mitochondrial genome of the male sterile which are either different or absent in the normal plants. Recent reports that T-DNA can be found in the chloroplasts of some plants transformed with *Agrobacterium tumefaciens* suggests further possibilities of targeting transforming DNA to organelle genomes.

The catalogue of genes that are potentially useful in agriculture, that are being isolated and cloned, is steadily increasing. It includes genes for storage proteins in maize, wheat, peas and beans; genes for the large subunit and the small subunits of rubisco from a range of plants including algae, blue-green algae and several crop plants; genes for a variety of enzymes expressed in plants, including alcohol dehydrogenase in maize, amylase in wheat, sucrose synthetase in maize, and genes involved in phenol propanoid synthesis in Antirrhinum and other plants. Genes of bacterial origin that are of interest include those for nitrogenase and other processes involved in nitrogen fixation, and a gene for the insecticidal endotoxin polypeptide produced by *Bacillus thuringiensis*. Genes for antibiotic resistance, such as chloramphenicol transacetylase and neomycin phosphotransferase are useful markers for selecting transformed cells on the basis of their resistance.

A Shopping List of Genes

While RFLPs could one day greatly simplify the selection of polygenic characters, most breeders would probably agree that relatively simple probing methods to discriminate between different nuclear or cytoplasmic genotypes will be useful.

ORGANELLAR DNA AND CROP IMPROVEMENT

Bowman and Bingham demonstrated and compared organellar DNA restriction digests to show that the wheat variety Rendezvous, with resistance to eyespot *(Pseudocercosporella herpotrichoides)* derived from *Aegilops ventricosa,* has a chloroplast genome derived from *Aegilops* and not from *Triticum.* This information is very useful to the breeder concerned with cytoplasmic or organellar contributions to the phenotypes of his breeding lines. Probes that detect viral RNA in potato, chromosome IB with a short-arm from rye in European wheats, rapid methods for discriminating between mitochondrial DNAs and to identify cytoplasmic male sterile lines of maize provide further examples.

Nasrallah (1985) has shown that a cDNA clone prepared to stigma mRNA of *Brassica oleracea* can be used as a probe on DNA digests of segregating populations to recognise the *S alleles* that govern sporophytic self-incompatibility in *Brassica.* This work should soon lead to the transformation of *Brassica* with genomic DNA for specific S alleles and may better facilitate the production of hybrid seed in *Brassica* by using self-incompatibility.

GENES FOR DISEASE RESISTANCE

Genes for disease resistance so far remain elusive. There is still insufficient information on the nature of their products and mode of action to design direct methods. However, many microorganisms and certain plants produce proteins that are toxic to specific plant pathogens, both microbial pathogens and insect that feeds on plants.

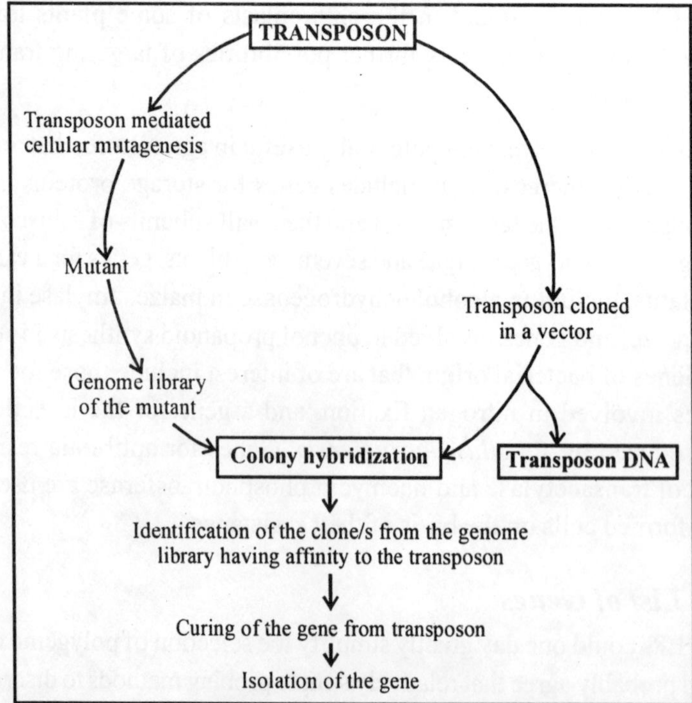

Diagram 10.1 Use of transposon mutagenesis for identification and isolation of a gene.

One aim of plant genetic engineering is to transfer the genes encoding these proteins toxic to economical plants with the hope that expression of toxic genes in these plants will provide biological control of atleast some serious plant diseases and insect pests. At present, the various diseases and the insect pests are controlled by the use of broad based chemicals, pesticides like insecticides and fungicides. However, they are causing damage to ecosystem and pollution to not only ground water but various fauna and flora.

A second approach is to explore the co-ordinated control of protein biosynthesis that occurs either when resistance is evoked by an avirulent pathogen or when a similar reaction is evoked by a chemical or physical treatment.

In investigating resistance genes, one can also follow another method in which the protein specific to resistance are sought for, the mRNA that is responsible having then to be found. Once in the possession of this a cDNA can be set up by means of the reverse transcriptease enzyme. This cDNA will then have in its nucleotide sequence the genetic information for the resistance that is being sought.

Transposon

A third approach is to use the method of transposon tagging. In brief this makes use of a mobile DNA element called a transposon. When the transposon jumps into an otherwise functional gene it has the effect of blocking correct transcription and produces a mutation of that gene. If the DNA sequence of the transposon is known and it is present in the plant in low copy number, probing DNA restriction fragments of the mutant will recover those which carry the transposon. Some of these will be flanked by DNA sequences of the mutated gene. Experiments are in progress in several laboratories to use maize transposons in an attempt to find mutations from resistance to susceptibility to several maize pathogens.

Transformation or DNA Delivery System of Agrobacterium tumefaciens

It is now common knowledge that the DNA delivery system *of Agrobacterium tumefaciens* can be harnessed to introduce foreign genetic information into those plants that are natural hosts of this bacterial pathogen. These plants are all dicotyledonous, but recent claims for transforming *Asparagus* suggest that progress may be made with other monocots including the cereals. Although naked DNA can be prepared and used to transform protoplasts, either by natural uptake methods or by electroporation, our present inability to regenerate and recover whole plants from cereal protoplasts is, for the time being, an obstacle (Flavell and Mathias, 1984).

Site Specific Mutagenesis

Plant breeders have attempted for several decades to develop varieties with increased lysine, tryptophan and methionine content particularly in cereal seeds loking to the importance as human and animal food. In corn, mutants such as *opaque-2, sugary- 1 , flory-2* have increased amount of lysine and/or methionine in seeds, but these mutant strains have undesirable soft kernels and produced lower yields. Several maize genes encoding Zeins have now been cloned and sequenced by site-specific mutagenesis for more lysine codon into Zein sequences. These high-lysine Zein coding sequences could be joined to strong promoters such as the CaMV35S promoter and reintroduced into maize plants by transformation by means of **electroporation or a microprojectile gun.**

Breakthrough in the Development of Neurotoxin-Free *Lathyrus sativus*

Lathyrus sativus, a legume, has unique features of tolerance to drought, and flood conditions. Cultivation of this crop is banned because of the presence of toxin BOAA or ODAP. The plant biotechnological methods could not be applied earlier in the absence of in vitro regeneration methods. A gene has been isolated from *Pseudomonas* strain which can degrade toxin BOAA.

Development of Salt Tolerant Varieties

Salinity is a problem in a very large area of the world. In order to extend intensive cultivation to marginal environments, the problem is to be faced and it has been reiterated time and again that solution shall come from an integrated or interdisciplinary approach. In that pursuit information has been generated through working with prokaryotic unicells, eukaryotic algae and higher plants. Osmoregulation with organic and inorganic solutes, cellular comparrmentation, vacuolar sequestering of ions, patch clamp studies, acidic proteins and genetic engineering of plants are some of the dimensions which represent an interface of basic and applied knowledge. The achievements which mainly concern functional mechanisms of salt tolerance, and their possible exploitation through conventional as well as modern biotechnological tools for crop improvement under saline conditions are of paramount importance. Table 10.2. presents the list of salt tolerant agricultural crops.

Applications of Tissue culture in Increasing Salt Tolerance

In the context of inducing or increasing salt tolerance of crop plants it would be appropriate to mention that tissue culture has been employed as a method to select salt tolerant lines which may not be an induction of salt tolerance *per se* but may represent an identification of salt tolerance genes in the form of isolation and characterisation of mutant cell lines and plants.

Table 10.2 Salt tolerance of agricultural crops (Maas, 1984).

Crop	Maximum soil salinity without yield loss (threshold) dS/m	% Decrease in yield at soil salinities above the threshold, % per dS/m
Sensitive Crops		
Bean	1.0	19
Carrot	1.0	14
Strawberry	1.0	33
Onion	1.2	16

Crop	Maximum soil salinity without yield loss (threshold) dS/m	% Decrease in yield at soil salinities above the threshold, % per dS/m
Almond	1.5	19
Blackberry	1.5	22
Boysenberry	1.5	22
Plum: prunet	1.5	18
Apricot	1.6	24
Orange	1.7	16
Peach	1.7	21
Grape fruit	1.8	16
Moderately Sensitive Crops		
Turnip	0.9	9.0
Radish	1.2	13
Lettuce	1.3	13
Clover, berseem	1.5	5.7
Clover, strawberry	1.5	12
Clover, red	1.5	12
Clover, alsike	1.5	12
Clover, ladino	1.5	12
Foxtail, meadow	1.5	9.6
Grape	1.5	9.6
Orchard grass	1.5	6.2
Pepper	1.5	14
Sweet potato	1.5	11
Broad bean	1.6	9.6
Com	1.7	12
Flax	1.7	12
Potato	1.7	12
Sugarcane	1.7	5.9
Cabbage	1.8	9.7
Celery	1.8	6.2
Com (forage)	1.8	7.4
Alfalfa	2.0	7.3
Spinach	2.0	7.6
Trefoil, big	2.0	19
Cowpea (forage)	2.5	11
Cucumber	2.5	13
Tomato	2.5	9.9
Broccoli	2.8	9.2
Vetch, common	3.0	11
Rice	3.0	12
Squash, scallop	3.2	16
Moderately Tolerant Crops		
Wild rye, beardless	2.7	6.0
Sudan grass	2.8	4.3

Crop	Maximum soil salinity without yield loss (threshold) dS/m	% Decrease in yield at soil salinities above the threshold, % per dS/m
Wheat grass, std. crested	3.5	4.0
Fescue, tall	3.9	5.3
Beet, red	4.0	9.0
Harding grass	4.6	7.6
Squash, zucchini	4.7	9.4
Cowpea	4.9	12
Soybean	5.0	20
Trefoil, birdsfoot	5.0	10
Ryegrass, perennial	5.6	7.6
Wheat, durum	5.7	5.4
Barley (forage)	6.0	7.1
Wheat	6.0	7.1
Sorghum	6.8	16
Tolerant Crops		
Date palm	4.0	3.6
Bermuda grass	6.9	66.4
Sugarbeet	7.0	5.9
Wheat grass, fairway crested	7.5	6.9
Wheat grass, tall	7.5	4.2
Cotton	7.7	5.2
Barley	8.0	5.0

Many terms have been used to described cell lines and regenerated plants isolated for salt tolerance, including adapted, resistant, selected and tolerant. The term *adapted* implies that the cells are able to grow in the stress environment but are not essentially genetically different. Generally used synonymously, the terms selected, tolerant and resistant imply that cells were exposed to stress and survived or performed better than the majority of the population, but again are not different genetically. Flick (1983) described a true mutant as a cell with a stable change in its genetic make-up, meeting the criteria of low frequency, stability in stress free environment and through regeneration and the heritability of mutation. Inheritance of a trait by progeny of regenerated plants is the most conclusive evidence of a true genetic change.

A shift toward halophytic nature has been suggested during the process of in vitro culture of salt sensitive plants (e.g. legumes) over medium containing high salt concentrations and their subsequent sub-culture over salt free medium. Such cells or lines which have been gradually adapted to the salt containing growth medium, have been found to retain their salt tolerance in salt free medium and to regenerate into salt tolerant plants. Besides, tissue culture techniques offer a unique opportunity to improve our understanding of the mechanisms and physiology of salt tolerance at the cellular level and to apply molecular techniques as a means to generate transgenic plants.

The genetic stability of the plants obtained through in vitro techniques has been referred as somaclonal variations. Such variations for salt tolerance have often been tried by plant breeders to select for attributes not expressed in the original material, but the success is delimited because one has to remain in the species specific genetic reaction norm, mutations included. Using tissue culture gene technology the limitation of species specific reaction norm will be annulated. Gene transfers not only from plant species to plant species but even from bacteria or animals to plants are possible which offer completely new facilities. Leaf disc transformation, protoplast based systems, direct and indirect pollen mediated transfers and methods to detect organ specific expressions of transplanted genes have been reviewed (Hess, 1990) in the context of transfer of agronomically useful genes like biological N_2 fixation, protein quality, virus cross protection, insect resistance and herbicide resistance. Drought and salt tolerance are the awaited candidates to be included in the list in order to realize a second green revolution. However, in the following paragraphs some cases are cited where an enhanced salt tolerance has been achieved through tissue culture.

Nabors et al. (1975) isolated cell lines of *Nicotiana tabacum* var. *Samsun* exhibiting continued growth in the presence of salt. Plants regenerated from cells tolerant to 6.4 gh^{-1} of NaCl also showed similar tolerance which was transmitted to next generation. Cell lines of *N. sylvestris* and *Capsicum annum* surviving in 10 and 20 gl^{-1} of NaCl were isolated. These cells lines retained their salt tolerance when sub-cultured for several passages in the absence of salt. Callus cultures of alfalfa selected for growth in presence of 10 gl^{-1} NaCl exhibited growth like a halophyte. A salt tolerant cell line isolated from haploid callus cultures of *Datura innoxia* was shown to retain its tolerance even after sub-culture in the absence of salt. Plantlets were regenerated from the tolerant cell line both in the presence and absence of salt. Several embryogenic cell lines of *Pennisetum americanum* were isolated which grew well in liquid suspension cultures containing 0.0058-1.16 per cent NaCl. These were maintained for a year through 50 passages in the presence of NaCl. Some of the cell lines retained their salt tolerance after sub-culture for several passages in the absence of salt. Hasegawa et al. (1980) reported that cells of *N. tabacum* Var. *Samsun* growing in presence of 10 gl^{-1} NaCl lost their salt tolerance on transferring to a salt free medium. It has also been reported that tobacco plants regenerated from salt tolerant cells had a higher survival rate in the presence of salt and the plants were reported to transmit tolerance to F_2 generation. Salt tolerant cell lines of pearl millet showed the ability to undergo somatic embryogenesis. Nabors et al. (1980) speculated that salt tolerance in regenerated tobacco was not carried by a single Mendelian gene.

Among physiological differences between salt tolerant and salt sensitive lines, both the cell types were similar in NaCl uptake to balance the osmotic challenge of salt in medium, but presumably tolerant cells excluded sodium from the cytoplasm. Contrastingly, level of Na^+ was found increasing in the callus of *Vigna aconitifolia* with increase in salt concentration of medium, with a concomitant decrease in K^+ in the callus.

A.few salt selected cell lines have been characterised as to the changes which occur in their ionic content under different levels of salt stress. Jai Ping et al. (1981) found that the Na^+-dependent soybean cell lines excluded Na^+ more efficiently than unselected lines. Dix and Street (1975) found that both selected and unselected lines of *Nicotiana sylvestris* took up Na^+ and Cl^- equally well. Pandey and Ganapathy (1984) found that difference in Na^+ and Cl^- uptake by selected and unselected cells of *Cicer arietinum* was noticeable only at higher salt concentration where selected lines absorbed more Na^+. CK Dix and Pearce (1981) found that both selected and unselected *N. sylvestris* cell lines accumulated proline on transfer from no salt medium to salt medium. Proline accumulated more rapidly in the unselected cell lines but was not sufficient to contribute in osmotic adjustment. Watad et al., (1983) found that proline increased in selected *N. tabacum* cells with increasing levels of salts and was proportionately reversible with decrease in salt concentration.

Stavarek and Rains (1985) studied some aspects of carbohydrate metabolism, respiration and energy costs of selected and unselected alfalfa cell lines. The non-selected cells show a considerable effect of salt stress on their metabolic functions, namely decrease in glucose uptake, respiration and the ability of cells to utilise glucose. In the NaCl-selected cells, the effect of salinity was not preceptible in metabolic parameters. The cells were able to maintain a level of productivity comparable to non-selected control cells.

Studies on the proteins produced during salt stress in selected and unselected tobacco cells have been conducted. Two protein bands were found which were more abundant in the selected cells than the unselected and another protein band that was unique to the selected cells. This particular selected cell line was not found stable and the tolerance was lost in salt free medium. Nevertheless, proteins also resumed to those found in the unselected cells.

These physiological mechanisms comparing NaCl-selected and non-selected cells suggest differences which could explain important components of survival and productivity of plants exposed to salt stress. A major limitation to yield in stress environment may be efficient use of metabolic energy.

While cells used in physiological studies are likely to provide information on cellular level functions, mechanisms that are dependent upon whole plant structures (e.g., root-shoot interactions) or whole plant functions (e.g., specialised xylem transfer cells) will be difficult to study in cell culture systems. In programmes focussed on the incorporation of potential genotypes into a breeding scheme, plants must be regenerated from cultured cells. In several species, the capacity to regenerate plants decreases with time in culture. This may be a technical problem which can be overcome by a better understanding of the process of regeneration and media sequence used to regenerate plants.

Thus, a more complete understanding is required of the mechanism of salt tolerance in in vitro conditions. It has been suggested that NaCl tolerant variants could arise by point or chromosomal mutation or by gene amplification. Gene amplification can result in either stable or unstable methotrexate-resistant Chinese hamster cell lines. Stable amplification is correlated with chromosome based gene duplications whereas, unstable amplification is correlated with the presence of small paired extra-chromosomal elements denoted as double minute chromosomes, Earlier, Biedler and Spengler (1976) have correlated loss of methotrexate resistance with siminution in size of homogenously staining regions within chromosome.

In addition to tobacco, considerable work has been done to select for salt tolerance in rice, wheat, pearl millet, porso millet and oats. In all these cereals numerous cell lines tolerant to 0.05, 0.10 and 0.15 M NaCl have been obtained and more than 2000 rice and oat plants and over 200 wheat, pearl millet and porso millet plants have been regenerated from salt tolerant cultures. All these endeavours have been summarised in tissue culture for crops project (TCPP) progress report Ketchum et al, 1987).

Future Trends in Plant Breeding

Plant breeding has benefited in many ways from the application of discoveries in science and engineering. The two that are exciting some interest at the present time are single seed descent and chemical hybridizing agents. Single seed descent is used to hasten progress towards homozygosity in small grains like wheat and barley that are self-pollinated. Chemical hybridizing agents, or gametocides, facilitate large-scale production of F_1 hybrid seed without the need of genetic male sterility or hand emasculation. These chemicals are applied before flowering and interfere with the production of viable pollen. The treated plants thus act as female parents and are wind pollinated by adjacent untreated plants. Provided the problems involved in producing large quantities of seed to meet appropriate purity standards can be overcome. F_1 hybrids could lead to fairly rapid increases in yield.

Possible Application of Biotechnology in Nematode Management

During the past decades, tremendous progress has been made in our knowledge of plant parasitic nematodes, the damage they do and in ways to manage them. However, there is much more that needs to be done if we are to improve our agricultural production efficiency by effectively managing plant parasitic nematodes. By seeking help from technological advances in chromatography. High performance liquid chromatography, Mass spectrophotometry. NMR spectrophotometry, electrophoresis, isoelectric focusing etc., the tools of biotechnology (classical breeding for resistance, biological control of plant parasitic nematodes, taxonomy, culturing of nematodes and the use of bioregulators such as allelochemics for management of nematodes) enable us to think in terms of single nematode, single cell and single gene analysis. Nematologists are utilizing these unique opportunities to advance our discipline further so as to make this great revolution in applied biology and agricultural sciences a success.

Non-recombinant DNA technologies, mutation techniques, an alternative method of combining genomes is by somatic hybridization through protoplast fusion (for example to combine the genome of the nematode resistant wild species *(Solanum sisymbriifolium)* with the susceptible eggplant *(Solanum melongena]*, embryo-rescue technique are certain important techniques which have been used to develop nematode resistant plants.

Embryo Rescue Technique

This technique involves hybridization between economically important species and a wild sexually incompatible relative species followed by excision and culturing of hybrid zygotic embryo on nutrient media) One of the classical examples of successful application of embryo rescue techniques in plant nematology is the synthesis of tomato cultivar resistant to root-knot nematode. Cultivated tomato species *(Lycopersicun esculentum)* is highly susceptible to *Meloidogyne* spp. But the related wild species *L. peruvianum* has a high degree of resistance against these nematodes. A cross between the two yielded a hybrid in which endosperm did not develop and the embryo was aborted. This problem can be solved by embryo culture. Thus, this technique provides a mechanism to transfer nematode resistance gene from *L. peruvianum* into cultivated species.

Protoplast Fusion

By employing conventional breeding techniques, it is not possible to obtain interspecific and intergeneric hybrids. The protoplast fusion technique provides an opportunity to overcome this problem in order to evolve interspecific and intergeneric hybrids which are sometimes absolutely essential in breeding involves fusion of somatic cell, it is also called parasexual or somatic cell, hybridization. This technique involves three steps:

1. *Isolation of protoplast:* Plant tissues from 2 different sources are subjected to a mixture of enzymes (cellulases, pectinases) in an osmotically stable solution such as sorbitol and mannitol.
2. Fusion of protoplast in presence of fusogen like polyethylene glycol.
3. Culture of hybrid plants to regenerate whole plant.

This technique involving interspecific crosses was perhaps successfully implemented by Fassuliotis *et al,* (1982) with an aim evolve root-knot resistant egg plants. Somatic hybrids were obtained by fusing protoplasts of *Solanum melongena* and *Solanum sisymbrifolim* (wild species) resistant to root-knot nematode. This trait i.e. resistance to the nematode was transferred from wild to cultivated species since no reproduction was seen in infectivity tests.

Recombinant DNA Technology

The pioneering work of Paul Berg, Herbert Boyer and Stauley Cohen in the early 1970's led to the development of recombinant DNA technology, which has revolutionized biochemistry. One of the major breakthroughs in this area of research in the fields of nematology is the identification of a dominant *Meloidogyne incognita* gene imparting resistance to the root-knot nematode and its linkage to an acid phosphatase gene (Rick and Fobes, 1974). The close association between an important phenotypic character a relatively easily identifiable electrophoretic marker could be exploited for faster and more reliable methods for nematode resistance.

Biocontrol Strategies

Opportunities in recombinant DNA technology may extent to produce more effective biological control agents for plant parasitic nematodes. The toxin gene from the insect pathogen *Bacillus thrunigiensis* has been transferred to the root colonizing bacterium *Pseudomonas flourescens* Migula. Similar techniques can be used to produce fungal strains with increased virulence, nematoxic or nematostatic metabolic production, lesser effects on non-target species.

The sophisticated sensory organs in nematodes provide a strong credence for the presence of well developed communication system in these animals. The intervention in the communication system of nematodes offer several possibilities of biocontrol. Carbohydrate moieties play a very important role in the host recognition. Laboratory studies have revealed that when lectine bind to these moieties, host recognition phenomenon is disturbed in nematodes reproducing amphimictically intervention in the pheromone communication system offers possibilities of some control strategies. Saturation of worm's environment with a synthetic pheromone emnating from various sources such as slow release capsules, feed additives or soil amendments would disrupt reproduction continually so that gradients useful to the pest are eliminated.

There are certain bacteria that attract 2nd stage *Meloidogyne incognita* while others repel them. Rhizosphere bacteria are positioned in the ideal location to intercept root feeding nematode. Seeds could be coated with an inoculum of strain that could colonize the root and produce nematode repellents bacteria that produce a nematode attractant, can be formulated into a nematicide pellet increasing its effectiveness.

Avermectins (AVM) are macrocyclic lactones derived from the mycelia of *Streptomyces avermitilis.* They have potent antihelminthic and insecticidal action. The gene responsible for the systhesis of avermectin from the genome of *Streptomyces avermitilis* can be isolated and cloned by r-DNA technology and introduced with nematode attracting bacteria.

Nematode Physiology

Sterols are major components of all cell membrane and metabolic precursors of steroid hormones and many other compounds. Although sterols are essential for growth and reproduction of nematodes, but these animals lack the ability to biosynthesize sterols *de novo.* Consequently interference with uptake, metabolism or utilisation of sterols offer a means for disruption of membrane function and hormonal regulation. Two such compounds are azasteroids and long chain altrylammes. Future investigation of nematode sterol metabolic pathways open new areas for nematode management.

Nematode Taxonomy

Restriction fragment analysis has received the most attention in nematology. This technique relies on use of endonuclease restriction enzymes to generate fragments of DNA which are subsequently separated electrophoretically on a gel. These enzymes recognize a district nucleotide sequence usually a series of 4-6 nucleotide base pairs and precisely but the genome everywhere that sequence exists. The fragmented DNA is separated by electrophoresis and the banding pattern is visualized either by fluorescent stain (ethidium-bromide) or by autoradiography. When applied to homologous regions of DNA of different species, restriction endonucleases can survey for nucleotide sequence alteration such as loss or addition of restriction sites or DNA insertions or deletions. Such alterations result in dissimilar patterns on gels and referred to as Restriction Fragment Length Polymorphisms (RELPs). RELPs have been used to differentiate between species and populations *of Meloidogyne, Romanomermis,* pathotypes of *Bursaphelenchus.* These polymorphism represents divergence among repetitive DNA sequences. For obtaining further information this technique is combined with DNA hybridization. The molecular basis of nucleic acid hybridization lies in the complementary of nucleotides in double stranded DNA and the precise hydrogen bonds that hold the strands together. In order to assess RELPs in homologous regions of the genome, DNA is but by restriction endonucleases the fragments electrophoretically separatd on a gel and the DNA is denatured and transferred to a nitrocelluloses filter. A second DNA fragment the 'probe' DNA is labeled *in vitro* either by radioactive compound or biotinylated nucleotide. Once labelled and thermally denatured, the probe is able to bind to any single stranded DNA molecule that shows nucleotide homology with the probe. Using a cloned alcohol dehydrogenase gene from *Drosophila,* Bolla (1987) reported difference between two strains *Bursaphelenchus xylophilus.*

Culturing of Plant Parasitic Nematodes

Plant parasitic nematodes are obligate parasites. Because of their requirement for living tissues as food sources, they have been particularly living host tissues as food sources, they been particularly difficult to bring into monospecific cultures under sterile conditions. Lack of cultures of living nematodes has seriously retarded the study to these organisms. The important area of phytopatholgical research is the pathgenicity to pursue when large population of microbiolgically sterile test organisms are not available. Advances in tissue culture technology if successfully applied to nematode culturing could held to overcome serious limitations culturing species of important genera such as *Meloidogyne* and *Heterodera* on callus.

Research on callus and plantlet development could influence our ability to manipulate nurse development in callus tissue substrates. Also, the use of new biotechnological techniques should enhance the possibility of determining the food requirements of different species the parasitic nematodes and ultimately the development of a chemically media for culturing some species.

□□□

CHAPTER-11

Transgenic Plants

Recent developments in biotechnology have revolutionized the way to introduce any new trait or characteristic by utilizing the technique of genetic engineering, which otherwise by conventional breeding involving, sexual hybridization, is a very lengthy procedure. Genetic engineering involves the successful introduction, integration and expression of a gene from its normal location into a cell of tissue that does not contain it. The plants obtained through genetic engineering contain a gene or genes usually from unrelated organism; such genes are called transgenes and the plants containing transgenes are known as transgenic plant.

Genetic engineering is a young branch of science. It was only in 1983 that chimeric genes were first expressed in genetically transformed plant tissue (Bevan *et al.*, 1983; Herrera-Esrrella *et al,* 1983). Only after this development, and the concomitant availability of selectable and non selectable marker genes which express in plant tissue, did work on plant genetic transformation systems begin in earnest. Since that time there has been an explosion in the literature of plant genetic transformation.

New emerging gene transfer technologies have enormous potential for plant improvement by introducing foreign genes in plant cells or tissues of both monocot and dicot plants. During the last 14 years, a number of transgenic plants have been produced in many important crops. Besides selectable marker genes, a number of other foreign genes conferring herbicide resistance, insect resistance, viral, bacterial and fungal resistance and genes related to increase shelf life of plants and plant products (via antisense RNA technology), have been transferred to plants from a wide range of plants, bacterial, and viral systems. In the majority of cases, foreign genes show-expression in transgenic plants and are stably inherited in the progeny without detrimental effects on the host plant. Moreover, transgenic plants under field conditions have also maintained increased levels of insect resistance, herbicide resistance and higher levels of virus protection. However, there is a need to have extensive laboratory and field testing of transgenic plants and their progenies before their usefulness can be realized on the commercial scale.

Genetic engineering of plants could ultimately become a reliable tool for crop improvement depending on cost effective production of transgenic plants and their products without compromise on quality, yield and nutrition. So far, no genetically engineered cultivar has been released to farmers and is delayed by 8-10 years. However, numerous, public institutions and private companies are conducting field trials of transgenic plants in different parts of the world. Flavr Savr transgenic tomato (improved shelf life) produced by Calgenes Inc. USA, will reach the market very soon. In future, faith in genetic engineering for crop improvement will be dependent on how, well consumer accepts transgenic tomato Flavr Savr which will be the criteria of its commercial success.

The term **transgenic** is used to describe plants and or animals which have had DNA introduced into them by means other than transfer of DNA from a sperm cell to an egg. The DNA transferred by other than sexual processes may be DNA from same species, from other plant or animal species or even from a prokaryote. Today many transgenic plants as well as animals have been engineered by the scientists with special beneficial properties.

History of Transgenic Plants

Earlier the conventional methods of plant breeding were used to produce insect resistant varieties, especially more recently in widecross programmes. But in some cases it has not been possible to introduce resistance by this means due to lack of variation in insect resistance in sexually crossable materials. The common crown gall bacterium, *Agrobacterium tumifaciens,* a ubiquitous soil organism with the remarkable ability to move a portion of its own DNA into a plant cell during infection. In the early 1980s number of groups in United States and Europe find ways to use this bacterium to insert other genes into plant cells, they were then able to recover whole plant from transformed cells (Bartn et al., 1987; Herrera-Estrella et al., 1983). The resulting plants were found to have incorporated the inserted genes into their own genomes and thus had acquired new, stable and inherited traits. This led to the use of bacterial genes for resistance to antibiotics and for the production of unusual amino acids.

The first successful use of this technology was reported by a Belgian biotechnology company, Plant Genetic Systems, in July of 1987 (Vaeck et al., 1987). Using a gene from *B. thuringiensis* this group developed tobacco plants that produced enough of the endotoxin to kill first instar Manduca sexta larvae. Insect placed on the leaves showed symptoms of feeding suppression after about 18 hrs and death within three days. Levels of endotoxin was as low as 30ng per gram of leaf provided complete protection against the nematodes and production of endotoxin was shown to be inherited like a simple dominant trait.

The first report was followed in august 1987 by similar results from a research group at the Monsanto Company using tomato. Shortly thereafter the Agracetus Company reported success in tobacco (Barton et al., 1987; Fischhoff et al., 1987). In late 1987 a group of scientists in United Kingdom reported success with a gene from the cowpea. This gene encodes a serine protease inhibitor, which is believed to impart resistance to a beetle herbivore of the cowpea by interfering with its digestion. When introduced in tobacco, it imparted resistance to Heliothis virescens. Growth reduction in the insect was correlated with the levels of inhibitions.

Up till 1980s, production of transgenic plants was restricted mainly to tobacco, petunia or tomato of the family solanaceae. However, now in more than 40 dicot species transgenic plants have been successfully produced (Table 11.1). The production of transgenic plants has now been extended to several monocotyledons like wheat, maize, rice and oats.

Table 11.1 A list of higher plants developed as transgenic plants using different methods.

Herbaceous dicotyledons	Nicotiana tobacum (tobacco), *Plumbaginifolia* (wild tobacco), *Petunia hybrida* (petunia), *Lycopersicon esculentum* (tomato), *Solenum tuberosum* (potato), *Solanum melongena* (eggplant), *Arabidopsis thaliana*, *Lactuca sativa* (lettuce), *Apium graveolens* (celery), *Helianthus annuus* (sunflower), *Linum usitatissimum* (flax), *Brassica napus* (oilseed rape: canola), *Brassica oleracea* (cauliflower), *Brassica oleracea var capitata* (cabbage), *Brassica rapa* (syn. B. campestris), Gossy*pium hirsutum* (cotton), *Beta vulgaris* (sugarbeet), *Glycine max* (soybean), *Pisum satium* (pea), *Medicago sativa* (alfalfa), *varia, Lotus conuculatum* (lotus), *Vigna aconitifolia*, *Cucumis sativus* (cucumber), *Cucumis melo* (muskmelon), *Cichorium intybus* (chicory), *Daucus carota* (carrot), *Armoracia* sp. (horse radish), *Glycorhiza globra* (licorice), *Digitalis purpurea* (foxglove), *Ipomoea batatas* (sweet potato), *Ipomoea purpurea* (morning glory), *Fragaria* sp. (strawberry), *Actinidia* sp. (kiwi), *Carica papaya* (papaya), *Vitis vinifera* (grape), *Vaccinium macrocarpon* (ctanberry), *Dianthus caryophyllus* (carnation), *Chrysanthemum* sp. (chrysanthemum), *rosa* sp. (rose).
Woody dicotyledons	*Populus* sp. (poplar), *Malus sylvestris* (apple), *Pyrus communis* (pear) *Azadirachta indica* (neem), *Juglans regia* (walnut).
Monocotyledons	*Asparagus* sp. (asparagus), *Dactylis glomerata* (orchard grape), *Secale cereale* (rye), *Oriza sativa* (rice), *Triticum aestivum* (wheat), *Zea mays* (corn), *Avena sativa* (oats), *Festuca arundinaccea* (tall fescue).
Gymmosperms (a conifer)	*Picca glauca* (white spruce)

How to Develop Transgenic Plants

Agrobacterium tumefaciens, A Remarkable Genetic System

Agrobacterium tumefaciens, the gram negative, rod shaped aerobic bacterium caused little interest when Smith and Townsend (1907) reported that it is the causative agent of a widespread neoplastic plant disease crown gall, which is characterized by the overproduction of tissues at the infection site. This pathogen adsorbs on the surface of cells exposed at wound sites, most frequently just below the soil surface at the root crown and hence named crown gall tumour. Later, a key discovery was made in 1974 by Zaenen et al where they found that *A. tumefaciens* contains a large extrachromosomal element (more than 200 kb) harbouring genes involved in crown gall induction. This tumour inducing plasmid was given the name Ti plasmid by van Larebecke *et al* (1975). However, this bacterium received extraordinary attention after the discovery that the bacteria is capable of modifying the genetic material of host cells by transferring part of its Ti plasmid which gets incorporated into the host cell DNA and expresses itself with the normal host DNA. With researchers analysing and unraveling the molecular mechanism underlying crown gall induction, the *Agrobacterium* gene transfer system has now become one of the most significant discoveries triggering a series of existing developments in molecular biology and genetic engineering.

The first transgenic plants expressing engineered foreign genes were tobacco plants produced by the use of *Agrobacterium tumefaciens* vectors (Horsch *et al,* 1984, DeBlock *et al.*, 1984). Transformation was confirmed by the presence of foreign DNA sequences in both primary transformants and their progeny and by an antibiotic resistance phenotype conferred by a chimeric neomycin phosphotransferase gene.

Agrobacterium constitutes an excellent system for introducing genes into plant cells, because of the following reasons:

1. DNA can be introduced into whole plant tissues, which by pass the need for protoplasts, and
2. The integration of T-DNA is a relatively precise process.

The region of DNA to be transferred is defined by the *border sequences,* occasional rearrangements do occur, but in most cases an intact T-DNA region is inserted into the plant genome. This contrasts with free DNA delivery systems in which the plasmids routinely undergo rearrangements and concentration reactions before insertion and, can lead to chromosomal rearrangements during insertion in both systems. Sequencing of insertion sites shows that only small duplications or other changes occur in flanking sequences during T-DNA integration. The stability of expression of most genes that are introduced by *Agrobacterium* appears to be excellent.

Survey of literature reveals that integrated T-DNAs give consistent genetic maps and appropriate segregation ratio. Introduced traits have been found to be stable over at least five generations during cross breeding and seed increase on in genetically engineered tomato and oilseed rape plants. This stability is critical to the commercialization of transgenic plants.

Agrobacterium-Mediated Gene Transfer

The utility of this bacterium as a gene transfer system was first recognised when it was demonstrated that the crown galls were actually produced as a result of the transfer and integration of genes from the bacterium into the genome of the plant cells. Virulent strains *of Agrobacterium* contain large Ti- (tumour inducing) plasmids, which are responsible for the DNA transfer and subsequent disease symptoms. Genetic and molecular analyses showed that Ti-plasmids contain two sets of sequences necessary for gene transfer to plants, one or more T-DNA (transferred DNA) regions that are transferred to the plant, and the *Vir* (Virulent) genes which are not, themselves, transferred during infection. The T-DNA regions are flanked by 25 base pair right and left border DNA sequences that were shown to be responsible for the T-DNA transfer to infected plant cell and that it can be deleted. **Phytohormones** synthesis interference with plant regeneration replaced with marker gene or desirable gene.The transferred genes have been shown to be stably incorporated into the plant genome, and the genetic analysis of their progeny have shown that the incorporated traits often display normal Mendelian inheritance.

Vectors for Agrobacterium Mediated Plant Transformation

Plant transformation vectors based on *Agrobacterium* can generally be divided into two categories: those that cointegrate into a resident Ti-plasmid **(cointegrate vectors)** and those that replicate autonomously **(the binary vectors).** Both the vectors have several common features imposed upon them by the requirements *of Agrobacterium.*

Cointegrating Vectors

Cointegrating transformation vectors must include a region of homology between the vector plasmid and the Ti-plasmid. This requirement for homology means that the vector is capable of integrating into a limited number of Ti plasmids. The first utilizes the disarmed *Agrobacterium* Ti-plasmid PG 3850. In this plasmid the **phytohormones gene** of the C 58 plasmid have been excised and replaced by pBR 322 sequence. Any plasmid containing the pBR 322 sequence homology can be cointegrated into the disarmed Ti-plasmid. The border sequences as well as a nopaline synthase gene are part of the Ti-plasmid, and the cointegration places the new sequences between the **T-DNA borders.**

Phytohormones Genes: According to Fraley *et al,* (1985) when the **right border** and all the phytohormone genes are removed from the Ti-plasmid, a **left border** and a small part of the original T-DNA, referred to as the **Limited Internal Homology (LIH),** remains intact. The vector to be introduced into *Agrobacterium* contains the LIH region for homologous recombination as well as a right border. The cointegrated DNA reconstructs a functional T-DNA with a righr border and left border. This system has been used extensively for introduction of many genes into plants. Once the cointegrate has been formed, the plasmid is stable in *Agrobacterium* and is virtually impossible to lose.

Binary vectors, on the other hand, are not completely stable in *Agrobacterium* in the absence of drug selection. There is also evidence that a cointegrating vector can transform tomato at a higher frequency than a binary vector.

Binary Vectors

These vectors are different from cointegrating vectors. Instead of a region of homology with the Ti-plasmid, they contain origins of replication from a broad host-range plasmid. These replication origins permit autonomous replication of the vector in *E. coli* and *Agrobacterium.* Since the plasmid does not need to form a cointegrate, these plasmids are considerably easier to introduce into *Agrobacterium.* A major advantage to binary vectors is their lack of dependence on a specific Ti-plasmid. The vector may be introduced into virtually any *Agrobacterium* host containing any Ti or Ri plasmid as long as the *vir* helper functions are provided. This may be important in the transformation of some plant species, since different *Agrobacterium* strains exhibit major differences in their abilities to infect different plant species.

Selectable Markers for Plant Transformation

Several selectable and non selectable marker genes viz. neomycin phospo-transferase, luciferase, streptomycin phosphotransferase, chloramphenicol acetyl transferase. hygromycin phosphotransferase, p-glucuronidase, and dihydrofolate reductase are widely available today for plant transformation and have also been transferred in large number of plant species (Table 11.2 to 11.5). Several requirements 9must be considered in the development of a truly useful selectable marker system.

It is most critical that the selective agent be inhibitory to plant cells. However, not all compounds toxic to plant cells are necessarily useful as selective agents. Cells that are not transformed can be killed in such a manner that they become toxic to adjacent transformed cells. This presumably happens because **of leakage** of toxic compounds, such as phenols, from the dying cells. If this occurs even high level expression of a resistance gene in the transformed cells is insufficient to rescue these cells. The best selective agents are compounds that arrest growth of non transformed cells or slowly kill them.

Table 11.2 Transgenic field crops

Plants	Method and Gene Transferred
Tobacco	At, NPT-II
Alfalfa	At, NPT-II
Cotton	At, NPT-II,
Sunflower	At, NPT-II
Moth Bean	At, NPT-II
Soybean	At, GUS
Safflowcr	At, GUS
Green Bean	At, GUS
Cowpea	EL* GUS
Chickpea	At, NPT-II

At: *Agrobacterium lumefaciens,* NPT-II: Neomycin phosphotransferase -II, GUS: β-glucuronidase, EL*: Electroporation.

Neomycin Phosphotransferase type II (NPT-II) Enzyme: It is most widely used selectable marker. It was originally isolated from the prokaryotic transposon Tn 5. It detoxifies aminoglycoside compounds such as kanamycin and G 418 by phosphorylation. This gene, fused to constitutive plant transcriptional promoters, has been used successfully to transform a large number of plant species (Table 11.2, 11.3, 11.4 and 11.5), and has been incorporated into numerous plant transformation on vectors.

Transformation of Plants

There are two basic approaches which have been used to obtain transgenic plants:

1. Co-cultivation of regenerating protoplasts
2. The leaf disc procedure

The production of transgenic plants can be divided into four main steps

1. Introduction of foreign genes into modified *Agrobacterium* strain
2. Cocultivation *of Agrobacterium* strains with protoplasts, plant cells or tissues
3. Selection and regeneration of transformants and
4. Analysis and verification of gene expression in transformed plants.

Table 11.3 Transgenic Vegetable Crops

Plants	Method and Gene Transferred
Carrot	Ar, NPT-II
Tomato	At, NPT-II
Cucumber	Ar, NPT-II
	At, Ar, NPT-II
Potato	At, NPT-II
Oil seed rape	At, NPT-II
Lettuce	At, NPT-II
Cauliflower	Ar, Mannopine
	At, NPT-II
Celery	At, NPT-II
Brinjal	At, NPT-II
Watermelon	At, NPT-II
Sugarbeet	At, NPT-II
Muskmelon	At, NPT-II
Bell pepper	At, GUS
Peas	At, NPT-II
Sweet potato	MB* GUS
Cabbage	At, GUS

At: *Agrobacterium tumefaciens,* Ar: *Agrobacterium rhizogenes,* NPT-II: Neomycin phosphotransferase-II, GUS: β glucuronidase, MB*: Micro-projectile bombardment

Table 11.4 Transgenic Fruit Crops

Plants	Method and Gene Transferred
Walnut	At, NPT-II
Apple	Ar, NPT-II
Peach	At, Octopine
Grapevine	At, NPT-II
Papaya	MB*, NPT-II, At, cp gene
Strawberry	At, GUS
Citrus spp.	At, NPT-II
Apricot	At, cp gene
Pecan	At, GUS,

At: *Agrobacterium tumefaciens,* cp gene: Coat protein gene, NPT-II: neomycin phosphotransferase-II, GUS: β-glucuronidase, MB*: Microprojectile bombardment

Table 11.5. Transgenic Forest Plants

Plants	Method and Gene Transferred
Populus hybrida (Poplar)	At, NPT-II, 'aroA'
Pseudotsuga menziesii (Douglas fir)	At, Octopine, NPT-II
Azadirachta indica (Neem)	At, Octopine, NPT-II
*Picea mariana** (Black spruce)	El*, CAT
*Pinus banksiana** (Jack Pine)	El*, CAT
Salix spp. (willows)	At, NPT-II
Catharanthus rosens (Madagascar)	Ar, Agropine

Plants	Method and Gene Transferred
Digitalis purpurea (Foxglove)	At, GUS
Picea rubens, P. Glauca (Red spruce, White spruce)	MB*, GUS, NPT-II
Larix X *eurolepis* (Larch)	MB*, GUS, NPT-II
Liquidambar styraciflua (Sweetgum)	At, NPT-II, Bt gene
Santalum album (Sandalwood)	At, NPT-II

At: *Agrobacterium tumefaciens,* aro A' gene: Glyphosate resistance, Bt gene: -endotoxm gene, NPT-II: neomycin phophotransferase-II, GUS: β-glucurnonidase, MB*: Microprojectile bombardment, CAT: Chloramphenicol acetyltransferase, El: Electroporation.

Most transgenic plants produced to date, were created through the use of the *Agrobacterium*-mediated gene transfer system. Other methods that have the potential to influence the production of transgenic plants include, microinjection, electroporation, direct injection into reproductive organs, PEG mediated gene transfer, particle gun and liposome mediated gene transfer.

Herbicide Resistance

Herbicide resistance engineering into crops represents a new alternative for conferring selectivity and enhancing crop safety from herbicide. Research has largely concentrated on those herbicides with properties such as high unit activity, low toxicity, low soil mobility, rapid biodegradation and broad spectrum activity against various weeds.

Engineered Resistance against Herbicide

The development of crop plants that are tolerant to herbicide is an important approach to control weeds. Most of the modern herbicides interfere with amino acid biosynthesis in plants. Once the mode of action of a herbicide is established, the transfer of resistance to the herbicide can be done by two different approaches: (1) the first approach is modification of the target side and, (2) detoxification of herbicide. Transgenic plants carrying the genes for herbicide tolerance has been produced in tobacco, petunia, tomato, potato, *Populus* etc.

Modification of the target

This approach has been made for developing resistance against atleast three herbicides viz. glyphosate, sulphonyl urea and inidazolinones.

Resistance against glyphosate

Glyphosate, an active components of round up herbicides, is used as non-selective post-emergence herbicides. It is suggested that glyphosate inhibits aromatic amino acid biosynthesis by competitive inhibitor of enzyme 5-enol-pyruvyl shikimate-3 phosphate synthase [EPSPS]. EPSPS catalyzes the synthesis of phenylalanine, tyrosine and tryptophan. The aro-A gene which codes EPSPS has been isolated and sequenced from *E.coli, Salmonella typhimurium* and *Arabidopsis thaliana.*

There are two possible methods for producing glyphosate resistant plants by genetic engineering:

1. transfer of a glyphosate sensitive EPSPS under the control of a powerful promoter causing over expression of the protein and/or
2. transfer of gene which codes a mutated, glyphosate resistant EPSPS.

cDNA of the EPSPS gene with 35 S cauliflower mosaic virus promoter was constructed and the transformed petunia cell lines with this construct exhibited 40-fold increase in EPSPS activity.

An alternative method is insertion and expression of mutated bacterial aro A gene coding for glyophosate resistant EPSPS. The source of the aro A gene was a mutagenized strain of *S. Typhimurium, E.coli* and was transferred to tomato and/or tobacco. Transgenic tobacco plants, showing tolerance to commercial levels of glyphosate are field tested in USA.

Resistance against sulfonylurea and imidazolinone herbicides

The herbicidal action of sulfonylureas and imidazolinomes is based on their ability to inhibit non-competitively an enzyme of branched-chain amino acid biosynthesis, acetolactate synthase (ALS). ALS is the first enzyme in the biosynthetic chain resulting in the synthesis of the branched chain amino acids valine, leucine and isoleucine.

Mutated strains rėsistant to sulfonylureas have been obtained from bacteria, fungi and plant cell cultures. The resistance of all these mutants forms of the ALS. Transgenic tobacco plants expressing a mutant ALS gene from tobacco or *Arabidopsis,* were produced that were tolerant to sulfonylurea herbicides.

Table 11.6. Mechanisms of action of different herbicides and basis of achieving resistance against them in transgenic plants.

Active principle of herbicide	Inhibited pathway	Target product	Use	Basis of resist
Amino acid biosynthesis inhibitors				
Glyphosate (Roundup)	Aromatic amino acid biosynthesis	EPSPS	Broad spectrum	Overexpression of EPSPS gene bacterial *aroA* gene
Sulphonylurea and Imidazolinones	Branched chain amino acids	ALS	Selected crops	Mutant ALS gene
Phosphinothricin (Basta)	Glutamine	GS	Broad spectrum	Gene amplification bar gene: detoxification
Photosynthesis inhibitors				
Atrazine (Lasso)	Photosystem II	Qb (32 kDa protein)	Selected crops	Mutant PsbA gene GST gene; detoxification
Bromoxynil (Bructril)	Photosynthesis		Selected crops	bxn gene; detoxification

Resistance against Phosphinothricin

L-Phosphinothricin (PPT) is a naturally occurring amino acid with herbicidal activity and is active ingredient of herbicide 'Basta'. PPT is a potent inhibitor of glutamine synthase and thus causes rapid increase in ammonia concentration in plants. This leads to death of plant cell, "bar" gene from *Streptomyces hygroscopicus* encodes phosphino thricin acetyl transferase enzyme that acetylates the free-NH_2 groups of phophinothricin and renders it inactivates the herbicide Basta.

This 'bar' gene with ca MV 35S promoter has been transferred to tobacco, tomato and potato plants using *Agrobacterium* Ti plasmid vector system. The transgenic plants have been tested for 'Basta' resistance.

Resistance against Atrazine

Atrazine is another herbicide, which inhibits photosynthesis. A 32 R Da protein, QB is responsible for herbicide binding and is encoded by P^{sb} A gene. The DNA sequences of P^{sb} A are highly conserved in different plants. More than 20 weed species have developed resistance to atrazine herbicides for example, *Amaranthus hybirds.*

Detoxification or degradation or herbicide

This approach is based on selective use of herbicide, which will kill the weeds and not the crop. A number of detoxifying enzymes have been identified in plants as well as microbes. Few examples are given here, This herbicide (Bromoxynil) inhibits photosynthesis (i) glutathione-S transferase or GST (in maize and other plants): It detoxifies the herbicide atrazine; (ii) nitrilase: It is coded by gene bxn in *Klebstella penumoniae,* and it detoxfies the herbicide bromoxynil, and (iii) phosphinothricin acetyl transferase or PAT (coded by *bar* gene in *Streptomyces* and transgenic plants using the *bxn* gene from *Streptomyces* spp. It detoxifies the herbicide PPT (L-phosphinothricin).

Transgenic tomato plants using the *bxn* gene from *Klebsiella ozaenae* and *bar gene* from *Streptomyces* and transgenic plants in potato, oilseed rape *(Brassica napus)* and sugarbeet using *bar* gene from *Streptomyces* have been obtained and were found to be herbicide resistant. Other crops for engineered herbicide tolerance include soybean, cotton and corn.

The development of crop plants that are resistant to such herbicide would provide more effective, and less costly weed control. Fillatti *et al,* (1987), Piruzian *et al.,* (1988) and DeBlock *et al,* (1989) have successfully transferred the herbicide resistant gene in tomato, poplar, tobacco, oil seed rape and cauliflower. Before commercialization of herbicide resistant crop plants, factors such as potential loss in vigour and yield, herbicide performance crop and chemical registration costs and potential for out crossing to weed species should be considered.

Engineered Resistance against Pest and Insects

Man has sought to alter the balance of nature to a great extent, by heavily depending upon chemical pesticides. Despite making every effort to control the pests rather manage them, 37 per cent of all crop production is lost world-wide to pests, of which 13 per cent is lost to insects alone. Therefore, insect pest constitutes a formidable group of crop destroyers. The main problem cropped up due to the development of insecticide resistant population. Now it is not uncommon, for examples of resistance in a major insect pest to be observed within the first year of field use. To worsen the situation, in the past indiscriminate use of chemical insecticides have multiplied the magnitude of damage caused by the insects-where elimination of a wide range of predators and parasitoids along with the primary pest have turned in secondary pest becoming with even more devastating effects.

Some of the most serious losses of crop production to insects may not be in economic terms especially in the third world countries. The recently developed insecticides are not available to poor farmers as they are costly and even if they are available the measures necessary for their safe use are lacking, resulting in serious consequences for the health of many agricultural workers.

Hence it is imperative to think and develop other avenues of pest management because the earlier concept of pest control is obsolete now. Breeding crops for resistance is one of the most important avenues to put a check on the non-judicious use of insecticides. In this context plant genetic engineering provides a handy tool for the development of better varieties. Transgenic plants offer lot of advantages over the chemical insecticides. The most obvious ones being:

1. The material is confined to the plant expressing it and, therefore, it does not leach into the environment.
2. The active factor is biodegradable and choice of suitable genes/gene products can ensure it is not toxic to man and animals.
3. Consumer acceptability: In recent years there has been much concern over the presence of pesticides residues in food crops; inherently resistant crops should offer the consumer the alternative of produce containing a well-defined and characterized gene product opposed to unspecified pesticide residues.

Many different strategies are under investigation to exploit the opportunities afforded by genetic engineering to control insect pests. One type of approach is to manipulate the organisms which are naturally pathogens of insects in order to enhance their efficacy as biocontrol agents. *Bacillus thuringiensis* is a bacterial pathgen of certain insects, the genes which encode the insect control proteins responsible for its insecticidal activity have been cloned and transformed to other bacterial host species in an attempt to improve on its poor persistence in the field. Another recently reported example is introduction of genes encoding the toxin from the venom of insectivorous spider/mites into the viral genome of the insect pathogenic baculoviruses in order to increase the rapidity with which the insect succumbs viral infection.

The production of insect resistant plant is another application of biotechnology with important implications for crop improvement and for both the seed and agrochemical industries. Pests and insects cause severe damage to our crops and thus use of insectides and pesticides is a common measure. With genetic engineering, crops can now be protected with insects without applying any synthetic chemical. More than 30 crop species have been successfully transformed including corn, cotton, soybean, and rice with insecticidal properties. Progress in engineering insect resistance in transgenic plants have been achieved through the use of the insect control protein genes of *Bacillus thuringiensis* is an entomocidal bacterium that produces an insect control protein which is lethal to lepidopteran larvae, although some strain with toxicity to coleopteran and dipteran larvae have been described. Some of the scientists have successfully introduced the insect resistant gene in tobacco, tomato and mustard.

For transgenic plant production with pest resistance properties, two major approaches have been developed:

1. the introduction of protein delta endo toxins from Bt and
2. introduction of protease inhibitors in plants.

Bacillus thuringiensis EndoToxins

The entomocidal bacterium *Bacillus thuringiensis,* upon sporulation normally produces a parasporal crystaline toxin. When ingested by a susceptible insect, a combination of the high pH and proteinases of the insects midgut solubiliges the crystal and yields active toxin. The effects of the toxin occur within minutes of indigestion, leading to diruption of insect's midgut cells. Bt to xin activity has been found against many species of insects within the orders of Lepidoptera, Diptera and Coleoptera.

These delta end a toxins are very specific in their action are safe insecticides, but their use is limited due to high production cost and due to instability of crystal proteins when exposed in the field.

The above toxin gene *(ht2)* from *B. thuringiensis* has been isolated and used for *Agrobacterium* plasmid mediated transformation of tobacco, cotton, potato and tomato plants. These transgenic plants were resistant to the feeding damage of *Manducta sexta* larvae (tobacco horn worm), a pest of tobacco. Experiments of feeding the leaves of these plants to larvae of *M. sexta,* showed 75%-100% mortality of the larvae, while the control plants carrying no transgenes were severely damaged.

Field tests using transgenic insect resistant plants were also conducted with tobacco and tomato and may be used for commercial cultivation in the near future.

Table 11.7. Insecticidal efficacy of CpTi in artificial diets and in transgenic plants

Insects killed by CpTi in artificial diets	Insects, against whom resistance noticed in transgenic tobacco plants (with CpTi)
Lepidoptera	
Heliothis verescens *H.zea* *Spodoptera hitoralis* *Chilo partellus*	*Helothis virescens* *H. zea* *Spodoptera litroralis* *Manducta sexta, Autographa gamma*
Coleoptera	
Callosobrichus maculatus, Anthonomus grandis, Diabrotica undecimpunctata, Tribolium confisum Costelytra zealandica	these insects do not attack control tobacco plants
Orthoptera	
Locusta migratoria	

Protease Inhibitors

Protease inhibitors are proteins with antimetabolic activity against wide range of insects. Such inhibitors are widely distributed within the plant kingdom and can accumulate to high levels in seeds and storage organs. Other major advantages of protease inhibitors is their inactivation with cooking.

In cowpea *(Vigna unguiculata),* trypsin inhibitor (CpTI) level was shown to be responsible for its resistance to attack by the major storage pest of its seeds (i.e. bruchid beetle = *Callosobruchus maculatus).* A variety of insects have shown to be toxic to CpTI. This CpTI gene was joined with CaMV 35S promoter, and one more marker genes and was used to infect tobacco leaf discs. *Agrobacterium* was used for transformation and the transgenic tobacco plants express a high level of CpTI. The CpTI gene in transgenic plants is stably inherited and there is no serious 'yield penatly'. Thus like Bt toxin, CpTI can also be used as a protectant against insect attack in transgenic plants. However, extensive field trials are necessary before releasing these transgenic plants to the farmers.

Gene for Other Insecticidal Secondary Metabolites

The production of secondary metabolites by plants can also implicate the resistance to insect attack. However, biosynthesis of each of these metabolites involves many steps, each controlled by a separate gene. Furthermore, these genes are tissue specific in expression. These features make the production of transgenic plants difficult in this case. Utilizing these facilities, transgenic plants with insect resistance due to secondary metabolites will be available in the near future.

If susceptible strain of a crop is inoculated with a mild strain of a virus, then the susceptible strain develops resistance against more virulent strains. The phenomenon is known as cross protection has been used to reduce yield losses in crops like tomatoes against tomato mosaic virus (TMV), in potato against potato spindle tuber viroid and in citrus against citrus tristeza virus. In most cases of crops protection, the symptoms of infection are delayed, and even the replication of virus is suppressed, although eventually the severe strain may be able to overcome the protection.

Table 11.8 Examples of secondary compounds from legume seeds with demonstrated insecticidal properties (after Gatehouse *et al.*, 1991).

Compounds	Bruchid (insect)	Legume
Non-protein antimetabolites		
Alkaloids		
2,5-dihydroxymethy 1-3,4-dihydroxypyrrolidine (DMDP)	*Callosobruchus maculatus*	*Lonchocarpus* sp.
Castanospermine	*Callosobruchus maculatus*	*Castamospermum austrole*
Non-protein amino acids		
p-Aminophenylalanine	*Collosobruchus maculantus* *Zabrotes subfasciatus*	*Vigna sp.*

Compounds	Bruchid (insect)	Legume
Rotenoids	*Cullosobruchus maculatus*	*Longiscorpus salvadoreensis*
Saponins	*Callosobruchus chinensis*	*Phaseolus vulgaris*
	Callosobruchus chinensis	*Glycine max*
Polysaccharides		
Pectosans	*Callosobruchus chinensis*	*Phaseoh s vulgaris*
Heteropolysaccharide	*Callosobruchus chinensis*	*Phaseolus vulgaris*
Polysaccharide (?)	*Callosobruches chinensis*	*Vigrna spp.*
Protein antimetabolites		
Lectins (phytohaemagglutinins)	*Callosobruchus maculatus*	*Phaseolus vulgaris*
α-Amylase inhibitors	*Zabrotes subfascitus*	*Phaseolus vulgaris*
	Callosobruchus chinensis	*Phaseolus vulgaris*
Protease inhibitors	*Collosobruchus maculatus*	*Vigna unguiculata*
Arcelin / LLP	*Zabrotes subfascitus*	*Phaseolus vulgaris*
	Zabrotes subfasciatus	*Phaseolus vulgaris*

Disease Resistance

Disease resistant gene can be categorised under three main heads i.e. virus, bacterial and fungal disease resistance. Of all plant pathogens, viruses are most intimately associated with their hosts, their genomes are relatively small and well characterized. Significant resistance to tobacco mosaic virus (TMV) infection termed coat protein-mediated protection has been achieved by expressing only the coat protein gene of TMV in transgenic plants. This approach produced similar result in transgenic tomato and potato plants against a broad spectrum of plant viruses including alfalfa mosaic virus, cucumber mosaic virus, potato virus X and potato virus Y. Transgenic tomatoes carrying the TMV coat protein gene have been evaluated in green house and field tests and shown to be highly resistant to viral infection. Likewise resistance has been created against various groups of viruses.

Recently, a new class of bacteriacidal proteins (lysozyme, cecropins and attacins) have been identified in the pupae of giant silk moth *Hyalophora cecropia.* These lytic proteins have also shown a potent *in vitro* antifungal activity against several pathogenic fungi, including *Phytophthora infestans.* Transgenic tobacco plants that express a barley ribosome-inactivating proteins, exhibited heightened protection against agronomically deleterious fungus *Rhizoctonia solani.* Similarly, tobacco pathogen-related proteins, namely PR-S and osmotin have been shown to be serologically related to seamatin, an antifungal protein of maize. Woloshuk et al., (1991) showed that Osmotin and related proteins from tomato had antifungal activity against *Phytophthora infestans.*

Post Harvest Management

Antisense RNA technology in combination gene transfer technology is being used to increase the shelf life of the plant and plant products. The recent approach to develop this technology aims to mimic mutation by reducing gene expression using RNA that contains the complementary sequences to a given RNA. Transgenic plants have been used to study the effects of antisense RNA on the expression of endogenous genes such as those encoding chalcone synthase Since chalcone synthase is the key enzyme in flavonoid biosynthesis inactivation of these genes by antisense RNA was scored on phenotypic basis as a change in flower pigmentation in petunia and tobacco. In transgenic tomato plants expressing antisense RNA of polygalacturonase, endogenous level of this enzyme activity was extremely reduced in ripening fruits.

Engineered Resistance against Viral Infection

Over seven hundred plant viruses are recognised causing various diseases and significant crop losses. Due to the lack of 'viricides', the practical control of viruses depends on methods that prevent or restrict virus infection. Various strategies have been applied for protecting viral infection, like:

Cross Protection

Various disadvantages of this practice includes (i) possibility of mutation in inducing mild virus strain (ii) possibility of synergism between inducing virus and another unrelated virus, (iii) possibility of unnecessary spread of mild virus causing threat for future yield losses and (iv) possibility of some yield losses due to mild strain also. Transgenic plants have been produced in tobacco, tomato, and potato with single gene using a broad spectrum of plant viruses.

Gene for virus coat

Coat protein - mediated resistance to viruses has been one of the successes of plant genetic engineering. Coat protein gene from tobacco mosaic virus (TMV), classified as a ositive strand RNA virus, has been transferred to genome of tobacco plants. In the transgenic plant, expression of coat protein (CP), as well as low and delayed infection was observed when inoculated with TMV. However, in such transgenic plants, coat protein gene should be constitutively expressed and may thus have effects on the nutritional value of plants. Subsequently Cp genes isolated from different plus strand RNA plant viruses have been cloned into plant expression vectors and transferred to various crop plants like tobacco, tomato and potato, using *Agrobacterium* mediated plant transformation system. Potatoes have been produced which have coat protein genes and are tolerant in field tests to both PVX and PVY. DNA coding for a component of TMV replicase enzyme, was also transferred to tobacco plants, conferring resistance to TMV.

Antisense RNA Mediated Protection

Antisense constructs also generates protection against RNA and DNA viruses. Transgenic tobacco plants expressing an tissue transcript (antisense - coat protein gene) were protected from CMV infection.

Transgenic tobacco plants with nucleocapsid (N) gene, CaMV 35S promoter and a leader sequence (to enhance expression) showed expression of N-gene. Such transgenic plants were resistant to TS WV TSWV or tomato spotted wilt virus is a negative strand RNA virus with RNA joined firmly to the nucleocapside protein.

Satellite RNA and its use for transformation.

Satellite RNAs are associated with specific strains of some plant RNA viruses, though not necessary for their replication. These extragenomic RNA species depend on a helper virus for replication. Some satellite RNAs are known to attenuate the symptoms of their helpervirus. Thus satellite RNA might be used to protect crop plants against the effect of CMV infection. The nature of protection in transgenic plants by satellite RNA is different from coat - protein protection.

Defective Interfering RNA/DNA Protection

Defective integering (DI) particles are deletion mutants of genomic viral sequences, which depend on their parent virus for replication. Like satellite RNAs, DI particles can attenuate the disease symptoms of their helper virus by interfering with its replications. Artificial DI molecules of TYMV, TMV and BMV have been constructed and inserted in plant protoplast e.g. deletion mutants of BMV RNA 2 exhibited reduced replication of MBV RNA 1 and 2 in barley protoplasts.

Replicase Mediated Protection

It has been discovered that transgenic tobacco expressing part of replicase gene of TMV strain was highly resistant to this and closely related strains.

Currently, atleast three approaches are available to broad spectrum resistance. The first is based on the replication and cell to cell movement of virus. Second approach relies on the transgenic expression of antivirus protein of non-plant origin including antibodies. The third approach is based on genetic manipulation of natural viral defences in plants.

Engineered Resistance Against Fungal Pathogen

Genetic engineering has opened vistas with isolation of resistance genes and gene transformation methods to develop transgenic plants. Resistance to fungal pathogen is achieved by isolating and transferring fungal resistance genes to the dressed susceptible crop. Compounds toxic to fungus can also be used to produce transgenic plants as done in case of tobacco. Antifungal proteins like **endo chitinase** and β-l,3-endo **glucanases** are thought to inhibit fungal growth. Transgenic tobacco plants with endochitinase CH *5* B gene with Ca MV 35S promoter exhibits reduced seedling mortality when grown in presence of *Rhizoctonia solani.*

Another method is using ribosome inactivating proteins (RIPs). These RIPs donot interfere with the functioning of self ribosomes but exert inhibitory activity towards fungal ribosomes. When RIPs are mixed with endo chitinase or β-l,3-glucanases, the antifungal activity exhibits synergistic increase in *in vitro* assay. Transgenic tobacco with barley type 1 RIP gene and potato Wun 1 promoter showed an increased resistance to *R. solaui.*

Table 11.9 Some pathogens for which resistance has been transferred in some crop plants.

Pathogen	Disease	Resistance gene	Source of gene	Transgent crop
Psedomones syringae	Wild fire	Acetyl	-	Tobacčo
Allernaria longipes	Brownspot	Chitinase gene	*Senatia marcescens* (soil bacterium)	Tobacco
Rhizoctonia solani		Chitinase	Bean	Tobacco
Phytophthora infestans	Late blight	Osmotin gene	Potato	Potato

Genetic Engineering and Plant Lipids

Plant lipids are major components of the human diet and are present in plants as storage oil. Engineered vegetable oils have been tested as field trials. A high stereat phenotype has been engineered in rapeseed by antisense approach. Stearoyl-acyl carrier protein (Steatroyl - ACP) desaturase is a chloroplast enzyme which catalyzes the first desaturation step in seed oil biosynthesis, converting stearoyl - ACP to oleayl-ACP. This coding region of the gene for rapeseed desaturate was fused in an inverted position to the napin storage protein promoter and to the promoter of seed-specific acyl carrier protein gene, and then introduced into rapeseed. This reduced desaturase activity in engineered plant resulting in increase in level of stearate in seeds (upto 40% of the total fatty acids), with corresponding decrease in oleate. Due to tissue specific expression of antisense RNA, integrity of membrane lipids in leaf remained unaffected.

Length of fatty acids can be manipulated with thioesterase enzyme, which is responsible for the accumulation of short chain fatty acids in triglycerides. The gene has been cloned and expressed in developing seed of *Arabidopsis,* where it caused the accumulation of C12 storage lipids. It is useful for application in confectionery production.

The field of storage lipid modification via the engineering of the fatty acid biosynthesis enzymes promises to be a very fertile one that will contribute to production of healthier foods as well as to production of chemical feed stocks.

Genetic Engineering and Storage proteins

Human nutrition requires a balanced source of amino acids and the amino acid balance of many plant products is unsatisfactory. Genetic engineering may be carried out in crops making following three approaches:

1. expressing a desirable, heterologous storage protein,
2. increasing the level of a desirable, but little expressed endogenions protein and,
3. suppressing the expression of antinutritional proteins.

A heterologous species of sulphur-rich storage protein in Brazil nut containing 18% methionine and 8% cysteine has been expressed in tobacco and rapeseed. In transgenic tobacco, phaseolin gene has been expressed, leading to accumulation of methionine rich phaseolin.

Another approach for increasing amino acid synthesis is by engineering of a rate-limiting enzyme. For example, in a tobacco mutant lysine overproduction was caused by a lysine-insensitive dehydro-picolinate synthase. Forty fold increase in free lysine has been obtained by expression of a bacterial dehydropicolinate synthase in plant chloroplasts.

Genomic clones encoding two subunit of soybean (3-conglycin (75 protein) genes were transferred into petunia plants. The temporally regulated expression of these two genes in transgenic plants mimics their temporal regulation during soybean embryo development. The gene encoding for 42 kd potato storage protein, patatin was introduced into tobacco. Stable synthesis of patatin protein was confirmed in the heterologons environment. Furthermore, transgenic tobacco possessing the 5' and 3' flanking sequences of the bean β-phaseolin gene fused with 15 kDa zein gene could synthesize up to 1.6% of total seed protein as zein. Thus, zein is deposited and accumulates in vacuolar protein bodies of the tobacco embryo and endosperm.

CHAPTER-12

Biotechnology and Crop Improvement in India

India's Efforts

Crops meet human requirments for food, medicines, spices and beverages. It is necessary to increase the output of crops several fold per unit area because of increasing human population with consequent decrease in land area.

In India, food production during 1994-95 exceeded our target and we have produced 191 million tonnes food grains, 67 million tonnes of vegetables, and over 21 million tonnes of oilseeds (Annual Report 1995-96, ICAR, New Delhi). The radical export growth has been registered in agricultural commodities during the current year. The biotechnological oriented research efforts have resulted in spectacular progress in overall food production including rice, oilseed, plantation crops, sugarcane, cotton etc. The credit goes to ICAR - an apex organization for agricultural research, education and front-line technology transfer in the country. It has provided a concrete base to Indian agriculture in realizing enhanced productivity with the record harvest. Efforts have also been made by ICAR, New Delhi to conserve and utilize agro-biodiversity by collecting exotic and indigenous valuable germplasm under plant improvement programmes of the National Agricultural Research System (NARS).

Conservation of Plant Genetic Resources

The National Bureau of Plant Genetic Resources (NBPGR) of ICAR is the prime organization which has its mandated activities, viz. exploration and germplasm collection, exchange, plant quarantine, characterization and preliminary evaluation and conservation in plant genetic resources.

Germplasm Collection

NBPGR is responsible for collection and conservation of fast-eroding plant genetic resources. Nine explorations were made during the period (April-December, 1995) for germplasm collection, and the accessions in 9 crops were collected from different agro-climatic regions of the country sigmoid and diversity collected in some of the crops included important land races of pointed gourd of parmal (Trikolwa, Dandal, Sautokhawa, Niwia, Hilli, Jhilli, Kalichak, Kelwa and Blokia) from eastern Uttar Pradesh and Bihar, showing variability in vine length, number of branches, length, volume and weight of fruits, *phalsa* from Telangana region of Andhra Pradesh, showing variability in plant habit, leaf size, stem colour, fruit taste, size and shape and seed; *khejra (Prosopis cineraria)* from Rajasthan having variation in plant canopy, branching, foliage density, maturity period and shape and size of pods or seeds; bael *(Aegle marmelos)* from eastern Uttar Pradesh and adjoining parts of Bihar, having diversity in fruit shape and size, shell thickness, pulp colour, taste, seed and mucilage content; jackfruit from Bihar and adjoining parts of Orissa, varying in fruit size, shape and taste; pomegranate from Himachal Pradesh, differing in fruit size, shape and maturity and having resistance to fruitfly; and Frenchbean from Himachal Pradesh, possessing variability in pod length, seed size, shape and bearing.

Germplasm Exchange

More than 57,700 accessions in different agri-horti-silvicultural crops including multiplication trials, screening nuisenes in wheat and rice were introduced from various countries. Some of the promising introductions were: high-yielding and cold-tolerant wheat lines from Mongolia; blast-resistant rice from the Philippines; sorghum lines having tolerance to *Striga* from the USA, high-yielding and early-maturing yellow sarson cultivars from Bangladesh, collection of different species of *Carthamus* from the USA, pea germplasm resistant to blight disease from the UK; cowpea lines resistant to various pests and diseases; white, brown bold-seeded, high-yielding, bushy plant type and vegetable type soybean germplasm from Taiwan; potato cultivars Mainechip and Prestile suitable for making chips and accessions including wild types with resistance to late blight. bacterial wilt and spindle tube viroid; hot pepper cultivars with high capsaicin content; early-maturing, black, long and round, purple brinjal lines; watermelon breeding lines resistant to races 0, 1 and 2 of *Fusarium* wilt from the USA; high-yielding ginger lines from Malaysia; high-yielding and weevil-resistant sweet potato cultivar Miramiyataka from Japan; good dessert type peach *(Prunus persica)* having yellow, juicy flesh and resistance to peach bacterial spot; 6 varieties of papaya with high yield and good fruit quality from the USA and 14 spp. of *Leucaena* suitable for paper industry from the UK. About 36,400 samples of Indian germplasm were exported to on request from research organizations in various countries. More than 45,600 germplasm samples of different crops were supplied to various indentors in the country, such as public research institutions or private seed industries and farmers, for utilization in crop-improvement programmes, basic research or for direct cultivation.

Long-term Conservation

Accessions conserved for long term included maize (190), barley (94), buckwheat (70), oat (262), rice (767), wheat (575), amaranth (133), sorghum (302), finger millet (24), kodomillet (4), Panicum (2), sesame (283), soybean (179), sunflower (259), castor (86), niger (57), groundnut (52), safflower (123), pigeonpea (255), lentil (65), *Lathyrus* (950), jute (695), Kenaf (290), bottlegourd (95), bittergourd (9), brinjal (87), tomato (95), Chinese cabbage (21), *kasuri methi* (48); Solanum spp. (S). spinach (7), sowa (4), chillies (193), coriander (36), medicinal and aromatic plains (16), and icleased varieties of other crops (43).

Germplasm Evaluation

Approximately 35.500 germplasm accessions of different agri-horticultural crops were grown for multiplication, characterization, preliminary evaluation and maintenance at NBPGR headquarters and its regional stations Promising accessions were identified for earliness, dwarfness and large head type in sunflower; high yield in mustard; and early, high yield in soybean, sesame, okra, *Medicago* and Frenchbean. Promising materials were also identified for their resistance or tolerance to various biotic stresses. These included wheat lines resistant to yellow, brown and black rusts and powdery mildew and wild relative *Triticum bioticum, T. dicoccoides, Aegilops biuncialis* and *Hordeum bulbosum* resistant to all 3 rusts, powdery mildew and smuts; lentil lines resistant to *Ascochyta, Uromyces, Erysiphe, Fusarium* and viruses; okra lines resistant to yellow-vein mosaic; *Oiyza officinals,* a wild species of rice, resistant to grassy stunt and several lines of rice showing field resistance to stem borer, blast, leaf-roller and gall-midge. In fruit crops, cultivurs Krassavica and Plum Beauty of pluma and Viva Gold and Nugget of apricot were found promising for yield and quality traits

Germplasm Conservation

A total of 6,383 accessions which met the gene bank standards were added to the base collection for the long-term conservation in the National Gene Bank. They included cereals, pseudocereals. oilseeds, pulses, fibres, vegetables, spices, medicinal and aromatic plants and released varieties of different crops in addition, 1,771 samples of lentil from the International Centre for Agricultural Research in Dry Areas (1CARDA) were conserved for duplicate safety. Samples of indigenous collections were stored as voucher specimen. More than 150,490 accessions were stored as base collection at -20°C

In-Vitro Conservation by Tissue Culture

In addition to maintaining the existing in-vitro collection of banana, sweet potato, yams, garlic, ginger, *Curcuma* and several medicinal and aromatic and endangered plant species, 15 new accessions of ginger, 21 morphotypes *of Curcuma* and 11 accessions of *Dioscorea* were added to the collection. *In-vuro* multiplication method was developed for *Coleus parviflorus* and 3 accessions were established in tissue culture. Shoot cultures of *Gentiana* could be maintained for 18 months on a modified medium at 25°C without subculturing.

Genetic diversity in 15 species of *Allium* and *Abelmoschus* was analysed by isozyme electrophoresis. Random amplified polymorphic DNA (RAPD) analysis revealed that intraspecific variation in *Allium* W as minimum but greater diversity was observed between species. Low polymorphism for isozyme and RAPDs is found in okra. An investigation with several accessions of *Solanium melongena, S. insanum* and *S. incanum* revealed that the karyotypes were nearly identical among the 3 taxa and meiosis was regular in their hybrids. These observations confirmed the earlier conclusion on the basis of allozyme and DNA studies that the 3 are very closely related genetically *(Annual Report- 1995-96, ICAR. New Delhi).*

Application of Biotechnology to Rice Production

In its study, Agriculture : Toward 2000, FAO (1979) has projected the need for the production of an additional 300 million tons of rice between 1974-1976 and the end of the century. According to Dr. M. S. Swaminathan, the rate of growth in world demand for rice during the 1980s is expected to be on the order of 2.9 pel cent a year. In contrast to the other cereals, the demand for rice would remain overwhelmingly for direct use as human food. If past trends in demand and production continue, the developing countries gross import requirement for rice would rise from 8.3 million tons in 1974-1976 to 33 million tons in 2000. Through expanded irrigation and improved productivity, it should be possible for developing countries to meet 200 million tons of the additional demand through greater home production. This will require an optimum blend of technology, services, and public policies.

Table 12.1. Following are the varieties of rice released by the Central Sub Committee on Crop Standards and release of varieties*

Variety	State
Rained, upland ecosystem	
Birsadhan 105, Birsadhan 106, Birsadhan 107, Birsadhan 201, Birsadhan 202, Turant Dhan	Bihar
IET7564	Karnataka
Rainfed, lowland ecosystem	
Vaidehi	
Irrigated ecosystem	
Ratnagiri 3, Karjat 2, Karjat 3	Maharashtra
Arvinda	Pondicherry
Gautham, Sakuntala	Bihar
Pant Dhan 12	Punjab
Amrut	Karnataka
Aromatic rice	
Taraori Basmati	Haryana
Ranbir Basmati	Jammu and Kashmir
High altitude	
Chenab, Jheelam	Jammu and Kashmir
Central varieties Semi-deep water	
Jitendra	West Bengal, Uttar Pradesh, Bihar
Purnendu	Orissa, West Bengal, Uttar Pradesh
Irrigated	
Pusa 834	Andhra Pradesh, Karnataka, Eastern Uttar Pradesh, Madhya Pradesh, Orissa Tripura
Rice hybrid	
KHRI	Karnataka

Source: Annual Report 1995-96, ICAR, New Delhi.

Improvement of rice variety and identification of suitable genotypes for various agroecological situations with emphasis on rainfed environments are on the way. Isolation of varieties with high degree of resistance to insects, pests and diseases for increased and stable yields received equal emphasis. Twenty live varieties have been notified and released for cultivation in different states, of which 7 were for rainfed, upland ecosystem, 1 each for lowland, semi-deep and deep water, 2 for high altitude or hilly regions and 10 for irrigated areas. Two rice varieties of export quality have also been released.

Several years of breeding and selection work are needed to incorporate useful genes from a suitable donor strain into a commercially popular variety. The breeding of the rice variety IR36, which now occupies over 10 million ha. in Asia, took about 7 years. Seed production on a scale necessary to cover large areas takes another 2-3 years. Thus about 10 years is needed from the time a cross is made until it makes a widespread impact, even when two or three crops can be grown in a year. Using such techniques as *rapid generation advance,* some strains of rice will produce four crops a year.

Another recent example of the time taken for transferring a desirable gene from one genetic background to another is the work on high-lysine corn. The opaque-2 gene discovered at Purdue University in the early 1960s has been associated with several undesirable traits. It is only after 12 years of patient research that scientists at the International Maize and Wheat Research Centre in Mexico (CIMMYT) have been able to combine the high-lysine character with other desirable traits.

Besides the 4 hybrids released in Andhra Pradesh, Karnataka and Tamil Nadu, 1 more hybrid CNRH 3 suitable for boro season has been released in West Bengal, Two hybrids 1 each from Directorate of Rice Research, Hyderabad (CRH 1), and Pioneer Overseas Corporation (PBH 1) were identified. More than 140 rice hybrid varieties were evaluated in multilocation trials. Five new cytoplasmic male-sterile (CMS) lines with stable male sterility and better out-crossing ability were developed. The CMS multiplication programme has shown a great success, the maximum seed yield being 2.8 tonnes/ha.

Identification of 3 indigenous sources of temperature-genic male sterility (TGMS) is an important step towards further strengthening the hybrid breeding in rice. The transfer of TGMS system into appropriate agronomic background would help to develop 2-line hybrids with higher yield and easy seed production.

Development of Molecular Tools in Rice

In rice, transformation experiments have been initiated using biolistic particle-delivery system. For successful transformation, protocols for production of embryogenic calli from different explants as well as preculturing or isolated immature embryos were standardized. Initial trials using *Act-Gus* plasmid construct showed *Gus-positive* transient expression. Establishment of embryogenic cell suspensions with high regeneration potential is a pre-requisite for successful protoplast culture. For developing suspension cell cultures callus from mature seed embryos, immature developing embryos and juvenile infloresr-nce are normally used. But in rice, new source of embryogenic calli from the rachilla of the floret were used for developing suspension cell cultures.

By using a virulence gene probe and in combination with traditional method of identifying pathotypes of bacterial leaf blight, several isolates were analysed. The molecular analysis of the diversity in pathogen populations indicates prevalence of 1 dominant pathotype (pathotype 1) characterized by distinct RFLP pattern, in many parts of the country.

Wheat

Five wheat varieties including 1 of emmer wheat have been identified for different zones. DDK 1001 is the first-ever dwarf, high-yielding emmer wheat.

The yield-maximization trials showed the importance of Zn and organic manure for stepping up the yield level. Except at Ludhiana, application of N, P and K @ 120, 60 and 40 kg/ha + $ZnSO_4$ @ 25 kg/ha + farmyard manure (FYM) @ 10 tonnes/ha at all sites gave significantly higher yield than conventional practice of NPK alone. Similarly, there was a significant increase in yield when NPK level was raised to 150, 75 and 50 kg/ha keeping the level of both $ZnSO_4$ and FYM constant. When nutrient level was further raised to 180, 90 and 60 kg/ha with $ZnSC_4$ and FYM the yield increased to 6.8 tonnes/ha at Hisar in WH 542, 6.5 tonnes/ha at Pantnagar in UP 2338 and 6.3 tonnes/ha at Karnal in PBW 343. Thus for achieving a productivity level of more than 6 tonnes/ha, it will be necessary to increase the tonnes/ha, it will be necessary to increase the NPK level to 180, 90 and 60 kg/ha + $ZnSO_4$ and FYM. This holds good for all varieties, as all the genotypes consistently behaved in the same manner at the 3 locations.

No-tillage has an edge over conventional tillage in wheat sowing because the sowing time is advanced by 5-6 days. Wheat yield of 4.95 tonnes/ha under no-tillage in dry-seeded plot of rice was significantly higher than wheat yield of 4.03 tonnes/ha obtained in plot of transplanted rice.

Table 12.2 Following are the wheat varieties identified for release for different production conditions

Variety/Production condition	Zone
PBW 373: Irrigated, late sown	North-western plains zone
HD 2643: Irrigated, late sown	North-western plains zone
HW 2004: Rainfed, timely sown	Central Zone
NIAW 34: Irrigated, late sown	Plains zone
DDK 1001 (emmer wheat): Irrigated, timely sown	Plains zone

* *Source :* Annual Report 1995-96, ICAR, New Delhi

Forage Crops

In India, twenty-two varieties have been released in different crops. Callus induction and regeneration of plants and their subsequent establishment in the field was successfully attempted in *Cenchrus ciliaris, Dichanthium annulatum* and *Panicum maximum* through various culture (solid/agar) media and also in suspension cultures using different sources of tissues or organs, viz young leaf base, nodal segments, immature inflorescence, embryonal axis from mature seeds etc. Several variants, such as albino plantlets, creeping (spreading or prostrate) and semi-erect type of plants among various somaclones in *D. annulatum* were found stable. The androgenic plants were successfully achieved in *Dichanthium annulatum.* The plants raised from anther cultures could also be successfully transferred to the field.

Oilseed Crops

Soybean

At National Research Centre (NRC) for Soybean, 2,500 exotic indigenous germplasm accessions of soybean have been maintained and evaluated. A defect-rectified, high-yielding (2.5-3.0 tonnes/ha) mutant NRC 2 (parent Bragg) with distinct improved seed longevity as well as early maturity has been identified. The variety yielded highest (3.23 tonnes/ha) among the 4 latest varieties in frontline demonstration at farmer's field in Malwa region. Variety NRC 12 yielding high (3.0-3.5 tonnes/ha) with resistance to pod-shattering as well as insect-pest is the first success of attaining resistance or tolerance against stemfly *(Melanagromyza sojae)* in soybean. NRC 1, a mutant of native black-seeded variety Bhatt, with exceptionally high germmability (85%) after 18 months of storage under ambient conditions and tolerance to water-logging, could be further improved for pod-shattering resistance. NRC 7, a recently developed variety with high degree of resistance to pod-shattering was highest oil-yielder with more than 2% higher oil than other .varieties. This variety shows high degree of drought tolerance, resulting in complete seed filling even under water stress. Stabilizing 4-seeded pods in segregants has been highly encouraging towards breakthrough for high yield. Of the 314 lines in M_4, 11 mutants of Gaurav showed higher protein content (43-45%) than the control (42.8%). Twentyflve mutants of Bragg had oil content above 20.5% and 3 had above 21% compared with the control (20.34%). Four mutants of PK 472 showed better seed viability than the control.

Groundnut

Germplasm-supply activity remained suspended to arrest the spread of PStV disease prevalent at Junagadh. A total of 1,500 accessions of PStV-free germplasm have been regenerated from satellite centre at Bhubaneshwar and 150 accessions have been supplied to each of the 5 leading centres to enrich the gene pool. Three advanced breeding lines, viz. IR 28, PBDR 6 and PBDR 13, were found to have multiple-disease resistance.

Applying chemical mutagen, a total of 89,762 M2 populations were screened and 1,285 individuals of 20 different types of mutant selected. A total of 76 promising selections were made in F_4 and F_7 generations. Fifty promising strains of Spanish and Virginia types and 10 HPS types have been identified for yield trials.

Rapeseed-Mustard

Yellow-seeded OO mustard has been developed through transfer of genes for low glucosinolate from *B. campestris* cv. Tobin to *B juncea* genotypes CM 8814, CM 9164 and WF 1. The genotype CVWF 1 resulted in the development of O glucosinolate genotypes CCWF 116 and CCWF 112. All the O glucosinolate genotypes had intermediate level of erucic acid. The evaluation of elite-breeding lines NDR 8501, CSTR 338-1 and CSTR 244-1 recorded high salt-tolerance index at Jodhpur. ICS 3-1 and CCM 2 gave higher seed yield in saline condition at Sampla: and Kranti gave the highest yield at Faizabad.

Germplasm lines, viz. C 2, C 4, C 5, C 6, C 9, C 10, C 11, C 13, C 14, C 15, C 16, S 71, S 117, Rabacca, Erucam, Kala Amb, CNDF 3, CNDF 6, CNDF 26, CE 7, HC 9001, Pusa Bold x RE 5, YST x ENMS 1, MPC 8 x NC 57367, PC 5, RL 91-64, from Ludhiana; PCR 3, RN 398 and DYS 25-9 from Bharatpur; and CSR 1086 and R 7006 from Bathinda, showed resistance to aphid infestation (*Annual Report 1995-96, ICAR, New Delhi).*

Sunflower

Method of inoculation for the downy-mildew disease has been standardized. Whole-seedling dip in spore suspension of 5 x 10^4 spores/ml was found the most effective. Study of anther culture indicated that uninucleate stage is the right phase for anther culture. The bud should be 2-3 mm with white petals having greenish tinge at the tip. Among hybrids developed through the male-sterile system, DSH 133, DSH 136, DSH 132, DSH 138 and DSH 126 were found superior in yield than the checks.

Ideal cultural practices for the yield maximization of the parental lines of hybrid have been standardized. Staggered sowing of male parent (6D-1) by 8 days earlier to female recorded significantly highest mean seed yield of 833 kg/ha. Blocking system of planting was found better than planting male and female in 1:3 row ratio. Staggering male 8 days earlier to female with recommended fertilizer dose (60, 90 and 60 kg/ha of N, P_2O_5 and K_2O) was found to be optimum for hybrid-seed production of KBSH 1. Seed treatment with Apron 35 SD had no deleterous effect on germination and seedling vigour of the treated seeds till 3 months after treatment. However, these parameters were affected adversely when stored for more than 3 months after treatments compared with the control and hence it is not desirable to store the treated seeds beyond 3 months.

Sesame

For the first time, inter-specific cross between *Sesamum mulayanum* and *S. indicum* was effected successfully.

Safflower

Three promising varieties AKSF 68, SSF 132 and JLSF 298, and 4 promising hybrids DSH 107, MKH 11, DSH 130 and DSH 129 have been developed for different situations.

Castor

SHB 145, a medium-duration hybrid of 180-240 days, has been identified for cultivation in castor-growing areas under irrigated conditions. It has the yield potential of 2,826 kg/ha and possesses tolerance to wilt. A castor hybrid DCH 30, developed by the Directorate of Oilseeds Research, Hyderabad, has been identified for rainfed tracts of south India. It matures earlier than the existing hybrid varieties. An inter-cropping of castor with pigeonpea showed the least incidence of *Spodoptera litura.* Neem-kernel extract resulted in the lowest pupal count per seedling, indicating its insecticidal effect on leaf-miner larvae.

Linseed

Out of 4,722 germplasm lines evaluated, 5 were found promising for early maturity, 5 for high oil (45-50%), 11 for powdery-mildew tolerance, 4 for *Alternaria-blight* tolerance, 4 for wilt resistance, and 8 for budfly resistance.

Niger

A total of 1,268 gemplasm lines have been evaluated at different centres. Few accessions, viz. Sagarbulk, No. 35, DHL 1, CHH 5, Comp II, 1C 11015, Phule 4, CHH 6, RCR 64, Poladish and 1C 74596, were found high yielding.

Commercial Crops

Sugarcane

The use of tissue culture technique in sugar cane improvement as an adjunct to the coventional system has been recently emphasised. Somaclonal variation has been well documented in sugarcane and report of somaclones with improvement in specific desirable traits are on the increase. The utility of the technique will depend upon the judicious selection of breeding objectives. Development of suitable screening techniques for isolating disease resistance at the cellular level will greatly help in utilising this technology for sugar cane improvement. It is possible to induce structural and numerical chromosomal variation by passing the material through a callus phase and other manipulations at the cellular level.

Cell and tissue culture techniques in sugarcane have taken two major thrusts:

1. related to modifications in plant's genome due to passage through the tissue culture system, and
2. concerned with the use of cultures for the elucidation of biochemical parameters.

It is proposed to use cell culture techniques on sugarcane to genetically improve the varieties specially with reference to sugar storage, disease resistance and adaptability to the physical environment. Much work has been done to understand the metabolism of sugarcane. It is noteworthy to mention here that commercial sugarcane is heterozygous, with germplasm from 3-5 *Saccharum* species. Through tissue culture techniques, high yielding disease and pest-resistant clones have been developed during the past 65 years for most of the sugarcane producing areas by conventional breeding and selection programmes with 10 to 15 years after identifying new clones. Once the clones are released, they can be vegetatively propagated as long as desired.

Two varieties, viz, Co 8011 and CoM 88121, qualified for zonal release in the Peninsular zone. Variety Co 86010 performed well in Tamil Nadu, where it recorded an increase of 20.8 tonnes/ha in cane yield and 1.57 tonnes/ha in sugar yield compared with the existing standard early variety CoC 671. The variety yielded 146.1 tonnes/ha, which was 14.6% higher than that of CoC 671 (127.5 tonnes/ha) and was found suitable for planting in interior and southern Tamil Nadu. The variety is moderately resistant to drought and shows vigorous early growth compared with many existing early varieties. The variety is moderately resistant to both red-rot and smut diseases and suitable for planting any time between October and February. Variety Co 86032 identified for Peninsular zone is spreading fast in Karnataka and Maharashtra. It has also entered into Madhya Pradesh.

Screening of somaclones produced through red-rot toxin-treated cultures yielded 11 clones MR types. The protein profiles of red-rot pathogen collected from different host clones or locations showed marked variations in their native protein constitution. Delay in flowering by 13-46 days in Co 1148, Co 62197, Co 87270, BO 91, CoLk 8102, CoH 15, DW 83-411 and DW 84-465 (early flowering) was attained under field condition and thus synthronization with late-flowering varieties could be achieved by altering photoperiod.

Involvement of sugarcane phytoalexins in red-rot resistance was observed. The phytoalexin compound was 3-deoxyanthocyanidin. Three phytoalexin fractions, viz. luteolinidin, apigeninidin and 3 caffel acid ester of 5-O apigeninidin, contribute to disease resistance in sugarcane against red rot. This is the first information in sugarcane about involvement of phytoalexin fractions, in disease resistance. Occurrence of sugarcane bacilliform virus (SCBV) on sugarcane genotypes was confirmed by enzyme-linked immunosorbent assay and electron microscopic studies. The virus particles were bacilliform in shape of about 108-118 mm x 20-21 mm in size. The virus was serologically closely related to another bananavirus, banana streak virus, which infests banana.

Cotton

Two *intra-hirsutum* hybrids, viz. Fateh for Punjab and TM 1312 (Surya) for Tamil Nadu and Andhra Pradesh, and 2 *G. hirsutum* varieties, viz. LK 861 and LAM 389, both for Andhra Pradesh were notified for commercial cultivation.

Ten hybrids, viz. HA 151, HA 175, HA 116, N 431, HA 149, HA 34, Omri, HA 195, HA 200 and Eldad, received from Israel, were put under extensive testing during *Kharif* 1995 at 11 main centres and 27 subcentres throughout the country, representing varied agro-climatic zones. All the hybrids were found highly susceptible to sucking pests at all the locations. They were also susceptible to cotton leaf-curl virus in the North zone. Hence none of them could express their yield potential under Indian conditions.

Jute and Allied Fibres

Tossa jute variety JRO 36 E having tolerance to water logging has been developed for north Bengal, Assam and north Bihar. Sunnhemp germplasm SUEX 015 has been identified as day—neutral, early-flowering type. Out of 200 white jute types, 32 sunnhemp types and 25 mesta types could be screened. CIJ 021 jute was found moderately susceptible to root- and stem-rot. In sunnhemp types, 3 were found resistant to wilt and 1 to anthracnose.

Plantation Crops

Plantation crops meet human requirements for food, timber, medicins, spices and beverages. It is necessary to increase the output of plantation crops several fold per unit area because of increasing human population associated with shrinking land area. Plantation crops present certain unique problems for the plant breeder in terms of their improvement. Tissue culture technique is an important tool for rapid multiplication of several economically important plantation crops which can be of immense use in the multiplication of true-to-type high yielding plants on a large scale for planting. This technique has been successfully applied in banana, cardamom, eucalyptus, and teak.

In India, there is an urgent need to increase productivity of plantation crops in order to meet the growing demands for inland consumption and export. Plantation crops have a vital role in the Indian economy. About 41 per cent of tea, 50 per cent of coffee, 57 per cent of cardamom and 77 per cent of pepper produced in India and is exported to other countries (Muliyar, 1983).

Cashew (Anacardium occidentale L.)

Cashew *(Anacardium occidentale* L.), a member of the family *Anacardiaceae* with the natural order Sapinadales, is an evergreen tree. It is a native of tropical Central and South America; but is now distributed all over the tropics and part of warm sub-tropics. However, colder sub-tropical areas are not suitable for cashew due to cold and frost; although the tree may grow and even bear a few flowers. However, very little efforts have been made to collect historical evidence of the cashew cultivation, except the first illustrative description of cashew was given by French naturalist, Thevet in 1558 AD. The country of origin is north Brazil from where it has been throughly dispersed throughout the tropcial low land of Mexico and West Indies. In the early stages, it was introduced into West and East Africa and India by the early Portuguese travellers in the 15th and 16th centuries, mainly for checking soil erosion on the coast. It is now extensively spread in the coastal areas of East Africa, Tanganyika, Kenya, Mozambique, Uganda, Gold Coast, Nigeria and Angola in Africa; Madagascar and the Malay Peninsula; Florida, Peru, Hawaii, Tahiti, Mauritius and Seychelles. It has obtained the greatest extension in Sri Lanka and along the coastal tracts of India. It was planted particularly along the shore and has become naturalised throughout the West coast (Kerala, Karnataka, Goa and Ratnagiri tract of Maharashtra); East Coast (Tamil Nadu, Andhra Pradesh, Orissa, West Bengal) and Andaman Islands. At present, both Malabar and South Karnataka contribute maximum cashew production in India. Indian cashew *(Anacardium occidentals* L.) account for 27 per cent in the world market. However, imports, about 300,000 tonnes of raw nuts from East African countries in order to meet the demand of the processing industry.

Cashew has been traditionally neglected as a wasteland crop with a very low average yield of less than one kg per tree. There are trees which are known to give very high yields of 20 kg per tree. Seed progeny shows marked variation in fruit and nut characters. If this can be doubled or tripled through somatic embryogenesis in a shorter period of time, the rate of multiplication of elite trees, can be greatly accelerated.

Table 12.3 Area and production of some plantation crops in India

Plant species	Area'000 ha	Production per annum'000 tonnes
Camellia sinensis	400	645
Cqffea arabica	220	190
Theobroma cacao	23	4
Elettaria cardamomum	100	3.5
Piper nigrum	110	30
Curcuma longa	88	200
Zingiber qfficinales	40	80
Anacardium occidentale	460	150
Hevea brasihensis	350	235
Cocos nucifera	1100	5700*

*million nuts

Standardization of Multilocation Technique for Cashew

For breeding of cashew, two objectives may be aimed:

1. The production of improved seedling strains and
2. The development of superior clones.

All the improved varieties of cashew have so far originated from seedlings which have to be multiplied vegetatively to obtain true-to-type planting material. The potentialities of the selected trees can easily be maintained through vegetative multiplication. The selected seedlings of F_1 generation may be multiplied vegetatively. In scion material, nut size can be a critical factor,too small a nut or very large nut can be a disadvantage, however, early fruiting forms might be of special significance.

To date no serious attempts have been made for in vitro embryogenesis of this crop, apart from a brief abstract in cashew. There is ample scope to develop technology for rapid multiplication of this important earner of foreign exchange.

Rubber (Hevea brasiliensis)

Rubber *(Hevea brasiliensis* Muell. Arg.) is the principal source of natural rubber covering over 350,000 ha land in India and producing about 200,000 tonnes of natural rubber.

Table 12.4. Tissue culture of *Hevea brasiliensis*

Explant source	Response	Authors
Zygotic embryos	Plantlets	Toruan and Suryatmana, 1977
Shoot apex		
Axillary buds	R	Carron and Enjalric, 1982
Stem	C,E	Wilson and Street, 1975
	SC	Wilson et al., 1976
Leaf	C-SE	Carron and Enjalric, 1982
Anthers	C-SE plantlets	Paranjothy, 1974 (but no survival)
	C-SE, 100 plantlets	Wang et al., 1980
	Haploid and Aneuploid E, haploid plantlets	Hu and Hao, 1980

*Abbreviations: C = Callus; E = Embryos; SC = Suspension culture; SE = Somatic embryos; SH = Shootlets

About 50,000 tonnes of rubber, natural and synthetic, are being imported to meet indigenous requirements. In the traditional rubber-growing areas, the scope for further expansion is extreamly limited. Clones suited to non-traditional areas will have to be selected and identified and such high yielders will have to be selectively propagated.

Genetic base available in *Hevea* is rather limited. Induced tetraploids, triploids and genetic dwarfs have been used in the breeding programme. Rapid multiplication of such types is possible through tissue culture and subsequent somatic embryogenesis would provide a method of multiplying it in commercial scale . Callus and organogenesis would be useful in getting more variants in order to enhance the crop's genetic base. Anther culture has been exploited in China to produce haploid and aneuploid plants. On the whole, attempts to exploit tissue culture for mass production of clonal material for improvement are still in their infancy.

Beverage Crops

Beverage crops like tea and coffee are important cash crops, earning sizeable foreign exchange of about Rs. 10 billion for India annually. At present tea *(Camellia sinensis)* is being grown on about 400,000 ha of land and the annual production is 645 M kg. In order to retain its 28 per cent share of international trade and to meet the increasing domestic demand, India needs to improve the productivity of tea substantially. This objective can be realised by replanting the old, less productive tea plantations with improved plant material.

Tea (Camellia sinensis)

Tea is an exclusively cross-pollinating, self-incompatible crop, polymorphic in origin, which are the characters that hamper both genetic studies and breeding as well. The intense heterogeneity in the seedling populations provide tremendous scope for clonal selection. In the tea plant, the differentiation from callus has been obtained only in a limited number of cases. Pollen pro-embryoids, anther callus and roots from callus have been reported. Kato (1982) was able to induce some shootlets from calli originated from epidermis and sub-epidermal tissues and only a few could be established in the soil. Most intensive efforts are needed towards this end.

Coffee (Coffea arabica)

Coffee *(Coffea arabica* L.) is the second important commercial crop. India accounts for only about 2-2.5 percent of world production and export about 50 per cent of its produce worth about Rs. 3 billion annually. Both *robusta* and *arabica* types of coffee are cultivated in almost equal quantities, in India. *C. conephora* var. *Robusta* shows great variability, since it is a cross-pollinating species. C. *arabica* is a tetraploid, self-pollinating species which can also be multiplied on a mass-scale through tissue culture. In coffee embryogenesis through callus was reported in the early 1970s. Callus cultures of several different varieties of coffee have been established without difficulty. Callus has been obtained from both seedling and mature leaf explants. High frequency somatic embryogenesis could be achieved and plantlets should obtained.

Table 12.5. Tissue culture of Coffee spp.

Explant source	Response
C. *arabica:*	
Stem	C
Endosperm	C
Various plant parts	R,E
Var. Bourbon	
Leaf	SE, plantlets
C conephora var. Robusta:	
Stem	C,SE
Leaf	C,SE
Anther culture	
C. arabica	C
C canephora	C,SE
C. excelsa	C,SE
C. liberica & C. arabica	Haploid, C, pro-embryos

Abbreviations'. C - Callus; E - Embryoid; R - Roots; SE - Somatic embryos

Shoot apices of *Coffea arabica* seedlings were cultured to produce multiple shootlets and plants. Undifferentiated haploid callus tissue was also obtained in C *arabica.* Therefore, tissue culture appears to have the scope for large-scale multiplication and would be helpful in yield stabilisation. A combined effort involving the use of new varieties (e.g., var. *Cauvery),* rapid clonal multiplication, mycorrhiza for better phosphorus utilisation and integrated pest management will be helpful in increasing productivity.

Spices and Condiments

Of India's trade to a large extent depends on spices and condiments since ancient times. Spices and condiments include cardamom, pepper, ginger, turmeric, nutmeg and mace, clove, cinnamom and vanilla. Unlike other economic crops, spices continue to be cultivated in the same way today as they were thousands of years ago. Tissue culture has considerable scope to modify and multiply these crops to increase their productivity.

Cardamom (Elettaria cardamommum Mation)

Cardamom *(Elettaria cardamommum* Mation) is one among them in India, earning sizeable foreign exchange (about Rs. 650 million in 1986). Its productivity has stagnated at around 63 kg/ha in India as compared to 250 kg/ha in Guatemala, the major competitor for Indian cardamom in the world market. The crop has also suffered extensive damage due to a disease called *Katte.* Tissue culture provides an efficient and rapid method of propagation of high yielding clones and virus-free plants and can be produced through meristem culture which is always free from any virus infections. Callus culture and regeneration in cardamom provides a good system for selecting variants with characters such as early flowering, bold capsules and resistance to diseases.

Apical shoot tip culture and panicle culture to produce multiple shoots and plantlets have been reported. Shoot buds and immature panicles were selected as starting material from elite clones identified in planters fields, that had traits such as compound panicles, bold capsules and high yield potential. The rate of multiplication achieved is about 5,000 plantlets per shoot explant in a year. The plantlets have been transferred to the field.

Pepper (Piper nigrum L.)

Pepper *(Piper nigrum* L.) is the most important foreign exchange earner among spices for India. Black pepper is considered to be superior in quality. The export earnings are reported to be about Rs. 2 billion per year. The present average yield is about 275 kg/ha^{-4} in India as against 340 kg/ha^{-4} in Brazil. Tissue culture can be of use in rapid multiplication of such high yielding material. Shoot tip culture and callus culture from various parts of the pepper plant have been established (Mathews and Rao, 1984).

Ginger (Zingiber officinals)

India is the leading ginger *(Zingiber officinals)* producing country and accounts for about 50 per cent of dry ginger output in the world. The ginger produced in Kerala is considered to be one of the best types in the world. It is grown in about 40,000 ha and the total yield is about 80,000 tonnes. It is exported in three forms: fresh, pickled or processed and dry. The average yield ha^{-4} is about 1.9 tonnes ha^{-4} which can be substantially improved through clonal propagation of high yielding types employing tissue culture techniques. Multiplication rates of over 15x10 plantlets per year from an initial bud of ginger have been estimated (Nadgauda et al, 1980).

Tuber Crops

Tropical tuber crops are used as a major source of carbohydrate in the diet of about 450 million people of 80 countries of the humid tropics. Among them, cassava (known as tapioca, manioc, Yucca and mandioca), perhaps the only starchy food crop that has the ability to grow under adverse conditions in the developing tropical countries with a good or poor harvest without any crop failure. It is an attractive crop of the people in the tropics, living even in subsistence level. who consume it either as main or secondary staple food. Its productivity appears to be significantly higher than that of other staple food crops. In terms of calories per unit land area per unit of time, Cassava can produce 250 x 10^3 calories per hectare per day, compared to 176 x 10^3 for rice and 110 x 10^3 for wheat, indicating its superiority over cereals in terms of biological value.

Sweet Potato **(Ipomoea** ***spp.)***

House (1908) grouped *Ipomoea batatas* with 25 wild species, included diploid, tetraploid and hexaploid species. The diploid (2n = 30) wild species include: */. trichocarpa, I. lacunosa, I. ramoni, I. triloba, I. setifera, I. tricolor, I. palmata, I. obscura, I. biloba, I. purpurea, I crassicanlis, I. paniculata* etc. The triploid (2n = 45) species are : */. gracilis* etc. The tetraploid (2n = 60) species are: */. tiliacea, I. littoralis, 1. autorescens, I. biloba* etc: The hexaploid (2n = 90) species include: */. trifida.*

Cytoiogical aspects: The important cytological aspects of sweet potato are:

1. incompatiblity,
2. Sterility in male and female phases and
3. Production of chromosomal races.

Crop Improvement Strategy: The crop improvement strategy is mostly confined to:

1. Establishment of indigenous breeding population through germplasm collection, conservation and evaluation,
2. Introduction of superior exotic genotypes, and
3. Selection and inter-varietal hybridization.

The aims of sweet potato improvement are: higher tuber yield, desirable plant type, resistance to diseases and pests, resistance to drought, shorter growing period, good cooking and keeping quality of tubers, photo-insensitivity, wider adaptability and high protein and carotene contents. Hence, improvement programme will be confined to the establishment of a breeding population having broad genetic base, introduction of exotic genotypes, selection, hybridization, polyploidy and mutation, etc.

At CTCRI, Trivandrum, the build up of germplasm bank, both indigenous and exotic collections, added to 609 in 1983. The germplasm has to be evaluated critically for various agronomic characters, compatibility groups and for disease and pest resistance, for further utilization in the breeding programme or for direct selection.

Tissue Culture Techniques: Ipomoea batatas Improvement

- *Meristem culture:* the best known tissue culture work on sweet potato is meristem culture by Mori (1971). Addition of auxins and cytokinins help in speedy regeneration of meristem and thus is found most suitable. Mori (1971) has been able to establish virus-free meristem culture of sweet potato, eliminating most prevalent virus from the crop. MS medium supplemented with adenine and kinetin are used for this purpose.
- *Embryo culture:* The embryo culture in sweet potato can help to recover rare hybrids, (i) MS medium supplemented with GA_3 and coconut milk is found essential for initial morphogenesis of the immature embryos and to enhance callus development, (ii) combintion of NAA, BAP and GA_3 can produce multiple seedlings, (iii) Sequence transfer of embryos to the above media is ideal for the recovery of seedling, rather than hormones in combination, (iv) For the development of mature embryos, auxins and cytokinins are not essential.

Casava *(Manihot Spp.)*

Cassava is botanically named as *Manihot esculanta* Crantz; the name *Manihot* comes from the Brazillian name of the plant (Marafioti, 1970); Crantz in 1966 first used the species epithet *esculanta* and thus the botanical name *Manihot esculanta* Crantz, is found in the nomenclature. Cassava has also substantial potential for use as livestock and poultry feed for which one-third of total cassava produced in the tropics, goes for compound animal feed in the form of chips and pellets in the international trade in the E.E.C. countries and also for local consumption. Further, since the cassava tubers are rich in starch, they are increasingly used as raw material for many industries in various forms for sizing the yarn, finishing of cloth; as a thickner for printing cloth; paper and food industries. Tapioca flour is used in various forms like sago, dextrine, glucose, core binder, alcohol etc. Thus, cassava has multifarious uses.

Cassava improvements are aimed at:

1. Evolving high yielding strains/hybrids.
2. Evolving early maturing varieties/hybrids.
3. Varieties resistance to pests and diseases.
4. Producing good quality tubers with high starch and low fibre content.
5. Possessing low levels of prussic acid (HCN).
6. Adopted to a wide range of environmental conditions.
7. Compact tuberous roots with short neck; easy harvest etc.

- *Micropropagation:* For micropropagation many elite clones need virus cleaning for use in disease-free areas. For this purpose, micropropagation systems have been developed to rapidly multiply the elite materials within a short time. CIAT (1982) reported that single node cuttings from meristem-tip culture are free from bacterial blight and frog-skin disease.

- *Cryopreservation:* Conservation and exchange of germplasm are hindered by perpetuation and spread of disease causing agents through planting material. Meristem tip cultures have been used for cryo-preservation of germplasm with 90% tissue survival and 10% plant regeneration; which are used in international exchange of germplasm. In vitro culture used in germplasm exchange, safe-guards against danger of pest and diseases dissemination.

Elephant-Foot-Yam *(Amorphophallus campanulatus [RoXb.] Blume)*

Elephant foot yam or white yam *(Amorphophallus campanulatus)* is an edible aroid, belongs to the family, Araceae. This is tropical tuber crop, belongs to the tribe, *Leiodieas* and the genus, *Amorphophallus*. This is a robust herbaceous plant, with an erect, solitary long pseudostem and bearing a tripartite leaf at the top and the underground corm is large, globulosey rounded and depressed. Elephant foot yam, locally known as, Ol, Suram or Chena, etc. in India, is a subsidiary food crop which contains sufficient qualities of starch. It is believed to be a native of South-East Asia. It is indigenous to tropical Asia and Africa. In India, it is extensively grown in West Bengal, Assam, Orissa, Kerala and some parts of Tamil Nadu and Andhra Pradesh. Eventhough this crop is of minor importance or considered as secondary food crop, it is becoming very much popular in different parts of India for its palatibility and better cooking quality.

The micropropagation techniques have not yet been well developed for its improvement (Mandal, 1993).

Aroids (Colocasia)

The improvement strategy of aroids is confined to the establishment of indigenous germplasm collection, conservation, evaluation, introduction of superier exotic genotypes and inter-varietal hybridization. In the germplasm, considerable variation is observed in shape, size, flesh colour and acridity of both corm and cormels. Since hybridization is restricted due to male sterility in cocoyams, more emphasis has been given to germplasm collection, critical screening and evaluation. At CTCRI, 300 ace. in taro and 65 ace. in fannia have been collected, maintained and catalogued on the basis of morphological and agronomical characters. This is to identify distinct groups and types of economic importance for direct utilization and for further genetic improvement of aroids.

Tissue Culture Technique for Aroids (Colocasia) Improvement

Virus Elimination Through Meristem Culture: Cormel tips of *Colocasia* plants are brought into culture in MS medium. Sodium hypochloride treatment is found effective in eliminating contamination, originating from sources within the explant. Dasheen mosaic symptoms used to get masked in the later part of the growing season and hence selection of planting material free from mosaic at the time of harvest, is a great problem. Cormel tip culture will be able to eliminate this dasheen mosaic and could be maintained symptom-free through many generations within the glasshouse.

- *Meristem culture:* Meristem culture for rapid multiplication of planting material, in special tissue culture, has been suggested by Mapes (1973). Meristems from the lateral buds of the corm are grown which proliferates to produce a mass of cells with the help of appropriate media used. Those cells can differentiate into small cocoyam plants.

Minor Tuber Crops

Yam-bean (Pachyrrhizus angulatus Rich; P. erosus L.)

Yam bean, *Pachyrrhizus angulatus,* belongs to the family of *Leguminaceae* and sub-family *Papilionaceae.* In India, this is locally called 'Mishri Kand' in Bihar and Uttar Pradesh; 'Shankalu' in West Bengal. This is a popular crop in the Northern and North-Eastern states. This is cultivated for its fleshy white tuberous roots which are eaten raw in salads and cooked as a vegetable. It is eaten directly after peeling. It is invariably used for table purpose without cooking or boiling. Yam bean being a legume, is particularly significant as the only tuberous crop plant with which nitrogen-fixing bacteria are associated.

The information available on the breeding of this crop is very much limited. Induced mutation breeding by gamma rays and EMS is found to be the effective tools in yam bean improvement. Induced tetraploidy showed increased gross morphology. Yam-bean being a tropical tuber crop, prefer a long moist climate and high intensity of sunlight. Due to poor and ineffective nodulation, this crop has positive response to nitrogen upto 40 kg/ha.

The micropropagation techniques have not yet been well developed for its improvement.

Chinese Potato (Coleus perviflorus Benth)

Coleus is the tuber crop of minor importance. *Coleus perviflorus* belongs to the family Labiatae, is commonly known as Chinese Potato or country potato. In Kerala (India), it is locally known as 'Koorka'. Nearly 200 species of *Coleus* are distributed in the tropical and subtropical regions of Asia, Africa and Pacific Islands; among which 8 species are recorded in India. *Coleus perviflorus* is one of the important tropical minor tuber crop, grown in Kerala and Tamil Nadu in India which is used as a vegetable having special flavour and taste. However, many species are ornamental and are of medicinal value. *Coleus* is grown in India, Sri Lanka, Java, Indo-China and tropical Africa for its edible tubers. In India, it is extensively cultivated in Kerala and also grown in Tamil Nadu and Karnataka. At CTCRI, 63 collections of *Coleus* are maintained for evaluation. Systematic evaluation of them has resulted in the identification of a superior selections, Cp-11. In *Coleus,* crop improvement through conventional breeding method is not fesiable, due to sterility problems.

It does not set seeds and hence is propagated through seed tubers and subsequently from stem cuttings to raise the main crop. Induction of variability is possible only through mutation or through the exploitation of somatic variation that may exist in the plant.

Breeding Constraints

1. Non-availability of genetically varient types in the germplasm and source of germplasm collection, is highly restricted.
2. It flowers profusely but due to completely sterile nature of pollen grains, hybridization and further selection has no scope. Highly irregular meiosis and occurrence of desynopsis might have resulted in complete sterility of this crop.

- *Tissue Culture:* It is possible to induce direct regeneration of plantlets from leaf lamina and petiole explants. Since direct regeneration from somatic tissues and regeneration of callus can bring out some clonal variability, this technique could be of practical value in the crop improvement programme.

FRUIT CROPS

India is an important fruit growing country in the world due to its varied climatic conditions still there is limited planting material to boost its production upto international standard.

Banana /**Musa paradisiaca** *L.)*

In India, banana *(Musa paradisiaca* L.) grows in about 270,000 ha of land. Numerous varieties exist with yields ranging from 26,000 to 55,000 kg per ha. Most of the commercial cultivars are triploid and thus seed sterile. Conventional vegetative methods are slow; therefore, attention has been turned to in vitro techniques. For in vitro propagation, all possible meristem containing plant parts, e.g., suckers, peepers, dormant eyes and the base of the parental pseudostem, have been used. Somatic embryogenesis has been reported but the embryos failed to germinate into plantlets. The process for scale up of clonal plantlets through meristem culture was developed and it is possible to produce 5,000 plantlets from a single shoot tip in a year. Meristem culture of banana enables: (1) rapid multiplication, and (2) elimination *of bunchy top* virus, disease.

Mango (Mengifera indica L.)

In India Mango *(Mengifera indica* L.) accounts for about half the total area under fruit plantations. During the year a new hybrid of mango Arka Neelkiran (IIHR 20-4) has given yield of 12 tonnes/ha. Indian cultivars are monoembryonic and are propagated vegetatively by grafting. There is an urgent need to multiply elite varieties at a faster rate to cover new areas, as well as to replace old plantations. Regeneration of somatic embryos from nucellus could be more readily induced from polyembryonic mango cultivars than the monoembryonic types. The Indian varieties such as *Dusheri* and *Alfonso* have also been amenable to somatic embryogenesis. Callus tissue and rooting has been reported from cotyledons.

Citrus

The total area under *Citrus* fruit in India exceed 68,000 ha. At present, yields are on the decline. Nucellar embryony plays an important role in natural and artificial selection in the evolution of *Citrus.* Nucellar embryos are virus-free and genetically uniform source for root stocks. Chaturvedi and Mitra (1974) obtained callus from stem and leaf segments of *C. maxima* (Pomelo) shoot in culture. It had the potential to produce 400 plants from each gram in five to six months. After a prolonged period of culture, this callus became embryogenic. At the National Botanical Research Institute, Lucknow, complete plants obtained through tissue culture in several *Citrus* species have been established successfully in soil.

Many species of citrus have been improved through tissue culture techniques including the understanding the physiological processes of citrus. Propagation of citrus and virus elimination have been obtained by the use of this technique. Nuclear cultures from fertilized and unfertilized ovules have provided a mean of eliminating virus diseases and rejuvenating old citrus clones which do not produce nucellar seedlings naturally. The technique also facilitate the safe and simple transfer of such clones from one country to another. The technique is also found effective in obtaining hybrid **Poncitrus** plantlets from polyembryonic citrus cultivars in which their development is usually inhibited. Mutation breeding of citrus holds great promise, particularly when single, embryonic cells are used. Work is on the way for production of haploid and somatic hybrids for the improvement of scion and rootstock cultivars.

❑❑❑

CHAPTER-13

Biotechnology in Forestry

The flora of a region is a dynamic entity. It is continuously evolving by speculation, migration and extinction of forms in response to environmental influences. Forest has attaracted of, all the various sections of society not of India but the world over. Dr. Richard Baker has said "Tree is Life" and some one has said that forest precedes mankind.

The huge Indian population and its fast growth rate demand food, power, fuel wood, timber, etc. every day. Its land area is shrinking due to fast increasing population growth. India has about 130 M ha of wastelands. Wastelands are lands that are degraded and are lying unutilised due to constraints such as salinity, alkalinity, water logging, wind and soil erosion, etc. and are unsuitable for agriculture. The need of the hour is a quantum jump in existing afforestation targets, reclamation of wastelands and protection of forest cover which will boost our forest products. Forests are the backbone of agriculture, animal husbandry, industrial development, etc. Large scale plantations requires huge amount of planting materials like seeds or seedlings. In order to achieve higher forest yields, it is necessary that the planting materials should be of perfect quality. Therefore, we have to develop high yielding novel varieties which can resist various pathogens, pests, herbicide and stress, etc. Such plants can be obtained by conventional methods of tree breeding and non conventional methods of genetic engineering and somatic hybridization.

Unlike annual crops, the breeding behavior of trees is complex and their life cycle is long. Therefore, one has to wait for many years to produce tree varieties with desirable traits. The modern non-conventional methods of genetic engineering, tissue culture and somatic hybridization have developed new hopes to enhance forest productivity with quick results. The non-conventional methods of plant production is very expensive, therefore, it is necessary to combine conventional methods with new biotechnological methods for high productivity, both qualitatively and quantitatively.

Selection of Trees for Indian Forestry

Forest cover provides healthy environment while deforestation causes environmental hazards, degradation, soil erosion, recurrent floods, drought and desertification. Table presents Statewise forest area and geographical distribution etc. With less than 2 per cent of the world's forest area, we are called upon to meet the requirements of nearly 15 per cent of the world's population. The annual target of foresting 5 million hectares of land set by the Government of India. To achieve this target efforts have to be made by the different agencies associated with the task like forest department, voluntary agencies, individuals, societies and industrial houses, etc. We to raise productive forests by identified species. Selection of suitable species for different problem-soils and agro-climatic conditions is absolutely essential for effective reclamation of wastelands. A list of trees and shrubs for various agro-climatic conditions of India is given below:

Table 13.1. Trees and shrubs of various climatic conditions of India

Name of Plants	Uses
Arid Tropical Regions	
Acacia catechu. Acacia nilotica, Acacia tortilis, Albizia lebbeck, Anogeissus latifolia, A, pendula ,Cassia siamea, Casuarina equisetifolia, Dalbergia sissoo, Diospyros melanoxylon, Eucalyptus sp., Gmelina robusta, Madhuca longifolia, Prosopis sp., Terminalia indica	Fuel wood
Acacia nilotica, Acacia tortilis, Ailanthus excelsa, Albizia lebbeck, Albizia procera, Areca catechu, Dalbergia sissoo, Prosopis sp., Terminalia sp.	Forage
Albizia lebbeck, Albizia procera, Dalbergia sissoo, Gmelina arborea, Gmelina robusta. Holoptclea sp., Pongamia pinnata, Shorea robusta, Stercularia sp., Tectona grandis	Timber
Bombax; cciba, Diospyros melanoxylon, Eucalyptus sp.	Pulp wood
Humid Tropical Regions	
Acacia catechu, Albizia lebbeck, Cassia siamea, Casuarina equisetifolia, Dalbergia sissoo, Hibiscus integrifolia, Melia azadirachta	Fuelwood
Accadia catechu, Dalbergia sissoo, Gmelina sp., Terminalia sp. Adina cordifolia	Forage
Albizia sp., Artocarpus chaplasha, Chukrasia velutina, Dalbergia latifolia, Dalbergia sisso, Gmelina sp., Kydia calycina, Michelia champaca, Phoebe attennata, Shorea robusta, Tectona grandis,	Timber
Eucalyptus sp., Populus sp.,	Pulp wood
Tropical Regions	
Acacia mearnsii, Acer campbellii, A. caesium, Alnus sp. Celtis australis, Quercus sp., Terminalia ciliata, Tsuga dumosa	Fuelwood
Acacia mearnsii, Grewia opliwa	Forage
Abies pindrow, Alnus sp., Cedrus deodara, Cryptomeria japonica, Juglans regia, Picea smithiana, Pinus sp., Salix alba	Timber
Eucalyptus globulus, Pinus patula	Pulp wood

Forest Resources

Tissue culture technology together with mycorrhizae, nitrogen fixing microorganisms and institutional support, etc. have been used to enhance the productivity of forest trees in the United States, Brazil, India, Japan and some West European countries. Success has also been obtained in the case of orchids, ornamentals and a number of vegetable and fruit species. Species of strawberry and apple are now produced commercially by tissue culture. In India, the technique of raising plants through tissue culture has been perfected for a few tree Species, e.g., *Bamboo, Dalbergia, Eucalyptus, Leucaena, Prosopis, Santalum, Sesbania* and *Tectona* etc. In order to achieve the quantum jump in the production of biomass for fuel, forage, timber and pulp (soft) wood using the new tools of biotechnology, the following species have been identified for mass propagation through tissue culture techniques (micropropagation); *Acacia nilotica, Alnus nepalensis, Harrrdwickia binata, Madhuca latifolia, Prosopis cineraria, Tamarindus indica,*

Dendrocalamus strictus, Bambusa arundinacea, Bambusa vulgaris, Tectona grandis, Shorea robusta, Dalbergia latifolia, Santalum album, Populus deltoides.

Mass propagation of elite plants through tissue culture, scaling up the production of artificial seeds, cryopreservation of the gene pools of elite trees, exploiting somaclonal variants for tree improvement, microspore and anther culture for fixation of heterosis, development, in some important tree species of a genetic transformation system using *Agrobacterium tumefaciens* and regeneration of transformed plantlets by tissue culture, genetic manipulation of trees through protoplast fusion or somatic hybridisation, large-scale field evaluation of tissue culture raised plants are some of the important ways to increase the forest production technology in present day.

The impact of biotechnology can be observed in the following manners:

1. enhancing the planting and survival of trees on sites that are unproductive environmentally.
2. increasing the yields of both traditional and new forest product.
3. use of disease-free stock.
4. improving the yield of wood-based products by efficient conversion processes including the use of microorganisms.
5. facilitating the identification and conservation of genetic resources.

Techniques used to achieve these goals are rejuvenation through clonal propagation including cuttings, micrografting and tissue culture, haploid generation, somatic embryogenesis, protoplast fusion, gene transfer, cryopreservation breakdown of wood components; and forest soil microbiological improvement and risk detection. The use of these technologies varies from species to species and their out put also depends upon plant species.

Specific Recent Technologies Relevant to Forestry

Clonal Propagation Including Tissue Culture

Considerable progress has been achieved in developing new clones of many industrial crops like tea, coffee, cocoa etc. using modern biotechnological methods but still we need substantial work in this direction in India. However, in recent years in Australia, Brazil, Canada, Europe, Japan and the USA *inter alia* with a wide range of species including some 44 angiosperms and 19 gymnosperms (Bajaj, 1986; Dodds, 1983); rejuvenation of tropical pines and temperate oaks and valuable ornamental woody plants is in progress through micropropagation. In Britain, clonal propagules of Si ka spruce may exceed a million in 1987. Propagation of 100-year-old trees of redwood and teak has been achieved (Boulay et al.,

1979; Gupta et al., 1980) but generally, juvenile material has been used. Micrografting is a technique that inserts that apical dome of a selected tree on to the tip of a rootstock and it encourages rejuvenation; it has been successful with apple, cherry and citrus species, rubber and eucalyptus (Jonard, 1986).

Clonal propagation through tissue culture offers the possibility of juvenile screening for growth rate and disease susceptibility. The recent work in this direction includes the development of model in vitro systems to examine the disease defense and resistance mechanisms of tree species. These will then be extended to evaluate all selected genotypes before bulking and distribution into national breeding populations. Tissue cultures and micropropagules, will also permit the evaluation of trees for tolerance of difficult soils (acid, alkaline, saline or affected by heavy metals), adverse climatic factors (extremes of temperature and rainfall) or aerial pollutants. The latter is of prime concern to temperate, industrialized countries.

However, the evaluation of trees, for difficult environmental conditions is a major imperative for tropical countries where billions of hectares are below optimum potential productivity because of soil reaction or nutrient status; in tissue culture screening identifies disease resistance, it is a means of propagating virus-free planting stock. Meristem culture is well established for agricultural and horticultural species and have applications in tree species such as poplar, that are commonly affected by virus diseases. The following three techniques have relatively of high potentiality to develop new variants which have already discussed in detail.

1. Somatic embryogenesis
2. Generation of haploids and
3. Protoplast fusion

Mierobial Breakdown of Cell wall Components

Paper industries and farmers face the problems of disposal of lignin in wood and straw, respectively. Lignin inhibits straw breakdown in soil and must be removed during pulp manufacture to facilitate fibre separation and subsequent bonding in the paper. Various wood-rot fungi have been shown to produce lignase enzymes which can destroy lignin upto 3 to 4 per cent. The US Forest Products Laboratory has examined the mutants and genetically engineered variants of *Phanerochaete chrysosporhim.*

Fermentation of xylose from hemicellulose in wood by *Candida tropicalis* is an important aspect of wood technology; this leaves the cellulose intact and the fermentation products including ethanol, acetone, glycerol and butanediol (precursors of synthetic rubber). This technology is useful to develop clean environment and requires less energy than that of the traditional wood processing methods.

Anaerobic digestion of plant biomass and sewage sludge in simple biogas generators is cheap and provides low purity methane for energy while leaving mineral nutrients and organic structure for soil improvement. On the large scale, higher purities of methane (90 per cent) may be obtained with thermophilic digesters packed with immobilized organisms, thus eliminating the need for purification of the gas.

RAPID PROPAGATION OF SOME FOREST TREES

Rapid Propagation of Eucalyptus tereticornis Sm.

Eucalyptus offers an ideal industrial feed stock for pulp and paper due to its short rotation cycle, by virtue of which it provides a competitive advantage against the increasing prices of superior quality, long duration timbers, such as *Tectona grandis, Dalbergia sissoo, Shorea robusta* etc. *Eucalyptus* timber is also suitable for various types of construction (for example door and window frames). To meet the increasing demand for timber in the future, fast growing *Eucalyptus* plantations are expected to make a great contribution.

For the production of uniform plant material *Eucalyptus,* being cross-,pollinated, needs to be multiplied vegetatively. This is necessary in view of the heterogeneity observed in plantations raised from seed. In recent years, tissue culture technology has been successfully used in some *Eucalyptus* species viz., E. *citriodora, E. resinifera* and *E. maculaia, E. sideroxylon, Eucalyptus* species and hybrids, *E. camaldulensis* and *E. grandis* for the production of clonally uniform plantlets from the desirable germplasm. However, the success of exploiting this technique commercially has not been achieved in all the *Eucalyptus* species of commercial value. Dr. R. I. S. Gill and his associates of Forestry Deptt, Punjab Agricultural University, Ludhiana worked on the tissue culture aspects *of Eucalyptus* spp. The data presented here are based on their studies.

Methods for Tissue Culture: The mature trees *of Eucalyptus tereticornis* (6-8 years) are to be cut in February The coppice shoots developed within two months. The twigs (10-15 cm) comprising 5-8 nodes are then excised from the stump. The leaves are pinched off and the twigs having nodal buds are first washed thoroughly in distilled water to remove the dust particles and then cleaned with a swab of cotton dipped in alcohol (absolute). The twigs are then cut into stem segments (1.5 cm). The segments consisting of leaf petiole and nodal bud are surface sterilized (0.1% Mercuric chloride, 8 minutes) and washed thrice with sterile water. The material left in water at the final washing. The nodal segments are cultured on different media modifications based on Murashige and Skoog (1962) medium and the cultures are incubated at 25 ± 2°C under 16 h light and 8 h dark conditions.

Table 13.2. *In vitro* shoot bud regeneration (%) from nodal segment explant of *Eucalyptus* cultured on different media compositions*.

Kind of growth regulator (mg/1) in MS media	Weeks after culturing		Mean
	3 WAC	6 WAC	
Basal (MS)	31.66 (19/60) 34.13	35.0 (21/60) 36.22	33.33 (20/60) 35.17
BAP(1.0)	36.66 (22/60) 37.24	40.0 (24/60) 39.19	38.33 (23/60) 38.21
BAP (0.5)	16.66 (10/60) 24.03	23.33 (14/60) 28.76	20.0 (12/60) 26.40
BAP(1.0) + IBA(0.5)	43.33 (26/60)	46.66 (28/60)	45.0 (27/60)

Kind of growth regulator (mg/1) in MS media	Weeks after culturing		Mean
	3 WAC	6 WAC	
BAP(1.O) + IBA(0.2)	41.14 25.0 (15/60) 29.91	43.05 28.33 (17/60) 32.12	42.10 26.66 (16/60) 31.02

L.S.D (P = 0.05) Media = 3.61, Weeks after culturing = 2.28, Interaction (Media x WAC) = NS

Figures in parenthesis indicate the number of responsive cultures (numerator)/ total number of cultures (denominator).. Figures given in italics are Arc Sine transformed values.

*(Based on R. I. S. Gill, S. S.,Gill and S. S. Gosal, 1997, In: Trends in Plant Tissue Culture and Biotechnology, Agro Botanica, Bikaner)

Tissue culture *of Eucalyptus* undergoes 4 main stages: (1) initiation (2) multiplication (3) rooting and (4) establishment. The shoot bud initiation takes 3-4 months for establishment of cultures. The multiplication of shoot buds. can be carried out in high cytokinin medium. In this medium multiplication rate is increased upto 3 to 4 times, when the cultures are transferred to low cytokinin medium during subcultured period of 4 weeks. For shoot elongation, the subdivided shoot but clumps are transferred on half-strength MS medium having activated charcoal (0.2%). The elongated shoots (3-4 cm) are then transferred to rooting medium. Rooting occurs within 10 to 15 days on both solid as well as liquid medium. The rooted plantlets thus, obtained may be hardened and established in open conditions. This method of shoot proliferation from mature trees would allow selection and multiplication of superior genotypes under field conditions.

Table 13.3 *In vitro* rooting (%) of tissue culture derived shoots of *Eucalyptus* in liquid media.

Kind of auxin (mg/1) in ½ MS media	Days after culturing		Mean
	10 DAC	20 DAC	
IBA (0.5)	61.66 (37/60) 51.73	85.0 (51/60) 67.38	73.33 (44/60) 59.55
IBA(1.0)	43.33 (26/60) 41.15	63.33 (38/60) 52.77	53.33 (32/60) 46.96
IBA (1.0) ± NAA(0.2)	20.0 (12/60) 26.25	20.0 (12/60) 26.25	20.0 (12/60) 26.25
IBA (0.5) ± NAA(0.2)	23.33 (14/60) 28.77	26.66 (16/60) 30.93	25.0 (15/60) 29.85

L.S.D. (P = 0.05), Media = 5.68, Days after culturing = 4.01, Interaction (Media x DAC) = 8.03

Figures in parenthesis indicate the number of responding cultures (numerator)/ total number of explants cultured (denominator). Figures given in italics are Arc Sine transformed values,

The data given in Table (13.2 and 13.3) are means of three teplicate batches of 20 cultures each. All the results were analysed statistically using ANOVA and means were tested for significance using least significance difference (LSD).

Direct Somatic Embryogenesis From Immature Seeds of Rosewood (Dalbergia latifolia Roxb.)

Biotechnological approaches of tree tissue culture lag behind those of herbaceous crop species mainly because tree tissues or at least those from mature individuals, cannot be similarly manipulated in culture to yield high frequency of regeneration of plantlets from any desired starting cell, cellular aggregates or organ (Haissig *et al,* 1987).

Rosewood *(Dalbergia latifolia}* belonging to the family Leguminosae commonly known as East Indian Rosewood or Black wood is an important timber yielding tree which is of great commercial value. Bulk of this commodity in the world is met by India and is a good source of foreign exchange to the country.

In view of the continuous felling and lack of cultivated plantations, Rosewood has become a fearer commodity and production has substantially gone down. The gap between supply demand had widened resulting in a near crisis situation in the world market. Due to the constraints involved in the conventional propagation methods like poor seed viability, there is shortage of planting stocks. This could be overcome by *in vitro* mass production. Dr. G. Laxmi Sita and M. Murahdhara Rao, Indian Institute of Science, Bangalore have undertaken a programme on tissue culture of Rosewood with a view to develop viable tissue culture methods for mass propagation. They earlier reported mass propagation fro.n callus culture by organogenesis and enhanced axillary shoot multiplication. It is well recognised that the ultimate goal for mass propagation is by inducing somatic embryogenesis from suspension cultures. They found that the age of the explant plays an important role in the induction of somatic embryogenesis.

Maximum number of somatic embryos (32) was observed from 90 day old cotyledons where somatic embryos started emerging within 7-1P days whereas from 120 day old explants the induction was 20-25 days after transfer. Fully mature cotyledons, excised from the dry seeds, did not show any embryogenesis but callused profusely regardless of the media composition. Sixty day old immature seeds did not respond and only turned brown. For further development of these embryos into complete plantlets, embryos with well developed cotyledons were transferred to growth regulator free medium with full, half strength MS major salts. Approximately 30% of the embryos of normal shape germinated within 20-30 days after transfer.

Dr. G. Laxmi Sita and M. Muralidhara Rao are of opinion that cotyledons from immature zygotic embryos provide the best source for somatic embryogenesis in Rosewood. Cotyledons from 90 day old pods give the highest frequency of direct somatic embryogenesis. Direct regeneration from cotyledonary explants offers an ideal system for co-cultivation using *Agrobacterium* mediated genetic transformation. Work on these lines using marker genes is in progress.

Micropropagation of Butea monosperma (Lam) Taub.

Butea monosperma Lam *(Butea frondosa* Koen) is a deciduous tree belonging to the family Leguminosae (Papilionaceae). In summer the gorgeous canopy of scarlet flowers look like flames, giving the tree its characteristics name the flame of the forest. *Butea* is a frost hardy, drought -resistant tree. Due to heavy deforestation, the trees are facing extinction especially along the west coast of India. *Butea* would be very useful for afforestation of eroded, degraded lands and has a great potential as an ornamental and shade tree in parks and gardens. Leguminous trees besides befog able to meet the demands for fodder, fuel and timber hold some promise as a means of maintaining soil nitrogen without the use of fertilizers. International bodies like FAO, NAS, ICRAF are actively promoting the planting of woody legumes.

Kavitha R. Kulkarni, Smitha Hegde and L. D. Souza (1997) reported that multiple shoots can be initiated from axillary buds of mature trees of *Butea* on WP medium with BAP. These were limited in number, only few shoots can be obtained from an axillary bud. The shoot tips turned brown and became necrotic. A large number of multiple buds can be obtained from cotyledonary nodes on half strength WP medium supplemented with BAP.

Cotyledon segments gives rise to white friable callus on MS medium with 5 mg/1 kinetin. The callus gives rise to shoots some of which produced roots. Green nodular callus are formed from the cotyledon segments on MS medium with 5 mg/1 BAP. The nodular callus gives rise to embryoids which germinated when subcultured on the same medium giving rise to plantlets with weak roots.

In nature root tubers are produced in *Butea* which give rise to new plants. Experiments were conducted by Kavitha, R. Kulkarni and his associates to induce such tubers *in vitro.* According to them the hypocotyl often day old seedlings was cut off just below the cotyledon and the seedlings were cultured on MS medium with Kin and phloroglucinol. The cut surface of the hypocotyl produced a mass of cells which developed into globular tuber like structures. These tubers when subcultured on WP medium supplemented with BAP, phloroglucinol, gave rise to shoots. Secondary tubers were formed from the primary tubers. Shoots from axillary buds as well from tubers do not root easily. Roots could be induced in a few shoots by planting them in moist sand supplemented with $MnSO_4$ and IBA. Rooted plants have been transferred to pots in the green house and also to the field.

Applications of Biotechnologies to Forest Soil Microbiology

The genetic manipulations of nitrogen fixing relationships between plants and rhizobia and actinomycetes reported by Postgate (1987). The principles clearly apply to the many leguminous trees now is used, particularly in the tropics and to the non-leguminous, nitrogen fixing species such as alder, *Casuarina* and some elms. Enhancement of the amount of nitrogen fixed has obvious benefits to rural populations in the Third World who cannot afford or are unable to import fertilizers. This is one of the major aims of agro forestry systems for adverse conditions such as high temperature, low rainfall, and extreme soil pH; for any condition there is potential for selection of fungal strain and for synthesis of tree-fungal association genotypes. This may be achieved in vitro and mass production of fungal spores for nursery inoculation in culture. One of the common problems in establishing tree plantations is the existence in the soil infected with fungal mycelia. It is hard to determine the risk of infection by normal sampling and microscopic examination because of the difficulty of species identification. The studies on the feasibility of using monoclonal antibody techniques to overcome this problem is in progress.

CHAPTER-14

Biotechnology in Relation to Nitrogen Fixation and Plant Productivity

Nitrogen is one of the chief and important constituents of biotic and abiotic structures. The biotic structures include protein and nucleic acid molecules that play the basic role in cell structure, metabolism, growth, reproduction and transmission of heritable characters. Therefore, in absence of constant supply of this unavoidable element, life can not exist.

Abiotic nitrogen component includes gaseous nitrogen, the most abundant element of atmosphere (78.08% v/v), and seems to have a highly complex nutrient cycle in the terrestrial and aquatic ecosystems. There is an interrelationship between different elements required for plant growth in environment. These all are affected by microbes. Atmospheric nitrogen is chemically inert. Therefore, it can be used, as such, by most of the living organisms. The common source of nitrogen in the nature is the nitrate ions (NO_3^-). The nitrate ion is often absorbed by the plants as a mineral raw material from the environment. The absorbed nitrate ions are then converted into amino groups (NH_2^-). The conversion of NO_3^- to NH_2 group is only possible by plants and not by animals. Minerals accumulation in soil, dissolution of rock by water, addition of fertilizers in soil are also the sources of NO_3^- in the soil.

Apart from these sources, atmospheric nitrogen and oxygen combine together in presence of lightening which come down with the help of rain-water to the ground, and serve as NO_3^- source on the earth. Absorbed nitrogen (in the form of NO_3^-) incorporates into organic structures of plants which remain intact until death. Animals also obtain nitrogen by eating plants. Finally, both plants and animals die or decay. Through gradual decay of all nitrogenous compound of dead plants and animals, it gets converted into ammonia (NH_3^+). The so forming NH_3 is then utilized as a source of nutrition for nitrifying bacteria. Usually, these bacteria are of two kinds:

1. Those bacteria which absorb ammonia and convert it into nitrite ions i.e., $NH_3 \rightarrow NO_2^-$
2. Those bacteria which absorb nitrite ions and convert it into nitrate i.e., $NO_2 \rightarrow NO_3^-$

Thus, the environment again gets NO_3^- by the activities of these nitrifying bacteria. The available NO_3^- in the environment acts upon denitrifying bacteria which converts it into molecular, atmospheric nitrogen. The free molecular nitrogen is then, utilized by nitrogen fixing organisms in which few species of bacteria and blue green algae have been included.

They are often present in soil and water. They absorb nitrogen and incorporate it into ammonia and protein. The process is called **Nitrogen fixation**.

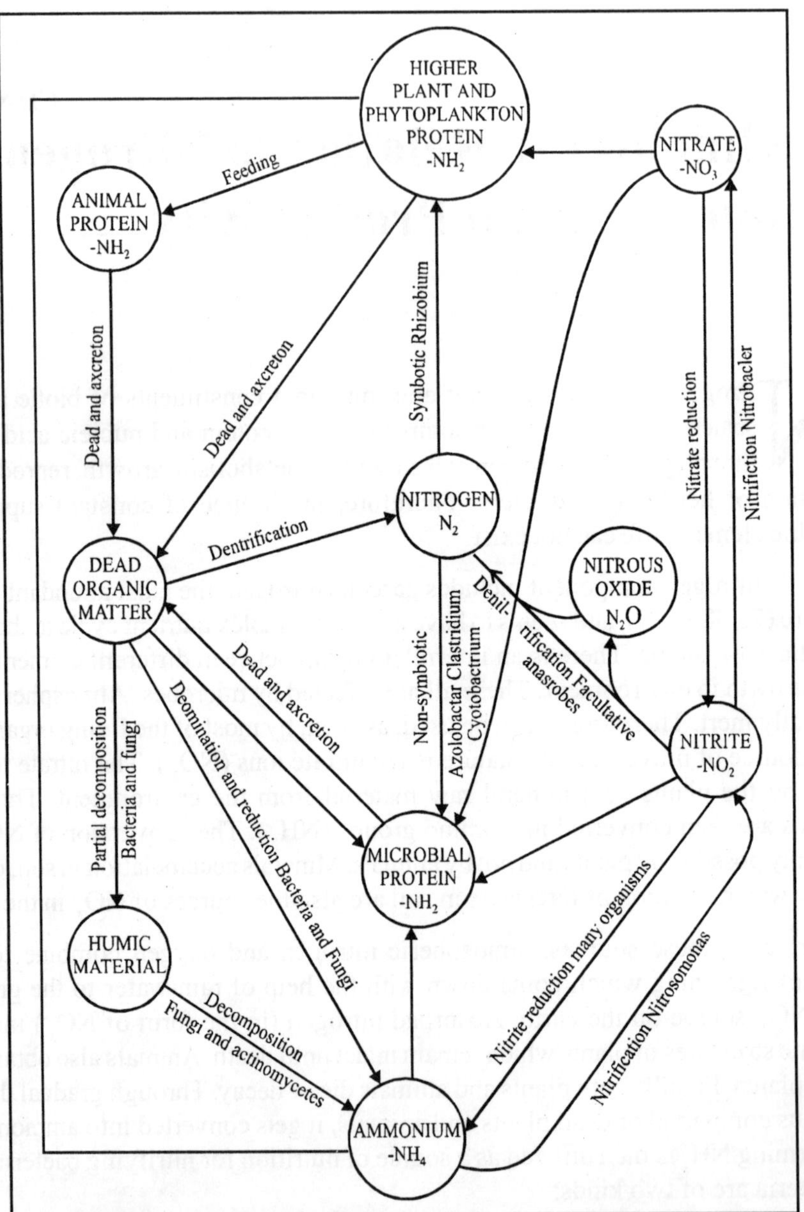

Diagram 14.1 The chemical conversions in the nitrogen cycle.

Nitrogen is the mineral nutrient most needed by plants and it often limits plant growth. Some plifit species have formed mutualistic symbiosis with nitrogen fixing *prokaryotes* and some *eukaryotes*. Those organisms that can directly utilise atmospheric nitrogen as a nitrogen source, are called *diazotrophs,* and belong to the kingdoms, *eubacteria* and *archaebacteria.* They are able to live independently on soil N for growth. Inside the root nodules, the bacteria on the other hand are supplied with carbon from the host and are sheltered from competition with other organisms. The most important mutualistic symbiosis are root nodule symbioses, mainly formed by members of the genera *Rhizobium, Bradyrhizobium* and *Azorhizobium* on legumes, and by the genus *Frankia* on some non-legumes. The reduction of atmospheric dinitrogen (N_2) to ammonium (NH^+_4) which occurs in nitrogen fixing organisms is catalysed by an enzyme complex called *nitrogenase.* A minimum 16 ATP and $8e^-$ are needed for a reduction of one molecule of N_2. It is estimated that nitrogen fixation could range from one to a few kg/ha/year in lichens and free living bacteria, and up to a hundred or possibly a few hundred kg / ha year in legumes and actinorhizal plants.

Nitrogen fixation has always been a subject of great importance in biology, but since two decades the recent biotechnological development has lent its practical urgency. Nitrogenous fertilizers have become more and more expensive in terms of fossil energy, and hard cash; the population pressure have increased the need for high protein plant food. Biotechnologists are actively engaged in improving the efficiency of microorganisms to fix atmospheric nitrogen in fields. This will improve the functioning of the nitrogen cycle in biosphere which will ultimatly increase plant productivity. The programme of improvement also includes the creation of nitrogen-fixing symbiotic association other than those existing between leguminous plants and bacteria of the genus *Rhizobium.* This will help to improve agricultural production.

Diazotrophic Microorganisms

The fixation of atmospheric nitrogen by prokaryotic organisms is known as diazotrophy. The phenomenon has also been observed in eukaryotic unicellular green alga, isolated from hot spring, by Yanada and Sakaguchi (1980).

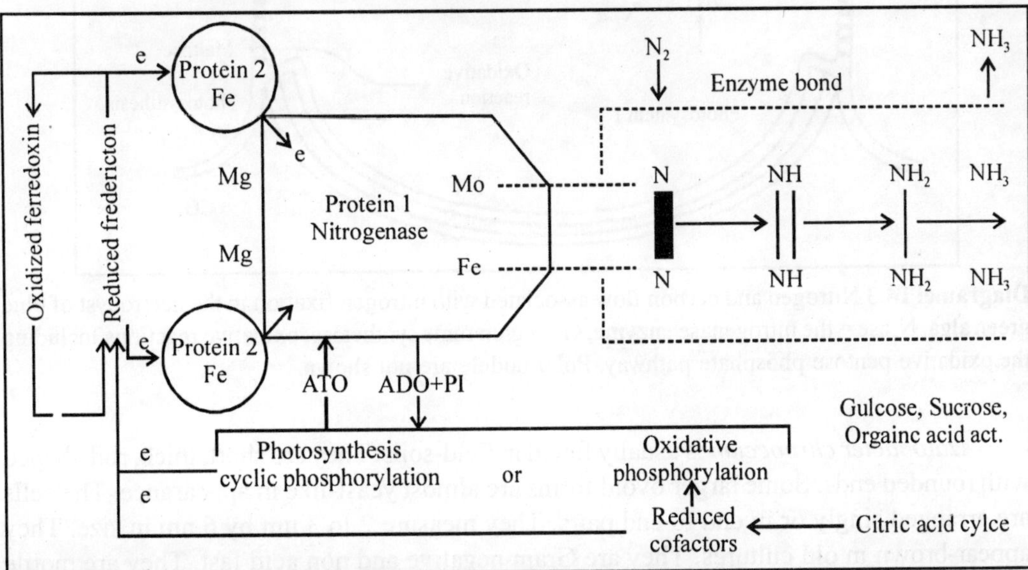

Diagrame. 14.2 A generalised scheme for the action of nitrogenase (incorporating results from studies on various organisms).

Biological Nitrogen Fixation

This process is brought about by nitrogen-fixing microorganisms. These are soil inhibitants e.g. bacteria, blue green algae, molds, etc. Those microorganisms which utilize gaseous nitrogen directly and independently in the soil are known as asymbiotic or non-symbiotic nitrogen fixing organisms and the process by which they brought about is referred as non-symbiotic nitrogen fixation. Those microorganisms which utilize gaseous nitrogen indirectly through the mediation of other living organisms in the soil are called symbiotic nitrogen fixing bacteria and the process brought about by them is referred as symbiotic nitrogen fixation.

Asymbiotic Nitrogen Fixation

Microorganisms

Bacteria: As described earlier, soil contains free-living bacteria that are capable of fixing molecular nitrogen into its compounds. This process was first recognized in bacterium *Clostridium pasturianum* which is a Gram-positive, spore forming rod-shaped bacterium. After seven years of the isolation of *C: pasturianum,* two more important free-living nitrogen-fixiti bacteria were isolated by Beijernick (1901). These newly identified bacteria were *Azatobactor chroocum* and *A. Agile.*

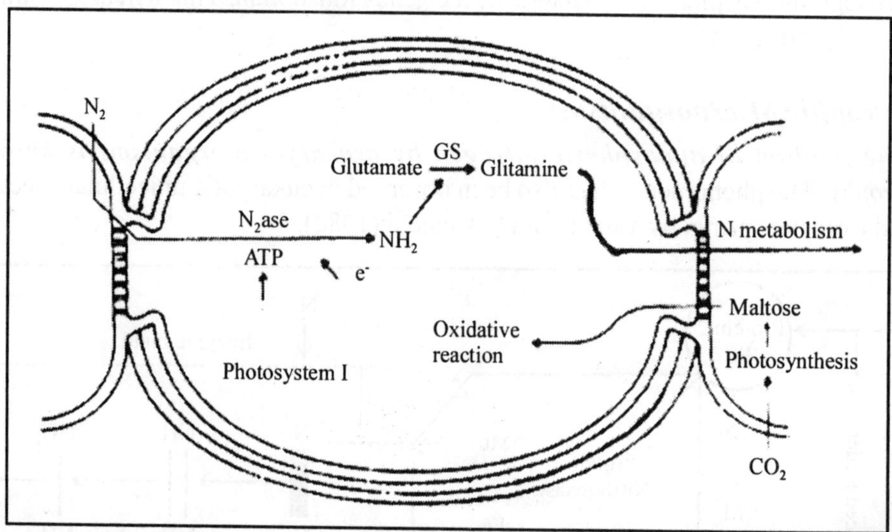

Diagrame. 14.3 Nitrogen and carbon flow associated with nitrogen fixation in the heterocyst of blue green alga. N_2ase = the nitrogenase enzyme; GS = glutamine synthetase, oxidative reactions,including the.oxidative pentose phosphate pathway. Polar nodule are not shown.

Azatobacter chrooccum is usually found in field-soils. They are short, thick, rod-shaped with rounded ends. Some larger ovoid forms are almost yeast-like in appearance. The cells are arranged singly or in end to end pairs. They measure 2 to 3 nm by 6 nm in size. They appear-brown in old cultures. They are Gram-negative and non acid fast. They are motile and the motility is due to the presence of polar flagellum. Physiologically, they are aerobic and remain active in between 25° C. They fix atmospheric nitrogen only after utilizing carbohydrate (dextrose, maltose, lactate, etc.) and give carbon dioxide as a by product. *A. glile* also actively fixes atmospheric nitrogen and produces carbon dioxide.

Apart from *Azatobacter,* other non-symbiotic bacteria like *Rhodospirillum pneumoiniae, Rhodopseudomonas, Chlorobium, Diplococcus pneumoiniae, Azatobacter aerogenes. Micrococcus sulfurens* have also been known to fix atmospheric nitrogen asymbiotically.

The ability to fix nitrogen by microorganisms was confirmed by means of the technique known as *acetylene reduction* to *ethylene* by diazotrophic microorganisms. The mechanism of this conversion is controlled by an enzymatic complex (or nitrogenous enzyme) which can reduce gaseous nitrogen to ammonia.

The enzyme nitrogenase is sensitive towards oxygen. There are reports which reveal that several groups of microorganisms fix nitrogen in presence of minute quantities of oxygen. Such microorganisms are known as **microaerobic fixers.** They oxidize methane. The examples are bacteria, Spirilla *(Aquaspirillum and Azosprillium), Xanthobacter autorophicus* and S. *flavus, Thiobacillus ferrooxidans and* non-heterocystous microaerobic cyanobacteria *(Oscillatoria, Plectonema).* Certain coliform bacteria also fix nitrogen only in aerobic conditions (e.g., *Citrobactor, Enterobacter, Erwinia).* A comprehensive lists of microorganisms have been presented in the book, but in near future they may need revision in light of 'findings of biochemical and physiological researches'. It is established that oxygen not only inhibits the activity of nitrogenase enzyme but also regulates its biosynthesis, so that in conditions that are physiologically hardly favourable to nitrogen fixation, the enzyme is not synthesized.

Mechanism of Asymbiotic Nitrogen Fixation

The knowledge of asymbiotic nitrogen fixation is still insufficient. It was previously believed that *Azatobacter* fixes atmospheric nitrogen with the help of enzyme system called azatase and nitrogenase is only one of the compounds of this system. Beijerinck (1901) indicated that fixation of atmospheric nitrogen took place at the nitrous acid stage, but it was not proved. According to one more concept, nitrogen is fixed through the stage of hydrolysis to hydroxylamine or to ammonium nitrate which is produced from nitrogen and water as shown below:

$$N_2 + 2H_2O \rightarrow NH_4NO_2$$

But this concept has not been proved. The hydrolysis of nitrogen is very difficult under normal conditions.

It was further proposed that *Azatobacter* fixes nitrogen after its reduction by the active hydrogen. However, this hypothesis was also not proved.

At last the mechanism of nitrogen fixation was described on the physiological basis. It was proved that the protoplasm of *Azatobacter* produces special catalysts which take part in fixing free nitrogen. The catalysts contain carboxyl and amino group as an important parts which do not take any active part in fixing atmospheric nitrogen. Phenyl hydrogen and hydroxylamine are also present in the composition of the catalysts. It was presumed that molecular nitrogen reacts with catalysts of nitrogen fixation through oxygen of carboxyl group and produces hydrazine as the first product of the reaction. The hydrazine, is then, further reduced by active hydrogen and is transformed into amino acids. They finally enter the proteins of the cell protoplasm. This incorporation of amino acid into protoplasms helps into the growth of a bacterium.

Requirements of Nitrogen Fixation

1. Oxygen and carboxyl group of the catalysts play an important role in nitrogen fixation.
2. A very small amount of chemical energy (ATP) is used up during nitrogen fixation. About 1 kilo calories fix 2 mg of atmospheric nitrogen. It shows the relationship between respiration and nitrogen fixation.

Some strict anaerobic bacteria are more sensitive to oxygen. *Clostridium pasteurianum* is an obligate anaerobe, therefore, the mechanism of nitrogen fixation of such organism is similar to that *of Azatobacter.* because they produce large amount of hydrogen during their metabolic reaction. The various possible pathways of nitrogen fixation have been shown in different illustrations.

Symbiotic Nitrogen Fixation

Agriculturists have noted that a relatively large group of plants, the legumes, are capable of fixing atmospheric nitrogen through a symbiotic association with soil bacteria called *Rhizobium.* These bacteria or nodule bacteria, usually dwell in the roots of various beans, peas, lupines, cloves and other leguminous plants.

Hellriegal and Wilfarth (1888) first of all demonstrated that an increase in nitrogen content in soil is due to the presence of small tumor-like outgrowth on the roots of leguminous plants. It was found that in absence of root nodules, growth of succeding plant-crop was retarded. The results of various experiments reveal that root nodule forming plants can feed upon atmospheric nitrogen and this capacity of plants can be obserbed only in those plants which are growing in non-sterilized or bacteria containing soils. These symbiotic bacteria e.g., *Rhizobium* utilize the free atmospheric nitrogen and synthesize it into new nitrogenous compounds, which are utilized by plants for their growth; and bacteria get their food from these plants. Such a mutual beneficial association of bacteria and plants is referred to symbiosis.

Symbiosis was originally defined by Bary in 1879. Symbiosis is the living together of any two organisms. Mutulism is a relationship where both partners benefit, though it greads into those where the advantage or disadvantage to one of the partner may not always be obvious. Commensal relationships are usually rather loose association of the benefit of another without any effect on itself.

Formation of Nodule in Leguminous Plants

Roots of most of the leguminous plants possess circular out-growth or swelling called nodules or tubercules, which are formed after infection of certain types of nitrogen fixing bacteria particularly the different species *of Rhizobium.*

These bacteria have the capacity to fix free atmospheric nitrogen of soil in nodule. They infect or get their enterance in root through the soft-hair or other epidermal cells by damaging them. In the first stage of the infection, bacteria grow very profusely at the tip of the root hair and form a long filament in the root hair- called *infection thread.* This thread reaches the endodermis and pericycle area through cortex tissue. Cells of this area (cortex) go on dividing and form a young nodule. The newly formed young nodule pushes the overlying parenchyma and the epidermal tissue towards outside and produce a small swelling on the surface of the root. Anatomically, a nodule is made of this walled parenchmatous cells which is filled up with the nitrogen fixing organisms. The size and shape of the root nodule varies according to plants in which occur. It is not necessary that all the bacteria which infect the root, produce nodule.

According to Wipf and Cooper (1939) the root nodule always contains double number of chromosomes as against of normal somatic tissue. If the root lacks such cells (cells with double chromosome number), there will be no formation of root nodule, even after the formation of infection thread. Those plants which do not bear root-nodules will never be able to fix atmospheric nitrogen in the plants. It has been observed by Alien and Alien (1947) that out of 429 genera of family Leguminosae about 179 genera containing 949 species form nodules.

Nitrogen Fixing Organisms found in Nodules

Rhinibium : The presence of these bacteria in the root nodule can be seen under high power microscope by crushing a washed nodule between two glass slider. Frank (1877) and Beijernick 1888) discovered *Rhizobium* a free living or symbiotic bacteria of leguminous as well as non-leguminous plants. This kind of bacteria, preferentially, infect legumes as compared to non-legumes. Structurally, *Rhizobium* are rod shaped but great variation can be observed during their life cycle. These are coccoid. very small, highly motile and ellipsoidal forms. The bacteroids are usually irregular with X, Y, star and club-shaped forms. They are 0.5 to 0.9 micron by 1.2 to 3.0 micron in size. All the species are non-sporing and non-acid fast. These symbiotic bacteria *Rhuzobium* are difficult to cultivate in ordinary culture media but they grow on manitol agar.

Rhizobia are susceptible to antibiotics (produced by other microorganisms of soil) and also to the action of bacteriophages. Fungicides, herbicides and other plant protectants are also toxic to *Rhizobia'*. Occurrence of different species of *Rhizobium* in leguminous plants can be enumerated as follows:

Agrobacterium (A. radiobacter, A. tumefacions) : It has also been included in the species: *'Rhizobium radiobacter,* but this splitting of *Rhizobium* at the genetic level into *Agrobacterium* is not accepted by most of the microbiologists.

One more kind of slow growing asymbiotic bacteria of lupin, soybean, cowpea have been identified and placed under a separate genus *Phytomyxa (P. japonicum).* In short, the root nodule genera can be delimited into two genera: *Rhizobium* and *Phytomyxa:* the former comprises of three species: *R. leguminosarum, R. meliloti* and *R. radiobacter,* while the latter bears only one species *P. japonicum.*

Table 14.1 List of some plants in which *Rhizobium* is found

Name of the plants	*Rhizobium* spp.
Clover	*R. trifolii*
Bean	*R. phaseoli*
Lupinus	*R. lupini*
Ornithopus	*R. lupini*
Soybean	*R. japonicum*
Melilotus, Trigonella	*R. meliloti*
Pea	*R. legeumimnosarum*

According to Elken (1969) *Rhizobium* can also be classified on the basis of DNA composition, for example guanine and cytosine composition of *R. leguminosarum* and *R. meliloti* is 56.6 to 63.1 per cent while G+C composition of *P. japonicum* is 62.8 to 65.5 per cent.

Structure and Function of Nodule

The outermost layer of the nodule constitutes the bacteroid zone which is enclosed by several layers of cortical cells. The rate of nitrogen fixation of nodule is directly proportional to the volume of the effective nodule. Sometimes ineffective strains produce ineffective nodules. These nodules are relatively small and contain poorly developed tissue associated with morphological abnormalities.

Leghaemoglobin

The effective nodules are usually larger and pink in colour. This is due to the presence of red coloured leghaemoglobin (leg indicating its presence in legume root nodule). This pigment is similar to haemoglobin of blood and is found in nodules between bacteroids and membrane envelops, enclosing them. Chemically, it is haem-protein and it contains a haem meoiety attached to a peptide chain and represents the globin part of the molecule. The molecular weight of leghaemoglobin is 16,000 - 17,000 daltons. The amount of leghaemoglobin in nodules has the direct relationship between amount of atmospheric nitrogen fixed by legumes. The functions of leghaemoglobin are as follows:

1. It represents an active site of nitrogen absorption and reduction.
2. It acts as a specific electron carrier in nitrogen fixation.
3. It regulates the oxygen-supply in the nodule.
4. It acts as an oxygen-carrier.

According to Scholander (1960), leghaemoglobin acts as a biological-valve in regulating the supply of oxygen to bacteroids in nodules.

Site of Nitrogen Fixation in Nodules

According to Bergersen (1969) the membrane envelops surrounding the groups of bacteroids are the site of nitrogen fixation, but in recent years it has been proved that bacteroids are the seat of nitrogen fixation.

Mechanism of Nitrogen Fixation in Legumes

The mechanism of nitrogen fixation in legumes has not yet been understood. However, the basic prerequisite for nitrogen fixation in extracts of nodule-bacteriods are similar to those that have been described for asymbiotic bacteria such as *Azatobacter*. It has not been well established that how in both asymbiotic and symbiotic system 15N is converted into NH_3 and then to amino acids, because the intermediate products between 15N to NH3 have not been determined.

A good number of theories have been proposed by various workers. Two of them are important, which are as follows:

Theory of Virtanen

Nitrogen fixation in root appears immediately after nodule formation. Virtanen (1948) proposed the following scheme of nitrogen fixation in root nodules:

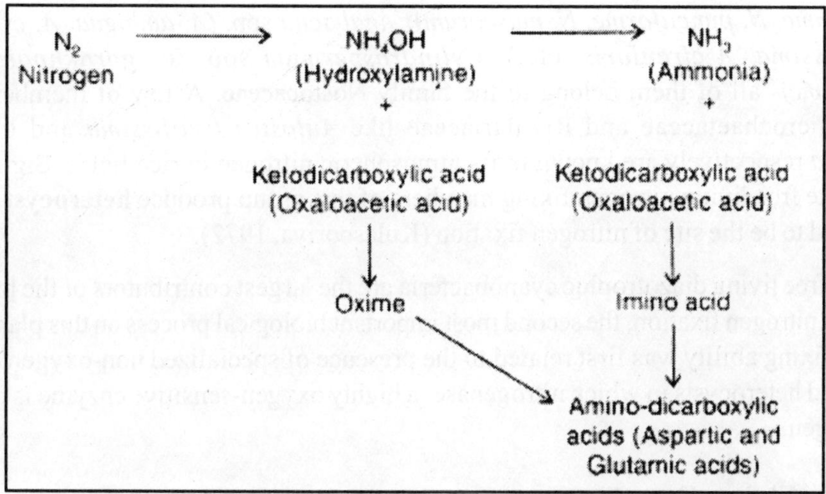

Diagrame 14.5 Mechanism of nitrogen fixation (Virtanen, 1948)

According to him young plants fix more amount of nitrogen than the old plants. A great part of nitrogen is converted into L-aspartic acid and L-glutamic acid. Apart from these, α-alanine is also present in the nodule which is produced from L-aspartic acid by decarboxylation. In addition to this, small amount of Oxime-N and Nitrite-N are also present.

Theory of Burri's and Wilson

According to this theory hydroxylamine is the central compound of nitrogen fixation, from which ammonia is formed through reduction or it can also directly react to the keto acid.

Both these theories reveal the formation of amino dicarboxylic acid as the primary products in biological nitrogen fixation.

NITROGEN CONVERTERS IN THE SOIL

The oxidation of ammonia to nitrate in the soil is one of the important phases in nitrogen cycle. This is brought about by two groups of bacteria- *Nitrosomonas* and *Nitrobacter.* The required energy for the growth of these bacteria is obtained through the oxidation of ammonia or nitrate. Winogradsky (1891) first isolated these autotrophic bacteria and described the conversion of ammonia to nitrite, and nitrite to nitrate. This is called **nitrification.** The conversion of nitrate to nitrous oxide and nitrogen gas by various kinds of soil organisms is called **denitrification.** The process releases nitrogen gas to the atmosphere and completes the complex nitrogen cycle in nature.

Cyanobacteria and Nitrogen fixation

Frank (1889) first reported the ability of nitrogen fixation by blue green algae. 'According to Drevves (1928) *Nostoc'punctiforme* and *Anabaena variabilis* have the ability to fix nitrogen in the soil. The principal genera of this group known to fix nitrogen are *Nostoc* spp. *(N. cumnnme. N. punctiforme, N. muscorum); Anabaena* spp. *(A. ambigua. A. cylindrica, A, fertilissima. A circularis,* etc.), *Cylindrospermum* spp. *(C. gorakhporense, C. licheniforme)*- all of them belong to the family Nostocaceae. A few of members of the family Microchaetaceae and Rivulariaceae like *Aulosira fertilissima* and *Calothrix brevissima* respectively are known to fix atmospheric nitrogen in rice fields. Significantly most of the free living nitrogen fixing members of this group produce **heterocyst** which is considered to be the site of nitrogen fixation (Kulasooriya, 1972).

The free living diazotrophic cyanobacteria are the largest contributors of the process of biological nitrogen fixation, the second most important biological process on this planet. Their nitrogen fixing ability was first related to the presence of specialized non-oxygen evolving cells called heterocysts in which nitrogenase; a highly oxygen-sensitive enzyme is produced from oxygen.

Nitrogen Fixing Fungi

The fungi also independently play an important role in fixing free nitrogen in soil. The fungi that are responsible for this function are of species of *Mactosporium commune, Cladosporium herbarum, Phoma, Alternaria tenuiis, Rhodotorula* spp. etc. The occurrence of these fungi has been noted in forest-land.

BIOCHEMICAL ASPECTS OF DIAZOTROPHY

An American company (DuPont) in 1960 started the fundamental research on the biochemistry of nitrogen fixation specially on nitrogenase activity in cell free extracts of *Clostridium pasteurianum.* The experiments were conducted in absence of air because of the very great oxygen sensitivity of nitrogenase. The techniques like chromatography, electrophoresis and fractionation under anaerobic conditions, helped the researchers to isolate nitrogenase from *Clostridium pasteurianum.* Thereafter, nitrogenase was easily isolated from other anaerobic nitrogen fixers while this extraction was more difficult with *Azotobacter vinelandii.* The so isolated enzyme has the property to accept electorn from sodium dithionite and was oxygen-tolerant and evolved hydrogen in the absence of nitrogen. In 1981, nitrogenase in the form of crude preparations was isolated from about 30 microbes and out of which about *6* of them were purified.

Biochemically, nitrogenase is a binary enzyme and consists of two brown metalloproteins. of which the joint activity is essential to reduce nitrogen to ammonia.

During nitrogen conversion, nitrogenase binds to molybdoferroprotein (MoFe-protein) which was named **disnitrogenase.** When nitrogen binds at the molybdenum level in a separable fragment of protein, called **FeMoco.** The two componcnt proteins of the enzyme nitrogenas are quickly and irreversibly destroyed by exposure to oxygen.

Nitrogenase is also able to reduce other substrates in the place of nitrogen, including hydrogen ions and, in this case it evolves hydrogen.

Nitrogenase Producing E. coli *Cells*

Because of the close relationship between *E. coli* and *Klebsiella pneumoniae, E. coli* is specially qualified for the investigation of the nitrogenase system. With *E. coli* C-M74 cells, high nitrogenase activity can be attained in culture medium, supplemented with 50mg/ l ammonium sulfate, 1% aspartic acid, and 0.1% yeast extract.

During reduction of nitrogen to ammonia, electrons, are required which are transferred to nitrogenase by a specific transport proteins; the *Ferrodoxins* or the *Flavodoxins.* ATP serves as energy source. The consumption of ATP molecules is 12 to 15 per nitrogen molecule. The reason for this high consumption of ATP has not been entirely elucidated.

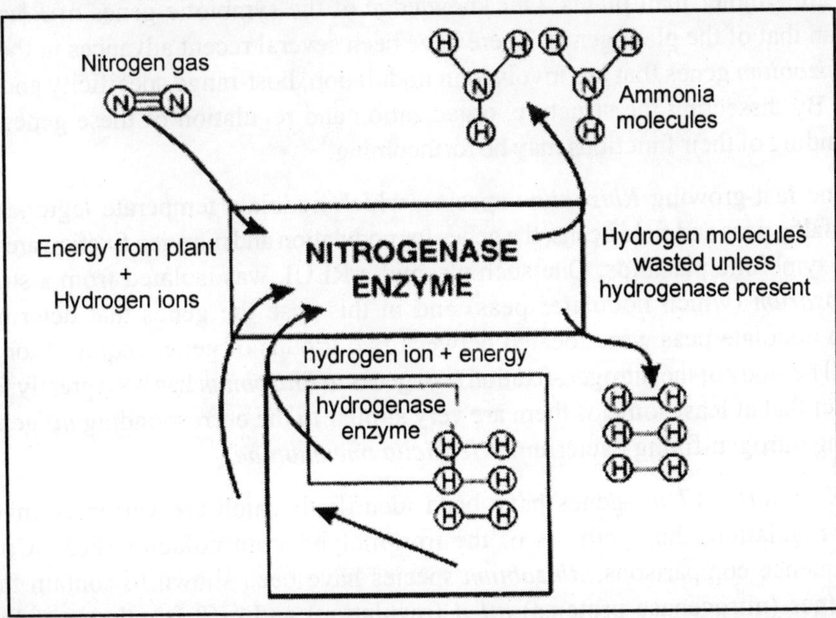

Diagrame 14.7 Representation of the mechanism of nitrogen fixation and the action of hydrogenase.

As evident from the reactions that hydrogen is produced during nitrogen fixation. Certain nitrogen-fixing aerobic bacteria possess hydrogenase enzyme which has the ability to recycle hydrogen produced by nitrogenase in vitro. Therefore, the re-utilization of this hydrogen generates more ATP (by oxidative phosphorylation) and consequently improves the efficiency of nitrogen fixation.

Rhizobium is also associated with soybean plants which contain hydrogen uptake (or *hup)* genes which have the ability to recycle hydrogen and back into the nitrogenase system and fix nitrogen. Through this mechanism the energy is harvested by plants which is otherwise lost.

The nitrogenase enzyme of bacteria fix nitrogen by converting nitrogen gas into ammonia using hydrogen ions and energy by host plant. During the process there is a production of energy rich hydrogen gas. Some bacteria possess hydrogenase enzyme which can capture hydrogen gas and converts it into hydrogen ions and release energy which can feed back into nitrogenase enzyme.

Genetics of Free-living and Symbiotic Diazotrophs

Organization of Nitrogen-fixation Genes

In molecular terms, very little is known of the means by which the prokaryote specifically recognizes and invades its respective host-plant. However, there has been a growing interest in approaching this problem by identifying and analysing the relevant symbiotic genes and their gene products. Because of the agricultural importance of the *Rhizobium-legume* interaction, the most detailed studies have been conducted on this system. Also, because of bacteria are simpler than plants. Our knowledge of the symbiotic genes of *Rhizobium* is better than that of the plant genes. There have been several recent advances in the analysis of the *Rhizobium* genes that are involved in nodulation, host-range specificity and nitrogen fixation. By dissecting the structure, organization and regulation of these genes, a better understanding of their functions may be forthcoming.

In the fast-growing *Rhizobium* species, which nodulate temperate legumes such as clover, alfalfa, peas and field beans, the genes for nodulation and nitrogen fixation are clustered on large symbiotic plasmids. One such plasmid, pRLUI, was isolated from a strain of *R. leguminosarum* (which nodulates peas) and in this case the genes that determined the ability to nodulate peas were located between two groups of genes required for nitrogen fixation. The study of the nitrogen-fixation *(nif)* genes in *Rhizobium* has been greatly facilitated by the fact that at least some of them are very similar to the corresponding *nif* genes in the free-living nitrogen-fixing bacterium, *Klebsiella pneumoniae.*

In *Klebsiella,* 17 *nif* genes have been identified which are variously involved in *nif* gene regulation, the synthesis of the iron-molybdenum cofactor (FeMoCo). Using DNA sequence comparisons, *Rhizobium* species have been shown to contain the *nifHD* and *K* genes (nitrogenase proteins) *nif A* (regulation) and *nifB* (synthesis of FeMoCo). Further studies may well demonstrate similarities between other *Rhizobium* and *Klebsiella nif genes.*

The study of nitrogen fixation genes was first carried out in *Klebsiella pneumoniae* (strain M5al); and spectacular progress achieved in 1970s. Although the first report came from the mutants of nitrogen fixing bacteria which was isolated and biochemically characterized as early in 1959. The genetic research by Streicher et al. (1971) of U.S.A. and Dixon of Agric. Research Council, Unit of Nitrogen fixation (U.K.) added more knowledge in this field. Dixon and Postgate (1971) successfully transferred the nitrogen fixation genes, called *nif,* to *Klebsiella pneumoniae.* Later on they could able to transfer these genes to *Escherichia coli.* This work encouraged the researchers of this field specially for *Rhizobium* and the presence of *nif* genes in this microorganism. More than a dozen genes, termed the *nif* genes are involved in the assembly of nitrogen-fixation apparatus. The *nif genes* are not found in a free state but are clustered in one region (they are not scattered throughout the bacterial DNA). Therefore, it is much easier to stretch of DNA in a *Rhizobium* chromosome and insert the whole batch into another organism. This has also been done in *Agrobacterium tumefaciens* (Ti plasmids). In 1981, the number *of nif* genes was 17 and they were organized in eight transcription units (or operons), one of which was monocistronic. Brill (1980) also included two other genes: *nif-R* between L and F, and *nif-W* between F and M.

Nod-Genes for Nodulation

Concerning the genes (called *nod* genes) for nodulation and host-range specificity, an important observation is that, despite the morphological complexity of the infection process, only a few genes on the symbiotic plasmid are required for nodulation. In the *R. leguminosarum* symbiotic plasmid pRL1 JI, less than 10kb appears to be required for nodulation and for the determination of host-range specificity for peas. The reason has been analysed by DNA sequencing (and thus identifying genes as long open reading frames), isolation and characterization of mutants and by studying the regulation of the *nod* genes.

The organiztion of *nif* genes of other free-living diazotrophs from which *nif* mutants were isolated *(Azatobacter vinelandii, Azospirillum lipoferum, Clostridium pasteurianum, Rhodopseudomonas capsulata* and a few cyanobacteria) was as like to that of *Klebsiella pneumoniae.*

REGULATION OF NITROGEN-FIXATION GENES

The regulation of *nif* genes is rather complex. Ammonia and nitrogenous compounds (like nitrates and amino acids) inhibit the expression of nitrogenase activity. Excess of aeration in the culture medium also inhibits the expression of *nif*-genes. For induction of *nif* genes, the presence of nitrogen is not necessary, because these genes can express themselves in cells cultivated in an argon atmosphere. In fact, the *nif-genes* expression is the result of a derepression rather than that of an induction

Eady et al. (1978), Roberts et al. (1978) reported that all the products of *nif*-genes are not present in the cultures carried out in the presence of ammonia. When ammonia is added to a completely derepressed cultures, a rapid repression of the *nif* phenotype occurs and nitrogenase activity as well as the enzymic proteins disappear within 45 minutes.

The presence of ammonia in the medium plays very important role in the synthesis of various biomolecules, e.g., in presence of ammonia, glutamine synthetase reacts with ATP and seems to lose the enzymic and regulatory functions of *nif* genes.

The presence of oxygen in the culture medium represses the biosynthesis of nitrogenase enzyme in *Klebsiella pneumonia.* Similar results have also been observed in aerobic and anaerobic bacteria. However, the mechanism of oxygen repression seems to be independent from that of ammonia in such microorganisms.

Molybdate does not have any direct effect on the regulation of *b-nif* genes, however, molybdoprotein is probably involved in the regulation of at least one of the *Klebsiella nif* operons (m/YKDH). The presence of molybdenum transport and storage proteins is known in *Clostridium pasteurianum* and other diazotrophs.

It is now established, after number of current reports that the regulation of the w/-genes of *Klebsiella pneumoniae* is carried out by both *nif* A and *nifL* genes as well as by genes that are remote from the *nif* cluster *(gln* genes). However, there are other genes which are involved in the expression of *nif-genes,* they are as follows:

nar D = which is involved in molybdenum processing.

unc and *a* gene in his operon = which influence ATP supply.

nim gene = gene of the uncertain function near tip.

Genetics of Symbiotic Diazotrophs

Much works has been done on the fast growing nitrogen-fixing *bacteria-Rhizobium.* The genetic map of *Rhizobium* is now established. The studies reveal that *R.leguniosarum, R. phaseoli* and *R. trifolii* have very similar gene organization. The genes for the component protein subunits of nitrogenase (molybdoferroprotein and ferroprotein) are formed by a single operon in *Rhizobium leguminisarum.* Similar situation has been reported in *Azatobactor vinelandii* and in *Klebsiella pneumoniae*

Transfer of *Nif*-genes to Microorganisms

As mentioned earlier that nif-genes are not found in a free state but are clustered in one region (they are scattered throughout the bacterial DNA). Therefore, it is much easy to cut out the relavent stretch of DNA in a *Rhizobium* chromosome and insert the whole batch into another organism. The construction of self-transmissible plasmids (Ti-plasmids) bearing *nif*-genes made it possible to shift the nitrogen-fixing capacity to non-fixing species or to non-fixing mutant of diazotrophic species like *E. coli (E. coli* C-M74) , *Salmonella typhimurium, Serratia marcescens, Pseudomonas flurescence,* and *nif*-mutant of *Azatobacter vinelandii.* It is quite easy to transfer *nif*-genes from one species to another, or form one microbial genus to another under laboratory conditions. This may be done by three methods of gene-transfer: conjugation, transduction and transformation. Therefore, the transfer of *nif*-genes between the plasmid and bacterial chromosomes can take place under laboratory as well as under natural conditions.

Transfer of *Nif*-genes and Development of New Nitrogen-fixing plants

It is easy to transfer grouped genes and the *nif*-gene cluster. Plant viruses could be used as vectors for this purpose. The cauliflower mosaic virus is a deoxyribo virus to which it is possible to join certain *nif*-genes which could become integrated into the plant genome or which could be transcribed into the plant cell at the same time as those of the virus when the latter's replication takes place.

Agrobacterium tumefaciens is a pathogenic bacterium, which contains a large plasmid, Ti and is responsible for the induction of tumours in plant tissues on the site of the infection by the bacterium. These plasmids are of 150 to 230 kilobases and can able to develop'tumours, therefore, they are called Ti or **tumour inducing plasmids.** *nif*-plasmids are introduced by either transformation or by conjugation. The DNA of *nif*-plasmids have two following regions:

1. The larger region contains information that makes it possible for the bacterium to catabolize derivatives of the basic amino acids, or opines, synthesized by the rumour cells.
2. It corresponds to the T-DNA and is incorporated into the genome of the plant cell. This T-DNA includes the ability of utilization of the latter as well the transformation of its metabolism in order to produce opines. The molecular weight of T-DNA is 15 x 10^6 daltons.

The development of tumours in plant tissue by *Ti -plasmid* of *Agrobacterium tumefaciens* is brougnt about by the transfer of a segment of bacterial DNA (from a prokaryotic) and its integration into the genome of eukaryotic cells. This introduction of *Ti -plasmid* was successfully made by Schell and his coworkers. Out of the seven genes of T-DNA of *Ti -plasmid*, they found that five of them are responsible for tumorization, that is in blocking cell differentiation. It is of interest to mention that *Agrobacterium tumefaciens* infects only dicotyledonous plants, however, it is not easy to introduce Ti plasmids into the cells of monocotyledonous plants like cereals.

Kemp (1981) used nucleic acid recombination and cloning techniques to transfer genes to plant cells. He has transferred genes for the bean protein- **phaseolin** to sunflower plants.

This transfer was followed by the synthesis of mRNA of the storage protein in the sunflower cells, made tumoral by the T-DNA *of Agrobacterium tumefaciens* which was used as a vector for the gene.

The development of plants which can able to fix atmospheric nitrogen independently of symbiotic microbes remains a difficult and complex task. However, in the present situation, it is possible to develop and gain better knowledge of the factors that induce nitrogen-fixing symbiosis, with a view of improving their efficiency.

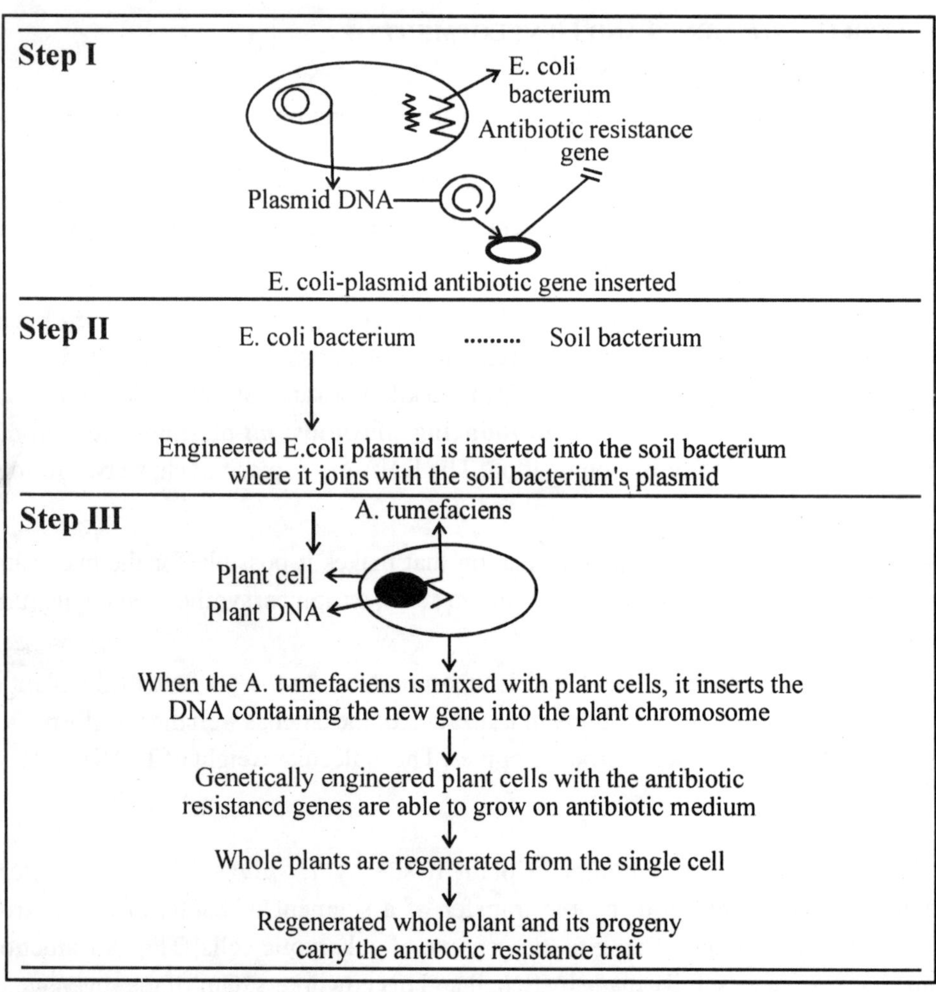

Diagrame 14.8 Schematic presentation of plant genetic engineering with Ti plasmids from *Agrobacterium tumefaciens.*

Nitrogen Fixing Trees and Forest Management

To maintain the sustained productivity of forest ecosystems, nitrogen input to the soil from both biological and non-biological sources is essential. Natural forest ecosystems contain climax vegetation in which the soil has been developed for a long period by the interaction of microorganisms and higher plants. Nitrogen fixation is one of the main processes in which nitrogen is brought into the soil.

Future intensive forest cultivation, with consequent short rotations, will require replacement of nitrogen reserves at a greater rate than the present secondary succession leading to old-growth forests. With the increased use of trees genetically selected for increased yield, it will be necessary to maximize the potential for biological nitrogen fixation in future forest management, so as to maintain the long-term productivity of forest lands without significantly accelerating rates of nutrient depletion.

Nitrogen fixing microorganisms which grow symbiotically with C-autotrophic higher plants, can indirectly utilize much more energy than free-living N-fixing microorganisms, and contribute the majority of the N_2 input in forest ecosystems.

Woody nitrogen-fixing plants have a much greater influence on the forest ecosystem than herbaceous ones. A total of 997 nitrogen fixing woody species were included in a masterlist from a database maintained by the University of Hawaii NiFTAL Project (Halliday and Nakao, 1982); even this masterlist is not complete. Brewbaker et al. (1983) recommended 44 species of 22 genera as economically important nitrogen fixing trees.

The following categories of nitrogen-fixing tree/soil/micro-organism associations have been studied: (a) root-nodule symbiosis with *Rhizobium,* (b) root-nodule symbiosis with actinomycetes, (c) mycorrhizal symbiosis, (d) rhizosphere systems, and (e) root-pathogen complexes.

Nitrogen Transfer Between Nitrogen Fixing and Non-Nitrogen Fixing Mycorrhizal Plants

General Description of Symbiosis

The term *symbiosis* was first used by Frank in 1877 to describe the regular coexistence of different organisms such as fungi and algae in lichens. It was used as a neutral term that did not imply parasitism. A decade later De Bary (1887) used symbiosis to include parasite as well. But the meaning of the terms *symbiosis* and *parasite* has been changed later on. Symbiosis was used more and more for mutually beneficial associations between dissimilar organisms, and parasite and parasitism came to be almost synonymous with pathogen and pathogenesis (Smith and Read, 1997).

In recent years and symbiosis is defined as *the living together of differently named organisms.* This definition includes all associations ranging from *mutualistic,* in which all organisms involved are believed to derive benefit; to *parasitic,* in which one organism benefits to the disadvantage of other member of the association. A more precise definition of mutualism is that associations are *mutualistic* if the fitness (offspring produced) of the associating organisms is greater than they are living apart. In a symbiosis, the organism with the large size is the host and the smaller is the symbiont. The symbiont can either be external to the host (ectosymbiotic) or within it (endosymbiotic). The symbiosis is *obligate* for an organism which is unable to survive and reproduce in the absence of its living partner, and *facultative* if it is able to do so even if the partner does during the association.

In a dual association, symbiosis is wide-spread especially between higher plants and bacteria in N- fixing symbiosis; between higher plants and fungi in most mycorrhizae; between algae and fungi in lichens; between algae and coelenterate in corals and so on. However, it is not currently known whether symbiosis occurs between higher plants. In addition, the tripartite symbiosis of plants, N- fixing bacteria and mycorrhizae (either vesicular-arbuscular mycorrhiza or ectomycorrhiza) has also been found in some *Acacia, Albizia, Casuarina* and *Leucaena* species. Symbioses are often found in nutrient limiting conditions and nutritional interactions play a key role in most symbioses. There are two important types of symbioses: mycorrhizal symbiosis and N-fixing symbiosis.

Mycorrhizal Symbiosis

The word *mycorrhiza* is derived from the Greek *mykes* (fungus) and *rhiza* (root). Mycorrhizae are highly evolved, mutualistic symbioses or associations between soil fungi and plant roots. The partners in this association are members of the fungus kingdom *(Basidiomycetes, Ascomycetes* and *Zygomycetes)* and most vascular plants. Mycorrhizal plants have been found in every continent and in every major vegetation type. Depending on the plant and fungal species involved as well as distinct morphological patterns, at least seven different types of mycorrhizal associations have been recognised. They are: vesicular-arbuscular mycorrhiza (VAM); ectomycorrhiza (ECM); orchid mycorrhiza; ericoid mycorrhiza; ectendomycorrhiza; arbutoid mycorrhiza and monotropoid mycorrhiza. In addition, dual associations of both ECM and VAM have been found in some trees and shrubs such as *Acacia, Casuarina, Eucalyptus* etc.

Mycorrhizal fungi generally benefit their host plants by:

1. increasing the physiologically absorbing surface area of the root system;
2. increasing the ability of plants to capture water and nutrients such as nitrogen, phosphorus, or other essential elements from the soil;
3. increasing the tolerance of plants to drought, high soil temperature, and extremes of soil acidity caused by high levels of metals such as sulfur, manganese, and aluminium;
4. providing protection from certain plant pathogenic fungi and nematodes that attack roots; and
5. modifying the transpiration rates and the composition of rhizosphere microflora by excretion of chelating compounds or ectoenzymes

In return for these benefits, the fungus partner receives carbohydrates, vitamins and other nutrients supplied by the plant partner (for further details, refer to Smith and Read, Mycorrhizal Symbiosis, 2nd ed., Academic Press, 1997).

Vesicular-arbuscular mycorrhiza (VAM): Where the *Zygomycete* fungi form arbuscules and/or vesicles and external hyphal networks in the soil and grow extensively within the cells of.the cortex, are formed by nearly all vascular plants.

Ectomycorrhizae (ECM): They are characterised by dense mycelial sheaths around the roots and a Hartig net between root cells, are limited to mostly temperate forest trees and *Basidiomycetes* and some other fungi. Mycorrhizae involve an intimate association of the host plant's root tissue and the fungus, but there is also fungal tissue that extends into the soil as individual hyphae and in some species as more complex strands. The external hyphae can take up mineral nutrients from the soil and transport them into the host root. Mycorrhizal associations can therefore be of great potential benefit to the host plant in nutrient limited systems.

Leguminous Plants / Rhizobiaceae Symbiosis

Symbiotic N_2 fixation in agriculture can be attributed mainly to legumes - the plants in the *Leguminosae.* It is estimated that *Leguminosae* contains more than 200 genera, and 20,000 species, which ranges from small plants, such as the clover, to the large trees such as *Acacia* species. Approximately 90% of them ran fix nitrogen from the atmosphere with *Rhizobiaceae,* either with *Azorhizobium, Bradyrhizobium, Rhizobium* or with *Sinorhizobium* in root nodules. The important agricultural legumes can be divided into three groups: crops that grown for their commodities (e.g. grain legumes); forage legumes that may be grazed or harvested for animal fodder; and trees or shrubs in agroforestry systems. Worldwide approximately 1.5 million km of land are cultivated with grain legume plants, mainly in *Glycine max, Phaseolus vulgaris, Pisum sativum, Arachix hypogaea, Cajanus cajan* etc.. The annual harvest of grain legumes are about 200 million tonnes, which provides the plant protein source for human and animal consumption, or vegetable oil and other raw materials. The areas partially covered by the forage legumes mainly in *Trifolium, Lotus, Medicago, Stylosanthes, Macroptilium* and *Mimosa* etc., is even much larger, about 30 million km^2 of grassland in the five continents. The third group consists of the genera *Acacia, Albizia, Alnus, Leucaena, Robinia* etc., which are mostly used as timber, fuelwood or craftwood, and in the pharmaceutical industry as antibacterial and antifungal agents, or food additives and any other functions.

There are five genera of the *Rhizobiaceae: Azorhizobium, Bradyrhizobium, Rhizobium, Sinorhizobium* and *Photorhizobium.* All members of *Rhizobiaceae* are characterised by a gram-negative cell wall structure. Cells are generally rod-shaped, non spore-forming and motile with differently arranged flagella. All rhizobia are aerobic bacteria that persist saprophytically in the soil until they infect a root hair cell. The enzymes from the bacteria degrade part of the cell wall and allow bacteria entry into the root-hair cell itself, which-lead the root hair to produce a threadlike structure called the infection thread. The bacteria multiply extensively inside the thread, which extends inwardly and penetrates through/ between the cortex cells. In the inner cortex cells the bacteria are released into the cytoplasm and stimulate some cells (especially tetraploid cells) to divide. Each enlarged, non-motile bacterium is referred to as a bacteroid. These result in a proliferation of tissues, eventually forming a mature root nodule. A typical root nodule cell contains several thousand bacteroids. The nodule contains a protein called leghemoglobin, which gives legume nodules a pink colour due to its prosthetic heme group. Leghemoglobin is thought to help transport O_2 into the bacteroids, which is essential for bacteroid respiration. Nitrogen fixation in root nodules occurs directly within the bacteroids. The host plant provides bacteroids with carbohydrates, which they oxidise and from which they obtain energy. These carbohydrates are first formed in leaves during photosynthesis and then are translocated through phloem to the root nodules. Sucrose is the most abundant carbohydrate translocated, at least in legumes. Some of electrons and ATP obtained during oxidation by the bacteroids are used to reduce N_2 to NH_4^+, which is catalysed by nitrogenase.

Non-leguminous Plants / Frankiaceae symbiosis

About 200 plant species from 25 genera, 8 families and 7 orders, all non-leguminous angiosperms, have been found to form nodule symbiosis with N_2fixing *actinomycetes* belonging to the genus of *Frankiae.* All of these species are perennial dicots and, with few exceptions, woody shrubs or trees. They are named actinorhizal plants from *actino* in *actinomycete,* and from *rhiza* in the Greek word for root. Actinorhizal plants can be found both on every continent except Antarctica and in most climatic zones. They typically grow on disturbed marginal soils and are pioneer species early in successional plant community development, such as *Dryas* species in arctic tundra; *Caxiuirina, Hippophae, Myrica* and *Elaeagnus* species in coastal dunes; *Alnus* and *Myrica* species in riparian; *Alnus* and *Dryas* species in glacial till; *Casuarina , Purshia, Ceanothus Cercocarpus, Comptonia* and *Cowania* species in chaparral and xeric; *Alnus* species in alpine; *Alnus, Casuarina, Coriaria* and *Shepherdia* species in forest. Globally, especially in wherever indigenous legumes are absent or rare, actinorhizal plants have potential applications in soil amelioration and reforestation, as fuelwood, timber and pulp, and as windbreak or even for addressing **pyro-de-nitrification.**

The symbiont *Frankia* is a gram-positive, filamentous bacterium belonging to the family *Frankiaceae* within the order *Actinomycetes.* Speciation in *Frankia* is not yet clear and within the genus different isolates are classified. Almost *all of Frankia* are characterised by three structural forms of hyphae, sporangia and vesicles. The hyphae are branched with a diameter of 0.5 to 1.5 μm and the mature vesicle is spherical with a diameter of 2 to 4 μm. Both the hyphae and the mature vesicles are septate. When actinorhizal plants, such as *Alnus* and *Casuarina* seedlings are excavated from soil containing *Frankia,* numerous small, multilobed, coralloid-type, amber or whitish nodules are found on their root systems. The oldest and biggest nodules are close to the stem base and the youngest are on the distal parts of the root system. Some *Casuarina* nodules may reach a size of around 5-10 mm in diameter and a weight of about 1 g in dry matter. It has been reported that *Frankia* produced sporangia and vesicles as soon as the microorganism escaped from the mother nodule. This indicates that the formation *of Frankia* structures inside the nodule might be under host control. In general, the nitrogen fixation rates of *Frankia*-actinorhizal plant symbioses are comparable to that of *Rhizobia-legume* symbioses.

Benefits From Nitrogen Fixing Plants To Non-Nitrogen Fixing Plants

The nitrogen fixing plants can directly fertilise the soil and indirectly neighbouring plants through above- and below-ground litter, through root exudates and leakage from leaves and roots, and through the excreta of grazing animals. When nitrogen fixing plants are grown in mixed plantations with non-nitrogen fixing plants, an increase in both growth and yield of the non-nitrogen fixing plants have often been found. Moreover, the legume with the non-legume intercrops, on average, yield more efficiently per unit land with higher area- x -time equivalence ratios than either intercrops with two legume species or two non-legume species.

The nitrogen fixing plants (donor) and non-nitrogen fixing plants (receiver) could provide a good system for investigating nitrogen transfer, which was often termed the process of **nitrogen decomposition** and uptake in a neighbouring plant, where neighbouring plants might have differing nitrogen status.

Nitrogen transfer could lead to increase productivity of intercrops and forests without increasing the use of nitrogenous fertilisers if properly managed and quantified. However, the roles of mycorrhizal hyphae in direct N transfer were somewhat inconclusive, and it is not yet clear? if the transfer can be large enough to contribute significantly to the N status of the receiver in agricultural or natural ecosystems.

Nitrogen Transfer In Mycorrhizal Plants

The idea of nitrogen fixed by a legume component may be available to its associated non-legume plant originated from earlier studies in the late. In several experiments it was found that as much as 10 to 30% of the total N fixed in field pea was deposited in the growth media of sand, predominantly as amino acids, and a major part was taken up by an associated cereal. In general, the process of N-deposition and uptake in a neighbouring plant is often termed N-transfer. Since, then substantial N-transfer have been observed in several legume/crop intercropping systems such as soybean or cowpea/maize or sorghum, cowpea/rice, gram clover/wheat, barley or oat/vetch or lupin etc., although some experiments showed little or no N-transfer between them

Compared to pure stand cultivation, possible benefits obtained by non-legumes or trees may partly be due to transfer of the symbiotically fixed N by legumes, either through directly release from nodulated roots or through decomposition of dead nodule and root tissue in the soil or through a common mycorrhizal network or even enhanced by mycorrhizal hyphae. However, the mechanism of N transfer and the role of mycorrhizal hyphae in the direct transfer of nitrogen are not well established. Therefore, studies in identifying the pathways of N transfer and whether mycorrhizal hyphae play a direct role in facilitating N transfer are needed.

Nitrogen Nutrition in Mycorrhizal Plants

Forms of Nitrogen Used by Mycorrhizal Associations: The two most available forms of inorganic N for plants and both VA mycorrhizal and ectomycorrhizal fungi, are NO_3^- and NH_4^+. Originally, NH_4^+ is mainly derived from the reduction of atmospheric dinitrogen symbtotically fixed by nitrogen fixing plants, either by *Rhizobia* in legumes or by *Frankia* in actinorhizal plants. In general, NO_3^- is the dominant form of nitrogen-available to plants and fungi in almost all agricultural soils due to the rapid nitrification of NH_4^+. In contrast, NH_4^+, which is released from soil humus and other organic N sources by ammonifying organisms, predominates and NO_3^- may be almost entirely absent in many undisturbed or very acidic soils. NO_3^- is highly mobile and is readily transported towards the plant roots by mass flow and diffusion. NH_4^+ is absorbed to negatively charged soil particles and transported towards the plant roots mainly by diffusion.

Mycorrhizal Effects on Nodulation and Nitrogen Fixation: The possibility of a direct involvement of ectomycorrhizal fungi in N acquisition by plants was first suggested by Frank in 1894 (Smith and Read, 1997). Then after 50 years of Frank's work, the probably first observation of growth, nodulation and mycorrhizal status of a large number of legumes was made by Asai in 1944. In most cases, improved nodulation and N fixation in mycorrhizal plants appears to be the result of relief from P stress and possibly uptake of some other essential micronutrients, which result in both a general improvement in growth and yield and indirect effects upon the N-fixing system. The differences between mycorrhizal and non-mycorrhizal plants usually disappear if the latter are supplied with a readily available P source.

It was known that cereals intercropped with grain legumes generally benefit from the association in terms of increased grain and nitrogen yields per unit area compared with monocropped cereals. This indicates that N-transfer from the nitrogen fixing plants (donor) to an associated non-nitrogen fixing plant (receiver) is most likely to happen in nature either via the interception and uptake of released fixed nitrogen from donor plant by the roots of receiver plant or through the root mycorrhizal between donor and receiver plants (Newman, 1988; Newman *et al.,* 1992), or even enhanced by mycorrhizal hyphae. Therefore, in agricultural and/or natural ecological communities, mycorrhizal associations may be important factors influencing the performance of both donor and receiver plants through the acquisition and translocation of nitrogen by the mycorrhizal fungus, particularly when the relatively immobile NH_4^+-N rather than the mobile NO^-_3 -N is the major source of plant available N.

It is therefore, concluded that Nitrogen fixing and non-nitrogen fixing plants can provide a good example for investigating the N-transfer between the plants where neighbouring plants may have differing nitrogen status. In general, with or without a split root system, combined with the fine nylon mesh to allow the direct mycorrhizal link but not root contact, together with ^{15}N-isotope labelling technique to enrich the N in the donor plant, some experiments have demonstrated a below-ground N-transfer whereas others found no evidence for transfer of nitrogen between donor and receiver plants via mycorrhizal hyphae (Smith and Read, 1997). Furthermore, there is controversy about the extent to which direct N-transfer is actually being facilitated by mycorrhizal hyphae or by some other indirect means, and whether it is of agricultural or ecological significance if N-transfer do occur between these plants through mycorrhizal hypha.

The following questions still remain to be answered:

1. Does N transfer occurs between mycorrhizal plants at all? If so, how much N is transferred and how much N-transferred is enhanced by mycorrhizal hyphae?
2. How much net N transfer is from nitrogen fixing to non-nitrogen fixing mycorrhizal plants and *vice versa?*

Nitrogen Fixation Research in India

The word on bio-chemistry of nitrogen fixation has mainly been done at Haryana Agriculture University, Indian Agriculture Research Institute, Punjab Agriculture University and BARC. At HAU, a pathway for ureide biogenesis in nodules of pigeonpea has been proposed. The further metabolism of ureides in aerial parts particularly leaves and developing pods has also been worked out. Detailed biochemical and physiological studies explaining nodule senescence have also been carried out at IARI. Attempts have been made to understand the regulation of uptake hydrogenase so as to reduce the energy cost of biological nitrogen fixation. In addition, studies are concentrated on increasing nitrogen fixing efficiency by insertion of uptake hydrogenase genes or by increasing nitrogenase activity in nodules.

At BARC, early interactions in legume- *Rhizobium* symbiosis have been investigated, according to which, host-induced alterations of capsular polysaccharides and de novo synthesis of specific proteins by both the partners in response to each other, appear to be among the first steps in the series of signal response interactions. At PAU, the role of extracellular invertase of *Rhizobium* in free sugar metabolism in developing root nodules of legumes has been ascertained (Adopted from Dr. Randhir Singh, In: Souvenir, Int. Cong. Plant Physiology, Feb. 15-20, 1988, New Delhi).

According to H.N. Krishnamoorty (In: Souvenir, Int. Cong. of Plant Physiology, Feb. 15-20, 1988, New Delhi) the activity of nodule is also controlled by nitrogenase enzyme. This enzyme is about 75% efficient. This is because, when soil aeration is poor and nitrogen availability is limited, electrons instead of being used to reduce nitrogen, now combine with protons, i.e.hydrogen. As a result, nodules evolve hydrogen instead of fixing nitrogen. This is a wasteful process. Sufficient progress is being made to identify strains of *Rhizobia* which evolve less hydrogen and fix more nitrogen.

The most ideal situation would be the one in which the higher plant is able to fix nitrogen without the help of the bacteria. Towards this end, attempts have been made to transfer nitrogen fixing genes (*nif*-genes) to higher plants from *Rhizobium.* However, it appears to be a long way before it is achieved.

In India, intensive forest cultivation of nitrogen fixing trees, with consequent short rotation is needed. This will require replacement of nitrogen reserves at a greater rate than the present secondary succession leading to old-growth forests. With the increased use of trees, genetically selected, it will be necessary to maximize the potential for biological nitrogen fixation in furure forest management, so as to maintain the long term productivity of forest lands without significantly accelerating the rate of nutrient depletion. India is in urgent need of additional professional training and better transfer of technology in this context.

Further, improved strains and new species of nitrogen-fixing microorganisms should be investigated/developed in order to reduce the use of chemical fertilizers in the fields.

❑❑❑

CHAPTER-15

Genetic Engineering

INTRODUCTION

In nature genetic recombination brings about natural evolutionary changes but now a days it is possible to employ DNA recombination artificially, to direct or engineer the evolution of new micro organisms. The intentional recombination of genes from different sources by artifical means (recombinant DNA or rDNA- technology) is the basis for genetic engineering i.e. the creation of new genetic varieties of organisms using recombinant DNA technology.

Genetic engineering constitutes one of the basic foundations of modern biotechnology because genetic modification or manipulation of useful microorganisms is vital for their profitable utilization in the production of useful, high-value products. Besides microbes, cell cultures of plants and animals can be genetically manipulated to advantage; this is possible in view of our ability in many cases to raise protoplast cultures (and fuse them) which can be handled like microbes in genetic experiments.

Techniques of genetic engineering have made it possible in some cases to transfer genes from one organism to another by overcoming the species barrier.

Gene manipulation experiments require the use of certain enzymes concerned with nucleic acid metabolism. These enzymes make it possible to manipulate DNA *in vitro.* The most important of these enzymes are those called *restriction endonucleases,* which are part of the armoury that bacteria have 'evolved as defence mechanisms against invasion of alien genetic elements such as bacteriophages and plasmids. The nucleases hydrolyze nucleic acids into nucleotides. An exonuclease chews the DNA strand between two intercalary nucleotides. Restriction endonucleases cut DNA at specific sequences. The bacteria which produce these restriction endonucleases protect the specific sequences in their own DNA by masking them with certain chemical groups with the consequence that a restriction endonuclease cannot cut the specific sequence in the bacterium's own DNA but can attack only the foreign DNA.

When the DNA is cut specifically by an endonuclease, this cleavage often gives rise to DNA fragments with single-stranded sticky ends. These sticky ends may be rejoined by means of another kind of enzyme called ligase. Still other enzymes are available for cutting out and processing the required gene fragments from a donor genome.

Usually, the substrates that restriction endonucleases attack are palindromic DNA sequences that read the same both backwards and forwards (e.g., the word MADAM).

Over 100 different endonucleases have so far been isolated and characterized. Perhaps the most popular example is *EcoR* 1, produced by *Escherichia coli.* This *EcoRl* attacks the sequence

5'...GAATTC...3'

3'...CTTAAG...5'.

It may be noted that this sequence has a symmetry around its centre. The enzyme produces single-strand cuts or "nicks" between A and G:

3'...CTTAAG...5'.

Cleavage of the foregoing palindrome by *EcoRl* is diagrammatically illustrated in diagrame 15.1.

Endonucleases can be used to make new recombinants from virtually any type of DNA, because each endonuclease produces identical, complementary ends on any DNA molecule it attacks.

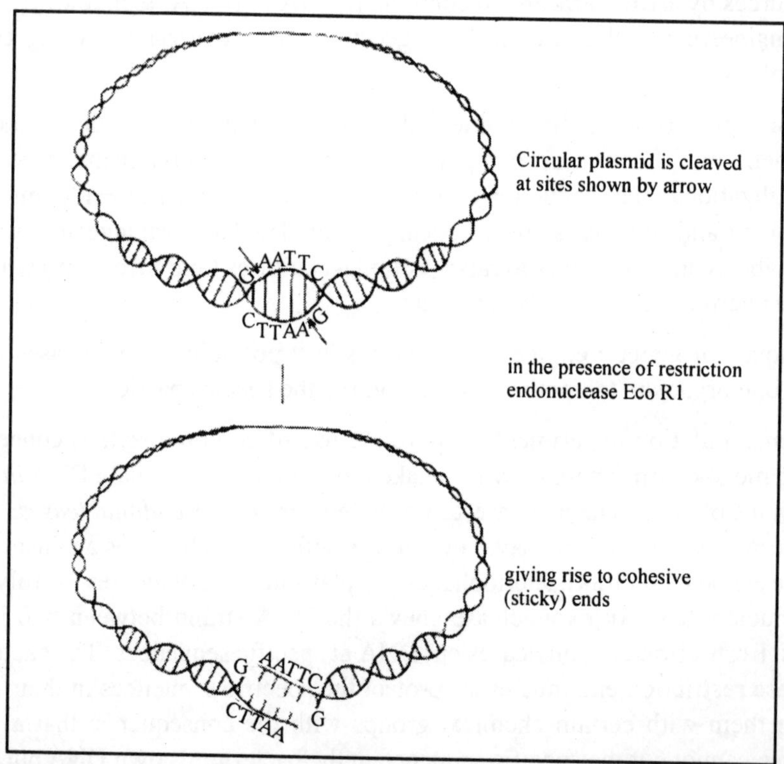

Figure 5.1 Action of *EcoRl* on plasmid DNA, producing cohesive sticky ends.

Diagram 15.2 illustrates how an endonuclease called *BamH*1 cuts palindromic sequences, forming sticky ends. When two such fragments of DNA_1 and DNA_2 come together, they join by the formation of complementary base pairs.

Certain DNA molecules can act as gene vectors. These vectors carry nucleic acid from one cell to another. Plasmids and transposons are examples of these vectors and are now described in more detail.

Plasmids and temperate bacteriophages constitute important tools of genetic manipulation. A certain gene sequence is first enzymatically incorporated into a suitable plasmid or phage (Diag. 15.3). The modified plasmid so produced then enters a bacterial cell where it multiplies repeatedly, producing large numbers of its copies, each of which carries the incorporated gene sequence. When such transformed bacteria are plated on an agar medium, the desired clone may be isolated by usual microbiological selection procedures. This whole process is called gene cloning. Gene cloning essentially means the isolation and selective replication of some specific DNA sequence. The sequence to be cloned may originate from any source or could even be man-made. Customarily, it is replicated inside a bacterium.

Diagram 15.4 shows some preferred routes for gene cloning. Diagram 15.5 illustrates the process of molecular cloning of genomic DNA into plasmid DNA.

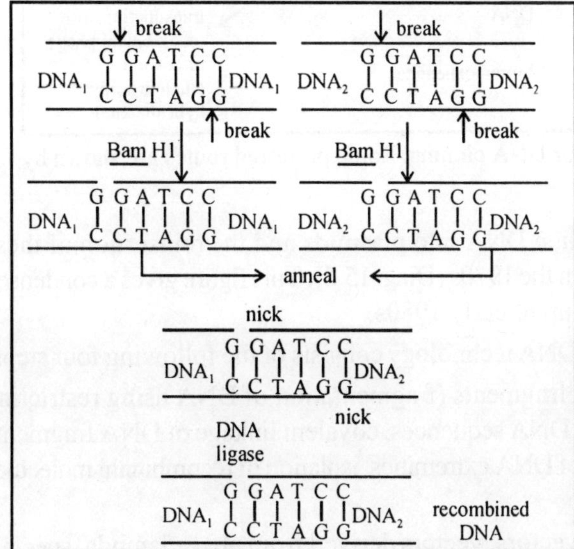

Diagram 15.2 Breakage and rejoining of two different DNA fragments, producing recombinant DNA.

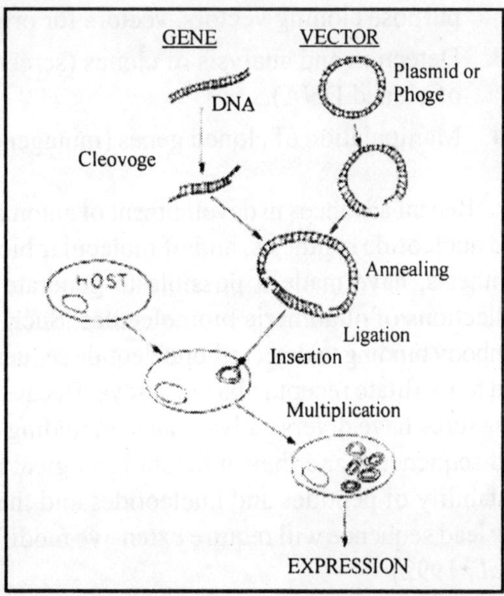

Diagram 15.3 Steps in the insertion of a new gene into a host organism using a vector.

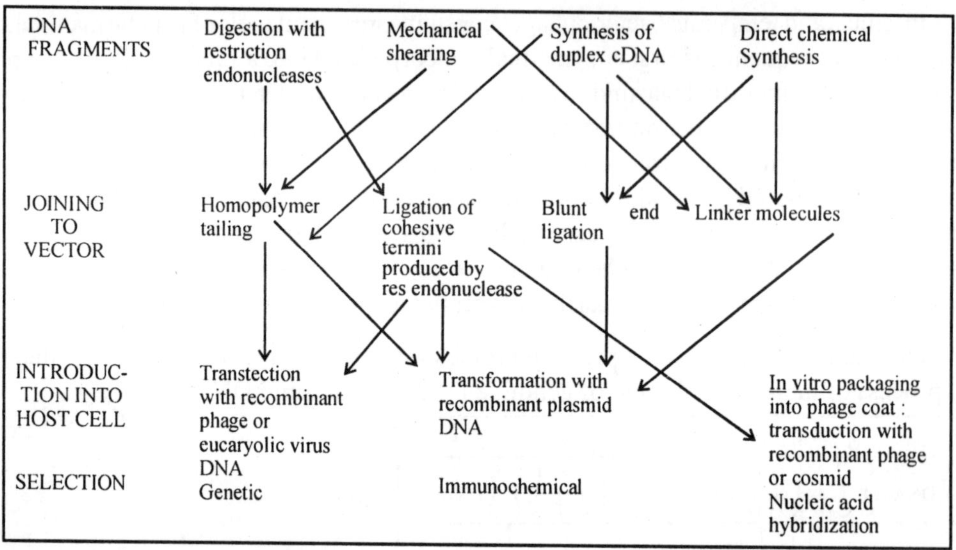

Diagram 15.4 A generalized scheme for DNA cloning. Some preferred routes are shown by arrows. (After Old and Primrose, 1981).

The first reports of insertion of alien DNA into plasmids and the reinsertion of these modified plasmids into *E. coli* appeared in the 1970s (Diag. 15.6). This figure gives a condensed history of recombinant DNA research up to early 1980s.

In summary, *in vitro* recombinant DNA technology consists of the following four steps:

1. Generation and cloning of DNA fragments (fragmentation of DNA using restriction enzymes, enrichment for specific DNA sequences, covalent linkage of DNA fragments to vector molecules, modification of DNA extremities, isolation of recombinant molecules and interspecies DNA transfer).
2. Use of cloning vectors (plasmid vectors, vectors derived from phage lambda, special-purpose cloning vectors, vectors for organisms other than *E. coli).*
3. Detection and analysis of clones (screening of recombinant clones, characterization of cloned DNA).
4. Manipulation of cloned genes (mutagenesis. efficient expression of cloned genes).

Recent advances in development of automated chemical methods for solid-phase peptide and nucleotide synthesis, and of molecular biological methods for protein and nucleic acid synthesis, have made it possible to generate new kinds of compound libraries, namely, collections of oligomeric biomolecules. Such libraries have been used to map epitopes for antibody binding to detect ribonucleotide sequences with specific binding or catalytic activity, and to facilitate receptor-based assays. Because of their modular structure these oligomeric structures have diverse advantages, including the ease with which they can be synthesized and sequenced, and their inherent biological relevance. On the other hand, the metabolic instability of peptides and nucleotides and their poor absorption characteristics mean that any lead sequence will require extensive modification before it could be used *in vivo* (Simon *et al.,* 1992)

Simon *et al,* (1992) have described the development of oligomeric N-substituted glycines as a motif for the generation of chemically diverse libraries of novel molecules. These oligomers are called "peptoids".

For designing an alternative to the natural polymers, Simon *et al.* postulated five desirable attributes for any modular system: (1) the monomers should be straightforward to synthesize in large amounts, (2) the monomers should have a wide variety of functional groups presented as side chains off of the oligomeric backbone, (3) the linking chemistry should be high yielding and amenable to automation, (4) the linkage should be resistant to hydrolytic enzymes, and (5) the monomers should be achiral. Diagram 15.7 compares peptides and peptoids (Diag. 15.8) revealing the similarities in the spacing of the side chains and the carbonyl groups, and the differences in the chirality of the two monomers. Though these peptoids are simply isomers of peptides this should not obscure the important differences in stereochemical and conformational characteristics of the two oligomers.

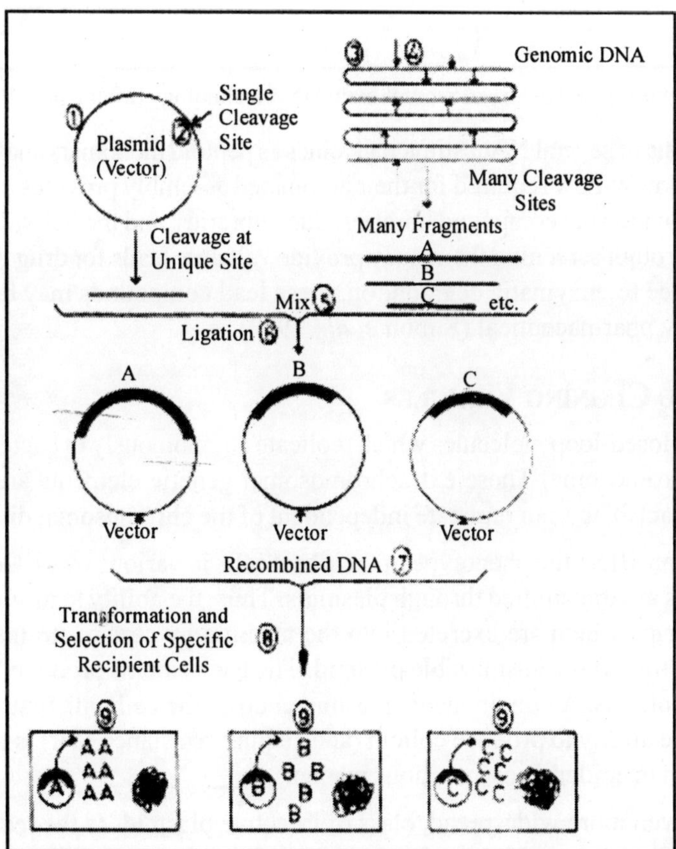

Diagram 15.5 General sketch for molecular cloning of genomic DNA into plasmid DNA. Plasmid DNA (1) with a unique restriction site (2) is cleaved. Genomic DNA (3) is also digested by the same restriction enzyme (4). The two DNA digests are mixed (5) and treated with ligase (6) to form hybrid DNA molecules. A set of hybrid DNA molecules (7) is produced. Recipient cells are transformed with the recombined DNA (8). Transformants that express a specific product are selected and cloned (9). (After Pasternak and Click, 1987.)

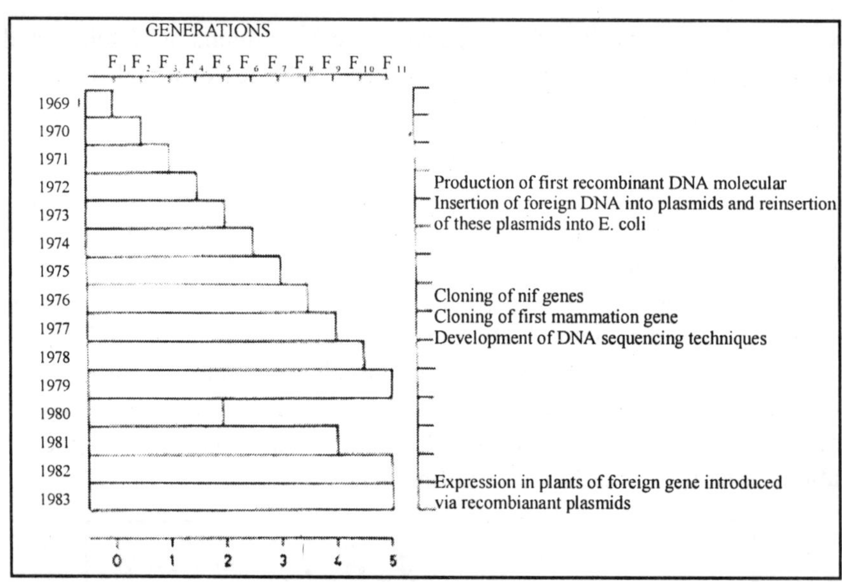

Diagram 15.6 History of recombinant DNA technology (after Riley, 1984).

The synthesis of several N-substituted glycines as peptoid monomers and the development of the chemical know-how required for their automated assembly provides a modular system for the creation of unusual compound libraries. Such libraries, and the individual peptoids that are identified through screening them, may provide valuable leads for drug design. By virtue of their resistance to enzymatic degradation, these lead compounds may be well along the way toward new pharmaceutical (Simon *et al.,* 1992).

Plasmids and Cloning Vehicles

These are closed-loop molecules which replicate autonomously in bacterial cells outside the bacterial chromosome. These extrachromosomal genetic elements are transmitted by cell-to-cell contact. They can replicate independent of the chromosomal division.

Plasmids can affect the phenotype of their host cells in various ways. Certain properties of bacterial cells are transmitted through plasmids. Thus, the ability to produce bacteriocins (antibiotic proteins which are excreted into the medium by certain strains and which kill other bacterial strains) is transmissible plasmid. The bacteriocins produced by *Escherichia coli* are called colicins. A colicinogenic plasmid endows the cell with resistance against its own colicin, The ability to produce colicin (and acquire resistance to it) is transmitted from cell to cell, often in epidemic or infectious manner.

Another, even more widespread, class of infective plasmids is the resistance-transfer, or R-plasmids. These are infectious determinants of resistance to antibiotics or drugs. One plasmid quite commonly confers resistance to several different drugs. For instance, Rl isolated from *Salmonella paratyphi* (and transmissible to *E. coli)* harbours determinants of resistance to ampicillin, chloramphenicol, kanamycin, streptomycin, spectinomycin, and sulphonamide. Another similar plasmid is R6 which includes neomycin and tetracycline resistance characters but is sensitive to ampicillin and spectinomycin.

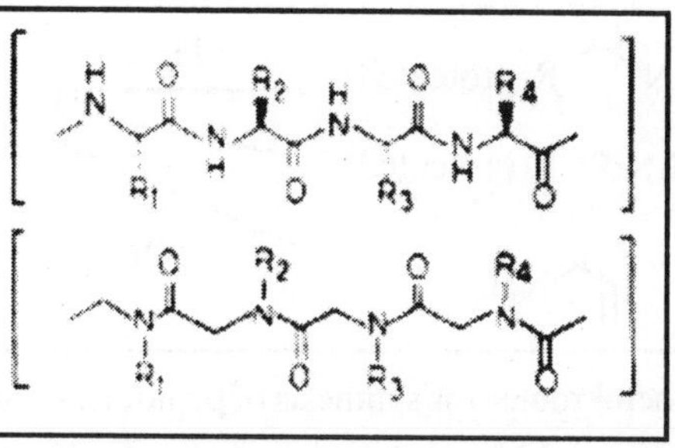

Diagram 15.7 Comparison between peptides and peptoids showing the similarity of spacing of the side chains, and the lack of stereochemistry of the peptoid monomers. As for any chemically synthesized modular system, the choice of functional groups is virtually limitless. (After Simon *et al.*, 1992.)

The best-studied class of transmissible plasmids is the F-factor (for fertility) of *E. coli* strain K12. Those cells which contain F-factor behave as males whereas those which lack F-factor are females.

Some plasmids are not infectious by themselves. For their passage, they require the assistance of another, self-transmissible, plasmid. ColEl is a good example of the non-self-transmissible plasmids. This determines colicin production and has become important for the construction of cloning vehicles.

The bacterial cells which harbour the aforestated plasmids contain two types of DNA. One is the main or chromosomal DNA. The second is a minor DNA component consisting of relatively small, covalently-closed, supercoiled molecules. The minor component (i.e., plasmid DNA) can be separated from the chromosomal DNA by various methods, e.g., by sedimentation equilibrium in a caesium chloride density gradient.

The plasmid molecules separated from the chromosomal DNA can also be seen under the electron microscope.

The sizes of various plasmids differ greatly. Generally, non-transmissible plasmids are smaller than transmissible plasmids. Also, the number of copies per cell varies over a broad range. The larger plasmids, e.g., F-factor, tend to occur in one or a few copies per cell. Smaller plasmids, e.g., ColEl, can occur in large numbers per cell.

Those *E. coli* cells which contain transferable plasmids characteristically bear distinctive surface filaments called pili which are essential for cell-to-cell plasmid transfer. Plasmid transfer involves DNA synthesis and only a single DNA strand is transferred, which strand is always the same strand. After entry into the recipient host cell, the single strand synthesizes its complementary strand.

Some types of pili (e.g., those produced by cells having F- or Col-plasmids) contain specific adsorption sites through which bacteriophages attack the *E. coli* cell. Some of the phages are quite specific: thus, those which attach to 1-pili do not attach to F-pili, and vice versa.

Another property of several plasmids is their mutual incompatibility. For instance, two different plasmids may not be able to replicate in the same host cell.

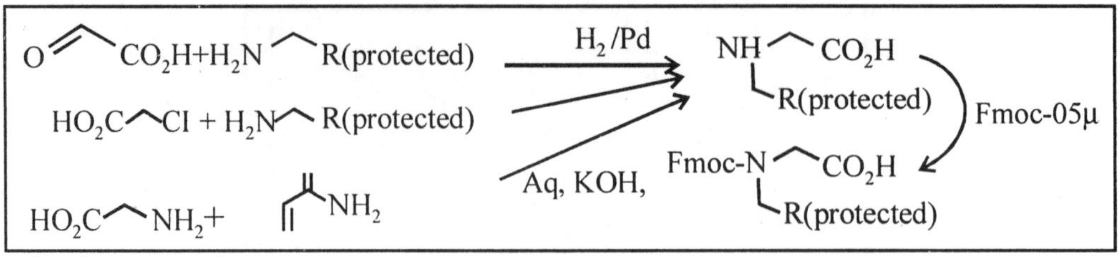

O CO$_2$H+H$_2$N R(protected) — H$_2$/Pd → NH CO$_2$H R(protected)

HO$_2$C Cl + H$_2$N R(protected) → Fmoc-05µ

HO$_2$C NH$_2$+ NH$_2$ — Aq, KOH, → Fmoc-N CO$_2$H R(protected)

Diagram 5.8 General routes for synthesis of peptoid monomers (after Simon *et al.*, 1992).

The importance of genetic stability in the scaling-up of recombinant DNA technology cannot be overemphasized. In general, most organisms harbouring recombinant plasmids turn out to be less fit than their plasmid-free counterparts under non-selective conditions.

Entirely synthetic plasmids have been made; these contain a convenient arrangement of restriction enzyme cleavage sites, a promoter site, and antibiotic-resistance genes. These synthetic plasmids are commonly named after their inventor, for instance, p^{SC101} denotes plasmid 101 designed by Stanley Cohen.

If bacterial cells having plasmids are cultured without selection for plasmid retention, the plasmids are liable to be lost unless the plasmid codes for an active partition function to ensure distribution to each progeny at cell division. This is especially true for several chimaeric plasmids that incorporate alien genes into vector derived from Coli-like plasmids. For example, the plasmids which encode human insulin chains can be lost from the population when *E. coli* KI2 strains containing these plasmids are propagated under non-selective conditions.

The techniques *of disparate cloning* and *operon fusions* (Franklin, 1978) have made it possible for eucaryotic structural genes to function by fusion to the regulators of procaryotic operons. Certain plasmids of *E. coli* can be redesigned in such a manner that insertions of foreign DNA may be selected and so that such DNA may use prokaryotic promoters for expression. The *in vitro* engineering of the mammalian somatostatin gene into the *lac* operon of *E. coli* constitutes an elegant example of the success of disparate cloning methods (Itakura *et al.*, 1977). Somatostatin is naturally produced in the brain, but recombinant DNA techniques have made it possible to produce large quantities of this hormone through bacteria. Somatostatin is used in the treatment of certain diseases including diabetes.

Bacteriophage lambda can be adapted to serve as a safe, lytic cloning vehicle. Blattner *et al.* (1977) constructed particularly valuable strains called "charon phages", which have been developed as cloning vectors.

By means of plasmids it is possible to amplify and purify DNA sequences. The F-factor of E. *coli* exemplifies a naturally-occurring cloning vector of bacterial genes because these plasmids act as vehicles for the selective amplification of short segments of the *E. coli* chromosome that they incorporate.

A good cloning vehicle is one which has only a single site for cutting by a restriction endonuclease. For instance, ColEl contains a single site for *EcoR*$_I$ action. When *EcoR*$_I$ attacks ColEl, a linear molecule, with complementary single-strand tails 5'-AATT... at its ends, is generated. The tail ends are sticky and can also be reannealed or rejoined together by complementary base-pairing, even with the identical ends of any other $EcoR_I$ restriction fragment, and this is where the importance and value of plasmids lies. It gives us a tool whereby we may reanneal a mixture of ColEl molecules, that have been cut up with $EcoR_I$, with $EcoR_I$ fragments of some alien DNA, and thereby reconstitute a circular molecule in which the alien DNA segments have become inserted between the ColEl cut ends. This kind of insertion engineering can be further sealed by treatment of the system with DNA ligase, the enzyme which mediates joining of the juxtaposed 3'-OH end with the 5'-phosphate end of the single-strand termini.

The foregoing protocol is only one example. In place of *Eco*Rl, we can use *Hind*III or similar other endonucleases, each of which cuts at some particular, specific site in the base sequence.

The last two decades have witnessed great strides in plasmid biology. People no longer prefer to use naturally-occurring plasmids now, but are going in for elaborately constructed compound plasmids as cloning vehicles. This kind of plasmid is designed in such a manner that it carries markers to signal both its presence in the host cell and to indicate that it harbours a DNA insert. One of the best examples of such an artificial compound plasmids is pBR322. It carries genes for ampicillin resistance and tetracycline resistance and can replicate repeatedly in the presence of chloramphenicol. Sequences for cloning can be successfully inserted in either of four unique restriction sites for *Eco*Rl, *Hind*III, *Bam*H1. and *Pst*1.

Debabov (1982) has used pBR322 as the cloning vector to make strains of E. *coli* which produce 55 g/ litre of L-threonine, in a system where the conversion of carbohydrate into amino acids was more than 40%.

A number of hybrid vectors for gene cloning in various hosts have been, developed during the last decade. This has become possible because of the availability of (1) microbial plasmids with self-replicating mechanism and high copy number, and (2) restriction endonucleases with high degree of DNA sequence specificity. These vectors can serve to transfer genes not only between closely-related organisms but also between organisms belonging to different genera, families, or even classes.

Plasmid vectors constitute the central foundation of recombinant DNA technology and genetic engineering. Some of the more important vectors available for cloning in various hosts are now briefly described (these vectors are some natural bacteriophages and plasmids of *Escherichia coli):*

1. *Broad host range vectors:* Many large plasmids of the P-group (e.g., RP4), originally isolated from *Pseudomonas,* are conjugatively transferred to *Agrobacterium tumefaciens, Rhizobium,* and *Azotobacter.* Tables 15.1 and 15.2 list such vectors.
2. *Bifunctional Bacillus-Escherichia vectors: B. subtilis* is the bacterium of choice for large-scale production of diverse metabolites. It is a non-pathogenic bacterium whose genetic map is well-deciphered. Some vectors available for cloning in *Bacillus* are listed in Table 15.3.

3. *Bacteriophages as vectors of E. coli:* An example of this kind is the M_{13} system in E. *coli.* Table 15.4 gives a list of M_{13} vectors and their derivatives.

Table 15.1 Some general-purpose plasmid vectors in *E. coli* (after Thompson, 1982)

Plasmid	Genetic marker (s)	Cloning site (s) / Phenotype (s)
pMB9	Colicin immunity	*Eco*R1, *Sma*1, *Hpa*1/None, *Hind*III, *Bam*H1/Tc-S
pBR322	Tc-r, ampicillin resistance (Ap-r)	*Eco*R1, *Bal*1, *Pvu*II/None, *Hind*III, *Bam*H1, *Sal*1, *Sph*1/Tc -S, *Ava*1, *Pst*1, *Pvu*1/Ap-S
pBR325	Tc-r, Ap-r, Cm-r	*Eco*R1/Cm-S, *Pst*1, *Pvu*1/Ap-S, *Hind*III, *Bam*H1, *Sal*1, *Sph*1/Tc-S, *Ava*1/None
pKC7	Ap-r, kanamycin resistance (Km-r)	*Bg*lI, *Bcl*1/Km-S, *Pvu*1/Ap-S, *Eco*R1, *Hind*III, *Sma*1, *Xho*1, *Bst*EII, *Bam*H1/None
pACYC184	Tc-r, Cm-r	*Hind*III, *Bam*H1, *Sal*1/Tc-S, *Eco*R1/Cm-S
pAC105	Colicin immunity	*Eco*R1
pMK16	Tc-r, Km-r, Colicin immunity	*Bam*H1, *Sal*1, *Hinc*II/Tc-S, *Xho*1, *Sam*1/Km-S, *Eco*R1/None

Table 15.2 Some special-purpose cloning vectors in *E.coli* (after Thompson, 1982)

Plasmid	Genetic marker(s)	Cloning sites/Phenotypes	Number of copies per chromosome
pMF3	Ap-r	*Eco*R1, *Hind*III, *Bam*H1/None	1-2
pDF41	Trp.E+	*Eco*R1, *Hind*III, *Bam*H1, *Sal*1/None	1-2
pRK2501	Tc-r, Km-r	*Sal*/Tc-S, *Hind*III, *Xho*1/Km-S, *Eco*R1, *Bg*/1/None	2-4
pBRH1	Ap-r	Promoter having *Eco*R1 can express Tc-r on cloning	14
pGA39	Cm-r	*Hind*III, *Xma*1, *Pst*1 or blunt-ended fragments having promoters can express Tc-r upon cloning	15
pKY2289	Ap-r	DNA insertion at *Eco*R1 or *Xma*1 site allows colonies to grow on plates containing mitomycin C	17

Table 15.3 *Bacillus* plasmids suitable for cloning (after Subbaiah *et al.*, 1985)

Plasmid	Genetic marker(s)	Cloning site(s)
pUB110	Km-r	*Eco*R1
pHV33	Ap-r, Tc-r, Cm-r	*Eco*R1, *Hind*III, *Bam*H1, *Sal*1, *Pvu*1, *Ara*1
pBD6	Km-r, Sm-r	*Bam*H1
pBC16	Tc-r	*Eco*R1
pBC16-1	Tc-r	*Eco*R1, *Hind*III
pGY31	Tc-r	*Eco*R1

Table 15.4 Some M_{13} vectors (after Bibb et.al., 1980)

Phage	Phenotypic marker(s)	Cloning sites/Phenotypes
M13 mp 2	Blue plaques	*Eco*R1/white plaques
M13 mp 5	Blue plaques	*Hind*III/white plaques
fd 101	Ap-r, Km-r	*Pst*1/Ap-S, *Hind*III, *Sma*1/Km-S
fd 107	Ap-r	*Pst*1/Ap-S, *Sal*1, *Hind*III, *Eco*R1/None
fd Tet.	Tc-r	*Eco*R1, *Hind*III, *Xba*1, *Ava*I/None

Once a gene has been cloned and purified, one may use the technique of genetic transformation to reintroduce the purified, cloned. DNA sequence and establish it in a suitable, competent, recipient bacterial cell. Competence, or receptivity, in bacterial cells can often be induced by exposing them to low concentrations of calcium chloride. One can now synthesize or engineer recombinant plasmids in cell-free systems and then introduce them into *E. coli* cells by the calcium chloride treatment. Following successful transformation in this way, the plasmids go on replicating autonomously and indefinitely in the host cells.

A frequent impediment in gene cloning work that has been observed in *Bacillus subtilis* (Ehrlich *et al.,* 1986) and some other microbes is the structural plasmid instability or plasmid rearrangements. These rearrangements are caused by illegitimate recombination which occurs much more frequently in plasmids than in the chromosome. Thus, directly-repeated sequences 4-kb long recombine with a frequency of about 10% per cell generation when carried on a plasmid, but the same sequences recombine some 1000 times less frequently when carried on the chromosome of this bacterium (Ehrlich *et al.,* 1986).

In some cases, the cloning of eucaryotic genes in microbial hosts leads to the irretrievable loss of valuable information about the state of cytosine methylation and the interaction of regulatory proteins with their respective recognition sequences *in vivo.* Besides, studies of naturally-occurring mutations have required that each mutant sequence be cloned before being sequenced. Another problem is that occasionally cloning procedures themselves introduce artefacts (e.g., deletion mutations) into the DNA fragments of interest. Many such problems can be overcome by resorting to direct genomic sequencing of native, uncloned DNA (Saluz and Jost. 1987).

Strategies for Improving Plasmid Stability in Genetically Modified Bacteria in Bioreactors

The rDNA technology provides a direct approach for the production of a wide range of biochemical products from industrial microorganisms, as well as from mammalian and plant cells.

The organisms most commonly exploited for industrial production purposes are *Escherichia coli, Bacillus subtilis,* and *Saccharomyces cerevisiae,* primarily for the production of recombinant proteins. To manufacture such proteins, DNA sequences encoding the protein must be transferred into the cells. This is usually achieved by the introduction of extrachromosomal DNA in the form of plasmids or phages. The cells which carry these recombinant genes can be cultured in special bioreactors for the large-scale production of proteins including a high proportion of specific product (Kumar *et al.,* 1991). Another application of these microorganisms is the introduction of new enzymatic activities which may generate greater quantities of enzymes, to deregulate existing metabolic pathways or to create new ones. This metabolic design is used to produce higher levels of amino acids and to increase the substrate spectrum of microorganisms. The creation of such modified systems is achieved by sequential steps, including identification and isolation of the required genes, construction of the expression vectors and screening for the optimal host system. Generating the genetic equipment in a desired host is followed by the optimization of the bioprocess, which involves scale-up the technology for the separation of the product (Kumar *et al.,* 1991).

Plasmids are widely used as vectors in genetic engineering for the expression of foreign genes in procaryotic or eucaryotic (yeast) cells. A variety of plasmids are known but not all are useful in commercial bioprocessing. Desirable characteristics for useful vectors include (1) high copy number, (2) possession of several unique restriction endonuclease cleavage sites, (3) small size, (4) genetic stability, (5) screening markers (e.g., antibiotic resistance gene encoded by the plasmid), and (6) simple procedures for transfer into the host. High plasmid copy number leads to higher concentration of transcription template. To use any plasmid as a vector it should bear several restriction sites for inserting DNA fragments. A small plasmid imposes less of a metabolic burden on the host and also facilitates transformation. A most important parameter is that the expression plasmid should be stably maintained in the host for several generations (Kumar *et al.,* 1991).

The productivity of a bioreactor employing recombinant strains is largely affected by the degree to which the plasmid-free (P^-) cells are generated and propagated. This phenomenon can complicate scaling-up operations. The P^- cells are generated from plasmid-harbouring (P^+) cells by segregational instability which is caused by defective partitioning of the plasmids between the daughter cells during cell division. The resulting cells (with plasmid absent or structurally altered) are non-productive.

The P^+ cells usually grow more slowly than the P^- cells because the P^+ cells have to synthesize more DNA, mRNA, and protein. This leads to a lower maximum growth rate for P^+ cells. The growth rate of P^+ cells also depends upon the toxicity of the coded proteins and strength of the promoters. Thus, once generated in the reactor, P^- cells may propagate rapidly, leading to a mixed population with the P^- proportion of population increasing and leading to poor economics of the total bioprocess.

Although continuous systems are highly productive for many microbial production processes, their application is more limited for recombinant organisms because of plasmid instability. Most recombinant proteins are therefore produced by either batch or fed-batch techniques. The generation of P^- cells is a common phenomenon in both continuous and batch culture. However, the situation is more complicated in continuous processes where the cells go through a greater number of generations than in batch processes (Kumar *et al.,* 1991).

The strategies to improve plasmid stability can be categorized into selective and non-selective methods. The former include maintaining selection for antibiotic resistance by use of antibiotics in the growth medium, complementation of host auxotrophy by incorporating auxotrophic markers on plasmid vectors, lysogenic phage repression and incorporation of suicide proteins and RNA whose synthesis is repressed in the presence of the plasmid. Non-selective methods include incorporation of partition loci to obtain controlled partitioning of the plasmid to the daughter cells during cell division, compensation of auxotrophic defect of the host coded on the plasmid and application of specific culture conditions (Kumar *et al,* 1991).

Plasmid stability is a must for expression of heterologous proteins. The strategies to achieve this have been classified into either cellular/molecular or bioprocess strategies (Kumar *et al.,* 1991). The cellular/ molecular strategies include, for example, modulating genes at the segregational step and post-se'gregational effects. These methods may prevent the formation of P cells in the reactor, or kill or inhibit proliferation of the P~ cells after the segregational step. Chromosomal integration is another strategy. Bioprocess strategies also include inhibition or separation of P" cells from the mixed population in a bioreactor. Only a few strategies such as two-stage cultivation, recycling conditions (between different dilution rates or different substrate concentrations) can nullify the growth advantage of P" cells.

Extant strategies are designed mainly to overcome the segregational instability of plasmids and to nullify or eliminate the growth advantage of P" cells.

The presence and transcription of specific sequences located on plasmid expression vectors can lead to plasmid instability. Both the size of the inserted DNA and the act of introducing foreign DNA are factors which can affect genetic stability.

Earlier, it was proposed that 'illegitimate' recombination might be a consequence of errors of the DNA-modifying enzymes during rearrangement or replication procedures, which affect stability of the plasmids. Smaller-size plasmids (<10 kb), used as vectors for Gram-positive bacteria, are replicated by the 'rolling-circle replication'. By this mode of replication single-stranded DNA (ssDNA) is generated and s.uch plasmids are therefore known as ssDNA plasmids. These plasmids are liable to suffer frequent errors during replication. In some Gram-positive bacteria the plasmids replicate (size >10 kb) with low error levels. These plasmids use the mechanism of 'theta replication' instead of 'rolling-circle', suggesting that the former replication might be less error-prone than the latter one. Using these plasmids*, large DNA fragments (up to 40 kb) can be efficiently cloned and maintained for 1 50 generations (Kumar *et al.* 1991).

Optimizing environmental conditions for the culture of organisms is an important consideration for successful operation of any bioprocess. However, for culturing recombinants, there is the additional necessity of plasmid stability. Therefore, in bioprocesses using recombinant organisms, the objectives are high plasmid stability, high volumetric productivity, high yield coefficients, and low costs of ingredients and energy. For recombinants, plasmid stability during cell culture is a primary consideration, with other factors which contribute to improved productivity assuming secondary importance. In general, use of a complex growth medium has been found to stabilize some plasmids during cultivation.

Recombinant microbes tend to accumulate high intracellular concentrations of the proteins expressed from the introduced genes. In addition, they are grown to high cell densities, essential for increasing productivity in bioprocesses. For large-scale culture in particular, this leads to the need for an inexpensive, simple C-source substrate such as glucose. Overfeeding with glucose, however, leads to a bacterial Crabtree effect (the inhibition of oxygen consumption in cellular respiration that is produced by increasing concentrations of glucose) under aerobic conditions in which acetate and CO_2 formation occurs; high levels of these metabolites in the culture broth can inhibit further growth. To overcome these limitations, a fed-batck culture strategy using different feeding policies has been developed for *E. coli* and *Saccharomyces cerevisiae.*

One problem for scaling-up bioprocesses is the construction of reactors which allow a high level of oxygen input and in which the medium is well mixed. Micromixing problems in industrial scale bioreactors need to be solved for the culture of recombinant systems (Kumar *et at.* 1991).

Strain Construction

Mutation and selection have traditionally played the major role in the development of many currently-used organisms for industrial production of amino acids and nucleotides. The starting point is organisms having some capacity to synthesize the desired product but which require several mutations leading to deregulation in the biosynthetic pathway so as to permit better product yield. The multiple mutations enable channelization of carbon sources to the desired products and minimize loss of carbon in byproducts or its diversion to less important products.

It is now possible to prepare semisynthetic genes by substituting a synthetic oligodeoxyribonucleotide segment containing desired changes in its nucleotide sequence into the total DNA gene coding for a cloned protein. This has given us the tools for systematic genetic manipulation of the primary structures of proteins. By using the technique of oligonucleotide site-directed mutagenesis, it is possible to change a single amino acid in the primary sequence of a protein. In this way, one can engineer proteins and enzymes to modify their behaviour.

The site-directed mutagenesis of any protein (or its gene) involves (1) cloning of the concerned gene in some suitable vector,.(2) expression of the cloned gene, (3) determination of the sequence of the gene, (4) synthesis of oligodeoxynucleotide, (5) oligonucleotide-directed *in vitro* mutagenesis, and (6) identification and isolation of mutant clones. A plasmid vector is nicked with a restriction enzyme (Diag.15.9). Exonuclease III is used to digest away a part of the coding strand at the 3'-end. A 16-19 base long synthetic oligonucleotide is made sufficiently complementary to the template so as to hybridize with the appropriate sequence. DNA polymerase and DNA ligase are used as shown (Diag.15.9). A heteroduplex is formed. Transformation and *in vivo* replication then yield homoduplexes whose sequences are either the same as the sequence of the wild-type DNA or the sequence of the synthetic oligonucleotide containing the desired mutation. Colonies can then be screened using the synthetic oligonucleotide primer labelled with ^{32}P, as a hybridization probe, to permit detection of the desired mutant.

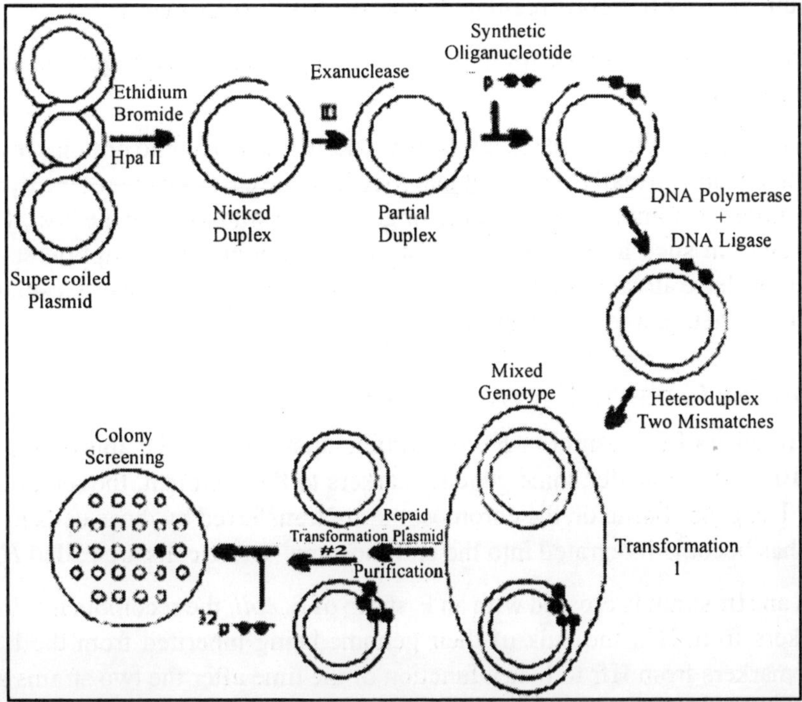

Diagram 15.9 Strategy for oligodeoxynucleotide site-directed mutagenesis of a protein involving *a* plasmid vector (after Dalbadie-McFarland *et al.*, 1982).

In recent years, genetic manipulation techniques have been used for developing strains that overproduce metabolites. Here, the starting strains are usually those which have not previously been subjected to mutagenesis (Old and Primrose, 1981). The chief idea is to increase the number of copies of structural genes coding for the relevant enzymes and also to increase the frequency of transcription which, of course, is related to the frequency of binding of RNA polymerase to the promoter region. The biosynthetic genes may often be incorporated *in vitro* into a plasmid such as pBR322 which, upon entry into the cell by genetic transformation, will produce multiple copies of the genes. The transcription frequency may be increased by constructing a hybrid plasmid *in vitro* which harbours the structural genes of the biosynthetic enzymes but at the same time lacks the regulatory genes (i.e., promoter and operator). In this hybrid, the structural genes are in fact attached next to an efficiently and frequently-read promoter and operator, and are then subject to regulation by these genes.

Expression Vectors

These exemplify special-purpose cloning vectors. Once a gene has been cloned and identified, subsequent steps may involve its recloning into some secondary vector so that its transcription is directed by a suitable vector-associated promoter. Some promoters that have been employed for this purpose are *lac* and *trp* of *E. coli* and the phage lambda P_L promoters. Hallewell and Emtage (1980) developed certain plasmid vectors containing the *trp* operon promoter suitable for efficiently regulated expression of foreign genes.

This work exemplifies a vector in which one may insert DNA fragments downstream from the promoter.

In some cases, the cloned genes fail to be properly translated into proteins, even though transcription occurs normally. This problem may in some cases be overcome by fusing the gene to the amino-terminal protein of a vector gene that is translated in the host. The vector-associated gene in such a case provides both translation and transcription start signals. Suitable vectors have already been made to permit fusion of a cloned gene to β-galactosidase anthranilate synthetase and β-lactamase.

F-Plasmid and Genetic Recombination

The presence of F-plasmid in *E. coli* confers upon the cell the ability to act as male (Diag. 15.10) and to transfer some genetic markers to the recipient, though at a very low frequency. The genes borne on the chromosome are transferred by those cells in which the F-plasmid has become integrated into the chromosome. These cells are called *Hfr.*

When an Hfr strain is crossed with an F^- strain of E. *coli,* the recombinants inherit only a few markers from Hfr, the bulk of their genome being inherited from the F^- cell. The transfer of markers from Hfr to F^- is a function of the time after the two strains are mixed. The mating between the Hfr and F^- strains can be interrupted at known intervals of time by

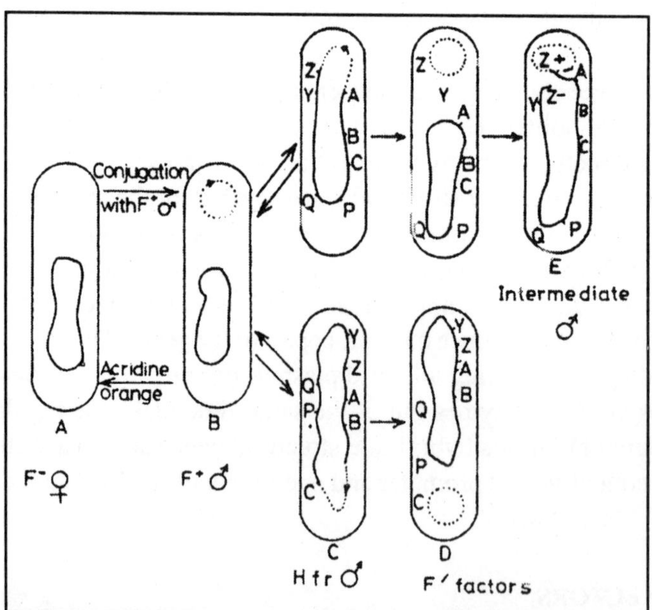

Diagram 15.10 Characteristics and interrelations of F+, Hfr, and F^- strains of E.coli (dotted lines show the sex factor; arrows show the polarity of transfer; letters A-Z, represent position of the bacterial genes).

subjecting the conjugants to mechanical agitation in a blender. This makes it possible to map the gene order (and distance) by analysis of the recombinants from crosses involving different strains of Hfr and F⁻, in terms of time units. A circular composite genome map can be drawn in this way (Diag. 15.11).

Unlike the F^+ cells, the Hfr strains do not, as a rule, transfer the F^- factor from cell to cell. Secondly, the vast majority of recombinants produced in Hfr x F^- crosses are F^-, not Hfr. Both these findings suggest that the F-plasmid becomes integrated into the chromosome, in the case of the Hfr strain.

However, after a certain period, the integrated F-plasmid can also get detached from its chromosomal location and once again assume an independent, autonomous, extrachromosomal state; inother words, Hfr reverts back into F^+. In some cases, during this process, the F-factor also carries a small segment of the bacterial chromosome (containing 1, 2 or a few chromosomal genes) along with it. These modified plasmids are called F^1 (F-prime) to distinguish them from the original F-plasmid. Those strains which harbour the F^1-plasmids (contain some extra genome) are called incomplete diploids, or merozygotes.

Plasmids

The term phasmid is used to designate a hybrid between a plasmid and a phage. Whereas plasmids are restricted to an intracellular state, phage particles can exist extracellularly as infectious virus particles. Kahn and Helinski (1978) have succeeded in reconstructing the plasmid ColEl, by means of appropriate recombinant DNA techniques, with a view to allowing it to be packaged *in vivo* into bacteriophage particles. This kind of plasmid packaged in the form of phage particles may then be injected into a suitable recipient bacterium. In these attempts, phage P4 is generally used (Diag. 15.12). The phasmid so produced has been designated P420 and is readily interconvertible between the phage and the plasmid states.

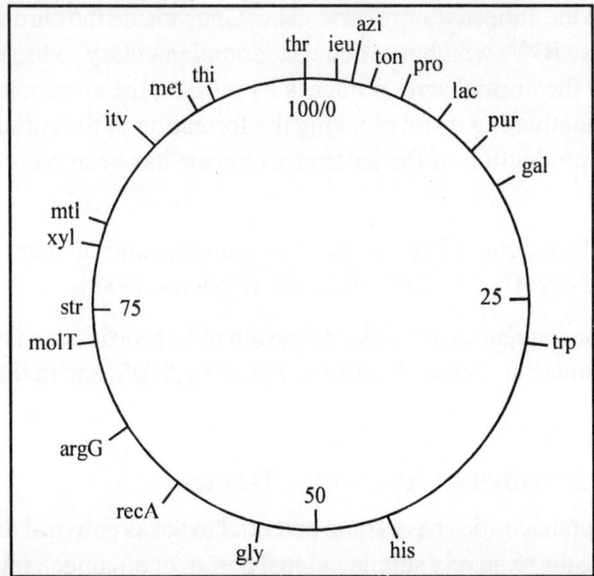

Diagram 15.11 Temporal map of *E. coli* genome, calibrated from 0 to 100 minutes.

The next decade is likely to see increasing use of phasmids in various research experiments designed to study the molecular and genetic basis of the extrachromosomal state of DNA in bacteria and also the replication of various other circular DNA molecules found in eucaryotes as well as procaryotes (Helinski, 1979).

"ANTISENSE" RNA TECHNOLOGY

A major recent breakthrough in genetic engineering is what has been termed the "antisense technique". This technique involves putting a gene into a cell "backwards", i.e., in the reverse orientation, so as to regulate its expression. In this technique, an 'antisense' sequence complementary to the coding strand is used to specifically block expression of the gene. The binding of the antisense sequence to the coding RNA by base-pairing interferes with its translation into protein, thereby reducing the amount of protein produced. This kind of antisense mechanism has been known to operate naturally in bacteria where it controls gene expression. Several attempts have been made in recent years to adapt the bacterial mechanism for engineering the cells of plants. Researchers at the Free University of Amsterdam have now used, for the first time, antisense RNA to turn off an endogenous plant gene—the gene determining the flower colour *of Petunia.* The same approach has been used to genetically engineer better tomatoes which are bruise-proof and which can be left to ripen on the plants (instead of plucking them green and then exposing to ethylene to elicit the colour). The objective was to turn down the production of the key enzyme polygalacturonase, which is responsible for fruit softening. This enzyme is switched on only when the fruit ripens, and it then lyses the cell walls. Researchers have first cloned the polygalacturonase gene-and then hooked it up, backwards, to a regulatory sequence that ensures continuous switching on of the antisense gene following its insertion back into the plant. The Ti-plasmid vector has been used for carrying the gene back into the plant chromosome. Accordingly, the transgenic tomato contains two versions of the gene, namely, the normal gene and the antisense or reverse gene. Once inside the chromosome, the reverse gene makes antisense RNA which is, of course, complementary to the mRNA made by the normal gene. When the normal gene produces its mRNA, the antisense RNA is thought to bind to it, thereby inactivating it and blocking the formation of the softening enzyme. Some 90% decline in the production of the softening enzyme has been recorded, and the trait is genetically stable.

This antisense technique is likely to have wide applications in plant biotechnology such as the production of decaffeinated coffee beans (Roberts, 1988).

Use of antisense sequences to block expression of a specific gene is also being explored in the treatment of cancer and viral infections, including AIDS, and in the control of genes of trypanosomes.

ANTISENSE NUCLEOTIDES AS ANTIVIRAL DRUGS

Antisense oligonucleotides have some potential to act as antiviral drugs (Cohen, 1989). A major advantage is the relatively simple rational design of oligonucleotides which bind only to specific nucleic acid sequences, compared with conventional drugs which are frequently targeted against sites of unknown structure in proteins. In Watson-Crick base pairing, heterocyclic bases of an antisense oligonucleotide form hydrogen bonds with the heterocyclic bases of target single-stranded nucleic acids (RNA or single-stranded DNA). In Hoogsteen base pairing, the heterocyclic bases of target are double-stranded nucleic acids (i.e., double-stranded DNA). Both these models of binding by antisense oligonucleotides can potentially regulate gene expression, and may have possible use in modulating some human and plant diseases (Agrawal, 1992).

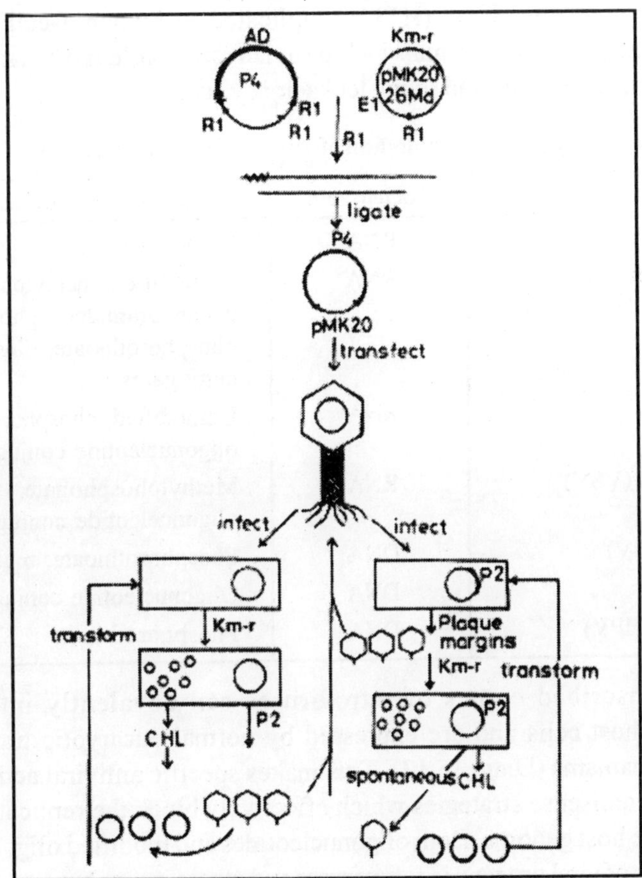

Diagram 15.12 A hybrid ColEI-P4 phasmid. Plasmid pMK20 is derived from ColEl resistant to kanamycin. (P2. bacteriophage P2; CHL, cells incubated in presence of chloramphenicol.) (Based on the data of Kahn and Helinski, 1978.)

It has been known since 1977 that cell-free translation of mRNA is in-hibited by the binding of an oligonucleotide complementary to a short segment of the mRNA to form a duplex. The first example of specific inhibition of gene expression *in vivo* by a synthetic oligodeoxynucleotide was the inhibition of Rous sarcoma virus replication in infected chicken fibroblasts by a synthetic oligodeoxynucleotide complementary to a part of the viral genome (Agrawal, 1992). There was also inhibition of transformation of primary chick fibroblasts into malignant sarcoma cells.

One good advantage of antisense oligonucleotides as chemotherapeutic agents compared with conventional drugs is the possibility of designing an oligonucleotide which binds specifically to its target nucleic acid sequence. Another merit is the relative ease of design and synthesis.

Antisense oligonucleotides have been designed for use against many different viruses (Table 15.5). Retroviruses have been a major target in studies,of antiviral oligonucleotide strategies because they are of much medical and veterinary concern. Human retroviruses include human immunodeficiency virus (HIV, associated with AIDS; Wickstrom, 1991) and human T-cell lymphotropic virus (HTLV, implicated in human T-cell leukaemia); and retroviruses infecting domestic animals include avian sarcoma-leukosis viruses (ASV-ALV) (affecting poultry), and bovine and feline leukaemia viruses.

Table 15.5 Antisense oligonucleotide inhibition of viruses (from Agrawal, 1992)

Virus	Nucleic acid	Comments/Conditions
Rous sarcoma (RSV)	RNA	Unmodified
Human immunodeficiency (HIV)	RNA	Unmodified, methylphosphonate, phosphoramidate, phosphorselenoate, phosphorothioate, oligonucleotide conjugates
Influenza	RNA	Unmodified, phosphorothioate, oligonucleotide conjugates
Vesicular stomatitis (VSV)	RNA	Methylphosphonate, oligonucleotide conjugates
Herpes simplex (HSV)	DNA	Phosphorothioate, methylphosphonate
Simian (SV40)	DNA	Oligonucleotide conjugates
Human papilloma (HPV)	DNA	Phosphorothioate

Reverse-transcribed copies of retroviruses can covalently integrate into the chromosomes of host cells and are expressed by normal eucaryotic transcriptional and translational mechanisms (Diag. 15.13). This makes specific antiviral action difficult, and recourse is had to antisense strategies which effectively block the replication cycle before integration into the host genome. Both oligonucleotides and modified oligonucleotides have blocked production of viral progeny in retroviruses and other viruses, but preventing integration appears to be the most desirable approach for retroviruses.

The targets for antisense oligonucleotides are, in general, single-stranded linear nucleic acid molecules, folded back on themselves, to form secondary and tertiary structures, thereby minimizing the possibility of antisense oligonucleotide binding to the target. Antisense oligonucleotides inhibit viral replication with molecules designed to target individual genes as well as splice-donor and splice-acceptor sites; thus blocking the expression of integrated proviral genes. Some phosphorothioate oligonucleotides can cause selective but non-sequence specific inhibition of the reverse transcriptase of HIV.

Translation of mRNA may be blocked by the binding of a complementary oligonucleotide. There are two possible mechanisms by which this can occur: (1) by base-specific hybridization, thus preventing access by the translation machinery, i.e., 'hybridization arrest'; or (2) by forming an RNA-DNA duplex which is recognized by the intracellular nuclease RNase H, specific for digesting RNA in an RNA-DNA duplex. Various chemical modifications of the oligonucleotides can result in three different modes of action (see A, B, C in Diag. 15.13). The antisense oligonucleotide (in A) binds the target by base-specific hybridization, causing both hybridization arrest and RNase H activation. Degradation of mRNA by RNase

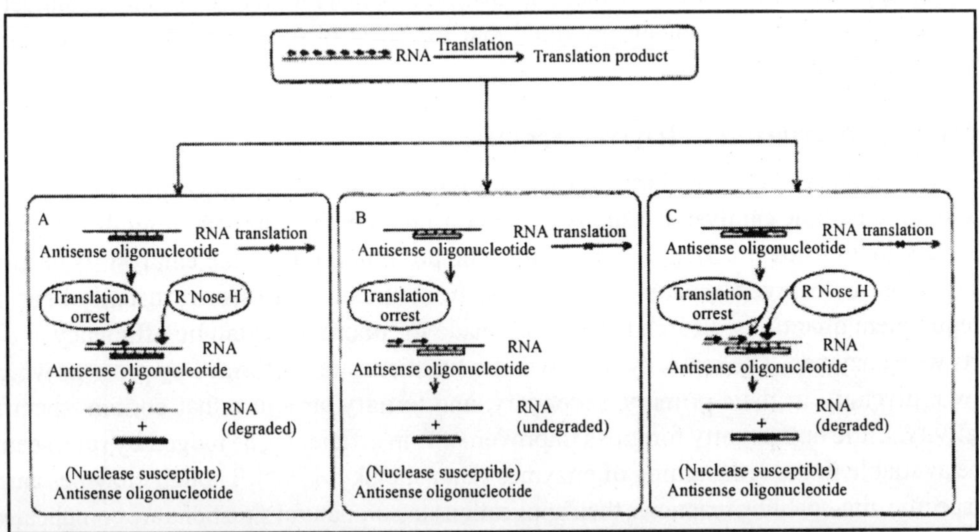

Diagram 15.13 Blockage of mRNA translation by a complementary oligonucleotide. A brief account is given here (Source: Agrawal, 1992.)

RNaseH releases the oligonucleotide, which can then bind to other copies of the target mRNA. The susceptibility of the oligonucleotide to cleavage by other nucleases (i.e., the *in vivo* half-life of the antisense oligonucleotide) is therefore a major parameter affecting this 'catalytic' mode of degradation. Unmodified phosphodiester oligonucleotides and their phosphorothioate analogs fall into this category. The antisense oligonucleotide (in B) binds to the target by base-specific hybridization causing translation arrest, but doe's not activate RNase H. Oligoribonucleotides and their analogs, and oligodeoxyribonucleotides containing methylphosphonate, phosphoramidate, and various non-phosphate internucleotide linkages fall in this category. These oligonucleotides are nuclease resistant, and are effective in inhibiting translation, but are generally required in higher molar concentrations than those which activate RNase H. The oligonucleotides in this group (C) combine the features of (A) and (B): the oligonucleotide contains internucleotide linkages which activate RNase H (e.g., phosphodiester, phosphorothioate), flanked by nuclease-resistant internucleotide linkages which do not activate RNase H (e.g., methylphosphonate, phosphoramidates', non-phosphate internucleotide linkages etc.). Digestion of the mRNA target in the RNA-DNA duplex releases the oligonucleotide, which, because of its nuclease resistance and, hence, longer *in vivo* half-life, is more effective than oligonucleotides in category (A). Some oligonucleotides in category (C) are hybrids of oligodeoxyribonucleotides (central region) and either Oligoribonucleotides or their analogs (flanking regions).

The antiviral activity of an antisense oligonucleotide depends usually on the binding of the oligonucleotide to the target nucleic acid, thereby disrupting the function of.the target, either by hybridization arrest (Diag. 15.13 and 15.14) or by destruction of the target (RNA) via RNase H (Table 15.5; Diag. 15.13).

The antiviral properties of antisense oligonucleotides and their various analogs, together with their apparently low toxicity in mice and rats, indicate that they may be suitable candidates for pharmaceutical development, provided that the costs of producing them are brought down to reasonable levels.

Catalytic Antibodies (Ribozymes)

Many industrial chemical transformations require either a promoter or a catalyst. The promoter, unlike the catalyst, is usually either consumed in the reaction, or tends to show a relatively low turnover efficiency. Therefore the promoter, unlike the catalyst, is used in stoichiometric (or even greater) proportions. In contrast, catalysts are usually used in subequivalent quantities. Search is therefore made for improved catalytic efficiency and it starts with enzymes. The majority of enzymatic functions are performed by proteins which show a diversity in their primary, secondary, and tertiary structure that confers specific reactivity. Little opportunity for gross improvements in efficiency, as judged by rate, seems to be available in the redesigning of enzymes (Danishefsky, 1993). Though enzymes carry out specific, life-enabling reactions, their applications to non-natural situations are complicated. We might distinguish between two kinds of non-natural settings. One is that of a natural type of reaction (such as oxidation, reduction, or aldol condensation) with non-natural substrates. Considerable scope exists for achieving enzymatic modulation of artificial substrates undergoing natural reactions. It has been less easy to adapt protein-based enzymes to *catalyze* reaction types not included in their original "capability". Consequently the protein structure of enzyme has shown little adaptability in acquiring wholly unanticipated reactions.

Two main complications hinder constructing *de novo* protein based catalysts (that is, artificial enzymes) to accommodate unnatural reactions. It is difficult to interrelate the active site structure of the proposed protein with its ability for catalysis. For the moment the capacity to obtain peptides and proteins of defined amino acid sequence by fully synthetic or recombinant means has not much helped in obtaining valuable *de novo* fashioned artificial enzymes.

It is for this reason that catalytic antibodies have attracted attention. The massive power of the immune system is directed toward producing antibodies whose combining sites are complementary to antigen which is so designed as to simulate the proposed transition state for a chemical reaction. When exposed to reaction substrates, the binding forces of the antibody accelerate the reaction. Virtually the full binding force of the antibody is, in principle, applicable for purposes of catalysis. Catalytic antibodies can catalyze known but non-natural chemical reactions quite well. Hitherto, they have been used to catalyze "typical" organic reactions. But, recently, catalysis by an antibody has also been achieved in an otherwise disfavoured reaction (Danishefsky, 1993).

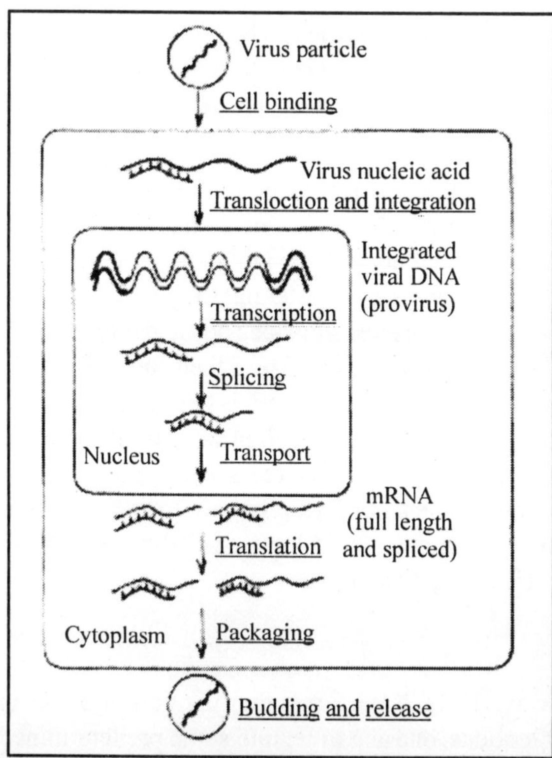

Diagram 15.14 A retrovirus replication cycle and possible target sites of sequence-specific action for antisense oligonucleotides. The first step in the virus replication cycle is binding to the cell surface. This is followed by entry, uncoating, and release of viral genetic material. An antisense oligonucleotide could hybridize with viral nucleic acid and inhibit processing required for translocation and integration. Integrated viral genome remains in a double-stranded form called provirus. At the provirus stage, antisense oligonucleotides could inhibit transcription by formation of a triple helix. Antisense oligonucleotides could also inhibit RNA polymerase activity by binding to locally opened DNA or to nascent RNA hybridization. After the provirus stage, the steps which could be inhibited by antisense oligonucleotides are: splicing (through hybridization at intron-exon junctions or the lariat branch site); transport of mRNA from nucleus to cytoplasm; and finally translation of mRNA. Translation could be inhibited by targeting antisense oligonucleotides to, for example, the cap site, primer binding site, initiation sites, polyadenylation site, packaging site, and protease cleavage site. (Agrawal, 1992.)

Ribozymes destroy RNA that carries the message of disease, and now are on the verge of leaving the lab for the clinic. The ability to target ribozymes to cleave viral RNA *in vitro* has generated much speculation about their potential therapeutic value as antiviral agents *in vitro* (Sullenger and Cech, 1993).

Already, attempts are being made to insert the gene for a specific RNA sequence into the cells of HIV-infected patients (Dropulic *et al,* 1993). A specially designed ribozyme—an RNA molecule engineered to seek out and destroy the RNA genome of HIV by cutting it in two—is being tried with the expectation that lymphocytes containing the ribozyme gene will have a better chance of surviving HIV infection.

RNA enzymes have potential as therapies for diseases such as AIDS, cancer, and chronic hepatitis.

Many ribozymes were dubbed with terms such as hammerhead, hairpin, and axehead, inspired by their three-dimensional shapes (Symons, 1992). The key to their unique activity lies in their structure: they contain stretches of nucleotides that base-pair with a complementary RNA region, and they have a catalytic section, like the active site of a protein enzyme, that chops the bound RNA while the base-pairing holds it in place (Barinaga, 1993). These features make catalytic RNAs ideal material for bioengineering: a ribozyme can be custom-designed to recognize and base-pair with a specific cellular RNA that a researcher would like to eliminate. Once designed, the ribozyme can be then turned loose in the cell to kill its target. There are plenty of clinical situations where physicians would like to target and destroy a "bad RNA", chronic viral diseases, cancers initiated by a mutated oncogene. HIV is another tempting target.

Attempts are underway to develop a way to target a ribozyme to the site in the cell where the HIV RNA accumulates, thereby improving the ribozyme's chances of hitting home. Delivering the goods to the right site in the cell and to the right cell is an important challenge for ribozyme therapy. For example, ribozymes are a potential therapy for chronic hepatitis B. But unlike white blood cells, which can be removed from the body and reintroduced for treating HIV, the liver obviously cannot be temporarily removed, and therefore ribozymes need to be delivered to the organ while it's still in the body. Some work is being planned with a ribozyme taken from a liver-infecting viroid called delta that often tags along with the hepatitis virus. Delta might be able to be exploited not only as a source of ribozymes, but also as a delivery vehicle. If one could engineer delta to be a non-symptomatic form that would deliver and express certain sequences to liver cells, we would have a self-limiting vector delivery system that could combat hepatitis (Barinaga, 1993).

Gene Machine

A computerized gene machine is now available. It performs the function of building DNA sequences to order by combining nucleotides, one at a time, into some predetermined sequence. For instance, if a G (guanine) is needed, the growing sequence is placed in a solution containing modified G nucleotides. The purpose of the modification is to allow only one G to be added, in order to prevent several G's from attaching to the sequence. Following the incorporation of one G, the chemical block (modification) has to be removed before a different nucleotide (or even another G) can be added. Diagram 15.15 gives an outline of; a gene machine.

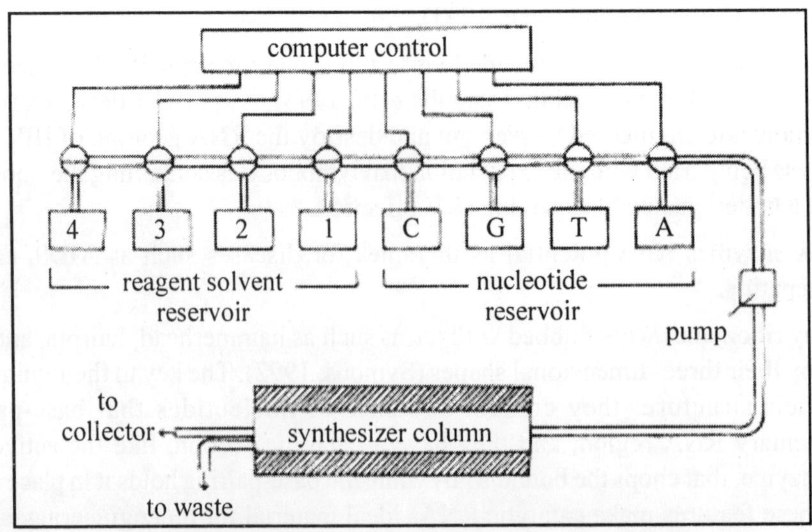

Diagram 15.15 Design of a computerized gene machine.

One such machine has a column reactor in which oligonucleotides are grown on solid support beads packed into cassettes. These prepacked cassettes can be colour-coded according to the first 3'-terminal in the required gene sequence. The appropriate cassette is inserted into one of the column reactors before the start of the synthesis. At the end of the process, the cassette is removed from the reactor, releasing the oligonucleotide. There is no need to remove the solid support. The solid support beads are made from an inert polymer.

The machine has a peristaltic pump and nine glass bottles used for washing solutions, oxidation, capping, and detritylation operations. Some bottles serve to contain waste solutions whereas others can store modified nucleotides or additional reagents.

The gene assembly process involves detritylation, activation, coupling, oxidation, and capping operations, which are controlled by microprocessor. Several intervening washes are also involved.

Although gene machines presently offer the quickest way of synthesizing predetermined nucleotide sequences, their limitation is that only short DNA sequences can be built.

Genetic Engineering of Animals

Much of the knowledge currently available for laboratory animals such as *Drosophila* and mouse is being extrapolated to higher animals with a view to transferring today's gene-splicing and recombinant DNA technology to domestic animals. In these attempts, it is desirable to integrate the exotic genes with the endemic genome, especially with reference to those genes which influence the various economic traits. Seller and Beckman (1982) considered the possibility of using recombinant DNA probes to identify specific segments of chromosomal DNA produced by using restriction endonucleases to break DNA in the genome. The technique enables one to characterize an animal's genotype at a number of locations. A more immediate possibility involves,the techniques of freezing and cleaving embryos, in conjunction with embryo transfer programmes.

Animal breeders have sought major genes as a quick means of making significant genetic changes in a population. In fact, the genes now being exploited commercially are dwarfing genes in chicken, muscular hypertrophy genes in cattle, and genes determining meat quality in pigs. In sheep, a gene which dramatically affects the ovulation rate has been reported (Cunningham, 1984).

DNA Hybridization

An inherent part of several genetic manipulation programmes is DNA hybridization. The rate at which two polynucleotide strands join together *in vitro* depends on the extent to which their nucleotide sequences complement each other. The relative ease with which single-stranded cDNA can be produced has catalyzed new possibilities. Often, it can be made radioactive to facilitate its proper identification, and it can then be employed as a DNA probe for locating complementary sequences elsewhere.

For instance, in chromosome mapping, cDNA binds to the precise position on a chromosome where a gene is located. Again, the genes of different organisms can often be intercompared by their ability to bind to a specific sequence of cDNA, so that the relationships between them may be elucidated.

One of the most useful applications of recombinant DNA technology has been the synthesis of insulin genes. The steps involved in the process, leading to the formation of active human insulin, are shown in Diag. 15.16.

Pulido *et al.* (1987) have used the 15-bp synthetic oligonucleotide 5'-GCTGTGAGGAA ATAC-3' as a probe to screen a human DNA library and detect recombinant phages containing genes of the interferon alpha family. One of the clones was found to carry the interferon alpha 2 gene, which was processed to substitute the sequence encoding the leader peptide by an ATG initiating codon. They then placed this construction under the direction of suitable promoters of an expression vector generating a plasmid (pIN89) which expressed substantial amounts of the mature form of interferon alpha 2 gene in *E. coli.*

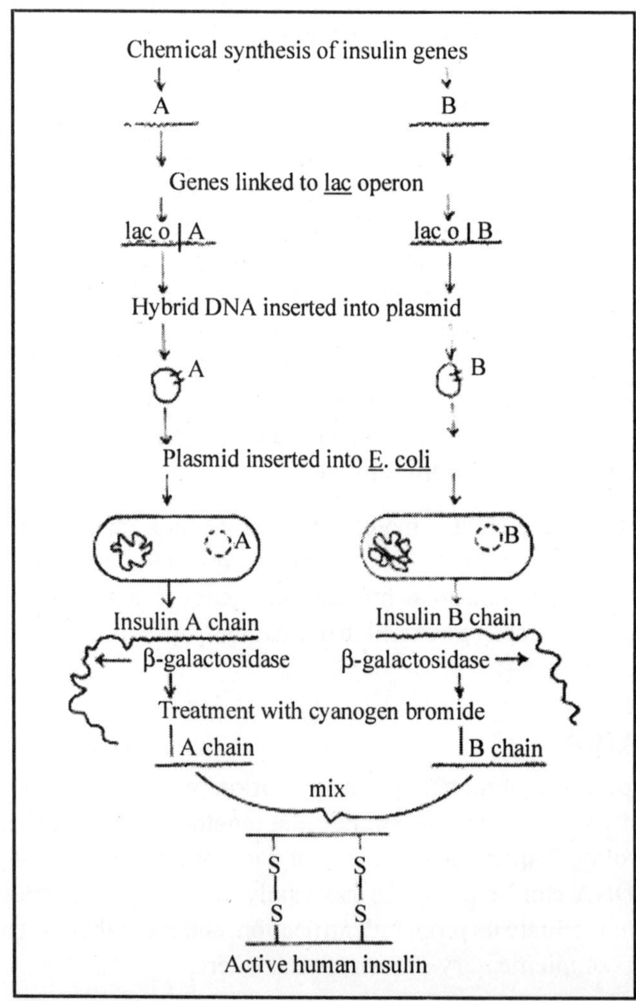

Diagram 15.16 Use of recombinant DNA technology, including DNA hybridization and gene cloning, for production of insulin genes.

DNA FINGERPRINTING

Forensic DNA fingerprinting is a powerful technology that gives us very reliable tool able to link the blood, semen, or hair left at the scene of a crime to a suspect's DNA. This new technology allows us to match two DNA samples. Successful application of the technology depends on several factors, e.g., reliable statistical analysis and very precise, high quality DNA analysis.

There has been much controversy on the statistics. At issue are just how accurate the estimated probabilities are—and how accurate they need to bc. The question arises once a crime lab determines that two DNA samples match. This is done by examining the DNA at several sites where its sequence is known to vary. If all the sites match, it's "strong evidence" that both samples came from the same person. But to estimate the strength, the lab calculates the frequency with which each sequence variation, or allele, occurs in the population to which the suspect belongs by examining, say, the Caucasian database. Then using what is known as the multiplication or product rule, the frequencies of the individual alleles are multiplied to calculate the frequency with which the complete pattern occurs in that population, often resulting in vanishingly small numbers (Roberts, 1992).

But several leading population geneticists feel that the numbers generated by this procedure are misleading and are based on misapprehension of population genetics theory. Populations contain subgroups in which the frequencies of the markers used in DNA fingerprinting vary dramatically from their frequencies in the population at large. That means the likelihood of a match between samples may be grossly over- or underestimated.

Other experts concede that population substructure exists, but insist that current procedures are conservative enough to compensate for it.

A committee constituted by the National Academy of Sciences, USA, does assume that population substructure exists, but they devised a practical and sound approach for accounting for it: using the multiplication rule, but in combination with what they call the

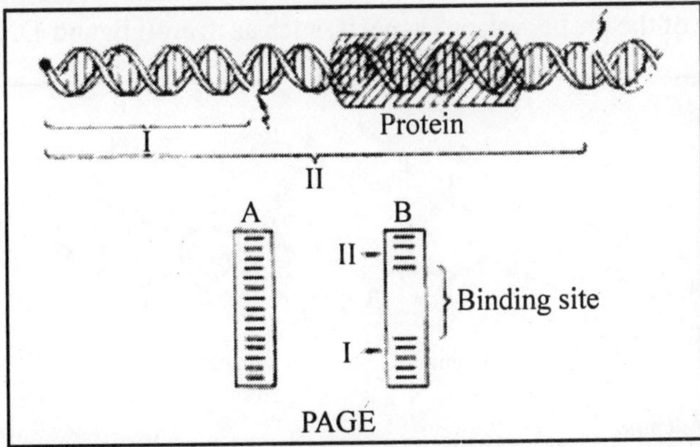

Diagram 15.17 Concept of a footprinting experiment. The DNA molecule is labelled on one strand at one end (•), and cleaved (I) at a density of one cleavage per DNA molecule. Each cleavage point thus corresponds to a specific length of labelled DNA fragment as analyzed by gel electrophoresis in polyacrylamide (PAGE). (A, control; B, in the presence of protein (ligand). (After Nielsen, 1991.)

"ceiling principle". This may ensure that the frequency estimates are biased in favour of the suspect. First, crime labs have to establish the ceiling, or upper bound, frequency for each allele at each site in 15-20 genetically homogeneous populations, such as English, German, Russian, etc. This would be done by collecting blood samples and establishing cell lines from 100 individuals in each population. At the time of calculating the odds of a match, the lab would use the highest frequency found in any of the populations, or 5%, whichever is higher. The end result is that the most "extravagant" probability estimates will be replaced with numbers in the range of 1 in several hundred thousand or a million.

DNA Footprinting

Ligand-DNA interactions play a fundamental role in the biological functions of DNA. Regulation of gene expression is largely based on sequence-specific, protein-DNA interactions in terms of binding of transcription factors and repressers to the regulatory regions (promoters, enhancers) of genes. The binding of such proteins to their recognition sequences of the DNA either promotes (in the case of transcription factors) or restricts (in the case of repressers) binding of RNA polymerase to the gene promoter region. Similarly, many drugs bind to DNA in a sequence selective manner and thus interfere with DNA function.

The structural analysis of ligand-DNA complexes can often be obtained by rather simple footprinting experiments. Diagram 15.17 illustrates the concept of a typical footprinting experiment. A DNA fragment containing the base sequence of interest is radioisotope-labelled on one strand at one end (usually with ^{32}P). The DNA is then treated with the footprinting probe under conditions that result in one modification per DNA fragment on average. The modification usually is directly, or can be converted to, a scission of the DNA backbone, and the position of this modification can be determined by subsequent analysis of the DNA by high-resolution gel electrophoretic analysis, since each cleavage position will correspond to a band in the gel (Nielsen, 1991). By comparing the cleavage pattern in the absence and presence of DNA binding ligand, regions and single base positions which are protected from modification by the ligand can be identified as a "footprint" in the cleavage pattern. By choice of the probe, various aspects, such as overall ligand-DNA contact region,

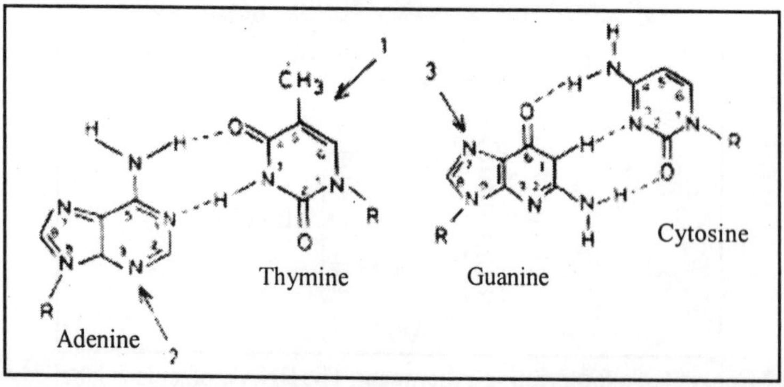

Diagram 15.18 Watson-Crick base pairs of double helical B-DNA. Positions attacked by dimethyl sulphate (1 and 2) and $OsO_4/KMnO_4$ and psoralen (3) are indicated by arrows. The major groove is facing upwards and the minor groove downwards.

contacts with the DNA bases, or contacts with deoxyribose or the phosphates of the DNA backbone, can be analyzed. Thus, the choice of footprinting probe is an essential part of a footprinting experiment (Nielsen, 1991). Some examples of these probes include DNase I, dimethyl sulphate (Diag. 15.18), Fe(II)EDTA, UV-B radiation and psoralens/UV-A radiation (Nielsen, 1991); for *in vivo* footprinting analysis, dimethyl sulphate is a very useful probe but enzymatic probes are unsuitable.

Genetic Diseases

Over 3000 different diseases are transmitted from parent to offspring. Out of these, specific defects in DNA in some twelve inherited diseases (including Huntington's disease, sickle-cell anaemia) have so far been successfully identified. About a dozen genes probably involved in cancers have been discovered. Over 50 human genes had been sequenced (Watson *et al.,* 1983) by the early 1980s. Over 1600 human genes have already been mapped.

DNA diagnostics involve the analysis of disease at the nucleic acid level. These diagnostics may provide rapid, automated analyses for nucleic acid sequences associated with genetic diseases. DNA diagnostics are also believed to facilitate the identification of disease-associated genes at birth, thus creating new opportunities for preventive medicine (Landegren *et al.,* 1988).

The human genome contains about 10^5 genes, encoded by about 3-5% of the total 3 x 10^9 base pairs of DNA. This DNA is distributed on 24 different chromosomes. Each person inherits a complete set of 22 autosomes plus one X or *Y* sex chromosome from each parent. This means that each autosomal gene is present in two copies. Genes contain exons (protein-coding regions) and introns (non-coding regions). Individual genes may be made of up to about,2 million base pairs. The ultimate tool of detecting DNA sequence variants is DNA sequence analysis for which some sequencers are now available. Two commonly-used methods for such an analysis are the chemical degradation technique and the chain termination technique (Landegren *et al.,* 1988).

Diagram 15.19 illustrates the idea of DNA diagnostics in relation to human genome. In recent years, recombinant DNA methods have become available which can readily detect DNA mutations and identify molecular defects in man that account for heritable diseases, somatic mutations associated with neoplasia, and also acquired infectious diseases. Advances in recombinant DNA technology have enabled us to diagnose several diseases, and now occupy a prestigious place in the repertoire of clinical physicians.

No doubt, the detection of a new mutation at a gene locus is a difficult, hard challenge. New mutational events Polyacrylamide gel electrophoresis occur frequently in X-linked recessive diseases, e.g., Lesch-Nyhan syndrome (deficiency of hypoxanthine-guanine phosphoribosyl transferase enzyme), urea cycle defect resulting from a deficiency in ornithine carbamylase, and Duchenne muscular dystrophy. Certain novel methods have now been developed to tackle the aforestated diagnostic challenge. One of these methods, called the ribonuclease A cleavage method, can detect single base mismatches between the radioactive RNA probe (normal) and the patient's genome or transcribed sequences (Caskey, 1987).

It is now possible by site-directed mutagenesis to change essentially any amino acid in any protein.

Neoplasia means acquired genetic alteration. Oncogenes can be activated by point mutations or by chromosomal rearrangements. Likewise, T-cell leukaemias can also be activated by chromosomal changes.

Direct detection of various genetic disorders at the DNA level is now possible using cloned gene or oligonucleotide probes. Further, the use of restriction fragment length polymorphisms associated with 'linked DNA segments can permit not only the diagnosis of hitherto undetectable diseases but also the chromosomal localization of the loci involved.

Current estimates (Cooper and Schmidtke, 1987) indicate that about 3% of all new borns are afflicted by genetic disorders. Over 3000 genetic traits have so far been implicated in the pathology of human inherited disease. Recombinant DNA technology makes it possible to isolate and amplify any segment of the human genome by molecular cloning. Table 15.6 lists several cases where genetic diseases have been successfully analyzed directly,by using gene probes to detect intragenic defects. We also know that there are several human diseases in which the genetic defect is expressed in cultured cells. Table 15.7 gives some specific examples.

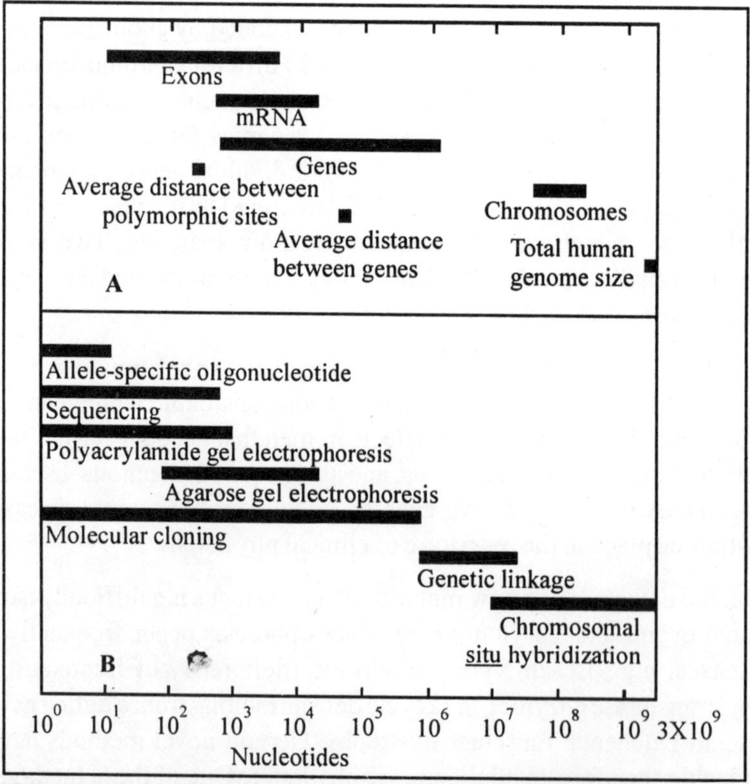

Diagram 15.19 Diagnostic techniques and the human genome. A, size ranges of some informational units of the human genome; B, size ranges over which various diagnostic techniques may be useful. (Condensed from Landegren *et al.*, 1988.)

Table 15.6 Direct analysis of some genetic diseases using gene probes to detect intragenic defects (condensed from Cooper and Schmidtke, 1987)

Disease(s)	Probe
Achondroplasia	Collagen
Adrenal hyperplasia	Steroid 21 hydroxylase
Atherosclerosis	Apolipoprotein A-1
Diabetes mellitus	Insulin
Hypercholesterolaemia	Low-density lipoprotein receptor
Leukaemia and lymphoma	T-cell antigen receptor
Phenylketonuria	Phenylalanine hydroxylase
Sickle-cell anaemia	b-globin synthetic oligonucleotide

Table 15.7 Selected human diseases in which the genetic defect is expressed in cultured cells

Disease / Syndrome	Molecules affected
Tay-Sach	Hexosaminidase A (absent)
Sandhoff	Hexosaminidase A and B (absent)
Galactosaemia	UDP-galactose transferase
Gaucher	Glucocerebrosidase
Hunter	Alpha-L-Iduronidase
Lesch-Nyhan	Hypoxanthine-guanine phosphoribosyl transferase
Xeroderma pigmentosum	DNA repair enzymes

For a proper understanding of molecular pathogenesis, it is necessary to find the gene responsible for the pathological disorder. This task is going to become easier as the human genome is mapped. However, a single clinical entity (such as Friedrich's ataxia) may have more than one cause while different clinical entities (such as multiple endocrine neoplasias and Hirschsprung's disease) can stem from mutations in a single gene, in this case the oncogene is *ret.* Many common conditions, including some major killers, are multifactorial and pose even more intractable problems. In Europe, a polymorphism in or near the gene for angiotensin converting enzyme appears to be responsible for as many cases of coronary artery disease as smoking.

Though HI V-1 was first isolated over a decade ago, it is still not clear exactly what features of an immune response confer protection against AIDS, so that designing an effective vaccine has proved extremely difficult. But a protective response is possible: rhesus monkeys infected with S1V bearing a deletion in the nef-gene resist subsequent challenge with the native virus.

Once the pathogenesis of a condition is known, the therapeutic intervention becomes possible, though there are pitfalls. The structure of a protein complexed with a potential drug may have to be determined several times before rational design can hone the drug to perfection. Similarly, the regions of a mouse monoclonal antibody that determine its complementarity with an antigen may not be the only parts that must be retained when it is "humanized".

While researches on other organisms provide useful models applicable to man, the non-human systems are sometimes not adequately extrapolated to the humans. Thus, the discovery of the giant Duchenne muscular dystrophy gene of man could not have been predicted from work on other systems, and so also is the case with the identification of recessive cancer genes.

Several Mendelian disorders in human patients have been recorded in which identified biochemical abnormalities have led to gene cloning in recent years. These are listed in Table 15.8.

Table 15.8 Some disorders in which identified biochemical abnormalities have led to gene cloning (after White and Caskey, 1988)

Disorder	Abnormality	Deficiency	Method for gene cloning
Lesch-Nyhan	X-linked hyperuricemia	Hypoxanthine-guanine phosphoribosyl transferase	Isolation of genomic fragments after nuclear DNA gene transfer Isolation of cDNA from cells in which the gene is amplified
Citrullinemia	Elevated serum/urinary citrulline	Arginosuccinase synthetase	Isolation of cDNA from cells with abnormal gene regulation
Phenylketonuria	Elevated serum/urinary phenylalanine	Phenylalanine hydroxylase	Antibody enrichment of polysomes containing phenylalanine
Tay-Sach	Excess GM_2 ganglioside in postmortem brain tissue of children	Hexosaminidase A, or its subunit	Antibody identification of a cDNA in a bacterial expression vector

The Human Genome Project

The genetic material in human egg and sperm cells (i.e., germ cells) contains 3 x 10^9 base pairs (bp) of DNA (National Research Council, 1988). Given the four-letter alphabet of DNA symbolized with the letters G, A, T, and C the sequence of 3 x 10^9 bp corresponds to 750 megabytes of information (Olson, 1993). If the sequence of the human genome could be determined, it could be stored on a desktop computer. However, it is extremely difficult to sequence DNA on this scale. A landmark advance in DNA analysis occurred in 1970 with the discovery of site-specific restriction enzymes, which can scan any source of DNA for every occurrence of a particular string of bases (for example, the enzyme *EcoR* 1 recognizes the string GAATTC). Restriction enzymes cut both strands of the double helix at their recognition sites. They enable us to develop precise physical maps of DNA simply by determining the coordinates in base pairs of the sites at which particular enzymes cleave. These maps derive their utility through annotation: mapped landmarks provide reference points relative to which functional DNA sequences such as genes can be localized. Restriction enzymes also facilitate a key step in the cut-and-splice procedures by which recombinant-DNA molecules (i.e., DNA clones) are constructed (Olson, 1993).

Recombinant DNA (Watson *et al.,* 1992) technology has two dimensions: synthetic and analytical. The former is important as it enables us to'design and construct a DNA molecule that programs a bacterium to synthesize a mammalian protein and provides a means to large amounts of the pure protein. The ability to alter the structure of the protein through site-directed mutagenesis lends genuine novelty to the resultant biosynthetic opportunities (Olson, 1993). In contrast the importance of recombinant DNA technology in making the Human Genome Project feasible from its analytical dimensions. Cloning makes it possible to purify individual recombinant DNA molecules from complex mixtures and then to prepare biochemically useful amounts of the molecules by culturing the microbial strains into which they have been introduced (Olson, 1993).

Genetic mapping requires an ability to distinguish between the two copies of the genome present in the somatic cells from which the germ cells are derived. Subtle differences in the base sequence of different instances of the human genome sometimes alter restriction sites and, hence, restriction fragment sizes. These alterations are detectable even in complex genomes by a method known as gel transfer hybridization, developed in 1975. In 1987, the first global human genetic map, based on "restriction fragment length polymorphism", was published; fairly good methods for the actual determination of DNA sequence became available in 1977 (Watson *et al.,* 1992).

Subsequent technical advances have made it possible to determine individual DNA sequences of 10^5 bp.

Three broad aims are involved in the specific mapping and sequencing objectives of the Human Genome Project:

1. To improve the research infrastructure of human genetics.
2. To help establish DNA sequence as the primary interface between knowledge of human biology and knowledge of the biology of model organisms.
3. To launch an open-ended effort to improve the analytical biochemistry of DNA.

Analysis of human genetics is limited to the examination of individuals, families, and populations in contemporary society. Much progress has been made possible by the development of the polymerase chain reaction (PCR). PCR amplification depends on a pair of short, synthetic "primers" (i.e., single-stranded DNA molecules whose ends can be extended by DNA polymerase under the direction of template molecules). The test sample contains the template molecules, and the primers direct the amplification to a particular segment of the template DNA, commonly a region only a few hundred base pairs in length. Starting with a minute sample of total human DNA, one may amplify any such region 1 billionfold while leaving the rest of the genome at its original concentration (Olson, 1993). The PCR has made a profound effect on physical mapping. One other new development that has also improved the prospects for the construction of large-scale physical maps, has been the introduction of the yeast artificial chromosome (YAC) cloning system, first described in 1987 (Watson *et al.,* 1992). YACs allow large segments of DNA to be cloned as linear, artificial chromosomes into the yeast host *Saccharomyces cerevisiae.* Even some of the earliest YAC clones were 10 times the size of the largest clones that had been constructed previously. Furthermore, the YAC system appears capable of cloning a higher proportion of the genomic DNA of many organisms than could be recovered using earlier systems (Olson, 1993; Watson *et al.*, 1992).

The YAC technology has already evolved to the point where specific segments of the human genome could be recovered efficiently in YAC clones, and in 1992, complete YAC based physical maps of human chromosome 21 and the human Y chromosome were published (Olson, 1993; Watson *et al.*, 1992).

Another important advance in physical mapping has been the development of fluorescence *in situ* hybridization (FISH) which uses DNA probes that can detect segments of the human genome by DNA-DNA hybridization on samples of lysed metaphase cells. Attachment of fluorescent molecules to the probe : DNA allows visualization in the light microscope of the position on a chromosome to which the probe binds.

While the PCR, together with such new techniques as YAC cloning and FISH, has been highly successful in physical mapping, PCR-based methods have also transformed genetic mapping. In particular, the PCR has allowed development of a new class of genetic markers that have a particularly high probability of existing in alternate forms in different instances of the human genome (Olson, 1993; Watson, *et al,* 1992). These markers are based on short, repetitive DNA sequences that are widely distributed in the human genome.

Human Gene Therapy

Gene therapy can be defined as the direct modification of the content, organization, or expression of defective genetic information in cells or organisms to provide functional genes and gene products (Friedmann, 1983). It may be thought of in two main categories, somatic and germline. Somatic cell therapy aims at correcting some grave disease by repairing the concerned gene which is responsible for the disease. The genetic therapy of certain severe immunodeficiency diseases is a good example of somatic therapy. Because of a defective gene, the body is unable to produce a protein that is essential for normal immune function. By altering or repairing this gene, it is theoretically possible to cure the disorder. In this i| therapy, germline cells are not involved. In contrast, germline therapy aims at correcting defects in reproductive cells, thereby not only mitigating disease but also alleviating it in a manner that the corrected genes would be transmitted to the progeny.

Yet another type of gene therapy that may be practised within the next decade is the so-called "enhancement therapy". In this, the idea is to alter a gene in order to affect some feature such as eye colour and height.

Recent advances in molecular biology have brought these manipulations within reach of at least a few medical experts. The ideal approach aims at site-specific insertion of functional genes into their normal locations in the genome. Unfortunately, we know little, as yet, about any effective systems for demonstrating such events in higher eucaryotes. Because of this limitation, we are left with the alternative strategy of random insertion of additional foreign genetic material into mutant genomes as the method of, providing previously missing genetic information to mutant cells.

Some methods of inserting foreign genetic material into mammalian cells include (1) microinjection, (2) uptake or transfection of naked DNA, (3) fusion with liposomes or bacterial protoplasts, and (4) use of viral vectors.

One useful model for gene therapy at present involves the removal of suitable target cells from the body, followed by their genetic transformation *in vitro,* and their reintroduction into the recipient. Another approach involves the introduction, via retroviral vectors, of a foreign gene into accessible target cells (e.g., bone marrow stem cells) *in vitro,* followed by the reinsertion of the altered cells into the patient (Friedmann, 1985).

Retroviruses are useful tools for gene delivery into other systems. These viruses are infectious cancer-causing agents whose RNA genome is reverse-transcribed into DNA, the resulting DNA being orderly integrated into host chromosomes. An interesting property of retroviral genome is that it consists ofai complêx of *two* identical chains of RNA, i.e., is diploid. Retroviral oncogenesis usually depends on transduction or insertional activation of cellular genes, which can be isolated and studied. Retroviruses include several veterinary pathogens and also two important human pathogens, the causal agents of the acquired immunodeficiency syndrome (AIDS) and adult T-cell lymphoma/leukaemia. As retroviruses are natural genetic vectors, they may be modified to serve as genetic vectors for experimental and therapeutic use. Furthermore insertion of retroviral DNA (by reverse transcription) into host chromosomes can be used to mark cell lines and to produce developmental mutants (Varmus, 1988).

The first clinical gene transfer (albeit only a marker gene) in an approved protocol took place on 22 May 1989 and the first approved gene therapy protocol for correction of adenosine deaminase (ADA) deficiency began on 14 September 1990. By now there are 11 active clinical protocols underway on three continents (Anderson, 1992).

In 1980 an unsuccessful attempt was made to carry out gene therapy for beta-thalassemia with the use of calcium phosphate-mediated DNA transfer. Retroviral-mediated gene transfer was developed in the early 1980s in animal models. Some alternate gene delivery techniques are reviewed in Lindsten and Patterson (1992).

The first Government approved human genetic engineering experiment, initiated in the USA in 1989, was for the transfer of gene-marked immune cells (specifically, tumour-infiltrating lymphocytes, TI L) into patients having advanced cancer. The protocol had two primary objectives: (1) to demonstrate that an exogenous gene could be safely transferred into a patient, and (2) to show that the gene could be detected in cells taken back out of the same patient. However, this procedure is expensive and clinically difficult. Over 60% of the patients fail to respond to this treatment; even those who do respond often fail within a year. It is likely that only a few of the heterologous cells administered to a patient are really effective in killing cancer cells *in vivo.*

The first cancer gene therapy protocol was a direct outgrowth from the TIL gene marker protocol. Once it was shown that gene-modified TIL could be safely given to patients, a new protocol was started in which the gene for tumour necrosis factor (TNF) was added to the vector. Here the idea was to make the TIL more effective against advanced malignant melanoma (Anderson, 1992).

TNF itself is a potent anticancer agent in mice. In humans, however, its.toxic effects are strong. By putting a TNF gene into TIL and then letting the TIL "home" in to tumour deposits, it may be possible to develop effectively high doses of TNF in tumour sites and avoid systemic side effects. However, because the bulk of TIL cells are probably destroyed in liver, spleen, and lungs, and also because the production of TNF from the exogenous gene occurs from a heterologous promoter, there is some possibility of the production of large amounts of ectopic TNF with toxic effects. Accordingly, a safety trial was needed to determine if toxic concentrations of TNF might develop in the liver or some other organ. The first patient began treatment in January 1991 and a number of patients are currently under treatment. So far there have been no side effects from the gene transfer and no apparent organ toxicity from secreted TNF (Anderson, 1992).

There appear to be at least two ways to improve TIL immunotherapy by gene transfer: either add a gene to the TIL or tumour-specific T-cells to make them more effective, or add a gene to the tumour cells with a view to inducing thc body's immune system to make more effective TIL.

Two other gene therapy protocols have also been approved. The first, from the University of Michigan, plans to insert a low density lipoprotein (LDL) receptor gene into hepatocytes obtained from patients suffering from familial hypercholesterolaemia; this disease results from a defective LDL receptor gene. The gene-corrected hepatocytes are proposed to be injected back into the portal circulation of the patient. The second, from the University of Washington, involves cell therapy for a complication of acquired immunodeficiency syndrome (AIDS) in which a suicide gene is inserted into the therapeutic cytotoxic T-cells to confer protection in case the T-cells happen to become too toxic.

Apart from the United States, some work along the above lines is also being planned in China. This Chinese work relates to haemophilia B. A retroviral vector containing a factor IX gene has been used to transduce autologous skin fibroblasts growing in culture. The factor IX-secreting autologous fibroblasts are then injected subcutaneously into the patients.

The observations of retroviral mediated gene transfer on 100 monkey years and 20 patients- years have shown no side effects and any pathology. As a result of a replication of defective retroviral vector no malignancy was observed (Anderson, 1992). However investigators at National Institute of Health in USA have now described three monkeys who developed malignant T-cell lymphomas after a bone marrow transplantation and gene transfer protocol with a helper virus contaminated retroviral vector preparation. The helper virus may possibly have been directly responsible for these lymphomas. This result reaffirms the necessity for clinical protocols to use vector preparations that are free of helper virus, as is indeed required for all approved protocols.

After long debates on somatic cell gene therapy and ethical and social implications, it has immerged that somatic cell therapy for the purpose of treating a serious disease is an ethical therapeutic option. But regarding germline gene therapy controversy still persist i.e. ethical or not.

There is some concern about using gene transfer to insert genes into humans for the purpose of enhancement i.e. attempting to improve desired characteristics. We have too little understanding of what normal function is to attempt to improve upon what we think is normal (Anderson, 1992). Correction of the genetic defect is one thing, but to attempt to alter a characteristic such as size is quite different. The area is further complicated by major social implications as well as by the problem of how to define when a given gene is being used for treatment and when it is being used for inhancement.

So long cells are removed from a patient, the desired gene is inserted, and the gene modified cells are returned back to the same patient, gene therapy is not likely to produce any substantial medical impact because these techniques are expansive and require advanced medical expertise. But a number of applications of gene transfer are being devised to ensure that gene therapy may be applied to a wide range of diseases in near future.

Gene therapy is likely to have its major impact on health care only after vectors have been developed that can be efficiently and directly injected into patient in the same way as insulin is administered today. Vectors needs to be designed that will target specific target cell, insert their genetic information into a safe site in the genome, and be regulated by normal physiological signals. When efficient retroviral, viral and synthetic vectors of this type are produced, then gene therapy may have a profound impact on medicine. After the discovery of Human Genome Project, which is a library of genetic information in our cells, then gene therapy is likely to be used extensively not only to cure certain diseases but also to prevent many diseases by providing protective genes before the disease become manifest. However, there is the possibility of misuse of genetic engineering technology looms large, and society must ensure that gene therapy is used only for the treatment of diseases (Anderson, 1992).

Out of two inherited diseases, the adenosine deaminase (ADA) deficiency is some what controlled by drug therapy with the missing enzyme, while the bone marrow transplants represent a long term treatment with 50% cure rate,since bone marrow is a suitable target organ for gene transfer. In contrast other disease Duchenne muscular destrophy is caused by a deficiency of dystrophin. The exact function of dystrophin is not known and no effective treatment is known. Any gene transfer protocol would need to be targeted at muscle cells which are difficult to transfer and do not seem to have a stem cell population as in bone marrow cells.

Gene therapy protocols aimed at correcting ADA deficiency are already being tested in humans. The general strategy has been to remove T-lymphocytes, infect them with a retrovirus or an adenovirus construct containing a functional ADA gene and then return the transformed cells to the patient. A gene expression on level of 10 – 20% of normal ADA activity can restore immune competence. Future trials are aimed at isolating stem cell populations from bone marrow which will avoid the need to reintroduce transfected cells at regular intervals in order to maintain ADA gene expression.

In case of gene therapy in muscular destrophy faces many problems, since muscles represent approximately 75% of the body mass of an individual and so the target tissue is very large. In addition myoblasts, the appropriate cell type for transfection , are not migratory and large areas of muscle might to be transplanted or transfected in situ.

There are about 15 human diseases that may be treatable via gene therapy. These are all diseases which involve the haematopoietic system and which can be cured at present via bone marrow transplantation. Other inborn defect, such as Lesch-Nyhan syndrome(a deficiency of hypoxanthine-guanine phosphoribosyl-transferase) might be amenable to gene therapy. Particular attention is needed if the target cells resided in the central nervous system. Gene therapy by organ transplantation might be a new route for correcting some diseases.

Genetic Marker Systems

DNA sequence polymorphisms between individuals can be used for genetic mapping. Several marker systems that promise to meet requirements of an automated genetic diagnostic assay have become available (Table 15.9). Many of these assays are based on DNA amplification.

Restriction Fragment Length Polymorphism (RFLP)

RELP markers are codominant (hetrozygotes can be distinguished from either homozygote) and provide complete genetic information at a single locus. The amount of DNA required for RFLP analyses is relatively large (5 - 10μg). Multiple southern blots (Diagram 15.20), corresponding to hundreds of individuals, can be probed simultaneously. New genetic markers or genes can easily be located within the context of an exciting RFLP map, but very little is known about the distribution of markers in the germplasm.

Diagram 15.21 illustrates the principles of RFLP. The two alleles of a gene X are flanked by cleavage sites for restriction enzyme A and B on both chromosomes. Though the position of the two A sites are identical; the two alleles differ with regard to the position of one of the B sites. The B site at the right hand end of gene X is absent in one allele, but another B site (bold type) is present further to the right. If cellular DNA is cut either with enzyme A or B, a Southern bolt analysis (Diagram 15.20) reveals only one band in DNA cut with B after hybridization with a radioactively labeled probe of gene X (Botstein et. al., 1980).

An alternative to one of the disadvantages of RFLP markers, the need for radioactive probes, has become available with the availability of sensitive non-radioactive detection systems. Automation of RFLP mapping is difficult, and it may be more practical to turn to one of the DNA-amplification based marker systems to provide an automated genotype assay (Rafalski and Tingey, 1993).

Random Amplification of Polymorphic DNA (RAPD)

Technology for the amplification of discrete loci with single, random-sequence, oligonucleotide primers is simple and easy to use. The RAPD amplification reaction is performed on a genomic DNA template and primed by an arbitrary oligonucleotide primer, resulting in the amplification of several discrete DNA products. These are usually separated on agarose gels and visualized by ethidium bromide staining. Each amplification product is derived from a region of the genome that contains two short DNA segments with some

Table 15.9 Comparison of two systems for generating genetic markers (condensed from Rafalski and Tingey, 1993)

	RELP	RAPD
Principle	Endonuclease restriction	DNA amplification with random primers
Genomic abundance	Southern blotting, hybridzation. High	Primers Very high
Dominance	Codominant	Dominant
Amount of DNA required	2-10 μg	10-25 ng
Development costs	Medium	Low
Start-up costs	Medium/High	Low

Sequence information is not required in either of the 2 systems. Types of polymorphism for both systems are single-base changes. insertions and deletions.

sequence homology to the primer; these segments must be present on opposite DNA strands, and be sufficiently close to each other to allow DNA amplification to occur. The polymorphisms between individuals result from sequence differences in one or both of the primer binding sites, and become manifest as the presence or absence of a particular RAPD band. Such polymorphisms behave as dominant genetic markers. Analysis of RAPD markers lends itself to automated breeding applications because it requires only small amounts of DNA (15-25 ng), a non-radioactive assay that can be performed in several hours, and a simple experimental set-up (Rafalski and Tingey, 1993).

RAPD technology enables researchers to screen for DNA sequence-based polymorphisms at a large number of loci. Sets of short primers (usually 10 mers) suitable for RAPD amplification are available commercially or may be readily synthesized. RAPD markers are dominant (profiles are scored for the to presence or absence of a single allele) known about marker linked to a trait of interest is available, it becomes easy to turn the RAPD assay into a dystrophin. These PCR-type assay based on secondary DNA sequence, by use of allele-specific PCR (AS-transfer protocol wouldgation, or a sequence-characterized amplified region (SCAR) assay (see Rafalski to have a stem cell population.

Aerosol Gene Delivery

Molecular cloning techniques have brought within our potential reach the identification and isolation of an increasing variety of genes with mutations responsible for important human diseases. To date, attempts to replace absent or mutated genes in human patients have had to rely on *ex vivo* techniques because methods for safe and effective *in vivo* gene delivery are not available. Retroviruses, adenoviruses, and liposomes have been used in animal model studies in attempts to increase the efficiency of gene transfer. DNA has been introduced into animals by intratracheal, intravenous, intramuscular, and intraarterial injections (Stribling *et* al., 1992). The lung is a particularly attractive organ for *in vivo* gene therapy because of its direct accessibility via the airway. Introduced genes have shown expression in the lungs of rodents after intratracheal instillation. Aerosol delivery is a good method as it results in deep penetration of material into the lungs, and can deposit aerosolized material throughout the airways and alveoli of healthy individuals. Aerosol administration has delivered biologically active macromolecules to the lungs of humans. Stribling *et al.* (1992) have shown that aerosol delivery of a chloramphenicol acetyltransferase (CAT) reporter gene complexed to a cationic liposome carrier can produce generalized CAT gene expression in mouse lungs *in vivo.* The ability to express transgenes selectively within the lung is likely to greatly facilitate the development of gene therapy for a variety of human diseases.

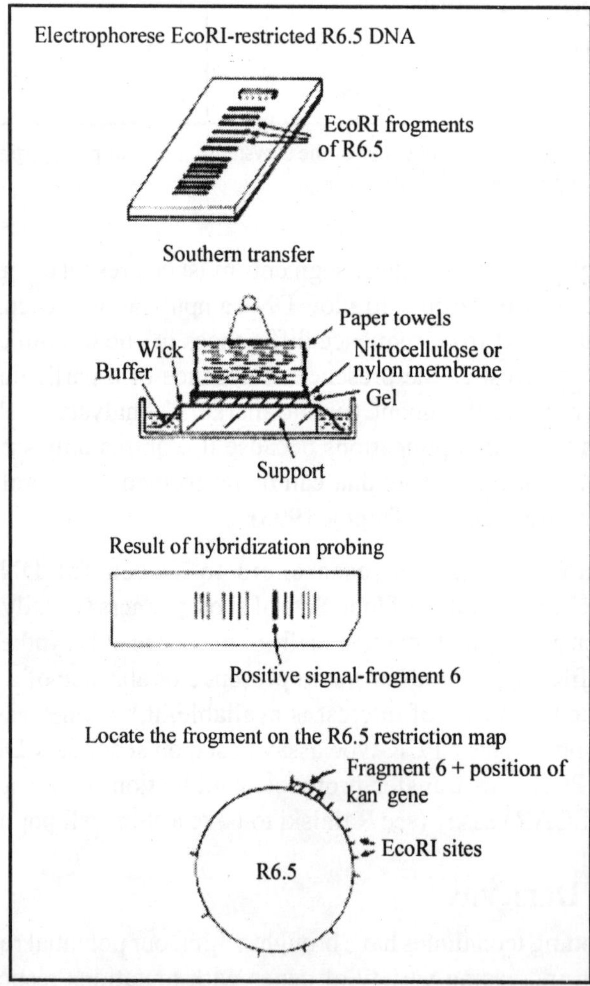

Diagram 15.20 Southern hybridization.

Biolistic Missiles

Although it is fairly easy to transform the nuclear genes of diverse plants and animals, there is a notable lack of suitable methods for introducing genes into mitochondria or chloroplasts. Recently, microprojectiles (missiles) coated with DNA have been used to introduce genes into these organelles. Boynton et al. (1988) and Johnston et al. (1988) have succeeded in shooting missing segments of DNA into chloroplasts of *Chlamydomonas* and mitochondria of yeast. The "bullets" used were tungsten microprojectiles. Both the photosynthetic capacity of a *Chlamydomonas* mutant and the respiratory capacity of defective yeast were restored. It was also demonstrated that the inserted cloned DNA directly replaced the defective or missing DNA in the organelle. As a result of the introduction of the DNA, restored functioning was inherited by the daughter cells, showing that the shot-in DNA had become integrated and expressed in the progeny cells.

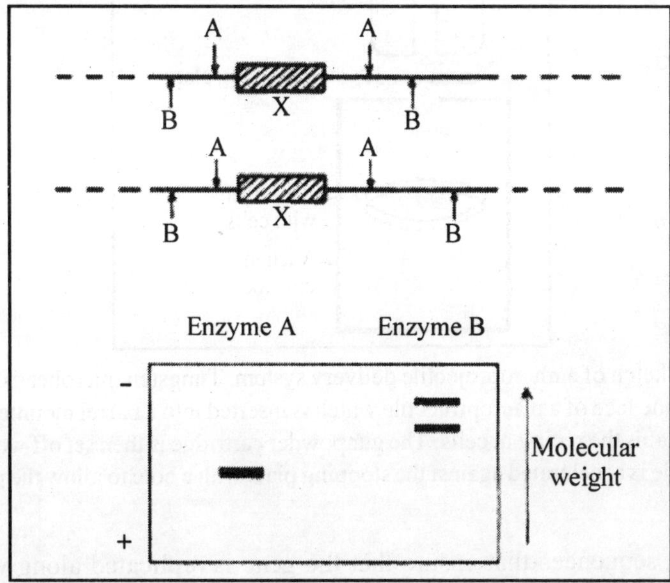

Diagram 15.21 Principles of RFLP (after Botstein *et al.*, 1980)

The foregoing bombardment technique (Diag. 15.22) opens up exciting possibilities for manipulating organelle genomes by molecular genetic techniques in the same way as nuclear genomes.

Molecular Engineering

Molecular engineering makes it possible to remove, insert, or substitute nucleotide sequences in target genes. The target gene is cloned. This is followed by site-specific deletion, substitution, or insertion of DNA obtained from other genes or synthesized *in vitro*. It is possible to construct promoters that ensure constitutive expression in specific hosts or tissues. Genetic engineering technology also has the potential to design a gene that is expressed to some predetermined level in a specific subset of animal or even human cells (Roizman, 1988). Molecular engineering also permits the tailoring of gene products required for specified needs. The engineered genes can be made to express by introducing the genes into cells either by themselves or in a vector. Some vectors allow the engineered gene to be inserted into a specific site in the target genome. Other vectors carry a gene that imparts to the recipient cell a selective advantage for growth in special media; these vectors furthermore

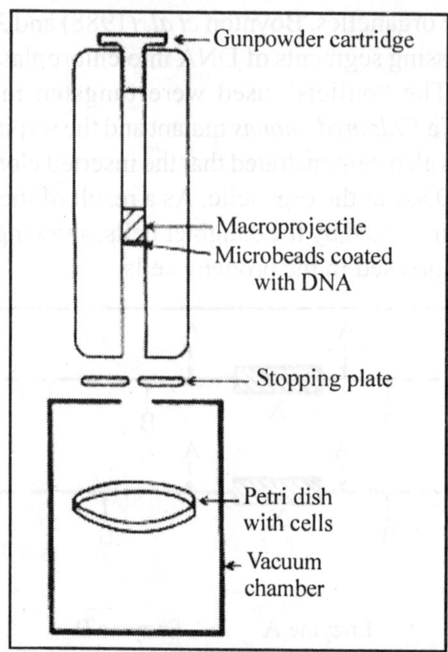

Diagram 15.22 Sketch of a microprojectile delivery system. Tungsten microbeads coated with DNA are deposited on one face of a macroprojectile which is inserted into a barrel mounted above a vacuum chamber that contains the recipient cells. The gunpowder cartridge is then set off with a firing pin, and the macroprojectile is accelerated against the stopping plate with a hole to allow the pellets to bombard the cells.

may have DNA sequences that ensure that the gene is replicated along with the cellular genome. A third category of vectors (most viruses) are designed to introduce the engineered gene efficiently and simultaneously into a large number of cells.

As compared to molecular engineering, genetic engineering also involves specific deletion, replacement, or insertion of DNA, but by homologous recombination in cells rather than by construction *in vitro*. For instance, insertional substitution or deletion proceeds by a double recombination event through the homologous flanking regions. From this, it follows that viable recombinant genomes can arise only if the genome segments deleted or interrupted are dispensable with respect to replication and function.

Polymerase Chain Reaction

Polymerase Chain Reaction (PCR) is an extremely useful technique (Diag. 15.23) with many applications in molecular biology (Ehrlich *et al.*, 1989). Because of the potential to select and amplify sequences of DNA, starting with extremely small amounts of DNA, the technique can revolutionize the manner in which molecular biology experiments are carried out. The PCR can be used in clinical genetics (including phenylketonuria screening, cystic fibrosis, Duchenne muscular dystrophy, and Von Willebrand's disease). It also finds application in the study of highly polymorphic regions of the genome and for the detection of rare sequences. Diagram 15.23 illustrates the basic PCR technique.

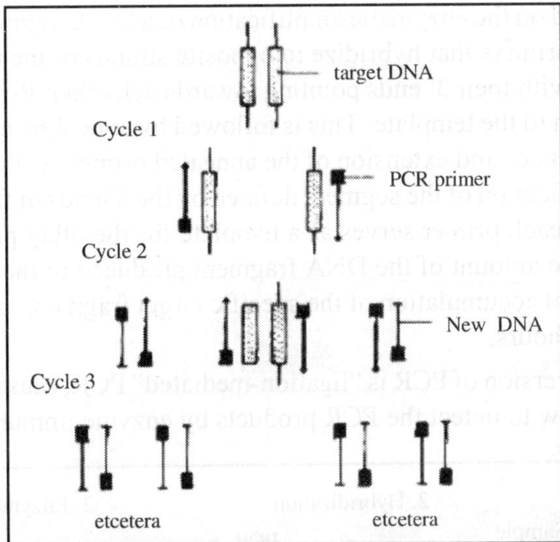

Diagram 15.23 Illustration of the basic polymerase chain reaction technique. In the first cycle, the target DNA is denatured, the specific primers anneal to the single-stranded target DNA and the strand is copied by the DNA polymerase. Similarly, in the second cycle also the DNA is again denatured, the primers anneal and the DNA is copied by the DNA polymerase.The third cycle consists of the same procedures and at this stage a population of DNA molecules, which are flanked by the specific primers, is produced. During the fourth and subsequent cycles these molecules are further amplified and eventually become the predominant DNA species within the mixture. (After Hide and Tail, 1991.)

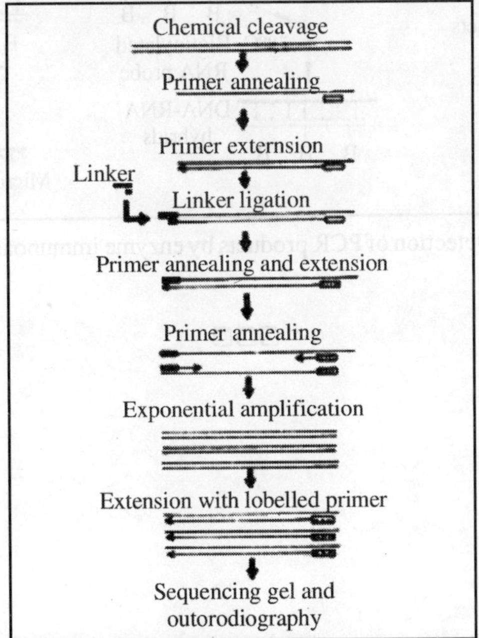

Diagram 15.24 Vertically striped box is the second gene-specific primer positioned with its extending 3'-end to that of the first primer so as to increase specificity. Box with wavy lines represents a radioactively end-labelled primer to visualize the sequence upon electrophoresis and autoradiography.

The PCR is based on the enzymatic amplification of a DNA fragment that is flanked by two oligonucleotide primers that hybridize to opposite strands of the target sequence. The primers are oriented with their 3'-ends pointing toward each other. Repeated cycles of heat denaturation are given to the template. This is followed by annealing of the primers to their complementary sequences and extension of the annealed primers with a DNA polymerase, resulting in the amplification of the segment defined by the 5'-ends of the PCR primers. The extension product of each primer serves as a template for the other primer; consequently, each cycle doubles the amount of the DNA fragment produced in the previous cycle. The result is an exponential accumulation of the specific target fragment, up to several million-fold within just a few hours.

A more modern version of PCR is "ligation-mediated" PCR, illustrated in Diag. 15.24. Diag. 15.25 shows how to detect the PCR products by enzyme immunoassay.

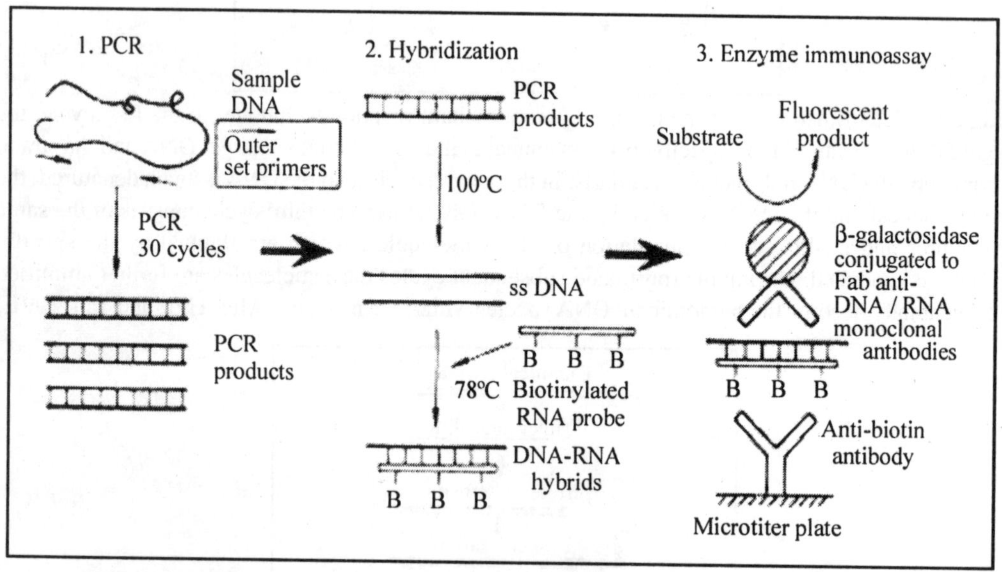

Diagram 15.25 Detection of PCR products by enzyme immunoassay (PCR-EIA).

□□□

CHAPTER-16

Synthetic Seeds

The Natural Seed

Seeds are the dormant stage of spermatophytic plants life-cycle. At this stage a germplasm can be stored for many years and on providing favourable conditions, these seeds germinate to give rise to new plants.

The seed stage represents a unique developmental phase of the spermatophyte seed plants life-cycle, and as such involves structures, not characteristic of other stages of development. The essential structure of seed is defined as a *ripened ovule consisting of an embryo surrounded by its coats.* Anatomically a seed consists of some old or parental sporophyte tissue viz. the seed coats, which are derived from the integument's and nuecllus, in some gives endosperm, which may be either sporophytic tissue or fertilized triploid tissue, and the egg cell gives embryo i.e. the new young sporophyte. The normal seed contains storage food materials which it utilizes during the process of its germination. These substances are frequently found in the cotyledons or endosperm. Thus endosperm may contain variety of stored materials such as starch, oils, proteins etc.

Development of the Concepts of Tissue Culture and Artificial Seeds

P. R. White is acknowledged as the father of tissue culture in the United States. He was the first to grow excised root tips of tomatoes *(Lycopersicon* sps.*)* in continuous culture. When new material is started in culture, grown in vitro (literally, in glass), it develops very small juvenile shoots, which are reminiscent of seedlings. A plantlet continues to produce and maintain small stems and leaves throughout its duration in culture. This is fortunate because most mature material would be too unwieldy for micropropagation to succeed in a test tube. After multiplication in culture and when transferred to soil outside the laboratory, the plantlets will produce leaves of normal size and assume the mature features of the plants from which they originated.

When plants are multiplied vegetatively as distinguished from those grown from seeds-whether by tissue culture or by cuttings, all the offspring from a single plant can be classified as a clone. This means that the genetic make-up of each offspring is identical to that of all the other offspring and to that of the single parent. On the other hand, plants propagated by seed, resulting from sexual reproduction, are not clones because each seed and the resultant plant has a unique genetic make-up a mixture from 2 parents, different from either parent and different from one seed to another. The term *cloning,* with respect to tissue culture, refers to the process of propagating in culture large numbers of selected plants with the same genotype (the same genes or hereditary factors) as their respective parent plant.

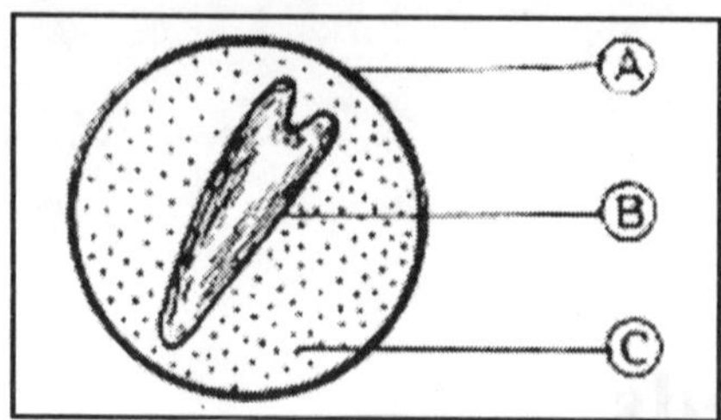

Diagram 16.1 The concept of artificial seed. Capsule gel with hydrophobic membrane, the (A) artificial seed coat, (B) somatic embryo, (C) artificial endosperms

The liquid cell suspension cultures have particular significance for the mass production of cells. One common source of cells for cell suspension is from friable callus, although specific cells, such as from leaf mesophyll (the thin, soft tissue between the upper and lower epidermis of the leaf), are also grown in suspension. Cells in suspension can form embryoids (somatic embryos) in the process of embryogenesis. Embryos may multiply and/or be induced to form plantlets in the process of morphogenesis. Many hybrid plants produce embryos that do not mature to viable seeds. These embryos can be rescued, removed from the seed at immature stage, and then grown in culture.

Suspension cultures have been enhanced by new methods, that can continuously introduce fresh medium into the suspension culture, thereby, enabling the production of thousands of cells or embryos in a single container with a minimum of manual transfer. This is one way that tissue cultur can compete with the plentiful seed production in nature.

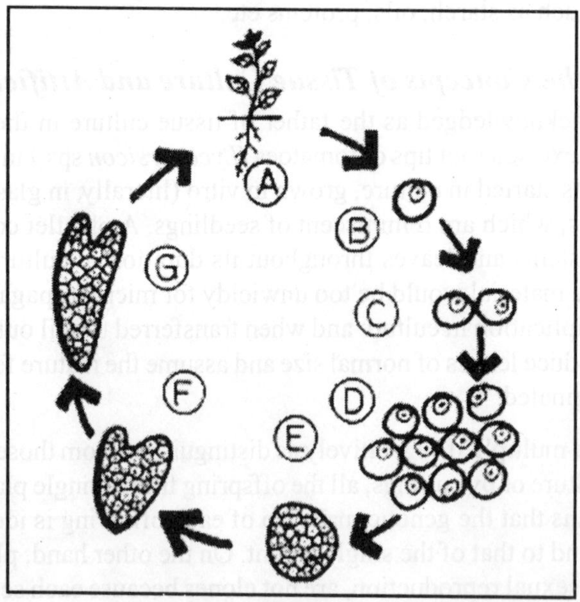

Diagram 16.2 Different stages of somatic embryogenesis. (A) a young plant, (B) isolated single cells from leaves , only one cell is shown. (C) doublet formation after first mitosis in culture. (D) cell colony (proembryo mass) following many mitosis. (E) Globular embryo. (F) heart shape embryo. (G) torpedo stage embryo.

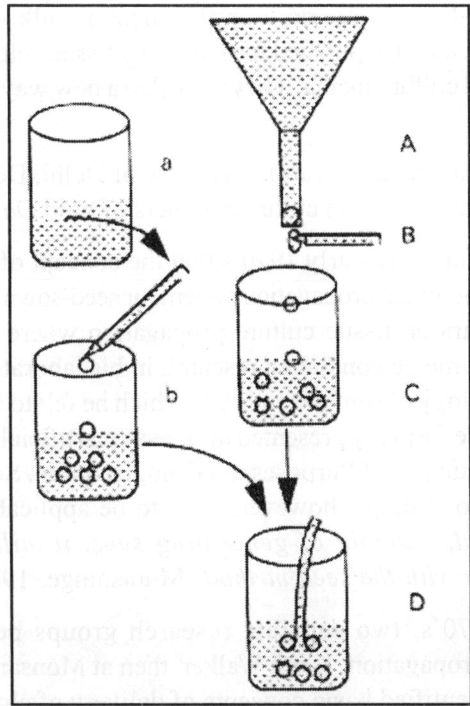

Diagram 16.3 Preparation of artificial seeds via encapsulation with calcium alginate beads using a draping method. Alginate (2%) is either mixed with somatic embryos (a) or poured in a separatory funnel (A), droping of alginate beads along with embryo into a bath of calcium nitrate (100 mM) solution (b) or a single embryo is inserted into alignate drop using a plastic pipet having 4mm inside diameter (B), falling of alginate beads into a bath of calcium nitrate (C) and washing of the capsules in water (D). Diargram modified from Redembaugh et al., 1991.

Interest in anther and pollen culture; the tissue culturing of anthers or pollen to obtain haploid (cells with half the normal number of chromosomes of vegetative cells) clones- is spurred by the practical applications of such haploid cultures. Haploid (n) plants are sterile, but if the chromosomes duplicate, either spontaneously or by induction, the plants will be diploid (2n, which is normal for the vegetative state), and their progeny will be true to form. Considering the fact that it takes several generations of inbreeding to obtain a pure line by conventional means, it is little wonder that plant breeders are interested in anther culture.

Discovery of Synthetic Seeds

The origin of the idea of an artificial seed is difficult to determine. Certainly, those who first produced somatic embryos may have considered such an application (Steward, et al., 1958 and Reinert, 1958). The discovery of somatic embryogenesis in carrot in the year 1958 almost simultaneously by F. C. Steward (USA) and J. Reinert (Germany). F. C. Steward, a renowned plant physiologist, at Cornell University in New York, was so impressed by the dramatic effects of coconut milk in carrot culture media that he set beside his other objectives in order to dedicate himself to the study of growth factors in this and other liquid endosperms.

Among the active materials he extracted from the coconut milk were several ingredients that are now commonly included in purified form in many tissue culture media. Coconut milk is still used in some orchid culture media. They provided a new way of propagating the plant species.

Later, S. Guha and S.C. Maheshwari (University of Delhi, Delhi) in 1964 discovered the formation of pollen embryos from cultured anthers of wild *Datura innoxia.*

However, it was not until the early 1970's that the concept of using somatic embryos began to be presented as a potential propagation system for seed-sown crops. Toshio Murashige gave a number of seminars on tissue culture propagation where he concluded with this concept. For a period of time he conducted research in his laboratory that was focused on the developmental physiology of somatic embryos which he felt to be the limiting factor for large-scale propagation. He formally presented his ideas on artificial seeds at the Symposium on Tissue Culture for Horticultural Purposes in Ghent, Belgium, September 6-9, 1977. His terse comments in the proceedings, however, were to be applicable, *the cloning method must be extremely rapid, capable of generating several million plants daily, and competitive economically with the seed method* (Murashinge, 1977).

During the mid-1970's, two separate research groups began work on somatic embryogenesis for crop propagation. Keith Walker, then at Monsanto Company, directed a group of scientists that identified basic concepts of delivery of cloned, agricultural crops. Since the focus was to develop thrifty somatic embryo systems that would recapitulate zygotic embryogenesis, their choice was the advanced system developed for *Medicago sativa* L. (alfalfa) using a line (Regen S) identified by Bingham, et al. (1975). Soybean and vegetable crops were also of interest to them. Walker cited two reports that had a strong impact on their thinking about the use of somatic embryos for crop propagation. Early in 1980, Walker moved to Plant Genetics, Inc. where Redenbaugh, et al. (1984 and 1986) discovered that hydrogels such as sodium alginate which could be used to produce single-embryo artificial seeds. In a few experiments, the artificial seeds were planted in the greenhouse with plant production (7% for alfalfa and 10% for celerly).

Street (1977) advocated the problem of reliability in embryogenesis. According to him morphogenic competence is determined from the time of culture initiation, such that there is the need to have an initiation medium that will ensure that the competent cells are involved in callus formation. Sunderland (1977) demonstrated that the production of hundreds of morphologically uniform embryos from *Datura* and *Nicotiana* pollen.

Robert Lawrence (of Union Carbide) started to develop various methods for cloning forest trees. It was difficult for him to produce hybrids for crops such as celery and lettuce. This group focused on delivery of somatic embryos using fluid drilling technology (Lawrence, 1981) and using polyoxyethylene to form seed tapes or sheets. Lawrence and Walker's groups came together at a symposium workshop, Advances in Methods of In Vitro Cloning for Large Scale Propagation of Plants, held September 21-22, 1981 at the W. Alton Jones Cell Science Center in Lake Placid, New York. They discuss the various concepts like how low-cost, high-volume propagation system can be developed for vegetable and agronomic crops using somatic embryos and delivered by fluid drilling (in a seed tape, or as an artificial seed). The Lawrence group (then at Agrigenetics)' subsequently began to use alginate for encapsulating carrot and celery somatic embryos (Lutz, et al. 1985). Because their encapsulation results with alginate were similar to those at Plant Genetics, they focused on the problems of somatic embryo physiology. In this way, they were successful in obtaining germination of carrot somatic embryos in vermiculite in a growth chamber.

Drew (1979) was active in developing methods to commercially propagate crops using somatic embryos. He suggested delivering carrot somatic embryos in a fluid drilling system, but was able to produce only three plants from carrot embryos on a carbohydrate-free medium. He could not get success in producing many plants through this system. He faced a crucial problem and found the very slow rate of development of plantlets derived from culture.

Kitto and Janick (1982) coated clumps of carrot embryos, roots, and callus with polyoxyethylene. Some embryos survived the coating process as well as a desiccation step (Kitto and Janick, 1985a and 1985b).

The early assessments of Murashige and Street (1977) on the difficulty of somatic embryogeny are still valid today. The quality and fidelity of somatic embryos are the limiting factors for development and scale-up of artificial seeds.

Interestingly, artificial seeds prepared from shoot buds can also be used for plant propagation, and this was reported by P.S. Rao's group from BARC, Bombay. Rice is the world's most important food crop and a primary food source for more than one third of the World's population. This crop has received considerable attention in biotechnological research programmes. Research on artificial seeds in rice is still in infancy and this technology through somatic embryogenesis would offer a great scope for large scale propagation of superior, elite hybrids (Brar and Khush, 1994). P. S. Rao and his associates have reported high frequency somatic embryogenesis from indica rice cultivars (Suprasanna *et al,* 1995) and utilized this embryogenic system for the production of artificial seeds.

Table 16.1 Important crop plants in which artificial seed production and plant conversion has been demonstrated

***In vitro* propagules for encapsulation**	**Crop**
Somatic embryos	Alfalfa, Celery, Brinjal, Carrot, Brassica, Lettuce, Sandalwood, Rice, Horseradish
Axillary buds / Adventitions buds	Mulberry, Eucalyptus, Vitis
Shoot tips	Banana, Cardamom, *Carum carvi*

Since then the induction of somatic and/or pollen embryogenesis has been reported in a wide array of plants, including several crop plants such as rice, wheat, triticale, maize, pearl-millet, sorghum, sugarcane, potato, sweet potato, eggplant, lettuce, carrot, alfalfa, soybean, cucumber, *Brassica* species, asparagus, coffee, tobacco and cotton. However, the production of high quality and uniform embryos (which is very important for the preparation of artificial seeds) has been limited to only certain crops like carrot and alfalfa.

Somatic embryogenesis has, also been obtained from non-zygotic explant tissue (coffee leaf), suggesting that somatic embryogenesis may be obtained from sexually immature tree, zygotic embryo or ovule tissue.

Difficulties in developing somatic embryogenesis systems for tree species are similar to those for herbaceous species. However, tree species may also exhibit a unique set of tissue culture-dependent variabilities.

Uses and Limitations of Artificial Seeds

Plants are traditionally propagated either by seeds or the vegetative propagules like stem cutting. Now most plants can be multiplied through tissue culture techniques, particularly shoot tip culture (clonal micropropagation).

Micropropagation through artificial seeds may be commercially exploited on a large scale, generating millions of plants in a few days, and this may become a profitable multibillion rupees industry in near future. This technology would be feasible and even competitive economically with the seed method.

Artificial seeds offer the possibility of a low-cost, high-volume propagation system that will compete with true seeds and transplants. Most likely, the technology will first be used with hybrid vegetable crops such as celery or with high-value flower and ornamental species. Because of the relative ease of producing large numbers of somatic embryos, artificial seeds will be applicable for monoculture as well as mixed genotype methods of agriculture. The artificial seed coating also has the potential to hold and deliver beneficial adjuvants such as growth-promoting rhizobacteria, plant nutrients and growth control agents, and pesticides.

Potential Uses of Artificial Seeds

1. Delivery Systems.
2. Reduced costs of transplants.
3. Direct greenhouse and field delivery of elite, select genotypes, hand-pollinated hybrids, genetically engineered plants, sterile and unstable genotypes, large-scale monocultures, mixed-genotype plantations.
4. Carrier for adjuvants such as microorganisms, plant growth regulators, and pesticides Protection of meiotically-unstable, elite genotypes.
5. Analytical Tools.
6. Comparative aid for zygotic embryogeny.
7. Production of large numbers of identical embryos.
8. Determine role of endosperm in embryo development and germination.
9. Study of seed coat formation.
10. The synthetic seeds so developed are breed true.
11. There are potential advantages of artificial seed technology specially for tree genetic engineering.

12. The artificial production of seeds has already been obtained successfully in *Zea mays, Apium graveolens, Daucus carota, Lactuca sativa, Madicago sativa, Brassica* spp., *Gossypium hirsutum, Velerina* sp., *Santalum* spp., etc.
13. The encapsulation of somatic embryos (hydrated or desiccated) provides a potential method to combine the advantages of clonal.
14. Propagation with the low-cost, high-volume capabilities of seed propagation.
15. These seeds can be produced within a short time (one month) whereas natural seeds are the end product of complex reproductive process and breeders have to wait for a long time for development of new varieties.
16. Artificial seeds can be produced at any time and in any season of a year.
17. Dormancy is the common feature of natural seeds, but by means of artificial seeds the dormancy period can be reduced to a great extent, thereby shortning the life cycle of a plant.
18. They are useful in preserving germplasm.
19. Synthetic seeds are applicable for large scale monocultures as well as mixed genotype plantations.
20. Such seeds give the protection of meiotically unstable, elite genotypes.
21. The synthetic seeds provide us knowledge to understand the development, anatomical characteristics of endosperm and seed coat formation.

Both these propagation methods have certain limitations such as the need of intensive labour, rooting of regenerated shoots and transplantation, slow and small scale multiplication. Micro-propagation has some additional problems like the need of acclimatization of tissue culture derived plants before they are transferred into field conditions (hardening) because of their tenderness due to the absence of lignification and low cuticle formation. By contrast, plant propagation via artificial seeds has several advantages over classical methods as well as micropropagation (with shoot tip culture). The advantages of this technology include the rapid and large-scale multiplication, minimal labour and low cost propagation. In addition, artificial seeds can be directly delivered to the field, thus eliminating transplantation and tissue hardening steps. They can also be provided with various kinds of adjuvants like plant growth regulators, useful microorganisms and pesticides to tailor a field-specific, plantable unit for a desired crop. However, the genetic uniformity is maintained in all these propagation methods.

Artificial seed technology can be very useful for the propagation of a variety of crop plants, especially crops for which true seeds are not used or readily available for multiplication (e.g. potato) or the true seeds are expensive (e.g. cucumber and geraniums), hybrid plants (e.g. hybrid rice) and many vegetatively propagated plants which are more prone to infections (e.g. day lily, garlic, potato, sugarcane, sweet potato, grape and mango).

This newly emerging technology would also be useful for multiplying genetically engineered plants (transgenic plants), somatic and cytoplasmic hybrids (obtained through protoplast fusion techniques), sterile and unstable genotypes. Besides, artificial seeds would be useful material for preservation of desirable elite genotypes (cryopreservation). They would also be valuable tools in experimental research to study the process of zygotic embryogenesis and understanding the role of endosperm in normal embryo development and germination.

Synthetic seed technology offers many useful advantages on a commercial scale. The resultant plant population from the synthetic seed will be uniform and the direct delivery of somatic embryos will save many subcultures to obtain plantlets from regenerated embryos. The encapsulated embryos could also be packed with pesticides, fertilizers, nitrogen fixing bacteria and even microscopic destroying worms.

At the Biotechnology Division of BARC, research on the development of protocols for synthetic seeds using somatic embryos, axillary buds and shoot tips is in progress in five economically important plants, sandalwood, rice, mulberry, banana and cardamom. The following pages describe the results obtained in this direction.

The artificial seed systems coupled with artificial intelligence and microcomputer systems like the most advanced robots which can mimic the motions and functions of a living being (i.e., automated encapsulation) would tremendously increase the efficiency of encapsulation and production of artificial seeds, and revolutionize the plant propagation method in the years ahead. This technology is gradually moving towards the commercial propagation of high value crops. However, there is a great need for refinement of this technology by the tackling of certain technical problems such as the need to produce high-quality and high-fidelity somatic embryos, and to avoid the genetic instability and variability of tissue culture derived plants (somaclonal variation, which is not preferred for crops where true-to type plants are important). These problems can be overcome if the process and regulation of somatic embryogenesis and origin of somaclonal variation are well understood. Intensive research is being carried out in several laboratories to address these vital issues. Further, the understanding about the storage, transport, handling, growth habit and harvest index of artificial seeds is essential. Similarly, the efforts to increase the output of embryos/plants per gram callus tissue of input (mass balance) and conversion frequency of artificial seeds are also needed. If all these problems are rectified with technical progress, no doubt this novel method can become valuable tool in agriculture to propagate crop species.

Although synthetic seed research is being caried out in many laboratories, the principle limitation for commercialization has been the somatic embryogenesis. Despite the fact that somatic embryogenesis has been achieved for many species, during the last decade for many of the important, high valued genotypes, it has not been reported. Therefore there is a need to shift the focus on somatic embryogenesis in valuable crop plants. Research on encapsulation of axillary buds, shoot tips or any other propagules should be taken up. Equally important is the need to develop new synthetic seed coatings for encapsulating embryos and other vegetative propagules. Synthetic seed technology in years to come would certainly find its application in plant propagation and delivery of tissue cultured plants.

Production of Synthetic Seeds

Synthetic or artificial seeds are the living seed-like structure derived from somatic embryoids in vitro culture after encapsulation by a hydrogel. *The preserved embryoids are termed as synthetic seeds.* Somatic embryoids are identical with zygotic embryos and give rise to plants only under controlled laboratory conditions. Somatic embryoids are without seed coats. In vitro embryoid develops from callus tissue and their induction is initiated by somatic embryogenesis supplimenting the medium with auxin and cytokinins in proper ratio.

Such seeds are contaminated with microbes and desiccate quickly when they are subjected to field conditions. Therefore, to get rid from this problem, they are encapsulated by a protective gel like calcium alginate. These encapsulated embryoids can resist unfavourable field conditions without desiccation. These seeds so developed behave like a true seed and are used as a substitute of natural seeds. They can also be sown directly in the greenhouse or in fields.

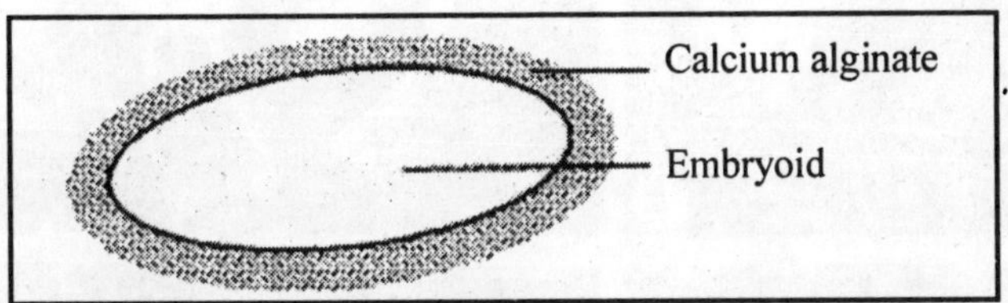

Diagram 16.4 Cross section of synthetic seed

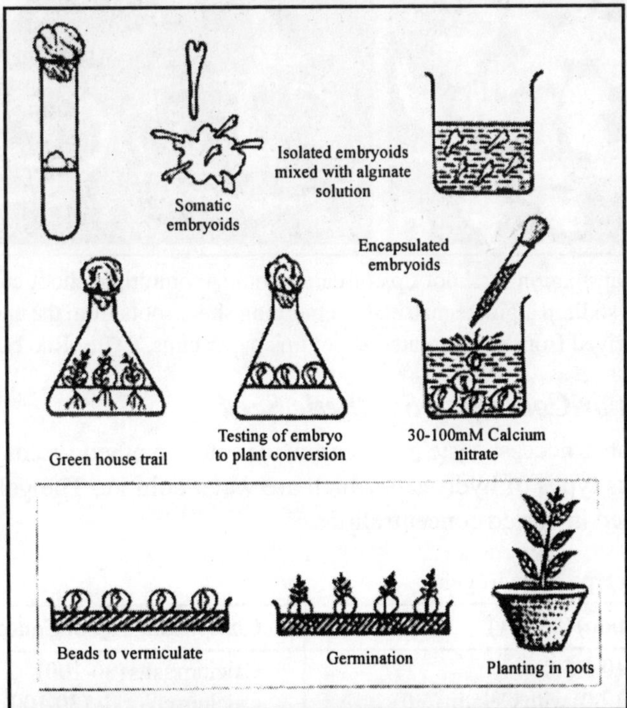

Diagram 16.5 Flow diagram presenting the procedure of synthetic seed production: Establishment of Callus Culture

Induction of Somatic Embryogenesis
⇓
Maturation of Somatic Embryos
⇓
Encapsulation of Somatic Embryos
⇓
Evaluation of Embryoid and Plant Conversion
⇓
Planting in Fields/green House

Diagram 16.6 Schematic presentation of steps of synthetic seed production

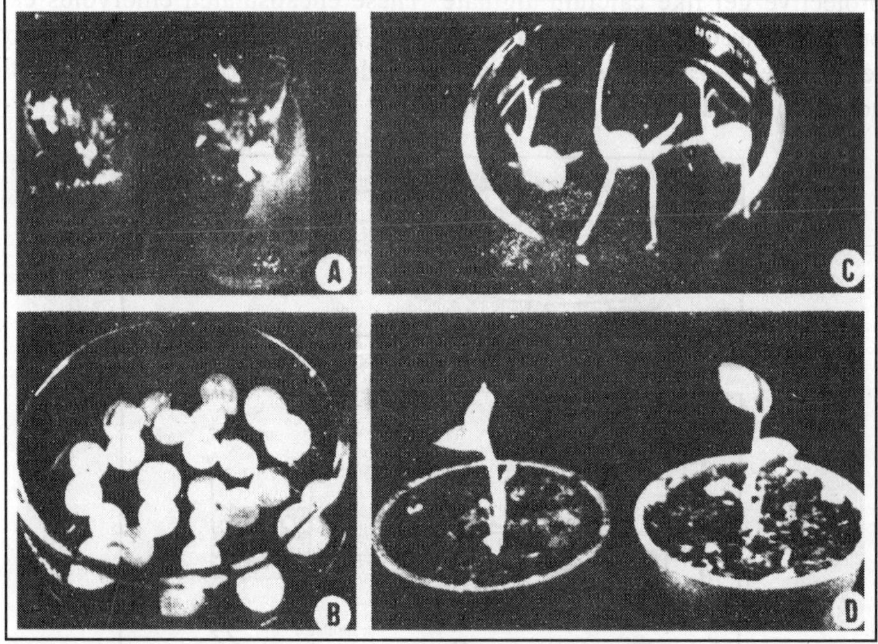

Diagram 16.7 Encapsulation of shoot tips of cardamom. A - multiple shoot cultures. B -Shoot tips encapsulated in 3% sodium alginate matrix, C - emerging shoot roots from the encapsulated shoot tips and D- plantlets derived from encapsulated shoot tips paper cups. (After Rao et. al, 1997).

Encapsulation or Coating of Synthetic Seed

Encapsulation is necessasry to produce and to protect synthetic seeds. The encapsulation is done by various types of hydrogels which are water soluble. The gel has a complexing agent which is used in varied concentrations.

Table 16.2 Various type of hydrogels.

Gel (Concentration) (% w/v)	Complexing Agent Concentration (μM)
Sodium alginate (0.5-5.0)	Calcium salts (30-100)
Sodium alginate (2.0) with Gelatin (5.0)	Calcium chloride (30-100)
Carragenan (0.2-0.8)	Potassium chloride
Locust Beam Gum (0.4-1.0)	Ammonium chloride (500)

The encapsulation or coating of carrot and celery somatic embryos is done by polyoxyethylene (Polyox) and dried the embryo in polyox mixture.

Two standard methods have been used for encapsulation of somatic embryos.

Gel Complexation via a Dropping Procedure

The most useful encapsulation system is to drip two per cent LF60 sodium alginate from a separatory funnel into a 100mM calcium nitrate solution. As the sodium alginate drops from at the tip of the funnel, the somatic embryos are inserted. The encapsulated embryos complex in calcium salt for 20 min, after which they are rinsed in water and then stored in a air tight container otherwise the capsule will dry out within 24 hour. It is a slow method of seeds production.

Automate Encapsulation Process

Automate encapsulation process is the recent and quick method of artificial seed production. An encapsulation machine can be used successfully to encapsulate somatic embryos, e.g., for alfalfa. Blank alginate capsules were planted in Speeding™ trays using a vaccum seeder. The blank capsules are planted in the field using a Stanhay planter. However, because for the rapid drying and the thickness of the alginate capsules, a hydrophobic coating is required for mechanical handling. For coating, an Elvax 4260 Copolymer™ (Dupont), is suitable for producing a slow-drying, non-tacky coating which allows embryo conversion. Based on this system of production of artificial seeds the cost factor is not high for most of the cash crops.

Alginate or carrageenan is also used for artificial seed production, e.g., of carrot, asparagus, Norway spruce, etc. Mascarenhas (India) reported the encapsulation *of Eucalyptus* somatic embryos. He obtained 50 per cent germination from such seeds. Alginate artificial seeds are spherical and transparent. The alginate capsule is generally non-inhibitory.

Mass Balance Concept

An additional concept that has greatly aided the improvement of artificial seed performance is *mass balance.* Mass balance considers the amount of tissue at the beginning of the experiment (or production run) and the number of high quality plants produced at the end of it. Simply emphasising the number of embryos per gram of fresh callus or the number of embryo-producing calli is not adequate. In fact, treatments that lead to higher numbers of embryos may actually produce fewer superior quality of embryos than another protocol. At this time, the artificial seed package, consisting of a calcium alginate bead coated with a hydrophobic Elvax polymer, appears to be sufficient.

Steps of Commercial Artificial Seed Production

The following steps are needed for commercial synthetic seed production:

1. Production of embryogenic tissue from transformed cells or tissue.
2. Large-scale production of synchronous somatic embryos.
3. Maturation of somatic embryos.

4. Non-toxic encapsulation/coating process.
5. Artificial endosperm/megagametophyte, depending on species.
6. Storage capability of artificial seeds.
7. High frequency, direct green house/nursery field conversion, depending on production requirements.
8. Low genetic and epigenetic variation.
9. Appropriate expression of engineered trait.

Artificial Seed Propagation

Artificial seed propagation could potentially reduce the time needed to insert a desirable gene into a productive forest, as compared to using seed as trie propagation method. A considerable advantage would be to eliminate or minimise the requirements for seed production using the following process:

1. Production of large-scale embryogenic tissue from genetically engineerecd cells;
2. Concurrent plant regeneration, confirmation of transformation, and *progeny* testing;
3. Cryogenic storage of potential superior lines;
4. Scale-up production and maturation of somatic embryos;
5. Encapsulation of somatic embryos as artificial seeds;
6. Either greehouse/nursery establishment, growth, and transplanting into the field or direct seedling.
7. The final stage will be the evaluation of production plantations for increased yield/ performance due to engineered trait.

Unlike herbaceous species, the growth and maturity of trees far exceed the time required for tissue culture manipulations. If the engineered trait is expressed only in the mature tree and is not stable during meiosis, then mass clonal propagation would be accomplished only through tissue culture methods (or, by traditional relatively low-volume ramet production).

Cryopreservation of desirable genotypes is one of the key components for artificial seed propagation of tree species. It retains the genetic gains from genetically engineered tree species without having to establish clonal orchards. Once the progeny tests are completed, then embryogenic tissue corresponding to the superior clones can be thawed for rapid scale-up production via somatic embryogenesis.

Although the use of artificial seed technology should be extremely valuable for the rapid introduction of genetically engineered material into production of forests.

Hydrogel Encapsulation of Artificial Seeds

Water soluble hydrogels have been found suitable for making artificial seeds. Two methods have been used to coat somatic embryos: gel complexation via a dropping procedure and molding. Redenbaugh *et al.* (1986) mixed alfalfa somatic embryos with sodium alginate (2% w/v) and dropped them into a calcium nitrate solution (100 mM). Surface complexation began immediately and the drops were gelled completely in 30 minutes. Alternatively, the embryos could be mixed in a temperature-dependent gel such as Gel-rite™, placed in the well of a microfilter plate, and gelled as the temperature was lowered. Somatic embryos from several crops have been encapsulated in alginate with plants recovered *in vitro*.

Table 16.3 Useful gels for encapsulation of somatic embryos

Gel	Conc. % w/v	Complexing Agent	Concentration (mM)
Sodium Alginate	0.5-5.0	Calcium Salts	30-100
Sodium Aginate with Gelatin	2.0	Calcium Chloride	30-100
Carrageenan	0.2-0.8	Potassium	
Locust Bean Gum	0.4-1.0	or Ammonium Chloride	500
Gelrite	0.25	Temperature lowered	

Table 16.4 Crops encapsulated in calcium alginate beads

Species	Common Name
Apium graveolens L.	Celery
Brassica species	Rapid-cycling Brassica
Daucus carota L.	Carrot
Gossypium hirsutum L.	Cotton
Lactuca sativa L.	Lettuce
Medicago sativa L.	Alfalfa
Zea mays L.	Corn

Initially, the effect of encapsulation was difficult to assess because of the overall poor quality of the somatic embryos. Although visually normal embryos were produced (i.e. bipolar, root and shoot axis, cotyledons), the germination and continued development of the embryos was very inconsistent. In fact, the use of germination (root elongation and emergence) as an efficacy assay was found to be unsuitable when shoot production and further growth was not observed concomitantly. Consequently, the approach for developing artificial seeds was shifted away from one focused on somatic embryogenesis (initiation of embryo formation) to one of somatic embryogeny (initiation, development, and maturation of embryos).

The concept of somatic embryogeny and the production of high quality embryos is not one that is widely followed by many researchers who have either focused on the production of a somatic embryogenesis system that results in some plant recovery or who have interest only in the study of the early stages of embryogenesis. This focus separated from an emphasis on producing mature, true-to-type, high quality embryos can possibly lead to conclusions based on abnormal somatic embryogenesis. To overcome this problem and to achieve high quality embryo production for scale-up of artificial seeds, measure the embryo response in terms of embryo-to-plant development or *conversion.* Essentially, embryo conversion frequency is the per cent of the somatic embryos that produce green plants having a normal phenotype. This assay has been critical for developing conditions and media that select for uniform plant production. Following are the events which are associated with the process of embryo-to-plant conversion.

1. Germination (radicle elongation)
2. Development of a vigorous root system
3. Growth and development of the shoot meristem
4. Production of true leaves
5. A direct shoot-to-root connection
6. Absence of hypertrophy of the hypocotyls.
7. Minimization of callus growth in the hypocotyl
8. A green plant with a normal phenotype.

Synthetic Seed and Forest Trees

The use of biotechnological approaches in forestry may be greatly enhanced and considerable time could be saved by using artificial seed technology. Genetic engineering in forestry will be similar to that for field and horticultural crops. Desirable genes should be identified, cloned and inserted into the tissue (protoplasts, cells-pollen, zygotic embryos, needle tissue, etc.). The putatively transformed tissue will be regenerated to plants and tested for expression of the genes. With annuals, the transformed individual plants can then be backcrossed with the original population for large-scale production of transformed seed within one to few years. However, for most tree species, after adequate gene expression is confirmed, scale-up production of transformed seeds deviates significantly at this point from that of annual and biennial crops because of the very long generation time for trees, particularly conifers.

Plain steps in genetic engineering of tree species focus on transformation, followed by traditional forest seed production and tree evaluation are as under:

1. regeneration of genetically engineered tissue;
2. confirming presence and expression of engineered gene;
3. bulking up propagules through vegetative propagation or seed production;
4. greenhouse/nursery establishment and growth;
5. seed orchard establishment with concurrent progeny testing, including evaluation of the engineered trait;
6. seed orchard roguing;
7. seed production in seed orchards and;
8. either nursery production with transplanting into the field or direct seedling.
9. the final stage will be the evaluation of plantations for increased performance due to the engineered trait.

□□□

CHAPTER-17

Environment and Energy

INTRODUCTION

Energy is an important input for development. It aims to human welfare covering household, agriculture, transport and industrial complexes. Countries all over the world engage in making strategy or policy on energy and look into a possibility of having energy systems with no or every limited environmental impacts. The fossil fuels exhaust one day. The energy crisis has shown that for sustainable development in energy sector, we must conserve the non-renewable sources and also replace/supplement them by non-pollunting renewable sources. The renewable ones are more or less pollution free, environmentally clean, and socially relevant. Moreover, no nation can afford to depend on only one form of energy. It shall have a mix of at least seven forms (biomass, solar, coal, petroleum, natural gas, hydro and nuclear).

The production of wastes as agricultural and industrial byproducts is a necessary consequence of modern civilization. The byproducts of activities in agriculture, forestry, dairying, and food industries can be used for various purposes and the resulting pollution can be minimized. These wastes or byproducts may be degraded into fermentative products by suitable microbes or may be transformed into proteins. For instance, the algae can be grown on wastewater to obtain protein-rich phycomass, while at the same time cleaning up the water itself.

Biomass is defined as the living matter or its residues and is a renewable or perpetual source. The common examples of biomass are wood, grass, herbage, grains, and bagasse.

In tropical countries such as India, biomass has an immense potential of significantly supplementing the meagre fossil fuel supplies. Some areas in which biomass can play an important role as an alternative source of energy are thermal applications (boilers, furnace kilns), shaft power applications (internal combustion engines, spark ignition, and compression ignition), and production of fuels. In all these applications, the first step involves the conversion of biomass into gaseous or liquid form.Thermochemical gasification of biomass constitutes important aspects of biomass utilization. Gasification by the fermentation route employing microorganisms is another notable alternative.

Huge amounts of agricultural residues, such as cereal straw and byproducts of corn or beet cultivation, are produced annually and can be converted into useful products. Blanch and Sciamanna (1980) have reviewed the composition of many cellulosic feedstock materials. These materials are softwoods or woody agricultural residues made up of celluloses, hemicellulose, and lignin. Table 17.1 shows the compositions of the raw materials used for ethanol production.

Table 17.1 Approximate composition (%) of cellulosic raw materials used for ethanol production (after Blanch and Sciamanna, 1980)

	Corn stover	Wheat straw	Rice straw and hulls	Bagasse	Cotton gin trash
Glucose	39	37	36-41	38	20
Mannose	0.3	0.8	2-3		2
Galactose	0.8	2.5	0.1-0.5	1	0.1
Xylose	15	19	14-15	23	4-5
Arabinose	3	2	2-5	2-3	2
Lignin	15	14	10-20	18	18
Ash	4	10	12-20	3	15
Protein	4	3		3	3

The types of biomass available for conversion into energy are region-dependent. Thus, sugarcane and cassava are suitable for hotter climates, cellulose for temperate areas, and hydrocarbon shrubs for arid or semiarid regions.

Biotechnological processes relating to the conversion of varied organic substances by fermentation (Diag. 17.1) and by microbial metabolism often cause serious environmental pollution. A single brewery can sometimes generate 10,000 m^3/day of effluent with a biological oxygen demand similar to that of the sewage from a city of some 200,000 people.

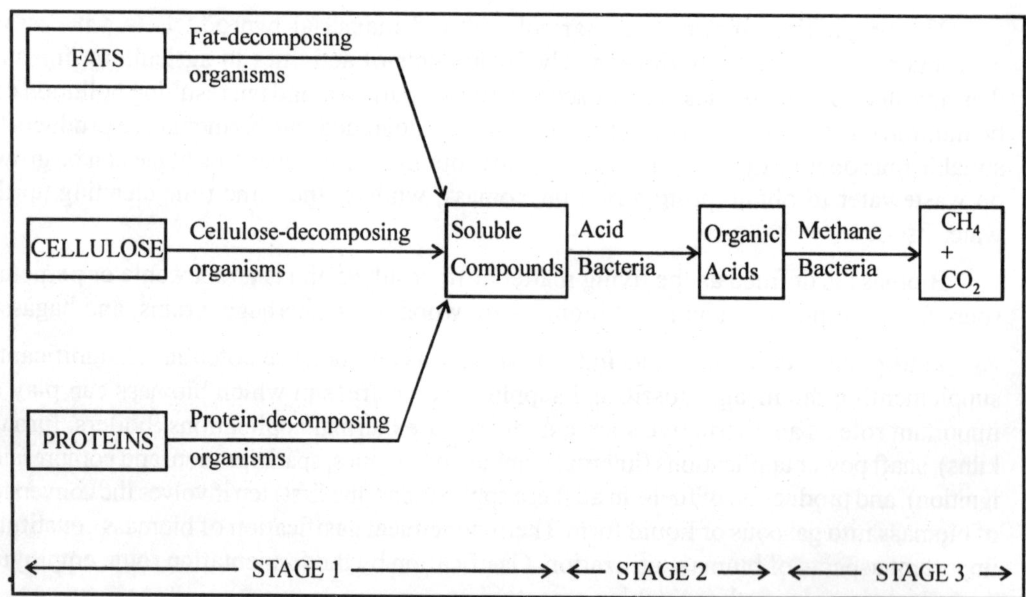

Diagram 17.1 Three stages in the anaerobic fermentation of organic materials.

Aspergillus flavus and *A. parasiticus* produce aflatoxins which are some of the most potent carcinogenic natural substances known. These species thus cause severe economic and health-related problems worldwide by infesting edible or useful plants and by growing on stored foods or feedstuffs. These problems are particularly acute in many developing countries.

Methanol is a fairly cheap chemical. It is a useful feedstock for the production of single cell protein biomass and also for industrial fermentations involving methylotrophic microbes. The facultative methylotrophs *Hyphomicrobium* spp. are frequently used in these fermentations.

Biomass Production

Several countries have launched vigorous programmes of biomass production. In New Zealand, radiata pine *(Pinus radiata)* constitutes some 75% of the annual timber cut in sawmills. Grafting and root cuttings of this tree are now being supplemented with tissue culture methods (Thorpe *et al.*, 1986). Micropropagation studies have been made with excised mature embryos. Currently, shoot initiation on excised embryos or cotyledons, elongation, rooting, and plantlet hardening are being attempted. These methods can produce over 200 shoots/clone and over 75% rooting. A somaclonal selection procedure for improvement of biomass energy crops is shown in Diag. 17.2.

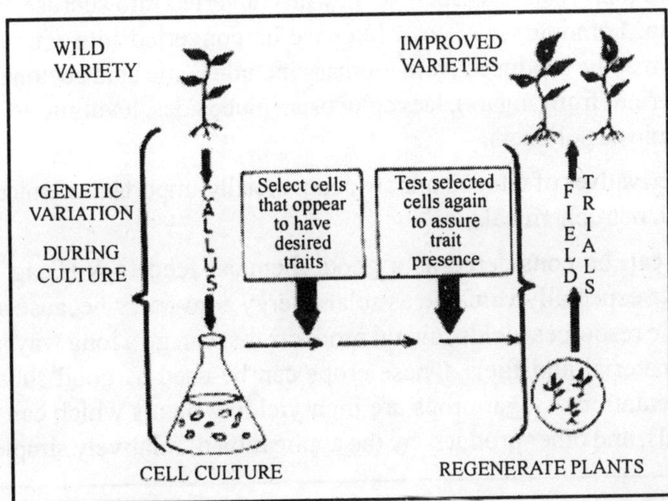

Diagram 17.2 Somaclonal selection scheme for improvement of biomass energy crops.

A serious thought is now being given to the idea of clonal propagation of selected, superior (elite) trees as a better alternative to the rather slow process of breeding. The clonal propagation of superior trees has the advantages that (1) it produces fast and immediate gains, and (2) favourable gene combinations can be transmitted intact to the propagules. However, old or mature conifers are rather difficult to clone, but explants from mature trees can sometimes form adventitious shoots, some of which successfully root.

There are several constraints in conventional tree breeding that hinder progress in developing high-yielding varieties. The recently-developed techniques of molecular biology can remove some of the crossing barriers and obstacles in breeding and cloning. In this context, the techniques of *in vitro* pollination and fertilization, *in ovulo* embryo culture and embryo rescue, protoplast fusion, dihaploids, somaclonal variants, and genetic manipulation in tree breeding for biomass production are potentially useful.

Certain species *of Euphorbia* are potent renewable resources for hydrocarbon (energy) production. These species bear latex and yield about 35% of their dry weight as simple organic extractables (Calvin *et al.,* 1982). Chemical analyses of the extracts of *E. lathyris* show that 5% of the dry weight is a mixture of reduced terpenoids, and 20% is a simple sugar (hexose). The terpenoids can be converted into a gasoline-like product and the hexoses may be fermented to ethanol. The conversion of certain biomass-derived gasoline-like materials into high-quality transportation fuels has already been demonstrated by Weisz *et al.* (1979). Calvin *et al.* (1982) have estimated that the total energy as liquid fuels obtainable from *E. lathyris,* assuming a biomass yield of 25 dry tons/ha/yr, is about 48 MJ ha/yr, out of which some 26 MJ is in the form of hydrocarbons and 22 MJ is in the form of ethanol. Attempts are now being made to further increase the hydrocarbon yield of this species.

Biomass is not only a source of fuels but also of chemicals. The scale of production of chemicals is lower but their prices are higher than those of fuels. Many useful chemicals are produced by traditional chemical reactions applied to biomass materials such as field and forest crops and their residues. Sucrose may be converted into sucrose acetate butyrate for use as a plastic. Lemongrass oil can likewise be converted into vitamin A. Some other chemicals that may be produced from biomass include lactic acid acetone, butanol, ethanol, ethylene, glycerine (from sugars), laevoglucosan, glucosides. levulinic acid, xylitol, furfural, lignin, and cellulosic polymers.

Several derivatives of fats and oils are commercially-important chemicals that are strongly competing with petrochemicals.

Biomass can be considered as a good chemical feedstock (Diag. 17.3). Sugar and starch crops are especially valuable as solar energy converters because an effective use of these renewable resources yields several products that can go a long way in ameliorating the scarcities of material and fuels. These crops can be used as good substrates for diverse classical fermentations. Sugar crops are high-yielding plants which can be converted into fuels, chemicals, and other products by the application of relatively simple technology.

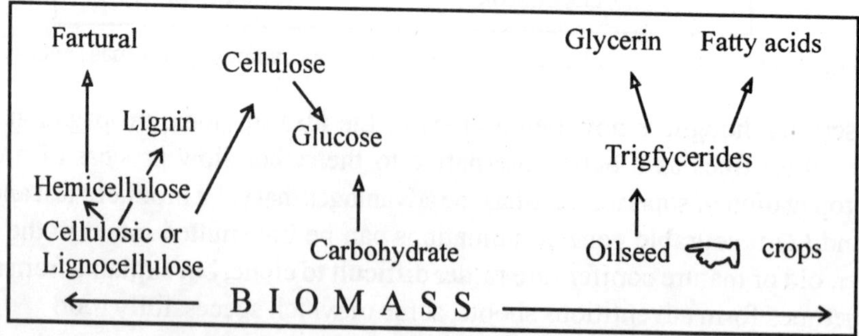

Diagram 17.3 Some feedstocks and primary chemicals derived from common biomass types (single arrow, feedstock; double arrow, primary chemical). (Modified from Lipinsky, 1981).

The two major sugar crops are sugarcane *(Saccharum officinarum)* and sugar beet *(Beta vulgaris).* The former is grown mostly in tropical countries whereas the latter is a temperate plant. Both these plants not only produce sugar but also yield several valuable byproducts such as fibre and bagasse (Diag. 17.4). The sugars can be fermented to ethanol.

The ability of some fungi to convert cellulose has been commercially exploited for producing edible mushrooms for human food. Three of the widely-used fungi are *Volvariella volvacea, Lentinus edodes,* and *Pleurotus* sp. All three produce human food (mushrooms) or may be fed to animals (Chang and Hayes, 1978).

V. volvacea (and some species of the genus) are cultivated on rice straw or other similar materials in Africa and the Far East. They can also be grown on water hyacinth, cotton, or banana leaves.

Lentinus edodes is popular as human food in China and Japan. It can be cultivated on wood, mainly oak. It effectively converts the lignocellulosic material of wood into fungal protein.

Pleurotus spp. preferentially decompose lignin but also utilize cellulose and other carbohydrates in wood. These fungi can convert sawmill residue into protein-rich food (Zadrazil, 1976).

Usually, the upgrading of materials such as wood and straw fodder requires some pretreatment with acid, alkali, or some other chemical. The objective of the pretreatment is to loosen up the structure, making the loosened material more amenable to attack by the degrading microbes, either in the digestive tract of herbivores or by enzymes that can break

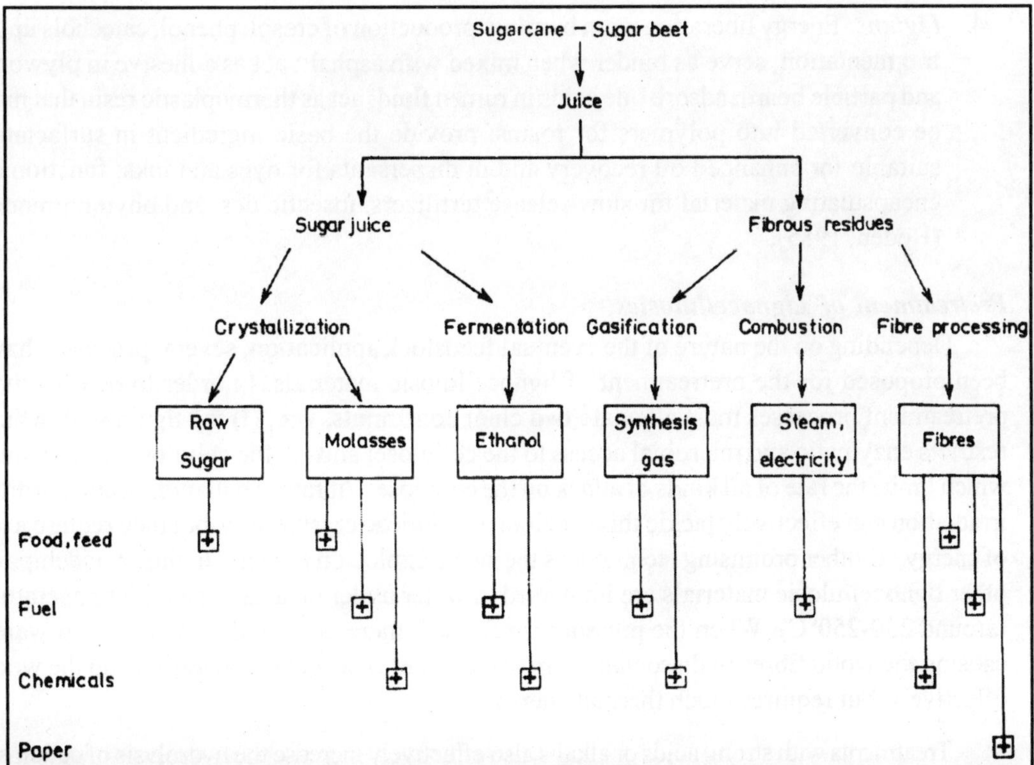

Diagram 17.4 General overview of sugar crop processing systems, showing the various byproducts obtained.

up cellulose and hemicellulose into more easily utilized sugars. The drawbacks of the pretreatment are, firstly, it causes some undesirable side reactions, and, secondly, it generates waste byproducts. A newer approach is to use better chemical (extraction with alcohol, phenol, or formic acid) or physical (explosive steam decompression) fractionation methods which often yield useful byproducts.

Fractionated lignocellulose yields cellulose fibres, microcrystalline cellulose, hemicellulose, and lignin. The possible uses of these products are as follows:

1. *Cellulose fibres:* Making paper, enzymatic production of sugar syrups which may be fermented (Diag. 17.4) to alcohols, polyols (glycerol, propylene glycol), ketones or acids; production of protein-rich animal feed through microbes.
2. *Microcrystalline cellulose:* Improving the printing quality of paper; making of suitable powders to be used as carriers for aromatic oils; manufacture of food-grade and pharmaceutical gels that resist freezing; use as carrier for enzymes; and provision of a large surface for chemical grafting (e.g., nitrification for explosives).
3. *Hemicellulose:* After hydrolysis can be used for microbial production of protein-rich animal feed; conversion to xylose after hydrolysis; production of xylitol by hydrogenation of xylose; production of furfural by dehydration of the pentoses; and making of ethanol from xylose by means of yeast strains.
4. *Lignins:* Energy liberation upon burning; production of cresol, phenol, catechols upon fragmentation; serve as binder when mixed with asphalt; act as adhesive in plywood and particle board; adsorb bile acids in rumen fluid; act as thermoplastic resin that may be converted into polymers for foams; provide the basic ingredient in surfactants suitable for enhanced oil recovery and in dispersants for dyes and inks; function as encapsulating material for slow-release fertilizers, insecticides, and phytohormones (Heden, 1985).

Pretreatment of Lignocellulosics

Depending on the nature of the eventual feedstock application, several processes have been proposed for the pretreatment of lignocellulosic materials. In order to be effective, pretreatment processes must alleviate two chief constraints, viz., (I) the lignin seal, which restricts enzymatic and microbial access to the cellulose; and (2) the cellulose crystallinity, which limits the rate of all kinds of attack on the cellulose. Vibratory ball milling and electron irradiation can effectively tackle this problem of cellulose crystallinity, but they require a lot of energy. Another promising technique is the steam explosion process. In this, woodchips or other lignocellulosic materials are immersed in water under pleasure at a high temperature (around 230-250°C). When the pressure is released, there is a rapid evaporation of water, causing the wood fibres to dissociate from one another. The technique loosens up the wood effectively, but requires much thermal energy.

Treatments with strong acids or alkalis also effectively increase the hydrolysis of cellulose, but these corrosive materials must then be completely removed, as otherwise the subsequent microbial growth will be affected or inhibited.

Forestry Biotechnology

Each year more than 3000 million cubic metres of wood are harvested, half of which is used as fuel wood. Forests also provide a variety of non-wood products that not only meet needs in food, fodder, and building materials in developing countries but also form one of the mainstays of the modern pharmaceutical industry. At present, forest production as well as its environmental functions in water regulation, soil holding, source of genetic diversity and provision of clean air, are under strain. Biotechnology can potentially make a significant contribution to reforestation and a more sustainable exploitation of forests.

In most Third World countries the forest cover is shrinking but in many rich countries, it is expanding.

Interest in wood production for energy purposes is increasing. In developing countries most people depend on gathering wood for meeting fuel needs. For rural population forests are also a major source of food, fodder, building materials, and medicines. In many countries a considerable number of small family enterprises are based on forestry products, their aggregate employment exceeding that of large-scale forest industry. The latter is often dominated by transnational companies that are much less oriented to local needs.

Forestry Research

Traditionally, research into sylviculture and forest management has tended to concentrate on wood processing and paper manufacturing, the overall objective being to produce more wood at less cost (see *Biotech. Develop. Monitor,* No. 5, Dec. 1990). Only recently has more attention been paid to the significance of shade and fodder trees for pastoralists, the many types of agroforestry adaptable to successful peasant farming on different types of soil, the accelerating fuelwood crisis and the role of forestry in rehabilitating marginal lands.

Biotechnology Potential

Germplasm Storage

Knowledge and availability of genetic resources are an essential input for applications of biotechnology. In developing countries exchange of germplasm is hindered by a lack of availability of reproductive materials and the absence of an exchange network in many multipurpose trees and other perennial crops. Deforestation, the rising emphasis on artificially set up forests and the increasing adoption of high-yielding varieties by subsistence farmers are narrowing the genetic base of important tree crops.

In the past two decades indigenous tree crops in South East Asia have been abandoned after the introduction of species *of Eucalyptus, Casuarina,* and *Acacia* from Australia. Tissue culture may be a promising method of preserving germplasm in addition to traditional *in situ* and *ex situ* conservation methods. However, at present genetic change in unorganized tissue or cells is still a problem and many species cannot yet be regenerated from cultured tissue.

Micropropagation

Breeding of trees is a time-consuming activity because their maturation takes over-ten years. Biotechnology may decrease the time required to identify and propagate superior trees. The most common biotechnologies applied include tissue and organ culture, somatic embryogenesis, and micrografting.

Additional advantages over root cuttings, the currently practised way of vegetatively producing trees, have the much higher-multiplication rates, a greater degree of control, and the smaller space requirement. Tamarind trees grow in small bushes, 3-4 metres high, when grown from tissue culture. Grown from seed, they reach a height of 10-12 metres. In this case tissue culture simplifies harvesting procedures.

For several forest trees, it is difficult to rejuvenate mature tissues. Some species produce exudates that inhibit growth *in vitro.* Consequently, regeneration of whole plants from cells or tissues is still unachievable for many species.

Tissue cultures could provide a base for improving trees by such advanced biotechnologies as protoplast fusion and multiple gene transfer. However, the application of these technologies is still very limited and expensive.

Compared to industrialized countries the countries of the South, face many additional problems. Most applications of forest biotechnology have been achieved with species that are not on their priority list. There is paucity of funds to conduct the needed research on fast-growing multipurpose trees that are important for them.

BIOFERTILIZATION

Environmental concerns, financial reasons and the need to reforestate marginal lands motivated research into nitrogen-fixing trees, *mycorrhizae* and biostimulants. Some tree species have a symbiotic relation with *Frankia* bacteria, just as some legumes have with *Rhizobium.* These microorganisms fix nitrogen from the air and transform it into ammoniacal form that can be absorbed by the trees as a nutrient. In this way these trees are particularly suitable for being grown on nitrogen-poor soils and as pioneer trees. Their leaves can form a base for a nitrogen-rich humus able to enrich the soil. Some nitrogen-fixing trees, e.g., *Casuarina,* are resistant to drought and salinity. These are being used in reforestation programmes and in fixing sandy soils in Egypt, China, and India. Intercropping with nitrogen-fixing trees enhances site productivity both by recycling organic matter and nutrients, and by improving soil texture and rainfall infiltration. In order to improve nitrogen-fixing ability, inoculation with nitrogen-fixing strains is already practised in many countries. Research is directed towards optimal combinations of tree species and microbial strains, production of non-contaminated inoculants, scaling-up of inoculant production, improvement of pure *Frankia* strains and selection and improvement of host-trees by breeding hybrids and improving tolerance to such stress factors as drought and salinity.

MYCORRHIZAE

These *are* root-fungus structures formed by symbiotic fungi that colonize the roots of most vascular plants. They enhance nutrient uptake, especially of phosphorus, and seem to increase disease resistance and tolerance to stresses like drought, salt, toxicants, and pH-extremes. Mycorrhizae are useful for nutrient-uptake in stress situations. As many tropical soils are characterized by phosphorus deficiency, inoculation programmes could be interesting especially for developing countries. Experiments on inoculation of trees with selected strains have dramatically increased survival and growth on adverse sites. Commercial production of fungus for large-scale nursery inoculant is still in a starting phase.

Biostimulants

Research into biostimulants is oriented foremost to finding non-polluting alternatives for chemical fertilizers. In forestry this kind of research is still very limited. A new biostimulant has been developed that consists of humic acids, marine algae extracts, a non-hormonal reductant plant metabolite, and B-vitamins. This blend appears to increase root and top growth of plants while greatly reducing fertilizer requirements, in pines and *Alnus*. it also increases resistance to stresses such as low soil water potential and possibly residual herbicides in soil (*Biotech. Develop. Monitor.* No. 5, Dec. 1990).

Biological Control of Pests

Biotechnology may contribute to the breeding of insect-resistant varieties by transfer of resistance properties into otherwise susceptible hosts, to the improvement of biological pest control methods using insect pathogenic microorganisms, and to disease detection.

Breeding pest resistant trees by genetic manipulation seems to be an elegant way of replacing chemical insecticides. However, because of the long life tune of trees compared to those of insects, there is a good chance that insects develop resistance against the built-in genes before the manipulated trees reach maturity. Careful action is also needed because alteration of relationships between the target-tree and specific insects can result in unforeseen changes in their ecosystem, causing even higher damage than the controlled insects ever do.

An alternative may be the use of insect pathogenic microorganisms. The rhinoceros beetle, for example, is a serious damaging agent of coconut and oil palms. Studies with *baculovirus* showed that infection and release of virus infected adults in coconut plantations in Malaysia could effectively control the beetle. Biotechnology may be applied to identify other biologically active agents. Genetic engineering could eventually introduce new properties into biological control agents, such as enhanced virulence, broader host specificity and longer shelf life.

Early disease detection can be achieved by the use of monoclonal and polyclonal antibodies. These techniques are already being applied for several tree crops.

Processing of Forest Products

The applications of biotechnology may change raw materials such as fodder, fibre, and fruits in such ways that harvesting and processing methods need to be altered. But biotechnology also offers possibilities for improved processing of wood residues and tree chemicals such as resins, phenolics, enzymes, waxes, flavourings, and pharmaceuticals. Some important tree crops in the chemical industry include *Pinus* spp. (resin), *Elaeis guineensia* (palm oil), and *Hevea brasiliensis* (rubber).

In industrialized countries, interest in new products from biomass resources, such as wood, cellulose, and lignin, is growing. Composites of conventional plastics with lignocellulosic non-woven mats can be pressed into rigid shapes to form doors, walls and auto body parts.

Optimization of fungal strains,and environmental conditions of fungi that are able to delignify wood partially may dramatically reduce the energy required for mechanical refining of wood. Use of these fungi may enable small-scale biological pulping by farmers in developing countries.

Perspectives for the South

Biotechnology can be applied in three major 'sub-sectors': tree farms or plantations, reforestation of natural forests and its sustainable exploitation, and small-scale agroforestry.

The traditional bias in forestry research and management and current financial investments in forestry biotechnology research indicate that the greatest attention is directed to large-scale plantations.

Eucalyptus species are fast growing trees, especially suitable for providing raw material for pulp and paper industries. *Leucaena* trees fix nitrogen and provide a wide range of services and products to small-scale farmers. They are not widely grown in plantations for sale.

Overexploitation of natural forests is mostly caused by local people who consider forest products as common goods. Commercial forestry activities are to blame here as far as they, by fair means or foul, drive out local people from better soils in order to make way for plantations. Commercial forestry will not be able to help these local rural people, who.are, often without any formal land rights, used to collect their fuel wood and lack the financial means for buying it. Their way out of fuelwood shortages could be to treat it like food and grow it as a subsistence crop through employing simple agroforestry techniques.

Plantations that usually consist of only one or a few species cannot replace the tremendous diversity of natural forests. Ecological relations are so complex that for some regions it is questionable whether satisfying results can be achieved before the forest will have totally disappeared.

Developing countries that earn foreign exchange by exporting wood or other forest products that can be produced in large-scale plantations, could benefit from current mainstream biotechnology research. However, most of them would be better off if greater attention were given to domestication, improvement, and genetic conservation of locally important multipurpose species.

Reforestation (Biomass Regeneration)

In many tropical countries and arid or semiarid zones, widespread destruction of forests in the last few decades has underlined the necessity to launch reforestation programmes so as to replenish some of the lost forest cover. Restoration of forest cover can, of course, be done by means of existing planting methods by planting tree saplings raised in, nurseries. But these existing practices are not sufficient and have to be supplemented with quicker and better methods. Direct seeding or broadcast seeding is one such method. In this case, the seeds are first coated with suitable chemicals to repel birds, rodents, and insects; and then sown in the area to be forested.

Another popular planting technique makes use of helicopters and aeroplanes to broadcast the seeds over a vast area. This practice is quite popular in Australia, New Zealand, Canada, and the USA. The technique has not yet gained popularity in the tropics. Aerial seeding from planes or helicopters has been used for sowing pastures as well as agricultural crops such as soybean, wheat, and rice. The technique is well-suited for reforesting sites having rough terrain, debris, or difficult access. Table 17.2 lists some species that have been successfully sown from the air. Table 17.3 lists some promising candidates for aerial seed lings in developing countries.

Table 17.2 Some plants that have been successfully sown by aerial sowing (condensed from NAP, 1981)

Species	Location
Acacia auriculiformis	Indonesia
Calliandra calothyrsus	Indonesia
Sesbania grandiflora	Indonesia
Eucalyptus spp.	Australia
Leucaena leucocephala	Pacific Islands
Liriodendron tulipifera	USA
Picea mariana	Canada
Pinus spp.	USA, New Zealand
Populus spp.	USA
Spathodia campanulata	USA

Table 17.3 Some possible candidates for aerial seeding in developing countries (condensed from NAP, 1981)

Humid tropics	Semiarid areas	Tropical highlands
Acacia spp., *Albizia lebbek, Avicennia* spp., *Calliandra calothyrsus, Cassia* spp., *Casuarina* spp., *Derris indica* (= *Pongamia glabra*), *Ficus* spp., *Leucaena leucocephala, Melia azedarach, Syzygium cumini, Terminalia catappa*	*Acacia nilotica, A. senegal, Anacardium occidentale, Azadirachta indica, Eucalyptus citriodora, Haloxylon* spp., *Prosopis* spp., *Zizyphus* spp.	*Alnus acuminata, A. nepalensis, A. rubra, Callitirs* spp., *Eucalyptus globulus, Grevillea robusta, Inga* spp., *Robinia pseudoacacia*

Advanced Materials

Though non-living, many materials and advanced composite material systems have diverse applications in biotechnology. Materials account for nearly 60% of the manufacturing cost of industrial products. Successful application of new materials is vital for many industrial sectors.

Modern engineering materials fall into three main categories: (1) metals, (2) ceramics, and (3) polymers. Metals are well established in engineering applications but advances in processing and alloying technology are continuing to improve the performance of metallic materials to their limit and provide fresh scope. Compared with metals, ceramics have superior wear resistance, chemical stability, high temperature strength, and low thermal conductivity, but suffer from brittleness. The thrust in ceramics development is in processing and control of defects to increase product reliability. Innovations in the polymer industry are helping to extend temperature tolerance capability and physical, chemical and mechanical performance. All three classes of materials can be either "functional" (e.g., special magnetic/ optical properties) or "structural", i.e., load-bearing in nature (Hossain, 1992).

In the development of structural materials, "composites" (Diagram 17.5) represent a cornerstone for progress. These materials consist of fibrous or paniculate reinforcements held together by a common matrix and have properties superior to those of the constituents. They may be divided into the following:

1. Metal-matrix composites (MMC).
2. Ceramic-matrix composites (CMC).
3. Polymer-matrix composites (PMC).

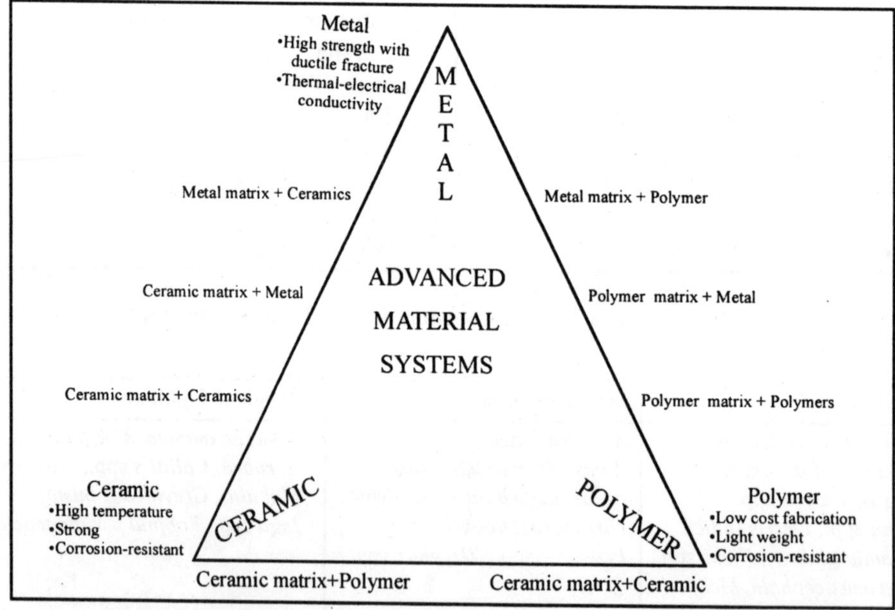

Diagram 17.5 Some advanced material systems and their properties.

Materials having properties, such as high specific stiffness, high temperature strength and high environmental resistance that are significantly better than those of more conventional materials such as steel and aluminium, are designated as advanced materials. Advanced materials can be tailor-made to have properties for specific applications. For this reason they are also known as "engineered materials". Some examples of typical advanced materials are:

Metals	:	superalloys, shape memory alloys, rapidly solidified materials;
Polymers	:	thermoplastics, polymer blends, elastomers, adhesives, inorganic polymers;
Ceramics	:	alumina, zirconia, silicon nitride, silicon carbide, coatings;
Composites	:	aramid fibre composites, s-glass composites, carbon-fibre composites, SiC reinforced aluminium or titanium.

Advanced materials are much more costly than conventional materials. This means that exploitation depends upon the relative importance of cost and performance. Aerospace, automobile, and transport industries act as catalysts in the development and wider diffusion of these materials.

Over the past decade various market projections have shown a strong growth potential in advanced ceramics. Several countries have initiated standardization activities for structural applications, particularly in the areas of materials analysis, characterization, and mechanical testing.

Polymers have been in wide use for many years. The volume of plastics on the worldwide market now exceeds that of metals. Current developments centre on engineering polymers with improved mechanical and thermal properties. Polymer-matrix composites fall into two main categories:

1. glass reinforced plastics typically based on thermosetting resins with low stiffness glass-fibres. These have been in use for 30-40 years in transport, marine and leisure-goods industries; and
2. advanced composites based on epoxies reinforced with fibres of high-stiffness glass (s-glass), graphite, aramid or other organic fibres are used in high-value added products for aerospace, -sports equipment and engineering and automotive sectors.

Currently, advanced composites represent only about 5% of the overall market but they are expected to grow at high rate in the near future.

In some advanced industrialized countries, significant investments are being made to develop MMCs because the demand for these materials is expected to grow in airframes, reciprocating parts in automobiles, leisure goods, and various other industrial applications. Designers need good reliable design data to realize the market potential but at present there is a serious lack of proven or standardized test methods needed for generating the data.

Standardization of advanced materials is still at an early stage. Such materials are usually first used by the aerospace industry where cost is less important than performance. A standard developed by the aerospace industry for its own use may not be suitable for the engineering industry in general and more work is often needed to translate the industry-specific standards into broader ones.

By its nature materials industry is an enabling technology and the users occur in different sectors. The materials supply industry tends to dominate standardization activities whilst users do not become sufficiently involved. The problem is not easy to solve because user industries have other pressures to cope with and they are content to leave standardization to suppliers. However, efforts must be made to attract users into standards-related activities whenever possible (Hossain, 1992).

Trade in materials is international in character. A material developed in one country can be produced in another and subsequently incorporated in industrial products in other countries. It is important that specifications, codes of practice, and standards are developed on an international basis.

Biodegradable Materials

Polymeric materials occurring naturally or produced from renewable resources are extremely useful as alternatives to petrochemical-based polymers and plastics.

Meanwhile, there is growing concern about the disposal, of plastics and their environmental impact. Industry is looking for ways to minimize the unnecessary use of plastics to complement recycling and reuse programmes. Others are working on new materials or modifications to old ones to reduce the environmental impact of plastics. Current efforts are concentrated on developing a family of plastic materials which are produced from renewable resources while being completely biodegradable. PHBV (polyhydroxybutyrate-polyhydroxyvalerate) is a good example of biodegradable plastics. These thermoplastic materials are a family of PHBV copolymers which fit neatly into our ecosystem.

Manufacturing Process

Polyhydroxybutyrate (PHB) homopolymer is produced in nature by a wide variety of bacteria which store it as a ready source of carbon and energy. PHB is useful but its brittleness, deficiencies in thermal stability, and difficult processability limit its usefulness. The addition of polyhydroxyvalerate to the polymer chain (Table 17.4) overcomes these limitations (Luzier, 1992).

Table 17.4 Typical properties of PHBV.

Property	**HV content (mol %)**		
	0	**10**	**20**
Melting point (0° C)	177	140	130
Crystallinity (%)	80	60	35
Tensile strength (MPa)	40	25	20
Flexural modulus (GPa)	3.5	1.5	0.8
Extension at break (%)	8	20	50
Notched izod impact strength (J/m)	60	110	350

PHBV copolymers are thermoplastic polyesters. They are composed of hydroxybutyrate (HB) units with between 0 and 24% of hydroxyvalerate (HV) units appearing randomly throughout the polymer chain (Diagram 17.6).

ICI produces PHBV by fermentation, using *Alcaligenes.* These bacteria can grow on a wide range of carbon sources in both aerobic and anaerobic conditions. Current production uses *Alcaligenes eutrophus.* This strain grows very efficiently on glucose, and can be safely handled in large quantities.

To begin the fermentation process, *A. eutrophus* is inoculated into a fed-batch reactor containing a balanced glucose medium. All nutrients are in excess except phosphorus. The medium's phosphate content is limited to support only a certain amount of cell growth.

The phosphate content decreases as the culture grows such that the culture eventually reaches phosphate starvation. Up to this point in the fermentation, very little PHB has accumulated in the cells. But in stage two of the process in which glucose is added, the cells

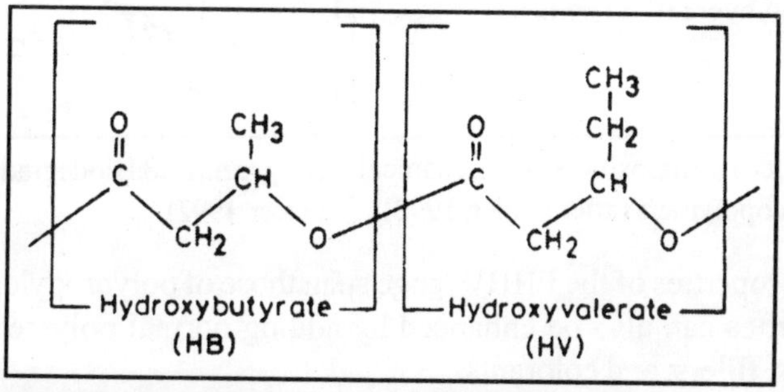

Hydroxybutyrate (HB)
Hydroxyvalerate (HV)

Diagram 17.6 Chemical composition of hydroxybutyrate-hydroxyvalerate copolymer.

cannot convert the glucose to amino acids/proteins because of the low phosphate availability. Consequently, the dry weight of the biomass rises significantly as the cells convert the glucose feed to PHB, causing massive amounts of PHB to accumulate in the cells. The PHB concentration can account for up to 80% of the biomass's total dry weight at the end of the fermentation process (Luzier, 1992).

Carbon sources other than glucose can be used. Some work is underway to assess the use of various agricultural byproducts (e.g., molasses and sugar beets).

A. eutrophus produces PHB homopolymer when fed exclusively glucose under the appropriate conditions. However, PHBV is formed if in addition to glucose the bacteria are fed a controlled amount of propionic acid during the second stage of fermentation. *A. eutrophus* incorporates a predictable amount of HV units randomly with the HB segments to form the PHBV copolymer. Therefore, a family of polymers with specific HV contents having a range of different properties can be biologically manufactured.

The last stage of PHBV production involves separating the polymer from the cells. This is done by aqueous extraction, where the cell walls are broken and the polymer is extracted and purified.

PHBV Properties

PHB is brittle and difficult to process as it decomposes above its 177°C melting point. Adding HV to the polymer leads to several improvements (Diag. 17.7 and 17.8), including drop in melting point, reduction in average crystallinity, and increased flexibility and toughness.

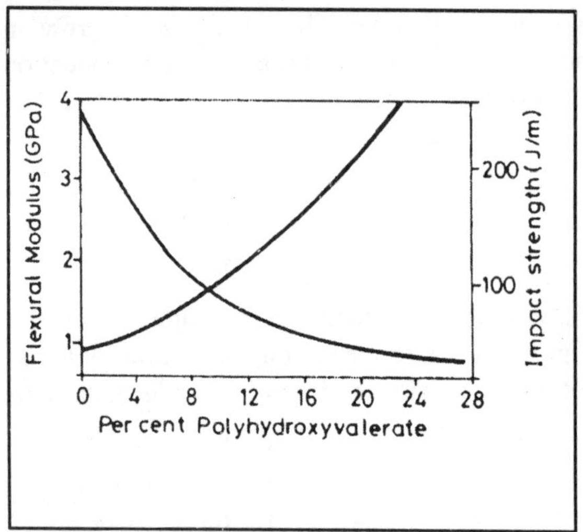

Diagram 17.7 Effect of composition on mechanical properties of PHBV copolymers (after Luzier, 1992).

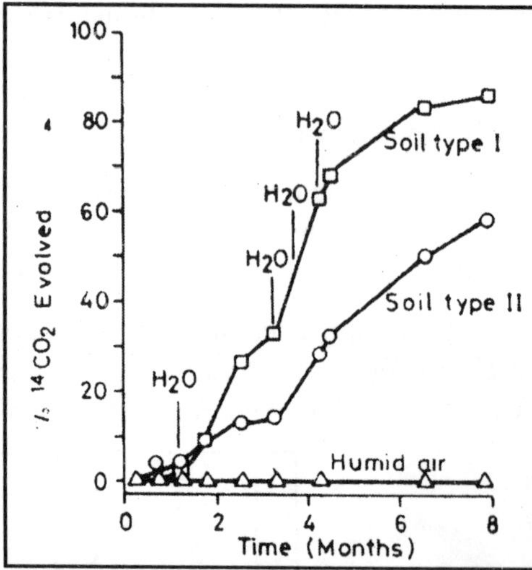

Diagram 17.8 Biodegradation of PHBV (after Luzier, 1992).

Some of the properties of the PHBV range span those of polypropylene to polyethylene. But PHBV properties can also be enhanced by adding normal polymer additives such as natural plasticizers, fillers, and colorants.

PHBV copolymers are naturally produced by bacteria from agricultural raw materials, and they can be processed to make a variety of useful products, where their biodegradability (Table 17.5) and naturalness are quite beneficial. PHBV copolymers are still in the first stage of commercialization. Pellets or powder of PHBV are currently to produce injection-molded articles, blow-molded bottles, extruded sheet, film, paper coatings, and fibres.

Biodegradability

Microorganisms use PHBV as an energy source and degrade it by secreting enzymes into HB and HV segments. These fragments are used by the cells as a carbon source for growth.

Biodegradation rates depend on surface area, microbial activity of the disposal environment, pH. temperature, moisture level, and the presence of other nutrient materials. PHBV is not affected by moisture alone. The environment must be microbially active. No

Table 17.5 PHBV biodegradable polyester biodegradation

Environment	1 - mm molding	
	100% weight loss (weeks)	**Surface erosion (µm / week)**
Anaerobic sewage	6	100
Estuarine sediment	40	10
Aerobic sewage	60	7
Soil	75	3
Seawater	350	1

degradation occurs under normal storage conditions, and the material is indefinitely stable in air (Luzier, 1992).

PHBV degrades in a wide range of environments. Degradation occurs most rapidly in anaerobic sewage and slowest in seawater. No harmful intermediates are produced during degradation in a simulated landfill environment, PHBV showed about a 60% weight loss after 50 weeks. Shredded municipal waste was used at 35°C with percolating water to neutralize and accelerate the system (Luzier, 1992).

APPLICATIONS

PHBV's key properties are biodegradability, biocompatibility, and its manufacture from renewable resources. Primary application areas in which these features meet some market needs are (1) disposable personal hygiene, (2) packaging, and (3) medical: PHBV's biocompatibility coupled with its slow hydrolytic degradation have potential in reconstructive surgery and controlled release fields (Luzier, 1992).

PHBV fits quite neatly into the ecosystem (Diag. 17.9). It is a polymer which is naturally produced by bacteria from agricultural raw materials.

PHBV is still in the early stages of commercial development. It is an excellent example of how new technology can help meet society's needs for plastic materials and clean environment (Luzier, 1992).

BIOPOLYMER PRODUCTION BY AN *ALCALIGENES* SP. FOR BIODEGRADABLE PLASTICS

There is currently much interest in certain polymers designed to replace some of the industrial polymers which are petrochemical based. Among these biopolymers is poly-beta-hydroxybutyrate (PHB) which is a biodegradable, biocompatible, thermoplastic produced by various microorganisms. The material can be made into films, fibres, and sheets, and moulded into shapes and bottles. According to Byrom (1987), PHB and its copolymer with hydroxy valeric acid (PHV) are being developed for a variety of applications.

PHB is an intracellular storage compound that acts as a reserve of carbon and energy (Anderson and Dawes, 1990). The polymer accumulates as distinct granules in the cell, and has been reported to accumulate up to 70-80% of cell dry weight for strains of *Alcaligenes eutrophus,* under conditions of nitrogen or phosphate limitation and excess of carbon source (Shimizu *et al.,* 1992). The cells are centrifuged, ruptured, and treated with enzymes to solubilize the non-PHB components. After washing and flocculation, PHB is recovered as a white powder (Rhee *et al.,* 1992). The main problem limiting the widespread use of PHB and associated copolymers is the relatively high cost of fermentation substrates and the product recovery costs compared to petroleum-derived raw materials (Byrom, 1987).

Certain strains of *Alcaligenes* can form more than 80% cell dry weight as PHB, when the cells are grown under nitrogen-limitation in fed-batch cultures (Shimizu *et al.,* 1992).

Attention is also being given to the possible use of recombinant strains for PHB production. The operon responsible for the production of PHB in *Alcaligenes eutrophus* has been cloned into *E. coli* (Slater *et al.*, 1988; Peoples and Sinskey, 1989). Since growth and biomass productivities with *E. coli* are high, and the level of PHB accumulation has been reported to be about 90% (Slater *et al.*, 1988), the use of recombinant strains could improve the economic viability of a PHB process.

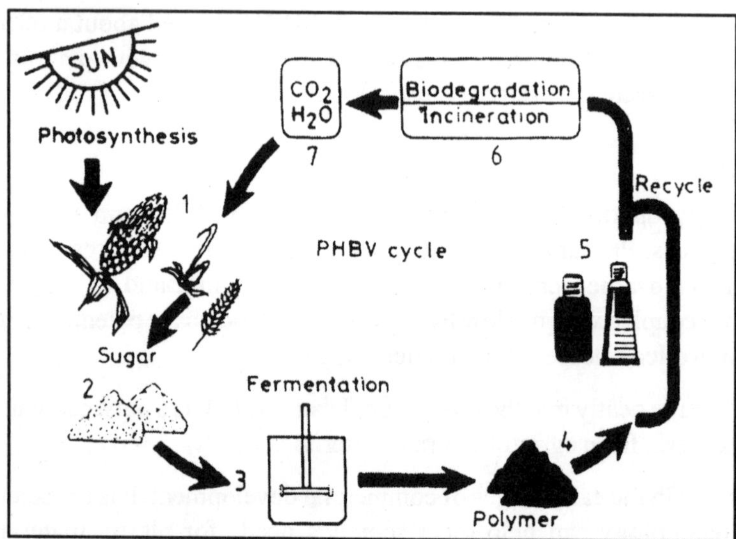

Diagram 17.9 PHBV cycle (I. carbohydrate from photosynthesis; 2, sugar feedstock; 3. fermentation process; 4, extracted polymer; 5, plastic product; 6, disposal options; 7, end products return to cycle). (After Luzier, 1992).

The expression of PHB in transgenic plants has kindled great interest with a report that genes from *A. eutrophus* which encode the two enzymes required to convert aceto-acetyl-coenzyme A to PHB had been placed under the transcriptional control of the cauliflower mosaic virus 35S promoter and introduced into *Arabidopsis thaliana* (Poirier *et al.*, 1992). Transgenic plant lines that contain both genes accumulate PHB as electron lucent granules in the cytoplasm, nucleus and vacuole; the size and appearance of these granules being similar to the PHB granules that accumulate in bacteria.

Strain SH-69 of *Alcaligenes* sp. can accumulate poly-beta-hydroxyalkanoates (PHAs) from a range of carbon sources. In batch culture SH-69 can produce copolyesters consisting of 3-hydroxybutyrale (3HB) and 3-hydroxyvalerate (3HV) from simple carbohydrates that are not generally considered as precursors of 3HV monomer units. The content of PHA and the proportions of monomer units vary depending on'the carbon and nitrogen sources used.

In Table 17.6, the influence of various carbon and nitrogen sources on the PHA content and composition is shown.

Anderson and Dawes (1990) have reviewed the growing interest in the commercial development of biodegradable alternatives to petrochemical plastics. The first PHA consumer products were launched in 1990 with more recent test marketing of hair care products in bottles manufactured from ICI's Biopol. a 3HB-CO-3HV copolymer.

Table 17.6 PHA production by *Alcaligenes* sp. strain SH-69 after 24 hr of batch culture with different carbon and nitrogen sources (source: Kim *et al.*, 1992)

Carbon source (0.11 M)	**Nitrogen source***	**Dry cell weight (g/1)**	**PHA content (wt %)**	**PHA composition (mol %)**	
				3HB	**3HV**
Glucose	Yeast extract	3.2	45.2	98.4	1.6
	Urea	1.4	14.8	85.6	14.4
	Ammonium sulphate	1.8	29.3	93.3	6.7
Sucrose	Yeast extract	1.5	18.9	98.3	1.7
	Urea	1.3	4.0	93.5	6.5
	Ammonium sulphate	1.2	15.1	92.0	8.0
Sorbitol	Yeast extract	3.1	44.8	93.1	6.9
	Urea	1.8	37.2	85.7	14.3
	Ammonium sulphate	1.7	28.1	93.5	6.5
Mannitol	Yeast extract	3.4	58.7	94.1	5.9
	Urea	2.2	18.2	92.5	7.5
	Ammonium sulphate	1.8	29.0	93.3	6.7
Sodium	Yeast extract	2.3	34.5	91.9	8.1
	Urea	1.6	5.3	78.1	21.9
	Ammonium sulphate	2.7	41.1	86.7	13.3

3HB. 3-hydroxybutyrate. 3HV. 3-hydroxyvalerate
The concentrations of nitrogen sources were as follows
yeast extract *(2* g / l). urea (1 g/l). ammonium sulphate (2.2 g/l)

Future research is likely to focus on raw material cost reductions as well as the development of recombinant microbial strains and transgenic plants for PHA production (Rhee *et al.*, 1992).

BIOENERGY

For most of the world's people, biomass, rather than oil, is the major energy source. The developing countries obtain over 40% of their energy from wood, crops, crop residues, and human, animal, and industrial wastes. Bioenergy fuels are produced from wood, crops, forest residues, and agricultural residues. These materials are subjected to physical (e.g., chipping, compacting, drying), chemical (e.g., gasification, liquefaction), and biological (fermentation digestion) treatments to produce biofuels which include woodfuel. charcoal, vegetable oil, alcohols, and biogas. The end-use processes can be combustion (simple or advanced) or engines (e.g.. diescl, steam). The final products are low-grade or high-grade heat, power, and transport.

Although much of the alcohol being produced at present is synthetic (non-microbial). The rising costs of petroleum have rekindled fresh interest in producing ethanol by fermentation for use as fuel. Likewise, biogas is being produced as a source of energy in many countries, in fact, the microbial generation of biogas is the most practical process to produce fuel for farm and community use (Table 17.7). This is because ethanol for fuel requires a much greater capital investment than is the case for biogas.

Table 17.7 Some fuels from microbial processes

Fuel	Typical source(s)	Process / Remark
Ethanol	Molasses, grains, phytomass	Much capital needed
Methanol	Methane	None commercially available as yet
Methane	Waste materials	Practically feasible for farm and community use
Hydrogen	Algae	None commercially available as yet

The continuing concern for the long-term consequances of large-scale use of fossil and nuclear fuels has generated much interest in the potential of several bioenergy systems (Diag. 17.10).

Rising population pressures have forced the issue of whether to use some piece of land for food plants or fuel purposes; for wastelands or marginal land, the choice is easily in favour of fuel or energy cropping. Land availability for biomass crops varies from country to country. In the future, areas now growing food or feed crop surpluses (as in many developed countries) may become available for energy crops, but such availability is also intimately linked to the prices of petroleum and fossil fuels. The cost of energy from biomass may best be made more competitive in two ways, viz., (1) by increasing productivity relative to energy and other inputs, and (2) by substantially improving the efficiency of the conversion process. Accordingly, if one has to produce energy from biomass, one must develop the crops and the technology to convert them appropriate to a bioenergy industry (not the food, feed, fibre, chemicals, or waste management industries). The development of these technologies is bound to have important environmental consequences, both negative and positive. Examples of the likely adverse effects include soil erosion, nutrient depletion, waste disposal, degradation of water quality, and air pollution. The likely environmental improvements may be associated with:

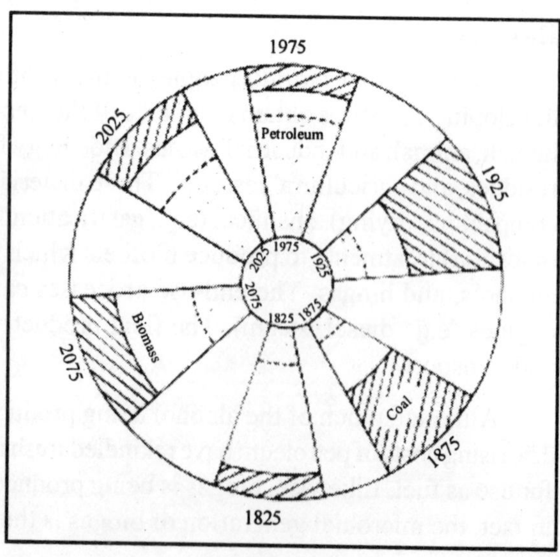

Diagram 17.10 A hypothetical and speculative representation of the relative importance of petroleum, coal, and biomass as industrial chemical feedstocks for the 250-year period 1825-2075 A.D. The amounts shown (not to scale) are estimated percentages of total feedstocks utilized. Perhaps by 2075, the cycle may be back to almost complete dependence on biomass. (After Goheen, 1981).

1. Anaerobic digestion of byproducts of food and beverage industries. This will minimize or eliminate pollution. Diagram 17.11 illustrates a development programme for non-conventional anaerobic treatment technology.
2. Renovation of wastewater streams and municipal sludge by producing biomass and useful energy. This technology is potentially applicable to eutrophic lake restoration.
3. Biomass fuel utilization. This adds little to the acid rain problem. Also, because biomass recycles $C0_2$. it does not enrich the atmosphere with CO_2.
4. Recycling of sludges and fly ash through energy crops. These materials are not tolerated in food chains (Smith, 1987).
5. Substitution of ethanol for lead in gasoline. This minimizes the hazards from various enhancers, such as benzene, which act as pollutants.

Ethanol can be generated from residues having a high sugar content, such as molasses and corn. The alcohol yield depends on the amount of starch or fermentable sugars present in the substrate. Diagram 17.12 shows the design of a fermentation system for producing ethanol from molasses. This fermentation occurs at the normal atmospheric pressure (Faith *et al.*, 1974).

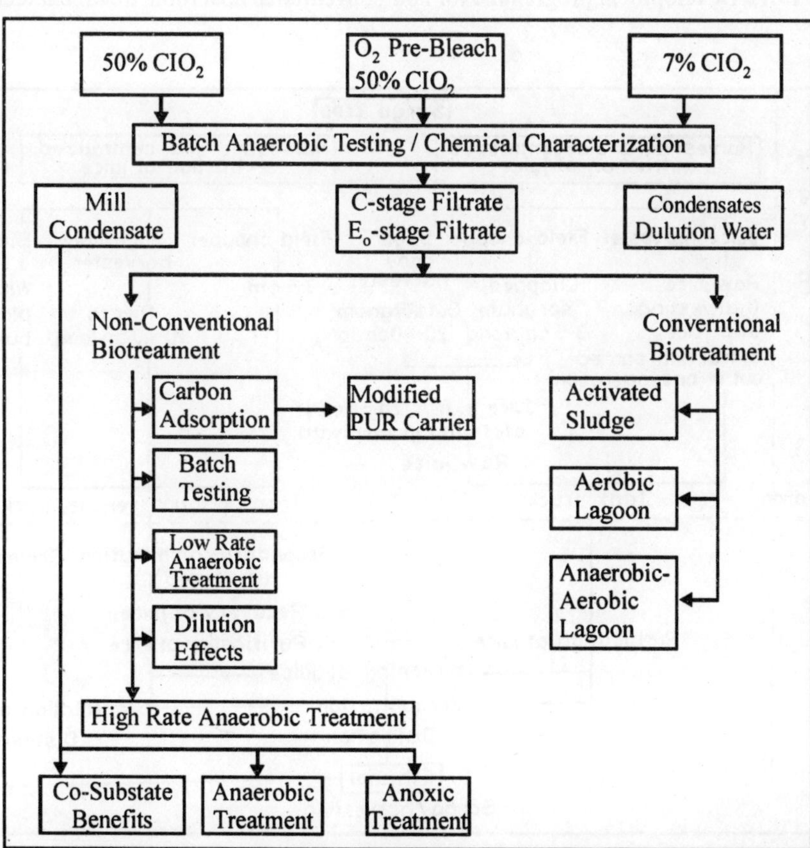

Diagram 17.11 Development programme for non-conventional anaerobic treatment technology.

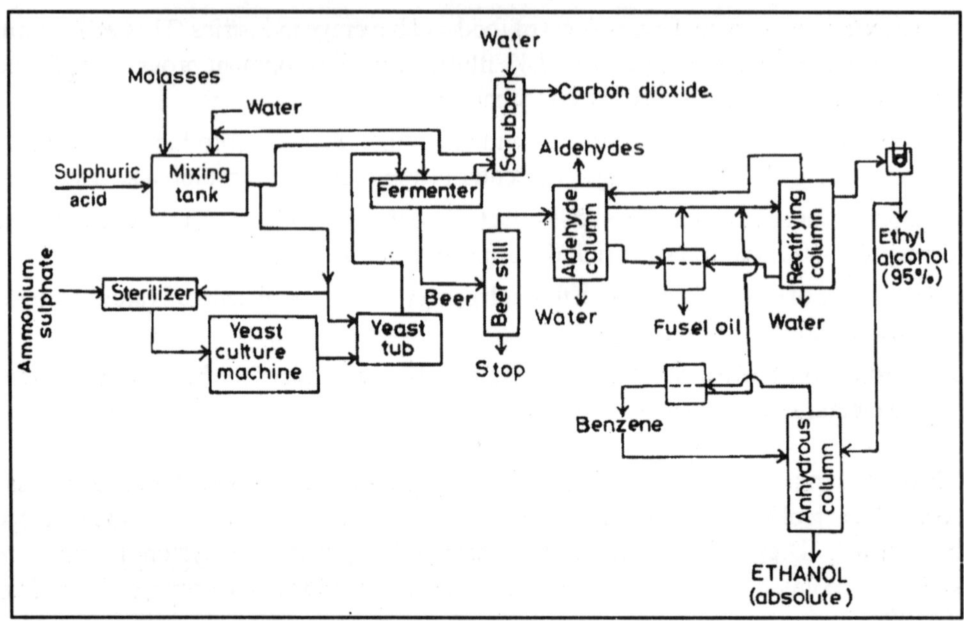

Diagram 17.12 Development programme for non-conventional anaerobic treatment technology.

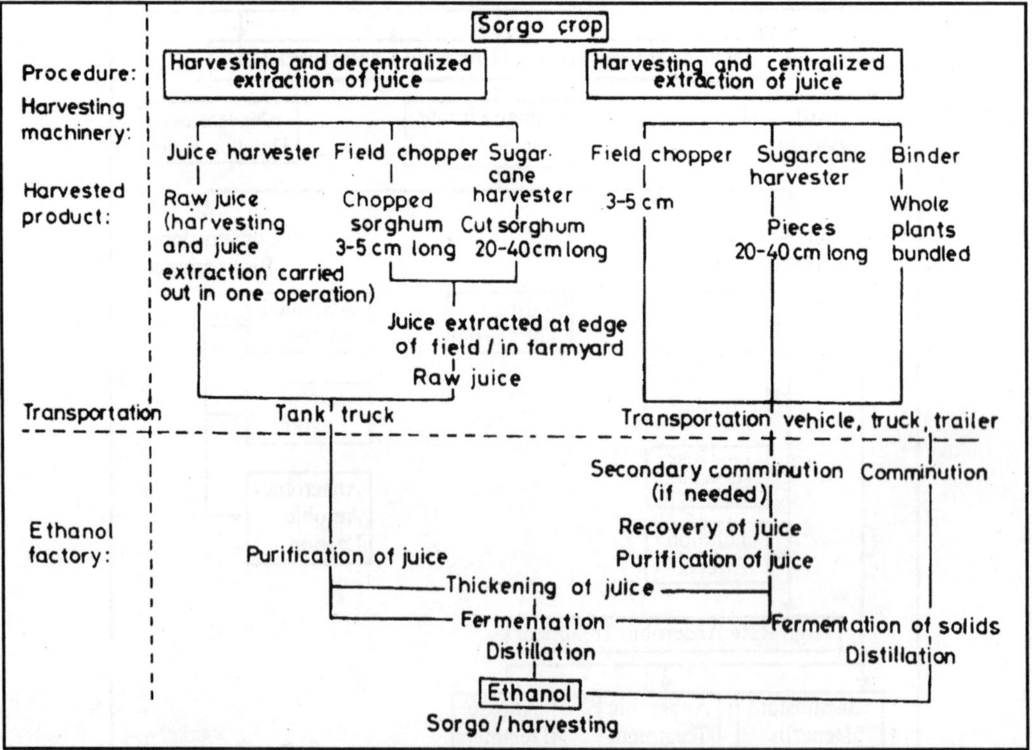

Diagram 17.13 Development programme for non-conventional anaerobic treatment technology.

Some increases in ethanol yields can be obtained by resorting to rapid fermentations, using a vacuum, and by recycling the microbial cells.

Bioenergy from wood burning has, of course, been used by man since time immemorial, but in many parts of the world, various types of energy plantations are now being practised as an adjunct and as important contributors to our future energy needs. Sugarcane, molasses (Diag. 17.12), sugar beet, surplus grains, and other renewable sources are potential substrates for making industrial alcohol. Vegetable oils may find application as diesel oil extenders or substitutes.

Some exciting new approaches have recently become available for saving some of the energy (up to 50% of the heat of combustion of ethanol) that is needed to make anhydrous alcohol by distillation. These include membrane separation, liquid/liquid extraction, and vapour phase adsorption. Equally promising is vacuum fermentation with suitable thermophilic anaerobes which can tolerate over 5% alcohol (Bungay 1983).

However, bioenergy will quite certainly be in strong competition with energy from fossil fuels, and this will necessitate due consideration of alternative uses for the biomass produced (Heden, 1985).

One of the earliest countries to exploit non-conventional energy sources for biomass production was Brazil. Brazil is one of the leading countries to produce ethanol from sugarcane and use the ethanol so produced as a substitute for fossil energy sources, particularly gasoline. Brazil has also made rapid progress in developing a sound technology for ethanol production.

Many of the state-owned cars in Brazil are being modified and redesigned to run on alcohol instead of gasoline. The advantage of using alcohol is its lower price in Brazil, about 25% lower than that of gasoline.

Though Brazil is perhaps the only country where fuel alcohol is used on a large scale, in several other countries also pilot projects have been started to produce ethanol as a substitute for gasoline. In some countries, the feasibility of wood gasification for methanol production is being carefully examined. Ethanol is more promising, and economically cheaper than methanol for use as fuel. A judicious use of ethanol can provide farm power. Some practical biological systems to convert lignocellulose into sugars for ethanol are now being developed. When operational, they are likely to substantially increase the economic attractiveness of this unconventional fuel, and may cause a profound impact on the economy of the developing countries in the tropics. Diagram 17.13 shows some possible ways of recovering ethanol from sweet sorghum.

Ethanol has proved suitable as fuel for lighting and cooking, and it can be used alone to power vehicles or mixed with gasoline as an octane booster. The stillage wastes produced as byproduct of ethanol production are used for feeding animals or as fertilizer.

The technology for ethanol production is based on the proven superiority of sugarcane as the major raw material. Three other efficient ethanol-producing materials are molasses (Diag. 17.12), cassava, and corn. Table 17.8 compares the yields of ethanol from various materials.

The yeast cells have been widely used in alcoholic fermentation processes worldwide. Diagram 17.14 illustrates the relations between yeast growth and alcoholic fermentation under different conditions.

It is a general observation that cell immobilization results in a decrease of cellular activity in the reactor. On the other hand, beneficial effects in terms of activity, physiological stability, and increased product yield are often encountered. For ethanol production, the yeast cells have been immobilized by entrapment within polysaccharide gels, particularly calcium alginate.

The worldwide annual production of sacchariferous byproducts is much lesser tnan that of amylaceous residues. Starch is thus an important raw material for ethanol production. However, because starchy materials must first be converted into sugary materials, the preparation of mashes from starchy residues is quite expensive energetically.

Table 17.8 Approximate yields of ethanol production from different biomass materials (after World Bank, 1980)

	Sugarcane	Molasses	Cassava	Corn
Ethanol/ton of biomass (litres/ton)	70	270	180	370
Biomass/ha of land (tons/ha)	50		12	6
Ethanol/ha of land (litres/ha)	3500		2160	2220

Another alternative being actively considered is to produce ethanol from lignocellulosic materials.

Diagram 17.4 and 17.12 outline the steps involved in ethanol production from sugarcane.

Upon hydrolysis, hemicellulose yields xylose, glucose, and other constituents (hemicellulose is typically made of xylan, araban, glucan, galactan, and mannan; these macromolecules are polymers of such simpler sugars as xylose, arabinose, glucose, galactose, and mannose; hemicellulose also contains uronic acid). Of these products, xylose can yield ethanol after isomerization (Diag. 17.15). Xylose is first isomerized to xylulose which is fermented to ethanol via the pentose phosphate metabolic pathway. In this process, *Saccharomyces cerevisiae* is used, along with exogenous addition of glucose isomerase. to convert xylose into ethanol. (The yeast cells lack the enzyme glucose isomerase which is needed to convert xylose into xylulose.)

Zymomonus mobilis is a particularly valuable microorganism for producing industrial ethanol. Table 17.9 shows the superiority of this microbe *over Saccharomyces carlsbergensis*. Further, since less substrate carbon is incorporated into biomass, somewhat higher product yields are obtained with *Zymomonas* fermentation (Bringer and Sahm, 1984).

Zymomonas cells are better producers of ethanol than the yeast cells, because this bacterium shows higher sugar consumption rate and lower growth rate as compared to the yeast cells. Also, it grows anaerobically — a special merit for use in immobilized state. By immobilizing *Zymomonas mobilis* cells, the cell density can be greatly increased and a continuous operation at a high dilution rate without wash out can be achieved; this, of course, is conducive to higher reactor efficiency. Whereas for yeast a maximum reactor productivity of about 30 g/l/hr is achievable, the corresponding value for *Zymomonas mobilis* can be as high as around 60 g/l/ hr.

Immobilized cell technology makes it possible to achieve high production rates with low ratės of cell growth.

Table 17.10 gives current estimates of worldwide availability of renewable agricultural resources that may be utilized for producing ethanol.

glucose concentration →

oxygen tension ↑	high	low
high	1 aerobic growth + aerobic alcohol fermentation	2 aerobic growth no alcohol
	CRABTREE EFFECT	PASTEUR EFFECT
low	3 anaerobiosis no growth alcohol formation	4 anaerobiosis no growth alcohol formation

Diagram 17.14 Yeast growth and alcoholic fermentation (Wohner et al., 1984)

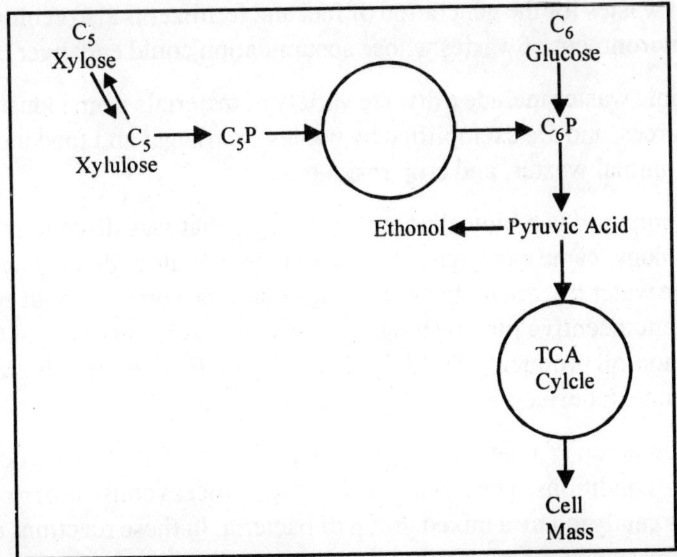

Diagram 17.15 Production of ethanol from xylose (or hemicellulose hydrolysate) (Tsao, 1986).

Table 17.9 Comparison of fermentations by *Zymomonas mobilis* and *Saccharomyces carlsbergensis* Initial sugar concentrations 100 g glucose/1

	Zymomonas	**Saccharomyces**
Specific growth rate μ (h^{-1})	0.276	0.123
Specific ethanol productivity	5.44	0.82
Cell yield	0.03	0.04
Ethanol yield (%)*	95.00	90.00

* 100% = 0.511 $g.g^{-1}$

Table 17.10 Rough estimates of annual global production/consumption of some renewable agricultural resources (after Wohner *et al.,* 1984)

	Quantity (tons dry matter)
Biomass	1.2×10^{11}
Utilizable wood production	1.3×10^{10}
Starch production	1.1×10^{9}
Sugar production	1.2×10^{8}
Crude oil consumption	3.0×10^{9}
Lactose waste in whey	1.0×10^{6}

Biogas

The rapidly-dwindling reserves of fossil fuels in recent years have stimulated a great interest in exploring the alternative sources of renewable energy such as solar energy and solid organic wastes. The technology for biogas production from organic wastes has received a tremendous boost in many Third World countries.

The use of wastes for the generation of fuel and fertilizer is also ecologically important as it rids the environment of wastes whose accumulation could endanger public health.

Solid organic wastes include a diverse variety of materials from industrial, agricultural, or domestic sources, and are exemplified by wastes from sugar and food industries, garbage, human refuse, animal wastes, and crop residues.

Biogas production is a biotechnological process that was discovered long before the word 'biotechnology' came into vogue. In industrialized countries biogas technology is mainly applied in waste water treatment. In developing countries, concern about energy supply has been an important incentive for new biogas programmes. The oil crises of the seventies and eighties jolted non-oil producing Third World countries. Besides, the shortages of fuelwood and the environmental effects of wood collection are grave.

Biogas production is a naturally occurring process that starts off when organic matter enters anaerobic conditions. The anaerobic digestion process consists of a complex series of reactions that is catalyzed by a mixed group of bacteria. In these reactions organic matter is converted step by step to mainly methane and carbon dioxide.

Polymers such as cellulose, hemicellulose, pectin, and starch are hydrolyzed to oligomers or monomers, which are then metabolized by fermentative bacteria, resulting in the production of hydrogen, carbon dioxide and volatile organic acids such as acetate, propionate, and butyrate. Finally, methanogenic bacteria produce methane from acetate, hydrogen, and carbon dioxide.

Key factors in the digestion process are the exclusion of air and light and a temperature close to 35 degrees Celsius. These conditions can be met in a hole in the ground, lined with brick or cement to keep the mixture of bacteria, water and feedstock ('slurry') from leaking out. A suitable cover excludes air and light and also collects the gas. In tropical and subtropical areas the ambient temperature is usually about right for production of biogas during most of the year.

The gas, stocked in the top of the dome in the most simple models, is piped off for use. After the feedstock is exhausted, it is pumped out and the residue is used as fertilizer. Anaerobic digestion not only breaks down organic materials into biogas, it also releases plant nutrients such as nitrogen, potassium, and phosphorus and converts them into a form that can be easily absorbed by plants.

The efficiency of the biodegradation process is determined by the proportion of different microbial strains and the extent to which conditions allow them to grow. It may be beneficial to add certain strains, especially in the starting phase, to rapidly stabilize the fermentation.

In upstream and downstream of the digester also, some improvements have been proposed. Some workers add a conditioning tank to prepare the feedstock before it enters the reactor. In other cases measures are taken to clean up the biogas, for instance to separate carbon dioxide from methane.

Anaerobic digesters can be fed with a range of substrates, including:

1. Domestic wastewaters, sewage sludges, and municipal solid wastes.
2. Agroindustrial wastewater, sludges and more solid materials.
3. Agricultural plant wastes and animal wastes.
4. Energy crops.

In industrialized countries suitable equipment is available for treating agroindustrial wastewaters. The primary objective is pollution control, but increasingly the produced biogas is recycled mainly for heating purposes. Not only agroindustrial wastes originating from food processing industries, but also the effluent from pharmaceutical, chemical, petrochemical and coal gasification plants are being considered as feedstocks for biomethanation.

Anaerobic treatment of domestic sewage sludge is widely practised as well. Domestic solid wastes in landfills are increasingly used for energy recovery. Landfills themselves behave like gigantic digesters. Pipes are installed in these landfills to collect the biogas. The fermentation in the landfills takes place under dry conditions, but is slow and not very efficient. Nevertheless, the biogas extraction from landfills is progressively and steadily spreading in some countries. Three strong trends in the development of biogas technology in developing countries need to be highlighted:

1. The increasing introduction of integrated biogas farming, involving polycultures, completed by livestock and fish-breeding and waste recycling processes.
2. Dry methane digestion for rural family use is more and more considered as an alternative for conventional (wet) fermentation processes.
3. Improved designs of digesters for industrial use are being increasingly imported from developed countries.

Most of the biogas plants in developing countries are situated in rural areas, often for small-scale treatment of domestic wastes. The number of industrial installations is growing. Researchers emphasize the aptness of biomethanation for treating municipal solid waste which is a major problem in Third World cities.

Municipal solid waste in developing countries is usually better suited to anaerobic digestion than in industrialized countries, due to its higher content of organic matter.

Among developing countries. China and India are well experienced in biogas technology. Historically, emphasis has been on small-scale domestic digesters. Both Chinese and Indian governments provide some incentives.

The early motivation for building methane digesters was mainly to improve sanitation and to recycle organic fertilizer, rather than the production of energy. More recently, motivation has shifted, and today the production of energy ranks first. Millions of households operate a small-scale digester and use biogas for cooking and lighting. These digesters are mainly fed with animal dung and night soil. By reducing the amount of pathogenic bacteria and viruses they have had a marked effect on the improvement of sanitary conditions in rural areas.

Stimulating biogas is desirable in view of the actual costs of conventional energy sources, the dependence on energy imports, and reduction of environmental pollution. Wider application of biogas in developing countries, however, depends on governmental policies.

The dissemination of biogas technology has proved to be not only a question of technological development: Just as important is the knowledge of how to manage the digester system, and insight in environmental and sanitary effects of diverse energy sources. Therefore, a well-organized extension service is necessary to emphasize both energy, sanitation, and fertilizer aspects as well as to provide training in integrating biogas technology in fanning systems or industries.

Further, people must be enabled to buy digesters. In India, small-scale biogas technology only reached those farmers who could afford initial investment. In contrast, due to governmental subsidy, the prices of digesters in China remain so low that even poor people can afford one.

Biogas has been utilized in China since the early years of this century. Millions of biogas digesters are now being used. In these, mostly crop stalks are used as the substrate. The digesters are almost entirely buried underground, with a fixed dome that serves as the gas holder. Another type of digester has a floating cover and is made of steel, plastic, concrete, or bamboo frame covered with asphalt. These digesters are of the high pressure type and require gastight joints, walls, and pipelines.

The following are some of the methods commonly employed to produce fuel from solid organic wastes:

1. Anaerobic fermentation of animal, human or agricultural wastes produces methane.
2. Treatment of domestic wastes, crop residues, and agro forestry byproducts with carbon monoxide and water produces fuel oil.
3. Pyrolysis of municipal wastes gives fuel gas, oil, and char.
4. Treatment of wastes with hydrogen gives substituted natural gas.

Out of these, method (1) is the simplest and most popular in several Asian countries such as China and India.

Raw Materials and Substrates

The naturally-occurring organic material of plant, animal, or human origin serves as the feed material for biogas production. The more biodegradable the material, the faster the digestion process. Table 17.11 lists some important organic materials used for biogas production.

Table 17.11 Some organic materials routinely used for methane generation

Type of waste	Examples
Crop residues	Sugarcane bagasse, weeds, corn stubble, straw, spoiled fodder Cattle dung, urine, poultry droppings, sheep and goat droppings, fishery wastes, blood and meat
Human	Faeces, urine, refuse
Agroindustrial	Oil cake, rice bran, wastes from fruit and vegetable processing
Forest litter	Twigs, barks, branches, leaves
From aquatic habitats	Water hyacinth, macrophytes, seaweeds

The most commonly-used feed materials are those of animal origin such as cattle dung. These require no special treatment as they have already undergone mechanical and biochemical treatment *by* the animals (in their guts). Some plant-derived materials are not so suitable in view of their high lignin content. Lignin tends to retard bacterial decomposition of the plant material, thereby slowing down the rate of gas generation.

In biogas plants, grass and cabbage wastes can be fermented efficiently if some (about *25%)* sewage or liquid manure is added or, alternatively, the acetic acid produced is neutralized by adding some alkali, e.g., NaOH and NH_4OH.

The sugar beet pulp is a valuable animal feed. It can also be fermented anaerobically to yield biogas (Diag. 17.16). Stoppok and Buchholz (1984) used a two-step anaerobic digestion of sugar beet pulp and, with retention times of 16-32 hr, obtained a biogas yield of about 80% of the theoretical value for total carbon conversion.

As a first step in the treatment of heavily-polluted industrial wastewater, microorganisms are employed to convert the organic waste into methane and CO_2. This conversion is catalyzed sequentially by different groups of microbes. These various microbes act together as a bioenergetic symbiotic team. Diagram 17.17 shows the design of a two-stage fermenter for microbial production of methane.

Diagram 17.18 outlines the process used for methane production from manure; the same can also be applied to pulpmill sludge.

Straw is a lignocellulosic waste material. The white rot fungus *Pleurotus* is the only microorganism known that can degrade lignin completely. The growth of this fungus on straw removes the lignin, and the remaining straw pulp can be advantageously used in a biogas fermenter to yield biogas.

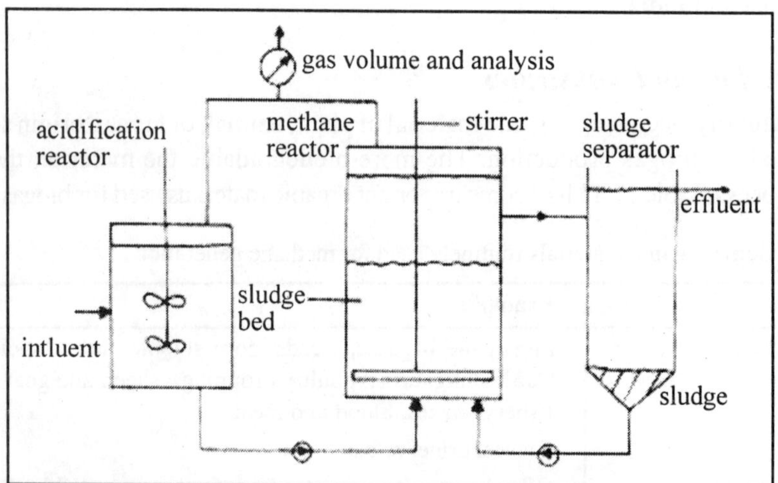

Diagram 17.16 Sketch of the reactor system for biogas generation from sugar beet pulp (after Stoppok and Buchholz, 1984).

One of the most abundant carbohydrates derived from plant biomass and wood is D-xylose. During paper manufacture, D-xylose is generated as a waste byproduct from hydrolysis of xylan (xylan is the chief constituent of hemicellulose).

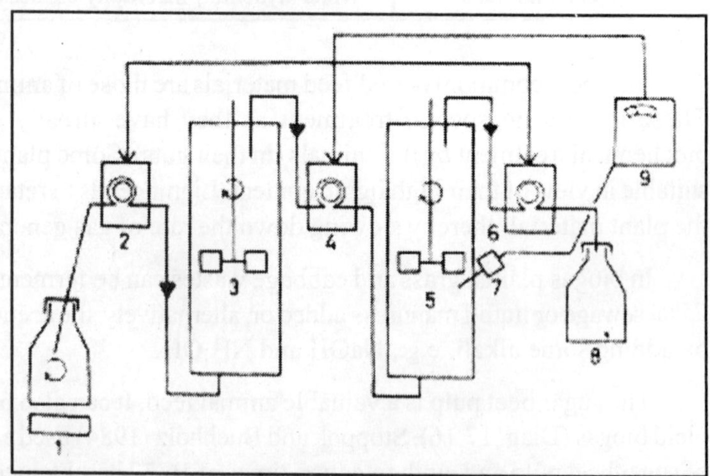

Diagram 17.17 Sketch of a two-stage fermenter for microbial generation of methane (1. mixed substrate storage container; 2. feed pump; 3. first-stage reactor; 4. transfer pump; 5. second-stage reactor; 6. discharge pump; 7. pH electrode; 8. receiver; 9. pH meter and controller for synchronized operation of all pumps). (After Trosch *et al*.. 1984.)

Diverse workers have attempted to convert xylose into ethanol or some other useful product by employing the yeast cells. However, yeast has been found not to ferment xylose to ethanol possibly because of NADH accumulation under the prevailing anaerobiosis. Only a few species of fungi can ferment it to ethanol and that too very slowly. By employing glucose isomerase, xylose can be converted into xylulose, which can be easily fermented by several species of yeasts. Another approach to this problem is genetic manipulation. Hollenberg and Wilhelm (1984) have introduced a xylose-isomerase gene in *Saccharomyces cerevisiae,* thereby enabling the cells to convert xylose directly into xylulose. The isomerase gene was isolated from *Bacillus subtilis*.

The suitability of any anaerobic digester system for a particular situation may be judged in terms of the following important requirements (Bu' Lock, 1986):

1. Maximizing the solids loading capacity of the digester so as to handle the waste without added water.
2. Matching the residence times of the solid and liquid wastes to their differential biodegradabilities.
3. Ensuring maximum retention times and suitable optimal conditions for the microbes.
4. Matching the operating temperature to the available low-grade heat supplies, including the heat content of the incoming waste (such as stillage).

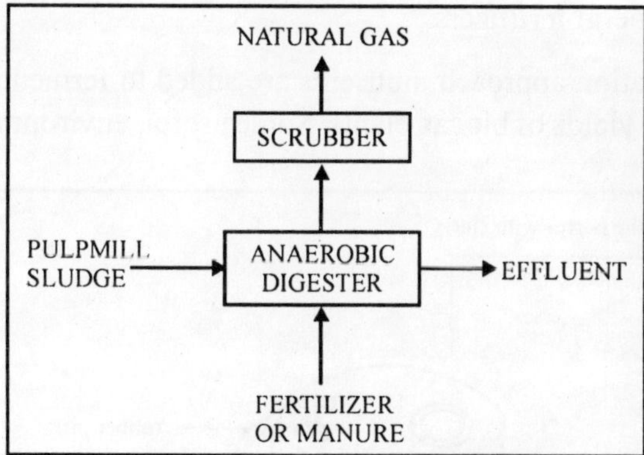

Diagram 17.18 Outline of process used for methane production from pulpmill sludge.

Technology

"Biogas" comprises a mixture of methane, carbon dioxide, hydrogen sulphide, and ammonia. These gases are produced during anaerobic digestion of organic wastes. The digestion is a two-stage process, each stage being catalyzed by a specific group of microbes. The acid-forming bacteria first break down the cellulosic material into simpler organic compounds such as acetic and propionic acids, CO_2, and some ammonia. In the second stage, the methane-forming bacteria break down these acids into methane and CO_2. A balanced cooperation between the acid formers and the biogas plant improves the efficiency of the biogas plant, for whose maintenance and efficient operation the following conditions must be maintained:

1. A proper temperature range, depending on the temperature tolerance of the acid-forming bacteria. Usually, it is 30-40°C (for mesophilic bacteria) and 50-60°C (for thermophiles).
2. A suitable pH, usually 6.6-7.5. The methane formers function best at pH 7-7.2. Lime may be added to buffer the system at this pH.
3. A solids concentration of about 10-12% in the slurry. This seems best for the digestion process.
4. Slow digestion under stagnant situations. Agitation or stirring increases the rate of digestion, leading to an increase in gas production.

Diagram 17.19 shows the design of a laboratory-scale biogas plant designed in Ghana.

Biogas for Municipal Planning

Biogas is not only of interest to individual users but is also important for holistic municipal planning. Biogas technology can benefit municipalities in various ways such as pollution avoidance, e.g., by reducing emissions of methane and ammonia, and the use of digested sludge as a substitute for chemical fertilizers. Groundwater pollution is also considerably reduced. For an individual farmer, biogas means less work and digested sludge not only hinders germination of weed seeds but also stimulates the growth of the crop plants to a greater extent than mineral fertilizers.

In the cofermentation approach, nutrients are added to fermentable substrates with a view to improving the yields of biogas plants. Some useful, environment-friendly additives

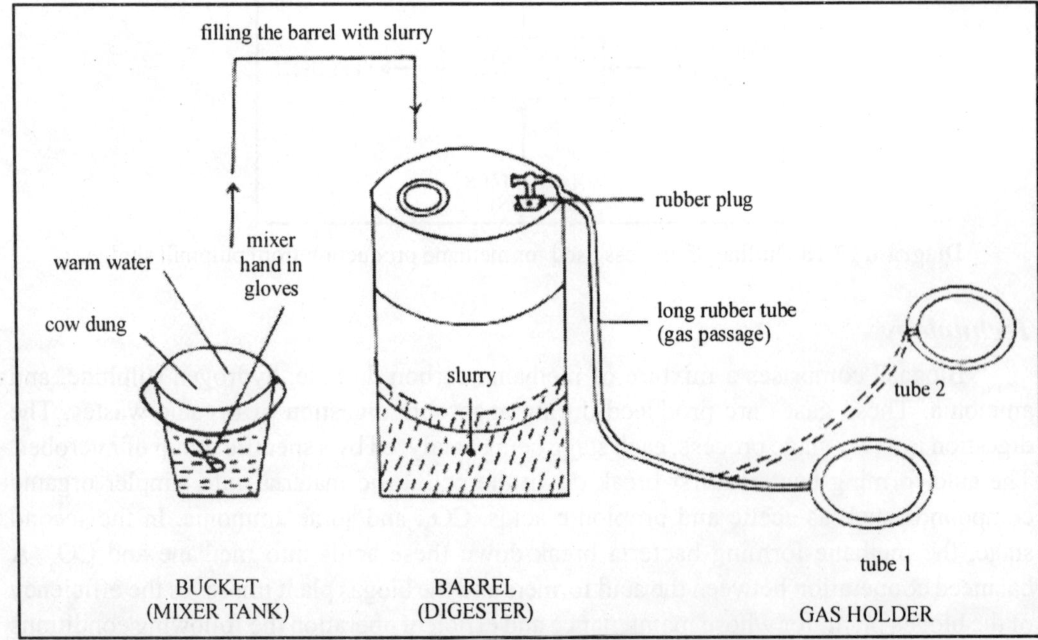

Diagram 17.19 Sketch of a batch biogas plant (after Abbam, 1985).

include lawn cuttings, maize silage, used cooking oil, brewery waste, and household waste. As the composting of household wastes involves heavy costs for municipalities, this alternative disposal via biogas plants is an attractive proposition.

Another useful approach is the solid manure technology in which plants with steel tanks and concrete slurry pits are used for housing solid manure and bedding straw with a view to obtaining high energy yields. As solid manure has much higher content of organic dry matter, the gas yield can often be tripled as compared to the yield from the more liquid dung-urine substrate whose organic matter percentage is comparatively lower. The solid components need to be thoroughly pulverized, first in the influent collecting tank and then in the digester.

Two types of biogas plant which optimize gas production using a stirrer are in vogue: the concrete pit plant (biogas storage plant, Diag. 17.20) and the steel tank plant (through-flow type, Diag. 17.21). With the concrete pit plant, a concrete liquid manure tank with a concrete cover is expanded to convert it into a biogas plant. Storage and digestion occur in the tank (Kellner and Neumann, 1992). The gas formed in the digesting chamber is collected in the chamber itself, in a bag made of plastic sheet. Also, open manure pits may be covered with double plastic sheet of which the outer, fabric-reinforced one imparts shape whereas the lower one rises and falls depending on gas generation or consumption (Diag. 17.20). This type of plant is more compact and cheaper than a steel tank plant.

Unlike the above storage-type plant, the steel tank plant has been used since long as a through-flow type (Diag. 17.21) even with problematic liquid manures. This plant has a horizontal steel tank with a paddle-type stirrer. The gas is stored in the tank. The plant copes up well with floating scum and sediment layers even with such problematic manures as liquefied solid manure with high straw content and pig manure. However, this plant requires much space and is limited by weather changes, etc.

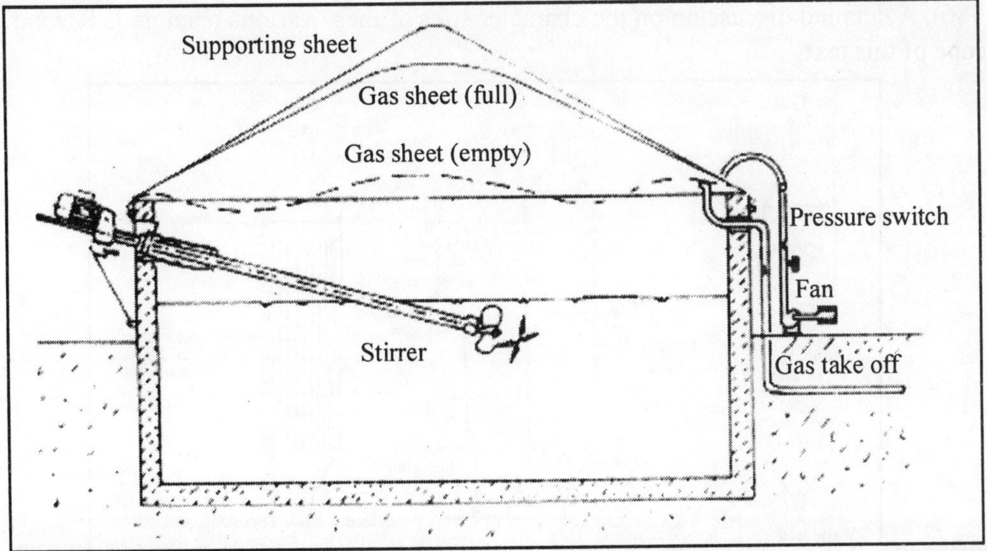

Diagram 17.20 Storage type biogas plant with double-skin pit cover and swivel-mounted gaslight stirrer (after Kellner and Neumann, 1992).

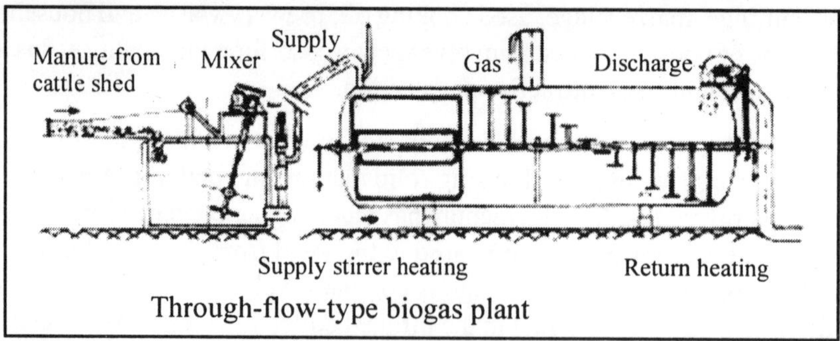

Diagram 17.21 A through-flow type biogas plant (Kellner and Neumann, 1992).

BIOGAS FROM WASTEWATER

During the last two decades; industrial wastewaters have been increasingly used in fermentation systems for the production of methane. In these systems, the methanogenic bacteria are retained within the bioreactor. The bacteria convert acetic acid into methane. Retention is achieved by flocculation and settling or by attachment to stationary support surfaces.

The rate of methane generation of a reactor is proportional to the concentration of organic material in the substrate solution, and the fraction of this material actually converted into methane. The rate is inversely proportional to the hydraulic residence time and the oxygen content of the organic material converted into methane.

Examples of the different types of advanced reactors used for methane production from wastewater are (1) anaerobic contact reactor, (2) anaerobic filter, (3) upflow anaerobic sludge bed reactor (Diag. 17.22 A), (4) anaerobic fluidized and expanded bed reactor (Diag. 17.22 B), and (5) downflow stationary fixed film reactor (Diag. 17.22 C) (Van den Berg, 1986). A detailed discussion on the characteristics of these various reactors is beyond the scope of this text.

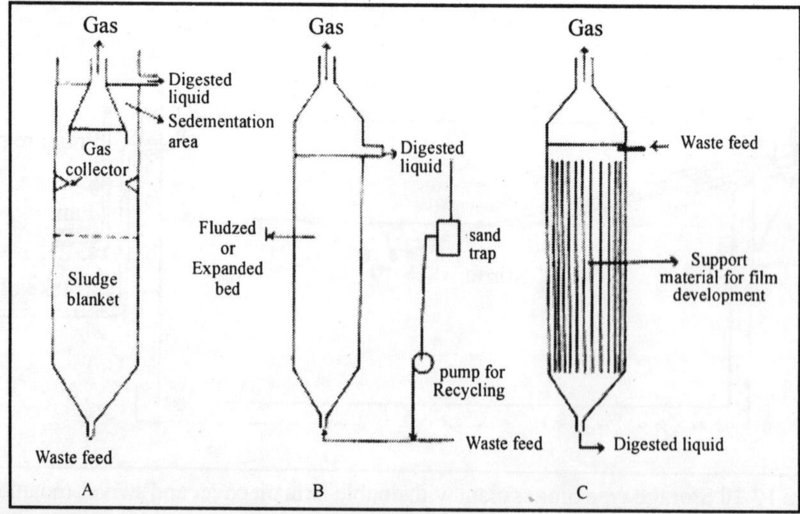

Diagram 17.22 Designs of the upflow anaerobic sludge bed reactor (after Van den Berg, 1986).

Producer Gas

Although ethanol has attracted much attention as a non-petroleum fuel for motor transport, there exist some alternatives that have a similar potential. These include hydrogen, methanol, and liquid fuel from coal. Vegetable oil and oil from tar sands and shale also have such potential. Producer gas is another such alternative.

Producer gas is generated from such solid fuels as wood, charcoal, coal, peat, and agricultural residues. It can be used to power internal combustion engines. It is made when a stream of air passes through a bed of glowing coal. This coal may come from the burning of wood, charcoal, coke, peat, or from waste materials such as corncobs, peanut shells, bagasse, straw, and paper.

The gas generation occurs in a gasifier which is a metal tank with a firebox, a grate, air inlets, and an outlet for the gas produced. Mainly carbon monoxide and hydrogen are generated. These gases are combustible when mixed with air. In the cylinder of a spark-ignition gasoline engine, the gas-air mixture ignites with spark plug. In diesel engines, however, producer gas does not ignite on its own, but it is possible to operate diesel equipment on producer gas. The latter is mixed with the combustion air and then a small volume of diesel fuel is injected into the cylinders to provide ignition.

The producer gas-generation equipment has four basic components, viz., (1) a generator, which makes gas from the solid fuel; (2) a cleaner, to filter soot and ash from the hot gas; (3) a cooler, to condense tars and other impurities; and (4) a valve, that mixes the gas with air; also, a throttle valve, to meter the mixture into the engine intake manifold.

The generator is typically a cylindrical or rectangular metal tank housing fuel, a firebox, and an ash pit. The fuel falls into the combustion chamber and keeps burning in the air blowing through this firebox. A red-hot bed of charcoal is thus produced. The three commonly used types of combustion chambers (Diagram 17.23) differ in the relative positions of the air inlet and gas outlet.

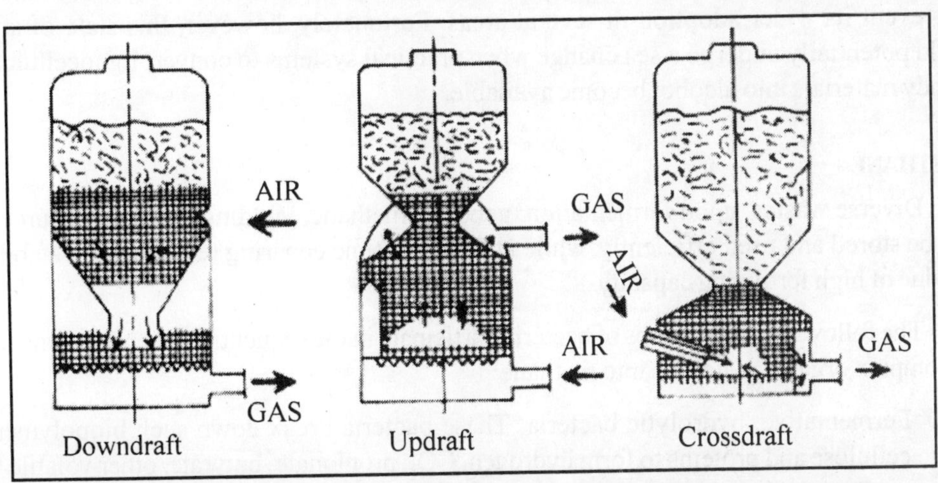

Diagram 17.23 Three types of generator used for producer gas generation.

Some advantages and disadvantages of the producer gas are as follows:

1. It is a practical and proven fuel.
2. Its generators are fairly simple to make.
3. It can be used to fuel cars, trucks, boats, trains, trolleys, and motorcycles.
4. It requires no significant modification of existing engine design.
5. The equipment required to produce it can use renewable fuels, e.g., biomass.

The limitations of the producer gas are:

1. It generates less power than petroleum (when used in an internal combustion engine).
2. Vehicles run on it are cumbersome and clumsy as the generator has to be carried on a trailer.
3. It is bulky and difficult to store and handle.
4. Its carbon monoxide component can be hazardous.
5. Extensive deforestation if excessive use of wood fuel or biomass is made to produce it. Also, the use of wood in gas generators means that much less wood is available to the poor for fuelwood use.

In many developing countries, the process of petroleum fuels has risen above the cost of biomass-based energy. This has stimulated the use of the cheaper biomass-based energy sources. An increased production of biomass is essential both to meet the energy requirements in developing countries and to check the menace of deforestation. For a balanced development, it is also necessary to give a greater attention to the relationships amongst biomass resource exhaustion, agriculture, and energy problems. Of the energy technologies currently available, agroforestry and improved charcoal production appear to be the most promising. Biogas generation requires integrated resource management, which is commonly found only in better organized communities, not among the poorest of the poor. As regards fuel alcohol, it has become clear that small-scale fuel alcohol production is not yet economically feasible. The potential competition for arable land associated with large-scale alcohol fuel production tends to prevent its wider adoption in several areas. Fortunately, however, this state of affairs could potentially undergo a sea change when practical systems to convert lignocellulose in woody materials into alcohol become available.

Methane

Diverse wastes, upon fermentation, generate methane, yielding an energy source that can be stored and used efficiently, while at the same time ensuring retention of a stabilized residue of high fertilizing capacity.

The following three groups of bacteria participate in the sequential anaerobic conversion of complex organic materials into methane:

1. Fermentative, hydrolytic bacteria: These bacteria break down such biopolymers as cellulose and proteins to form hydrogen, CO_2, propionate, butyrate, other volatile fatty acids, and ethanol.

2. Acetogenic bacteria: These bacteria convert most of the products stated in (1) into acetate.
3. Methanogenic bacteria: They use acetate and H_2/CO_2 or formate and H_2/CO_2 to produce methane.

Examples of category (1) include *Clostridium, Bacteroides, Ruminococcus, Escherichia coli,* and *Bacillus* spp. Those belonging to group (2) include *Syntrophomonas, Desulphovibrio,* and others. Examples of group (3) are *Methanococcus, Methanoth'rix,* and *Methanosarcina.* These and other methanogens can derive their energy from the production of methane.

A slurry of waste organic matter is fed to an enclosed biogas plant or digester in which the gas formed is trapped by an inverted drum which covers the surface of the liquid. As gas is produced, the drum rises, acting as a gas-storage chamber. From this chamber, the gas may be drawn off as per requirement. The gas contains about 60% methane, about 35% CO_2, and small amounts of H_2S, H_2, and N_2.

A large number of designs for biogas plants are now available and in use in several countries. Diagram 17.24 shows one designed by the Indian Agricultural Research Institute, New Delhi.

Water hyacinth *(Eicchornia crassipes)* is a highly-productive and widespread, notorious aquatic weed endowed with rapid proliferation in diverse kinds of water bodies. It is especially suited for methane generation because of its chemical composition (Widyanto *et al.*, 1979). Its carbon : nitrogen ratio, soft organic matter, high moisture and low lignin contents are all conducive to good biomethanation. The carbon : nitrogen ratio (29) is particularly well-suited for anaerobic digestion.

Experimentally, fermentation of cellulose to methane has been estimated to achieve efficiencies of up to about 75%, depending on loading concentrations and other factors (Daniels, 1984).

Methane can be generated by anaerobic digestion of pulpmill sludge which is either supplemented with fertilizer or manure. This process requires a cleaning or scrubbing operation in order to increase the quality of the biogas produced to the level of a regular natural, commercial gas supply.

METHANOL

Methanol provides an attractive alternative fuel to gasoline. It can be made from gas, coal, or wood. It can be stored and used in existing equipments. Up to 15% of methanol can be added to commercial gasoline in the car's engine. In fact, the methanol-gasoline mixture results in improved economy, lower exhaust temperature, reduced emissions, and better performance as compared to gasoline alone. Methanol is particularly well-suited for use in fuel cell's for producing electricity. Diagram 17.25 outlines the sources, distribution, and uses of methanol.

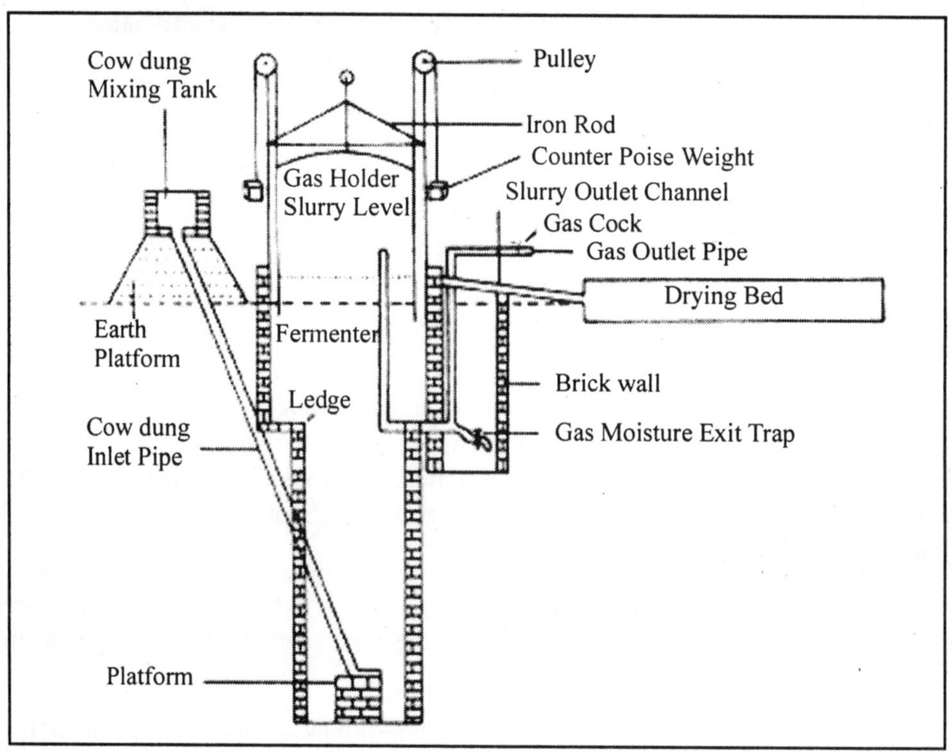

Diagram 17.24 Design of a gobar gas plant developed at Indian Agricultural Research Institute, New Delhi.

Methanol is manufactured from carbon monoxide and hydrogen, both of which can be obtained by incomplete oxidation of any carbonaceous fuel with oxygen or water. It is usually obtained from methane by partial oxidation with water. Methane gas is also produced biologically by the breakdown of natural wastes, refuse, garbage, pig and chicken manure, and sewage. Such methane can be used for powering automobiles. Diagram 17.26 shows the design of an oxygen refuse converter. In this, we can dispose off our municipal garbage while simultaneously generating useful energy.

Microbial Degradation and Conversion

Wastes rich in carbohydrates can be transformed by conventional microbial fermentations (Diag. 17.27) or by biotechnological processes involving industrial microbiology. Diagram 17.27 illustrates how microbial cellulolytic enzymes hydrolyze cellulose.

Maize, wheat, cassava, and potato are rich sources of starch. Upon hydrolysis (acid, or enzymatic), starch yields glucose and dextrins. The resulting glucose is used as substrate for production of ethanol by fermentation.

Hemicelluloses are associated with cellulose in the cell walls of plants, making up 10% of the wood mass of coniferous trees and about 20% of that of broad-leaved trees. Their content goes up to about 30% in straw and in maize cobs. Upon hydrolysis, hemicelluloses

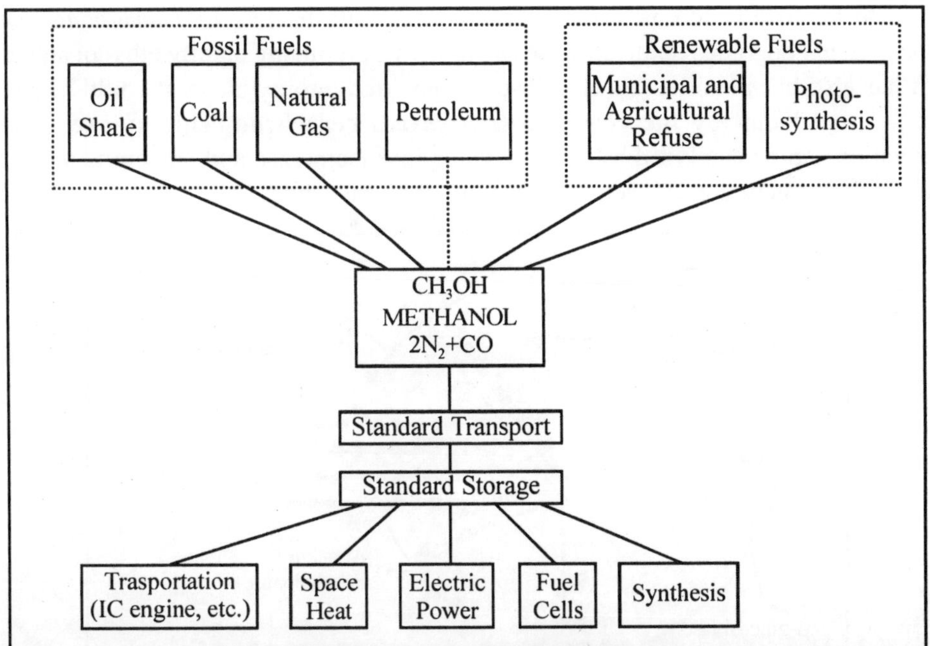

Diagram 17.25 Sources, transport, and applications of methanol (after Reed and Lerner, 1974).

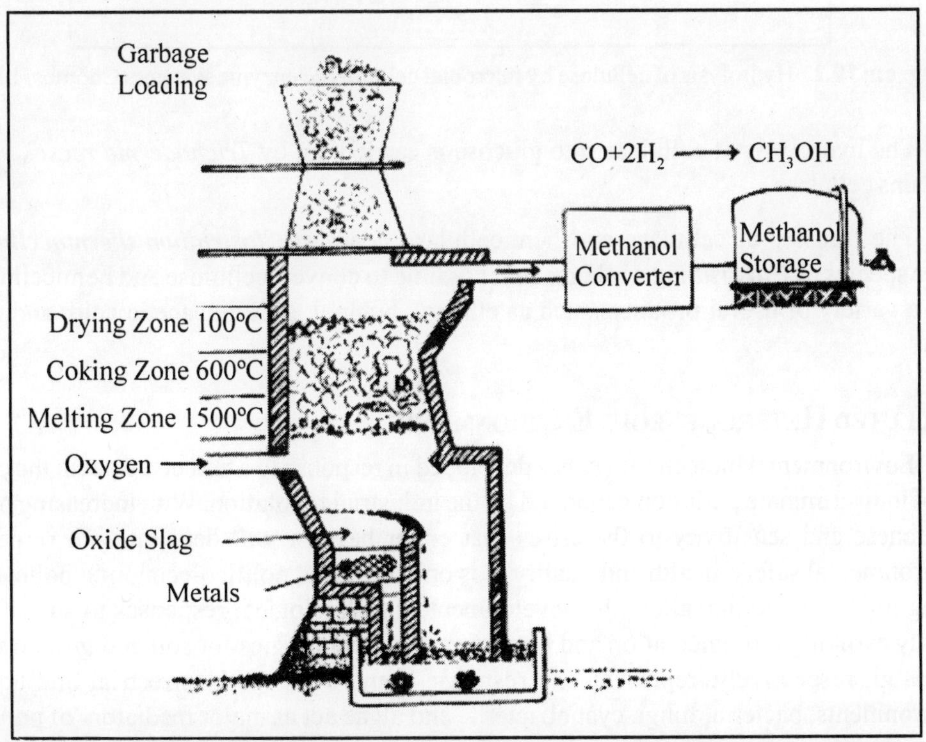

Diagram 17.26 Sketch of an oxygen refuse converter for conversion of municipal garbage into CO_2 and H_2 or methanol.

give rise to pentoses such as xylose. The thermophilic bacterium *Thermobacteroides saccharolyticum* degrades hemicelluloses at 40-42°C, producing chiefly ethanol and lactic acid. It has also been shown that if xylose is converted into xylulose, it could be readily fermented to ethanol by the yeast *Saccharomyces cerevisiae.* This xylose-xylulose isomerization can be catalyzed by the enzyme xylose isomerase and one could attempt to incorporate the 1 gene for xylose isomerase into the yeast cells.

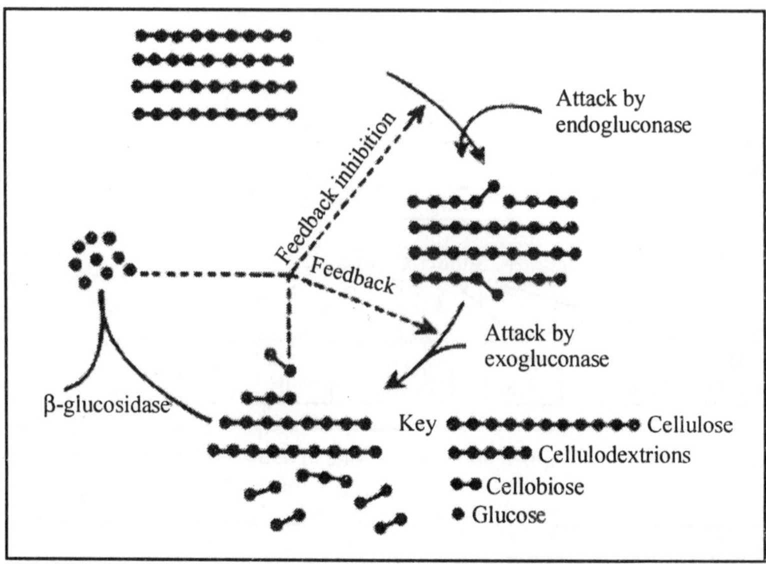

Diagram 17.27 Hydrolysis of cellulose by microbial cellulolytic enzyme systems (Coombs, 1987).

The hydrolysis of cellulose into glucose is carried out by *Trichoderma reesei,* which contains cellulase.

The transfer of cellulase and hemicellulase genes *rf Clostridium thermocellum* to other species of *Clostridium* could make it possible to convert cellulose and hemicelluloses into a variety of useful products such as ethanol, butanol, acetone, acetic acid, and lactic acid.

Polluted Heterogeneous Environments

Environmental biotechnology has developed in response to a perceived need: the clean-up of indiscriminate pollution generated by the industrial revolution. With increasing public awareness and sensitivity to the crises that either have or will develop with respect to environmental safety, health and quality, it is only now that politico-economic policies are being implemented that allow the development of technological responses to such crises. Newly evolving bioremediation and biorestoration technologies for soil and ground water clean-up, respectively, represent one response scenario. In both terrestrial and aquatic environments, bacteria, fungi, cyanobacteria, and algae act as major mediators of pollutant fluxes by changing driving forces between sources and sinks and by changing both the chemical and physical status of environments subjected to macroscale pollutant fluxes. The fundamental problems in assessing levels of mediation in any polluted environment arise from an inability to either characterize or quantify these communities on the one hand, and to differentiate between perturbed and non-perturbed states within such communities, on the other hand (Hamer and Heitzer, 1991).

Three of the most important constraints in the consideration of any microbial system are that (1) no microbial system is ever in equilibrium (steady state), but changes with respect to time; (2) gradients exist in all microbial systems thereby creating conditions at reaction sites that are entirely different from those measured by either probes inserted into or samples removed from systems; and (3) when the physiological states of microbes differ, their dynamic response to identical perturbations will also differ (Hamer, 1990). According to Hamer and Heitzer (1991): "Until the enunciation of the "Superfund Strategies" some five years ago (Hirschhorn, 1988) the major emphasis of pollution control for environmental quality maintenance for the previous 25 years had been efforts to reduce the effects of polluting discharges, particularly gaseous and liquid effluent streams, by reducing bulk pollutant concentrations in such streams.

This was in marked contrast to earlier philosophies where, on the assumption of a virtually unlimited capacity of receiving environments to accept pollutants, pollution management was, almost exclusively, based on dispersal and dilution in receiving environments rather than on treatment. Essentially, apparent amelioration by dilution was sought in order to avoid incurring costs for treatment. As far as dispersal and dilution policies were concerned, waste solids and sludges were largely ignored, as were residues from repetitive and frequently excessive applications of toxic chemicals to land.

Since long concentrated solid and semisolid wastes have been disposed in landfills. Erroneously, such dumping has been thought to imply immobilization. This ignored the well established precesses for pollutant transport and transfer between various environmental compartments. Waste dumping (disposal) and gross environmental abuse are by no means the prerogatives of primary industries concerned with natural resource exploitation, but have been widely performed by many secondary and tertiary industries, as well as by municipalities. In recent years, a critical point with respect to pollutant accumulation and adverse effects from pollutant transport has been reached, and it is no longer possible for the frequently surreptitious adverse effects of pollutant dumping and/or release to remain unnoticed and unattended.

Biodegradation is a well-known, but incompletely understood phenomenon. It occurs in virtually all aquatic and terrestrial environments. Qualitative methods are available for assessing both biodegradation and biodegradability, with particular reference to aquatic environments. It is usually not possible to ascertain pollutant biodegradation and mobilization quantitatively in complex matrices such as soils and sediments (Hamer and Heitzer, 1991).

Subsurface Heterogeneity

Both macroscopic and microscopic heterogeneities are clearly evident in subsurface environments. Typical soil microenvironments comprise three principal inorganic particulates: sand, silt, and clay; a physically-and chemically-diverse organic fraction; an aqueous phase containing unevenly distributed organic and inorganic matter; a gaseous phase of markedly different composition from the overlying atmosphere; and, of course, a diversity of procaryotic and eucaryotic organisms (Burns, 1980).

Because of their extensive surface areas and their ion exchange properties the soil clays are generally considered to be the most influential inorganic particles affecting biological activity. Sand and silt are comparatively inert in comparison to clays. Although they tend to retain neither water nor films of humic matter, they mediate both gas and water diffusion, as well as aggregate formation. Soil organic matter comprises several fractions: biotic debris, breakdown products, and varied range of colloidal and polymeric humic substances. Heavily polluted soils further have mixtures of organic and/or inorganic chemicals with which they have been polluted and their partial breakdown products (Hamer and Heitzer, 1991). The physical properties of the microbes present in soil are largely determined by their surface charge, but their metabolic activity is mostly determined by their physiological state and prevailing environmental conditions.

According to Marshall (1980), the habitats available to soil microbes are shaped by the structural organization of the various paniculate fractions present. Fluctuating wet and dry conditions influence the availability of organic matter, the degree of weathering, the leaching and redeposition of soluble materials and the translocation of smaller paniculate fractions, including colloids, in any soil.

All microbes require water for growth, but despite this they can grow in environments showing a wide range of water availabilities. Water availability is a complex function of both absorption and solution factors and is expressed either as water activity or water potential. When water availability is affected by absorption, the effect is said to be matric and when affected by solute interaction, is said to be osmotic. In simple systems where changes result because of changing solute concentrations, it can be treated conveniently by-considering water activity. In soils, where changes in temperature, solute concentrations and interactions between the volumetric water content and the geometrical and physico-chemical properties of solid matrices are involved, water potential forms a more convenient basis (Griffin and Luard, 1979) because it incorporates all the variations in matric potential derived from liquid-solid and liquid-gas interfaces involving capillarity, repulsion between charged colloidal particles and absorption by surfaces, membranes, and macromolecules. The water content of soil and the matric potential both depend on the pore size distribution. Water potential is a selective factor with respect to microbes in soils. For movement in a soil matrix unicellular microbes require a continuum of water-filled pores of requisite diameter to provide appropriate pathways, whereas most filamentous fungi have the ability to bridge air-filled pores and to penetrate semisolid matter. These facts have clear implications for soil systems where it is proposed to enhance natural microbial activity in order to accelerate pollutant biodegradation (Hamer and Heitzer, 1991).

Microbes and Environmental Clean-up

One of the means available for the treatment of contaminated environments is the diverse flora of the microbial world, which offers some hope for a reversal of environmental degradation. The effective cleanup of chemically-contaminated sites in the environment, notably water, soil, and sediments, requires a multidisciplinary effort. To achieve this, the environmental biotechnologist has a range of physical, chemical, and microbiological methods available as options to carry out the restoration of contaminated sites. The selection of the appropriate strategy depends on a detailed understanding of the abiotic and biotic characteristics of the contaminated system and a thorough understanding of the processes involved.

Biological treatment strategies have advantages over incineration and solvent extraction in being cheaper and in the possibility that formation of toxic compounds or derivatives is considerably reduced. Microorganisms are important in nutrient and element cycling in the biosphere and their biodegradative potential has been harnessed and to some extent optimized, for example, by the wastewater treatment industry.

Bioremediation

Bioremediation is the process whereby the degradation of polluting compounds occurs as a result of biochemical activity of organisms. Two options are available to utilize the biodegradative potential of microorganisms in polluted environments. These are:

1. *In situ* Biodegradation, whereby the activity of microorganisms already present in the particular environment is targeted. More effective biodegradation results from enhanced activity of such microbes, either by increasing their activity, e.g., by the addition of suitable additional nutrients which were otherwise limiting their activity, and/or by increasing their number so that their activity can be more effectively manifested in the particular system (Mason *et al.,* 1992).
2. Bioaugmentation, whereby specific microorganisms which possess a known specific biodegradative or other potential activity are introduced into the affected environment. The basis of this technique is to target the contaminating chemical pollutant with a microorganism able to degrade it. The strategy of introducing microorganisms into environments alien to those where they are usually found is surprisingly common in environmental biology and parallels can be seen, for example, in the use of biological pest control agents, in composting, and in biofertilizers.

In bioremediation, the contaminant is made an integral part of the microbial food chain. The end result is an organic rich, non-toxic compost. It is being applied for petroleum and coal derived wastes using a process in which microbial interaction and subsequent degradation of contaminant hydrocarbons is optimized and oil release techniques developed for improved oil recovery (Sheehy, 1992).

Sites contaminated with a variety of substances including petroleum and petroleum derivatives, coal tars, nitroaromatics, byproducts of agriculture and aquaculture, cyanides, pesticides and herbicides can be restored by bioremediation.

There is need for developing a comprehensive range of technical and scientific skills in bioremediation including isolation and selection of appropriate microorganisms, treatability studies and process simulation testwork at scales ranging from the laboratory to pilot plant.

Bioremediation technology can successfully be applied commercially for the treatment of contaminated soils, groundwater, sludges and ongoing waste streams. Bioremediation has also been used for several land based fuel spills.

Bioremediation uses naturally-occurring microorganisms to degrade organic chemical contaminants and can be applied to many industrial situations. The contaminants which can be treated include petroleum hydrocarbons, polyaromatic hydrocarbons, chlorinated phenols, petrochemicals and some pesticides such as phenoxy-herbicides. Bioremediation processes can operate in soil or bioreactors, in one of several configurations including land treatment, *in situ* bioremediation, soil slurry bioreactors and fixed film or stirred tank bioreactors.

Bioremediation for Marine Oil Spills

Among the many countermeasures that are available for use during marine oil spills, bioremediation has received increasing attention during recent years.

Biodegradation is the natural process whereby bacteria or other microorganisms alter and break down organic molecules into other substances, such as fatty acids and carbon dioxide. Bioremediation is the act of adding fertilizers or other materials to contaminated environments, such as oil spill sites, to accelerate the natural biodegradation process. Bioremediation is also used in other applications, including sewage treatment, terrestrial oil spills, and experimentally for hazardous wastes.

Three main types of bioremediation technologies are currently being developed or used for treatment of oil spills: (1) addition of nutrients to oiled shorelines, (2) addition of microbegto oiled shorelines, and (3) addition of nutrients and/or microbes to open water oil slicks. All three technologies attempt to accelerate biodegradation of oil.

Biodegradation is one of the main ways in which spilled oil is weathered. It occurs in most environments at varying rates, depending on localized environmental conditions and on the composition of the oil (for example, heavier oils are more resistant to biodegradation than lighter oils). Among the many environmental factors that will affect biodegradation rates, oxygen, nutrients, and temperature are probably the most important (Atlas, 1981; DeFlaun and Mayer, 1983).

Simply adding oil to an environment will stimulate growth of indigenous microbes, since the oil provides increased amounts of carbon, the microbes food source (Lee and Levy, 1991).- There is a lag period before indigenous microbial communities begin to degrade oil. This may fee due to the fact that oil is initially toxic to microbial organisms, and the most toxic fractions must be weathered before microbes can grow; this can take a period of several days to several weeks (Lee and Levy, 1989).

The primary processes of microbial degradation are aerobic though some anaerobic degradation may also occur. Low-energy, sheltered environments probably have the lowest rates of biodegradation, especially in subsurface sediments. Oil in anaerobic sediments in marshes or other environments may degrade very little, with oil persisting in some cases for several years (Lee and Levy, 1991). High-energy environments (more exposed, with greater impact from waves) usually show rapid biodegradation, in part because of physical weathering, but also because wave action supplies oxygen and nutrients to the microbial communities, facilitating biodegradation (Hoff. 1992).

Microbial populations that undergo rapid growth in the presence of spilled oil may become limited by inadequate amounts of nitrogen and phosphorus. Nutrients are more likely to be limiting to the biodegradation on oiled shorelines or oil slicks, than for degradation of suspended oil particles in the water column (Atlas, 1981).

At extremely high salinities, biodegradation is inhibited but this is not likely to be a problem in the normal range of salinities usually encountered in marine and coastal environments.

Types of Bioremediation

Nutrient addition: The theory behind bioremediation by nutrient addition is simple: microbes already living on an impacted shoreline have a sudden new source of food—carbon compounds in the spilled oil. After the initial toxicity of the oil decreases through evaporation of the volatile compounds and after indigenous species of hydrocarbon-degrading microbes become acclimated, they begin to break down the oil, and their population grows. At this point, the sudden increase in numbers of microbes may deplete existing supplies of nutrients and this may limit further growth of the microbial population. The microbial population can continue to increase with added nutrients, and degrade oil at a faster rate, than it could without the supplemental nutrients (Hoff, 1992).

This technique appears promising for use on oiled shorelines. The potential advantages of any bioremediation technique must be balanced against possible detrimental environmental effects, including introduction of contaminants, toxicity to aquatic organisms, and physical impacts. Some fertilizer products, whose primary use is in a terrestrial setting, may contain trace elements as micronutrients (e.g., copper or mercury) that would be introduced into an aquatic environment with potentially much more significant toxicological effects (Mearns, 1991). Others may produce byproducts such as ammonia and nitrates that may be toxic.

Nutrient addition can include a variety of application techniques as well as numerous commercial products, usually fertilizers. These products can be grouped into three basic categories: soluble inorganic nutrients, oleophilic formulations, and slow release formulations.

Inorganic nutrients include a wide variety of water-soluble garden or agricultural fertilizers that can be mixed with seawater and sprayed on shorelines. These fertilizers can be formulated with different ratios of nitrogen and phosphorus and usually include small quantities of trace elements. Some advantages of inorganic nutrients are that they are readily available, inexpensive, and usually consist of compounds with well-known properties. However, since these formulations are water-soluble, they may be washed off the shoreline by tidal action, requiring frequent, repeated applications.

Oleophilic formulations were developed to solve the problem of solutions washing off rocks or beaches, and to provide nutrients at the oil-water interface, where bacteria will be metabolizing the oil. Oleophilic products are chemically "sticky", adhere to oil on rocks or other substrates, and are designed to remain at the oil-water interface and to be readily accessible to oil-degrading microbes.

Slow-release formulations release quantities of nutrients over a longer period of time, and to remain in the area where they are applied. They include various products with mixes of nitrogen, phosphorus and other compounds, packaged in dissolvable capsules or briquettes.

Microbial Addition

Adding microbes to contaminated areas, also known as "seeding", is conducted with the aim of enhancing biodegradation of an oil-impacted area with selected strains of microbes that are capable of degrading hydrocarbons. However, the effectiveness of this technique is not well supported in the scientific literature (Atlas, 1981). In fact, studies indicate that addition of microbes to an open environment may not increase biodegradation because "foreign" strains of bacteria are frequently out-competed by indigenous species, and thus disappear quickly from the microbial community (Lee and Levy, 1989).

No strain of bacteria, whether indigenous or from an outside source, is likely to degrade oil actively until after the most toxic components of the oil have evaporated. Therefore, claims of "instant success" from products containing microbes should be regarded with scepticism (Hoff, 1992).

As in 1991 no genetically-engineered microorganisms are being considered for use in bioremediation.

Open-Water Bioremediation

To date, no studies appear to have evaluated use of bioremediation in an open ocean situation.

Biodegradation in the water probably occurs at the water surface. Therefore, any product or nutrient added would need to stay at this interface and follow the oil slick as it moves. For bioremediation to be successful on open water, the nutrients or microbes would have to remain with the oil slick for the time it takes microbes to become acclimated to the oil and begin biodegrading. There have been concerns at potential toxicity in application of nutrient additions and microbes in open water. The dilution factor is likely to be much greater on open water, however, and this is likely to lessen the risk from direct toxic effects.

Monitoring

There is no single measure that will accurately measure effectiveness or toxicity of a bioremediation application. Most of the larger bioremediation monitoring programs that have been undertaken (Table 17.12) have used a combination of the techniques discussed, depending on their specific concerns and objectives (Prince *et al.,* 1990). As a minimum, a monitoring plan at a bioremediation field test or application should include at least the following end points (Hoff, 1992):

1. To measure effectiveness, track changes in indicator hydrocarbon compounds using gas chromatography/mass spectroscopy (GC/MS). As a minimum, collect samples at the beginning and end of the application period at control and treated sites.
2. Conduct toxicity testing using bioassays to determine acute and/or chronic toxicity to aquatic organisms. Include sediment bioassays if the bioremediation compounds are likely to lodge in sediments. Sample at treated and control sites.
3. Monitor environmental impact on aquatic habitats through chemical analysis of sediments or water for potentially toxic compounds (such as heavy metals) that may be, part of a bioremediation product. Collect samples at the beginning and end of the application period at control and treated sites (Hoff, 1992).

Genetically-Engineered Microbes for Bioremediation

There has been much debate concerning the potential to apply genetically-engineered microorganisms into the environment. It is often very difficult to isolate a single microorganism able to completely mineralize a particular pollutant to CO_2 and H_2O. It is more likely that the necessary enzymes to achieve the breakdown of a compound are found distributed in the metabolic potential of different microorganism (Mason *et al.,* 1992). One proposal that has been made is to pool all of the biodegradative enzymes of interest together into a single "superbug". Complete biodegradation can then be achieved by engineering the biochemical capability of a particular carrier microorganism. Many of the genes which encode the necessary enzymes for biodegradation of some of the more persistent pollutants, are not found on the bacterial chromosome but are carried on plasmids. These plasmids are much more mobile than the main chromosome and their natural transfer is a regular feature of genetic exchange in the environment. Various natural mechanisms also exist by which chromosomal genes can be transferred in the natural environment. The current technological status in molecular genetics has rendered these genes readily amenable to laboratory manipulation, and transfer and construction of strains of bacteria with additional biodegradative phenotypes is feasible. Genetically-engineered microorganisms (GEMs) are already being tested in several countries to assess their potential for enhanced bioremediation (Mason *et al.,* 1992).

Microorganisms, Consortia, and Communities

Microorganisms are rarely, if ever, found in pure cultures. Microbial community structure in the natural environment is complex and unfortunately little understood. During the last few years the nature of community reactions has become a focal point of research. The reasons for this are that biodegradation and nutrient cycling reactions are mediated by interacting groups of highly specialized microorganisms operating together in concert to carry out the microbial transformations observed on the macroscopic level (Hamer and Heitzer, 1991). The biochemical flexibility of a microbial community in a particular environment arises from the interactions among the various microbial consortia present. That biodegradation of contaminated environments should be a community level process rather than exclusively concerning the ability of an individual microorganism is both advantageous and

Table 17.12 Some bioremediation case histories and treatment history monitoring (after Hoff, 1992)

Incident	Location/ Substrate	Type of oil	Type of bioreme-diation	Products	Days/Hours monitored	End points measured	Application effective?
Exxon Valdez	Prince William Sound, Alaska	Prudhoe	Fertilizer	Inipol	1989: 99 days	Microbial counts, respirometry	Yes, partially
	Shorelines	Bay crude		Customblen	1990: 55 days	Water quality, bioassays	
Prall's Island	Arthur Kill, New Jersey	Fuel oil	Fertilizer	Customblen	92 days	Water quality, microbial counts	Inconclusive
	Gravel beach						
Apex Barges	Galveston Bay Texas	Partially refined	Microbial	Alpha BioSea with Miracle-Gro	9 days	Bioassays (acute)	No
	Marsh					Per cent oil in mousse Fatty acids	
Mega Borg	Gulf of Mexico	Angolan crude	Microbial	Alpha BioSea	7 hours	Bioassays (acute)	Inconclusive
	Open water					Per cent oil in mousse	

disadvantageous. From the negative perspective it is unfortunate that, under most conditions, no single microorganism can be added for achieving complete mineralization. In nature, the need for a single microorganism able to biodegrade a wide spectrum of problem pollutants (the so-called "superbug") is an unrealistic prospect: in nature, the biochemical diversities of microbial communities are quite remarkable. When conditions in a particular environment change slightly, then the microbial community is able to accommodate this change with either slight or dramatic shifts in the balance between different species of the community. If a localized environment is severely contaminated with a particular substance that has been shown to be biodegradable in laboratory studies, this implies one of three possible causes: (I) that the substrate is unavailable in the system due to either physical partitioning into an inaccessible site or because of chemical interactions binding the substrate to the surface, (2) that a key stage in the biodegradation process is missing from the biochemical potential of the indigenous community, or (3) that the appropriate organisms have become restricted as a result of the lack or limitation of some essential nutrient. This explains why natural *in situ* bioremediation might not work without first amending the nutrient requirements of the indigenous population and how direct manipulation of the community can allow a change in the microbial balance thereby enabling biodegradation to occur through the concerted effort of new microbial consortia (Mason *et. al.,* 1992).

Survival and Stability

Irrespective of the nature of its origin, the fact that a particular microorganism can be selected for or maybe engineered is only the first step in the overall bioremediation process. The second stage is to introduce that organism into the new environment in such a way that (1) it is able to establish itself without being out-competed, and (2) will remain active in the environment. There is already some information available concerning the survival and maintenance of introduced selected bacterial strains. In several instances it has been shown that, particularly in environments with low levels of hazardous pollutants, the introduced strain is unable to establish itself unless introduced at very high concentrations (Ramadan *et al*., 1990). This is frequently an unrealistic strategy. The main reasons for the failure of successful establishment include predation by protozoa, infection by bacteriophage or non-availability of nutrients as a result of competition by other microorganisms with higher substrate/ nutrient affinities. This scenario is particularly applicable to contaminated lakes and aquifers. Generalizations are difficult partly due to the enormous spectrum of microorganisms both from the points of view of taxonomic and biochemical diversity as well as a result of the different physical and chemical parameters which prevail in different environments and their influence on survival characteristics. How introduced strains interact with indigenous strains particularly with respect to genetic exchange and how the structure and function of the consortia are affected is also poorly understood.

Molecular Ecology and Detection Technology

Molecular ecology is concerned with the molecular mechanisms involved in the interaction between a microorganism and its environment. This entails study of what regulates the expression of particular genes and how microorganisms respond, at the genetic level upon encountering unfavourable physical conditions or when they are exposed to chemical pollutants.

Molecular ecology is also concerned with the development and application of new techniques for the more precise enumeration of specific bacteria even in a background of a diverse range of numerous other species. It is now possible to carefully monitor the fate of a particular microorganism much more precisely than hitherto. One important feature of the newly developed methods for molecular ecology is that they are based not only at the level of the whole microorganism but also at the level of single genes and therefore allow one to follow the fate of particular genes in a system; these are essential in the context of the question of environmental release (Mason *et al.,* 1992).

Environmental Complexity

Polluted environments can be broadly categorized into three groups: (1) those polluted by complex mixtures, e.g., crude oil, fuel oil, kerosene, coal tars; (2) those polluted by simple mixtures, e.g., simple solvents, explosives; and (3) those polluted by diverse wastes and ill-defined/modified chemicals, e.g., fire-damaged products. General solutions are possible for the first two categories but the complex and uncertain nature of the third category imposes the requirement to examine each on an individual basis. It is therefore not surprising that the pure culture monosubstrate approach cannot be applied in practice. Thus the problem facing the environmental biotechnologist is not only to develop strategies for successful application of a single microbial species but to define the mechanisms by which introduced populations consisting of several stable consortia will interact and compete in the natural environment (Mason *et ai, 1992).*

Bioaugmentation Strategies

Oil is one common pollutant which has both quantitative and qualitative effects on the natural environment. Permanent damage to many biological processes can result from contamination with some of the hydrocarbons in oil resulting in mutagenesis even when present at low concentrations. Complex transportation networks are a major source for oil pollution.

Bioaugmentation programmes have been employed for the purification of contaminated coastal areas and in wastewater treatment plants. In many instances the results have not been very good. However, it is important that these results are not interpreted so as to imply that bioaugmentation is an inefficient process.

Any uncontrolled addition of microorganisms to natural systems is unlikely to result in the achievement of the treatment objective. When sufficient knowledge and careful control of both the system and the system parameters are available, bioaugmentation can be a highly effective tool to combat environmental contamination problems (Mason *et al.,* 1992). How efficiently a specialized microorganism will carry out its desired function in a natural system depends on the local environmental conditions and on the process control. For the success of a bioaugmentation program it is necessary that the introduced microorganisms can establish themselves on the one hand, and on the other hand the metabolic pathway(s) for the biodegradation of the problem compound(s) must be active under the prevailing environmental conditions. One particular strategy for bioaugmentation uses the technique of immobilization. The immobilized cells of microorganisms are quite active and beneficial as far as the biodegradative capacity is concerned. In addition, an immobilized system need not be subject to washout and competition events and is now an area of increased interest for treatment programmes (Mason *et al.,* 1992).

USE OF MICROORGANISMS IN POLLUTION CONTROL

Sewage treatment is carried out by aerobic and anaerobic microorganisms. The effluents from yeast, oil. and cider breweries, from dairies, and from potato starch factories can be efficiently processed by an anaerobic process in which the active biological compound is recycled and which produces less residual sludge as well as less offensive odours.

Bacteria, algae, protozoa, and other microbes cooperate in sewage ponds to break down the organic wastes. Diagram 17.28 shows some generalized removal curves for biochemical oxygen demand (BOD), helminth eggs, excreted bacteria, and viruses *in* waste stabilization ponds at temperatures above 20°C. Diagram 17.29 and 17.30 show two sludge digesters, one standard rate and the other high rate.

Strains of *Pseudomonas* contain certain enzymes which can degrade diverse hydrocarbon molecules and toxic aromatic compounds such as benzene, toluene, and xylene. *P. putida* has plasmids carrying the genes coding for such enzymes. Four such plasmids are (1) OCI (degrades octane, hexane, decane); (2) XYL (degrades xylene, toluene); (3) CAM (degrades camphor); and (4) NAH (degrades naphthalene). The last two are self-transmissible by promoting mating between the bacterial cells. By hybridization, a composite strain containing the XYL and NAH plasmids, and parts of CAM and OCT was produced and found to be capable of growing on crude oil.

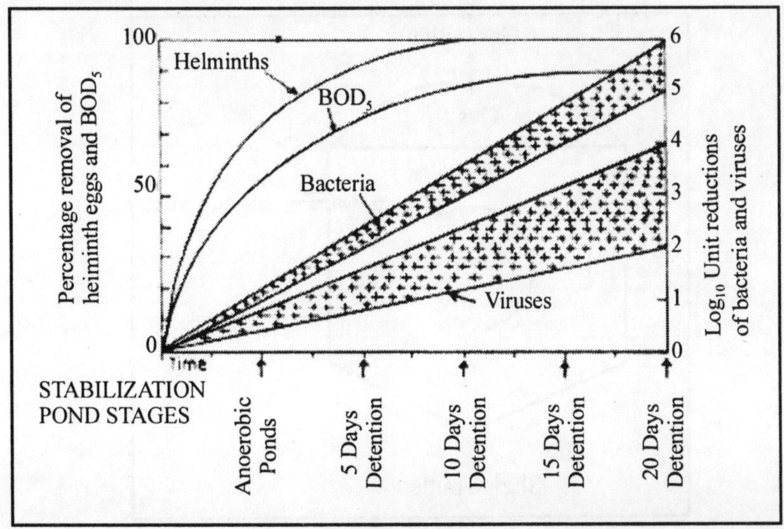

Diagram 17.28 Generalized removal curves for BOD in waste ponds (after Shuval *et al.,* 1985).

Certain microbes, which cannot by themselves degrade a given molecule, often modify the molecule in such a way that it will then be degraded by other microbes. This is known as cometabolism. The powerful insecticide parathion is decomposed by cometabolism *of Pseudomonas aeruginose* and *P. stutzeri.*

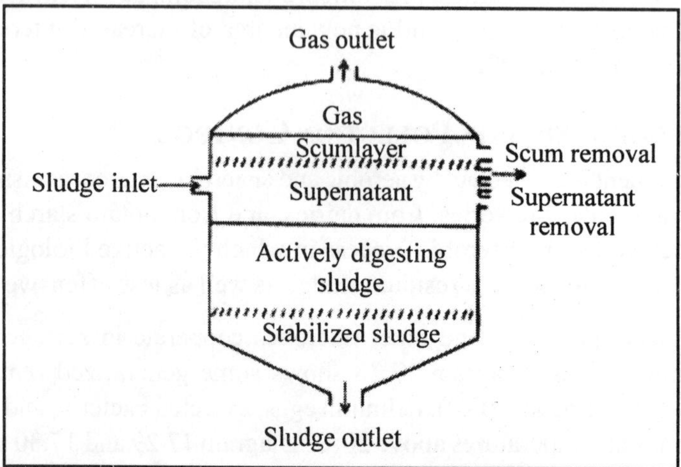

Diagram 17.29 Standard rate (stratified) digester.

Vermicomposting

A variety of rural wastes are available in our country but their nutrient content is usually low and they mineralize rather poorly. The nutrient status of these wastes can be improved through vermicomposting brought about by the activities of earthworms. The earthworms ingest the soil, partially break down the organic matter, mix the two fractions, and then eject the digested matter which is rich in nitrogenous compounds.

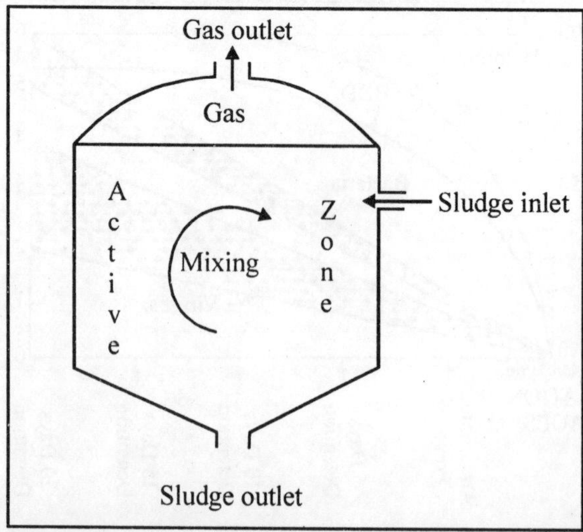

Diagram 17.30 High rate (complete mix digester).

HYDROGEN

Many microalgae and cyanobacteria can metabolize molecular hydrogen. The nitrogenase of cyanobacteria is not only involved in the conversion of N_2 into NH_3 but also reduces protons to hydrogen. Many nitrogen-fixing cyanobacteria have active hydrogenases which recycle the hydrogen gas, thus preventing it from being released in free form. However, if the hydrogenase action is inhibited by some means, then the hydrogen is released. When the cyanobacterial cells are incubated in atmospheres depleted in oxygen and nitrogen, their nitrogenase reduces protons as the exclusive substrate, and the hydrogen gas is evolved.

Diagram 17.31 brings out some likely phases of development of hydrogen technology and solar technology in the coming decades.

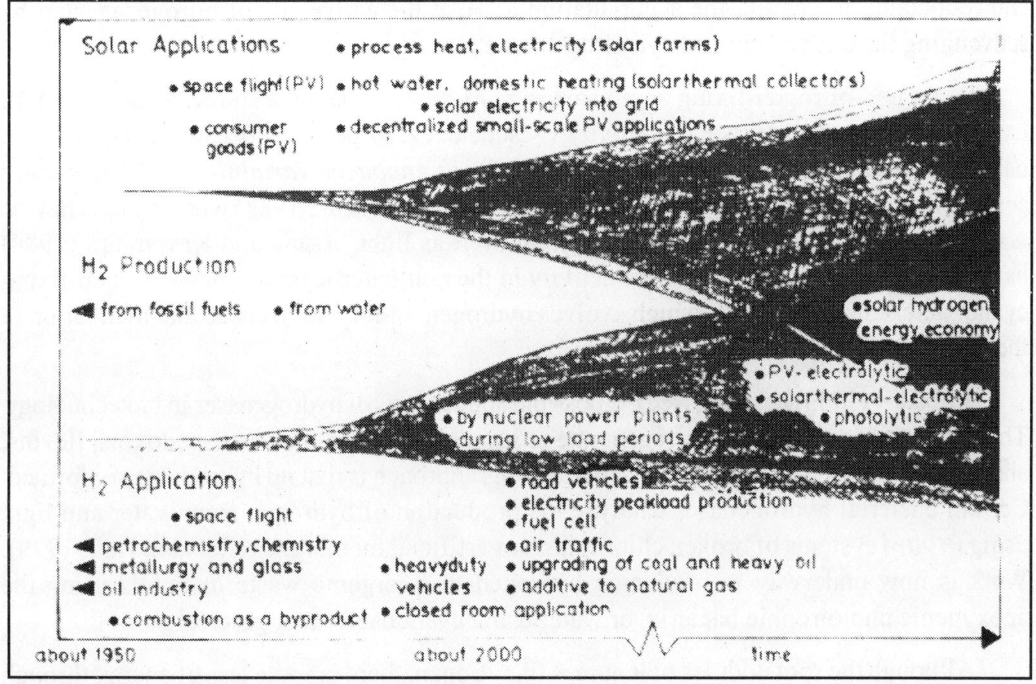

Diagram 17.31 The development phases of solar and hydrogen technology in future decades.

Hydrogen gas is a renewable, clean or non-polluting fuel (Diag. 17.31). Some green algae such as *Scenedesmus* and *Chlorella* also evolve hydrogen; such hydrogen evolution is inhibited by oxygen because the process is dependent on a hydrogenase which is sensitive to oxygen. Cyanobacterial systems are better because here the hydrogen evolution is dependent on nitrogenase which is relatively less sensitive to oxygen as compared to hydrogenase; another advantage of most cyanobacteria is that in their heterocysts the oxygen is quite low or absent. In the heterocystous cyanobacteria, the nitrogenase activity is maximal in the light but these cells also often show some activity in the dark (Asada *et al.*, 1985). In both light and dark, most reductant for nitrogenase comes from the oxidative pentose phosphate pathway (Bothe *et al.*, 1984).

Light-dependent hydrogen evolution by the nitrogen-fixing cyanobacteria can continue for several days (Asada *et al.,* 1985) in some strains of *Anabaena.*

Hydrogen release is an energy-wasting process (so far as the producing organism is concerned) and hence its significance to the alga must be as an electron sink. It provides a means for removing reducing equivalents when they are produced in excess of the organism's needs.

Many cyanobacteria can consume hydrogen gas via hydrogenase when it is supplied exogenously; hydrogen produced by nitrogenase is usually recycled instead of being released in free (storable) form (Lambert and Smith, 1981; Houchins, 1984). Such recycling of H_2 is thought to enhance the efficiency of nitrogen fixation (and hence growth) by recouping ATP and reductant, preventing the accumulation of H_2 at the active site of nitrogenase and by scavenging the oxygen via an oxyhydrogen reaction.

Some non-nitrogen-fixing organisms, e.g., *Microcystis* (Asada and Kawamura, 1984), have also retained hydrogenase activity (Peschek, 1979) possibly because they occur in habitats in which H_2 is present. In vegetative cells *of Anabaena variabilis,* the hydrogenase activity may be subject to a kind of "redox control" (Spiller *et al.,* 1983) whereby the enzyme is only activated upon removal of oxygen as well as light. Asada and Kawamura (1984) have demonstrated the hydrogenase activity in the non-heterocystous, non-nitrogen-fixing cyanobacterium *Microcystis* which evolves hydrogen rapidly upon anaerobic incubation in the dark.

Klibanov (1983) has reviewed the possible applications of hydrogenases in biotechnology. These enzymes may be used to regenerate reduced coenzymes such as ferredoxins, flavins, and nicotinamide dinucleotides. The merit *of* this approach is that no byproducts are formed. Certain bacterial hydrogenases catalyze the production of hydrogen from water and light using *in vitro* systems of broken chloroplasts or artificial membranes (Rao and Hall, 1979). Work is now underway in producing hydrogen from organic waste materials using the anoxygenic phototrophic bacteria, or water using cyanobacteria or green algae.

Although the root nodules of legumes fix nitrogen, there is some less of energy through hydrogen evolution from these nodules'. In fact, the H_2 evolution from many legumes may represent a significant proportion of the energy supplied to the nitrogenase system. This fact has necessitated studies of the relationship between H_2 metabolism and N_2 fixation in nodulated symbionts.

H_2 is a specific inhibitor of nitrogen fixation in clover plants, and some nodules on the roots of pea show the hydrogenase activity. This hydrogen seems to be produced via the N_2-fixing system. The nitrogenase enzyme requires a source of reductant for its activity and this reductant comes from the oxidation of carbon compounds. The ATP needed for the nitrogenase action is supplied by oxidative phosphorylation in the case of such aerobic organisms as *Rhizobium.* Diagram 17.32 illustrates the *in vitro* nitrogenase-catalyzed reactions under three different gases, namely, (1) N_2 (both protons and N_2 are reduced), (2) $N_2 + C_2H_2$ or argon + C_2H_2 (C_2H_2 is reduced to ethylene), and (3) argon (protons are reduced to H_2).

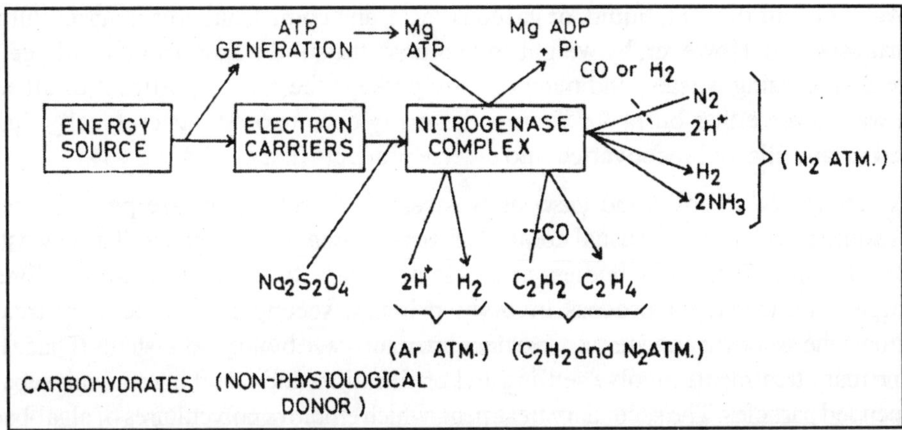

Diagram 17.32 Energy requirements for nitrogenase-catalyzed reactions.

The three types of H_2 reactions that can occur in the nitrogen-fixing organisms are (1) ATP-dependent H_2 evolution catalyzed by nitrogenase; (2) reaction between two protons and two electrons catalyzed by classical or reversible hydrogenase, yielding molecular H_2 (e.g., *Clostridium);* and (3) H_2 uptake catalyzed by an uptake hydrogenase; this uptake hydrogenase occurs in several cyanobacteria, *Azotobacter,* and legume root nodules.

Biomonitoring of Wastewater Toxics

The two basic biological assessment approaches used in abatement of wastewater toxicants are toxicity testing and biomonitoring. The toxicity test is a standardized laboratory method of measuring the response of aquatic organisms exposed to samples of wastewater. The test organism responses are typically classified into acute (i.e., quick or rapid effect) or chronic (slow, longer-term effect). The acute toxicity procedure commonly used is the 96-hour rainbow trout (a fish) lethality test; the endpoint of an acute toxicity test is death of the test organism. In contrast, the chronic toxicity tests usually measure some non-lethal but damaging effects such as fish reproduction failure and reduced growth. The duration of chronic tests is often longer than 96 hours.

Biomonitoring involves the use of aquatic organisms in assessing the impact of wastewater discharges on rivers or lakes. Unlike the toxicity tests, biomonitoring is conducted in the environment, not in the laboratory. Exposing fish (placed in cages) at various distances downstream of a wastewater outfall or collecting minnows or clams from the vicinity of a wastewater discharge for contaminant uptake analysis are good examples of biomonitoring.

WASTE TREATMENT

As waste materials usually occur in highly-diversified and time-variable compositions, their treatment is a challenging task. Waste treatment processes are usually operated in a non-aseptic environment, with mixed cultures of diverse microorganisms. It is a challenging task to stabilize the system and design a process that can ensure the required degree of reliability and product specification (Humphrey, 1986).

Wastes occur in solid, liquid, or gaseous form, and come from urban, agricultural, and industrial sources. However, by weight and volume, the solid wastes (urban garbage, crops and food processing wastes, and manures) are perhaps the most significant of all wastes. These wastes are either burnt for energy, thermally decomposed, anaerobically digested, dumped in landfills, or bioconverted into diverse products (Diag. 17.33).

As compared to solid and gaseous wastes, the liquid wastes are perhaps the most troublesome in view of their usual content of non-retractable substances. These wastes are usually discharged in water bodies, and it is necessary to treat the wastes before such discharge. The treatment process involves primary, secondary, and tertiary treatment operations; the secondary and tertiary portions are multistage biological systems (Diag. 17.34). Most primary treatments involve settling and/or filtration, and result in removal/ separation of suspended particles. The secondary treatment, which employs polycultures of algal-bacterial species, results in reduction of the BOD of the wastewater. The tertiary treatment phase is designed for denitrification or phosphate removal. The impact of the newer advances in biotechnology is likely to be felt mostly on the secondary and tertiary treatment phases.

The sketch of a bench-scale reactor for treatment of sludge is shown in Diag. 17.35.

Most biodegradable wastes are of organic origin and are largely cellulosic. However, non-cellulosic wastes are by no means insignificant. In fact, quantitatively, the non-cellulosic wastes are quite significant in many situations, and qualitatively they are extremely heterogeneous.

Returning Wastes to Land

The natural cycling of organic wastes involves nutrient absorption by plants, eating of plants by animals and man, and the return of the nutrients (as waste products) to soil. Human activities during the last few decades have altered this cycle materially, and the organic wastes, instead of being returned to soil, are now usually diverted to aquatic bodies such as rivers and lakes. Since only a small portion of the human food supply comes from aquatic life, it is much more prudent to return the nutrients to the land, thereby ensuring continued fertility and productivity of the soil.

Solid Wastes

Solid wastes are usually treated by composting, before being discharged onto land. Composting is an aerobic biological process which converts agricultural and human wastes into humus; humus is highly effective for soil improvement.

The conversion of wastes into humus is brought about by the cooperative interactions and activities of diverse soil organisms (Diag. 17.36), including microorganisms, invertebrates, and others. Diagram 17.37 outlines the composting process.

A simple form of compost pile can be made by alternately placing layers of soil, manure, and plant residues in a crude bin. Depending on the prevailing temperature, the pile heats up from microbial action within a few days. The composting process can be stimulated or accelerated by periodical wetting and turning, and is usually completed within 2-3 months.

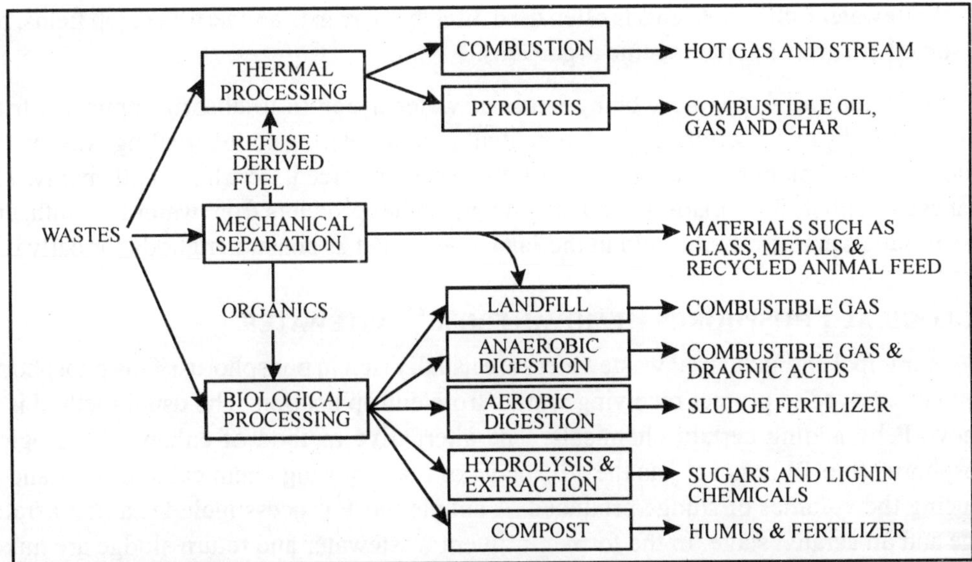

Diagram 17.33 General waste treatment processes (Humphrey, 1986).

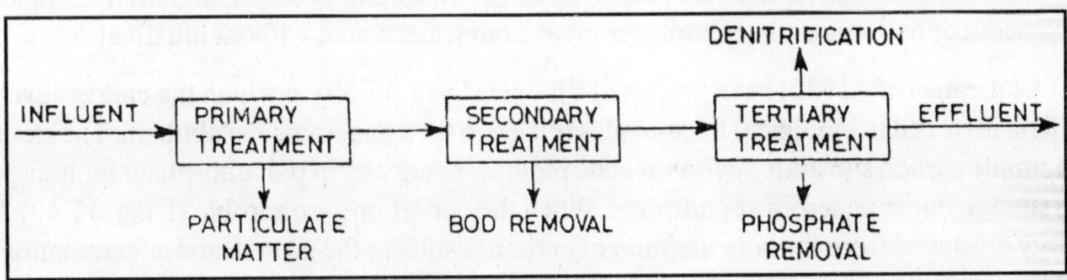

Diagram 17.34 The three stages involved in an aqueous waste treatment system.

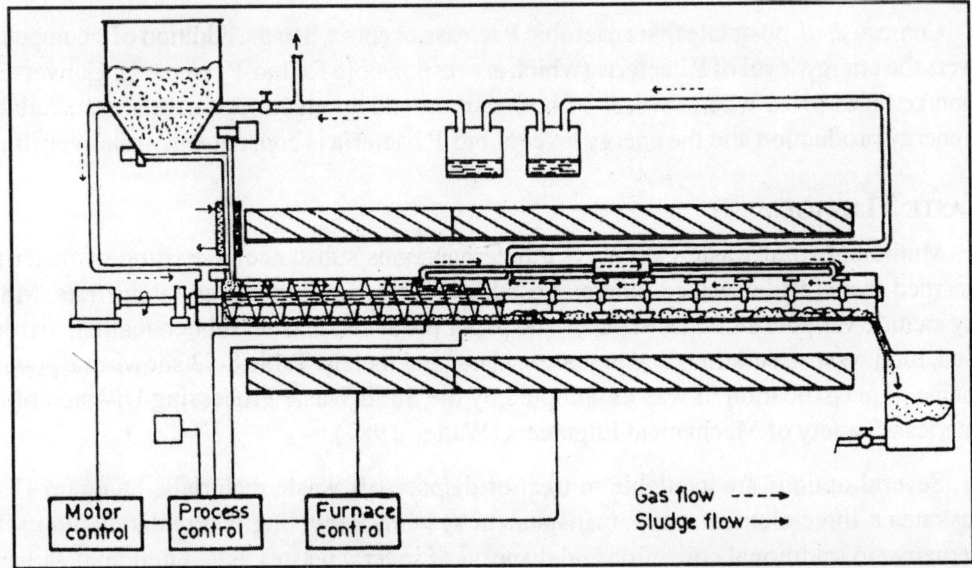

Diagram 17.35 Sketch of a bench-scale reactor (Campbell and Bridle, 1988).

Wastewater rich in nutrients is often used directly to irrigate and fertilize crop fields, but can sometimes transmit pathogenic organisms.

An increasing use is now being made of water hyacinth to absorb impurities from water. Water hyacinth offers a great potential as a simple means of treating wastewater while producing plant biomass that can be converted into feed,, fertilizer, and energy. The Japanese scientists have made alcohol and diverse other products from water hyacinth, and these products are now being sold in the Japanese market as consumer goods for daily use.

Biological Phosphorus Removal from Wastewater

Municipal and industrial wastewaters are usually rich in phosphorus. This phosphorus must be removed to protect receiving waters from eutrophication. The usual method is to remove P by adding certain chemicals. The alternative method of enhanced biological phosphorus (bio-P) removal has the advantage of not requiring chemical additions and of reducing the volumes of sludge produced. A simple bio-P process includes a non-aerated stage and an aerated stage. In the former, influent wastewater and return sludge are mixed under non-aerated conditions, leading to the release of phosphorus from the biomass. In the second stage, nitrification and phosphorus uptake from solution occur. Diagram 17.38 shows the process of biological phosphorus removal from wastewater, without nitrification.

Comeau *el al.* (1985) have proposed a model (Diag. 17.39) in which the energy level of bacteria involved in biological P removal is related to their anaerobic metabolism. The storage of a simple carbon substrate such as acetate requires energy from polyphosphate for transport and storage under anaerobic conditions. When the conditions are aerobic (Diag. 17.40), the energy produced from the consumption of carbon results in the growth and accumulation of P by the phosphorus bacteria. Another alternative possibility involves the use of nitrate by denitrifying phosphorus bacteria in the absence of oxygen.

Comeau *et al.* postulate that anaerobic P release occurs when the addition of a compound lowers the energy level of P bacteria (which are responsible for bio-P removal). Conversely, P uptake takes place when sources of both carbon and an electron acceptor are available for energy production and the energy level of bio-P bacteria is consequently relatively high.

Waste Management

Municipal solid waste (MSW) is a heterogeneous substance consisting of materials discarded from residences, commercial establishments, institutions, and industries. MSW may include variable amounts of paper and paper products, plastic, rubber, leather, textiles, wood, food wastes, ceramics and potteries, glass, and metals. Table 8-13 shows one popular scheme of classification of wastes, adopted by the Solid Waste Processing Division of the American Society of Mechanical Engineers (Walter, 1987).

Several options are available to treat or dispose off waste materials. Diagram 17.41 illustrates a three-element waste management system. Recycling is a realistic and useful alternative to traditional collection and disposal of several wastes, e.g., municipal garbage. The wastes remaining after source reduction/ recycling can be processed in some centralized

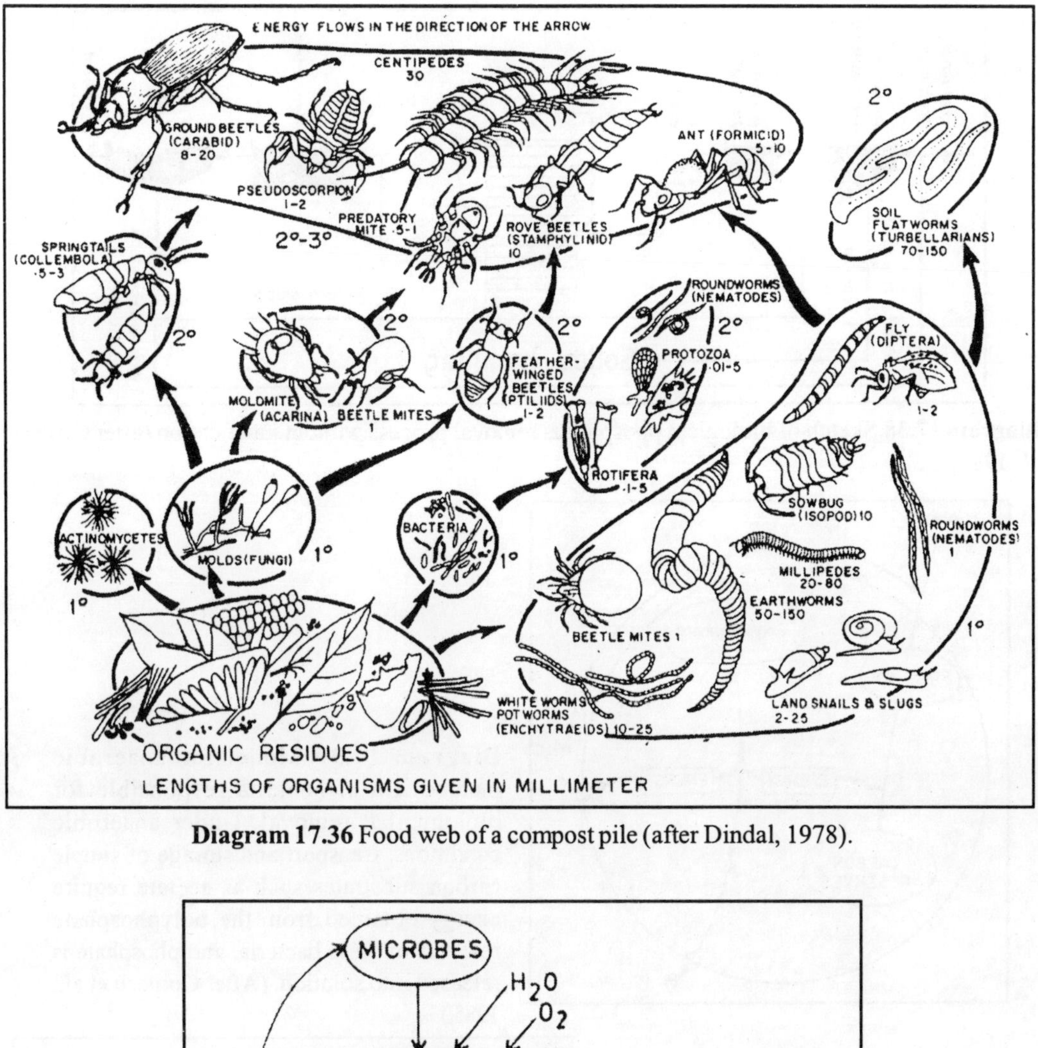

ENERGY FLOWS IN THE DIRECTION OF THE ARROW
CENTIPEDES 30
GROUND BEETLES (CARABID) 8-20
PSEUDOSCORPION 1-2
PREDATORY MITE .5-1
ROVE BEETLES (STAMPHYLINID) 10
ANT (FORMICID) 5-10
2°
SOIL FLATWORMS (TURBELLARIANS) 70-150
SPRINGTAILS (COLLEMBOLA) .5-3
2°-3°
2°
MOLDMITE (ACARINA)
BEETLE MITES 1
2°
FEATHER-WINGED BEETLES (PTILIIDS) 1-2
2°
ROUNDWORMS (NEMATODES)
2°
PROTOZOA .01-5
FLY (DIPTERA) 1-2
ROTIFERA .1-5
SOWBUG (ISOPOD) 10
ROUNDWORMS (NEMATODES) 1
MILLIPEDES 20-80
EARTHWORMS 50-150
ACTINOMYCETES
MOLDS (FUNGI) 1°
BACTERIA 1°
1°
BEETLE MITES 1
1°
WHITE WORMS POT WORMS (ENCHYTRAEIDS) 10-25
LAND SNAILS & SLUGS 2-25
ORGANIC RESIDUES
LENGTHS OF ORGANISMS GIVEN IN MILLIMETER

Diagram 17.36 Food web of a compost pile (after Dindal, 1978).

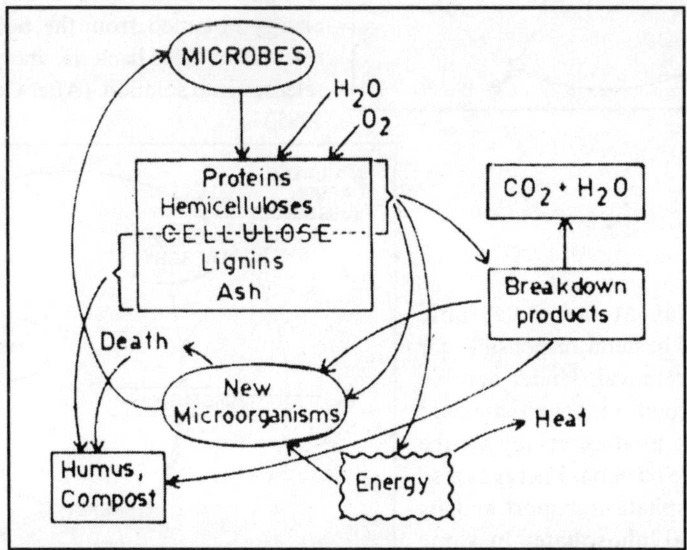

Diagram 17.37 Schematic sketch of the composting process.

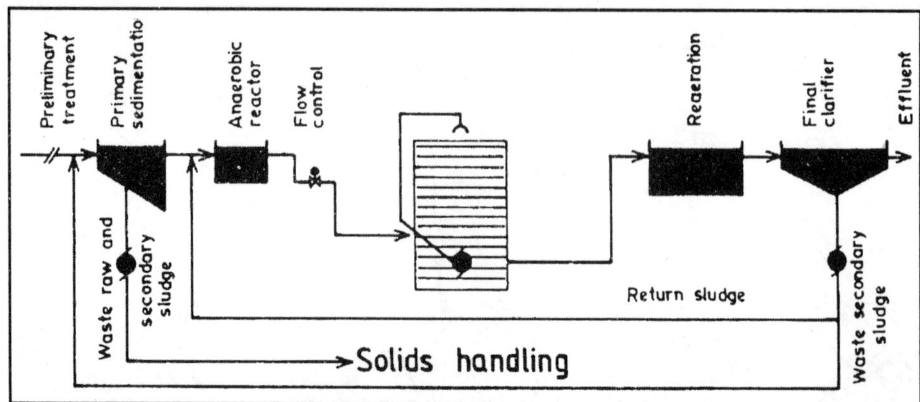

Diagram 17.38 Sketch of biological phosphorus removal process without nitrification (after Gibb *et al.*, 1989).

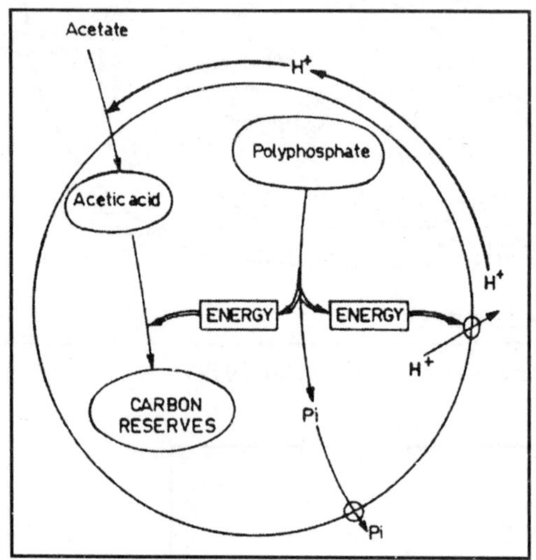

Diagram 17.39 Model for anaerobic metabolism of bacteria responsible for biological P removal. Under anaerobic conditions, transport and storage of simple carbon substrates such as acetate require energy obtained from the polyphosphate reserves of these bacteria, and phosphate is released into solution. (After Comeau et al., 1985.)

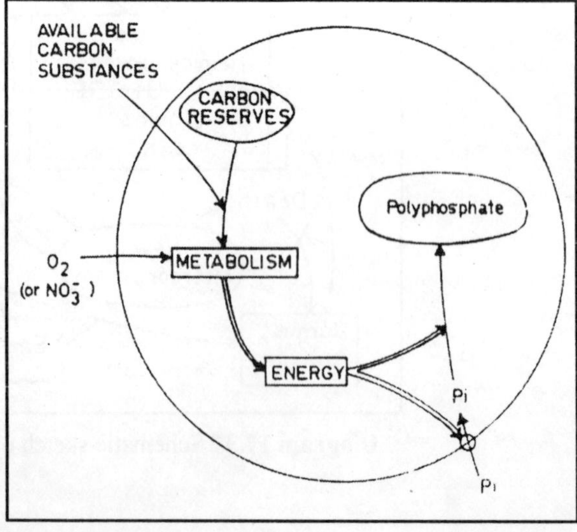

Diagram 17.40 Model for aerobic metabolism of bacteria responsible for biological P removal. Under aerobic conditions, carbon compounds are used with oxygen to produce energy for the growth of these bacteria. Energy is also used for phosphate transport and its storage as polyphosphate. In some cases, nitrate may be used instead of oxygen for the production of energy. (After Comeau *et al.*, 1985.)

plant to further reduce landfill requirements. Technological options include energy from waste, composting, refuse-derived fuels, or transfer stations. Based on these four technology options, six possible processing alternatives can be considered (Diag. 17.41), and some of these can even be combined.

Biosensors employing immobilized whole cells act as broad-spectrum sensors useful in water-monitoring to combat the increasing number of pollutants finding their way into the groundwater systems and from there into the drinking water. Table 17.14 lists four such potential biosensors; out of these, the BOD biosensor is already available in the UK market (Gronow *et al.*, 1985).

BIOLEACHING

Bacterial leaching is a process by which microorganisms found in the acidic waters of mines dissolve normally soluble sulphide ores to release their mineral content as an effluent. Once the effluent has been collected, the metal can be extracted easily. Similarly, bacteria can be used in the extraction of gold and silver from refractory ores. Refractory ores do not usually react to conventional treatment processes and it has been discovered that gold can be leached from refractory ores through normal leaching methods, after pretreating the ore with acidic solution containing bacteria.

Table 17.13 Classification of wastes (Walter, 1987)

Waste	Content	Composition	Moisture content (approx. %)
Trash	Highly combustible waste, e.g., paper, wood, plastics	100% trash	5
Rubbish	Combustible waste, e.g., paper, rags, wood, from domestic commercial, and industrial sources	Rubbish : garbage 80 : 20	25
Refuse	Rubbish and garbage from residential sources	Rubbish : garbage 50 : 50	50
Garbage	Animal and vegetable matter	Garbage : rubbish 65 : 35	70
Organic animal	Carcasses, organs, solid organic wastes	Animal and human tissue 100%	85
Gaseous, liquid or semiliquid	Industrial wastes	Variable	Variable
Semisolid or solid	Combustibles requiring hearth retort or grate burning equipment	Variable	Variable

Bacterial leaching occurs naturally. The process can be optimized by identifying the bacteria best suited to each mine site and the conditions needed for effective leaching such as acidity, temperature, and oxygen requirements (Acharya and Spencer, 1991). This new technology offers a number of advantages over conventional mining techniques:

1. Because it is known to recover metals from low grade ores, bacterial leaching can be applied to dumps of 'waste' abandoned at mine sites that are uneconomic to process using conventional technology.
2. The bioleaching of existing waste dumps eliminates the cost of mining the ore. This property was especially useful when metal prices were low.
3. Bioleaching can produce refined metal. This is especially relevant to many developing countries which have to send their mineral concentrates to advanced countries for refining.
4. New-environmental regulations in many developed countries make it difficult to use smelting and other technologies cost-effectively. Unlike smelters, bacterial leaching does not pollute air, and careful collection of the effluent minimies groundwater pollution. In fact, since the process occurs naturally, it is in the interest of mining companies to prevent the effluent from seeping into the groundwater around their mine sites.

The world's first commercial concentrate bacterial oxidation plant has been operating in South Africa since 1986 and has demonstrated greater efficiency (gold recovery averages 94 per cent through biooxidation compared to roaster recovery of 90 per cent). The same process is currently being used for refractory gold ores in Brazil.

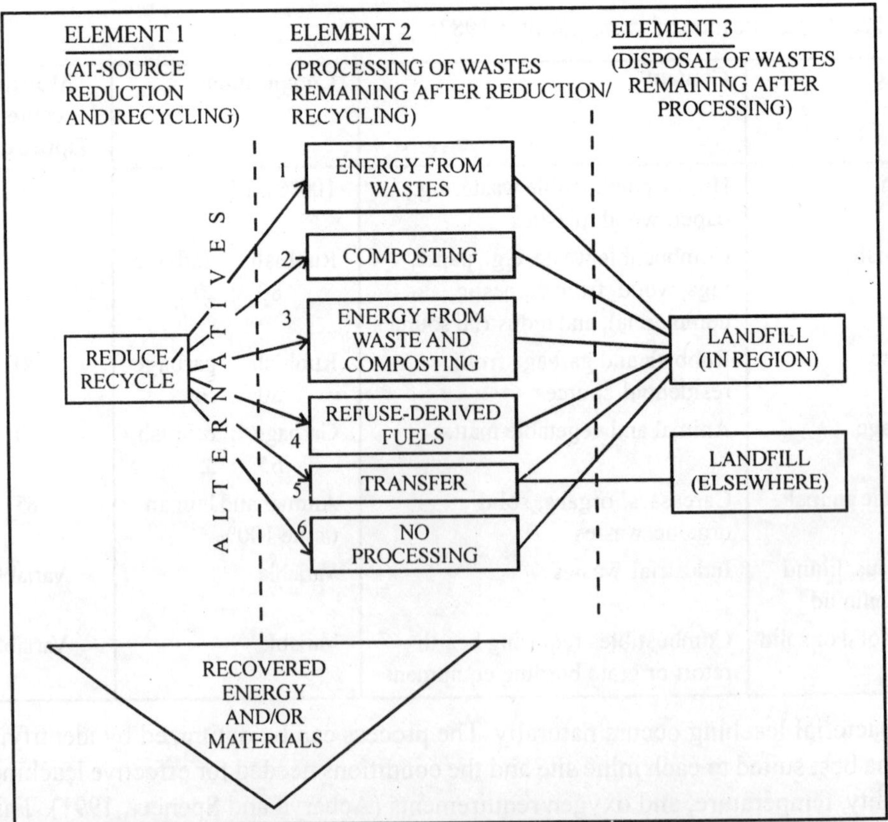

Diagram 17.41 Some alternatives for management/disposal of wastes.

Table 17.14 Four potential microbial biosensors for environmental monitoring (Gronow *et al.*, 1985)

Microbe	Sensor for	Transducer	Range (approx.)	Stability (approx. days)
Trichosporon cutaneum	BOD	O_2	1-40mg/1	15
Methylomonas flagellata	Methane	O_2	6mM	20
Azotobacter vinelandii	Nitrate, nitrite	NH_3	10^{-5} M	14
Bacillus subtilis	Mutagen	O_2	1.5 µg/cm^3	

It is for developing countries that bacterial leaching offers the greatest potential.

1. Since average ore grades are higher in developing country deposits, leach solutions contain a high metal content.
2. Cut-off grades have been higher in developing countries because of less sophisticated technology. The waste dumps therefore contain a higher grade of metal.

Among developing countries, Chile is one of the most advanced users of bacterial leaching. It plans to introduce copper bioleaching at most of its mine sites. In contrast, many other developing countries, especially large producers in Africa, have not used the potential of this new technology to develop their large copper deposits.

This uneven development is largely due to the general state of African economies with respect to foreign exchange problems, the reduced participation by foreign multinationals and the different attitudes of mining authorities to innovation and new technologies. In Chile, for example, the liberalization which accompanied 15 years of military dictatorship, has ensured a perfect environment for foreign investment. Chile's mining code grants virtual property rights over mining concessions; foreign company profits can be shipped abroad and capital can be pulled out after five years of investment. Labour unions have been destroyed with industry-wide stoppages forbidden and the hiring of temporary workers in the case of strikes.

Bioleaching provides advantages in terms of environmental safety, reduced overall costs of metal recovery, economic recovery of metal from lower grade ores and increased value added for those developing countries without conventional refining capabilities.

Technology has been a major factor not only in changing production processes but also enabling producers to market previously unmarketable products. The change takes place because of uneven access to the new technology. Biotechnology differs from previous technologies in that it is not altogether inaccessible to developing countries. The exploitation of this new technology has been successfully demonstrated by Chile (see Warhurst, 1985; Acharya and Spencer, 1991) in the production of copper and a number of other countries in the production of both base and precious metals.

MICROBIOLOGY IN WASTE MANAGEMENT

Until a century ago, the waste products from human activities used to be returned into the environment and underwent the biosphere's natural elimination processes without there being any long-term charge on the environment. During the last century, the increase in the amount of refuse has been accompanied by a decrease in its quality, mainly due to the production and dispersal of heavy metals and xenobibtic compounds.

In the last century the natural equilibrium has been upset by three causes:

1. Increase in population.
2. Widespread utilization, followed by diffusion into the environment, of toxic metals previously kept out of the biosphere in view of their concentration in ores.
3. Increase in the production and dispersion of xenobiotic compounds which are biodegradable at best with difficulty, often not at all (Gandolla and Aragno, 1992).

The amount of waste has grown but its return to the environment has decreased considerably (Diag. 17.42).

Waste disposal into the environment occurs in two ways: either by dispersal of the derivatives into the biosphere (sediments, soil, water, air), or by concentration (e.g., in landfills), in order to exclude them from the biosphere. Waste treatment has a double aim: to produce derivatives whose dispersal is acceptable (e.g., composts, certain gases), and to concentrate the dangerous compounds (e.g., heavy metals) and isolate them more or less indefinitely from the biosphere (Gandolla and Aragno, 1992). Over 90% of the mass of urban waste is deposited in landfills and less than 10% is incinerated. Incineration involves the dumping of the residue, which amounts to up to 25% of the initial waste volume.

Waste management technology involves three types of procedures: physical (sorting, compacting); chemical (combustion, chemical treatment of and liquid emissions); and biological anaerobic digestion, biofiltration). Biological processes can either be undesirable, and have to be controlled and minimized, or they are necessary, and should be used and optimized (Gandolla and Aragno, 1992).

A landfill containing organic material of biological origin (paper, cardboard,, domestic, agricultural, and some types of industrial wastes) can be compared to a huge bioreactor, in which biological degradations will occur, either aerobically or anaerobically depending"on the way the dumping is conducted (Baccini, 1989). In modern, compacted landfills which are like anaerobic bioreactor or methanogenic microbes convert the anaerobically degradable materials into a mixture of methane ; carbon dioxide. The functioning of the methanogenic microflora needs to be optimized with a view accelerating the stabilization of the waste mass, and to avoid emission into the percolating water and the atmosphere of the low molecular weight organic intermediates characteristic of incomplete degradations. In normally managed landfills, the biological activity is usually not optimal, owing to the coarse heterogeneity of the material deposited, the scarcity of water, and the lack of available nitrogen and phosphorus compounds.

The surface of the landfill should act as an aerobic biofilter and should oxidize methane and vola compounds diffusing from the inside of the landfill. Sometimes, hazardous, thermogenic, aerobic processes occur spontaneously at the periphery of the landfill, either

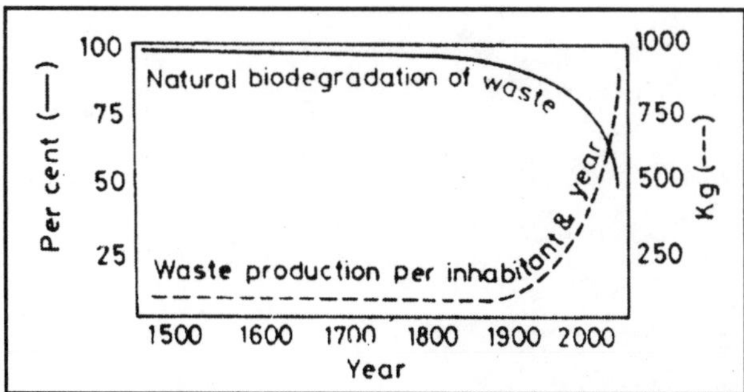

Diagram 17.42 Trends in the evolution of the amount biodegradability of waste produced by mankind.

due to composting of organic material follow contact with air, or to biological oxidation of methane air mixtures.

Certain dangerous compounds, e.g., antibiotics, must not be deposited in a landfill even if they biodegradable because they can lead to the selection and spread of resistant bacterial strains which might transfer their resistance to potential pathogens.

The classical composting system in heaps is simple and economical; it can be used on a small scale within the community. However, it can lead to environmental problems, such as groundwater pollution if not well managed, to the emission of noxious smells; at temperatures of 40-55°C, there is good growth of thermophilic fungi, including the cellulolytic *Aspergillus fumigatus*—a powerful allergen. Although it requires a more sophisticated technology and cannot be applied to small-scale plants, composting bioreactor minimizes some of the problems caused by heap-composting.

Organic wastes with a relatively low content of ligneous material are treated by a biomethanization process (Wise, 1987). This procedure should take place in a liquid suspension, at medium (35°C) or high (60°C) temperatures. A high quality biogas is produced by this method. Biomethanization involves use of thermally regulated biodigesters. The substrate composition should be optimized and the volatile fatty acids and gas composition monitored (Gandolla and Aragno, 1992).

□□□

CHAPTER-18

Biotechnology in Relation to Human and Animal Health

Historical Background

The branch of science dealing with human health is not very new. The human conciseness regarding their health problems started simultaneously with the apperance of diseases. This laid the foundation stone of immunological study in microbiology a branch of bioscience, which deals witn the study of immunity to infections. As early as 1000 AD Chinese physician Yo-Meishan successfully inoculated the emperor's grandson with dried crust of small pox to render him immune from a serious attack of the dreadful disease. Edward Jenner (1798) learnt from milk maid, infected with cowpox, that developed cowpox, were immune to small pox. Pasture (1822-1895) developed attenuated vaccines for anthrax and rabies. The classical work of Russian scientist E. Metchinikoff (1845-1916) on the biological theory of immunity marked a new stage in the history of immunobiology. He discovered phagocytosis and intracellular digestion in mesodermal cells of some animals. The phagocytic theory of immunity was expounded in 1883 at VII Congress of Russian Naturalists and Physicians in Odessa.

Paul Ehrlich (1854-1915) gave a theory of humoral immunity according to which certain substances in the blood serum, secreted by special cells under the influence of microbes and their toxins play an important role in the defence reactions of the body. The ability of developing specific immunity to invading or infective agent is a unique property of vertebrates.

Von Pirquet (1906) given the concept of allergy as his studies with the individual after exposure to an antigen demonstrated changed reactivity - in one case recognized as immunity in another as hypersensitivity. He coined the term allergy for this changed reactivity. Pirquet's concept was the fundamental one as it recognised behind all reactions to antigenic exposure no matter what their clinical outcome, a common biological process of specific sensitization. During the first four to five decades of this century investigations were mainly concerned with chemistry of antigen and antibodies. Modem immunological studies cover defence against infection, prevention of disease by immunisation, blood banking, and hypersensitivity including autoimmunity. Immunologiqal techniques are used to measure immune responses for diagnosis and progresses of certain diseases, hormones and drugs.

What is an immune System?

In body, the nonspecific immune mechanisms are usually inadequate to cope with foreign agents or substances, mainly when a particular virulent microorganism is involved and is able to evade phagocytic mechanisms. In vertebrate host, the specific immune system is endowed with three characteristics: specificity, memory and recognition of self-antigens.

Specificity: The immune system has the property to recognize foreign substances i.e., antigens or substances that can stimulate the formation of antibodies. After establishment of the contact, products of the immune system are elaborated and interact with antigen.

Memory: During response to foreign substances some lymphocytes give rise to memory cells and are permitted to act with speed and vigour the next time the foreign agent is encountered. The capacity of this memory makes feasible the process of vaccination. The first time the foreign agent is encountered, there is short lag time before the immune system can generate enough immune products to overcome such an infectious agent.

Self-recognition: There is an interaction of many foreign substances, with immune system. The immune system, however, can discriminate between the foreign substance and self substances. This is called self-recognition.

There are many terms which are often used in immunological studies. The definitions of various terms are as follows:

Antigen (Ag): A substance that initiates the immune system to form immune products specific for the substance. Chemically it may be protein, polysaccharides, or nucleic acids, that are either soluble or particulate. The antigens not only cause the formation of immune products but interact with them as well. They are also known as immunogens when the emphasis is on their ability to incite the formation of immune products (immunogenicity).

Antigenic determinant: Small chemical groups on the surface of antigens with which immune products interact.

Partial antigen or Hapten: It is the substance which can interact with specific antibody combining groups on an antibody molecule but which fails by itself to elicit the formation of a detectable amount of antibody (Hapten means *to grasp).*

Immunoglobulins (Ig): These are proteins which have demonstrable antibody activity and/or share a common antigenic specificity with any known antibody and are produced by cells that form antibody. Thus proteins like myeloma proteins, Bence-Jones proteins and subunits of antibodies are also known as immunoglobulins. Functionally they are two types: surface immunoglobulins: they are present on the surface of the lymphocytes where they act as specific receptor (recognition molecules) for the antigen, and secreted immunoglobulins these are the products of B lymphocytes and appear in the body fluids (humors as antibodies).

Antibody (Ab): An antibody is a immunoglobulin whose formation is induced by the introduction of an antigen in an animal body. It reacts with the corresponding antigen specifically in some observable way, that each antibody has binding sites for an identified antigen.

B lymphocytes (B Cell): A major class of lymphocytes that produce immunoglobulins and are primarily involved in humoral immunity, that is the production of antibodies.

T lymphocytes (T Cell): A major class of lymphocytes that are thymus dependent and form effector T lymphocytes on stimulation by- antigens. They also produce lymphokines, which are non-antibody mediators associated with inflammatory events and also associated with immunoregulation.

Humoral immunity: It is the immunity provided by antibodies, B cell immunity, antibody-mediated immunity.

Cell mediated immunity (CM): It is the immunity provided by T cells: T-cell immunity.

Complement: The term complement is applied to a group of co-factors occurring in fresh normal blood, serum and some body fluids that are activated characteristically by antigen-antibody interactions and subsequently mediated certain biological events of immunological reactions.

Immunological tolerance: Immunological tolerance is a central failure of responsiveness of immune system brought about by appropriate exposure to an antigen, in which immunologically competent cells fail to respond to that antigen.

Immune system: It is the system of the body which is responsible for all types of immune responses. Essentially it is constituted by the lymphoid organs and cells and is divisible into T cell division and B cell division.

Immunocyte: A mature immunologically competent cell is known as immunocyte.

Mycelomas: They are the cells of certain malignant tumours of bone marrow. They produce large quantities of abnormal immunoglobulines (antibodies) and can be grown in vitro indefinetly. Immunoglobulines produced by mycelomas (clones) in vitro have an identical structure. They are, in fact, monoclonal antibodies.

Principles of Immunology

Immunology is the study of immunity to infectious diseases in organisms. There are three types of protections against any infectious diseases. They are as follows:

- Nonsusceptibility: This is species characteristic of the host and gives complete protection against a particular microorganism.
- Natural resistance: This is the natural available capacity present in an organism and is due to physical and chemical characteristics of the host. The resistance of an individual varies according to time.
- Natural immunity: This is basically dependent upon the natural antibodies (modified blood globulins) which can able to react with antigens.

Immunoglobulins

These are protein molecules with demonstrable antibody activity. They are made of heterogenous group of proteins accounting for about 20% of total plasma proteins. They are mainly γ - globulins but a few are β-globulins.

Structure: They are glycoproteins and are composed of 82-96% polypeptide and 4-18% carbohydrate. The polypeptide part possesses the biological properties of antibodies.

Basic Unit: Each molecule of immunoglobulin has at least one basic unit or monomer which contains four polypeptide chains in two pairs, and each one have similar chains. One of the pairs have two or more than 2 amino acids. The larger pair with more amino acid is called heavy chain while with smaller one is called light chain, they are also known as H and L chains respectively. Every chain possesses an amino terminal portion which shows marked heterogeneity or variable region or V region and the carboxy terminal portion with a similar type of amino acid residue called constant region or C region. The 'V region'of both the chains are of equal length. The antigen combining site is formed by the 'V region of 'H' chain and 'L' chain. Therefore, a monomeric immunoglobulin (Ig) molecule has two antigen combining sites.

When polypeptide chains are folded three dimensionally by disulphide linkage the structure is called domains. Such domain in 'H' chains are known as VH, CH1, CH2 and CH3 and those in 'L' chain VL and CL. Each such segment is made of about 110 amino acid molecules. The 'H' and 'L' chain are linked with a disulphide linkage, and the region is known as hinge region of the molecule. On the basis of inter-chain disulphide linkage, the Ig may be of four types: IgGl, IgG2, IgG3 and IgG4. In Ig, a glycopolypeptide chain (like the size of' L' chain) is also found (with mol. wt. 15,000) and called joining chain or T chain.

Carbohydrate moieties: This part of the Ig is found in secretary component, T chain and constant region of the 'H' chain. It is not found in 'L' chain and any of the 'V region. Its function is still not certain, however, it is suggested that it plays a role in secretion of Ig by plasma cells.

Immunoglublin Classes and Subclasses

Based on the antigenic character of heavy chains the immunoglobulins are classified into various types, the important are: Immunoglobulin G (IgG), Immunoglobulin M (IgM), Immunoglobulin D (IgD), Immunoglobulin A (IgA), Immunoglobulin E (IgE).

IgG is the major immunoglobulin in human serum, accounting for 70 to 75% of the immunoglobulin pool. IgG is present in blood plasma and tissue fluids. The IgG class acts against bacteria and viruses by opsonizing the invaders and neutralizing toxins. It is also one of the two immunoglobulin classes that activate complement by the classical pathway. IgG is the only immunoglobulin molecule able to cross the placenta and provide naturally acquired immunity for the newborn.

There are four IgG subclasses (IgGl, IgG2, IgG3, and IgG4) that vary chemically in their chain composition and the number and arrangement of interchain disulfide bonds. About 65% of the total serum IgG is IgGl, and 23% is IgG2. Differences in biological function have been noted in these subclasses. For example, IgG2 antibodies are opsonic and develop in response to antitoxins. Anti-Rh antibodies are of the IgG1 or IgG3 subclass. IgG1 and IgG3 also bind best to monocytes and macrophages and activate complement most effectively. The IgG4 antibodies function as skinp sensitizing immunoglobulins.

IgM accounts for about 10% of the immunoglobulin pool. It is usually a polymer (pentamer) of five monomeric untis, each composed of two heavy chains and two light chains. The monomers are arranged in a pinwheel array with the Fc ends in the center, held together by a special J (joining) chain. IgM is the first immunoglobulin made during B-cell maturation and the first secreted into serum during primary antibody response. Since IgM is so large, it does not leave the bloodstream or cross the placenta. IgM agglutinates bacteria, activates complement by the classical pathway, and enhances the ingestion of pathogens by phagocytic cells. This class also contain special antibodies such as red blood cell agglutinins and heterophile antibodies.

Although most IgM appears to be pentameric, around 5% or less of human serum IgM exists in a hexameric form. This molecule contains six monomeric units but seems to lack a J chain. Hexameric IgM activates complement up to twentyfold more effectively than does the normal pentameric form. It has been suggested that bacterial cell wall antigens such as gram-negative lipopolysaccharides may directly stimulate B cells to form hexameric IgM without a J chain. If this is the case, the immunoglobulins formed during primary immune responses are less homogenous than previously throught.

IgA account for about 15% of the immunoglobulin pool. Some IgA is present in the serum as a monomer of two heavy and two light chains. Most IgA, however, occurs in the serum as a held toeether by a J chain. IgA has special features that are associated with secretory mucosal surfaces. IgA, when transported from the mucosa-associated lymphoid tissue to mucosal surfaces, acquires a protein termed the secretory component.

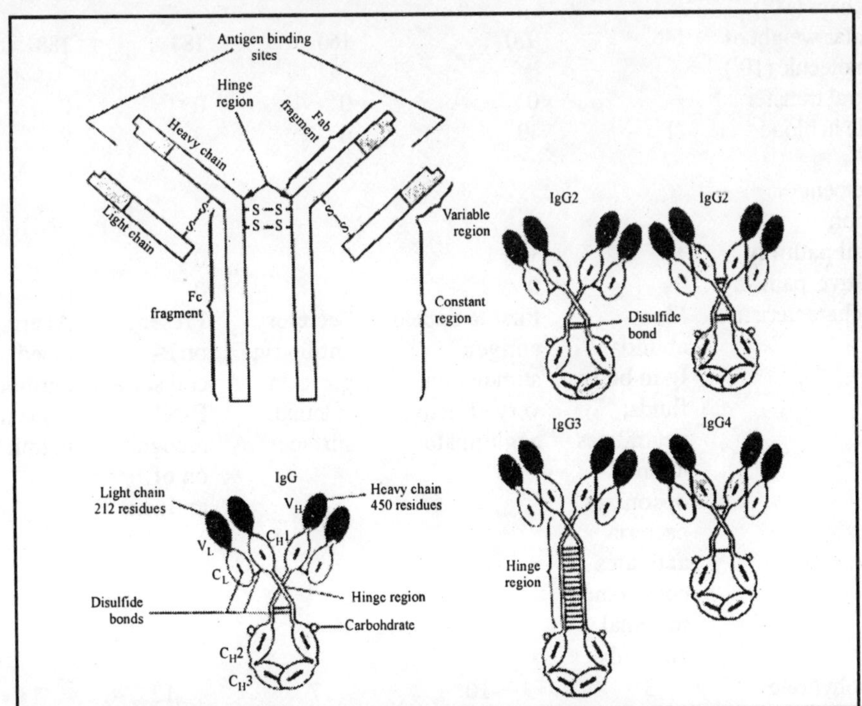

Diagram 18.1 (A) Basic structure of Immunogloublin (B) Different types of Ig. Note different types of inter chain disulphides linkage.

Secretory IgA (sLgA), as the modified molecule is now called, is the primary immunoglobulin of the secretory immune system. This system is found in the gastrointestinal tract, upper and lower respiratory tracts, and genitourinary system. Secretory IgA is also found in saliva, tears, and breast milk. In these fluids arid related body areas, sLgA plays a major role in protecting surface tissues against infectious microorganisms by the formation of an immune barrier. For example, in breast milk sLgA helps protect nursing newborns. In the intestine, sLgA attaches to viruses, bacteria, and protozoan parasites such as *Entamoeba histolytica.* This prevents pathogen adherence to mucosal surfaces and invasion of host tissues, a phenomenon known as immune exclusion. In addition, sLgA binds to antigens within the muscosal lamina propria, and the antigen-sLgA complexes are excreted through the adjacent epithelium into the gut lumen. This rids the body of locally formed immune complexes and decreases their access to the ciruculatory system. Secretory IgA also may neutralize viruses and other intracellular pathogens that reside within epithelial cells. Secretory IgA also plays a role in the alternate complement pathway.

Table 18.1 Physiochemical Properties of Human Immunoglobulin Classes

Property	Immunoglobulin Classes				
Heavy chain O_1	μ	A_1	-	-	
Mean serum concentration (mg/ml)	9	1.5	3.0	0.03	0.00005
Valency	2	5(10)	(2-4)	2	2
Molecular weight of heavy chain (10^3)	51	65	56	70	72
Molecular weight of entire molecule (10^3)	146	970	160[c]	184	188
Placentral transfer	+	0	0	0	0
Half-life in blood (days)[d]	21	10	6	3	2
Complement activation					
Classical pathway	++	+++	0	0	0
Alternative pathway	0	0	+	0	0
Major characteristics	Most abundant Ig in body fluids; neutralizes toxins, opsonizes bacteria, activates complement, maternal antibody	First to appear antigen stimulation; very effective agglutinator	Secretory antibody; protects external surfaces	Present on B-cell surface; B-cell recogniti on of antigen	Anaphylacti cmediating antibody: resistance to helminths
% carbohydrate	3	7-10	7	12	11

[a]Properties of IgG subclass 1., [b]Properties of IgA subclass 1., [c]sLgA = 360 - 400 kDa, [d]Time required for half of the antibodies to disappear.

IgD is an immunogiobulin found in trace amounts in the blood serum. It has a monomer structure similar to IgG. IgD antibodies do not fix complement and cannot cross the placenta, but they are abundant on the surface of B cells and bind antigens, thus signaling the B cell to start antibody production.

IgE makes up only 0.00005% of the total immunobulin pool. The classic skin-sensitizing and anaphylactic antibodies belong to this class. IgE molecules have four constant regiondomains ($C_\varepsilon 1$, $C_\varepsilon 2$, $C_\varepsilon 3$ and $C_\varepsilon 4$) and two light chains.

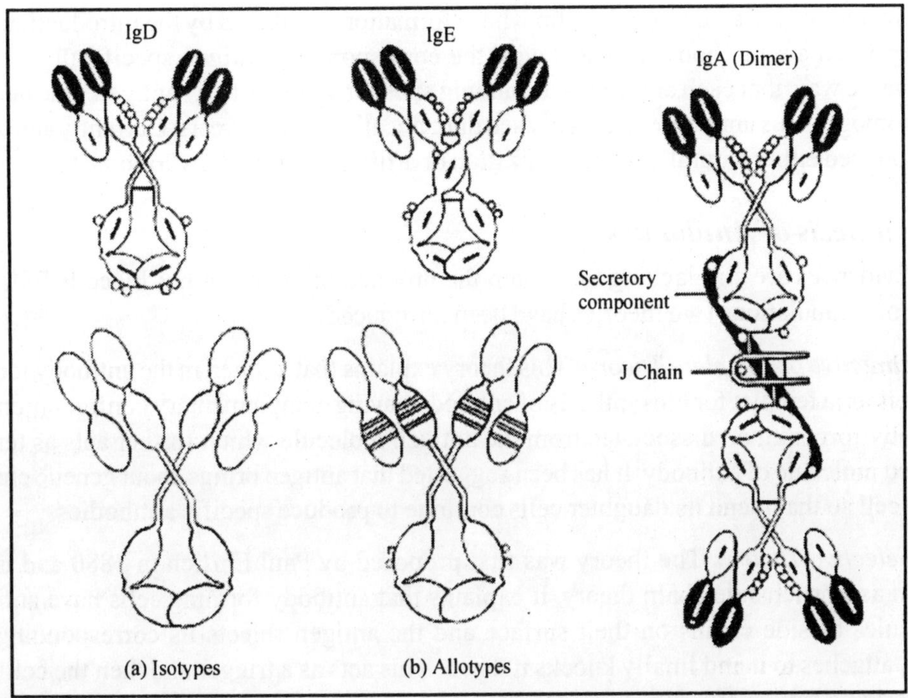

Diagram 18.2 Immunoglobulines: (a) Basic structure, (b) IgD: The structure of human Igd. The disulfide bonds linking protein chains are shown, (c) IgE: structure of human IgE.

The Fc portion of the $C_E 4$ chain can bind to special Fc receptors on mast cells, and basophils. When two IgE molecules on the surface of these cells are corss-linked by binding to the same antigen, the cells degranulate. This degranulation releases histamine and other pharmacological mediators of anaphylaxis. It also stimulates eosinophilia and gut hypermotility (increased rate of movement of the intestinal contents) that aid in the elimination of helminthic parasites. Thus, through IgE is present in small amounts, this class of antibodies has very potent biological capabilities.

Immunoglobulin Function

All immunoglobulin molecules are bifunctional. The Fab region is concerned with binding to antigen, whereas the Fc region mediates binding to host tissue, various cells of the immune system, some phagocytic cells, or the first component of the complement system. The binding of an antibody with an antigen usually does not cause destruction of the antigen or of the microorganism, cell or agent to which it is attached. Rather the anitbody serves to mark and identify the target for immunologic attack and to activate nonspecific immune responses that can destroy the target for phagocytosis by neutrophils and marcophages. This ability of an antibody to stimulate phagocytosis is termed opsonization. Immune destruction also is promoted by antibody-induced activation of the complement system.

Source of Antibodies

An antibody is a immunoglobulin whose formation is induced by the introduction of an antigen in an animal body. It reacts with the corresponding antigen specifically in some observable way, that each antibody has binding sites for an identified "antigen. The need for pure homogeneous antibodies has increased dramatically in recent years. Currently antibodies are produced either naturally by immunization or artifically through hybridoma formation.

Biosynthesis of Antibodies

Evidences are still lacking to explain the production of antibodies by cells following antigenic stimulation. Two theories have been introduced:

Directive or Template Theory: This theory explains that antigen in the antibody forming cells acts as a templet for biosynthesis of antibody having complementary configuration. The antibody formed than dissociates from the antigen molecule which further acts as templet for next molecule of antibody. It has been suggested that antigen brings about genetic changes in the cell so that it and its daughter cells continue to produce specific antibodies.

Selective Theory: The theory was first proposed by Paul Ehrlich in 1880 and is also known as Ehrlich side chain theory. It explains that antibody forming cells have antibody molecules as side chains on their surface and the antigen selects its corresponding side chain, attaches to it and finally knocks it down. This acts as a trigger and then the cell starts synthesizing similar side chains repeatedly which are dislodged to free circulation and unite with the antigen.

Diversity of Antibodies

One unique property of antibodies is their remarkable diversity. According to current estimates each human or mouse can synthesize more than 10 million different kinds of antibodies. How is this diversity generated? The answer is threefold:

1. rearrangement of antibody gene segments,
2. somatic mutations, and
3. generation of different codons during antibody gene splicing.

Immunoglobulin genes are split or interrupted genes with many exons. Embryonic B cells contain a small number of exons, close together on the same chromosome, that determine the constant (C) region of the light chains. Separated from them, but on the same chromosome, is a larger cluster of exons that determines the variable (V) region of the light chains. During

B-cell differentiation, one exon for the constant region is spliced exon for the variable region. This splicing produces a complete light-chain antibody gene. A similar splicing mechanism also occurs to join the constant and variable exons of the heavy chains.

Because the light-chain genes actually consist of three parts, and the heavy-chain genes consists of four, the formation of a finished antibody molecule is slightly more complicated than previously outlined. The germ line DNA for the light-chain gene contains multiple coding sequences called V and J (joining) regions. During the differentiation of a B cell, a deletion (which is variable in length) occurs that joins one V exon with one J exon. This DNA joining process is termed combinatorial joining since it can create many combinations of the V and J regions. When the light-chain gene is transcribed, transcription continues through the DNA region that encodes for the constant portion of the gene. RNA splicing subsequently joints the VJ and C regions creating mRNA.

Combinatorial joining in the formation of a heavy-chain gene occurs by means of DNA splicing of the heavy-chain counteparts of V and J along with a third set of D (diversity) sequences. Initially, all heavy chains have the u type of constant region. The corresponds to antibody class IgM. Another DNA splice joins the VDJ region with a different constant region that can subsequently change the class of antibody produced by the B cell.

In mouse the k light chains are formed from combinations of about 250-350 V_K and 4 J_K regions giving a maximum of approximately 1,400 different k chains. The o′ chains have their own vo′ and Jo′) regions but smaller in number than their k counterparts (6 different / chains). The heavy chains have approximately 250-1,000 V_H, 10-30 D, and 4 J_H regions giving a maximum 120,000 different combinations. Because any light chain can combine with any heavy, there will be a maximum of 2 x 10^8 possible k chain antibody types.

Table 18.2 Number of Antibodies Possible through the Combinatorial Joining of Mouse Germ Line Genesa

O light chains	**V regions = 2**
	J regions = 3 Combination = 2 x 3 = 6
k light chains	V_K regions = 250-350 J_K regions = 4 Combinations = 250 x 4 = 1,000 = 350 x 4 = 1,400
Heavy chains	V_H = 250-1,000 D = 10-30 J_H = 4 Combinations = 250 x 10 x 4 = 10,000 1,000 x 30 x 4 = 120,000
Diversity of antibodies	k-containing: 1000 x 10,000 = 107 1,400 x 120,000 = 2 x 108 O′-containing: 6 x 10,000 = 6 x 104 6 x 120,000 = 7 x 105

[a] *Approximate values.*

The value of 2×10^8 different anitbodies is actually an underestimate because antibody diversity is further aungmented by two processes:

1. The V regions of germ line DNA are susceptible to high rate of somatic mutation during B-cell development in the bone marrow. These mutation allow B-cell clones to produce different polypeptide sequences.
2. The junction for either VJ or VDJ splicing in combinatorial joining can occur between different nucleotides and thus one VJ splicing event can joint the V sequence CCTCCC with the J sequence TGGTGG in two ways:

   CCTCCC + TGGTGG = CCGTGG,

   which codes for the amino acids proline and tryptophan; and

   CCTCCC + TGGTGG = CCTCGG,

   which codes for proline and arginine. Thus the same VJ joining could produce polypeptides differing in a single amono acid.

Specificity of Antibodies

As noted previously, combinatorial joinings, somatic mutations, and variations in the splicing process generate the great variety of anitbodies produced by B cells. From a large, diverse B-cell pool, specific cells are stimulated by antigens to reproduce and form a B-cell clone that contains the same genetic information. This is known as the clonal selection theory, a hypothesis to explain immunologic specificity and memory.

The existence of a small B-cell clone (a population of cells derived asexually from a single parent) that can respond to one or a few antigens by producing the correct antibody is the first tenet of this theory. The lymphoid system is thus considered to contain many B-cell clones, each clone able to recognize a specific antigen. The antigen selects the appropriate clone of B cells (hence the pharse *clonal selection),* and the cells from the other clones are unaffected.

According to Secon Tenet, each B-cell clone is genetically programmed to respond to its own distinctive antigen before the antigen is introduced. The particular antibody for which an individual B cell is genetically competent is ingrated into the plasma membrane of that B cell and acts as a specific surface receptor for the corresponding antigen molecule. The reaction of the antibody and antigen initiates the differentiation and multiplication of the B cell to form two different cell populations: plasma cells and memory B cells.

Plasma cells are literally protein factories that produce about 2,000 antibodies per second in their brief five- to seven-day life span. Memory B cells can initiate the antibody-mediated immune response upon detecting the particular antigen molecule for wliich they are genetically programmed (i.e., they have specificty). These memory cells circulate more actively from blood to lymph and live much longer (years or even decades) than plasma cells. Memory cells are responsible for the immune system's rapid secondary antibody response to the same antigen.

Immunization

Specific antibodies can be produced naturally by the immunization of domestic animals or human volunteers. Whatever the source, purified antigen is injected into the host.

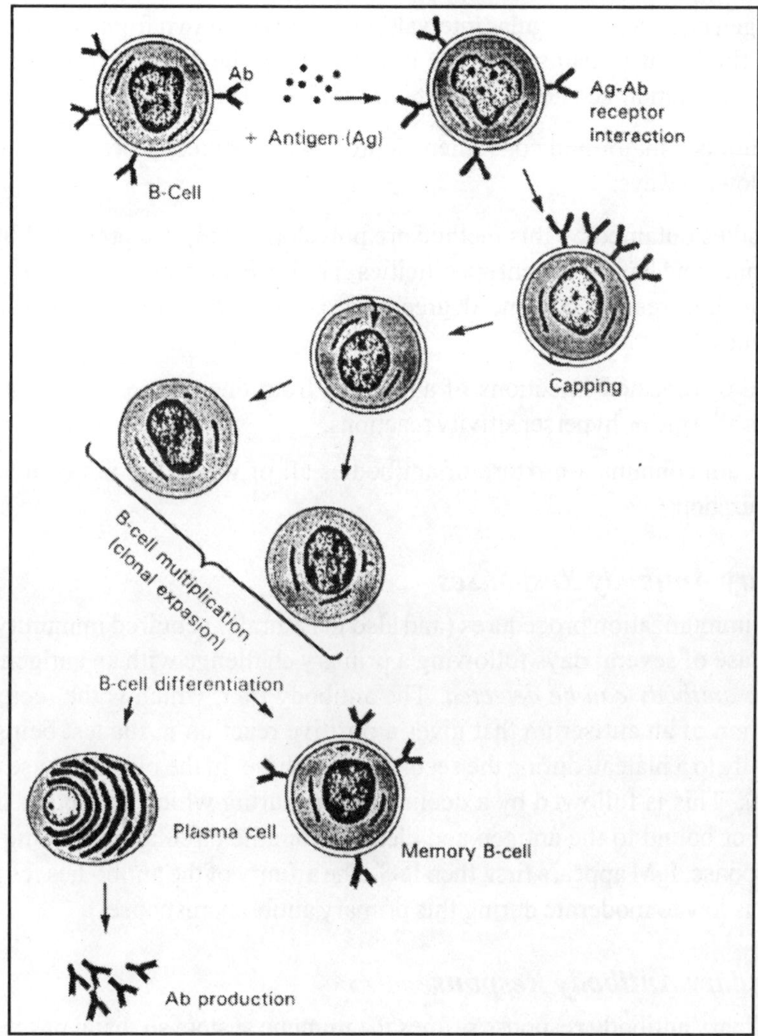

Diagram 18.3 Clonal Selection. It is through clonal selection that the .immune system can respond specifically to myriad of possible antigens, whether they are individual molecules or are attached to pathogens and abnormal cells such as cancer cells. B cells or B lymphocytes constantly roam the body, particularly the blood and lymphoid tissues. Each B cell synthesizes only one of the millions of possible antibodies and displays this antibody of the proper specificity (top left), it complexes with the antibody and capping occurs. (Capping is the regional aggregation of antibodies on the surface of the following Ag-Ab interaction.) The antigen is then internalized; the B cell swells and begins to divide rapidly, producing a B-cell clone. The activated B-cell clone differentiates into plasma cells and memory cells. Plasma cells form the specific antibody that immediately attack the antigen that provoked its formation. Memory B cells persist in the body and boost the immune system's readiness to eliminate the same antigen if it present itself in the future.

The host's immune system recognizes and responds to the antigen, and its B cells proliferate and differentiate to produce specific antibodies. To promote the efficiency of antigen stimulation of antibody production, the antigen is mixed with an adjuvant (Latin *adjuvans,* aiding), which enhances the rate and quantity of antibody produced. Following repeated antigen injections at regular intervals, blood is withdrawn from the post and allowed to clot. The fluid that remains after the blood clots is the serum. This serum has been obtained from an immunized host that contains the desired antibodies. It is called antiserum.

Antiserum is a major and convenient source of antibodies, however, its usefulness is limited in following ways:

1. Antibodies obtained by this method are polyclonal; they are produced by several B-cell clones and have different specificities. The decreases their sensitivity to particular antigens and results in some degree of cross-reaction with closed related antigen molecules.
2. Second or repeated injections of antiserum from one species to another can cause serious allergic or hypersensitivity reactions.
3. Antiserum contains a mixture of antibodies all of which are not of interest to give immunization.

The Primary Antibody Responses

During immunization procedures (and also in naturally acquired immunity) there is an initial lag phase of several days following a primary challenge with an antigen. During the *lag phase no antibody can be detected.* The antibody titer, which is the reciprocal of the highest dilution of an antiserum that gives a positive reaction in the test being used, rises logarithmically to a plateau during the second or log phase. In the plateau phase the antibody titer stablizes. This is followed by a decline phase, during which antibodies are naturally metabolized or bound to the antigen and cleared from the circulation. During the primary antibody response, IgM appears first, then IgG. The affinity of the antibodies for the antigen's determiants is low to moderate during this primary antibody response.

The Secondary Antibody Response

The primary antibody response primes the immune system so that it possesses specific immunologic memory through its clones of memory B cells. Upon secondary antigen challenge, the B cells mount a heightened or anamnestic [Greek *anamnesis,* remembrance] response to the same antigen. Compared to the primary antibody response, the secondary antibody response has a shorter lag phase, a more rapid log phase, persists for a longer plateau period, attains a higher IgG liter, and produces antibodies with a higher affinity for the antigen (affinity maturation).

Catalytic Antibodies

In the past several years immunologists have applied the principles of enzymology to create a new class of antibody molecules-catalytic antibodies. Catalytic antibodies accelerate specific chemical reactions by lowering the free energy of transition states. They accomplish this by binding reactants in a cleft or crevice on their surfaces and inducing structural changes in the substrate molecules. Because antibodies directed against a huge array of biopolymers, natural products, and synthetic molecules can be formed, catalytic antibodies offer a unique approach for generating tailor-made, enzyme-like catalysts.

Catalytic antibodies are made by coupling a transitionstate analogue to a carrier protein and injecting the combination into an experimental animal. Antibody-secreting spleen cells are taken from the animal and fused with myeloma cells. The hybrid antibody-secreting cells divide indefinitely and generate clone of cells, each hybrid clone secreting a monoclonal catalytic antibody with a unique antigen-binding pocket. A clone that makes catalytic antibody specific for the analogue is then selected.

Currently available catalytic antibodies transform relatively simple compounds. Much of the potential of catalytic antibodies for biotechnology and molecular biology depends on the development of catalytic antibodies able to act on proteins or nucleic acids. If this can be accomplished, catalytic antibodies could extend the immune system's innate capacity to defend the body. For example, one might stimulate the immune system of a patient with heart disease to produce antibodies that would break up the proteins in blood clots, forestalling heart attacks.

Antigens

The name antigens (Gk. *anti* = against, *genos* = genus) is given to organic substances of a colloid structure (proteins and different proteins complexes in combination with lipids of polysaccharides). It upon injection into the body (subcutaneously, intracutaneously, cutaneously, into the mucous membranes, intramuscularly, intravenously and orally) are capable of causing the production of antibodies and reacting specifically with them.

Proportion of Antigens

An antigen may be soluble substance such as horse serum proteins or a bacterial toxin, or it may be present on particulate matter like red blood cells, a bacterial cell, or a virus. An antigen is always a foreign substance for the host. An antigen must be capable of inducing an antibody response. The molecular weight of any antigen should be more than 6000 daltons. The portion of antigen that specifically combines with antibody is called its determinant group. Antigenic properties are pertinent to toxins of a plant origin (ricin, robin, abrin, cortin, etc.), toxins of an animal origin (toxins of snakes, spiders, scorpions etc.), enzymes, native foreign proteins, various cellular component, bacteria and their toxins and viruses etc.

Antigens are of two types: (a) Complete antigens and (b) Partial antigens:

Complete antigens: They cause the production of antibodies in the body, and react with them in vivo as well as in vitro.

Partial antigens: They are also known as haptens and do not cause the production of antibodies, but can react with them. Haptens includes lipids, complex carbohydrate and other substances.

The immunological specificity of antigens is linked with a determinant group found on the surface of the antigen as one or more active areas.

Isoantigens: These substances have antigenic properties and are contained in some individuals of a given species. Isoantigens are found in erythrocytes of animals and man.

Antigen-antibody Binding

Every antigen monomer has two similar antigen combining sites and each site is formed by 'V region of 'H' and 'L' chains. This helps the antibody molecule to link 2 similar antigens together. When many antibody and antigen unit join together, it results in the formation of a *lattice.* A large antigen and the cell surface of the microorganisms have several antigenetically active sites. A large number of antibody molecules are usually linked with antigenic sites and form an aggregates.

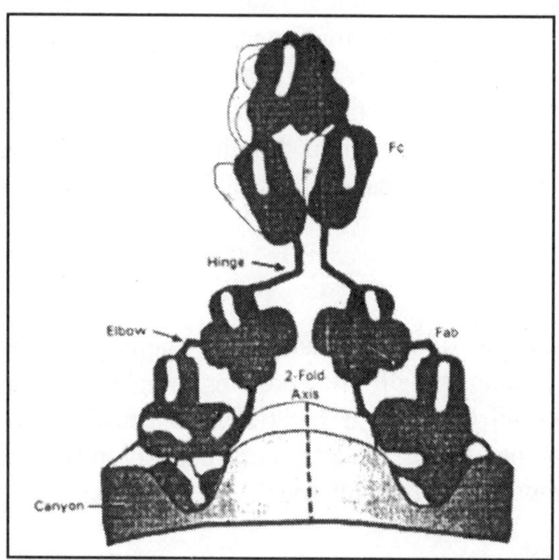

Diagram 18.4 Antigen-antibody binding: an example of antigen binding represented in the model.

The Immune Response

In immune mechanism there is an involvement of special types of cells called lymphocytes. The immunological response in an individual is brought about by the introduction of antigens or immunogens. The immune response mechanism is of two types: humoral immunity - involving production of antibodies, and cell-mediated immunity in which the lymphocytes react directly with foreign material. An immunologic response to an immunogen consists of one or more of the following components:

Antibody production: It involves 'B' cell division in immune system.

Development and specific cell mediated immunity which involves the 'T cell division of the immune system.

Immunological memory: Which involves one or both components of the immune system, is responsible for an accelerated immune response when the same antigen enters or is administered in the body after it has induced the first response (primary response).

Immunological tolerance: It is the specific immunological unresponsiveness brought about by certain antigens in certain circumstances.

Lymphocytes

Origin of Lymphocytes

Lymphocytes are small, nonphagocytic, mononuclear leukocytes that are immunologically competent, or are precursors of such cells. They lack stainable cytoplasmic granules and are formed in lymphatic tissues. Lymphocytes are the pivotal cells of the specific immunologic response. Undifferentiated lymphocytes are derived from bone marrow stem cells. They are produced in the bone marrow at a very high rate (ICr cells per day). Some lymphocytes migrate through the circulatory or lymphatic systems to the secondary lymphoid tissue (thymus, spllen, aggregated lymph nodules in the intestines, and lymph nodes) where they produce lymphocyte colonies.

T cells or T Lymphocytes

Thymus-dependent lymphocytes or T lymphocytes, migrate from thymus where they are influenced by the hormone- thymosin and become immunological competent. T lymphocytes are mainly involved in cellular type of immunological response that is with cellular immunity, such as rejection of foreign tissue.

Some T cells arc transported away from the thymus and enter the bloodstream where they comprise 70 to 80% of the circulating lymphocytes. Other T cells tend to reside in various organs of the lymphatic system, such as the lymph nodes and spleen. This thymus-depended differentiation of T cells (or theymoyctes) occurs during early childhood, and by adolescence the secondary lymphoid organs of the body generally contain a full complement of T cells.

Table 18.3 Classes of Lymphocytes

Lumphocyte	**Role**
T Cells	
T_H (helper) cells; also called CD4^1 cells	Provide assistance, or potenitiate expression of immune function by other lymphocytes
T_s (suppressor) cells; also called CD8 cells	Suppress or impair expression of immune function by other lymphocytes
T_c (cytotoxic) cells; also called CD8 cells	Bring about cytolysis and cell death of "targets"
T_D cells or T_{DTH} (delayed type hypresentivity) cells; also called CD4 cells	Recruit and regulate a variety of nonspecific blood cells and macrophages in expression or delayed (Type IV) hypresensitivity reactions
B Cells	
B lymphocytes	Proliferate and mature into antibody-producing cells
Plasma cells	Are mature, active antibody-producing cells
Null Cells	
Natural killer cells	Bring about cytolysis and death of target cells

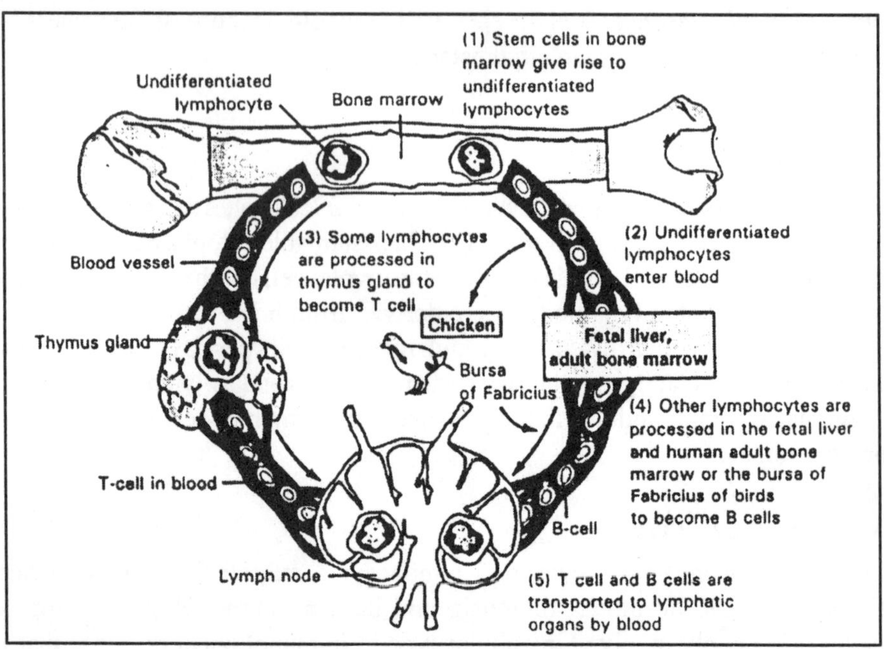

Diagram 18.5 Schematic presentation of lymphocyte development. Bone marrow releases undifferentiated lymphocytes which after processing become T and B cells.

B Cells or B Lymphocytes

B-lymphocytes: The origin of these lymphocytes is from the *bursa of Fabricus* of birds (a mass of lymphoid tissue near the cloaca). The letter B was originally derived from the *busa of Fabricius,* a specialized appendage of the cloaca of chickens where these lymphocytes differentiate. B cells are distributed by the bloom and make up 20 various lymphoid organs along with the T cells. These lymphocytes are mainly involved in the production of antibodies- humoral immunity. B lymphocytes differentiate the fetal liver and adult bone marrow.

Plasma Blasts: These cells are produced by B lymphocytes following antigenic stimulation. In fact, the lymphocytes change themselves into plasma blasts which multiply and differentiate into plasma cells. These are the large cells and contain relatively more basophilic cytoplasm, less developed endoplasmic reticulum, large nucleus with nucleoli and can multiply and differentiate themselves in plasma cells.

Plasma Cells: The plasma cells are uninucleated with amphophillic cytoplasm, rich in ribosomes, rough endoplasmic reticulum, with prominent Golgi bodies, acentric nucleus with cart wheel type chromatin.

Null Cell

There is a population of lymphoid cells that do not have characteristics of either T or B cells. These are called null cells because they lack the specific surface markers of B or T cells, and can be distinguished from them by the presence of cytoplasmic granules. It is

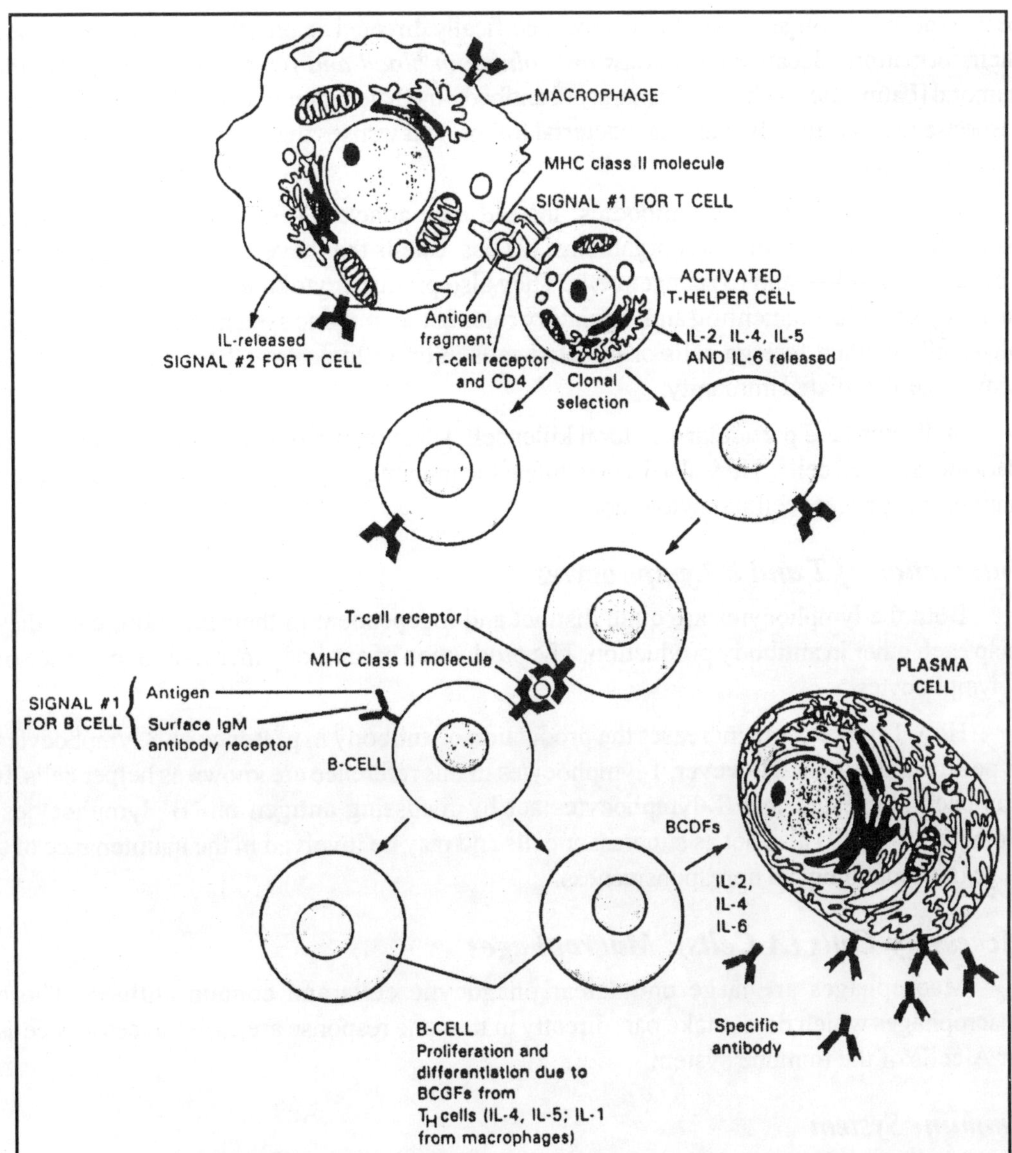

Diagram 18.6 T-dependent antigen triggering of a B cell. Schematic diagram of the events occurring in the interactions of macrophages, T-helper cells, and B cells that produce cell-mediated immunity.

currently believed that this population of cells contains most natural killer (NK) cells and antibody-dependent cytotoxic T cells. These cells are probably of bone marrow origin, however, their exact lineage is uncertain.

Function of Lymphocytes

Plasma cells are fully differentiated antibody-synthesizing cells that are derived from B lymphocytes. They respond to antigens by secreting antibodies into the blood and lymph. Antibodies are glycoproteins produced by plasma cells after the B cells in their lineage have been exposed to antigens. Antibodies are specifically directed against the antigen that caused their formation. Because *antibodies are soluble in blood and lymph fluids,* they provide humoral [Latin *humor,* a liquid] immunity or antibody-mediated immunity. The humoral immune response defends mostly bacteria, bacterial toxins, and viruses that enter the body's various fluid systems.

T cells do not secrete antibodies. Instead they attack (1) host cells that have been parasitized by viruses or microorganisms, (2) tissue cells that have been transplanted from one host to another, and (3) cancer cells. They also produce cytokines, chemical mediators that play specific augmenting and regulatory roles in the immune system. Since T cells must physically contact foreign cells or infected cells in order to destroy them, they are said to provide cell-mediated immunity.

Null cells (and particularly natural killer cells) destroy tumor cells and virus- and other parasite-infected cells. They also help regulate the immune response. Null cells often exhibit antibody-dependent cellular cytotoxicity.

Interaction of T and B Lymphocytes

Both the lymphocytes are quite distinct and independent in their function, even they help each other in antibody production. The production of antibody increases in presence of T lymphocytes.

How T lymphocyte increases the production of antibody in presence of B lymphocytes is poorly understood. However, T lymphocytes in this reference are known as helper cells. It has been suggested that T lymphocytes act by focussing antigen on 'B' lymphocytes. Sometimes T cells also act as suppressor cells and may be involved in the maintenance of a state of immunological unresponsiveness.

Accessory Cells (A Cells): Macrophages

Macrophages are large uninuclear phagocytic cells and contain antigen. Those macrophages which do not take part directly in immune response are called accessory cells or A-cells of the immune system.

Immune System

Immunological Tolerance

The important characteristics of immunological tolerance are: (i) it is induced by an antigen, (ii) it is specific, (iii) it is associated with immunological memory, and (iv) it is usually requires persistence of the antigen. Either both T and B cells or individual cells are involved in immunological tolerance.

Mechanisms of lmmunological Tolerance Development

These are as follows:

Clonal Abortion: When an antigen comes in contact with immature immunologically competent cells (T or B cells), maturation of these cells is aborted, which results in lack of corresponding active T and B cells to react with antigen in the later life.

Clonal Exhaustion: When an individual is challenged with repeated doses of T cells independent antigen, all the corresponding B cells are stimulated to differentiate into short lived antibody forming cells (plasma cells). Thereafter no competent cell is left to react with the antigen.

Functional Deletion: This has been explained on the basis of two aspects (i) by T cell dependent antigens, and (ii) by T cell independent antigen.

Antibody forming Cell Blockade: T cell independent antigen can also produce tolerance by blocking the antibody forming cells. It, however, requires lesser amount of antigen.

Generation of 'T' Suspensor Cells: Sometimes antigen may stimulate generation of 'T' suspensor cells which directly suppresses corresponding 'T' and B' cells to induce tolerance.

Factors Affecting Immunological Tolerance: There are two main factors which affect tolerance: (i) Age of an individual organism or the maturation of immune system, and (ii) nature of antigen, its dose and clearance in the organism. The tolerant-capacity of an individual persists so long as threshold amount of antigen persists in immune system, thus tolerance with a replicatied antigen (living cells) persists much longer than the foreign protein which is eliminated rapidly by catabolism.

Serology

Serology deals with the study of antigen, antibody reactions in vitro. It includes identification and quantitation of antigen or antibody using its known counterpart. The basis of these reactions and their application is the specificity of antigen antibody reactions.

Classification of Antigen - Antibody Reactions

The antigen-antibody reactions are termed on the basis of the observable effects of the reaction or the technique used. These include of:

1. Agglutination 2. Precipitation 3. Complement fixation 4. Opsonisation 5. Neutralisation 6. Immune cytolysis 7. Immune adherence 8. Immunofluorescence 9. Immunoelectrophoresis 10. Counter immune electrophoresis 11. Radio immunoassay 12. Enzyme linked immunosorbent assay and Immunoblotting.

Agglutination

When particulate form of antigen or antibody coated particles form clumps as a result of antigen-antibody interaction, the reaction is termed as agglutination reaction. Agglutination means clumping of the particles. In most agglutination reactions, the antigen is particulate and the antibody is in soluble form. However, if antibodies are coated on particles, agglutination of the latter can occur with soluble antigen. Agglutination occurs in two stages, the first stage- primary immunological reaction is the union between antigen and antibody and the second stage secondary reaction is the formation of visible clumps. The first stage is extremely rapid, it is completed in few seconds. It is not affected by temperature variation (0° to 40° C) and it does not require electrolytes. Higher concentration of ions may inhibit reaction by covering the oppositly charged antigen and antibody molecules. The second stage of the reaction is influenced by temperature and requires electrolytes. The ions reduce negative charges on particles which help in aggregation. Most reactions are accelerated by raising the temperature from 0° - 30° C. Some however, occur at low temperature only e.g., cold haemagglutination.

The agglutinate consists of a lattice of alternating antibody and antigen molecules. Immunoglobulin M is a potent agglutinating antibody. Because of its multiple valencies, one molecule of it can unite with 5 to 10 molecules of the antigen.

Precipitation

When soluble forms of antigen and antibody interact and form precipitate, the reaction is termed as precipitation reaction. Precipitation is an antigen antibody reaction in which the antigen is in soluble form. A precipitation reaction requires antibodies more than that for agglutination because with the decrease in size of particles, the total available surface of antigen increases. Though most precipitation reactions occur better at 37° C-45° C, more complete precipitation is frequently obtained at 0°-4° C. So quantitative precipitation tests are practically always refrigerated for an interval of one or more days . It is done by three tests: 1. Simple mixture. 2. Interfacial ring test. 3. Gel diffusion test.

Complement Fixation

Following antigen-antibody interaction, complement, if present, is activated and fixed to the antigen-antibody complex. The fixation of complement is detected by an indicator system – the sensitized or antibody coated sheep R.B.C. The test by which antigen or antibody is detected by activation of complement is known as complement fixation test.

Complement fixation test is based on the principle of fixation of complement factors to antigen antibody complexes which is detected by an indicator system consisting of sheep RBC and antibodies to sheep RBC. Unfixed complement causes haemolysis in indicator system. If complement is fixed to test antigen antibody complex, it is not available for the indicator system, hence haemolysis will not occur. If original test system is lacking in antigen or the corresponding antibody complement will remain free and haemolysis of sheep RBC occurs i.e. the test is negative (complement not fixed). Absence of haemolysis means complement is fixed to the test system i.e., the test is positive. In complement fixation test guinea-pig serum is usually used as source of complement and a calculated amount of complement is used which is just enough to be completely utilised by the test antigen antibody system. Further, the antigen and the serum may have anti-complementry activity. Therefore, first the titre of complement in guinea-pig serum is determined in presence of the test amount of antigen and the serum separately, and also in presence of pooled normal serum, and compared with the titre of complement obtained in absence of antigen and serum. Extra complement is used in the test system if antigen or serum possesses anti-complementry activity. In test system 5/4 of the titre of complement of pooled guinea-pig serum obtained in presence of 1 vol. of antigen and 1/5 vol normal serum, is used.

Complement fixation test is much more sensitive than agglutination and precipitation test but is more cumbersome to perform. At one time because of its sensitivity it was performed for detection of antigen or antibodies in diagnosis of a number of diseases and for the identification of the antigen. Complement fixation test has been used as Wassermann reaction in diagnosis of syphilis, and also for detection of specific antibodies to viruses, protozoa and rickettsia and a number of bacteria for diagnosis of the diseases caused by them e.g., diagnosis of deep seated gonococcol infections and amoebiasis, kala-azar, trypnosomal infections. It has also been used for detection of viruses grown in tissue culture or chick embryo. As a number of newer more sensitive and simple tests have been introduced in recent past, complement fixation test is now rarely performed.

Anti-complementry activity in serum develops on being kept at room temperature for some time. It can be eliminated by heating serum at 56° C for 30 minutes.

Opsonisation: Attachment of antibodies to particular antigen makes the latter easily phagocytosed by phagocytic cells. This enhanced phagocytosis by antigen-antibody interaction is known as opsonisation.

Neutralization: When attachment of antibodies to antigen neutralises the toxic effects of the antigen, the reaction is known as neutralisation e.g. Toxin-antitoxin interaction.

Immune Cytolysis: When antibodies and surface antigen of certain cells interact, it may cause cytolysis by complement activation.

Immune Adherence: Primate erythrocytes bear surface receptors for C 3. Therefore erythrocytes adhere to C3 attached to antigen antibody complexes. This is known as immune adherence. It is being utilised to detect antigen or antibodies as for complement fixation.

Immuno Fluorescence

When antigen or antibody detected by using fluorescent dye tagged antibody or antigen and fluorescent microscopy, the reaction/technique is known as immuno fluorescence.

Immunofluorescence technique was introduced following the use of fluorochrome labelled protein by Coon's and Kaplan (1950). It involves labelling of antibody with fluorescent dye followed by its use in detection or identification of antigen. It combines the sensitivity and specificity of immunology with precision of microscopy. The technique is more sensitive than agglutination, precipitation and complement fixation techniques. It can detect protein of the order of less than 1 μg/ml. of the body fluid.

Fluorescent dyes absorb ultraviolet light (between wave length 290 and 295 nm.) and emit light of longer wave length (525 nm) of visible spectrum. The fluorescent dyes in common use are Fluorescin isothiocyanate which emits green or apple green light and Lisiamine rhodamine B (RB 200) which emits orange light.

Immuno Electrophoresis: When a mixture of antigens are separated by electrophoresis and detected by immuno diffusion, the technique is termed as immuno electrophoresis.

Counter Immuno electrophoresis: When oppositly charged antigen and antibody molecules are subjected to migrate in an agar gel under the influence of electric current, a quick precipitation occurs where they meet in optimal concentration. This quick precipitation under the influence of electric current is known as counter immune electrophoresis.

Radio Immuno Assay (RIA): When antigen antibody reaction is detected by utilisation of radioactive labelled antigen or antibody and the technique is used to assay the antigen or the antibody in a given sample, the technique is known as Radio immuno assay.

This technique is used for the detection and quantitation of any substance that is antigenic and can be labelled with a radioactive isotope. In fact, this method is based on the competition between labelled (known) and unlabelled (unknown) antigen for the same antibody. The measurement of the amount of labelled antigen attached to antibody, it is essential to separate the antigen-antibody complexes from the mixture. Out of many available method, the most convenient method for such separation is solid phase radioimmunoassay where in the antibody is linked to an insoluble support e.g., agarose beads. The insoluble complex is then mixed with known and unknown antigen. The separation of antibody-bound labelled antigen from free-antigen is done by centrifugation and filtration. The radioactivity is then measured and percentage of labelled antigen bound to the antibody is calculated. Amount of unknown antigen is determined using reference standard curve.

Enzyme linked Immunosorbent Assay (ELISA): When the antigen antibody reaction is detected or one of the components is quantitated by enzyme labelled counter part and subsequent demonstration of fixed enzyme by its substrate, the reaction or the technique is known as enzyme linked immunosorbent assay.

ELISA, especially solid-phase ELISA is a improved diagnostic test since the technique readily lends itself to automation and is certainly feasible under field conditions. Although there are a number of different types of solid-phase EIA, the basic steps in each type are the same:

1. Attachment of the immunoreactant (generally antibody or antigen) to the solid phase to serve to capture the complementary reactant from the sample.
2. Incubation with the test sample so that the complementary reactants are always found in or compete for the second layer.
3. Amplification by, for example, enzyme-labelled antiglobulin.

The types of solid-phase EIA include: direct, indirect and bridge non-competitive ELSA in which antigen is immobilized on the solid phase; non-competitive EIA with antibody immobilized on the solid phase; and, competitive EIA with either antibody or antigen immobilized on the solid phase. The solid phase can be composed of a wide variety of materials including plastic, nitrocellulose membrane, paper, glass and cloth.

Immunoblotting: When antigens are first separated by poly-acrylamide gel electrophoresis and transferred on to nitro-cellulose paper strips, which are than used for detection of antibodies in the unknown samples.

It was first described by E. M. Southern in 1975. He devised a neat method for identification of DNA fragments from agarose-gels. The method involves the denaruration of DNA fragments to a nitrocellulose membrane, allowing the "probe" (radiolabelled RNA or complementary DNA) to hybridize with fragment of interest and finally detection of such fragment by autoradiography. This ingenious method is generally referred to as *Southern Blotting* after the name of its inventor. However, this method could not be applied directly to identify RNA fragments since RNA did not bind to nitrocellulose membrane. Alwine and his associates (1979) devised a method in which the nitrocellulose membrane was replaced by an RNA-binding paper (diazotized aminobenzyloxymethyl paper). This extension of Southern's method to RNA aquired the name *Northern Blotting.* Later, Thomas (1980) and others workers showed that RNA does, in fact, bind to nitrocellulose membrane under appropriate conditions. Towbin and his associates developed new method known as *Western Blot.*

Of the above reactions- agglutination, precipitation, complement fixation, neutralization, immunofluorescence, counter immuno electrophoresis, radioimmunoassay and enzyme linked immunosorbent assay are commonly used in serology laboratories.

Serological Tests

The serological diagnosis of an animal disease is a presumptive test which is usually then confirmed by direct culture of the causative virus or bacterium from excised tissues. Nucleic acid probes afford the opportunity to detect the organism directly in the tissue. Epizootic Haemorrhagic Disease Virus (EHD V) of deer has much in common with BTV.

Microbiological is also applied to: the genetic engineering of new bacteria (i.e. *Salmonella* and *Brucella)* for antigen and vaccine production; fermentation technology for antigen production; and, micromanipulation of embryos and in vitro fertilization for embryo transfer procedures which provides greater ability to control disease than ever before.

Biotechnology and Diagnoses of Animal Diseases

Animal Health

Animal health care covers many aspects including among others:

1. the improvement of animal productivity through either genetic engineering, animal embryo technology or growth hormones.
2. the improvement of animal nutrition.
3. the prevention and treatment of animal diseases by vaccines, monoclonal antibodies or interferons. Another extremely important aspect of animal health care is the diagnosis of animal disease. The scope of animal disease diagnosis is vast.

Objectives

1. To measures and to safeguard the Indian livestock population from the introduction of foreign animal diseases - for example, the import inspection system allows Indian producers access to genetic material such as semen and animal embryos from around the world.
2. To control and eradicate serious infectious and contagious indigenous diseases which threaten the economic viability of the livestock.

3. To ensure meat and animal product safety.
4. Development of research programme which promotes the effective achievement of these mandates.

Diagnostic tests: A wide variety of diagnostic tests should be performed. It is essential to have the capability to rapidly diagnose diseases like African Swine Fever and Bluetongue to Trichinosis and Vesicular Stomatitis to prevent their introduction in country. Therefore, it is essential to perform several tests like: simple agglutination procedures to highly specialized techniques, agar gel immunodiffusion (AGIO), buffered plate antigen test (BPAT), complement fixation (CF), culture, enzyme immunosorbent assay (E1A), fluorescent antibody (FA), histopathology, serum agglutination test (SAT), serum neutralization (SN) and tissue culture.

These classical diagnostic procedures reliable however, time-consuming, labour-intensive and expensive. There is obviously a continuing need to improve the capability for diagnosing animal diseases through the development of rapid, inexpensive, rugged and yet, of utmost importance, sensitive and specific diagnostic tests. A field test which is simple and rugged would allow veterinarians to quickly identify outbreaks of infectious disease, evaluate their spread and institute containment measures without the time required to send samples to a central testing laboratory.

The following two relatively recent biotechnologies provide the potential for development of improved diagnostic tests are: Monoclonal Antibody Production and Nucleic Acid Probe Technology.

Hybridomas Technology and Production of Monoclonal Antibodies

Hybridoma - a biotechnological tool- is the new path for achieving the goal of complete immunization of human body from infections. In 1975, a new era in the immunology was launched with the discovery of the hybridoma technique, a method of creating pure and uniform antibodies (immunoglobulins) against a specific target (antigen). To understand the basic concepts of hybridoma technology and its role in production of monoclonal antibodies.

Hybridomas are the hybrid cells of myceloma (cancer) cells with antibody producing cells (lymphocytes) from an immunized donor (animal). The hybrid cell or hybridoma resulting from this fusion has the ability to multiply rapidly and indefinitely in vitro and to produce an antibody of predetermined specificity, known as monoclonal antibody.

Somatic Cell Fusion

Barski et al. (1960) in laboratory of Virology and Tissue Culture of the Institute of Gaustave Roussy, at Villejuif (France); observed during their culture experiments of two stocks of tumour mouse cells that a new cell type was formed. This type had morphological characteristics and growth pattern that are different from those of parent cells. The nuclei of these new cells contained number of chromosomes equal to the sum total of chromosome numbers of the two parent cells. Unfortunately, the frequency of cell fusion was very low, between 10^{-4} to 10^{-6}.

Latter, the frequency of fusion of animal cells could increased by various workers using biotic and abiotic fusing agents. Okada (1962), succeeded in increasing the frequency of fusion between animal tumour cells by using the Japanese Haemagglutination Virus (JHV). He has also reported that JHV inactivated by ultraviolet radiation retained the ability to induce somatic cell fusion, so any complication resulting from using an infective virus could be avoided.

Harris and Watkins (1965), succeeded in somatic cell fusion between human HeLa cell and tumour mouse cells using inactivated Sendai virus as fusing agent. Weiss and Green (1967) noticed a peculiar feature in some somatic hybrid in which chromosomes of one of the parent cells gradually eliminated or lost. In somatic hybrid of human and mouse cells, human chromosomes generally lost after fusion while in somatic hybrid of monkey and mouse cells the chromosomes of monkey are eliminated. Davis (1981) used polyethylene glycol as fusing agent and could get somatic hybrids of animal and plant cells separately belonging to two different species and even hybrid of animal and plant cells.

Hybridoma Technology

The hybridoma technology for the production of standardized antibodies of a given class, specificity, and affinity has provided scientists with a tool that permits the analysis of vertually any antigenic molecule. Such reagents (antibodies) can be made in unlimited amounts whenever needed, thus, making them readily available to all investigators.

Kohler (1974) successfully produced a hybridoma by fusing a P3 myceloma cell (resistant to azaguanine) and a lymphocyte (from the spleen of a mouse) immunized against sheep-red blood cells. The experiment consisted of immunizing mice against red blood cells (the antibodies produced against this antigen are easily detected in the serum by means of an assay developed in 1963 by Jerne and Nordin), then mixing mouse myceloma P3 cells with spleen cells from immunized mice in the presence of polyethylene glycol. These chimeric cells, called hybridomas retained the property of immortality of the myceloma cells as well as that of secreting an antibody specific for a unknown antigen; resulting from the fusion of an antibody-secreting cell and of a tumour cell. They are capable of growing indefinitely in culture and of producing at the same time a particular species of antibody (a monoclonal antibody).

Production of Monoclonal Antibodies

The hybridoma technique seemed to be useful and applicable in producing monoclonal antibodies which is for affinity purification, tumor imaging, immunodiagnostics, and cancer treatment. They can also be useful in describing as well as predicting and optimizing secretion in antibody production system. The monoclonal antibody can be defined as a chemical reagent of known structure that can be reproduced at will, whereas conventional antiserum is a variable mixture of reagents that can never be reproduced, once the original supply is exhausted. The monoclonal antibody was purified by passage through chromatography columns containing the known antigen and, within 6 to 12 months, unlimited quantities of a single antibody could be obtained with a degree of purity and homogeneity. The clones selected can be stored by freezing technique.

It is now possible to obtain consistently large number of hybrids and one can select the lines producing the antibody of choice. The critical component in the production of hybridomas is the preparation of cells for fusion. Where rodents are used as a source of cell for fusion, immunization protocols must be deviced that optimize the proliferative response to antigen in the spleen. Plasmablasts appear to be more suitable as fusion partners than mature plasma cells.

The fusion and culture of hybridoma is quite straight forward. The scheme is equally useful when cultures of in vitro immunized cells are used as a source of plasmablasts. The production technique of monoclonal antibodies can be divided into three steps:

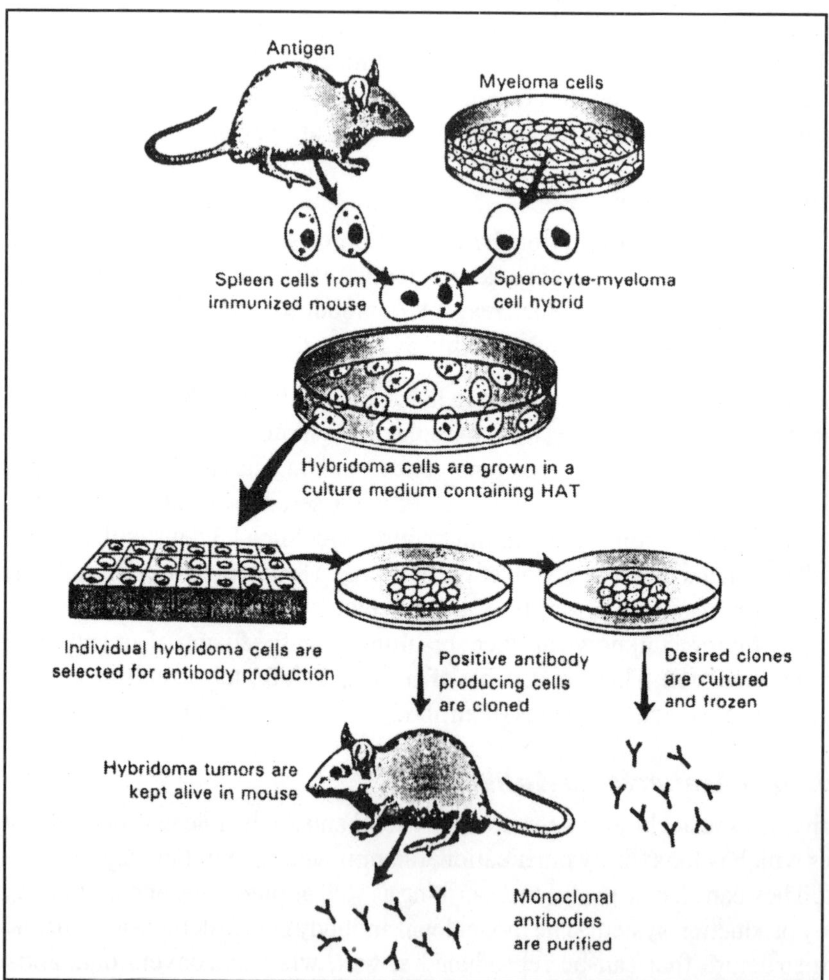

Diagram 18.7 Technique for the Production Monoclonal Antibodies. Antigen-stimulated spleen cells are fused with special mutant myeloma cells, yielding hybridomas. Each of them secretes a single, *Monoclonal* antibody. Once the hybridoma secreting the desired antigen is identified, it is cloned to generate many antibody-secreting cells that yield the huge quantity of a single antibody needed in medicine or science. Some hybridoma cells may be stored frozen and later cloned for antibody production or kept alive in laboratory animals.

Step I: Fusion and Culture of Hybridomas

1. Boost mice with an intravenous injection of antigen 72 hours before use. On the day of fusion, collect the spleen and make a free cell suspension by injecting and flushing the spleen with sterile serum-free medium. Centrifuge the cells and resuspend in lysing solution. When sterile water is used, suspend the cell in 1 ml of water and immediately dilute in 20 ml of medium. Any delay will result in the loss of plasmablasts and a consequent reduction in antibody-producing hybrids.

2. Maintain plasma cells in cell culture and feed into fresh flasks and medium for 16-24 hours before fusion to ensure that they are in early phase of growth at the time of fusion. At the time of use, collect the cells, centrifuge, and resuspend in serum-free medium.

3. Count both myceloma and spleen cells and then mix in the appropriate ratio. Depending on the properties of the tumour cells, the ratio of spleen to tumour cells may vary from 5:1 to 2:1.

4. Following mixing, the cells are centrifuged into a loose pellet by spinning at 1000 rpm for 10 to 15 minutes. Remove the supernatant and overlay the pellet with 1 ml of PEG. For 3 minutes, mix the PEG into the pellet. In doing so breakup the pellet into uniform small clumps.

5. Following fusion, dilute the cells in 30 ml of serum-free medium with the first 10ml medium being added and mixed at 1 ml per minute. Slow dilution reduces the risk of osmotic disruption of the fused cells. Centrifuge the cells and resuspend in complete medium containing HAT and then dispense into 96-well tissue culture plates, 10 plates for each 108 splenocytes used in the fusion. Added 106 thymocytes to each well to serve as feeder-cells. The latter step has proven critical for optimizing that outgrowth of newly formed hybrids. Feeder-cells and 2-ME in the medium exert a synergistic effect.

6. In carbon dioxide incubator with high humidity, incubate the cultures for 3-4 days with rapidly growing cultures between changes of culture medium, replace half the medium with fresh HAT medium every 4th day.

7. Identify and mark the wells containing hybrid colonies (hybridomas) on day 9 or 10 and then allow the colonies to grow to 500 or more cells. In rapidly growing cultures, supernatants can be collected and assayed for antibody activity by day 12-14. Where appropriate, collect and test the supernatants from the largest colonies first; test the remainder 2-4 days later.

8. After each test for antibody, transfer positive cell lines to 24 well plates. Add 3-5 x10^6 feeder cells to each well to promote rapid cell growth. Maintain cells in static culture for a minimum of 2 weeks by removing 50-75% of the hybrid cells after 2-3 days interval. The maneuver select for stable, rapidly growing, antibody-producing hybrids. Slow growing hybrids and hybrids that cease to synthesize antibody are eliminated.

9. After 2 weeks, make duplicate cultures and allow the cells to overgrow and die. Collect the supernatants and assay for the presence of antibody. Take cultures producing antibodies of the desired specificity from the master plate and expand into 6-well plates (2-3 wells for each cells lines). Harvest the cells twice for preservation in liquid nitrogen. Replenish the cultures with fresh medium and allow the cells to overgrow and die. Collect final supernatant for further analysis. The 6-well plates are essential for minimizing labour and time during this phase of hybridoma production.

Step II: Cloning and Preservation of Hybridomas

1. Following preliminary selection of hybrids, screen the final supernatants in detail to identify antibodies of immediate interest. Take the parent cell lines from the freezer and clone by limiting dilution. When viability is good, clone the cell lines immediately. Take the remainder of the excess cells and culture them in a T75 flask for 1-2 days. As soon as enough cells are present, harvest the cells and freeze one ampule to replace the ampule used for cloning. The culture should not be allowed to proliferate more than necessary to avoid change in composition of the cell line at this early stage of proccssing. As in the initial step, use feeder cells to promote growth.
2. At 6-8 days mark wells containing a single colony. At 12-14 days assay supernatants from the marked wells. Transfer 24-48 positive cultures to 24 well plates and maintain in static culture as described previously.
3. After noting which cloned lines are stable, expand 4-6 clones of each cell line into 6-well plates for cell preservation and production of antibody. Record which lines are stable and which are unstable.
4. Preserve 4-6 clones from each cell line in liquid nitrogen.

Step III: Production of Antibody

To produce antibody, culture the cloned cell lines in vitro or grow in ascites from the mice. After these steps, proper method is selected for assaying monoclonal antibodies in supernatants of hybridomas. Following methods are in use for assaying monoclonal antibodies:

(a) Enzyme-linked Immunosorbent assay (ELISA) (b) Radioimmunoassay (RIA). (c) Immunofluorescence (d) Cytotoxicity (e) Flow Cytometry, etc.

Application of Monoclonal Antibodies

Hybridomas can be stored by freezing technique. Hybridoma banks have been established in some institutes and laboratories in order to meet research needs. Many pharmaceutical firms are interested to the highest degree in the large-scale production of monoclonal antibodies from these hybridomas. Keeping in view their applications in various fields of medical science:

1. Dose determination of a medicine can be carried out by using the monoclonal antibodies of an animal, immunized against this particular medicine?

2. They are used to detect allergies, to carry out hormone tests, to diagnose viral diseases, to detect certain types of cancers, to monitor the presence or appearance of malignant cells after surgical or radio-therapeutic treatments.
3. The purification of complex mixtures or substances, the biological role which is important (proteins, hormones, toxins, etc.) could also be carried out with monoclonal antibodies.
4. The use of these antibodies was also envisaged for the labelling and precise identification of specialized cells such as neurones in order to gain better knowledge of the way in which these cells associate and operate.
5. Monoclonal antibody technique is also of great value in the area of the structure of cell membrane as membrane proteins are hard to purify.
6. In the field of direct therapy, serotherapy can be made more effective with the administration of a monoclonal antibody.
7. Monoclonal antibodies could also be used in the preparation of very specific vaccines, particularly against certain viral strains and against other parasites.
8. Monoclonal antibodies could also neutralize the action of lymphocytes responsible for the rejection of grafts and destroy the auto-antibodies produced in auto-immune diseases.
9. In association with medicinal substances, they could considerably increase the effectiveness of the latter on the target cells, while avoiding the serious side-effects of cancer therapies.

Many European and North-American firms are interested in the applications of monoclonal antibodies. In California, for-instance, certain companies are preparing diagnostic kits designed for the screening of certain lethal diseases. It is anticipated that the future support of some aspects of this hybridomas based monoclonal antibody technology will be tailored to the needs of each developing country in the world.

Immunotoxins

One result of hybridoma research is the production of immunotoxins. Immunotoxins are monoclonal antibodies that have been attached to a specific toxin or toxic agent (antibody + toxin = immunotoxin). Immunotoxins kill target cells and no others, because the antibody binds specifically to plasma membrane surface antigens found only on the target cells. This approach is being used to treat certain types of cancer.

In this procedure cancer cells from a person are injected into mice or rats to stimulate the production of specific antibodies against their plasma membrane antigens.Monoclonal antibodies are produced using hybridomas, purified,and attached to an agent toxic to the cancer cells.

When the immunotoxin is given to a cancer patient, it circulates through the body and binds only to the. cancer cells that have the appropriate surface antigens. After binding to the surface, the immunotoxin is taken into cancer cells by receptor-mediated endocytosis, and released inside. The immunotoxin then interfers with the metabolism of the target cells and kills them. Although this procedure is still experimental, it holds great promise in the treatment of certain types of cancer.

Monoclonal antibodies currently have following applications:

1. They are routinely used in the typing of tissue, in the identification and epidemiological study of infectious microorganisms.
2. They are used in the identification of tumor and other surface antigens
3. They are used in the classification of leukemias.
4. They are used in the identification of functional populations of different types T cells.

Anticipated future Monoclonal antibodies:

1. Passive immunizations against infectious agents and toxic drugs.
2. Tissue and organ graft protection.
3. Stimulation of tumor rejection and elimination.
4. Manipulation ol the immune response.
5. Preparation of more specific and sensitive diagnostic procedures.
6. Delivery of antitumor agents (immunotoxins) to tumor cells.

Nucleic Acid Probe Technology

The use of nucleic acid probes for diagnostic purposes is based on an entirely different set of principles, that is, the hybridization of complementary sequences of DNA or of DNA and RNA. The assumption is made that if specific DNA is present in a test sample then the organism must also be present. Whereas monoclonal antibodies (Mabs) can be used for both antigen and antibody detection (i.e. for scrodiagnosis), nucleic acid probes can only be used to detect *antigen* for want of a better word. Basically the steps of the procedure are as follows. The nucleic acid of the disease organism in question is extracted and bound to a membrane. DNA or RNA, a nucleotide sequence known to be unique to a region of the DNA or RNA of the disease organism is labelled (the *nucleic acid probe).* Conditions are created for the maximum binding of the probe to the DNA or RNA bound to the membrane. Unbound probe is washed off the membrane. Bound probe is then detected. The advantages afforded by nucleic acid probes over Mabs include; ease of test sample preparation (samples can actually be rather crude and include feces, tissue, blood, pus and other exudates); and, the ability to detect pathogenic determinants which would not be revealed immunologically.

Disadvantages

One of the main disadvantages to the use of nucleic acid probes is that the achievement of maximum sensitivity of detection still requires the use of radioactive labelling. A number of non radioactive detection systems such as biotin-avidin labelling, enzyme immunoassay, enzymic labelling, and fluorescence are being developed but there are still difficulties with high background in cruder sample preparations with these detection systems. In addition the. entire probe procedure is relatively time-consuming.

Enzyme-Linked Immunosorbent Assays (ELISA)

Microbial Research: Some Examples

The solid-phase ELSA can diagnose for a number of animal diseases including brucellosis, pseudorabies, bluetongue, paratuberculosis, infectious bovine, rhinorracheitis, trichinosis, epizootic haemorrhagic disease of deer, maedi-visna and rinderpest.

Brucellosis

Brucella abortus, the causative bacterium of brucellosis, causes uterine infections in cows which frequently result in abortions, and genital infections in bulls. It can also cause a persistent, latent infection in some animals. Brucellae are also highly infectious to humans causing a debilitating undulant fever. Intensive surveillance must be continued for a period to confirm that eradication is total and complete and to prevent reintroduction of the disease into the livestock. A highly standardized, automated, indirect solid-phase EIA technique should be used for detection of bovine antibody to *B abortus 2.*

Pseudorabies

Pseudorabies is a serious infectious viral disease of swine, cattle, sheep, dogs, cats and rats but is only naturally transmissible through swine. Pseudorabies virus (PRV) causes death of neonatal and weanling pigs. An indirect solid-phase EIA for the detection and quantitation of porcine antibody to PRV is in common use. It is faster and far more convenient than the standard serum neutralization (SN) test. A modified solid-phase EIA (dot-ELISA) in a dip-stick type of configuration has been developed which could have application as a rapid and economical field test for PRV diagnosis.

Bluetongue

Bluetongue is a viral disease of sheep and occasionally cattle. It is transmitted by insect vectors and is characterized by catarrhal stomatitis, rhinitis and enteritis and also by lameness. Both an indirect and competitive solid-phase EIA using a group-specific Mab can be used for detection of antibodies.

Biotechnology : Animal Vaccine Development and Production

Vaccine is an antigenic preparation administered with the object of stimulating the recipient's specific defence mechanisms in respect of given pathogen(s) or toxic agent(s). Some vaccines (e.g., the Sabin vaccine) are given orally while other (e.g., the Salk vaccine) are administered parentally. Vaccines are of following four main categories:

1. Inactivated vaccines e.g., vaccines against typhoid and cholera.
2. Those comprising suspensions of life (attenuated) pathogens, e.g., vaccines against yellow fever and tuberculosis.
3. Toxoids.
4. Those comprising a solution or suspension of the antigenic extracts of specific pathogens, e.g., the vaccine containing polysaccharide capsular material of *Streptococcus pneumoniae.*

Uses of Vaccines

The use of vaccines is extremely important for preventing various serious diseases. The development and production of these vaccines constitute an important function of the pharmaceutical industry. Vaccines are produced either by mutant strains of pathogens or by attenuating or inactivating virulent pathogens without removing the antigens necessary for eliciting the immune response.

Production of Vaccines

For the production of vaccines against viral diseases, strains of the virus are often grown by using embryonated eggs. Individuals who are allergic to eggs cannot be given such vaccine preparations. Viral vaccines are produced by using tissue culture. For example, the older rabies vaccine, which was produced in embryonated duck eggs and had painful side effects, has been replaced with a vaccine produced in human fibroblast tissue cultures which has far fewer side effects. The production of vaccines by bacteria, fungi, and protozoa generally involves growing the microbial strain on an artificial medium, which minimizes problems with allergic response. Vaccines should be tested and standardized before use.

Production of Vaccines

The majority of licensed vaccines for humans and animals presently in use are produced by conventional methods by manipulating the genes of bacteria causing diseases, and using the restructured genes for making antibodies. Another approach is to clone the genes for coat protein (antigens) of viruses in bacteria and preparing vaccines from bacteria. This is less hazardous than growing the whole virus in the laboratory. Some viruses like the smallpox and marburg virus are so dangerous that one does not actually want to propagate them in the laboratory. These include killed vaccines and live attenuated or inactivated vaccines. At present, a large number of viral vaccines are of the killed variety.

Advantages with Killed Vaccines

1. One of the major advantages of such vaccines is that they are relatively stable under environmental conditions, therefore, it is not as crucial to maintain a cold chain to ensure efficacy of the vaccines.
2. In specific disease situations such as rabies virus, clinicians are often reluctant to use live viral vaccines, because of the fear that they may inject themselves with the vaccine and there may be some adverse side effects. Although this possibility is extremely remote, the psychological trauma of injection with a virus such as rabies is sufficiently great to discourage some clinicians from using live virus vaccines.

Disadvantages with Killed Vaccines

1. The disadvantages of killed vaccines are that they do not replicate within the host and, therefore, large amounts of antigen are required for injection before induction of immunity. Since these vaccines are often produced in foreign tissue there is also the possibility of reactions developing against foreign proteins.

2. Since the vaccines are killed, they generally are injected intramuscularly.
3. A killed vaccine is more effective against systemic viruses than against viruses which replicate in local mucosal sites. This latter disadvantage has lead to the development of a large number of attenuated vaccines.

Advantages with Attenuated Vaccines

1. The main advantage of attenuated vaccines results primarily from their ability to replicate in the host.
2. Since their mode of action is similar to natural infections, immunity is generally of a broader spectrum than it is with killed virus vaccines. Thus they can induce a range of immune responses both locally, as well as systemically.
3. Attenuated vaccines develop immunity for longer duration than that of killed virus vaccines.
4. Since the virus replicates in the host and produces large quantities of proteins to which the host responds to the possibility of injecting foreign proteins is dramatically reduced with attenuated virus vaccines.

Disadvantages with Attenuated Vaccines

1. Since the vaccines are produced by passage in culture, to induce random mutation(s) or mutated with a specific agent and thereby reduce virulence, it is possible that passage in the natural host may result in reversion-back to virulence, e.g., attenuated polio virus. In the case of polio, reversion can occur within a few days of oral immunization.
2. Interference is also a potential problem. When viruses are grown in culture, there is a possibility to have other contaminating viruses present in it. For example the presence of BVD virus (a ubiquitous virus) in viral vaccines grown for immunizing cattle is very common. This virus is present in many of the cell lines and fatal bovine sera than are used for growing bovine viruses. Such interference may result in reduced replication of the attenuated virus vaccine and thus, reduced immunity.
3. Live attenuated virus vaccines are also extremely susceptible to environmental factors which may reduce their efficacy upon storage.
4. The attenuated virus vaccines induces latent infections and abortions if not administered properly or if administered at the wrong time in the animal's life.

Our Limitations

Killed and attenuated virus vaccines have not eliminated viral diseases (with the exception of small pox) completely. We still need to produce better vaccines that may be more efficacious and safer for use in human and animal medicine. There are a number of viruses for which we do not have vaccines due to the inability to grow virus in culture or in other economically acceptable culturing media. Some viruses may be of suppressive nature or impossible to attenuate by passage. In order to develop vaccines against exotic viruses, one requires excessive laboratory containment or restricts the use and application of such vaccines. Some of the newer technologies available would greatly eliminate some of these restrictions.

New Technologies

Table. 18.4 summarizes some of the newer technologies that are available and presently being used to produce new virus vaccines. These-technologies are based on classical approaches (reassortants, temperature sensitive or cold adapted, heterologous vaccines). These are based on the ability to manipulate the genetic material of the viruses in such a way as to either reduce the virulence of a specific virus in a specific way or identify the specific protective proteins and express them in a foreign host.

Table 18.4 Technologies for producing new vaccines.

Methodology	Example
1. Recombinant DNA - Expression of genes in foreign hosts	Influenza, AIDS, VSV
viruses (baculovirus, herpesvirus, adenovirus, vaccinia)	
bacteria (*Salmonella)*	Rotavirus
yeast	Hepatitis B
mammalian cells	Herpes
2. Reassortants	Influenza
3. Heterologous viruses	Rota
4. Genetic deletions	Herpes
5. Mutations	Polio
6. Antiidiotypes	Rabies

Steps of Sub-unit Vaccines Production

1. Identify protective proteins or epitopes on the proteins. Once this is done an individual can either produce a sub-unit vaccine by recombinant DNA technology or by synthetic peptide technology.
2. Identify gene coding for the protein.
3. Clone the gene coding for the specific protein and express it in a suitable expression system.
4. Purify the protective protein to homogeneity using bovine herpesvirus-1. BHV-1 has four major glycoproteins: GVP I, GVP II, GVP III and GVP IV.
5. Monoclonal antibodies immunosorbent columns are prepared and used for purification of large quantities of the BHV-1 glycoproteins. These glycoproteins are then mixed with the adjuvant avridine and used to immunize animals against BHV-1 virus.

Monoclonal Antibodies and Its Use in Lymphatic filariasis

Human lymphatic filariasis is most important vactor borne parasitic disease and is caused by two species of nematodes namely, *Wuchereria bancrofti* and *Brugia malayi.* This disease is endemic in India. A well known clinical symptom is swelling of legs, popularly known as elephantoid leg (elephant like). The methods developed to diagnosis the disease so far are not specific enough due to extensive sharing of antigen between helminths. Application of hybridoma-derived monoclonal antibodies to the diagnosis of lymphatic filariasis infection has offered a means for overcoming some of these limitations.

Vaccine Expression Systems

Once the specific protective proteins are identified, it is important to develop expression systems to produce large quantities of the specific protein in an economical fashion. At present, there are four different expression systems: (1) prokaryotic, (2) viruses, (3) eukaryotic and (4) mammalian.

Prokaryotic expression systems do not appear to be very useful for production of vaccines.

Bacterial expression systems is the viral protein produced in bacteria are often not folded properly for Induction of the desired immune response. A considerable amount of activity is being directed towards using other viruses such as vaccinia, herpesviruses or adenoviruses to express specific viral proteins in mammalian systems.

Virus expression or vaccinia expression: Another very popular expression system in the application of an insect virus, baculovirus, to produce high quantities of animal virus proteins in insect cells. Vaccinia virus will be used as an example where in a number of different viral proteins have been introduced into the vaccinia virus and are being used by a variety of different delivery systems to induce both local as well as systemic immunity. The advantages of vaccinia expression are that both a humoral, as well as a cellular immune response is ellicited. Even more attractive is that the vaccinia genome is very large and it is possible to delete large quantities of its genome and still maintain a viable virus. The non-essential vaccinia genes can be replaced with a number of genes coding for other proteins from other viruses. Expression of a number of genes in one virus would be much more economical to do than to culture each individual vaccine independently.

Vaccinia virus expression system is thermo stable: Finally, vaccinia can replicate in a wide variety of hosts, making it attractive for controlling infectious diseases in veterinary medicine and in human medicine. Furthermore, the thermal stability and its ability to replicate in a wide variety of hosts provides the opportunity to immunize wildlife against infectious diseases that can be transmitted to domestic livestock.

Vaccinia Virus Expression Systems for Wild Animals: Recently this system has been used is the case of wildlife rabies, where vaccinia vims containing the rabies virus glycoproteins is incorporated in bait and seeded in rural areas by dropping the bait from planes. Foxes and raccoons (a North-American nocturnal carnivore), which can be carriers of rabies virus, would eat the bait and be immunized against rabies virus. Using this approach, the number of animals that can be immunized, thereby reducing the epideminological spread of virus in the environment.

Peptide vaccines: The introduction of the specific protein into various expression systems, help us to identify the specific epitopes involved in inducing protective immunity and synthesis of the peptide. Using monoclonal antibodies against different proteins of bovine rotavirus, one can identify a immunodominant neutralizing epitope on the outer coat glycoprotein of bovine rotavirus. This epitope is identified by the ability of monoclonal antibody directed against it to neutralize virus in vitro as well as to prevent diarrhea in animals in vivo. The advantage of peptide vaccines is that it is possible to identify crucial epitopes on all viruses. Synthetic peptides generally are not very immunogenic unless they are linked to specific carriers and incorporated in strong adjuvants. In addition to incorporate adjuvants into peptide vaccines, the peptides can be engineered in such a way that they are linked to specific carriers which can act as ideal delivery systems for presenting the important epitopes on the peptide. Experience with virus carriers for synthetic peptides clearly indicates that the immunogenicity of these peptides linked to virus carriers approaches that of whole virus.

Expression of Viral Genes in Eukaryotic Cells System

It is the most natural method of producing non-infectious viral vaccines. The reason for the attractiveness of eukaryotic cells is that the proper level and degree of glycosylation and folding is more natural than in prokaryotic systems. At present a number of viral genes have been successfully expressed in yeast, mammalian cells, and more recently in green algae and filamentous water fungi. Green algae and filamentous water fungi provide large quantities of cheap proteins with the correct post-translational modification required for proper recognition of the host's immune system.

Use of Yeast as an Expression System

Yeast (S. *cerevisiae)* is used as an expression system because of our extensive experience with this organism and animals already have antibodies to yeast. Further, yeast do not have any oncogenes. This makes vaccines expressed in yeast potentially safer than vaccines produced in mammalian cells. Unfortunately, in some cases yeast may over-glycosylate proteins, which may influence immune response to that specific protein. Thus, the degree of glycosylation of highly glycosylated proteins may preclude its use in vaccine development.

Use of mammalian cells: The ultimate eukaryotic expression system is the use of mammalian cells for the continuous production and secretion of viral proteins and glycoproteins. However, the level of expression is relatively low and the high cost of cell culture is serious disadvantages in the use of mammalian cells for production of vaccines for veterinary use.

Microcarriers: For mammalian cells to be an economically viable vehicle. The development of microcarriers to produce large quantities of mammalian cells in a very concentrated environment, as well as strong promotors is in need of development. Extensive progress is being made in developing microcarriers to culture mammalian cells in a continuous fashion. In parallel, with the development of microcarrier systems are the requirement for new media and profusion of the bioreactors so that cells can grow continuously with minimal manipulations.

Control of Hepatitis-B Virus through Vaccines

Hepatitis-B is a major public health problem. An estimated 210 million Hepatitis-B virus carriers are in the world. India is known to be the second largest reservoir of this virus. The disease caused by this virus is characterised by higher morbidity, mortality and an amazing speed of spread of infection. About 80% liver cancer is attributed to Hepatitis-B infection. Till date no chemotherapeutic agents are available for its treatment. Prophylaxis against this has come in practice in 1981 with the introduction of a plasma derived vaccine. The scientists were fearing to introduce this vaccine because of the possible danger of spread of other infections including AIDS. In 1986, a genetically engineered yeast-derived vaccine has been introduced that is of non-plasma in its origin and has widely been accepted by the public.

Hepatitis-B: Plasma Derived Vaccine

The discovery of a unique HB antigen called Australia antigen (HBs Ag) in 1965 in the blood of infected persons. This has lead to the development of two effective vaccines against this dreaded diseases: a plasma derived vaccine, in 1981, and a more acceptable recombinant DNA (r-DNA) technology based yeast derived vaccine, in 1986. Dr. Baruch Blumberg received Nobel prize in 1976 for his pioneering contributions to HB.

The HBs Ag is present in the serum of infected people as small particles of 22 nm size. These nucleic acid-free particles have no infectious properties. When blood serum taken from an acute case of HB, contained enough HBs Ag to accord at least this partial protection. In 1981, through this concept, HB vaccine was made available by Merck & Co., USA and Institute Pasture Production, Paris as Hepatavax-B.

Hepatitis-B: Genetically Engineered Vaccine

A second generation of HB-vaccine has been developed by employing the techniques of genetic engineering. The fragments of DNA are joint to the DNA of a suitable vector. The vactor may be plasmid, phage or a cosmid. This is done by the use of restriction endonucleases. Artificial *linkers* containing restriction sites can be attached to the foreign DNA fragments to allow insertion into the vector. Once joined, the composite DNA is introduced into the bacterial or eukaryotic cell system by either transformation or transfection. After successful integration of the composite genome into that of the host cell, it affects the metabolism of the host cell. This further, directs and to translate the additional genetic information packed in vactor DNA; and consequently leads to, the synthesis of foreign proteins by the host cells. In this way host cell acts as mini-factory for the production of specific protein which is foreign to it.

Microencapsulation in Medicine: A New Approach

Living mammalian cells can be microencapsulated within a semipermeable membrane for the purpose of an artificial endocrine pancreas or for the large-scale growth of mammalian cells. This system has been used for pancreatic islets to enable transplantation of islets into diabetic rats and for hybridomas, lymphoblastoid cells, and fibroblasts for the production of human monoclonal antibodies and interferon.

Polymeric Devices

They are used for replacement of many types of diseased tissues and organs in humans, e.g., artificial joints, breast construction, large diameter vascular grafts, and controlled release of drugs. For these applications, insert surfaces, those do not interact with proteins or cells have been used to minimise long term tissue reactions. At the opposite extreme is the use of medical polymers as substrates for cell transplantation. In this case the ultimate replacement of the organ or tissue function depends primarily on the transplanted cells; the polymer plays an ancillary role as a substrate for adhesion of the cells and to give structure to the nascent tissue or organ.

Acquired Immuno Deficiency Syndrome (AIDS)

In the summer of 1980, in New York a patient died after a long illness caused by infection that the body can normally fight with or no problem. In June 1981, the centers for disease control (CDC) of USA reported that five young homosexual males in the Los Angeles area had contracted Pneumocystis Carinii, of which, two of the patients had died. This report signalled the beginning of an epidemic of retroviral disease characterized by profound immune-suppression associated with opportunistic infections, secondary neoplasms, and neurologic manifestations, which have come to be known as AIDS.

Etiology

AIDS is caused by HIV (Human Immuno Deficiency Virus). This human virus is a retrovirus belonging to the lentivirus family. Lentivirus family also includes feline immunodeficiency vims, simian immunodeficiency virus, visna virus of sheep, equine infectious anemia virus. In past, HIV was called by other names such as LAV (Lymphadenopathy associated virus), IDAV (Immune deficiency associated virus), HTLV-III (Human T-lymphotropic virus- type III).

Characteristics of HIV

1. a long incubation period, followed by a slowly progressive fatal outcome,
2. tropism for hematopoietic and nervous systems,
3. an ability to cause immuno-suppression,
4. cytopathic effects *in vitro.*

Structure of HIV

HIV is spherical in shape and has 0.1 micrometer diameter. It is differentiated into outer envelop, core shell and inner cone of RNA

Outer envelop: The outer envelop consists of proteins, which are distributed on the surface like a saucer ball made of 12 pentagons and 20 hexagons stiched together to make a sphere. A molecule of gp 120 proteins appears as a *knob* at the corners of the hexagons, with an extra molecule of the protein in the centre of each hexagon. Thus the total number of gp 120 molecules comes to 80.

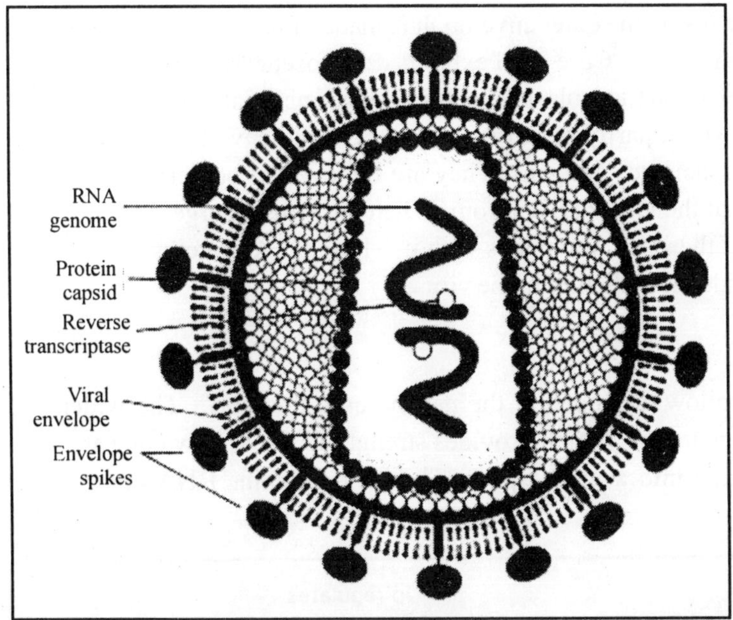

Diagram 18.8 (a) HIV virion. The virus particle is covered by a membrane that is derived from the host cell. Studding the membrane are viral glycoproteins, gp41 and gp120. Inside there is a core made up of proteins designated p18 and p24. The viral RNA and the enzyme reverse transcriptase are carried in the core.

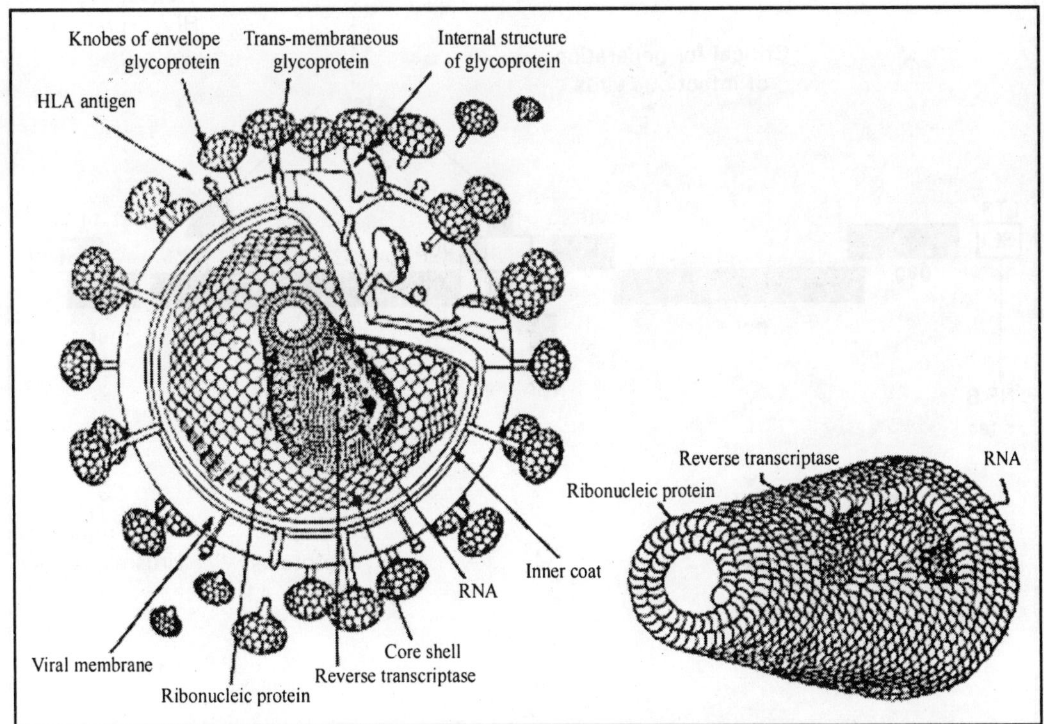

Diagram 18.8 (b) The latest model of the AIDS virus; (c) Enlarged view of the inner cone made of RNA molecule.

Core Shell

The core shell is found inside the outer envelop. It is made of proteins surrounding the centre of the core. It is a dense mass. Core shell exhibits deltacosehedron symmetry. It is a polygonal structure composed of 60 triangular elements forming a mix of alternating hexagonal and pentagonal structures which partly penetrate each other. The envelope also contains HLA (human-leucocyte- associated) antigens. They are believed to be derived from the membrane of human cell that the vims derives from. When the virus emerges or buds from these cells, it takes some of these HLA antigens with it. These HLA proteins do not form any set pattern. These provide individuality to the viruses.

Central or Inner Cone

The central cone is hollow and open at the narrow end, the "top". The wider end possesses a dimple- like indentation, which provides strength to the hollow cone and also accommodates more proteins into a given space. The cone contains RNA and reverse transcriptase enzyme.

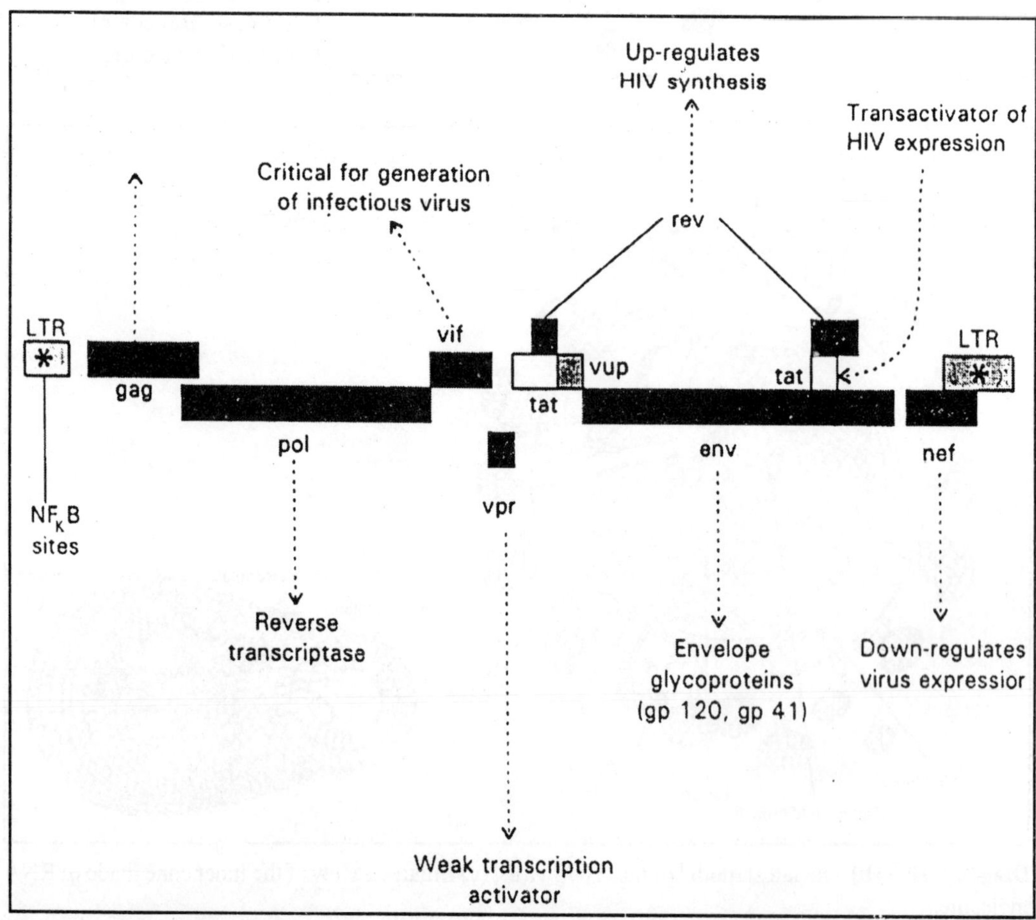

Diagram 18.9 HIV proviral genome. Several viral genes and their recognized functions are illustrated.

The HIV proviral genome contains following genes:

Structural genes	
gag gene	This gene codes for core proteins,
polgene	pol gene codes for reverse transcriptase.
env gene	This gene codes for envelope proteins.
Regulatory genes	
tat gene	tat gene is a transactivator gene regulating protein synthesis.
rev gene	It is the regulator of expression of virus.
vif gene	It is the viral infectivity factor and enhances viral replication.
nefgene	It is the negative factor and suppresses replication and may be responsible for latent period.

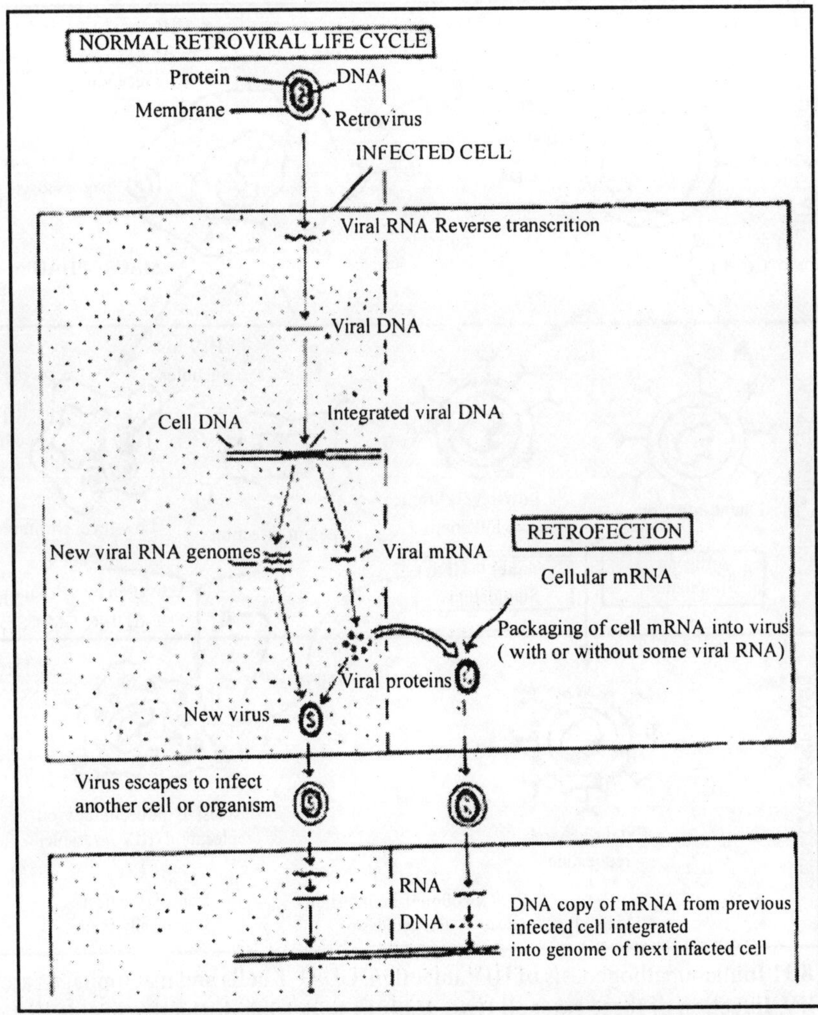

Diagram 18.10 Diagram showing the mechanism by which a retrovirus (including HIV) completes its life cycle and infects new cell.

Two genetically different but closely related forms of HIV, called HIV-1 and HIV-2, have been isolated from AIDS Patients. HIV-1 is the most common type associated with AIDS in USA, Europe, and central Africa, whereas, HIV-2 is common is West Africa.

Pathogenesis

A retro virus, including HIV, is a simple chemical package containing the viral RNA, along with a few molecules of reverse transcriptase enzyme, which copies the viral RNA into DNA as soon as the virus infects a new cell. This DNA then integrates into cellular DNA from where it orchestrates production of both the messenger RNA (mRNA). It codes for viral proteins, and new copies of the viral genome. Finally, newly formed viral genomes and proteins are assembled into new viruses, which escape from the cell.

There are two major targets of HIV: the immune system and the central nervous system.

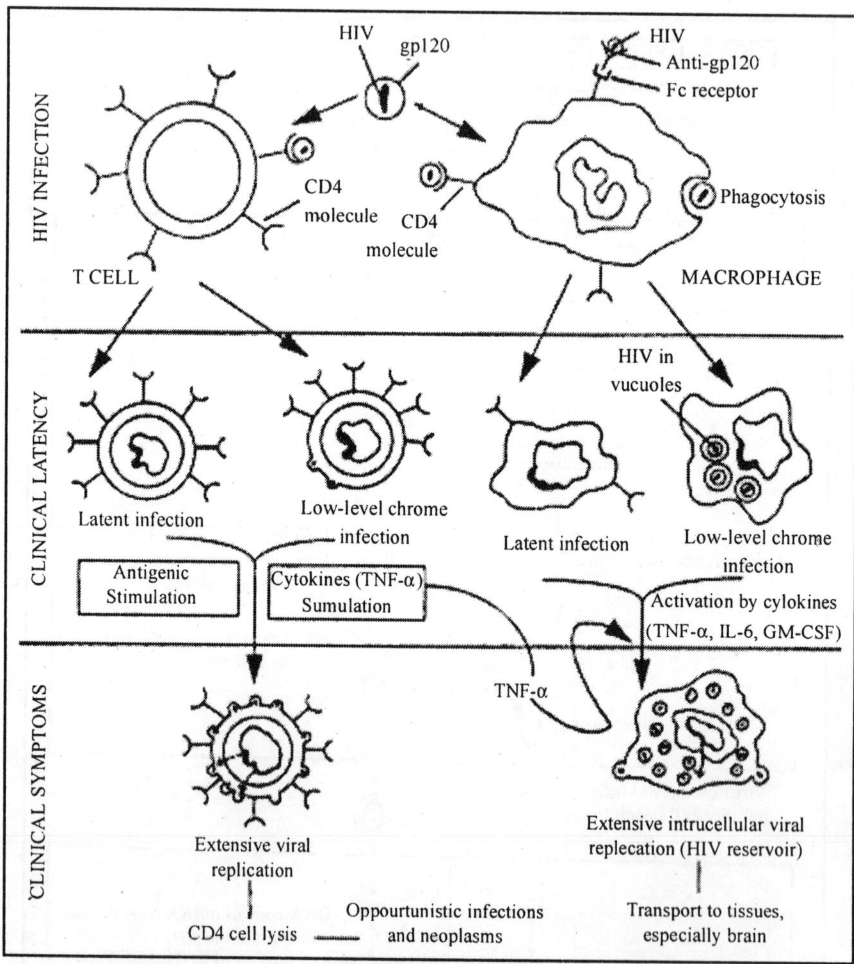

Diagram 18.11 Immunopathogenesis of HIV infection. CD4+ T cells and macrophages are the major targets of HIV. Infection of these two cell types leads to somewhat distinctive events that eventually lead to a marked loss of CD4+ T cells and dissemination of HIV into various tissues, especially the central nervous system.

Immunopathogenesis

Profound immunosuppression, primarily affecting cell-mediated immunity, is the hall mark of AIDS. This results chiefly from a severe loss of $CD4^+$ T cells as well as an impairment in the function of surviving helper T cells. HIV also infects marophages such as monocytes.

The CD4 molecule (present on the surface of T lymphocytes and macrophages) forms a high affinity receptor for HIV. This explains the selective tropism of the virus for $CD4^+$ T cells and its ability to infect other $CD4^+$ cells, particularly macrophages.

The various steps of infection of T cells by HIV are:

1. Binding of gp 120 envelop glycoprotein of HIV to CD4 molecules.
2. Binding is followed by fusion of the virus to the cell membrane and internalization. During this step viral gp 41 makes contact with some unidentified component of the cell membrane.
3. After internalization, the viral genome undergoes reverse transcription leading to formation of proviral DNA that is then integrated into host genome.
4. Integration of proviral DNA into host genome is followed by either of the two steps-

   (a) the provirus may remain locked into the chromosome for months or years and hence the infection may become latent, or

   (b) proviral DNA may be transcribed with the formation of complete viral particles that bud from the cell membrane leading to cell death.

Mechanisms other than direct cytolysis are involved in the causation of profound T-cell deficiency that characterizes late stages of HIV infection. These mechanisms may be-

1. Loss of immature precursors of $CD4^+$ T-cells, either by direct infection of thymic progenitor cells or by infection of accessory cells that secrete cytokines essential for $CD4^+$ T-cell proliferation.
2. Formation of syncytia (giant cells) by fusion of infected and uninfected cells. Fused cells develop *ballooning* and usually die within a few hours.
3. Autoimmune destruction of both infected and uninfected $CD4^+$ T-cells.

HIV also infect monocytes and macrophages. There are certain differences between HIV infection of T cells and macrophages, which may be-

1. Unlike T-cells, the majority of the macrophages that are infected by HIV are found in the tissues and not in peripheral blood. Infected macrophages are detected in tissues like brain, lymph nodes and lungs.
2. HIV may infect macrophages by the gp 120-CD4 pathway, and HIV may also enter macrophages by phagocytosis or by Fc receptor- mediated endocytosis of antibody-coated HIV particles.

3. Infected macrophages bud relatively small amounts of virus from the cell surface, but these cells contain large numbers of virus particles located exclusively in intracellular vacuoles.
4. Unlike CD4⁺ T- cells, macrophages are quite resistant to the cytopathic effects of HIV.

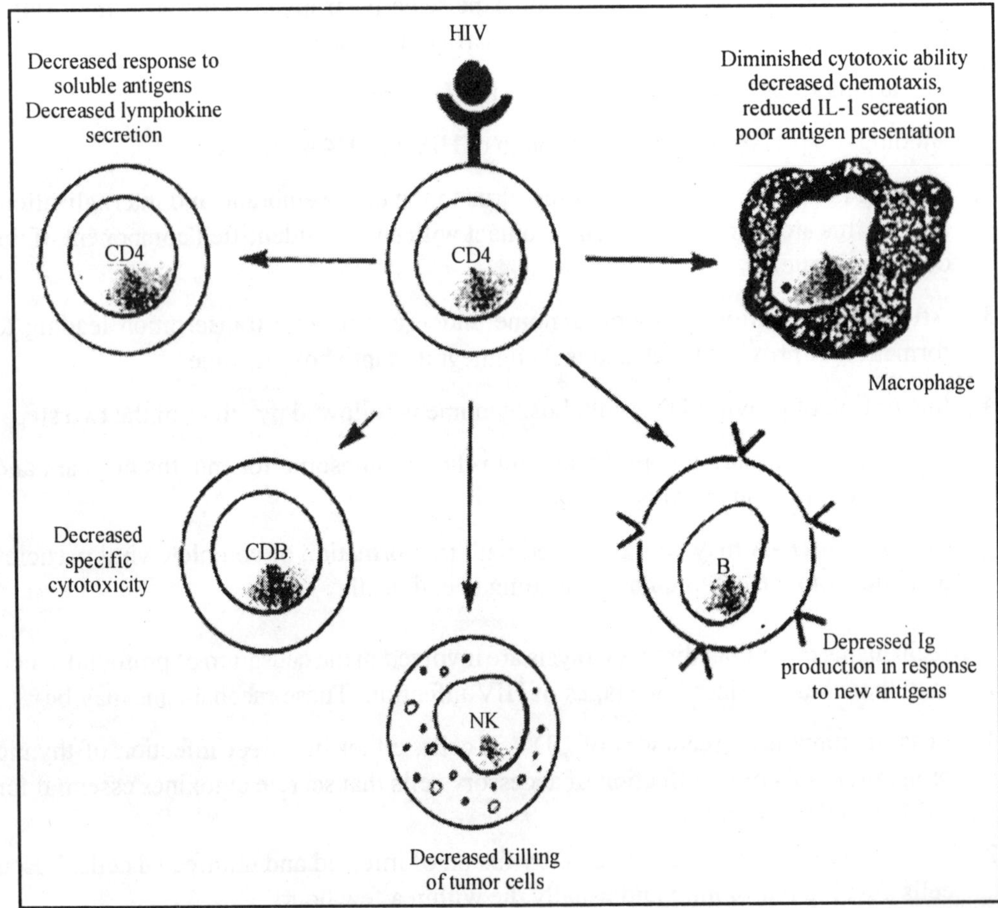

Diagram 18.12 The multiple effects of loss of CD4+ T cells by HIV infection.

HIV infection also causes profound abnormalities of D-ccll function. The gp 120 can promote B-cell growth and differentiation, and HIV- infected macrophages produce increased amounts of cytokine IL-6, which favours activation of cells. Despite the presence of activated B-cells, the AIDS patients are unable to mount an antibody response to a new antigen.

It must be recalled that the CD4⁺ T cells play a pivotal role in regulating the immune response: they produced a plethora of cytokines such as 1L-2, 1L4, IL-5, IFN-Y, macrophage chemotactic factors, and hematopoietic growth factors such as GM-CSF. Therefore, loss of this "master cell" has ripple effects on virtually every other cell of the immune system.

Major abnormalities of immune function in AIDS

Lymphopenia

Predominantly due to selective loss of the CD4+ helper inducer T-cell subset; inversion of CD4-CD8 ratio

Decreased T-cell Function in vivo

Susceptibility to opportunistic infections
Susceptibility to neoplasms
Decreased delayed-type hypersensitivity

Altered T-Cell Function in vitro

Decreased proliferative response to mitogens, alloantigens, and soluble antigens
Decreased specific cytotoxicity
Decreased helper function for pokeweed mitogen-induced B cell immunoglobulin production
Decreased IL-2 and IFN-Y production

Polyclonal B-Cell Activation

Hypergammaglobulinemia and circulating immune complexes
Inability to mount *de novo* antibody response to a new antigen
Refractoriness to the normal signals for B-cell activation *in vitro*

Altered Monocyte or Macrophage Functions

Decreased chemotaxis
Decreased HLA class II antigen expression

Pathogenesis of Central Nervous System Involvement

In addition to the lymphoid system, the nervous system is a major target of HIV infection. Macrophages and cells belonging to the monocyte and macrophage lineage (microglia) are the predominant cell types in the brain that are infected with'HIV. It is believed that HIV is carried into the brain by infected monocytes, hence neuronal damage may be secondary to release of cytokines or other toxic products from infected macrophages. HIV may also be found in brain in cell types, other than macrophages, including astrocytcs, oligodendrocytes, and endothelial cells.

About 60% AIDS patients suffer from dementia before they die; they will have problems with memory, thinking and behaviour. HIV infected brains exhibit shrinkage, small groups of inflammatory cells, and spaces in the *white matter.*

Symptoms of AIDS

Persons infected with HIV do not react in the same way. Some people remain symptomless for several years, whereas, other (60-70%) show symptoms earlier. The general symptoms associated with AIDS are:

1. Weight loss or abnormally slow growth in children.
2. Chronic diarrhoea for more than a month.

3. Prolonged fever for more than a month.
4. Persistent cough for more than a month.
5. Generalised puritic dermatitis.
6. An episode of herpes zoster (viral infection).
7. Oro-pharyngal candidiasis (fungal infection in mouth and throat).
8. Chronic progressive and disseminated herpes simplex infection.
9. Generalised enlargement of lymph glands (lymphadenopathy),
10. Repeated common infections such as otitis media (ear infection) or pharyngitis.
11. Lung diseases and skin tumors (Kaposis sarcoma).

Modes of HIV Infection

The common modes of HIV transmission may be as follows:

Sexual transmission

Male homosexual practice: HIV present in seminal fluid is transmitted to passive partner via anorectal abrasion.

Heterosexual transmission

1. Female to male: HIV infection present in female genital secretions and blood pass on to male partner via penis.
2. Male to female: From infected seminal fluid and blood, HIV is transmitted to female cervix.
3. In both sexes the risks are very considerably increased where there is genital ulceration or abrasion.
4. Artificial insemination: In Britain, women who were artificially inseminated with semen from symptomless carriers of HIV subsequently developed antibodies against virus.
5. Sexual partners of haemophiliacs infected with HIV have also developed the infection.

Blood transfusion

If HIV infected blood or blood products are transfused into a healthy man, the recipient develops AIDS. Haemophiliacs receiving HIV contaminated factor V1I1 develop AIDS.

Drug abuse

People who inject drugs intravenously can catch AIDS by sharing a needle or syringe with someone who is infected with HIV.

Mother to baby

HIV can pass from mother to child. An infected woman can pass the virus on to her child during pregnancy, at birth, or possibly, with her breast milk feeding the baby.

Biotechnology in the Pharmaceutical Industry

Recombinant DNA technology is also used for human and animal health. A recombinant DNA organism being made by combining an animal gene with a plasmid DNA, followed by introduction into a microorganism (Diag. 18.13). The recombinant DNA technology has been discussed in previous chapters. In the laboratory one would generate test-tube-scale cultures that contain the transformed cells that will now produce the protein coded by this animal gene. The next stages are development and production processes. This culture must be scaled up from the test-tube stage to the bioreactor, or fermenter, stage. The product must then be purified and packaged in suitable clinical form. Finally, before the product is ever subjected to clinical use, it is extensively tested in animal systems.

It is important to point out that the bulk of this overall process begins after the genetic engineer has completed his or her work. In a sense, the contribution of the molecular biologist, although crucial, is a small portion of the total process.

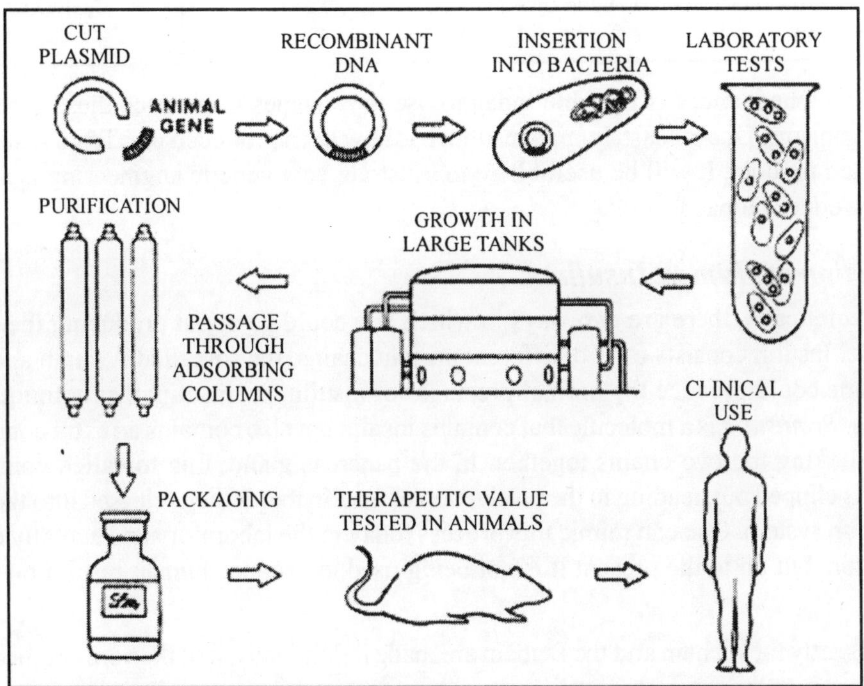

Diagram 18.13 Production of pharmaceuticals by recombinant DNA. Recombinant DNA can be used to add new gones to microorganisms, and these can be grown in fermentation tanks to produce proteins on a large scale. Purification and extensive testing in animals precede clinical application in human beings.

The general categories of these substances include hormones and growth factors, pain-relieving proteins, plasma proteins, enzymes, proteins in the immunology area, and possibly even new types of antibiotics.

The undermentioned table lists some of the growth factors and hormones that one might consider producing by this technology. The genes for almost all of these proteins have

Amino acid residues and molecular weight of human polypeptides potentially attractive for biosynthesis

Polypeptide	Amino acid residues	Molecular weight
Insulin	51	5,734
Proinsulin	82	-
Growth hormone	191	22,005
Calcitonin	32	3,421
Glucagon	29	3,483
Corticotropin (ACTH)	39	4,567
Prolactin	198	-
Placental lactogen	192	-
Parathyroid hormone	84	9,562
Nerve growth factor	118	13,000
Epidermal growth factors	-	6,100
Insulinlike growth factors (IGF-1 and IFG-2)	70,67	7,649,7,471
Thymopoietin	49	-

now been cloned, and it is possible today to use those genes to produce these proteins in microorganisms. One at least, human insulin, has now been produced on a large scale and is a marketed product. It will be useful here to illustrate how genetic engineering is actually used to produce human insulin.

Production of Human Insulin

Theoretically, there are two ways in which one could go about producing the insulin molecule. Insulin consists of two different protein chains, the so-called A chain and the B chain. One could produce the normal precursor of insulin, proinsulin, that is found in the pancreas. *Proinsulin* is a molecule that contains insulin but also contains an extra connecting peptide linking the two chains together. In the pancreas gland, this so-called connecting peptide is clipped out, leading to the production of insulin that is then released into the blood circulation system. One can mimic this process today in the laboratory and actually even in production, but up to the present it is not being used to produce human insulin on a large scale.

Presently the A chain and the B chain are made individually, and then are coupled in the plant to produce the bioactive insulin.

In separate plasmids the genes have been introduced individually for the A chain and B chain of insulin, and then these plasmids have been transformed into bacterial cells. The *Escherichia coli* are then grown in large fermentation facilities. The product that is initially made is a large chimeric protein consisting of the A chain or B chain attached to the end of a naturally occurring E *coli* protein.

This protein is subjected to a cleavage reaction in which the A chain and the B chain are chemically cleaved away from the rest of the chimeric molecule. Then, following several further purification steps, these two chains are combined, and the biosynthetic human insulin is recovered and purified.

Large amounts of the gene product of this plasmid accumulate in the *E. coli.* A thin-section electron micrograph of E. *coli* producing human insulin polypeptide shows dense areas, which are deposits of that protein within the cell. The protein is produced in very substantial amounts and can occupy a major portion of the cell.

This is one of the advantages of biotechnology today-by using appropriate control systems and regulatory systems on the plasmid being dealt with, one can make the protein of interest a major portion of the total protein of the microorganism. It can become a very efficient process.

It is a crystalline protein, and it has all of the characteristics of the insulin that is circulating in all of our bodies.

There are at least two advantages to being able to produce human insulin as opposed to continuing to use the pork and beef insulin that is currently used in many diabetic patients. First, the chemical structure of pork and beef insulin differs slightly from that of human insulin. Thus, there is the possibility of an improved therapy by using a molecule identical to the insulin that is already circulating in human bodies. The second advantage relates to the fact that currently produced pork and beef insulins are really by-products of the meat industry. Their production is subject to all of the economic pressures of the meat industry in terms of supply of pancreas glands. By production in microorganisms an essentially limitless supply to the particular pressures of the beef and pork markets.

Some of the plasma proteins that one might consider producing by this technology are albumin, globulms-α, β, γ, lipoprotems-α, β, plasminogen, fibrinogen, prothrombin and transferrin.

Albumin, for instance, is a protein that can now be manufactured using recombinant-DNA technology. At least one company is working to scale this process up to commercial levels. Many of the genes for other proteins in the plasma protein series have also been cloned.

□□□

CHAPTER-19

Biotechnology and Biodiversity

Biodiversity is a popular way of describing the diversity of life-forms on earth, it includes all life forms and the ecosystems of which they are a part. It forms the foundation for sustainable development, constitutes the basis for the environmental health of our planet, and is the source of economic and ecological security for future generations. Biological diversity may be defined by Rio de Janeiro (1992) as, "the variability among living organisms from all sources including, inter-alia, terrestrial, marine, and other aquatic ecosystems and the ecological complexes of which they are a part; this includes diversity within species and of ecosystems".

In other words biodiversity, may be defined as the *sum total of species richness, i.e. the number of species of plants, animals and micro-organisms occurring in a given habitat.*

Biodiversity refers to the variety and variability among living organisms and the ecosystem complexes in which they occur. It includes diversity of forms right from the molecular unit to the individual organism, and then on to the population, community, ecosystem, landscape and biospheric levels. In the simplest sense, 'Biodiversity' may be of following types:

Genetic diversity (diversity exist within species): It refers to the variation of genes within species. This constitutes distinct population of the same species or genetic variation within population or varieties within a species.

Species diversity (diversity exist within species): It refers to the variety of species within a region. Such diversity could be measured on the basis of number of species in a region.

Ecosystem diversity: In a ecosystem, there may exist different landforms, each of which supports different and specific vegetation. Ecosystem diversity in contrast to genetic and specific diversity is difficult to measure since the boundaries of the communities which constitute the various sub-ecosystems are elusive. Ecosystem diversity could best be understood if one studies the communities in various ecological niches within the given ecosystem; each community is associated with definite species complexes. These complexes are related to composition and structure of the biodiversity.

Diversity is studied under two parameters:

Point or α-diversity: It is represented by the number of species in a speecified areas. It increases with total number of individuals encompassed and thus with the increase in the area sampled wilh the productivity per unit area. It peaks along with gradient of productivity and declines in the most productive system

β-diversity: It is represented by the turnover of species across space. It depends on how large are species ranges.

Biodiversity at Global Level

At global level, it is estimated that there exists 5-30 million species of living forms on our earth. Of these, only 1.5 million have been identified. These include 3,00,000 species of green plants and fungi, 8,00,000 species of insects, 40,000 species of vertebrates, and 3,60,000 species of microorganisms.

The tropical forests are regarded as the richest in biodiversity. Scientists are of the opinion that whatever be the absolute numbers, more than half of the species on the earth live in moist tropical forests which is only 7% of the total land surface. Insects (80%) and primates (90%) make up most of the species. For instance, from a single tropical leguminous tree 43 ant species belonging to 26 genera have been retrieved. This approximately equals the ant diversity of all the British Isles. In 10 selected one hectare plots in Kalimanthan in Indonesia, Peter S. Aston of the Harvard University found more than 700 tree species, almost equal to the number of tree species native to all of North America.

The following explanations have been put forward with regard to the high species diversity in tropics:

1. In tropics, conditions for evolution were optimum and for extinction fewer;
2. In tropics, species diversity was conserved over geological time. This is because low rates of extinction prevailing there; and
3. Biological diversity is the result of interaction between climate, organisms, topography, parent soil materials, time and the heredity. The tropics is the ideal place for such an interaction.

However, these explanations need experimental observations and confirmation.

Biodiversity at Country Level

The Indian region (8°-30°N and 60°-97.5° E) having a geographical area of 329 million hectares is quite rich in biodiversity with a sizable percentage of endemic flora and fauna. This richness in biodiversity is due to immense variety of climatic and altitudinal conditions coupled with varied ecological habitats. These vary from the humid tropical Western Ghats to the hot desert of Rajasthan, from the cold desert of Ladakh and the icy mountain of Himalayas to the warm costs of peninsular India.

The country has over 1,15,000 species of plants and animals already identified and described. In addition, the country is very important Vavilonian Centre of biodiversity and origin of over 167 important cultivated plant species, and some domesticated animals. To name a few the following crops arose in the country and spread throughout the world: rice, sugarcane, Asiatic vignas, jute, mango, citrus, banana, several species of millets, several cucurbits, some ornamental orchids, several medicinal and aromatics. Infact, our country has been recognised as one of the world's top 12 megadiversity nations. This region is also a secondary centre of diversity for grain amaranthus, maize, red pepper, soyabean, potatoes and rubber plant.

There are several examples of plant germplasm from India making,significant contributions to plant improvement. Two cases are well known:

1. Nearly 20 cultivar of rice contain useful genes from wild rices of Kerala, which are responsible for resistance in cultivated rices; and
2. The contribution to the cantaloup (musk melon) industry (14,000 ha) of California by powdery mildew-resistant genes from the wild musk melon of our country.

In flora, the country can boast of 45,000 species which accounts for 15 per cent of the known world plants. Of the 15,000 species of flowering plants, 35 per cent are endemic and located in 26 endemic centres. Among the monocotyledons, out of 588 genera occurring in the country, 22 are strictly endemic. The family Poaceae has the highest endemism both by genera and species.

The North Eastern region could boast of being unique treasure house of orchids in the country, the abode of about 675 species out of 1,000 available in the Indian penninsula and against 17,000 species the world over. The important Indian orchids are: *Paphiopedilum fairieyanum* (Lindl) pfitz., *Cymbidium aloiflium* Sw., *Aerides crispum* Lindl., etc.

Animal Wealth

Our country is very rich in faunal wealth also. The country has nearly 75,000 animal species, about 80 per cent of which are insects. The distribution of major animal groups are shown in Table 19.1. In animals, the rate of endemism in reptiles is 33% and in amphibians 62%. Further there is wide diversity in domestic animals, such as buffaloes, goats, sheeps, pigs, poultry, horses, camels and yalks. Domesticated animals too have come from the same cradles of civilisation as the major crops.

There are no clear estimates about the marine biota though the coastline is 7,000 km long with a shelf zone of 4,52,460 sq km and extended economic zone of 20,13,410 sq km. There is an abundance of sea-weeds, fish, crustaceans, molluscs, corals, reptiles and mammals.

Information regarding other flora and fauna are patchy. Hundreds of new species may be present in our country awaiting discovery. The Western Ghats in Peninsular India, which

Table 19.1. Animal species

Group	Number of Species		Percentage of Endemism	World Percentage
	World	India		
Mammals	4,231	372	8	8.79
Birds	12,450	1,200	4	9.63
Reptiles	6,300	435	33	6.90
Amphibians	4,184	181	62	4.32
Fishes	23,000	2,000	-	8.69
Insects	8,00,000	60,000	-	7.5
Molluscs	1,00,000	5,000	-	0.5

extend in the southern states, are a treasure house of species diversity. Out of the described 15,000 species of the flowering plants in India about 5,000 species occur on the Western Ghats of Kerala; 235 are exclusive to this region. It is estimated that almost one-third of the animal varieties found in India have taken refuge in Western Ghats of Kerala alone.

Species Diversity and Ecosystem Stability

Species diversity or diversity of genes within species increases its ability to adapt to adverse environmental conditions. When these varieties or populations of these species are destroyed, the genetic diversity within the species is diminished. In many cases, habitat destruction has narrowed the genetic variability of species lowering the ability to adapt to changed environmental conditions. Finally, this diminished genetic variability may lead to ecosystem instability. In other words, the greater the variability of the species, the more is the ecosystem stability.

Ecosystem stability has been considered to be related to the cycling and recycling of nutrients, which in run, increases the efficiency of the resource use in the ecosystem. The high species diversity thus may be instrumental in cycling and recycling of nutrients and thereby achieving the stability of the ecosystem.

The survival and well being of the present day human population depends on several substances obtained from plants and animals. The nutritional needs of mankind are also met by wild and domesticated animal and plant species. Indeed, the biodiversity in wild and domesticated form, is the source for most of humanity's food, medicine, clothing and housing, much of the cultural diversity, and most of the intellectual and spiritual inspiration. It is, without doubt, the very basis of man's being. It is believed that 1/4th of the known global diversity, which might be useful to making in one way or other, is in serious risk of extinction. This calls for an integrated approach for conserving global biodiversity. Establishment of nature reserves or biospheres with lot of biophysical variability, maintenance of corridors with different nature reserves for the possible migration of the species in response to climate change, etc. are the immediate steps to be taken for conserving the very precious biological diversity on the earth planet. An international consensus on establishing global net-work for gene banks, microbiological resource centres, and marine parks is also important. At the same time conservation must be coupled with socio-economic development, especially in countries where population pressure threatens the national biotic resources.

Loss of Biodiversity: A Global Crisis

The loss of biological diversity is a global crisis. There is hardly any region on the Earth that is not facing ecological catastrophes. Of the 1.5 million species known to inhabit the Earth (humans are just one of them), one fourth to one third is likely to extinct within the next few decades. Biological extinction has been a natural phenomenon in geological history. But the rate of extinction was perhaps one species every 1000 years. But man's intervention has speeded up extinction rates all the more. Between 1600 and 1950, the rate of extinction went up to one species every 10 years. Currently it is perhaps one species every year.

The destruction of the world's tropical forests, which are disappearing at an alarming rate, is one of today's most urgent global environmental issues. A rich species diversity is slowly being lost for ever. Tropical forests are estimated to contain 50 to 90 per cent of the world's biodiversity.

Rainforest, the home to half of the world's life forms, continue to be destroyed at the rate of over 100,000 Km every year. This loss of biodiversity has immediate and long term effects on human spritual. The majority of the world's population still depends on wild plants and animals for their daily food, medicine, housing and household material, agriculture, fodder, fuel-wood, sprirual sustenance, and intellectual stimulation.

Table 19.2 Extent and loss of tropical forests in different ecosystems

Forest type	**Extent 1990 ('000 ha)**	**Annual decline 1981-90 (Per cent)**
Rain	713,790	06
Moist deciduous	591,779	09
Hill and mountain	201,417	1.1
Dry deciduous	178,579	0.9
Very dry	59,742	0.5
Desert	8,086	0.9

The loss is even more direct in the case of domesticated biodiversity. Traditional farmers of the world have developed an incredible variety of crops and livestock. This too has been eroded over the last few decades, as lakhs of traditional crop strains and hundreds of domesticated livestock breeds being replaced by a handful of laboratory-generated hybrids or by dominant cash crops.

The traditional diversity was bred to meet diverse human needs of nutrition, test, colour, ritual, smell, and to resist drought, flood and pests. It provided several kinds of insurance against crop failure to the farmer. Modern hybrids, on the other hand, while substantially increasing the grain yield and monetary profits, have forced the farmers to look elsewhere for their other daily needs, especially fodder.

India's biodiversity is one of the most significant in the world. As many as 45,000 species of wild plants and over 77,000 of wild animals have been recorded, which comprise about 6.5 per cent of the world's known wildlife. An assessment of wildlife habitat loss in tropical Asia in 1986 showed that the country had only 6,15,095 Km^2 out of its original wildlife habitat of 30,17,009 Km^2 i.e. loss of about 20 per cent. In the last few decades India has lost at least half of its forests, polluted over 70 per cent of its water bodies, built on or cultivated much of its grasslands, and degraded most of its coasts. Under such circumstances, none can say how many species have already lost.

The country has several problems such as overpopulation, large number of cattleheads, growing demand for land, energy, and water supply. Unplanned developmental works and overexploitation of resources have made its living resources most vulnerable. Of the world's 12 top priority biodiversity hot spots, India has two within its bounduaries. Overexploitation has not only resulted in shortages of various materials but also left our biodiversity exposed to various ecological threats. Over emphasis on timber logging has affected many animal species. Faunal losses have been mainly because of over-exploitation of certain species for trading purposes, habitat alteration and destruction; and pollution of streams, lakes and coastal zones.

Listing of Threatened Biodiversity

Red Data Book is the name given to the books dealing with threatened plants or animals of any region. Many countries have prepared their own Red Data Books (e.g. Britain, New Zealand, etc.). On the global level, the IUCN published Red Data Book in two volumes. Its opposite is the Green Book., which lists rare plants growing in protected areas like botanical gardens. A mimeographed Green Book for India has been brought about. It deals with about a hundred rare plant species growing in garden of Botanical Survey of India (BSI). The BSI has. also compiled three volumes of Red Data Book having information on endangered plant species. The UNEP has compiled endangered species of the world under the title, 'Blue Book'. The IUCN has defined Red Data Categories which specify the state of extinction process.

Vulnerable: These are the species whose population numbers are decreasing and are likely to become mote severely threatened with time and in near future, they may represent the category of endangered species, if unfavourable conditions in the environment continue to operate.

Endangered: Thc species with fewer individuals because of unfavourable environmental or human factors and that its natural regeneration is not able to keep place with exploitation or destruction by natural and unnatural means. If the same factors continue to operate as before the species would extinct soon, e.g., Indian Rhinoceros, Asiatic lion, and the great Indian Bustard.

Rare: The species (or texa) with small world population that are not at present endangered or vulnerable, but are at risk. Such species are usually localised within restricted geographical areas or habitats or are thinly scattered over a more extensive range. Rare species have a population of less than 20,000 individuals. Some species are naturally rare and have never occurred in greater numbers, yet they are able lo maintain these numbers. Other species become rare through man's action or other unnatural forces.

Extinct Species that are no longer known to exist in the wild but survive in cultivation. Generally, the term Extinct is used for the species that are no longer known to exist in the wild.

Threatened. It is a broader term that is used for species that fall into any of the above categories.

Threatened Animals

Though 1UCN in 1988 listed 23 species of mammals as endangered or vulnerable to extinction, 75 species are totally protected as listed under schedule I of the Wildlife (Protection) Act, 1972. Forest of the Western Ghats are famous for their endemic fauna. The lion-tailed macaque, the Nilgiri Langur, the Malabar large-spotted civet, the Nilgiri Tahri, etc. are some important endangered species. The Hoolock gibbon is the only ape found in the hilly forests of north-eastern India. Among the 19 primate species, 12 are endangered. The Manipur brown-antlered deer, once distributed in some parts of the northeast, is perhaps the most threatened species. There are some 50 to 100 of them in the Keibul Lamjao Sanctuary of Manipur. Over exploitation is one of the major causes for the present endangered status of the Himalayan musk deer. Human population has also caused depletion in the population of the Kashmir stag or Hangul.

Indian wild ass and buffalo are also among the threatened mammals.

The cheetah, lesser Indian rhinoceros have been extinct according to lUCN's Red Data Book. Other animals appear on the list are: Asiatic lion, snow leopard, swamp deer, elephant and tiger.

There are 1175 species of the birds, including 42 endemic, and 14 of these are considered threatened. N.J. Collar and P. Andrew in their book, 'Birds to Watch', have identified 70 species of threatened birds, 63 on the mainland, four on the Andaman Islands and three on the Nicobar Islands. A number of threatened endemic birds are from northeast. The wood-pigeon of the Western Ghats, the white-winged wood duck of northeast India and the forest owlet of the Satpura hills are among the threatened birds.

IUCN Red data book includes Bengal florican and cheer pheasant as endangered bird species.

The reptile fauna comprises 435 species including 238 species of snakes, about 30 kinds of turtles and many crocodiles and lizards. The tropical forests of Western Ghats.alone support some 20 species of snakes. In addition to their commercial exploitation for skin, the dwindling forest ecosystem also exposes these creatures to danger of extinction. The Indian python, a threatened species, is although widely distributed but its population has declined because of habitat destruction.

The four monitor lizards- the common Bengal monitor, the water monitor, the desert monitor and the yellow monitor- have been declared protected. Marine turtles; 'freshwater tortoises and crocodiles have also been exploited. The green turtle, the loggerhead and the Olive Ridley are the most widely consumed species. All the four species of sea turtles and five of mud turtles have been declared protected. The gharial, the saltwater crocodile and the marsh crocodile (mugger) are-exploited for their valuable skins. The three are included in IUCN list of threatened animals. But the population of crocodiles has been increased through breeding centres in Uttar Pradesh, Andhra Pradesh, Tamil Nadu, Orissa and West Bengal.

The amphibian fauna comprises 182 species, which includes salamanders, caecilians, frogs and toads. There is a high degree of endemism and out of 112 endemics, 84 occur in the Western Ghats and 20 in the northeast. Three of the Indian species, namely the Himalayan newt, *Tylatotriton verrucosus,* the Malabar tree-toad, *Pedostibes tuberculosus* and the Garo Hills tree-toad, *Pedostibes kempi;* have been identified as rare endangered by B.K. Tikader, in his book Threatened animals of India. The first mentioned is the only species of salamander known in the country and is distributed in Sikkim, north West Bengal, Arunachal Pradesh and Manipur. The Malabar tree-toad was recently rediscovered in the Silent Valley.

Table 19.3. Threatened Plants of India

Plants of Ornamental Value	*Aerides crispum Lindl, Cymbidium aloiflium* Sw. (Orchidaceae), *Diplorneris hirsula* (Orchidaceae), *Paphiopedilum fairieyanum* (Lindl.) Pfitz. (Orchidaceae) or 'Lady's Slipper Orchid', *Qaphiopedilum drurui* (Orchidaceae), *Rhododendron edgewortythii* Hk. f. (Ericaceae), *Symplocos chengapae* Raiz. & Sahni (Symplocaceae).
Plants of Medicinal Value	*Acorus calamus* (Araceae) or 'Safed Bach', *Atropa acuminata* Royle ex Lindl (Solanaceae) or 'Sag-angur', *Dioscorea deltoidea* (Dioscoreaceae), *Drosera* species (Droseraceac), *Podophyllum hexandrum* Royle (Podophyllaceae), *Rauwolfia serpentina* Benth. ex. Kurz (Apocyanaceae) or 'Sarpagandha', *Saussurea lappa* Clarke (Asteraceae) or 'Kuth'.
Plants of Scientific Value	*Balanophora involucrata* Hk. f. (Balanophoraccae), *Dischidia benghalensis* Coleb (Asclepiadaceae), *Nepenthes Kliasiana* Hk. f. (Nepenthaceae), *Sapria himalayana* Griff. (Rafflesiaceae).
Plants of Phytogeographic Significance	*Ceropegia jainii* Ansari & Kulkarni (Asclepiadaceae), *Frerea indica* Dalz. (Asclepiadaceae), *Glyphochola mysorensis* (Jain & Hem.) Clayton (Poaceae/Graminae), *Helicanthes danser* (Lorantahceae), *Jainia nicobarica* Balak (Rubiaceae), *Manisuris divergens* (Hack.) Ktze (Poaceae/Graminae), *Willisia warm* (Podostemaceae).
Trees of Forestry Importance	*Dysoxylum malabaricum* Hedd. (Meliaceae), *Pierocarpus santalinus* Linn. f. (Fabaceae/Papiolionaceae) or 'Rakta chandan'.
Other Economic Plants	*Decussocarpus wallichianus* (Presa) de Lauben (Podocarpaceae), *Phyllostachys bambusoides* Sieb. & Zucc. (Poaceae/Graminae), *Pinus gerardiana* Wall (Pinaceae) or Chilgoza, *Santalum album* Linn. (Santalaceae).

Causes for the Loss of Biodiversity

Proximate Causes

The important proximate causes for the loss of biodiversity are as follows:

1. Destruction of habitat: The natural habitat may be destroyed by man for his settlement, grazing grounds, agriculture, mining, industries, highway construction, drainage, dam building, etc. As a consequence of this, the species must either adapt to the changes, move elsewhere or may succump to predation, starvation or disease and eventually die. In our country, several rare butterfly species are facing extinction with the uncannity swift habitat destruction of the Western Ghats (of the 370 butterfly species available in the Ghats, up to 70 are at the bunk of extinction).

2. *Hunting:* From time immemorial, man has hunted for food. Commercially, wild animals are hunted for their products such as hides and skin, tusk, antlers, fur, meat, pharmaceuticals, perfumes, cosmetics and decoration purposes. In the country, rhino is hunted for its horns, tiger for bones and skin, musk deer for musk (have medicinal value), elephant for ivory. Gharial and crocodile for their skin, and jackal for thrivin fur trade in Kashmir. One of the most publicised commercial hunts is that of whale. The

whalebone or 'baleen' is used to make combs and other products. (ITES lists nine Indian animal species which have been severely depleted due to international trade). These are Fin Whale (*Balenoptera physalus*). Himalayan Musk deer (*Moschus moschiferus*), Green Turtle (*Chelonia mydas*). Hawksbill Turtle (*Eretmochelva imbrtcata*), Olive Ridley Turtle (*Dermochelys olivacea*), Salt-water Crocodile (*Crocodyhe porosus*), Desert Monitor Lizard (*Varanus griseus*), Yellow Monitor Lizard (*V. flavescens*), and Bengal Monitor Lizard (*V. bengalensis*). Hunting for sport is also a factor for loss of animal biodiversity.

3. *Over exploitation:* This is one of the main causes of the loss of not only economic species but also biological curiosities like the insectivorous and primitive species and other texa needed for teaching or laboratory work (like *Nepenthes, Gnelum, Psilotum,* etc.). Commercial exploitation of biodiversity has invariably meant its overuse and eventual destruction. This has been as true in the case of Indian wild mango trees which were turned into plywood as of the whales, that were hunted for tallow, in the oceans. Plants of medicinal value like *Podophyllum hexandrium. Coptis teeta, Aconitum, Discorea deltoidea, Rauvolfia serpentina, Quphiopedilum druri,* etc., and horticultural plants like orchids and rhododendrons come under the over-exploited category. Faunal losses have been mainly because of over-exploitation. For instance, excessively harvesting of marine organisms such as fish, molluscs, sea-cows and sea-turtles has resulted in extinction of these animals.

4. *Collection for zoo and research:* Animals and plants are collected throughout the world for zoos and biological laboratories for study and research in science and medicine. For example, primates such as monkeys and chimpanzees are sacrificed for research as they have anatomical, genetic and physiological similarities to human beings.

5. *Introduction of exotic species:* Native species are subjected to competition for food and space due to introduction of exotic species. For example, introduction of goats and rabbits in the Pacific and Indian regions has resulted in destruction of habitats of several plants, birds and reptiles.

6. *Control of pests and predators:* Predator and pest control measures, generally kill predators that are a component of balanced ecosystem and may also indiscriminately poison non-target species.

7. *Pollution:* Pollution alters the natural habitat. Water pollution especially injurious to the biotic components of estuary and coastal ecosystem. Toxic wastes entering the water bodies disturb the food chain, and so to the aquatic ecosystems. Insecticides, pesticides, sulphur and nitrogen oxides, acid rain, ozone depletion and global warming too, affect adversely the plant and animal species. The impact of coastal pollution is also very important. It is seen that coral reefs are being threatened by pollution from industrialisation along the coast, oil transport and offshore mining. Noise pollution is also the cause of wildlife extinction. This has been evidenced by the Canadian Wildlife Protection Fund. According to a study Arctic Whales are seen on the verge of extinction as a result of increasing noise of ships, particularly ice breakers and tankers.

8. *Deforestation:* One of the main causes for the loss of biodiversity is population explosion and resultant deforestation. Deforestation mainly results from population settlement, shifting cultivation, development projects, demand for fuel-wood, demand of wood for industry and other commercial purposes. In India, the rate of deforestation is 13,000 sq. km. annually. If this rate of deforestation continues, one can imagine the ultimate fate of our forest and biological richness. It is presumed that in coming years, the global loss of biodiversity from deforestation alone would be 100 species everyday.

9. *Other factors:* Other ecological factors that may also contribute to the extinction of plant and animal species are (a) Distribution range. The smaller the range of distribution, the greater the threat of extinction, (b) Degree of specialisation. The more specialised an organism is, the more vulnerable it is to extinction, (c) Position of the organism in the food chain. The higher the organism is in food chain, the more susceptible it becomes, (d) Reproductive rate. Large organisms tend to produce fewer off-springs at widely spaced intervals.

Biotechnology for Bio-conservation of Diversity

In the last twenty five years, all over the world, there has been a revolution in the field of Biotechnology- new discoveries and the inventions in the area of isolation and manipulation ot genes, better understanding of biological molecules and the advent of recombinant DNA technique. Biotechnologists all over the world have made efforts to create transgenic crops which will withstand the pests as also have enough resistance to withstand environmental stress. In fact. Biotechnology is inherently knowledge-intensive and having strong infrastructure would lead to value area of agriculture, animal husbandry, fisheries, forestry and medicine.

Biotechnology is not a miracle solution to the problem of biodiversity crisis. Rather, the use of biotechnology in the production of uniformity in plants and animals has threatened not only the life forms but also rendered entire community or ecosystem unstable. Further, indiscriminate and unregulated uses of genetically modified organisms pose a threat to mankind. In fact, uniformity (or homogeneity) in life forms accelerates the loss of biodiversity. The institutional structure that controls the biotechnology, therefore, should not overshadow those institutions that deal with conservation of biodiversity, and on no account ignore the rights and privileges of the local communities.

Biodiversity Conservation Methods

For much of the time man lived in a hunter-gather society and thus depended entirely on biodiversity for sustenance. But, with the increased dependence on agriculture and industrialisation, the emphasis on biodiversity has decreased. Indeed, the biodiversity, in wild and domesticated forms, is the source for most of humanity food, medicine, clothing and housing, much of the cultural diversity and most of the intellectual and spiritual inspiration. It is, without doubt, the very basis of life. Further that, a quarter of the earth's total biological 'diversity amounting to a million species, which might be useful to mankind in one way or other, is in serious risk of extinction over the next 2-3 decades. On realisation that the erosion of biodiversity may threaten the very extinction of life has awakened man to take steps to conserve it.

During the last seven years plans tor biodiversity conservation have been developed by the WRI and the IUCN with support from World Bank and other institutions. Basically, the conservation plan had an holistic approach and encompasses whole spectrum of biota and activities ranging from ecosystems at the macro level *(in situ* conservation) to DNA libraries at the molecular level *(ex-situ* conservation).

To conserve the biodiversity, the immediate task will be to devise and enforce time bound programme for saving plant and animal species as well as habitats of biological resources.

Conservation Methods

In-situ *Conservation:*

In-situ conservation refers to protection zones and areas of high biological diversity. These areas, described as natural ecosystems, will protect species with minimum human interference. The buffer zones or semi-natural ecosystems can allow for some human disturbance as long as the impact of humanity is not greater than any other factor.

For preservation of the endangered species, the only measure suggested is the strict protection against poaching of both vegetation as well as animal resources. Since most of the threatened organisms occur as components of biotic communities in open sites, restoring them in such habitats through judicious protection measures is required. For *in-situ* conservation, the biosphere reserve offers the best site of natural conservation of threatened flora. Today, India has 75 national parks and 421 wildlife sanctuaries covering an area of about 1.4 lakh km^2 constituting more than 4% of the total geographic area of the country, and one-fifth of the forest area. The protected area includes 23 tiger reserves as well as 14 biosphere reserves.

The Wildlife Institute of India has comprehensively reviewed the existing protected area network and highlighted the need to identify new protected areas in different parts of the country, in order to ensure representation of maximum wildlife habitats. It has made proposals to increase the existing network coverage in India to 147 national parks with an area of 49,435 km^2, and 519 wildlife sanctuaries with an area of 116,879 km^2 raising the coverage upto 5.06% of the total land area. The proposed network should be accepted.

The conservation efforts towards plant species have not been given adequate attention particularly of those which are of potential economic and scientific value.

Scientific studies in regard to *in situ* conservation should focus on the following lines where very little data are available on : (1) Applied research for conservation of living resources; (2) Interlinkages between plant and animal species, (3) Quantitative assessment of the conservation status of the species; (4) Successional status of key species in different ecosystems; (5) Multiplication and restoration of endangered, rare and endemic species using biotechnology; (6) Ecological restoration of degraded micro and macro-habitats; (7)

Identification of critical index species and their sensitive parameters; (8) Assessment of the impact of exotic, species on the ecosystem; (9) Determination of the impact on the ecosystem of various activities in the protected areas; (10) The possible climate change and its impact on biodiversity; (11) Hydrological changes including surface run-off and percolation in the protected areas; (12) Primary reduction and cycling of nutrients in the soil; (13) Studies on satellite mapping of all protected areas; and (14) Development of methodologies for classification of microhabitats.

The important point in *in-situ* conservation is that the forest trees, wild planrs, wild animals and micro-organisms all occur together in an ecosystem. Therefore, if an attempt is made to conserve and enrich the ecosystem, much can be achieved in a single step. This would be particularly advantageous in tropical forests where many species occur in low densities and have a high degree of endemism. Obviously there is an urgent need to coordinate efforts on the ground level. This would save not only time and effort but also the scarce fiscal resources and infrastructure.

To identify ecosystems that have been left out and are in urgent need of conservation, it is necessary to match the 12 bio-geographical provinces (viz. Ladakh, Himalayan highlands, Malabar rain forest, Bengal rain forest, Indus-Ganga monsoon forest, Assam-Burma monsoon (uicsl, Mahanadian, Coromandel, Decan thorn forest, Thai desert, Lakshadweep Islands, Andaman and Nicobar Islands) with the present day protected areas network. From such a study there will emerge the additional areas which are in need of conservation. The process of identification of the additional areas must be based, among other things, on the following-

1. Vavilonian centres of diversity of crop plants in the Indian region particularly with regard to wild ancestors of the crop plant genetic resources, non-crop plant genetic resources, and forest tree genetic resources;
2. Wild relatives of livestock;
3. Fish genetic resources;
4. Fresh water systems (rivers and lakes);
5. Marine fish and other economic sea animals;
6. Mangrove and coral systems;
7. Island ecosystems;
8. Threatened/ endangered biota including materials used for teaching; and
9. Unique and fragile ecosystems, including hot spots (NE Himalayas) and endemic areas.

So far, the approach has been restricted to fauna, especially large mammals and big wats in particular, because they are at the top of the food chain. Such a restricted view must change in favour of a holistic one in order to enable the country to save as much of the biological wealth as possible.

Ex-situ *Conservation*

India has done commendably well as far as *ex-siiu* conservation of crop genetic resources is concerned. It has also taken up such work on livestock, poultry and fish genetic resources. However, there is need to develop facilities for long and medium term conservation through.

1. Establishment of Genetic Enhancement Centres for producing good quality of seeds;
2. Enhancement in the existing zoos and botanical garden network;
3. Seed-gene banks
4. Tissue culture gene banks;
5. Pollen and spores banks.
6. Captive breeding in zoological gardens; and
7. in vivo and in vitro preservation

However both ex-situ and in-situ conservation of forest trees and micro-organisms (except nitrogen-fixing blue-green algae) have not received much attention.

Ex-situ and *in-situ* conservation should be given equal imponance as measures in biodiversity conservation. Release of genetically modified organisms should be regulated at national and international level, and there should be adequate dissemination of information about such release by the respective countries.

Biotechnology, Biodiversity and Intellectual Property Rights (IPR)

In the last twenty years, all over the world, there has been a revolution in the field of Biotechnology- new discoveries and the inventions in the area of isolation and manipulation of genes, better understanding of biological molecules and the advent of recombinanl DNA technique. Biotechnologists all over the world have made efforts to create transgenic crops which will withstand the pests as also have enough resistance to withstand environmental stress. In fact Biotechnology is inherently knowledge-intensive and having strong infrastructure would lead to value area of agriculture, animal husbandry, fisheries, forestry and medicine.

Biotechnology is not a miracle solution to the problem of biodiversity crisis. Rather, the use of biotechnology in the production of uniformity in plants and animals has threatened not only the life forms but also rendered entire community or ecosystem unstable. Further, indiscriminate and unregulated uses of genetically modified organisms pose a threat to mankind. In fact, uniformity (or homogeneity) in life forms accelerates the loss of biodiversity. The institutional structure that controls the biotechnology, therefore, should not overshadow those institutions that deal with conservation of biodiversity, and on no account ignore the rights and privileges of the local communities.

Productivity and Diversity

Productivity goes against diversity as it creates imperative for uniformity and homogenisation. This has generated the paradoxical situation in which modern plant improvement is based on the logic of uniformity and homogenisation. Green revolution, for instance is based on high productivity and low biodiversity. There is no need to combine high productivity and high genetic-diversity to enhance yield as well as to provide insulation against environmental stress and pollutants. Over the last few decades lakhs of traditional crop strains and hundreds of domesticated livestock breeds have been replaced by a handful of laboratory- generated hybrids or dominant cash crops. Similarly, forestry schemes introduced monocultures of commercial species like teak, eucalyptus and bamboo, and pushed into extinction diversity of local species. Agriculture modernization, fisheries, commercial forestry and animal husbandry thus produce uniform crops and domesticated livestocks and destroy the diversity of local species which fulfil local needs. Such a strategy of productivity increase based on the logic of destruction of biodiversity is no longer desirable as it will ultimately lead to loss of biodiversity.

Monocultures are ecologically unstable. Being genetically uniform, they invite diseases and pests; also vulnerable to environmental stress and pollutants. The technology for breeding high yielding varieties, indeed, a technology which breeds uniformity and at the same time threatens the biodiversity conservation and sustainability. If production continues to be based on the logic of uniformity and homogenisation, it will continue to displace diversity leading eventually to biodiversity erosion.

Biodiversity-Means of Production or Product

For peasants and forest-dwellers, biodiversity has been the source of sustenance for basic needs such as food, fiber, fodder, fuel, timber, shelter and medicine. The tribals and the farmers reproduce the necessary part of their means of livelihood by planting crop each year. The seed thus represents the capital with a simple biological barrier and would reproduce and multiply under suitable environmental conditions. New technologies by removing biological barrier transformed the means of production and product into mere 'raw material'. The cycle of regeneration of biodiversity is thus replaced by a linear flow of free germplasm from farms and forests into corporate labs and research stations, and the flow of modified uniform products as priced commodities from corporations to farms and forests. Through technological innovations, biodiversity is transformed from a renewable into non-renewable resource. It does not produce itself; it needs the help of inputs to produce. It is this shift from the biological processes of reproduction to the technological processes of production that underlies the problem of dispossession of farmers and tribals and the problem of erosion of biodiversity. The manufacture of the product in corporate labs is regarded as production. The reproduction of the raw material by nature and Third World Farmers and forest dwellers is more conservation. Biotechnology development thus leads to biodiversity erosion by way of converting the means of production or product into mere 'raw material'.

Politics of Patents and Intellectual Property Rights

Biotechnological processes use life forms or derivatives thereof, to make or modify products or processes for specific use. Under Intellectual Property Rights (IPRs), transformed microorganisms, plants and animals can be patented and become exclusive private property. The North has always used Third World Germplasm as a freely available resource and modified it. The issue of patent protection for modified life forms raises a number of unresolved political questions about the ownership and control of the genetic resource. By simply manipulating the life forms one does not acquires the patent or property right, because the modified life forms do not arise from nothing but from existing life-forms which belonging to others. Also, biotechnology does not create new genes, but merely relocates genes already existing in the organism. The advanced capitalist nations wish to retain free access to the developing world's storehouse of genetic diversity, while the South like to have the proprietory varieties of the North's industry declared a similarly public good. The North, however, resists this democracy. US has freely taken the biological diversity of Third World to earn millions of dollars of profits, none of which have been shared with Third World Countries, the original owners of the biological resource. For instance, an American Industry earned $8 million a year in 1962 simply by increasing the soluble solid contents of a wild tomato variety, *Lycopersicum chomrelewskii* taken from Peru. None of these profits or benefits were shared with Peru, the original contributor of the genetic material. The Convention on Biological Diversity is also not clear on this score. Industrialised countries, particularly the US interpreted key clauses of the treaty in a manner that would protect the interest of its own biotechnology industries. This is a clear set-back to the developing countries, who stand to lose the benefits due to them.

In absence of a proper biotech base, a developing coumry cannot match an industrial country although the former may be far richer in biodiversity. However, the Convention on Biodiversity, helped to place the subject matter of technology transfer and IPRs on the top of the agenda of policy and decision makers. Furthermore, access to genetic resources and transfer of technology are treated on the same plan.

On the issue of IPRs, the basic requirement of the Dunkel proposals is that inventions in all branches of technology shall be patentable, whether products or processes, if they meet the three tests of being new, involving innovative steps and being capable of industrial applications. It has also been provided that microorganisms will be patentable. In respect of plant varieties there is a separate obligation to provide them protection by patents or by an effective *sui-generis* system, or by a combination of the two. *Sui-generis* protection implies a system different from other categories of intellectual property protection (such as patents) and is a class by itself. Dunkel text, thus does not compel to patent seeds (i.e. plant varieties). So far we are concerned, seeds are also not patentable in India today, and we do not have any intension of changing this system. However, we will adopt our own system for the protection of plant varieties under which we may provide certificates for plant breeders right. The farmers rights include their using the seeds for their own needs or for exchange in the village community according to their traditional custom. Since farmers right will be fully safeguarded under system of protecting the plant breeders right, there is no truth in the allegation that the farmers will not be able to retain the seeds for their own use and that they will have to buy seeds every year from multinational companies. Furthermore, India is not in favour of the patenting naturally occurring life forms/ germplasm.

The extension of IPRs to plant varieties either in the form of patents or in the form of Plant breeder's Rights is bound to result in increase in prices of seeds, greater domination of agriculture by multinational companies and slower diffusion of new varieties. These would be in sharp contrast to the experience of the Green Revolution where the new varieties of seeds evolved by the government institutions percolated down to the fields in a short span of time with very little cost to the individual farmer. Our farmers will have to face great hardship due to the new regime. India calls for the removal of distortions in the IPRs regimes in areas related to prior existence of knowledge in nutrition. Mr. Kamal Nath, Environment and Forest Minister, referred to the harsh fact that IPRs regimes in these areas resulted in a virtual denial of benefit flows of financial return and markets to those very communities who had by their sustainable lifestyles preserved these systems and the natural resources on which they were based. He said that the market value of allopathic medicines derived from plants used in traditional remedies was calculated to be over $ 43 billion annually. Less than 0.01 per cent of profits had gone to the indigenous people who had led the researchers to them.

The South had not merely preserved much of its precious biogenetic resources, but also the knowledge and practices about their optimum and sustainable utilisation. Access to these resources have to be regulated and careful exercise, in keeping with the objectives of the convention and with due compensation to such people who have preserved their resources.

How to recognise and measure the value of indigenous knowledge, is one of the basic problems in deciding the compensation and for protection of farmers' IPRs. As a result of the persistent North-South split, the CSD could able to move forward on this contentious issue. However, the decision of the CSD to include in their medium-term programme, the knowledge, innovations and practices of indigenous and local communities is an important step in the direction of the protection of traditional knowledge and practices of the indigenous and local communities relevant to conservation and sustainable use of biological diversity.

Bio-Safety Protocol

India has demanded a clear comprehensive and legally binding international protocol on bio-safety under the convention on biodiversity. Addressing the first meeting of parties to the Convention on Biodiversity, at Nassau, Bahamas, on December 8, 1994 Mr. Kamal Nath demanded immediate and adequate safeguards against hasty experimentation and use of genetically modified organisms, since these have unimaginable repercussions. He feared that the developed world could become a playground for experimentation with such genetically modified organisms and it could only be checked through a legally binding agreement.

Socio-economic and Political Causes

The socio-economic and political causes of biodiversity loss vary from region to region, In recent times, they can be linked to governmental and international support for industrial forestry, agriculture and energy programmes. The enormous fires in the Amazon have been fueled by two main sources- State subsidies for the cattle industry (Caufield, 1984; Lutzenberger, 1987); and the taking over of the fertile lands in the North-East and South of Brazil by agribusiness operations to grow export crops. As Jose Lutzenberger noted in 1989:

□□□

CHAPTER-20

Enzymes Bioaccelerators

Enzymes are biological organic catalyst found within each living organism, which accelerates the biological reactions, do not affect the equilibrium constant and remains unchanged at the end of the reaction. Although it is produced by living cells but is itself not alive. The reactants bind to a specific site on the surface of enzyme molecule called active site. Enzymes show specificity for their substrate as well as for the reactions. Various factors such as temperature, pH, concentrations of enzymes and substrate affect the rate of enzyme catalysed reactions. Some enzymes require special additional factors for their normal activity. Allosteric enzymes have more than one active sites which may be located on the same subunit or on different subunits. There are some inhibitors which reduce the enzyme activity. Various enzyme-catalysed reactions characterize the metabolism of the cell. Regulation of these reactions is achieved by altering the enzyme activity.

The applications of enzymes (biological catalysts) that can change plants and animals in precise and often remarkably in dramatic fashion. In the hands of knowledgeable biotechnologists, enzymes become the tailor's scissors and the surgeon's scalpel. There is a strong dependence of advances in enzymology in the field of agriculture. Discovery of enzymes and elucidation of many of their properties were by agricultural chemists. The earliest enzymes studied were primarily from agriculturally important animals, plants and microorganisms. The fast developing wide field of biochemical engineering/biotechnology, and its tremendous promise for the future, have brought basic and applied scientists together in private sector companies, at universities, and in government and industrial laboratories.

Historical Background

The existence and power of enzymes were first recognized in the nineteenth century when reactions that has been considered to occur only in the presence of cells were found also to be mediated by cell-free extracts. It was the Buchner brothers, who showed that yeast, a living organism, yielded a non-living cell-free extract capable of fermenting sugars. Subsequent investigators of this field showed that this was due to the presence of many powerful catalysts in the extract, each one is highly specific for one of the steps in the breakdown of glucose. Similar catalysts, or enzymes, are found in all living things.

Enzyme (Gr. En = in zyma = yeast): Any substance, protein in whole or in part, that regulates the rate of a specific biochemical reaction in living organisms. It is capable of catalyzing a reaction in which substrate(s) are converted to product(s) through the formation of an intermediate enzyme -substrate complex. As with other catalysts, enzymes are responsible for accelerating the rate of a chemical reaction. On the basis of side of action enzymes are of two types:

Endoenzyme: The enzyme which acts within the cell in which they are synthesized inner core of or substrate molecule are known as endoenzyme.

Exoenzyme: The enzyme which act outside the cell in which they are synthesized or at the outer ends of substrate molecule are termed exoenzyme.

The chemical nature of enzymes remained in dispute for a long time. Willstatter, working with peroxidase, had chosen an enzyme of such high catalytic efficiency that he believed that his active preparations were protein-free. By contrast, Sumner's crystalline urease was of such relatively modest activity that his critics attributed the catalysis to a highly active trace contaminant rather than the purified protein itself. Improvements in fractionation procedures have since allowed the purification of many hundreds of enzymes from diverse sources, leading to the realization that all enzymes are proteins. On the basis of chemical composition enzymes are of two types:

1. Purely proteinaceous enzyme: These are made up of only proteins e.g., proteases that split protein and amylase that split starch.
2. Conjugate enzymes: These are made up of protein molecule, to which a non-protein group (prosthetic group or cofactor) is also attached e.g., oxidising enzymes.

It is essential to understand the difference between the following important terms - substrates, prosthetic group, apoenzyme, coenzymes and cofactor.

Substrate: The substance upon which an enzyme act is known as substrate. They are produced by the living protoplast of a cell.

Prosthetic group: It is the non-protein group of enzyme. The prosthetic group is firmly bound to protein component of the enzyme by chemical bonds and is not removed by dialysis. Haem, biotin and pyridoxal phosphate, like flavin,. usually function as prosthetic groups.

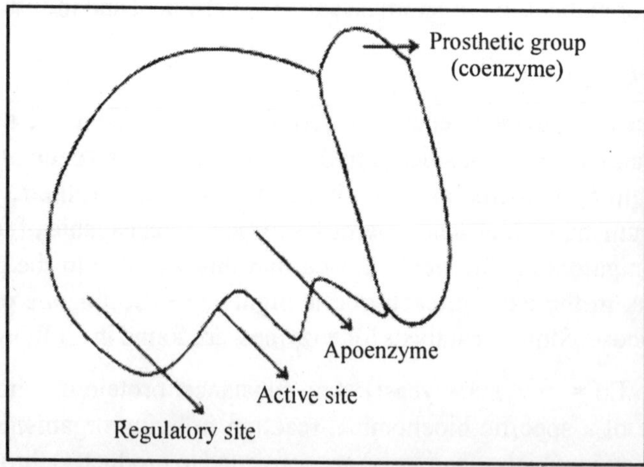

Diagram 20.1 Hypothetical morphology of an enzyme

Apoenzyme: If, however, the prosthetic group is removed, the remaining protein part of the enzyme is called as apoenzyme and it is inert.

Coenzyme: It is the co-worker of prosthetic group and apoenzyme because neither the apoenzyme nor the prosthetic group alone is enzymatically active. The prosthetic group, the coenzymes, as the name specifics, act as co-worker of an apoenzyme. They combine with the enzyme and leave it again in the course of a single catalytic cycle. The coenzymes are vitamins, DPN, NAD, tetrahydrofolate, Coenzyme A and adenosine phosphates, like $NAD(P)^+$.

Cofactor: When a prosthetic group consists of single atom of some metal like Mg^{++}, Fe^{++}, Cu^{++}, Mo^{+++}, Zn^{++}, then it is known as cofactor and can be easily separated from rest of the protein part. Many enzymes employ either metal ion cofactors or organic cofactors. Thus many oxidoreductases utilize haem, nicotinamide or flavin; the transferases use folate, coenzyme A, pyridoxal phosphate, thiamine and adenosine phosphates; some of the ligases use biotin etc.

Among these cofactors, some may be regarded as integral parts of their enzymes; they are blown as prosthetic groups.

There are several terms which are often used in biotechno-enzymology. These are as follows (arranged alphabatically):

— *Enzyme activation*: A mechanism in which the activity of an enzyme is increased by a direct effect on the enzyme, rather than due to new protein synthesis. It may be brought about through binding of an activator molecule at an allosteric site on the enzyme resulting in a change in the enzyme configuration, which leads to a change in the shape of the active site.

— *Enzyme activity*: An expression of the ability of a given enzyme preparation to catalyze a specific reaction. It may be defined in terms of the number of moles of substrate converted, or the number of moles of product produced, in unit time per unit weight of protein (e.g., micromoles per milligram protein per minute).

— *Enzyme amplification*: A technique used to visualize or quantify an immunoreaction in an assay procedure in which the enzyme label in the immnoassay is used to provide the trigger substance for a secondary system that can produce a large amount of a coloured product.

— *Enzyme analysis*: A technique in which a specific reaction catalyzed by an enzyme is used to determine either the amount of enzyme or substrate present in a sample that may be highly complexed.

— *Enzyme assay*: (i) A method for determining the activity of an enzyme sample, (ii) An assay used to determine the amount of a specific substance in a sample, where the means of detection is dependent on an enzyme-catalyzed reaction.

— *Enzyme turnover number*: The number of moles of substrate converted to product per minute per mole of enzyme. Related quantities are the catalytic central activity, which is the turnover number per active site of the enzyme protein, for enzymes with more than one active site. The molar catalytic activity is the turnover number in the units of sec^{-1}.

— *Enzyme equilibrium*: An expression which describes the state of an enzyme-catalyzed reaction in terms of the rates of forward and backward reactions. A reversible reaction is in equilibrium when the rates of the forward reaction and the reverse reaction are the same so that there is no net change in concentration of the reactants. All equilibria involving combination of enzymes with substrates or inhibitors are expressed in terms of dissociation constants rather than association constants; other equilibrium constants are written as association constants.

— *Enzyme extraction*: The removal of enzymes from contaminating materials in order to increase their specific activity. Techniques fall into two groups; those used to separate enzymes from solid substrate culture; those used to release enzymes from the interior of microbial cells.

— *Enzyme fermentation*: A process in which a microorganism is grown as a source of an industrial enzyme on a large scale. Industrial enzymes are produced by either solid substrate cultivations using fungal sources or conventional batch submerged culture techniques for bacterial source.

— *Enzyme immobilization*: The coversion of a soluble enzyme to a bound or insoluble form.

— *Enzyme induction:* The synthesis of an enzyme in response to an inducing agent which stimulate expression of the genes encoding the protein with a specific enzyme function.

— *Enzyme inhibition:* A mechanism whereby an enzyme is inactivated by a chemical agent. Enzymes are inhibited by binding of chemicals at either the active'site or control (allosteric) sites.

— *Enzyme isomerization:* The reversible changes in enzyme conformation in the course of a catalytic cycle.

— *Enzyme kinetic parameters, Enzyme parameters:* The parameters of the rate equations which remain constant, so long as temperature, pressure, pH value and buffer composition are constant. They are derived from the rate constants of the rate equations, and are frequently used to characterize the enzyme functionally.

— *Enzyme kinetics:* The study of the rates of enzyme-controlled reaction.

— *Enzyme production:* The processes whereby industrial enzymes are manufactured.

— *Enzyme reaction mechanism:* The basic priniples involved in the physical and chemical reactions associated with an enzyme catalyzed reaction.

— *Enzyme recovery:* The method used in the recovery of industrial enzymes.

— *Enzyme regulation:* Enzymes may be regulated at two levels (i) at the level of gene expression and protein synthesis through induction and repression; (ii) at the enzyme level through enzyme inhibition or activation as a result of the binding of effectors molecules at allosteric sites of the enzyme.

— *Enzyme repression:* A mechanism that prevents the synthesis of an enzyme by the formation of repressers that bind to DNA preventing transcription.

— *Enzyme separation:* Techniques used for the separation and purification of enzymes. These include solid separation, membrane separation, precipitation, absorption chromatography and gel filtration.

— *Enzyme species intermediates:* All covalent and noncovalent complexes between an enzyme and a substrate and or product or effector, and all the various enzyme isomers (enzyme isomerization). The concentration of the E.s. cannot be measured directly by means of steady state enzyme kinetics but their steady-state concentrations can be calculated from the kinetic equations for definite values of the concentration variables. In principle, the concentrations of the E.s. could be determined by the method of presteady-state kinetics.

— *Enzyme unit:* The amount of an enzyme that will catalyze the transformation of one micromole of substrate in a given time (e.g., one minute) under defined conditions of temperature, pH and substrate concentration.

— *Enzyme-linked immunosorbent assay (ELISA):* A sensitive analytical technique in which an enzyme is complexed to an antigen or antibody.

— *Killer enzymes or anti-enzymes:* The enzymes that inactivate other enzymes.

— *Enzymology:* A branch of science dealing with the chemical nature, biological activity, and biological significance of enzymes.

Properties of Enzymes

Enzymes are characterised by many bio-physical properties. Some common properties are given below:

1. *Proteinaceous nature:* Enzymes are made up of protein. They may be either purely made up of protein or associated with a non-protein part.
2. *Catalytic properties:* Enzymes act as catalysts and influence the speed of chemical reaction but themselves remain unchanged. A small quantity of enzyme can catalyse the transformation of a very large quantity of the substrate into end product e.g. the sucrase can hydrolyse 100,000 times of sucrose as compared with its own weight.
3. *Specificity of enzyme action:* The ability of an enzyme to catalyze one specific reaction and essentially no other is perhaps its most significant property. Close examination reveals that most enzymes can catalyze the same type of reaction.
4. *Reversibility of enzyme action:* Most of the reactions catalysed by enzymes are reversible and the enzyme can catalyze the reactions in both directions e.g. the lipase can catalyse not only the hydrolysis of fats into fatty acids and glycerol but also can synthesize fats from fatty acids and glycerol as shown below.

   Lipase

   Lipase Fats □ Fatty acids + glycerol
5. *Sensitivity of the enzymes:* The enzymes are sensitive to rise of temperature and are destroyed at higher temperature at 60-70° C. However, most of the enzymes act best between 20°C to 35° C. At 0° C or below, they are not destroyed but become inactivated.

Nomenclature of Enzymes

A commission has been established for deciding the name of enzyme and is known as Enzyme nomenclature commission. The terminology generally agreed for use in relation to enzymes. As in the example of urease, enzyme and classes of enzymes usually are named by appending the suffix -ase to the name or abbreviated name of a compound on which the enzyme acts. Enzymes that cause the cleavage of the polypeptide chain by reaction with water are referred to, as a class, as protemase. Naming of enzymes is known as nomenclature. There are various ways of naming enzymes. The principles of nomenclature are as follows:

1. The enzymes are named by adding the suffix-ase to the name of the substrate on which they act e.g. Proteinases, sucrase, nucleases, which break up proteins, sucrose and nucleic acids respectively.
2. The enzymes can be named according to the type of function they are performing e.g. dehydrogenases remove hydrogen, carboxylases help in adding CO_2, decarboxylases help in removal of CO_2, oxidases helping in oxidation and so on.
3. The enzymes are sometimes given double name, one after the nature of substrate upon which they act and second according to their function e.g. pyruvic decarboxylase catalyses the removal of CO_2 from pyruvic acid, alcoholic dehydrogenase catalyses the removal of hydrogen from alcohol.

Enzyme Classification

A system of rules by which enzymes are classified on the basis of the substrate they react with and the type of reaction catalyzed. This system provides both a systematic name and a four-part number code. The first number of the code place the enzyme into one of six groups indicating the type of reaction involved. The next two numbers indicate the groups involved in the reaction and the fourth number provides the absolute identification of the enzyme. The enzymes have been classified into following types:

1. Hydrolysing enzymes, 2. Oxidation-reduction enzymes, 3. Ligases, 4. Group transfer enzymes, 5. Desmolases, 6. Isomerizing enzymes, 7. Carboxylation enzymes.

Hydrolysing Enzymes

They catalyse the hydrolysis of complex big molecules into simple, smaller molecules. Depending upon the nature of food substance upon which the enzyme act, they can be further classified into following types.

1. *Carbohydrases:* They hydrolyse the complex polysaccharides into simpler monosaccharides e.g. sucrases, maltases, lactase, cellulase etc.
2. *Esterases:* They hydrolyse the substances containing ester linkage e.g. lipase, phosphatase.
3. *Proteolytic enzymes:* They hydrolyse the proteins into peptones, and amino acids e.g. pepsin, peptidases.

4. *Amidases:* They hydrolyse amides into ammonia and acids e.g. urease.
5. *Oxidation-Reduction Enzyme:* They catalyse oxidation and reduction reactions e.g. Oxidases, reductases, catalases, dehydrogenases.
6. *Ligases:* They bring linkage between two molecules with the simultaneous breakdown of ATP molecule which supplies energy e.g. synthetases which join amino acids to RNA.
7. *Group transfer enzymes:* The catalyse the transformation of a group from one kind of molecule to another or from one position of molecule to another position in the same molecule and are called as transferases.
8. *Desmolases:* They catalyse the reactions in which long carbon chain is broken or lengthened e.g. aldolase.
9. *Isomerizing enzymes:* They catalyse reactions in which an organic molecule is transformed into its isomeric form and are known as isomerases.
10. *Carboxylation enzymes:* They catalyse the reactions in which CO_2 is added or removed and are known as carboxylases or decarboxylases.

Enzymes as Chemo-thermo Regulators

The chemical reactions that take place within the living cell must be precisely controlled and coordinated. These reactions interlock somewhat as a multitude of assembly processes interlock on a modernday production line. For example glucose and other fuel molecules releases energy in a cells by burning. The cells could not tolerate and survive in such high temperatures nor effectively utilize energy that is released in short sudden bursts. Therefore, cells need a slow, steady release of energy that they must be able to regulate to meet metabolic energy requirements.

In cellular respiration, fuel molecules are slowly oxidized and energy is extracted in small amounts which includes sequences of 30 or more reactions (as many as 20 or 30 chemical transformations before it reaches some final state). Such chemical transformations require a system of flexible chemical control. The key elements of this control system are the remarkable enzymes.

What Enzymes Do?

An enzyme increases the speed of a chemical reaction without being consumed itself. An enzyme affects the rate of a reaction by lowering the energy needed to activate the reactions. Even a strongly exergonic reaction that releases more than enough energy as it proceeds is prevented from beginning by an energy barrier. This is because before new chemical bonds can-be formed, existing ones must be broken. The energy required to overcome this barrier and get the reaction going is called activation energy. An enzyme greatly reduces the activation energy necessary to initiate a chemical reaction. Enzyme promotes a chemical reaction. It does not influence the direction of a chemical reaction or the final concentrations of the molecules involved. They simply speed up reactions. For example carbonic anhydrase is an enzyme which promotes the interconversion of carbonic acid with carbon dioxide and water:

$$CO_2 + H_2O \xrightarrow{\text{Carbonic anhydrase}} \underset{\text{Carbonic acid}}{H_2CO_3}$$

This reaction occurs in animal cells and tissues (kidneys and red blood cells). Carbonic acid ionizes, forming bicarbonate ions, the forms in which most of the carbon dioxide in the blood is transported. The reaction shown above could proceed without the enzyme but would be very slow. However, in presence of carbonic anhydrase, the reaction proceeds about 10 million times faster. This reveals that a single molecule of this enzyme can promote the conversion of an estimated 600,000 molecules of carbon dioxide into carbonic acid each second.

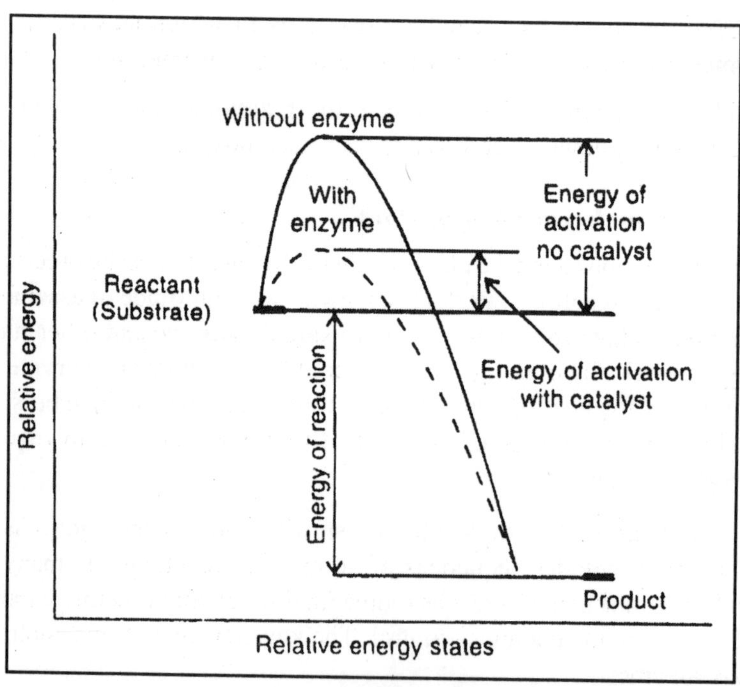

Diagram 20.2 An enzyme acclerates the rate of reaction through lowering activation energy

What is Enzyme Kinetics?

It is the study of enzymes in action. The extremely high rate of enzyme-catalysed reactions greatly facilitates this study. Consider, for example, the haem-containing proteins, haemoglobin and catalase. Haemoglobin binds oxygen. It may bind and release many oxygen molecules in the course of a minute but they remain oxygen molecules, and at any instant, only one is associated with each heam centre. Catalase, being an enzyme, has, a cumulative effect. Again no more than one H_2O_2 molecule will be bound per haem, but while it is bound it may react, and one therefore observes a rapid evolution of oxygen - about a million molecules per minute per enzyme molecule.

How Enzymes Work?

The Chemicals upon which an enzyme operates are referred to as its substrates. Enzymes are very specific. An enzyme will catalyze only a few closely related chemical reactions, or in many cases only one particular reaction. Enzymes form temporary chemical compounds with their substrates. Theses complexes then break up, releasing the product and regenerating the original enzyme molecules for reuse.

Enzyme + Substrate 1 + Substrate 2 ---> Enzyme Substrate Complex --->Enzyme + Product(s)

It is important to note that the enzyme itself is not permanently altered or consumed by the reaction.

Why does the enzyme-substrate complex break up into chemical products different from those that participated in its formation? As shown in Diag. 20.1, each enzyme has one or more regions called active sites, which in the case of a few enzymes have been shown to be actual indentations in the enzyme molecule. These active sites are located close to one another on the enzyme's surface, so during the course of a reaction, substrate molecules occupying these sites are temporarily brought together and react with one another. It is thought that when the enzyme and substrate bind together, the shape of the enzyme molecule changes slightly. This produces strain in critical bonds in the substrate molecules so that these bonds break. The new chemical compound thus formed has little affinity for the enzyme and moves away from it. An enzyme can be thought of as a molecular lock into which only specifically shaped molecular keys, the substrates, can fit.

Similar to a lock and key, however, the enzyme and its substrate seem not to be exactly complementary shapes. A recent model of enzyme action, known as the induced-fit model, is based on data indicating that the active sites of an enzyme are not rigid. When the substrate binds to the enzyme, it may induce a change in shape in the enzyme molecule, resulting in an optimum fit for the substrate-enzyme interaction. The change in shape of the enzyme molecule can put strain on the substrate. This stress may help bonds to break, thus promoting the reaction.

An organic, nonpolypeptide compound that serves as a cofactor is called a coenzyme. Many coenzymes are synthesized from vitamins, particularly from the B vitamins. The coenzyme serves as an adaptor, permitting the enzyme to accept a substrate for which it would, by itself, have little affinity.

Regulation of Enzymatic Action

The chemical reactions are regulated by enzymes, but what controls the enzymes?

1. *First:* The mechanism of enzyme control depends upon the amount of enzyme produced. The synthesis of each type of enzyme is directed by a specific gene. The gene, in turn, may be switched on by a signal from a hormone or by some other type of cellular product. When the gene is switched on, the enzyme is synthesized. Then the amount of enzyme present influences the rate of the reaction. Up to a maximum value, the

rate of an enzyme-dependent reaction increases as the concentration of the enzyme increases.

2. *Second:* Enzymatic control may depend upon the activation of enzyme molecules that are present in an inactive form in the cytoplasm. In the inactive form the active sites of the enzyme are inappropriately shaped, so the substrates do not fit. Among the factors that influence the shape (conformation) of the enzyme are acidity and alkalinity and the concentrations of certain salts. In a few enzymes activator site(s) (or an allosteric site), is also present. When a molecular activator, such as the substance cyclic AMP, occupies the activator site, the shape of the enzyme molecule changes, making the active sites better suited for binding with the substrate.

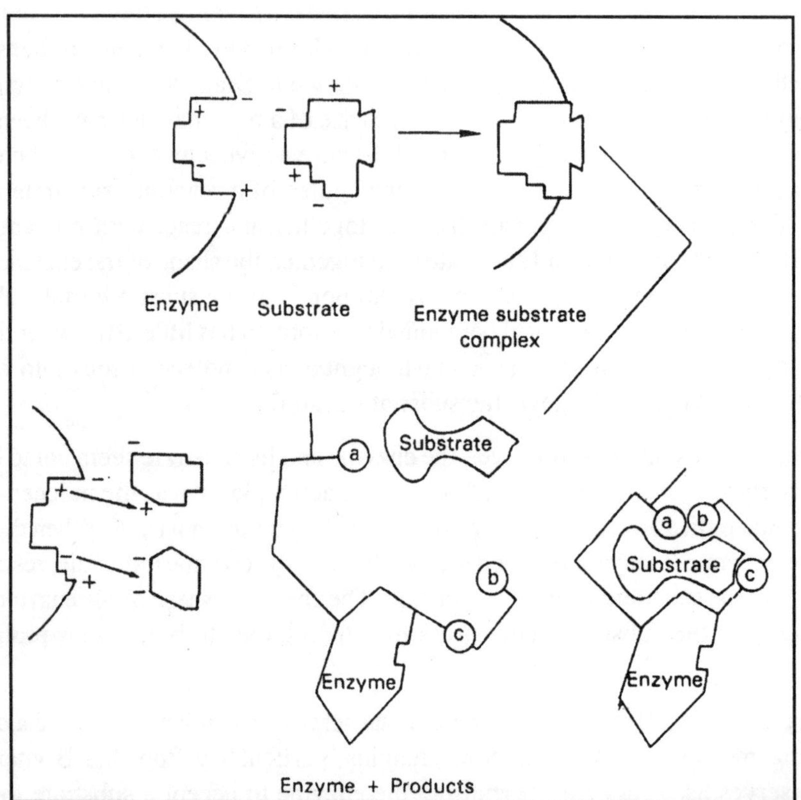

Diagram 20.3 (a) Enzyme-substrate complex formation, (Lock and key mechanism)

Enzyme Optima

Enzymes acts better under certain narrowly defined conditions known as optima. These conditions are optimum/appropriate temperature, pH, and salt concentration, etc. Enzyme pepsin (protein-digesting enzyme of the stomach) works best at the strongly acid pH of 2 whereas amylase (starch-digesting enzyme in saliva and pancreatic juice) works better at pH 8.5 (slightly alkaline). Strong acids or bases irreversibly in activate most enzymes. pH permanently changes the molecular conformation of enzymes. At low temperatures, enzymatic reactions occur very slowly or not at all, but their activity resumes when the temperature is raised to normal. The rates of most enzyme-regulated reactions increase with increasing temperature, within limits. Temperatures greater than 50°C or 60°C rapidly inactivate most enzymes by altering their secondary and tertiary level structures. The enzyme is said to be denatured at high temperature. This happens everyday while cooking an egg white; it changes in consistency as the protein is denatured.

Significance of Km value

The catalysis occurs on account of formation of transient enzyme-substrate (ES) complex. When substrate concentration is high, addition of enzyme can enhance the rate of reaction, till substrate concentration becomes limiting. Similarly, there is usually a direct proportionality between rate and substrate concentration until the enzyme concentration become limiting. The substrate concentration required to cause half the maximal reaction rate, a value named as Michaelis-Menten constant (Km). Km values are mostly constant and do not depend upon amount of enzymes. Usually the Km values for most of the enzymes studied, vary from 10^{-3} to 10^{-7}. Molar Km value can be implicated as indicator of substrate concentration, affinity of enzyme with its substrate and it partly indicates enzymes substrate concentration in the cellular compartment, where reaction occurs. It is the affinity part, for which Km values are used most. Km values and inversely proporsional to the affinity of enzyme for its substrate. Therefore, higher Km values suffers lower stability of ES complex.

Inhibitors, Activators and Inactivators of Enzymes

The rate of an enzyme-catalysed reaction may sometimes be altered in a specific manner by compounds other than the substrate(s). Activators increase the rate; inhibitors and inactivators decrease it. The study of such agents is of practical importance for several reasons:

1. Inhibition and activation of enzymes by key metabolites provides the normal means of metabolic fine control superimposed on the coarse control achieved by regulation of the synthesis and breakdown of active enzymes.
2. External interference with metabolism, whether by drugs, pesticides etc. or by undesirabale toxic agents, often depends on the inhibition of enzymes.
3. Inhibitors, and especially inactivators, provide a powerful tool for studying the chemical mechanisms of enzyme action.

Enzyme inhibition may be reversible or irreversible. Reversible inhibitors can be competitive or noncompetitive. In competitive inhibition the inhibitor completes with the normal substrate for the active site of the enzyme. A competitive inhibitor usually is chemically similar to the normal substrate and so fits the active site and binds with the enzyme. However, it is not similar enough to the normal substrate to take its place effectively. The enzyme cannot act upon it to form reaction products. A competitive inhibitor occupies the active site only temporarily and does not damage the enzyme irreversibly.

In noncompetitive inhibition the inhibitor binds with the enzyme at a site other than the active site. Such an inhibitor renders the enzyme inactive by altering its shape. Many important noncompetitive inhibitors are metabolic substances that help regulate enzyme activity by combining reversibly with the enzyme.

Many poisons are irreversible inhibitors that permanently inactivate or even destroy the enzyme. Nerve gases are irreversible inhibitors that poison the enzyme cholinesterase, essential in the normal function of nerves and muscles. A number of insecticides and drugs are irreversible inhibitors. The antibiotic penicillin and its chemical relatives inhibit a bacterial enzyme necessary for bacterial cell wall construction. Unable to produce new cell walls, susceptible bacteria cannot multiply effectively. Since human body cells do not have cell walls (and so do not employ the susceptible enzyme), penicillin is harmless to humans, except for the occasional allergic patient.

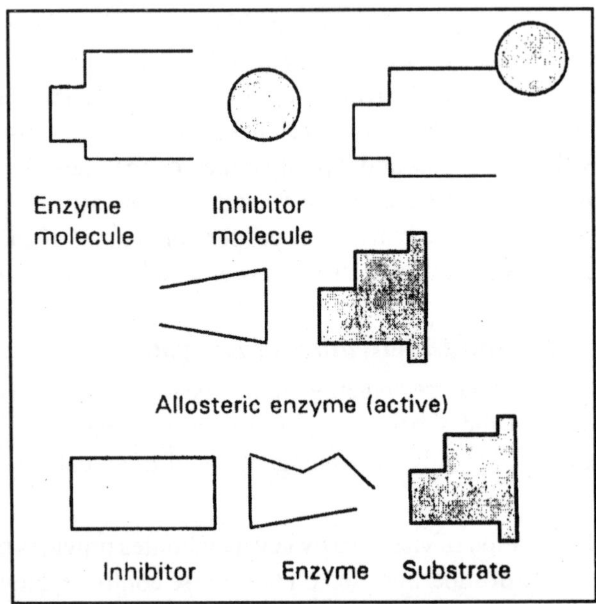

Diagram 20.4 Enzyme inhibition (a) non competitive, (b) Allosteric

Reversible and Irreversible Inhibition

Enzyme activity can be curtailed by various non-specific agents (acid or alkali, urea, detergents, proteases etc.) which disrupt protein structure. These agents interact with a protein at a small number of loci without markedly disrupting the three-dimensional structure. The two most important considerations in their classification are specificity and reversibility.

Multi-substrate Enzymes

The only true one-substrate enzymes are the isomerases, which catalyse reactions of the type A= B, and the lyases, which catalyse reactions of the type A - B + C, and are, therefore, one-substrate enzymes in one direction only. Hydrolases, catalysing reactions of the general type A-B + H_2O = A-OH + BH may be regarded as honorary one-substrate enzymes, since one substrate, water, is normally present at constant concentration.

The remaining groups are:

oxidoreductases:

oxidant + reductant = reduced product + oxidized product

transferases:

A + BX = AX + B

This large group includes methyl transferases, transaldolase, transketolase, acyl transferases, glycosyl transferases, transaminases, kinases etc.

ligases

X + Y + ATP = XY + ADP + P, (or AMP + ppi)

These three groups of enzymes by definition catalyse reactions involving more than one substrate.

Enzyme Secretion by Plant Cells

Enzyme secretion is a common feature of most of the living cells. By this process, cells may act on their environment by modifying its chemical composition and ensure their defence againt external agressions. In pluricellulaar organisms, groups of cells (glands) become specialized in the secretion of enzymes which are utilized in a function useful for the whole organism. The control of such a release of enzyme is often mediated by chemical substances which migrate from one tissue to another, one composed of secreting target cells. This coupling between stimulus and secretion is widely distributed in animal organisms.

The Mechanism of Protein Secretion

The synthesis, intracellular transport, and release of secretory proteins is a basic cellular function common to most eukaryotic cells (Chrispeels, 1976). In the pancreatic exocrine cell for example, the secretory proteins are the object of six steps or operations, which are: synthesis, seggregation, intracellular transport, concentration, intracellular storage and discharge (Palade, 1975). Proteins for export are generally synthesized on polysomes, attached to the membrane of the rough endoplasmic reticulum. In etiolated radish seedlings submitted to far red light, which is known to induce the synthesis of p-fructosidase, the enzyme activity is found associated to endoplasmic reticulum in a first time, then in Golgi apparatus and in cell wall. There is no evidence for the existence of a single intracellular route leading to enzyme secretion. Although the migration of proteins synthesized in endoplasmic reticulum, through Golgi towards the exterior of cell after exocytosis is likely to occur in some cases, direct transport from reticulum to plasmalemma and transfer of cytoplasmic enzymes accross plasmalernma cannot be excluded.

The Control of Secretion

The secretory process is a complex mechanism. It is known that calcium is essential for the control of the secretory process (Rubin, 1982). Calcium plays a critical role in the exocytosis process, which is a fusion-fission response involving the interaction of the plasmamembrane and secretory vesicle membrane. A second requirement for the occurrence of secretion is metabolic energy (Rubin, 1970). An adequate intracellular ionic equilibrium is also necessary. Ca^{2+} is considered as a second messenger in animal cells, able to trigger the discharge of secretory products by regulated cells. In addition, non-regulated cells often requires Ca^{2+} for their secretion. It would be, therefore, necessary to explore the knowledge concerning the regulation of Ca^{2+} in plant cells. It is known that Ca^{2+} is involved in the secretion of some plant enzymes, including peroxidase, α-amylase, and phosphatase. Recently, several articles have been published, which described modes of Ca^{2+} transport across plant membranes. An ATP-dependent Ca^{2+} uptake by isolated membranes vesicles was reported.

Some elements exist for substantiating the hypothesis that phytohormones regulate protein secretion through the mediation of second messenger such as c-AMP or Ca^2 Vcalmodulin However, the existence of an exchange mechanism between H^+ and Ca^{2+} suggests that plant growth regulators especially auxin could modify secretory process by a modification of the distribution of protons which indirectly affect Ca^{2+} compartmentation.

Hormonal Effects on Enzyme Secretion

There are a lot of works showing that treatment of whole plants, isolated plant organs or tissues with phytohormones of biological origin or with synthetic plant growth regulators results in changes in the activity of several enzymes (Barendse, 1983). A considerable number among these enzymes are of an exocellular nature but the process and control of their secretion has not been investigated in depth until now. On the other hand, there are much less papers establishing a correlation between endogenous hormonal status and enzyme levels or activities. It thus can be said that our knowledge of the hormonal control of enzyme secretion by plant cells is far from being well known.

Allostery and its Antecedents

Monod, Changeux and Jacob (1963) in their paper Allosteric Proteins and Cellular Control Systems proposed that in addition to active, substrate-specific sites, regulatory enzymes might possess separate allosteric sites specific for their regulator. Binding of a regulator molecule at such a site, they suggested, could influence events at the active site. The idea was not entirely new, but the paper drew together the information then available to provide a clear, persuasive argument for the widespread occurrence of regulation by this means. The name allosteric was intended to emphasize that the regulator need not bear any structural resemblance to the substrate.

Isoenzymes

Isoenzymes were discovered about 30 years ago and were at first regarded as interesting but of rare occurrences. Since then a wealth of information on enzyme heterogeneity has accrued and it now seems likely that at least half of all enzymes exist as isoenzymes. This is important in many areas of biological and medical science. Thus isoenzyme studies have provided the main experimental substance for the neutral drift controversy in genetics and evolution. Isoenzymes have greatly extended our understanding of metabolic regulation not only in animals but also in bacteria and plants; their existence has made available a multitude of highly sensitive markers for the study of differentiation and development, as well as providing indices of aberrant gene expression in carcinogenesis and other pathological processes. Isoenzymes are also being used increasingly in diagnostic clinical biochemistry. Isoenzymology crosses the traditional boundaries of the biological disciplines and it seems that advances and new applications in this field, are like to occur at a rapid pace.

The Cause of Enzyme Multiplicity

There are various causes of enzyme multiplicity and they may be divided into two categories. These are (a) genetic or primary causes, whereby the organism carries multiple genes each one encoding a different type of enzyme subunit; and (b) post-translational or secondary causes, whereby homogeneous enzyme subunits are modified differentially so as to produce a range of subunits from a single gene. There are, in turn, two types of genetic multiplicity; firstly, multiple alleles at a single genetic locus and secondly, multiple genetic loci.

Primary or Genetic Isoenzymes

Isoenzymes due to multiple alleles at a single genetic locus

In the diploid genome, each genetic locus is represented twice. For each locus, the individual will either be homozygous, possessing two identical alleles, or heterozygous, possessing two different alleles. Where the genetic locus encodes an enzyme subunit, the homozygous individual can only produce one type of subunit. However, the heterozygous individual with two different allelic variants will produce two different types of enzyme subunits. Within the individual the degree of enzyme multiplicity produced by multiple alleles is limited, as two different alleles per diploid locus is the maximum possible genetic variation of this type. However, from one individual to another, there may be considerable variation in the range of enzyme subunit types produced since there may be a variety of different alleles for the locus in the gene pool of the species.

The enzyme subunit types produced as a result of multiple alleles are likely to differ from each other only in minor ways, such as by individual amino acid substitutions caused by point mutations in the DNA nucleotide sequences. Where a genetic locus is active, both alleles will usually be expressed. Thus in the same individual, although the total activity of the enzyme may vary considerably between different kinds of cell, the isoenzyme profile will be constant throughout, homozygous individuals displaying only one subunit type, and heterozygotes possessing two different subunit types.

Isoenzymes due to Multiple Genetic Loci

Many enzymes are encoded at more than one genetic locus and, where this the case, each locus will produce a different type of enzyme subunit. As the expression of each genetic locus can be controlled independently, the organism may synthesize one type of enzyme subunit in a particular cell, and another enzyme subunit elsewhere. Furthermore, the expression of genetic loci may alter during the course of development, and therefore the type of enzyme subunit produced in a tissue may change. Thus multiple genetic loci permit differences in isoenzyme profile both from one tissue to another, and from one developmental stage to another even within the same tissue. Such variation in isoenzyme profile are not possible where the enzyme multiplicity is due to multiple alleles operating at only a single locus. However, as all members of the same species will possess the same genetic loci, multiple enzyme-encoding loci in the absence of multiple alleles cannot account for isoenzyme differences between one member of the species and the next. The isoenzyme subunits produced by multiple loci are likely to differ extensively as a result of numerous amino acid substitutions, and also deletions and additions of residues may have occurred resulting in small size differences between the subunits.

Secondary or Post-translational Isoenzymes

Proteins may be modified in numerous ways following their synthesis. Such possible modifications include the addition of carbohydrate, limited proteolysis, and the covalent modification of amino acid side chains. Post-synthetic alterations affecting only part of the enzyme subunit population, so that modified and unmodified subunits are found in the same organism, will result in isoenzymes. Where the post-translational modification process is very active in some tissues but not in others, the result will be a tissue-specific distribution pattern of the secondary isoenzymes, mimicking the effect of multiple genetic loci. An example of this is one particular pyruvate kinase subunit type. There are three major types of pyruvate kmase subunit in mammals, each coded for by an independent genetic locus.

Apparent Enzyme Multiplicity

It must be recognized that the term isoenzymes is often used loosely in an operational sense as it tends to be applied whenever enzyme multiplicity is observed. Multiple enzyme forms may therefore pass into the literature as isoenzymes when, for one reason or another, this term should not have been employed. The multiple forms may be artefacts resulting from laboratory manipulation of cells and cell extracts. Unphysiological ion concentrations may result in the nonspecific binding of ligands with consequent alteration of the apparent properties of the enzyme, or liberation of proteolytic enzymes on disrupting the cell may result in the partial degradation of the enzyme. In this and other ways enzyme multiplicity which was not present in the intact cell may be created. As an example, heterogeneity of the glycolytic enzyme phosphoglucose isomerase was eventually shown to be due to artefactual oxidation of enzyme sulphydryl groups.

Isoenzymes as Genetic Markers

Isoenzyme analysis is now an important experimental technique in genetics. Each genetically determined isoenzyme subunit type, by definition, is the result of a different gene, whether we are considering an enzyme encoded by multiple alleles at a single locus, or multiple loci. Therefore, wherever the isoenzyme subunit type is found, the gene coding for it is not only present, but is being expressed. In this way the isoenzyme subunit is the marker for its own gene. Theoretically, this is not restricted to enzymes; any polypeptide is a marker for its encoding gene. However, in practice, since enzymes are catalytically active they can be assayed specifically so that enzyme variants can be detected more readily than the genetic variants of non-enzymic proteins. It must be emphasized that only isoenzymes due to genetic multiplicity are of interest here. Use of isoenzymes as markers in modern biochemical genetics is so widespread that the hypothesis "one gene-one isoenzyme subunit" have been proposed (Rider and Taylor, 1980). The usefulness of isoenzymes as genetic markers is illustrated in the following experiment devised to test the Lyon Hypothesis. Not only was this a particularly elegant application of isoenzyme markers, but also was in its own right an important advance in our understanding of genetic regulation.

Enzymes in Food Industry

Applied enzymology has moved significantly beyond the early broad extra cellular hydrolytic enzyme preparation such as the bacterial α-amylase, papain, pectinase, and pancreatin where the commercial concentrate might contain as much as one per cent of the active labeled principle. Hydrolytic enzymes that modify starch, pectin, protein and fats to their component parts, and a few others like glucose isomerase and glucose oxidase, newly produced enzymes have become or are becoming commercially important in specific chemical transformation in food industry. The enzymes are commonly used in modification in flavour (cheese and butter flavour), in low, high, normal sweetness of sweetners, in corn wet milling, in feed, soy bean milk, baking, in modifying food gums, etc.

Table 20.1 Enzymes, coenzymes, and mineral activating substances.

Enzyme	Coenzyme or co-factor	Activating mineral elements
	In processes of glycolysis	
Phosphatase	Phosphate	Mg^{2+},
Phosphoglutamase		Mg^{2+}
Aldolase		Fe^{2+}, Zn^{2+}, Co^{2+}
Enolase		Mg^{2+}, Mn^{2+}, Zn^{2+}
Pyruvate kinase	ADP In the tricarboxylic cycle	Mg^{2+}, K^{+}, NH4 +
Decarboxylase and pyruvate dehydrase	Thiamine pyrophosphyte fatty acids, Co A	Mg^{2+}, Mn^{2+}
Isocitrate dehydrogenase	NADP	Mn^{2+}, Mg^{2+}
Succinate dehydrogenase	FAD	Ca^{2+}, Cr^{3+}, $A1^{3+}$
Aconitase		Fe^{2+}, -SH
Dehydrogenase		Mn^{2+}, K^{+}, Rb^{+}, NH_4^{+}
Pyruvic oxidase		Mg^{2+}, Mn^{2+}

Among the three most important functions namely, the mechanical, osmotic and chemical, the later depends solely upon the bio acceleration of metabolic reactions caused by enzymes

Enzymes and Heat Treatment for Preservation

Enzymes have pronounced effects on the colour, flavour, aroma, texture and nutritional quality of foods during growth and maturation, during harvest and post harvest storage and storage after processing. In tomato, the softening phenomenon in ripening is caused by its pectic enzymes. In peaches, apples, plums, grapes and avocados, the browning reaction is the result of polyphenol oxidase. In green leafy vegetables, lipoxygenase and other enzymes cause flavour and aroma deterioration, but other enzymes may also cause discolouration.

The frozen and food industry is based on the observations that sufficient heat treatment of vegetables and fruits to inactivate peroxidases, followed by freezing and frozen storage, could extend shelf-life from a few days or a few weeks to one or two years. The lipoxygenase is the major enzyme responsible for aroma deterioration in English green peas and green beans, while cystinelyase is responsible for aroma deterioration in broccoli and cauliflower

Enzymatic Modification of Proteins and Food industry

Proteolytic enzymes are used extensively for modifying proteins in various ways in food product and for waste management. These enzymes are used in baked and brewed products, cereals, cheese, chocolate/cocoa, eggs and egg products, feeds, fish, legumes, meat, milk, protein hydrolysates, wines, etc. For cross linking of two proteins of different properties, the enzymes like transglutamase, lipoxygenase, polyphenol oxidase and peroxidase are used. Proteolytic enzymes have long been used to produce protein hydrolysates for use in soups, bouillon, soysouce, tamaric sauce, etc.

Enzymes and Specialty Products

Enzymes, because of their high substrate specificity and stereo specificity, are ideal for producing special compounds required by the food and pharmaceutical industries. The conversion of corn starch produced by wet milling, to glucose and fructose has been the most successful commercial operation. Lipases are now being studied intensively to change triglyceride fatty acid composition.

The most successful commercial application of enzymes is in the amino acid industry. Amino acids for food and feed fortification, nutritional supplements, or as feed stock for down stream products have been made by fermentation processes, from protein hydrolysates or by chemical synthesis.

Enzyme and Plastein Reaction: This important reaction initiated by enzyme is being used successfully in Japan to produce phenylalanine-free peptide products for patients with phenylketonuria, surfactants for the cosmetic and food industry and antifreeze type compounds that have the capacity to prevent hard freezing of foods, blood and sperms.

Enzymes and Recombinant DNA Technology

Recombmant DNA Technology for all biotechnological purpose depends upon enzymes. The technology permits both an increase, as well as decrease, in the level of enzymes or other products in an organism and is being used in the following ways:

1. Understanding of the primary and secondary structures of DNA and RNA.
2. Sequencing of DNA and RNA.

3. The property to remove, via hydrolysis, specific nucleotide segments from plasmid (by use of restriction enzymes).
4. Ability to incorporate nucleotide sequences (by use of ligases) from other organisms or by chemical synthesis into the host organism.

The examples of this technology include chymopsin, insulin, bovine growth hormone, and proteolytic enzymes for use in detergents.

Bitterness in Orange and Grape Fruit Juices

The bitterness of orange and grapefruit juices at commercial level is a serious problem which develops due to the presence of limonin and naringin. Naringin is about 0.01 as bitter as limonin but is often produced in higher amount.

Bitterness caused by limonin (a triterpenoid) can be eliminated as follows.

1. By preventing its biosynthesis (probably synthesized .via the mevalonate pathway). This can be done by preharvest treatment with 1-naphthalene acetic acid.
2. By removing the rag and pulp from freshly expressed juice as soon as possible to prevent the precursor. The precursor of limonin is limonic acid, A-ring (mono) lactone.
3. By use of immobilized microbial cells which contain NADP-dependent limonin dehydrogenase. The enzyme on hydrolysis converts limonin to non-bitter products.
4. By use of unmmobilzed microbial cells which contain NADP-dependent limonin dehydrogenase. The enzyme on hydrolysis converts limonin to non-bitter products.

Bitterness caused by naringin can be eliminated as follows:

1. The bitterness caused by naringin can be eliminated by an enzyme, naringinase.
2. It can also be eliminated by using recombinant DNA techniques in the limonin biosynthetic pathway.

Elimination of Unwanted Compounds

Removal of cyanogenic glycosides from Cassava and lima beans

The food plants like cassava and lima beans contain toxic levels of cyanogenic glycosides. This compound can be eliminated from these vegetables by soaking them overnight which hydrolyses cyanogenic glycosides by specific glycosidases enzyme to glucose, HCN and acetone.

Removal of Pectic compounds (Cloud) from Orange juice

The freshely squeezed orange juice when allowed to stand separates and settles in the form of precipitation. The suspended pectic compound or cloud develops due to the presence of the enzyme pectin methylesterase. The problem can be overcome by heating the freshly squeezed juice which inactivates the enzyme. This process produces a cooked flavour.

Following are the areas of industrial use of microorganisms, and of application of major enzymes from microbes (main microbial species).

List of the microorganisms which produces enzymes used in Food Industry

Production of Fermented Beverages and Foods:

Dried bonito, soyasauce (*Aspergillus oryzae, pediococcus soyae, Saccharomyces rouxii, Torulopsis* spp.), vinegar (*Gluconobacter suboxidans*), pickled vegetables, cheeses (*Penicillium camembertii, P. roqueforti, Propionibacterium shermanii, Streptococcus* spp.), yoghurt (*Lactobacillus bulgaricus, Streptococcus thermophilus*), lactic acid (sour) drinks.

Alcoholic Fermentation

Bear (*Saccharomyces cerevisiae, S. carlsbergensis, S.uvarum*), cider (*S. cidri*), wine (*S. cerevisiae*), sake (*Aspergillus oryzae, Lactobacillus and Leuconostoc* spp., *S. cerevisiae*) and other fermentations of fruit juices, distilled sprits, etc.

Use of microbial cells and production of physiologically active substances in the food and pharmaceutical industries

1. Vaccine and microbial bioinsecticides (*Bacillus popilliae, B. thuringiensis*).
2. Baker's yeast (*Saccharomyces cerevisiae; Candida milleri*).
3. Fodder yeast (*Candida utilis, Saccharomycopsis lipolytica*).
4. *Spirulinas, chlorellas*, and other unicellular algae; single-cell proteins (*Methylophilus methylotrophus, Candida tropicalis, C. utilis, Saccharomycopsis lipolytica*).
5. Amino acids, mononucleotides (*Corynebacterium glutamicum*). Vitamins (riboflavin: *Eremothecium ashbyi*: Vitamin B_{12} *Propionibacterium* sp., *Pseudomonas denitrificans*).
6. Steroids (*Arthrobacter simplex, Mycobacterium* sp., *Rhizopus arrhizus, R. nigricans*).
7. Carotenoids (β-carotene: *Blakeslea trispora*; astaxanthin: *Phaffia rhodozyma*).
8. Gibberellins (*Gibberella fujikuroi*) and other plant growth hormones

Production of Industrial Solvents and Organic Acids for the Chemical Industry

Ethanol (*Kluyveniniyces fragilis, S. curevisiae, Zymomonas mobilis*) n-butanol and acetone (*Clostridium acetobutylicum, C. saccharoacetobutylicum*); acetic acid (*Acetobacterium woodii, Clostridium aceticum*); citric acid (*Aspergillus niger, Saccharomycopsis lipolytica*); fumaric and lactic acids (*Lactobacillus delbrueckii*).

Production of Polysaccharides for the Food Industry and for other Uses

Dextrans (*Leuconostoc mesenteroides*), levans, mannans; xanthans (*Xanthomonas campestris*).

Microorganisms used in cheese-making

Cottage cream: Streptococcuus lactis, S. cremoris, S. diacetilactis, Leuconostoc citrovorum

Soft Cheese

Brie (France): Streptococcus lactis, S.cremoris, Penicillium, camembertii, P.candidum, Brevibacterium linens.

Camembert (France): Streptococcus lactis, S. cremoris, Penicillium camembertii, P. candidum.

Semi-soft Cheese

Blue d' Auvergne (France): Streptococcus lactis, S.cremoris, Penicillium roqueforti or P. glaucum.

Gorgonzola (Italy): Streptococcus lactis, S. cremoris, Penicillium roqueforti or P. glaucum

Munster (Germany): Streptococcus lactis, S. cremoris, Brevibacterium linens

Roquefort (France): Streptococcus lactis, S. cremoris, Penicillium roqueforti or P. glaucum.

Hard cheeses

Cheddar: Streptococcus lactis, S.cremoris, S. durans (United Kingdom), Lactobacilluus case.

Colby (United States): Streptococcus lactis. S. cremoris, S. durans, Lactobacillus casei.

Edam (Netherlands): Streptococcus lactis, S. cremoris.

Gouda(Netherlands): Streptococcus lactis, S. cremoris.

Gruyere (Switzerland): Streptococcus lactis, S. thermophilus, Lactobacillus helveticus, Propiombacterium shermanii, or Lactobacillus bulgaricus and Propionibacterium freudenreichii.

Very hard cheeses

Parmesan (Italy): Streptococcus lactis, S.cremoris, S. thermophilus , Lactobacillus bulgaricus.

Production of antibiotics

Penicillins (Penicillium chrysogenum), cephalosporins (Cephalosporium acremonium), streptomycin, kanamycins, neomycins, tetracyclines, etc. (Strptomyces spp.), gramicidin-S (Bacillus brevis), polymyxin-B (Baacillus polymyxa), bacitracin (Bacillus subtilis).

Plant Enzymes in Resistance to Insects

During the last many decades, traditional approaches to host-plant resistance have shown that a number of plant natural products acts as constitutive bases of resistance against insects. It has also been realized that plant resistance against insects also consists of inducible and dynamic elements triggered by insect-feeding damage. These inducible elements of resistance have been found to be both plant enzymes and their end-products. Plant enzyme respond potentially at metabolic level of insect feeding, and that these induced responses can be of sufficient diversity and intensity to confer resistance.

Plant enzymes can confer resistance by perturbing the utilization of chemicals (anonists -kairomones) by an insect that are essential in utilizing its host plant. The perturbation arises from plant enzymes creating chemicals (antagonists - allomones) that are inhibitory to insects, or removing chemicals (agonists) that are useful. These perturbations can be caused by several general mechanisms that proceed when plant cells are broken by the feeding insects and during ensuing masticative and digestive processes. The plant enzymes may act in the following ways:

1. Direct production of antagonists: In this an enzyme converts a chemical to a more biologically active form. For example, glycosidases produced by plants (e.g., iridoids, saponins, phenolics, cardenolides, and alkaloids) potent influence upon insects and other animals. They are cyanogenic glucosides. Over 2000 species of plants are known to be cyanogenic. Cyanogenesis is probably a plant defence against unadapted, chewing insects. However, the mechanism of their action is not clear. Other enzymes like ureases (liberated from ammonia), polyphenol oxidases (produced from plant trichome as exudates of tomato and potato) also develop insect resistance in plants.
2. Indirect production of antagonists: In this an enzyme acting on a substrate to liberate a chemical messengers that trigger the de novo synthesis of a new and biologically more active chemicals.
3. Direct removal of agonists: In this an enzyme converts a chemical to a less biologically active form.
4. Indirect removal of agonists: In this an enzyme liberates a product which reacts with a second chemical rendering the last less biologically active.
5. Direct action of enzyme: In this an enzyme acts directly against the insect as a substrate.

Direct action of Enzyme

Insects act as a substrate for enzyme (like chitinase enzyme). Chitin is the major structural component of the peritrophic membrane of insects which is thus theoretically susceptible to attack by plant chitinases. One of the several functions of the peritrophic membrane of insects is to provide a physical barrier against the entry of pathogens or macromolecular toxins across the gut wall. Chitinase enhances larvel susceptibility to the bacterium Bacillus thuringinesis.

The Tomato Plant and Interaction of Plant Defences

The tomato plant contains number of enzymes which may be important mediators of resistance to insects (e.g., polyphenol oxidases, peroxidases, lipoxygenases, ureases, chitinases, phenylalanine, tyrosine ammonia lyases, endoploygalacturonases, etc.). The responses of the plant are complex and involve following three general modes:

1. The release of constitutive, non-activated defences;
2. the enzymatic activation of constitutive defences, and/or;
3. the induction of multiple enzyme systems leading to the de novo synthesis of many chemicals for defences (polyacetylenes, phenolics, proteinase inhibitors, etc.).

However, the impact and utility of these three general lines of defence upon insect pests are unclear.

There is considerable interest in utilizing proteinase inhibitor (PI) for host-plant resistance because of their adverse effects on insects and pathogens. Biotechnologists are actively working to know about PI genes or the genetic transfer of PI genes into different crop species.

Despite the limited research on the role of plant enzymes as mediators of resistance, opportunity knocks. Recent biotechnological techniques now make the interspecific transfer of genes fesible, and offer exciting, unique opportunities for developing crop resistance against a diversity of pests. Resistance conferred by the transfer of genes (i.e., for proteniase inhibitors), may be adversely affected by the inbuilt defence of the receiving plant. Gene transfers for herbicide resistance, aimed at altering the oxidative potential of foliage to enhance plant resistance to certain herbicides, may interfere with the defensive responses of the plant that are mediated by oxidative enzymes. We need joint efforts of molecular biologists, agronomists, food scientists, plant pathologists, weed scientists, and entomologists to achieve optimal success in crop improvements.

Integrated Pests Control and Enzymes

Pest control is being done by synthetic pesticides which have more restricted activity with shorter lifetimes in the environment. Enzyme and isoenzyme activities distinguished via separations by polyacrylamide gel electrophoresis and activity stain, are used commonly in determining changes in plants during breeding by conventional and non-conventional methods. Enzymes patterns established by the above techniques are ususlly related to pathogen and insect resistance, thus saving considerable time in analysis.

Abatement of Pollution

Treatment of industrial effluents, waste waters, disposal of wastes and garbage; recovery and reuse of biodegradable wastes. Microbial leaching of ores and recovery of mining and colliery wastes.

Enzymes and Fermentation

In typical enzyme fermentation, the organism is propagated through several stages of batch culture. Stock cultures from the research laboratories are generally preserved in freezedried ampules, periodically sampled and tested. Inoculum is usually tansferred from stage to stage in the proportion of 1 to 5 per cent by volume. There are two general procedures for culture: Surface culture (= solid-substrate cultures) and Submerged cultures for production of fermented beverages and foods.

- *Substrate Cultures:* The microorganisms are cultured on a basic substrate which contains high amount of nutrients and large surface, for example, wheat bran and/or rice bran and cereal meal. It also contains mineral substances and salts with low water content.

- *Submerged Cultures:* This type of culture is very common In principle, the same fermenters and general methods are used for the production of enzymes by the submerged process as for the production of antibiotic or single-cell protein. The production of enzyme is usually takes place in mechanically stirred tanks with the capacity of 10,000 to 100,000 liters in batch operation. This kind of fermentation lasts for 50 to 150 hours.

The scope of the book is limited, therefore, the descriptions of isolation of enzymes (disintegration of biological material, filtration, extraction, purification, concentration, ultracentrifugation, desalting etc.) upto the level of purified enzyme are not given.

Industrial Enzymes

There are about 3000 enzymes known and only few are manufactured at large scale. These enzymes are mainly extracellular hydrolytic enzymes which have the property to degrade naturally occuring polymers for example starch, proteins, pectins, and cellulose. Enzyme glucose isomerase is, however, an exception in this regard.

Industrial Enzyme Production

Enzymes are the directing and controlling biocatalysts present in all living matter which determine a particular chemical reaction. They accelerate the course of chemical reactions by several orders through a substantial decrease in the activation energy. The catalytic activity of the enzyme secreted by the cells or isolated from them is maintained under suitable conditions and permits the use of these enzymes as catalysts outside the cells.

1. Mutants that produce the enzyme constitutively, that is without an inducor;
2. The production of the desired enzyme in the mutants is not inhibited by repressers; and
3. The yield of enzyme is considerably increased by multiplication of the gene copies, and/or changes in the single sequences proceeding the structural gene to bring about high expression.

Role of Enzymes in Genetic Engineering

Fundamental knowledge of the structure, function and mechanism of DNA-modifying enzymes has been important not only in understanding how these enzymes catalyzes chemical reactions in vivo but also for the development of field of recombinant DNA technology. Several enzymes are involved in nucleic acid synthesis, especially when one considers the varied nature of enzymes in each group. For example with the polymerases, there are separate enzymes important in biosynthesis of DNA and RNA some with specificity for size of the chain length (gap) to be completed. Gene manipulation experiments require the use of certain enzymes concerned with nucleic acid metabolism. The most important of these enzymes are different kind of restriction endonucleases, exonucleases and ligases.

CHAPTER-21

Biotechnology and Agro-industrial Development

Agro-industry is one of the latest branch of applied bio-sciences which is, now a days, mainly related to biotechnology and/or genetic engineering. Genetic manipulation to bring about desired characteristics within a plant/crop has revolutionized this field. The use of microorganisms in different field of medicine and commodities production has also brought this branch of science in notice.

Agro-biotechnology is basically, a industry-based-process-biotechnology which is simply described as a discipline, to convert raw materials to final products when either the raw material and/or a stage in the production process involves biological entities. As the discovery, that waste products can be processed to manufacture/yield commercial valuable commodities, through the use of microorganisms, a number of industries came up in this field. This is most obvious in the production of fermented foods and drinks and microbiologically processed foods such as pickles and sauerkraut. In time, it was recognised that waste products could yield commercially valuable commodities when treated with specific microorganisms. Various commercial products, of important economic value, made by microorganisms are: (1) pharmaceuticals, including antibiotics, steroids, human protein, vaccines, and vitamins; (2) organic acids; (3) amino acids; (4) enzymes; (5) organic solvents; and (6) synthetic fuels. Many of these products can be produced both microbially and by chemical synthesis. The choice of which process to employ generally depends on economics, and it is not surprising that some products that have been produced by microorganisms, in ancient period are now produced chemically and vice-versa. Antibiotic and fermentation research was followed by the development of efficient industrial processes for the manufacture of vitamin, (riboflavin, cyanocobalamine), plant growth factors (gibberellins), enzymes (amylases, proteases, pectinases), amino acids (glutamate, lysine), flavour nucleotides (inosinate, guanylate), and polysaccharides (xanthanpolymer), etc.

Production of single-cell protein for feeding of animals and humans is another aspect of the beneficial application of the microbes. There is no doubt that, in the future, farming on land and sea will no longer be able to feed the world population even if our methods of food distribution improve tremendously. The logical answer will be single-cell protein grown on urban, agricultural, and industrial wastes (Demain, 1981).

The industrial process (technology), type of organisms, substrate utilized, and end products are varied. Many theoretical processes are possible but not practical. Technological problems are among the largest to be solved. To date, industrial microbiology is concerned at the level of the reaction between substrate and either bacteria or fungi. Algae or microzoans are not currently used. Several examples of industrial processes used in the production of specific substances are presented in this chapter.

The important characteristics that have made microbes useful in industries are:

1. their ability to grow on easily available and cheap raw materials,
2. their ability to maintain a physiological constancy, ;
3. their ability to bring about biochemical transformations under simple culture conditions. Microbes have tremendous capacity of carrying out a variety of reactions (especially secondary metabolism resulting in an inexhaustible supply of secondary metabolites available for commercial exploitation) and
4. high ratio of surface area to volume, which facilitates the rapid uptake of nutrients required to support high rates of metabolism and biosynthesis. These properties of microorganisms are highly preferable over synthetic processes. Presently, industrial microbiology has established as a strong arm of applied microbiology and it is unlikely that, at least in some industries, the microbes could be replaced by synthetic techniques in the near future.

The Fermentation Technology

In industrial microbiology the term fermentation is not used in its restricted scientific sense, referring to metabolic pathways that proceed by fermentation rather than respiration, but rather in a wider sense to include any chemical transformation of organic compounds carried out by using microorganisms and their enzymes. Industrial processes using microorganisms exploit the enzymatic activities of the microbes to produce substances of commercial value.

How and Why do Microorganisms Make Alcohol?

Energy conversion by living cell is the fundamental property. Living cells produce useful energy-ATP, which is regarded as the cells energy currency. Yeast has the property to maintain the stock of ATP, which is possible due to the consumption of sugars like glucose, and fructose. Sucrose is the main component of sugar-cane juice which consists of the glucose molecule attach to the one molecule of fructose. The first step of yeast's activity is to break apart the glucose and fructose units which enter the energy metabolization machinery to provide energy. If yeast grown in oxygenated medium, the sugar will be broken down step by step, into smaller and smaller molecules and at the end carbon dioxide is liberated. However, if there is little oxygen or no oxygen available to the yeast the series of chemical breakdown processes can not be completed and the sugar is broken down into ethanol, a fuel alcohol.

$$C_6H_{12}O_6 \rightarrow 2C_2H_5OH + 2CO_2$$

$$AG° = - 56Kcal$$

Glucose is splited up into two molecules of pyruvic acid via the reactions of glycolysis. Alcoholic fermentation and aerobic degradation follow the same reaction sequences up to this point. In fermentation pyruvic acid is degraded enzymetically to alcohol and carbon dioxide.

Organisms, such as yeast, which carry out alcoholic fermentation, contain the enzyme pyruvate decarboxylase (pyruvate decarboxylase-2-oxoacid carboxylase, E.G. 4.1.1.1) that catalyzes the decarboxylation of pyruvate to-acetaldehyde by an irreversible reaction. The enzyme has been found only in plant tissue upto now. In the final reaction of alcoholic fermentation, acetaldehyde is reduced to ethanol by NADH in the presence of alcohol dehydrogenase (alcohol: NAD oxidoreductase, E.G. 1.1.1.1). The enzyme is widely distributed and found in liver, retina and sera of animals, seeds and leaves of higher plants, and many microorganisms, including yeasts. Obviously the enzyme is not restricted to tissue which produce large amounts of ethanol.

Direct fermentation of cellulose to ethanol is of current interest. A thermophilic bacterium, *Clostridium thermocellum* and filamentous cellulolytic fungus, *Moniliu* sp. can produce ethanol directly from cellulose. However, in both cases, the fermentation rate is slow and final ethanol concentration remains low. In thermophilic pentose fermenting anaerobe, *Clostridium thennosaccharolvticum* is cultivated in combination with *C. thermocellum. This* mixture culture has been shown to ferment both Solka-Floc and corn stover to ethanol and also large quantities of acetic acid and lactic acid.

Petite yeasts yield upto twice as much alcohol as their normal relatives. If a normal yeast strain (IZ-1904) produces 41% of alcohol *then petite* verson of this yeast will produce 83%. *Zymomonus mobilis* - a bacterium - is the alternative of yeast. *Z. mobilis* is used as agave juice fermenter in Central America. It ferments sugar more efficiently to alcohol.

Fermentation Technology

The technology utilized in fermentation processes are designed to obtain maximum growth of an organism under the optimum physical conditions in a specific medium for the production of a desired end products. Three general types of growth environments are used: the flask, the shallow tray, and the vat or closed drum. Each of these procedures serves a specific need, and most fermentations use one of these basic procedures.

A fermenter or bioreactor is a container designed to provide an optimum environment in which microorganisms or enzymes can interact with a substrate and form the desired products. The fermenters are of two types:

1. *Open:* It allows continuous processing with substrate entering at one and end products leaving at the another.
2. *Closed:* In this type, the processing is done in the batches. This type of fermenter is used for the production of antibiotics. The microbes are grown on nutrients placed in the vessel at the start of the fermentation. The vessel is cooled by a water jacket. Air is pumped into bottom of the liquid, and acid or alkali added as necessary. A stirrer keeps the contents well mixed. Steam lines are provided so that the vessel can be sterilized after each fermentation batch.

During fermentation, it is necessary to regulate many factors within predetermined values: oxygen and carbon dioxide, pH, temperature and media concentration etc. It is also essential to maintain high degree of sterility within fermenter. The fermenter should be made of stainless steel or copper because such fermenters are resistant to steam sterilization.

Pre-treatment and Purification

Pre-treatment is always given to the raw substrate in addition to sterilization. For example molasses are treated for removal of iron salts from it. Molasses are the byproducts of sugarcane processing consisting of 50% sucrose. Similarly, starch in corn syrup is hydrolyzed to sugar before yeast can convert either to ethanol. Both the products are acidified and diluted.

Downstream Processing

At the end of the fermentation, the desired products may present in a very small quantities (just a few milligrams per cubic decimeter in case of pharmaceuticals). Therefore, it is necessary to give treatment to such undesirable waste products. The treatments of such products is collectivcly called downstream processing.

Purifying the Products

When fermentation process is completed, the vessel is full of a thick broth microbial cells, some unconsumed nutrients and dissolved products. It is necessary to clean the vessel after every operation.

Fermenter Design Considerations

Modern fermentation processes require a fermenter that should provide an environment suitable for the growth of a pure culture and/or a defined mixed culture, which can run free from contamination under controlled conditions. A well-designed vessel will also ensure that culture is contained with no aerosol leaks of the vessel contents, since repeated exposure to even a pathogen can in certain circumstances, be hazardous.

The design must incorporate a device for mixing the contents, an air supply for aerobic processes, probes to monitor the environment and regulators to control it. There must be provision for inoculation and sampling, as well as for charging and discharging the vessel. In continuous culture, it is necessary to monitor and control the flow rate of the medium as well as the culture volume and mass.

Incorporating all these features means that the construction has many potential sources for the entry of contamination. Good aseptic design at this point is crucial. The following design rules will apply:

1. There should be no direct connections between sterile and non-sterile parts of a system.
2. Minimize flange connections. These can move under vibration and heat and provide entry for contaminates.

3. Use all-welded construction if possible.
4. Avoid dead spaces and crevices, etc.
5. Various systems should be independently sterilizable.

Materials of the Bioreactor

It is essential that the bioreactor should be made of nontoxic materials, and able to withstand steam under pressure so that it can be sterilized and must be resistant to corrosive effects of sterilization and high or low pH. Pits and surface material can harbour microorganisms, therefore, the surface should be as smooth as possible. The material used should not affect, or to be affected by the environment.

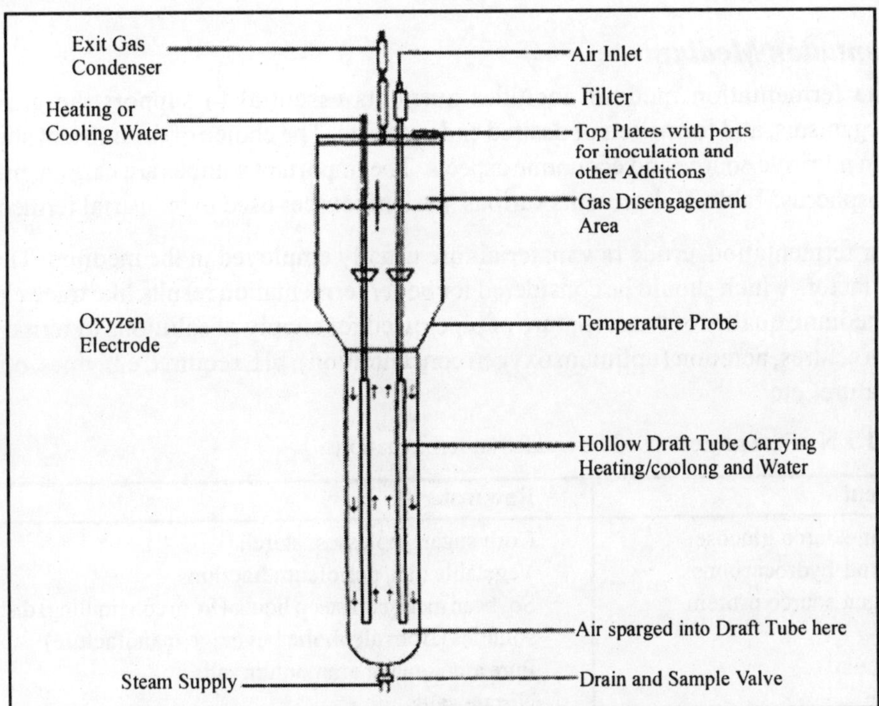

Diagram 21.1 Bioreactor

Selection of Microbes in Fermentation Industry

The choice of microorganisms for use in the fermentation industry begins with screening so as to find the right microorganism. Of the many species of microorganisms, relatively few posses the genetic information needed to produce economically useful products. The selection procedure employed in industry are designed to separate microorganisms that are potentially valuable in producing a commercially useful product from the rest. Selection of microbes that possess the potential for producing industrially important substances includes both naturally occurring microorganisms and genetic variants. The classic approach used to find new antibiotic-producing strains has been to screen large numbers of isolates from soil samples for microorganisms that naturally produce antimicrobial substances.

Microfilteration Membranes: Microfilteration protein containing solutions is a widely used processing step in the biotechnological industries. They are typically used including primary cell recovery, from fermentation broths and sterile filteration as the final steps in the production of protein products.

Genetic engineering has opened several new directions for employing microbes to produce economically important substances. Mutation and selection approach is hit or miss, whereas, the use of recombinant DNA technology permits the purposeful manipulation of genetic information to engineer a microorganism which can produce high yields of a variety of products. The use of biotechnology has great social consequences. The technique has the ability to modify the genetic composition of all organisms-*from microbes to humans.* However, genetic engineering raises serious ethical questions that society must now face.

Fermentation Medium

The fermentation medium includes nutrients essential to support the growth of microorganisms, and formation of desired end products. The choice of the nutrient should be based on nutritive source and economic aspects. The important sources are carbon, nitrogen, and phosphorus. Table 21.1 presents various nutrient sources used in industrial fermentation.

For fermentation, crude raw materials are usually employed in the medium. There are several factors which should be considered for better fermentation results like trace elements in the medium, quality of water, nature of pipes used for supply of solutions to fermentation reaction centres, aeration (optimum oxygen concentration), pH, required enzymes, optimum temperature, etc.

Table 21.1 Nutrient sources used for industrial fermentations.

Nutrient	Raw material
Carbon source glucose	Corn sugar, molasses, starch.
Fats and hydrocarbons	Vegetable oils, petroleum fractions
Nitrogen source protein	Soybean meal, cornsteep liquor (from corn milling) distiller's solubles (from alcoholic beverage manufacture)
Ammonia	Pure ammonia or ammonium salts
Nitrate	Nitrate salts
Nitrogen	Air (from nitrogen fixing organisms)
Phosphorus source	Phosphate salts

Regulatory Mechanisms and Industrial Fermentations

Several enzymes are involved in different metabolic pathways of a microorganism which act in an integrated manner. This is achieved by the regulatory mechanisms (e.g., induction, feedback regulation, catabolite regulation and energy charge regulation) which help the microbe to compete efficiently with other forms of life and to survive in nature. An ideal cell does not produce excess of metabolites, regardless of its environment. However, at industrial level, it is expected that microbes should produce excessive amount of metabolites which can be isolated. Organisms are screened for their ability to over produce the desired product from culture collections or from nature.

Enzyme Fermentation

Enzymes are the important sources of industrial products because of the following reasons:

1. Their rapidity and efficiency of action at low concentrations and under mild conditions of pH and temperature.
2. Their lack of toxicity.
3. The easy termination of their action by mild treatments.

The microbes are the major source of biological enzymes because of the reasons: (i) the case with which enzymes levels may be increased by environmental and genetic manipulations, (ii) Enzyme fermentations are economical on a large scale because of short fermentation cycles and inexpensive media, (iii) The screening procedures of microbes are very simple and a large number of culture samples can be examined in a short period. Once an active microbial strain is identified, fermentation parameters are optimized to maximize the growth.

Induction of Fermentation Enzymes

There are several catabolic enzymes which are of commercial importance and fall into the inducible category. The typical inducers are substrates e.g., starch for amylase, sucrose for invertase, lactose for p-galactosidase, etc. Certain analogues of the substrates are also known which act as potent enzyme inducers. There are reports which indicate that the products also act as inducers of enzymes.

The use of mutation techniques help us to eliminate the dependence of enzyme formation on inducer addition. Such technique of mutation is known as regulatory mutation, since its locus is not a structural but a regulatory gene. Mutants which produce a normally inducible enzyme without inducer are termed as constructive mutants.

Feedback Repression

During the fermentation process, microbes grow actively which leads to the repression of enzyme production. This occurs when pathway end-products build up in concentration or are added to growth media. These end-product are of low molecular weight (co-repressor) and can combine with an intercellular protein (aporepressor), coded by the regulatory gene, to produce a repressor. The repressor then stop the coding of structural gene for the production of enzyme. Such enzymes are known as repressible enzymes.

Table 21.2 Depression of biosynthetic enzymes by limited feeding of requirement to auxotrophs.

Auxotrophic requirement	Depressed enzymes	Increase
Leucine	Acetohydroxy acid synthetase	400-fold
Thiamine	4 enzymes of thiamine biosynthesis	upto 1,500-fold
Biotin	7-oxo-8-amino parlargenate	400-fold inotransferase

If internal accumulation of end-product (co-repressor) becomes limited, there is an increase in enzyme productions. This can be achieved by mixing an inhibitor of the pathway in the fermentation medium, and by limiting the growth factor supply fed to an auxotrophic mutant. This develops partial starvation for microbes, but the production of enzyme is increased.

Table 21.3 Carbon catabolic repression of enzymes.

Etzyme	Organism	Repressive carbon source
α-amylase	*Bacillus stearothermophilus*	Fructose
Cellulose	*Trichoderma viride*	Glucose, glycerol, starch, cellobiose
Protease		Glucose, *Bacillus megaterium*
Amyloglucosidase	*Endomycopsis bispora*	Starch, maltose, glucose, glycerol
Polygalacturonic	*Aeromonas liquefaciens*	Glucose, polygalacturonic acid, transeliminase acid
Invertase	*Neurospora crassa*	Mannose, glucose, fructose, xylose

Catabolic Repression

The catabolic inducible enzymes markedly repressed when cells are growing fast on a readily utilizable carbon source, many enzymes of industrial importance are subject to this type of regulation. If the use of repressing carbon sources is avoided in the fermentation medium the production of enzymes sensitive to catabolite repression is greatly stimulated. For example, use of mannose for growing *Pseudomonas fluorescens* var. *cellulosa* results in cell producing over 1,500 times as much cellulase as cells growing on galactose. Mutation can be used to obtain mutants resistant to carbon catabolic repression.

Hyperproduction by Deregulated Mutants - Gene Dosage

Regulatory mutants have often been obtained which produce more enzymes than their parents. For example, a constitutive polygalacturonase-producing mutant of *Acrocylindrium* was found to produce more enzyme than its inducible parent even in the presence of inducer. Sometimes hyperproducing mutants, contain multiple copies of structural gene coding for the enzyme. Increased number of copies of a gene increases production of its specific enzyme. In microorganisms, an increase in gene copies by genetic manipulation has been achieved by transferring plasmids or by the use of transducing phase.

Production of Alcoholic Beverages

The alcoholic beverages include beer, wine, whiskey, and other fermented beverages are economically important. The origins of bread and beverage making are lost in antiquity, but it is known that the Sumerians produced a beer consisting of moistened, fermented bread prior to 7,000 B.C. The Egyptians also made this bread beer, and regarded it as a holy gift from Osiris, the God of the death. They were also the first to make leavened bread. Both bread making and brewing were later refined by the Greeks and Romans, and numerous changes in the manner of making beer have been made since these early times.

Wine

The fermentation of the juice of grapes, cherries, and berries to produce wine is an old and well-established procedure. The characteristic and distinctive flavors of wine are partially determined by the variety of fruit used, the environmental conditions under which the fruits were grown, the parts of the fruit used, and the ability of the yeast to form flavored by-products (particularly esters). In this instance the sugar from these juices may be converted to alcohol via the Embden-Meyerhof pathway. Flavours and aromas occur in the wine as a result of the activities of chemical reactions other than sugars that proceed at the same time. The wine industry is basically the same the world over. Some of the differences that are found in wine originate in the type and quality of the grapes that are used. The physical environment of the grape plant, such as amounts of moisture and climatic temperature changes, as well as the general health of the plants, all reflect in the quality of the grape used in wine-making. Some have larger quantities of sugar, while others contain materials that are conducive in producing more desirable aroma, flavour, or other qualities in the wine.

The organisms used in the production of wine are also important. The type of organism used will influence the quality of the wine, and a variety of different species are used, some to produce especially dry or sweet wine. Traditionally, the yeasts used to ferment the juices were those occurring naturally on the surface of the fruit. Recently the trend has been utilize laboratory cultures of strains of *Saccharomyces cerevisiae* var. *ellipsoideus.* These strains are chosen for their ability to properly ferment the variety of grapes used and for their ability to impart flavors by formation of by-products which are characteristic for a particular wine.

The wine maker grow the yeasts in the winery by first inoculating about 1 to 5 lit of fruit pulp (must) from the laboratory culture slant and then serially transferring the inoculum to successively larger volumes of fresh must until a quantity suitable for bulk fermentation is obtained. The grape juice is treated with bisulfite to kill undesirable yeast and bacteria.

Fermentation is done in vats. They are usually made of wood or stone, and vary in capacity from a small keg to one that holds 190,000 lit or more The vat is covered to maintain anaerobic conditions that encourage fermentation and also discourage growth of acetic acid bacteria which could convert the vine into vinegar. The course of fermentation (particularly the nature of by-products) and the concentration of alcohol produced are influenced by the temperature, sugar concentration, acidity, and tannin content of the must and the amount of sulfite added.

Temperature is especially important as the yield of alcohol is higher at the lower temperatures and it also encourages the formation of pleasant flavors. At higher temperatures the alcohol resistance of the yeast is decreased. A favorable temperature for fermentation is in the vicinity of 10° C.

As the sugar content is over 30%, the fermentation is inefficient because alcohol accumulates and arrests the activity of the yeasts before all the sugar is converted to alcohol. A low acid content favors production of acetaldehyde, glycerol, and volatile and fixed acids, which have a cetrimental effect on the flavor, colour, and stability of the wine. In addition, a low acid content gives a lower yield of aromatic principles, which give the wine its bouquet.

After the partial fermentation, the racking operation begins and continues throughout the remainder of the fermentation period. In racking, wine is siphoned off from the sediment, which consists of dead yeast cells and other materials which settle at the bottom of the vat and is transferred to a fresh vat. This procedure is repeated for several times until the desired clarity of the wine is obtained. Before the last racking, a final clarification is performed by adding some substance (like casein, or bentonite) that carries additional sediment to the bottom,. Fermentation continues until the alcohol content is about 12%; and above this point the alcohol kills the yeast cells so that no more fermentation takes place. The wine is then stored in completely filled and sealed tanks for aging. Additional blending, fortification with additional alcohol, heat treatments, refregeration, or filtration may be required before the wine is bottled.

Beer

Beer is an alcoholic beverage prepared from fermented grains, usually barley. Several different starting materials may be used in the production of beer, but they all achieve the same end i.e. the production of a carbonated alcoholic drink. The top fermenting yeast, *Saccharomyces cerevisiae,* is the most widely used among all the yeasts. Strains should be chosen that are low-temperature-tolerant varieties. In this manner the low temperature favors the growth of the yeast and not bacteria, which may enter as contaminants on the starting materials. Not all beer is made from *Saccharomyces cerevisiae.* For example, *Saccharomyces carlbergensis* and *Saccharomyces monacensis* are used, especially in Holland. These yeasts grow at the bottom and are known as bottom fermenters.

Step I

In the initial steps of beer production, barley is malted through a controlled germination and grains are washed and soaked for 2 to 3 days to stimulate development of the embryo; the water is then drained off and the seeds germinated. The embryo is allowed to grow until the plumule attains a length equal to approximately three-fourths of the kernel length and is then dried to halt growth.

During the short period of germination, the embryo produces a number of digestive enzymes which begin to hydrolyze food reserves in the seed. From the brewer's standpoint, the most important of these are the carbohydrates which attack starch and sugars, and of these the most important are especially the amylases, which convert the starch into dextrins, maltose, and a little glucose. The maltose is the principal substratum for yeast fermentation.

Step II

Second step in beer production is known as mashing. The dried malt is ground and mixed with hot water. Most of the enzymatic conversion of starch to maltose by amylases takes place during mashing, which takes about 2 hours. The aqueous extract is separated from insoluble material and husks, and is then boiled with hops, the female inflorescence of the hop plant contaning essential oils and resins which impart characteristic flavors to the beer.

Step III

In this step the aqueous extract is cooled to a temperature between 8°C and 16°C (the temperature is partially determined by the type of beer being produced) and placed in fermentation vats. A selected strain of yeast is added. Traditionally *Saccharomyces carlsbergensis* is used to produce larger beer while *S. cerevisiae* is used to produce ales (in the United States, *S. carlsbergensis* may be used for both). During a fermentation period of about 5 to 14 days, the sugar is converted to alcohol. The yeasts settle to the bottom of the vat in a flocculent material in the production of the larger type of beer. In contrast, bubbles of carbon dioxide rise to the top of the vat and carry with them the yeast cells and dark flocculent materials from the liquid in the production of the ales. The resultant foamy scum must be periodically removed.

When the fermentation period is over, the newly formed beer is allowed to rest for a few days. This period allows the yeast cells to settle out and is then drawn off into weeks. The beer is clarified by addition of gelatinous materials in a manner similar to that in which wines are cleared and may also be filtered through diatomaceous earth. Antioxidants are added since beer changes flavour upon oxidation, and carbon dioxide is added either as the pure gas or by mixing in some freshly fermented beer. The beer is then bottled for sale.

Whiskey

They are made from fermented grains (corn, wheat, barley malt, or rye malt), which are mixed in varying proportions according to the type of whiskey being produced, and fermented by yeasts. The procedures are similar to those involved in beer production, the major exception is that the fermented grain broth is distilled in order to concentrate the alcohol.

Miso

This is an important commodity in Japan, made by fermentation of soybeans. It is sold as, pale-brownish paste, from which is made a pottage type of soup, serving as a regular morning dish. According to Sakaguchi (1961) the annual production of miso in Japan exceeds 50,000 tons. Cooked soybeans, *koji* made from rice, and 10 to 12 per cent of salt, are ground together ;along with a little miso from a previous batch. The mixture is packed into closed containers and allowed to ferment at 35 to 45° C, without aeration, for 3 to 5 months. It is then allowed to age from a month or more at room temperature before packaged for sale.

Organic Acid Fermentation

Citric Acid

Citric acid is the product of fermentation of numerous organisms. However, certain strains of the fungus *Aspergillus niger* produce commercially high yields of citric acid from a variety of 2-, 3-, 4-, 5-, 7-, or 12-carbon compounds. Four theories have been proposed for the mechanism of citric acid formation by *Aspergillus niger.* It would appear that metabolic shunt (induced by environment) is at work, since citric acid is usually not a metabolite that accumulates in the medium but is readily utilized as a carbon source. The culture conditions are minimal for nitrogen and carbon at low pH. Phosphate is usually deficient. *Aspergillus niger* appears to have more than one systems for the formation of citric acid since, it utilizes variety of different length carbon chains under special conditions.

Uses of Citric Acid

The important organic acid produced by fungi is citric acid which is used commercially as a flavoring ingredient in beverages and foods, especially in dry mixtures such as gelatins, and soft drink powders as tablets, and as the principal acid in the preparation of soft drinks, desserts, jams, jellies, candies, wines, and frozen fruits. About 90 % of the supply comes from *Aspergillus niger* fermantation. The acid is rapidly and almost completely metabolized in the human body and has wide pharmaceutical uses. Specially its incorporation is effervescent products and as citrates in blood transfusion. Citric acid is also used in astringent lotions to adjust the pH, in hair rinses and hair-setting preparations, in electroplating, in leather tanning, and in reactivating clogged with iron.

Citric acid was originally produced from calcium citrate obtained from cut lemons. Methods for producing citric acid from fungus mctabolism were introduced in the United States in 1923.

Mechanism of Citric Acid Formation

Citric acid is one of the principal organic acids produced in the citric acid (CA) cycle. Recent evidences indicates that the probable origin of citric acid is directly from the CA cycle. When mycelium actively grows, citric acid is produced as an intermediate in the CA cycle and is further diverted to growth-promoting biosynthesis or energy release. For commercial production of citric acid, culture conditions which are inhibitory to growth of the fungus are maintained and the CA cycle itself is inhibited or stopped by enzymatic inhibition of the low pH or by specific enzyme inhibitors, e.g., copper ions. During the production of citric acid, the activity of the condensing enzyme (operating in the condensation of acetyl CoA and oxaloacetic acid to citric) is increased, while the activities of the isocitric acid dehydrogenase and acotinase disappear. The enzyme acotinase is responsible to control the biosynthesis of isocitric acid from citric acid, and in turn isocitric dehydrogenase mediates in the hydrogen removal which yields oxalosuccinic acid from isocitric acid. Therefore, it has been suggested that inactivity of these latter enzymes is the reason for citric acid accumulation.

Commercial Production of Citric Acid

In commercial production of citric acid, *Aspergillus niger* may be cultured on the surface of the shallow pans or as submerged mycelium in aerated vats. Shallow-pan culture has been widely accepted and rapidly added by more efficient vat culture. The requirements and methodology for both the cultures are similar, although in actual plant operation many minute details are different. Beet molasses is the least expensive carbohydrate source, and has a high metallic ion content which may cause low yields of citric acid to be produced. Beet molasses has been widely used in the pan techniques, although pure sucrose or highly refined syrups give higher yields in submerged cultures in vats. Metallic cations are removed from the carbohydrate source by the use of cation-exchange processes and absorbents or (or in the case of beet molasses) by treatment with-ferricyanide. The carbohydrate is diluted to a concentration of about 20% to 25%. This high sugar-concentration inhibits formation of acids other than citric acid. The manufacture of citric acid needs a low pH, and hydrochloric acid is added to reduce the medium to the range of pH 2 to 5 when the fungal spores in the inoculum germinate. In the vat, the pH is kept below 3.5. Nitrogen is provided by adding ammonium salts. For vat cultures, copper or organic ions may be added as antagonists to iron and aid in the control of mycelial growth, which increases citric acid production. The prepared medium is added with spores from stock cultures. Highly aerobic conditions are needed, and submerged cultures are aerated with sterile air and agitated. Temperature must be maintained in the range of 25° to 30°C during incubation period (7 to 10 days). Therefore, the harvesting of the citric acid is started. Lime is added to the culture medium to precipitate any oxalic acid which formed, and the mycelium and calcium oxalate are filtered off. Additional filtration may be required to clarify the medium. A slurry of calcium citrate, which precipitates out. The calcium citrate is filtered out and then treated with sulfuric acid which precipitates the calcium. The dilute citric acid solution is purified by treatment with carbon and is demineralized by successive passage through cation- and anion-exchange resins. This purified citric acid solution is then evaporated, leaving behind citric acid crystals, which are further purified by recrystallizations from water.

Gluconic Acid

It is formed from sugars by the action of a large number of species of moulds, chiefly species of *Aspergillus* and *Penicillium.* The first fungus to be used was *Penicillium purpurogenum* var. *rubri-sclerotium,* and the fermentation was carried out in shallow pans of pure aluminium. Later it was found that certain strains of *P. chrysogenum* gave better results, and still later selected strains *of Aspergillus niger* were used.

D-Lactic Acid

Most of the work on the production of this acid has been done by American investigators. The fungus used is *Rhizopus oryzae.* Following a preliminary study of the physiology of the mould, a rapid process was worked out, using a rotary fermenter with forced aeration. The time for the fermentation was reduced to 30-35 hours, and yields of 70-75 per cent were claimed.

Gallic Acid

It was obtained by Calmette in 1902 by fermenting a clear extract of tannin by means of an organism which he named *Aspergillus gallomyces* (a strain of *A. niger),* the fungus being grown in a well-aerated and agitated liquid.

Fumaric Acid

It is produced, in yields of upwards of 65 per cent of the sugar consumed, when a strain *of Rhizopus arrhizus* is grown in shake-cultures in flasks. The sugar concentration was 10-16 per cent, and it was found to be essential to neutralize the acid continuously by means of calcium carbonate. The acid is required chiefly for the manufacture of plastics and varnishes.

Itaconic Acid

This substance was first obtained as a mould metabolic product of *Aspergillus itaconicus (A. glaucus* group). Of more interest in this connection was a paper by Calam, Oxford, and Raistrick (1939) describing the production of itaconic acid by a strain *of Aspergillus terreus.*

Glycerol

Glycerol is formed in small amounts during the normal alcoholic fermentation of sugar by yeasts, the maximum yield being about 3 per cent of the sugar consumed. It was found that the yield may be much improved by carrying out the fermentation in an alkaline medium, and still more so by adding sodium sulphite to the culture medium, which has the additional advantage that it inhibits the growth of bacteria without affecting the activity of the yeast. The yield of glycerol obtained was about 25 per cent of the sugar consumed.

Fermentation products have long been recognised as invaluable aids to the livestock producers of the world. As therapeutic antibiotics, such as penicillins and tetracyclines, they have been successfully used for individual animal treatment via injection, or for multiple animal treatments via drinking water or feed systems. These products were initially developed for animals as an extension to their use in human medicine. Over the past twenty years however, fermentation products have been introduced specifically foranimals as prophylactic and most importantly as growth promoting agents. These products promote the efficient production of meat under intensive conditions.

It is possible for medicinal agents to be added via drinking water systems, but this is not a convenient route except for short term medication of therapeutic or prophylactic medicines. All other methods of administration, however sophisticated in terms of multiple dosing, are on a single animal basis.

Vitamins

Vitamin is a generic term for a group of (unrelated) organic compounds-some or all of which are necessary, in small quantities for the normal metabolism and growth of microorganisms; they function as coenzymes or as components of coenzymes. Most microorganisms can synthesize the vitamins they require; those which cannot synthesize a particular vitamin must obtain it from the environment (e.g., growth medium). In some cases a precursor of the vitamin can replace the vitamin itself e.g., p-aminobenzoic acid can sometimes satisfy a requirement for folic acid. Microorganisms are important sources of certain vitamins e.g., vitamin B_{12} apparently can be synthesized only by certain microorganisms.

Vitamin B Complex

One of the best sources of the vitamin B complex is yeast, which is, in fact, one of the very few readily accessible foods containing the majority of the known substances comprising the vitamin B group. The increasing recognition of the importance of adequate supplies of these vitamins in the diet has led food manufacturers to put on the market a number of preparations of high potency, made from dried yeast, yeast extracts or autolysed yeasts. One of the B group, riboflavin, is now made in pure form by fermentation. Two closely related yeast-like organisms *Nematospora gossypii* and *Eremothecium ashbyi* are used in the production of riboflavin and by growing *E. ashbyi* on synthetic medium, the vitamin is readily isolated.

Several microorganisms produce vitamin B_{12} (cobalamin) by the fermentation of corn steep liquor and dextrose. In this instance the product (vitamin B_{12}) is contained within the cells and not in the supernatent liquid. The mycelium or bacterial cells are harvested and dried, and the vitamin is extracted in the appropriate manner. Another source of vitamins is found in whole yeast cells. Yeast cakes are prepared by growing large volumes of *Saccharomyces cerevisiae.* These cells are harvested and preserved into various sized lots. Vitamin B_{12} can also be produced commercially by direct fermentation, using *Propionibacterium shermanii* or *Pseudomonas denitrificans,* and these are the organisms used today for the production of this vitamin. *P. shermanii* can be grown in anaerobic culture for 3 days and in aerobic culture for 4 days to produce vitamin B_{12}. The growth medium for vitamin B_{12} production by these organisms contains glucose, corn steep liquor (a waste product of starch manufacture), and cobalt chloride. The medium is maintained at pH 7 by using ammonium hydroxide. *P. denitrificans* is grown for 2 days in aerated culture for vitamin B_{12} production, using a medium containing sucrose, glutamic acid, cobalt chloride, 5,6-dimethylbenzimidazole, and salts.

It is a by-product of acetone butanol fermentation and is produced by various *Clostridium* species. It is commercially produced by direct fermentation often uses the fungal species *Eremothecium ashbyii* and *Ashbya gossypii.*

Vitamin C

Vitamin C has been reported is the metabolic product of a strain of *Aspergillus niger.* This is more a curiosity than a possible means of manufacturing the vitamin by fermentation, since there are other easier methods of preparation.

Vitamin D

Ergosterol, the precursor of vitamin D, is synthesized by a number of moulds as well as by yeasts and there are a number of manufactured preparations of irradiated ergosterol, mostly made from yeasts.

Commercial Production of Antibiotics

In the original sense; any microbial product which, in low concentrations (of the order of micrograms/millilitre), is capable of inhibiting or killing (susceptible) microorganisms. The term was first used by Waksman in the 1940s; it distinguished the (then newly-discovered drug penicillin from the synthetic) sulphonamides which had been developed in the 1930s.

Difficulties in terminology become apparent when the penicillins (and other natural antimicrobial products) were subjected to chemical modification or synthesized de novo in the laboratory.

Currently, the term antibiotic is often extended to include drugs such as the Sulphonamides, Nalidixic Acid, and the semi-synthetic Penicillins. Many antibiotics are Chemotherapeutic agents but others are too highly toxic to be of any clinical value; certain antibiotics (e.g. colicins) are used for non-therapeutic purpose. Most of the antibiotics used as chemotherapeutic agents are active against bacterial pathogens; some (e.g., griseofulvin) are antifungal, others *(e.g.,fumagillin, suramin)* are antiprotozoal, while a few (e.g. the rifamycins) have limited antiviral activity.

The idea of using one microorganism to combat the evil effects of another is by no means a new one. Pasteur was probably the first to describe clearly the production of an antibiotic, and, during the intervening years, many claims have been made as to the therapeutic value of bacterial products.

Table 21.4 Some antibiotics produced by microorganisms

Antibiotics	Microorganisms
Amphotericin B	Streptomyces nodosus
Bacitracin	Bacillus licheniformis
Carbomycin	Streptomyces halstedii
Chloroterracycline	Streptomyces aureofaciens
Chloramphenicol	*Streptomyces venezuelae*
Erythromycin	*Streptomyces erythreus*
Fumagillin	*Aspergillus fumigatus*
Griseofulvin	*Penicillium griseofulvin, P. nigricans, P. urticae*
Kanamycin	*Streptomyces kanamyceticus*
Neomycin	*S.fradiae*
Novobiocin	*S. niveus, S. spheroides*
Nystatin	*S. noursei*
Oleandomycin	*S. antibioticus*
Oxytetracycline	*S. rimosus*
Penicillin	*Penicillium chrysogenum*
Polymyxin B	*Bacillus polymyxa*
Streptomycin	*S. griseus*
Tetracycline	Dechlorination and hydrogenation of chlorotetracycline; direct fermentation in dechlorinated medium
Vira A (Adenine arabinoside)	*S. antibioticus*

Production of antibiotics (antibacterial or antifungal compounds) is one of the largest and most important microbiological industries. Interest in utilization of antibiotics for therapy began in 1929 when Alexander Fleming found that a mold has contaminated his cultures of a pathogenic bacterium, *Staphylococcus aureus,* and had killed bacteria in its immediate vicinity leaving a clear ring of lysed bacteria. The mold was a *Penicillium* (later found to be *P. notatum).* Fleming named the unidentified lytic principle penicillin.

Screening of Antibiotic Producers

The search for antibiotics in the pharmaceutical industry presents a good example of how screening procedures are employed to select microorganisms for industrial applications. The discovery of new antibiotics results from laboriousre researches. Sample from many sources, including soil from around the world, are examined as potential sources of antibiotic-producing microorganisms; countless strains of microbial isolates are tested by pharmaceutical laboratories. One of the best penicillin-producing strains of *Penicillium* was isolated from an orange purchased at a road side fruit stand. Several antibiotic-actinomycetes were isolated from a manure-enriched pasture.

A useful antibiotic-producing strain must produce metabolites that inhibit the growth or reproduction of pathogens. This essential property can be assayed by using test strains and examining whether the isolate being screened produces substances that inhibit the growth of these test organisms. A positive result in such a primary screening procedure in no sense ensures the discovery of an industrially useful antibiotic-producing strain. It simply identifies those strains of microorganisms that have the potential for further development.

The search for new antibiotics still goes on, and new ones are frequently announced. Genetic engineering has opened up several new possibilities for employing microorganisms to produce economically important substances. The mutation and selection approach is hit or miss, whereas the use of recombinant DNA technology permits the purposeful manipulation of genetic information to engineer a microorganism which can produce high yields of a variety of products. The use of biotechnology has great social consequences.

Penicillin

Not only did penicillin developed interest in antibiotics and launch the antibiotic industry, but penicillin is still the most widely used antibiotic. It is the drug of choice when infection is caused by organisms susceptible to it. Penicillin is effective against Gram-positive bacteria and also against some of the larger viruses and rickettsia. Penicillin is a genetic term applied to an entire group of antibiotics which are closely related in structure. Most penicillins are produced by species of *Penicillium* while others are semisynthetic. The naturally occurring penicillins have the following basic structure and differ in the R groups.

Commercial Production of Penicillin

The original *P. notatum* Straub isolated by Fleming gave low yield of penicillin. Finally a strain of *P. chrysogenum* was isolated and was found to give superior yields of penicillin. Strains of *Penicillium* may produce more than 180 times as much penicillin as the original isolate with which the breeding program was initiated.

Penicillin is produced in submerged vat cultures using a selected strain of *P. chrysogenum.* Composition of the medium is of prime importance in obtaining a good yield. A medium giving an unusually good yield of penicillin is corn steep liquor, a cheap bulk material which is a by product of cornstarch manufacture. These nitrogenous compounds are of great importance giving a good yield of penicillin. The principal carbohydrate in corn steep liquor is lactose, which is the most favorable carbon source for penicillin production because it is assimilated slowly, giving the fungus a steady carbon source over a period of time and does not lead to the accumulation of organic acids. Precursor molecules (phenethylamine and phenylanine) needed for penicillin production are present in the corn steep liquor.

Procedure

About 30,000 liters of the medium are placed in a tank, sterilized and inoculated with an aqueous suspension of *P. chrysogenum* conidia. During the incubation period, the medium is aerated with sterile air and agitated. The tank is equipped with devices which allow the continuous addition of glucose syrups, sodium hydroxide, and sulphuric acid to maintain the pH between 6.8 and 7.4, cooling coils to maintain the temperature between 23° C and 25° C, devices for the introduction of antifoam agents, and pumps for the addition of the acyl or precursor.

The most commonly used precursor is phenylacetic acid. The other precursors in the form of a salt, amide, or ester of thc corresponding acid or amine may be added to yield penicillins bearing the desired acyl group. At first, the mycelium grows actively and utilizes lactic acid and organic nitrogen compounds as sources of carbon. Ammonia is formed during the breakdown of these nitrogenous compounds. Ammonia is now actively assimilated and the pH is lowered. Active the most commonly used precursor is phenylacetic acid. The other precursors in the form of a salt, amide, or ester of the corresponding acid or amine may be added to yield penicillins bearing the desired acyl group. At first, the mycelium grows actively and utilizes lactic acid and organic nitrogen compounds as sources of carbon. Ammonia is formed during the breakdown of these nitrogenous compounds. Ammonia is now actively assimilated and the pH is lowered. Active penicillin production is associated with this phase of lactose and ammonia utilization. Penicillin production ceases when the lactose is exhausted.

The duration of the incubation period is about 5 to 6 days. After incubation, the mycelium is filtered from the liquid medium which contains the penicillin. The penicillin is mixed with a solvent (amyl or butyl acetate) and the resulting emulsion is centrifuged to extract the acetate solvent which now contains the penicillin. A phosphate buffer is added to the acetate, and the penicillin is extracted with the phosphate buffer by cenrrifugation. This last step may be repeated, using successively smaller quantities of liquid. Butanol is added to the aqueous mixture, and the potassium salt of penicillin is crystallized from the solution. Potassium penicillin may be further purified and used in this form or it may be converted to procaine penicillin.

Penicillin Biosynthesis

Penicillin is produced along synthetic pathways not required for growth but superimposed upon those pathway required for maintenance. Key intermediates in the pathway are pyruvic acid, valine and cysteine. Penicillin production is at its maximum level when the citric acid cycle is not actively taking place or if a great deal of pyruvic acid is not being converted to acetyl CoA and then to fatty acid. The accumulating pyruvic acid can then be available for synthesis of L-valine, to form a dipeptide precursor of penicillin - L-cysteinyl-L-valine. This last intermediate is then combined with the acyl group from phenylacetic acid in the case of penicillin G or acyl groups from other specific precursors in the synthesis of the other penicillins. Many of the details concerning penicillin biosynthesis are still unknown.

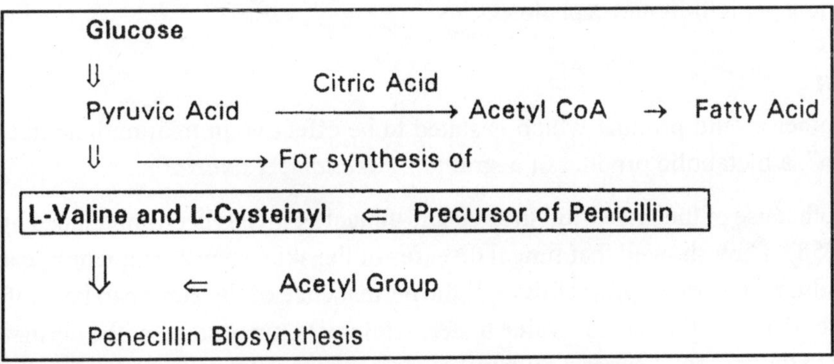

Diagram 21.2 Biosynthesis of Penicillin

Streptomycin

Streptomycin and various other antibiotics are produced using strains of *streptomyces griseus.* As in penicillin fermentation, spores of *S. griseus* are inoculated into a medium to establish a culture with a high mycelial biomass for introduction into an inoculum tank, with subsequent use for mycelial inoculum to initiate the fermentation process in a tank. The basic medium for the

1. Rapid growth of *S. griseus,* with production of mycelial biomass. Proteolytic enzymatic activity of *S. griseus* releases ammonia to the medium from the soybean meal, causing a rise in pH. During this initial fermentation phase there is little production of streptomycin.
2. A little additional production of mycelia. The glucose added in the medium and the ammonia released from the soybean meal are consumed during this phase. The pH remains fairly constant (7.6 to 8).
3. It is the final phase of the fermentation, after depletion of carbohydrates from the medium, streptomycin production ceases and the bacterial cells begin to lyse. There is a rapid increase in pH because of the release of ammonia from the lysed cells.
4. In the end of the fermentation, the mycelium is separated from the broth by filtration and the streptomycin is recovered. The purification consists of adsorbing the streptomycin onto activated charcoal and eluting with acid alcohol.

Cephalosporins

A cephalosporin C is made as the fermentation product of *Cephalosporium acremonium,* but this form of the antibiotic is not potent enough for clinical use. The cephalosporin C molecule, however, can be transformed by removal of an α-aminoadipic acid side chain to form 7-α-aminocephalosporanic acid, which can be further modified by adding side chains to form clinically useful products with relatively broad spectra of antibacterial action. We are at present into the so-called third-generation cephalosporins, such as moxalactam, which have been developed to combat bacteria that produce enzymes capable of degrading penicillins and cephalosporins.

Variotin

Another mould product which is stated to be effective in treating human mycoses is "Variotin", a metabolic product of a strain of *Paecilomyces variotin.*

Another use of the griseo-fulvin as an anti-fungal agent has been announced by Williams et al. (1958). They showed that fungal diseases of the skin, such as ringworm, can be cured by oral administration of griseofulvin. If the permanence of the cures can be confirmed, the antibiotic will be of the greatest value to dermatologists, since some of the dermatomycoses have, up to the present, been virtually incurable. Since then much work has been done on the subject and several papers have appeared in the medical press.

Animal Feed Production and Medication Types

Animal feed medication can be separated into the above three types and, although these appear to be distinctly different, there is appreciable overlap.

- *Veterinary therapeutics:* these are high potency antibiotics which are used in a manner similar to human pharmaceuticals.
- *Prophylactics:* these vary from high potency wide spectrum to relatively weak narrow spectrum antibiotics, they are given not only to prevent the development of diseases and growth of parasites, but also to avoid detrimental effects on animal growth. In the major group, the anticoccidial agents, the antimicrobial effect *is* incidental to the antiparasitic effect, but can provide some growth promotional effect.
- *Growth promoters :* the antimicrobial effects of these compounds alter the composition of the flora and fauna of the gut and hence promote better feed utilisation. In some cases this means faster growth but the main benefit is feed saving. *These two parameters are normally measured by the rate* of weight gain and feed conversion ratio of F.C.R. All of the current FSFA products are not used in human medicine and most have narrow ranges of activity.

The bulk of feed additives are used as growth promoters or prophylactic agents with over a thousand tones sold per annum in the U.K. alone. In terms of sales, the fermentation products outweight synthetic chemicals by a ratio of over ten to one.

The mentioned list gives some of the large number of additives in the market-place, covering the varying disease challenges, housing and feeding conditions - especially for growth promotion in pigs.

Table 21.5 Typical feed additives and their uses

Active Ingredient	Species	Use
Apramycin*	Pigs	Antibacterial, vs. enteritis
Avoparcin	Pigs, Broilers, Cattle	Growth Promotion
Bambermycin	Poultry, Pigs, Cattle	Growth Promotion
Chlortetracycline*	Calves, Pigs, Poultry	Antibacterial, wide spectrum
Levamisole	Cattle, Sheep	Anthelmintic
Lincomycin*	Pigs	Antibacterial, vs. dysentery
Olaquindox	Pigs	Growth Promotion
Monensin	Broilers, Cattle	Anticoccidial growth Promotion
Narasin	Broilers	Anticoccidial
Salinomycin	Broilers	Anticoccidial
Sulphamethazine*	Cattle, Sheep, Pigs,	Antibacterial, vs. atropic rhinitis
Tylosin	Pigs	Growth Promotion
Virginiamycin	Pigs, Broilers, Turkeys	Growth Promotion
Zinc Bacitracin	All	Growth Promotion

*Items which require veterinary prescription

Animal Feed Types

There are three broad types of feedstuff manufactured in the industry; mineral, vitamin supplements, protein concentrates and finished feeds; however, there is considerable variation within each type, especially in concentrates for cattle.

The animal vitamin supplement contains most of the essential minerals and vitamins. It is an integral part of the other two types of feed, and constitutes a 0.2 to 3 per cent of the final ration. Medicinal additives especially growth promoters and prophylactics tend to be added at this stage. The protein concentrate is a concentrated high protein radon which would form twenty to forty per cent of the complete feedingstuff.

Finished feed is a complete ration which can be fed throughout life. It contains all the essential minerals, vitamins, proteins and other nutrients. A typical composition is given by the following example of a pig starter feed:

Finished feeds and concentrates are often pelleted. This prevents ingredient segregation problems and gives a higher density diet. Considerable feed wastage by some animals is also avoided. Virtually all feeds for chickens are pelleted, ensuring that their strict nutrient requirements are met. About fifty per cent of pig rations are pelleted with about ten per cent for cattle. The pelleting process involves treatment of the feed with steam and then the resulting hot wet mash is forced through a circular die by one or more rollers. The pellets are then cooled/dried in a counter current of air.

Medicinal additives can be added to all the types of feed and they will undergo the various mixing, transportation, and other processes to which the feed is subjected.

Table 21.6. Composition of a pig starter feed.

Ingredient	**Percentage**
Barley meal	13
Soya meal	16
Wheat meal	31
Wheat feed	18
Maize meal	11
Fish meal	3
Meat and Bone meal	3
Fat	2
Limestone	1
Molasses	0.5
Mineral vitamin supplements	1.5
Declared analysis	
Oil	4.8
Protein	18.5
Fibre	4.6
Metabolisable energy	11.6MJ/kg

Incorporation of Feed Additives

The table below gives a breakdown of the chain of mixing steps which could be carried out involving feed additives.

Table 21.7. Feed Medication Routes

Feedmill	**Farm**
Feed Additive Product	Local Mixing
	Local Mixing
Mineral Vitamin Premix	Top Dressing
	Free Access
Protein Supplement	Local Mixing
Finished Feed	Free Access

The therapeutic drugs tend to be mixed directly into final feed, either on farm or at the feedmill. These drugs can only be used on prescription by a veterinary surgeon. The prophylactic and growth promoting agents tend to be mixed into the mineral vitamin supplements, which are then mixed into protein concentrates or finished feeds; normally carried out on different sites by different manufacturers.

Development of a New Active Ingredient

The identification and development of a new medicinal feed additive is similar to that of a new pharmaceutical drug. One major difference is that it is possible to test the drug in the target species at an earlier stage, and follow on to limited efficacy studies in that species, prior to obtaining an Animal Test Certificate. A second major difference is that more extensive toxicology studies are involved to ensure safety to the target species, human and to the environment.

There are again three main types of fermentation product developed for medication of animals via feeds, these are:

1. Purified fermentation products: Normally crystalline materials recovered by solvent extraction procedures, e.g. lincomycin.
2. Chemically modified fermentation products: Purified antibiotics, which have a side chain cleaved off and then replaced with another, thus giving modified or improved efficacy, e.g. procaine penicillin.
3. Dried broths: These contain essentially all the solids remaining from fermentation, e.g. monensin. The bulk of the water is removed by:
   (i) Filtration/centrifugation - this also removes aqueous soluble materials; and
   (ii) Total evaporation - using ageotropic, by reduced pressure or spray drying technique.

The fermentation stages for these products are very similar. Pure cultured organisms are grown under sterile conditions in large stirred tanks (these can have capacities of 20,000 gallons or more) containing various nutrients in an aerated aqueous medium. Strict controls on parameters such as air-supply, temperature and pH are made to ensure maximum rate of growth. When optimum growth has been achieved the fermentation process is terminated.

Medicinal Premix Product Development

Medicinal feed additives are normally marketed as premixes. These are simply a mixture of active ingredients and suitable edible inerts of cereal, vegetable or even mineral origin. (There are however examples of totally granular premixes, which are in effect standardised granulated raw materials). The following are typical examples of these edible inerts or premix diluents (sometimes known as carriers):

Limestone	Soya meal
Maize meal	Soya bean mill run
Maize gluten meal	Wheat middlings

These premix diluents vary widely in physical nature, nutritive value and cost. Not all diluents, however, are suitable for all products, or are universally acceptable to regulatory authorities. The following list gives the most important parameters used in assessing a new premix diluent, before any test is undertaken.

1. Must be acceptable to regulatory authorities
2. Cost - must be reasonably low
3. Particle size - must be within acceptable range
4. Long term availability/quality - suitable quality material in adequate quantities must be consistently available.

Distribution of Medicated Feed Additives

The feed additive manufacturer normally sales to major feed supplement manufacturers. These manufacturers then pass on either the premixes, or supplements containing these premixes to their own feed manufacturing plant, smaller feed compounders, large farmers or veterinary wholesalers.

The medicated premix can be incorporated into mineral/vitamin supplements or protein concentrates at different sites and by different companies, prior to incorporation into finished feeds.

Future Development

The medication of animals with fermentation products via feeds will continue to provide benefits in animal husbandry for some time ahead. In the existing range of products there will be an increased trend towards granulated products to reduce carryover active ingredients into other ratios within feedmills and also to reduce toxic hazards associated with dust. New fermentation and other products for existing and new uses will continue to be introduced despite increased regulatory and consumer pressure, group demands, and development costs. These new products will show new, broader, increased efficacy leading to increased cost benefits for the farmer.

The introduction of specific growth hormones and similar products, beginning in the mid 1990's have given benefits in addition lo those gains by current promoters and is thus unlikely to affect their sales. However, other developments of products such as monoclonal antibodies are likely to reduce prophylactic and therapeutic uses in the longer terms.

❑❑❑

CHAPTER-22

Biotechnology in Production of Secondary Plant Metabolites

The plants are the only natural primary producer of energy/some other metabolic products on earth. They are not only the primary producers of natural products such as food, fodder, fiber, oils, and wood but also, one of the important source of drugs, fragrances, food flavors, pigments, essential oils etc., as the secondary plant products or metabolites. As per UNESCO (1960) report, the different arid zone of vegetations are the hub of medicinal plants that can be exploited to extract natural chemicals, used as or in medicines.

The discovery of cell as fully independent body capable of doing all necessary living activity has revolutionized this field. This discovery was followed by development of tissue culture, protoplast fusion and genetic engineering technology. These technologies has inhanced the plant breeders capacity to grow a number of plants in a quick succession with desired characteristics. This has revolutionized the production of plants secondary metabolite products in the industrial unites. Some of the important secondary plant.metabolites include, proteins, flavonoides, rotenoides, alkaloids, steroids, glycosides, anti-pesticides, anti-microbials, pyrethrins etc.

The secondary metabolites have great economical and pharmocological importance and the industries are deeply interested in large variety of chemical substances being produced by plants due to their lesser toxicity. Though these substances are generally extracted from plant parts, the plant tissue culture technique has widened the scope and opened new vistas for the production of secondary metabolites.

Recently, Mitsui Petrochemical Industries (Tokyo) has achieved success in producing *shikonin* commercially in 750 lit fermentors from *Lithospermum* sp. plant cells. Shikonin is a valuable antibacterial dye. This is the only successful attempt so far acheived in the world by Japan. Thus the chances of exploitation of tissue culture technique for large scale production of useful metabolites are bright.

In India, the beginning of this work goes back to 1964 when Mitra and Kaul at NBRI (National Botanical Research Institute, Lucknow) showed the production of *reserpine* from *Rauwolfia serpentina* tissue culture. Later on work on various metabolites was curried out in other laboratories as well in India.

Tissue Culture and Secondary Plant Metabolites Production:

Plant cells in culture offer many advantages over intact plants for secondary metabolites production and their biosynthetic studies. These are as follows:

1. Plant cells are relatively easy to grow and can be kept under strict controlled nutritional and environmental conditions. Hence the uncertainities of climate and soil can be avoided.
2. Cells are cultured aseptically, devoid of many microorganisms or insects etc.
3. Suspension cultures, offers a very effective way of incorporating precursors which are often difficult to administer to the plant growing in nature.
4. The technology is now available for the relatively large scale production of plant suspensions, in batch cultures, closed continuous cultures system, open continuous culture system etc. and may also eventually provides an efficient means of producing commercially important plant products.

However, in order to realize the industrial applications of plant cell culture for medicinal compounds production, it is essential to satisfy the following conditions as minimum requirements:

1. The rate of cell growth and biosyntheis should be high enough to give a good production of the final product in a short period of time.
2. The cultured cells should be genetically stable to give a constant yield of the product.
3. The metabolites should be accumulated in cells without being catabolized rapidly or preferably, they should be relized into the liquid medium.
4. Production cost including the culture medium, precursor and chemical extraction should be low enough to be profitable to the industries.

As far as the first requirement is concerned experiments with various plant suspension cultures showed that the growth rate can be accelerated considerably by improving cultural conditions and selective breeding of cultured cells. Frozen storage technique (cryopreservation) may be used for the second requirement as suggested by Street (1975). For the above third condition, as plant cells generally tend to accumulate their secondary metabolites in vacuoles or the cytoplasm, it would be desirable to device a method for altering the permeability of plasma membrane. A surface active agent was used successfully by Tanaka et al. (1974). The forth requirement is related to the social demand of the product and production cost. The production cost depends upon the total cost of carbon source, cost of electricity necessary for aeration, stirring, temperature regulation etc. Therefore, it is advisable to use molasses, starch and alcohols as carbon source. Secondly, the introduction of continuous or semi-continuous culture in place of the batch culture system can reduce the cost for sterilization of the fermentor and the necessity for propagation of cells from the original stock.

Applications of New Culture Methods for the Producton of Secondary Metabolites

Recently, two new techniques which have enhanced the accumulation of secondary metabolites (phytochemical) in plant tissue culture have been developed: (i) Hairy root culture, and (ii) Elicitation of product accumulation. The simplicity of the procedures and outstanding results made the techniques most useful and have drawn attention of most of the tissue culturists.

Hairy Root Culture

Plants, mostly, growing in tropical areas are the major source of many of the fine chemicals (secondary metabolites). Their structural complexity makes their chemical synthesis uneconomic. Therefore, the biosynthesis of these plant products using plant cells in culture has long been accepted as a worthwhile objective. The production of red naphthoquinone, an anti-inflammatory drug, shikonin by suspension cultures of *Lithospermum erythrorhizon* is the important landmark in this direction.

Plant cell cultures proved to be more complex and very often the rate of production of desired compounds is also very low. Low productivity in cell suspension culture to some extent over come by:

1. manipulation in the composition of the medium.
2. tillering physical environment of cells, and
3. screening for clones of high producing cells.

The ability of selecting high-producing cells from cell suspension cultures is not an easy task. To maintain elivated productivity, it is necessary to screen repeatedly the desired clones. This inherent instability is associated with the changes at the genomic level, both inter-and intrachromosomal.

Diagram 22.1 A hairy root growth induced by *Agrobacterium rhizogenes.*

Recently, highly productive and stable hairy root culture has obtained by the genetic transformation of plant tissue by the pathogenic soil bacterium, *Agrobacterium rhizogenes*. The infection of dicotyledonous plants by *A. rhizogenes* causes roots to proliferate rapidly at the infection site. This phenotypic change results from the insertion into the plant genome of t-DNA (transfer DNA), carried on the bacterial Ri-plasmid, coding for auxin synthesis and other rhizogenic functions.

The hairy roots can be removed from the parent plant and can be cultured (definitely in simple defined media free of auxins and phytohormones. In this respect they differ from untransformed root cultures which are often dependent upon added auxins and phytohormones. Hairy root cultures are potentially applicable to the production of all root-derived secondary metabolites from dicotyledonous plants.

Many of the root synthesized compounds including tropane alkaloids, atropine and hyscyamine, steroidal precursors such as solasidine, and *Cathranthus* alkaloids are of sufficient high value (1000 US $/kg) to justify the exploitation of hairy root culture for their commercial production.

Establishment of Hairy Root Culture

For the establishment of hairy root cultures, the explant material is inoculated with a suspension of *A. rhizogenes* generated by growing bacteria in YMB medium for two days at 25° C with gyrator shaking, pelleting by centrigugation (5 x 10^3 rpm, 20 min) and resuspending the bacteria in YMB medium to form a fix suspension (approx. 10^{10} viable bacteria/ml). Transformation may be induced on aseptic plants grown from seeds, or on detached leaves, leaf-discs, petiols or stem segments from green-house plants following sterilization of the excised tissue with 10% (v/v) Domestos (Lever Bros.) for 20 min. Scratching the midrib of a leaf or the stem of a plantlet, with a needle of a hypodermis containing the thick bacterial suspension allows inoculation with small droplets containing *Agrobacterium rhizogenes*.

In some species, hairy roots may appear directly at the site of inoculation while in others a callus will form initially and hairy roots appear subsequently from it. In both cases hairy root appears within one to four weeks. Infection in resistant plant species might be increased by including 10M acetylsyringone in the medium in which the bacteria are suspended.

Solanum nigrum protoplasts have been cultured with *A. rhizogenes* and obtained transformed protoplast which subsequently developed calli. The calli spontaneously formed hairy roots when cultured on hormone-free media.

In most of the plant species, establishment of hairy root cultures in suspension is brought about by excising the transformed roots from the stockplant and placing them in growth medium with 30 g/1 sucrose as the energy source. However, some species, are more difficult to establish in independent culture. Various techniques, including suspending roots on the explant, until they are established, buffering the pH of the medium and/or using half of quarter strength, high phosphate or nitrate, or the addition of exogenous hormones, have been found to cause extensive callus formation.

Cultures may be cleared of bacteria by several passages in medium containing 500 mg/1 ampicillin, or 500 mg/1 ampicillin + 200 mg/1 cephalosporin C.

Characteristics of Hairy Root Cultures

When normal roots of dicotyledonous plants are infected with *A. rhizogenes* it is transformed into hairy roots which are characterized by:

1. High degree of lateral branching.
2. A profusion of root hairs.
3. Absence of geotropism.

Profuse branching, resulting in the formation of many meristems, is one of the reasons why such roots have the property of high growth-rate in culture. Both growth rate and extent of branching, me highly inter-related characteristics, vary between species and with the culture conditions. In *Nicotiana rustica* and *Datura stramonium* growth of cultures is influenced by the initial pH ol the culture medium. In *D. stramonium* the density of branching per unit length of primary root is also markedly affected by the ionic strength of the medium.

The property of hairy root culture which is most important from their commercial exploitation point of view is their high level production of secondary metabolites in absence of hormones in growth medium. Betalain synthesis in *Beta vulgaris* and hyoscyamine synthesis in *D. stramonium* increases two-folds and five-folds respectively in hairy root cultures.

Another key feature of hairy root culture is their productivity which is stable over many generations. In *Hyoscyamus muticus* and *N. rustica* hairy root cultures remained genetically constant for atleast one year. While in untransformed root cultures both genetic and chromosomal abnormalities are frequently observed. In hairy roots, the chromosome numbers resembles normal roots.

Genetics of Hairy Root Formation

The infection of plant root tissue with *A. rhizogenes* causes to pieces of t-DNA to be inserted into plant genome. The integration of foreign genome though is random, it alters the auxin metabolism of the transformed plant tissues in such a way that:

1. Hairy root morphology is expressed.
2. Amino acid metabolism alters to produce opines, and
3. Secondary metabolites synthesis is enhanced.

The synthesis of the opines (manopine or agropine) is the firm indication that hairy roots are indeed transformed. The strains of *Agrobacterium rhizogenes* which are used for transformation have good virulence. These are identified to have following plasmids, pRiA4, pRiHR[1], pRi1855 and pRi8196.

Manipulation of Hairy Root Cultures

Chemical or biochemical manipulation of hairy root cultures can enhance synthesis of biochemicals and/or a noval or uncharacteristic plant products. For this purpose natural precursors or matabolite analogues are used. When nicotinic acid was fed with *Nicotiana rustica* cultures it showed enhanced production of nicotine and anatabine. However, feeding of cadaverine to *Datura stramonium* cultures produces noval, as yet, to be unidentified alkaloids.

Hairy Root Cultures: Improvement of Secondary Metabolites Formation

Every plant species has its biosynthetic capabilities depending upon its genetic makeup and environmental conditions of its habitat. Therefore, some species are considered as high yielding for a particular fine chemical while others are made to be high-yielding by selective breeding programmes. For the production of a fine chemical, only high-yielding varieties are selected. In cultures these varieties are not genetically stable and thus their product formation is not always constant. Hairy roots have stable productive capabilities similar to the plants from the which they were derived as these remain genetically stable for a long time.

As plasmid (gene) transfer is involved in the formation of hairy roots, the system is highly amenable to manipulate at the genetic level. Recently foreign genes have also been inserted into plants during transformation with *Agrobacterium rhizogene* by either insertion into the Ri-plasmid. tDNA or in binary co-transformation using disarmed vactors derived from *Agrobacterium tumefaciens* (pBiN19). In the later case tDNA from both plasmids is inserted into the plant genome. Thus the manipulation of secondary metabolite by altering the expression of key genes in the pathway is now a reality in present context.

In presence of high levels of exogenous auxins, hairy root cultures spontaneously form callus from which suspension cultures can be obtained. However, such transformed suspension cultures are returned to hormone-free media resulting in roots regeneration. Therefore, commercial exploitation of hairy roots for secondary metabolite production in the high producing strains has two following advantages:

1. Regeneration of improved varieties of plants for growth in the fields, and
2. the exploitation of the secondary metabolite formation of the roots in bioreactor system.

A significant progress to both these ends is being made. However, no definite procedure of cell selection or genetic manipulation has been developed to improve valuable metabolite synthesis in a plant.

Elicitation of Product Accumulation

Elicitors, strictly speaking are compounds of biological origin involved in plant-microbe interaction. In the context of product accumulation by plant cells in suspension cultures elicitors are mediator compounds or stress agents. There are two kinds of elicitors, these are:

1. Abiotic elicitors
2. Biotic elicitors

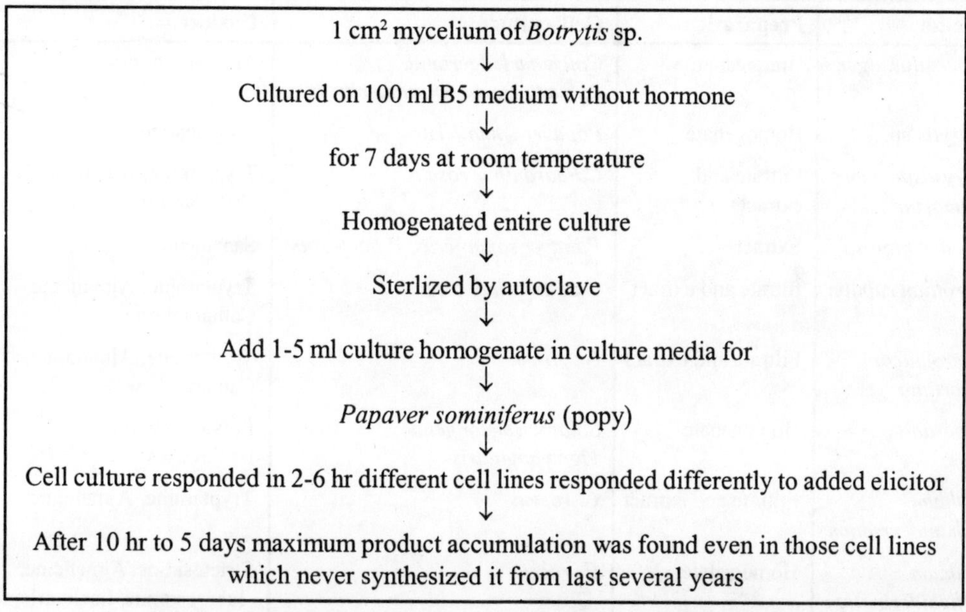

Diagram 22.2 Biotic elicitors of fungal or bacterial origin.

Abiotic elicitors are physical or chemical in nature e.g., ultra violet radiations, alkalinity, osmotic pressure or heavy metal ions etc. Biotic elicitors are complex culture homogenates of fungal or bacterial origin or fractions thereof. Both pathogenic *(Phytophthora, Botrytis, Verticillium* etc.) as well as non-pathogenic *(Aspergillus. Micromucor, Rhodotorula* etc.) microbes have been employed for this purpose. The chemical nature of the elicitors has identified as oligosaccharids, polysaccharids, glycoproteins and low molecular weight compounds like archidonic acid.

The products which accumulate in plant cell cultures due to elicitation may be antimicrobial in nature but they should not be confused with phytoalexins (a term used in plant pathology for pathogen-induced chemical defence in host plant), unless there is sufficient proof that the source plant responds to pathogens with rapid accumulation of the same product. Therefore, a new term has been coined for those compounds which in cell cultures are induceable by way of elicitation - Elicitation product or Elicitation metabolite.

Elicitors can be regarded as substitute of production media (that is optimum cultural conditions). Optimum employment of elicitors depends upon factors like: 1. Elicitor specificity, 2. Elicitor concentration, 3. Duration of elicitor contact, 4. Characters of cell line (clones), 5. Time course of elicitation, 6. Growth stage of culture, 7. Growth regulation and 8. Nutrient composition

A variety of elicitor preparations are generally tried for the yield and quality of a secondary metabolite for a cell line. For example *Botrytis* sp. preparation proved best with *Papaver sominiferus* (popy). The procedure followed in this experiment by Eilert et al. (1985) is as above.

Table 22.1 Some elicitor-induced products accumulation (Adopted from the abstracts of VI Congress of International Association of Plant Tissue Culture held at Minneapolis, 1986).

Elicitor	Preparation	Cell culture	Product
Aspergilus niger	Homogenate	*Cinchona ledgeriana, Rubia tinctoria, Morinda citrifolia*	Anthraquinones
Botrytis sp.	Homogenate	*Papaver somniferus*	Sanguinarine
Chrysosporium palmorum	Filtrate and extracts	*Catharanthus rosetis*	Tryptamine, Ajmalicine, Catharanthine
Dendrythion sp.	Extract	*Papaver somniferus, P. bacteatum*	Sanguinarine
Eurotium rubnim	filtrate and extract	*Catharanthus roseu*	Tryptamine, Ajmalicine, Catharanthine
Micromucor isabellina	Filtrate and extract	*C . roseus*	Tryptamine, Ajmalicine, Catharanthine
Nigeran sp.	Homogenate	*Solanum melongena* *Vigna angularis*	Polyacctylenes Isoflavoncs
Pythium aphanidermatum	Filtrate and extract	*C. roseus*	Tryptamine, Ajmalicine
Pythium aphanidcrinatum	Homogenate	*C. roseus*	Strictosidine, Ajmalicine, Taberosonine, Lachnuricine
Yeast	Carbohydrate preparation	*Glycine max*	Glyccolline

Elicitation of cell in suspension culture may react in the following ways:

1. In a given cell line different products may show highest levels of accumulation at different times.
2. Product accumulation may observed in cell lines which never synthesized it from last several years.
3. Elicitation will not cause an additive effect when applied to cells in production media, but may shorten the culture period required for maximum product accumulation.

Product accumulation due to elicitation has also been observed in growth media. Such occurrence may be due to excretion or leakage caused by cell breakdown.

Elicitation of secondary product accumulation in plant cell cultures would appear to develop into a powerful tool for biotechnologists and biochemists.

Biotransformation : Production of Phrmaceutical Compounds

One of the most promising fields in the biotechnological application of plant cell culture is the biotransformation of some less important substrates to medicinally useful products. The abiotic synthesis of these fine chemicals is very difficult and uneconomical also. Their synthesis by plant cell culture method is easier than by microorganisms. The importance of biotransformation technique has, therefore, very recently been realised by several biotechnologists. Detailed studies on biotransformation of steroids such as sex hormones and cardenolides have been undertaken by many workers in recent years.

Biotransformations in plant tissue cultures to transform steroids in a manner similar to that have widely been used with microbial system is very promising field. Stohs and Staba first time studied the biotransformation of steroids in which digitoxigenin was shown to convert (during a 7-day incubation with cultures of *D. lanata* and *D. purpurea)* in to a keddepositive compound, 3-dehydrodigitoxigenin. Furuya and his associates reported biotransformation of digitoxin to gitoxin purpurea glycoside-A and purpurea glycoside-B from suspension culture of D. *purpurea.*

Reinhard (1974) has studied the transformation capacity of *D. lanata* suspension culture transforming digitoxin to purpurea glycoside-A, digoxin deacetyl lanatoside-C, lanatoside-A, and lanatoside-C. Thus suspension culture of *D. lanata* are able to hydroxylate - the steroid nucleus of digitoxin at C-12, to link glucose to the carbohydrate side chain and to acetylate digitoxose.

Reinhard reported that both digitoxin and β-methyl digitoxin can be converted into medicinally more important cardiotonics by the specific hydroxylation at position 2 of the steroidal skeleton by suspension culture of *Digitalis lanata.* Feeding experiments showed that cultured cells of *Datura innoxia* possess a remarkably high capability for glycosylation of hydroquinone to form arbutin used as diuratic and urinary antiseptic.

It has been observed that anaerobic condition of cell physiology favours the biotransformation rate. Barz reported that anaerobic conditions which inhibits degrative reactions expidite glycosylation reaction of phenolic compounds nearly six-fold. Isomerization of D-tryptophan into a optically active form, L-tryptophan have been demonstrated into tobacco cell cultures using recemase enzyme.

Evaluation of Antimicrobial and Antifertility Activity

Screening of various cultures and extracts by bioassay for definite pharmacological activities may be an effective method for exploiting substances produced by cultured cells. Khanna screened aqueous and/or other tissue extracts of several plant species *viz., Atropa belladonna, Argemone mexicana, Artemisia scaparia, Agave weightii, Datura metel, D. tetula, Glycine max, Hyoscyamus niger, Lycopersicon esculentum, Solanum nigrun,* etc. and found antimicrobial activity against either Gram-positive or Gram-negative bacteria or both. She also isolated anit-microbial principles from tissues of these plant species. Quercetin obtained from *Agave weightii, Crotalaria burhia, C. juncea, D. metel, D. tatula, L. esculantum, Papaver rhoeas.* Whereas isorhamnetin isolated from tissue of *A. maxicana* was active against Gram-negative bacteria. She isolated kaempferol, a very strong flavonoid, from the tissue of *A. wightii, Pisum sativum, L. esculantum,* etc. and was found active against both bacteria and fungal human pathogen -*Candida albicans.* A-lignan and phyllemblin, isolated from tissue of *Emblica officinalis* also possess antibacterial and antifungal properties. She further extracted and screened kaempferol from various plant tissues pharmacologically against white male rats and observed antifertility activity (71 per cent) which had no side effects when tested by conventional methods. Nishi and Mitsuoka (1975) found that aqueous extracts of-callus and cell cultures of *Isodon japonies* inhibit the secretion of gastric juice in rats and promote the healing of acetic acid induced peptic ulcers on oral administration.

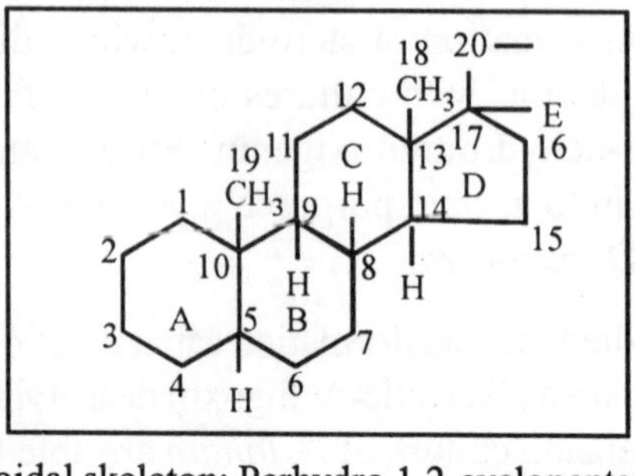

Diagram 22.3 Steroidal skeleton: Perhydro 1,2-cyclopentano phenenthrene ring.

Biological Control on Production of Secondary Metabolites

Active production of pharmacologically important compounds by plant cell cultures is mainly dependent upon following biological factors:

Growth Factor

Biosynthetic activity of cells in a batch culture depends on cell growth and substrate utilization. Not much is known about the correlation between the rate of secondary metabolite formation and age of individual cell in culture. However, production-growth pattern could be catagorise in three major types:

- *First type:* Product-production proceeds parallel with cell growth for example anthraquinones, morindone, nicotine, tropane alkaloids, etc.

- *Second type:* Product-production is delayed until cell growth declines or stops, for example polyphenol, shikonin, etc.

- *Third type:* Product-production declines as the cell growth increases for example, diosgenin, ascorbic acid, etc.

The growth phase of cells in cultures in many cases such as shikonin, anthocyanin and certain phenols production can be altered by auxin level in growth medium.

Morphogenic Differentiation

In nature, certain compounds are synthesized and stored up only in some specific plant parts such us essential oils in certain sex glands or duct; tropane alkaloids in roots of tobacco; and latex in laticifers ducts. Such compounds reported in intact plants can not be synthesized by cells in suspension cultures but if organogenesis is induced in cell cultures, these are synthesied in vitro. In *Scopolia parviflora* suspension culture, root initiation couplled with normal production of tropane alkaloids; alkaloid content increases many folds when organogenesis is induced in suspension culture of *D. innoxia*.

Shikonin, a derivative, found localized in cork cells only, has been produced exceptionally in suspension cultures. This finding has opened the path for in vitro production of certain monoterpenes, α-pinene, etc. without organogenesis in future.

Somaclonal Variation and Biosynthetic Activity

Cells grown in cultures for a long period undergo extensive genetic changes such as polyploidy, aneuploidy, anaphase bridge, fragmentation of chromosomes and gene mutation. Biosynthetic ability of cultured cells may be improved selecting high producing clones or by the artificial induction of genetic mutation.

The chemical mutagens tested for the induction of genetic variations in carrot cells are N-methyl-N-nitro-nitrosoquanidine. As a result, carotenoid yield in some improved clones increased three times in comparision to its wild types. Similarly, variants of sugarcane were isolated which showed appriciable increase in sucrose content in addition to disease resistance against Fiji virus, downy mildew, eye spot disease and smut, In some cases, the somaclones of sugarcane are unstable, while in other cases somaclones have been found stable when grown in four locations over five years. Somaclonal variation has also been reported in plants regenerated from protoplasts. Cellular variation, is therefore, another factor which can regulate secondary metabolism and has potential use in improving biosynthetic capabilities of culture strains. It is theoretically possible that the frequency of obtaining mutant cells may be markedly enhanced by mutagen treatment of haploid cells. However, mutation research on plant cell culture should be one of our major concerns for the near future.

Environmental Control on Production of Secondary Metabolites

Light is considered as an important environmental factor which controls the synthesis of most of the secondary metabolites in vitro. An appriciable quantitative changes have been observed in volatile oil content in cell cultures *of Ruta graveolens* when grown in light and dark separately.

Using Parsley cell culture, Hahlbrock et al. (1971) noted a marked increased in the accumulation of flavone and flavonol glycosides in illuminated cultured cells, specially with ultraviolet light. Stimulatory effects of light on the production of medicinal compounds have also been reported by many workers. Increased yield of polyphenols, plastoquinones, flavonoids and carotenoids has been observed when cultures were grown under light.

Inhibitory effects of light have also been observed specially with white and blue light on shikonin derivative formation when cell cultures of *Lithospermum* sp. However, in other cases no significant changes in the content of secondary metabolites could be observed when cell cultures were grown in light or dark.

Effects of Nutrients on Production of Secondary Metabolites

The addition of nutrients, growth hormones, vitamins, etc. in culture media is primarily aimed to increase cell growth in cultured condition. It has been reported that certain nutrients in culture media increase some secondary metabolites while others show an inhibitory effect.

Addition of sucrose in culture media above its ordinary level increases shikonin accumulation in cultured cells, lower concentration of sugar increases production of ubiquinone-10 in tobacco cell culture. Carbon-nitrogen ratio (C:N) plays a vital role in the increase production of catechol tannins in *Scynamore* sp. suspension culture. Unfortunatly, not much study has been made focused in this field.

Table 22.2 Some secondary metabolites detected in plant tissue culture.

Secondary Plant Products	Type of culture(s)	Stock plant
Flavones and Flavonoids		
Quercetin	Callus	*Agave wightii*
	(C)	*Datura metel*
		D. tatula
		Lycopersicon esculantum
		Solanum eviculare
		Petroselinum hortense
	Suspension	*Crotolaria juncea*
	Culture (SC)	*Artemisia scaparia*
Apigenin	(C)	*Trigonella corniculata*
		Solanum dulcamara
	(SC)	*Glycine max*
Lutcolin	(C)	*Datura pinnata*
	(SC)	*Papaver hortense*
Chrysocrion	(SC)	*P. hortense*
Scopolatin	(SC)	*S. luteum*
Negretin	(SC)	*Solanum tuberosum*
Isorhamnetin	(SC)	*Argemone maxicana*
		P. hortense
Kaempferol	(C)	*A. wightii*
	(C)	*C. cheiri*
	(C)	*P. sativum*
	(C)	*L. esculentum*
	(C)	*S. vaccaria*
	(C)	*S. dulcamara*
	(C)	*A. scoparia*
	(C)	*A. sativum*
	(C)	*G. max*
	(C)	*D. lablab*
Phyllemblin	(C)	*E. officinalis*
Artimissinine	(C)	*A. scaparia*
Vitexin	(C)	*V. decurrens*
Sinensetin	(C)	*Citrus aurantium*
Nobeletin	(C)	*C. aurantium*
Chalcones and Deoxyflavones		
Daidzein	(C & SC)	*P. aureus*
	(C & SC)	*G. max*

Secondary Plant Products	Type of culture(s)	Stock plant
2',4,4'-trihydroxy chalcone	(C)	*Phaseolus aureus*
Coumestanes and Coumarinochromans		
Pisatin	(C)	*Pisum sativum*
Soyagol	(C & SC)	*P. aureus*
Coumestrol	(C & SC)	*P. aureus*
Anthocyanins		
Cyanidin	(C)	*H. gracilis*
	(C)	*L. usitatisimum*
	(C)	*Dimorphothica auriculata*
Dophinidin	(C)	D. *auriculata*
Anthroquinones		
Rhein	(C)	*Cassia anqustifolia*
Digitoluutein	(C)	*Digitalis lanata*
4-hydroxydigitolutein	(C)	*D. lanata*
3-methyl purpurin	(C)	*D. lanata*
Alizartin	(C)	*Morinda citrifolia*
Tannins		
Catechin	(C)	*Camellia sinensis*
	(SC)	Paul's Sacrlet Rose
Epicatechin	(C)	*C. sinensis*
Naphthoquinones		
Plumbagin	(C)	*Plumbago zeylanica*
Sesquiterpenes		
Linderane	(C)	*Lindera strychinifolia*
Lindenenol	(C)	*L. strychinifolia*
Cryophyllene	(C)	*L. strychinifolia*
-bisabolene	(SC)	*Andrographis paniculata*
Sterols and Triterpenes		
Beta-silosterol	(C)	*N. tabacum*
Stigma sterol	(C)	*Dioscorea tokora*
Campesterol	(C)	*Withairia somnifera*
	(C)	Paul's Scarlet Rose
	(C)	*Tylophora indica*
	(C)	*Hslianthus annus*
	(C)	*D. metel*
	(C)	*T. foenumgracum*
Lanosterol	(C)	*S. nigrum*
	(C)	*S. elaegnifolium*
	(C)	*S. xanthocarpum*
	(C)	*S. carota*
	(C)	*D. pinnata*
Cholesterol	(C & SC)	*H. annus*
	(C & SC)	*S. indicum*

Secondary Plant Products	Type of culture(s)	Stock plant
Cycloartinol	(C & SC)	*N. tabacum*
Isofucosterol	(C & SC)	*H. annuus*
Obtusifoliol	(C & SC)	*N. tabacum*
Beta-amyrin	(C)	Paul's Scarlet Rose
	(C)	*T. indica*
Steroidal Alkaloids		
Solasonine	(C)	*S. xanthocarpum*
Tomatin	(C)	*L. esculentum*
carotenoids		
Voilaxanthin	(C)	Paul's Scarlet Rose
Zeaxanthin	(C)	*Ruta graveolens*
Neoxanthin		
Beta-carotene	(C)	*R. graveolens*
Lutein		
Lutein-5,6 epoxide		
Antheraxanthin		
Sapogenins		
Diosgenin	(C)	*S. nigrum*
	(C)	*S. xanthocarpum*
	(C)	*S. aviculare*
	(C)	*S. dulcamara*
Gitogenin	(C)	*A. wightii*
Hecogenin		
Tigogenin	(SC)	*S. nigrum*
Smilagenin	(C)	*Yucca aloifolia*
Sarsasapogenin		
Tiggenin		
Hecogenin		
Chlorogenin		

Precursors and Secondary Metabolites Production

When precursors or intermediate precursors are fed in culture medium they increase the synthesis of various phytochemicals in, in vitro. They are considered to be the limiting factors such as, cinnamic acid or phenyl alanin increases the synthesis of flavonoids and tropic acid for tropane alkaloid. High doses are proved to be lethal to tissue growth.

Shikonin production increases three-folds when *Lithospermum* sp. cultured cells are fed with L-phenylalanine, however, p-hydroxybenzoic acid is found ineffective. When *Datura* sp. cell suspension cultures are supplimented with hydroquinone, in traces, the arbutin synthesis increases considerably.

From economic point of view it is necessary that the precursor must be of low price than that of the ultimate product. Ogutuga and Northcote used inexpensive ammonia as nitrogen source and increased caffeine production nearly four times in tea suspension cultures.

Growth Regulators and Secondary Metabolites

Growth regulators affect growth and synthesis of secondary metabolites of cultured cells. Brain and Lockwood reported that the dose and nature of hormone supplimentation and the culture-age have very marked effect on steroidal levels while studying the cultured tissue of *Trigonella foenumgracum.* According to them certain combinations of auxins and cytokinins had synergistic effect while other had antagonastic influence on steroidal synthesis. The nicotine synthesis in the tobacco cell cultures is strongly inhibited by 2,4-D, whereas it is promoted by kinetin. The effect of growth regulators on secondary metabolism vary greatly depending upon kinds of metabolites.

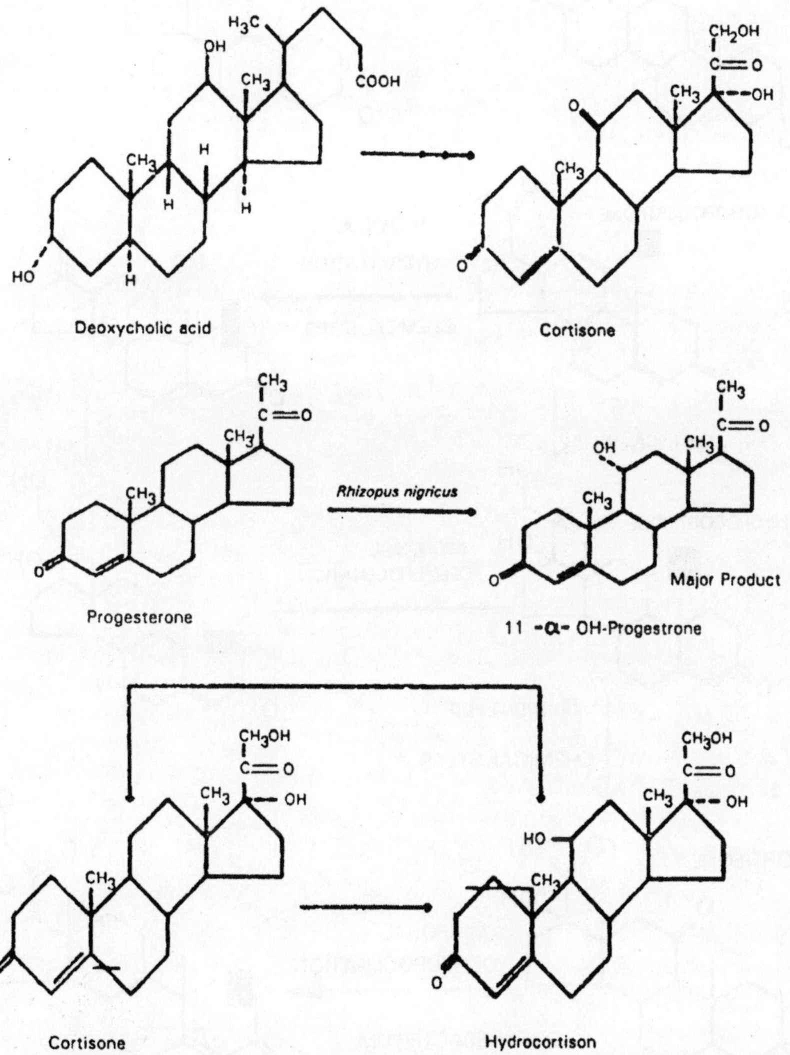

Deoxycholic acid → Cortisone

Progesterone —*Rhizopus nigricus*→ 11 -α- OH-Progestrone (Major Product)

Cortisone → Hydrocortison

Diagram 22.4 Some steroids produced from progesterone by transformation by several fungi, a Streptomycete *(Streptomyces lavendulae)* and Actinomycete. By chemical synthesis, the conversion of deoxycholic acid to cortisone requires 37 separate steps.

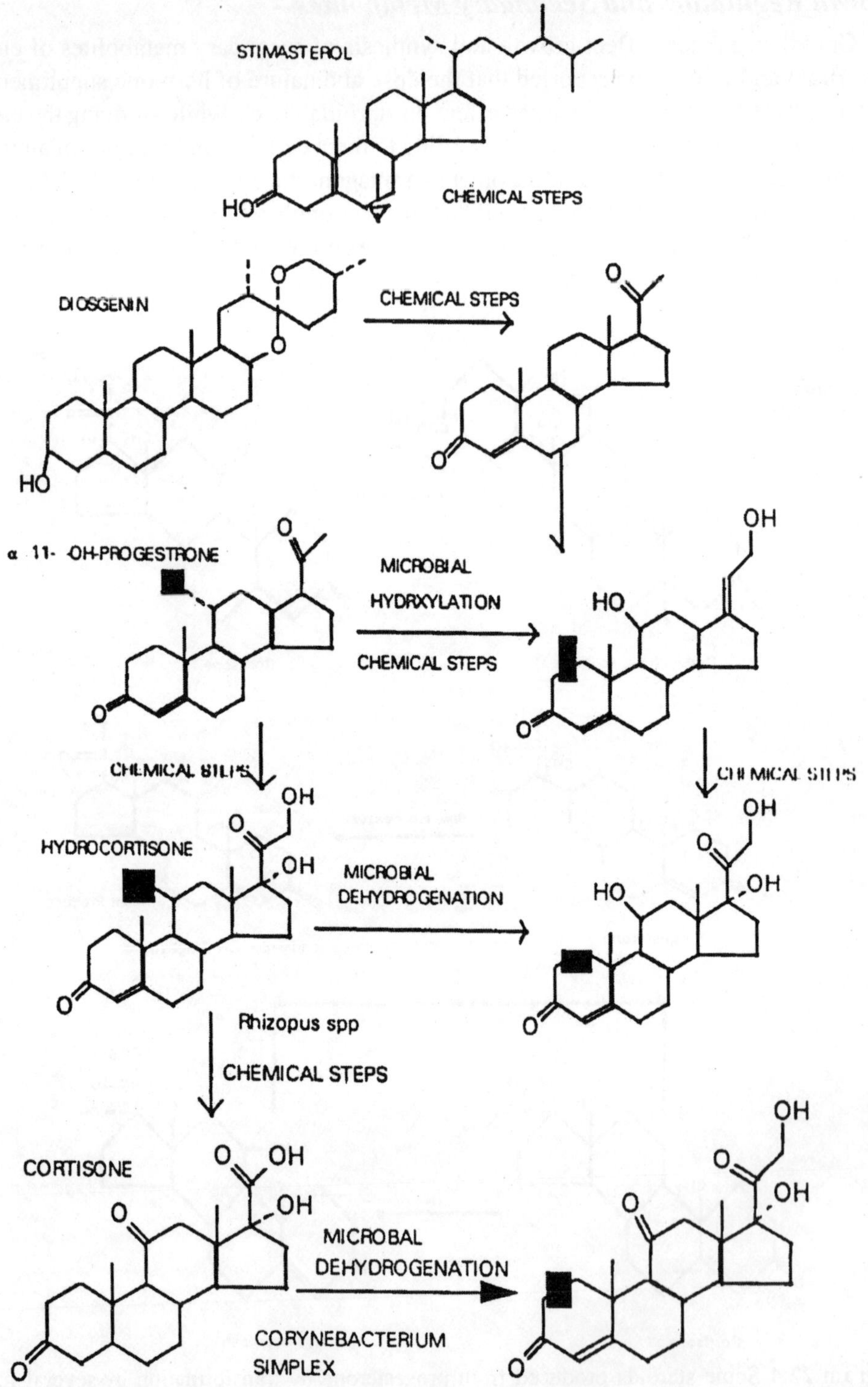

Diagram 22.4 Contd...

Tabata and his associates observed that the synthesis of shikonin is inhibited both by 2,4-D and NAA but not affected at all by IAA using the *Lithospermum* sp. cell cultures. In *Morinda* cultures, Zenk et al. (1975) observed that formation of anthraquinones takes place in presence of NAA but not in presence of 2,4-D. However, the synthesis of anthraquinones in *Cassia tora* remained ineffective by 2,4-D. Brain (1974) observed stimulatory effect of 2,4-D in L-DOPA synthesis. Although some promising information have already been obtained, more researches are needed for improving biosynthetic rate of secondary metabolites in, *in vitro* using biotechnological advances.

Steroids

Steroids are a group of organic compounds which have the four membered ring. They are biologically active, similar to hormones, produced by the testies, ovaries, adrenal cortex, and placenta. The steroids differ in the nature of their side groups or side chains, and these differences in structure confer different biological properties on the steroids. Steroids are widely used medically as anti-inflammatory agents, anesthetics, antifertility agents, and in the treatment of sterility.

Steroids are obtained directly from natural sources or can be synthesized. The steroid nucleus produced chemically after many transformations or additions in side chains. Microorganisms help in biotransformations of steroids are as follows:

1. *Rhizopus arrhizus* (fungus) hydroxylates progesterone forming another steroid, 11-α hydroxyprogesterone by introducing oxygen at position 11.
2. *Cunninghamella blakesleeana* (fungus) hydroxylates cortexolene to form hydroxygen, introduction of oxygen at the number 11 position.
3. *Rhizopus nigricans* (fungus) can also hydroxylate progesterone to produce 11-α-hydroxyprogesterone.
4. *Corynebacterium simplex* dehydrogenates cortisone to produce prednisone.
5. *Corynebacterium simplex* can also bring about dehydrogenation of hydrocortisone or cortisol to produce prednisolone.
6. *Nocardia restrictus* biotransforms δ^4-cholestene-19-hydroxy-3-one into estrone.
7. Androstenodione is converted into testosterone by yeast.

During a typical steroid transformation process, the microorganism, such as *Rhizopus nigricans,* is grown in a fermentation tank, using an appropriate growth medium and incubation conditions to achieve high biomass. In most cases, aeration and agitation are employed to achieve rapid growth. Therefore, steroid (e.g., progesterone) is added to a fermentor containing *R. nigricans.* The product is then recovered by extraction with methylene chloride or various other solvents; purified chromatographically, and recovered by crystallization.

Ergot, Ergotism and Ergot Alkaloids

Ergot is sclerotium of the pyrenomycete, *Claviceps purpurea* which infects rye and, less commonly, other grains. In infected plants, normal development of the grain is suppressed, and a hardened, purple-black sclerotium develops in place of the grain. (Diag. 22.5).

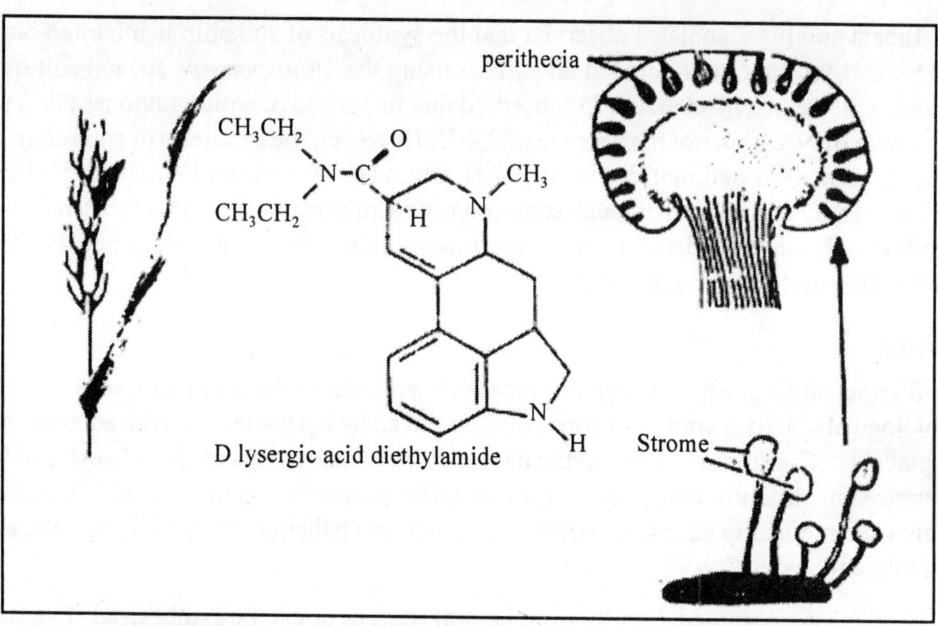

Diagram 22.5 Claviceps purpurea which causes ergot of grains.

Sometimes they are consumed in large quantities, as might happen if the sclerotia are included in milled flour, a disease known as ergotism (a condition of intoxication which follows the ingestion of excessive amounts of ergot alkaloids; symptoms may include vomiting, diarrhoea, thirst and convulsions, and gangrenous lesions may subsequently develop at the extremities). Cattle in the field may also be poisoned by the sclerotia if they eat infected pasture grasses. Ergotism in humans is marked by vomiting, feelings of intense heat or cold, pain in the muscles of the calf, a yellow colour in the face, lesions on the hands and feet, diarrhea, and an impairment of the mental functions.

The active principles in ergot that are responsible for ergotism are the alkaloids ergometrine, ergometrinine, ergotamine, and ergotaminine, which occur in the sclerotium. These alkaloids stimulate smooth muscle and selectively block the sympathetic nervous system. They have found its modern medical use in stimulating the uterus to contract to initiate childbirth, and also to hasten the return of the uterus to its normal size after childbirth. Until about 1950, efforts to produce the alkaloids in culture by conventional laboratory techniques had failed, and the only source of the alkaloids was the sclerotia, obtained from naturally infected rye. The search has continued for more reliable and productive methods of obtaining the ergot alkaloids.

□□□

CHAPTER-23

Biotechnology and Biomass Energy

The term energy derived from the Greek word '*energy*' meaning 'capacity to do work' is coined by Thomas Young (1773 - 1829); eighty years after the Newton's Classic concept, popularly known as kinetic energy. The behaviour of energy has been described by Laws of Thermodynamics. The first law of thermodynamics states that "no energy can be created or destroyed, but only can be transformed from one form to another. "The second law states that some energy is always lost into unavailable heat energy, no spontaneous transformation of energy from one form to other form is 100 per cent efficient".

Most organisms can synthesise molecules and macromolecules that serve as the structural and functional components of the cells. These components belong mostly to the classes of carbohydrates, proteins, fats, nucleic acids(DNA & RNA), vitamins, hormones etc. Among carbohydrates, glycogen and starch are especially important in energy metabolism. In a biosystem, in terms of energy, catabolic pathway is converging while the anabolic pathway is diverging. The stored micro- and macro-molecules liberates different form of energy by passing through a number of bio degradable processes.

The earth's planet cover is equivalent to, over 1,800 billion tones of dry matter, that is an energy equivalent of 30.10^{10} Joules, corresponding to the known reserves of fossil energy. Forests make up about 68 per cent of terrestrial biomass, grass ecosystems about 16 per cent, and cultivated lands only 8 per cent. For the earth as a whole, the 173 billion tones of dry matter produced every year by photosynthesis over 20 times the fossil energy consumed in the world, or again 200 times the energy contained in the food on the planet's for billion inhabitants.

The conventional energy sources of the world are dwindling fast. We have to come to realise that the world plenty of non-renewable natural sources does indeed have a bottom self. Bioenergy can play an important and vital role to meet energy crisis of the world. It has special relevance to India as its 80 per cent population resides in villages needing energy for cooking, lighting, and operating pumps for irrigation and drinking water, etc. India having largest population of cattle in the world and sizable forest potentialities, bioenergy can safely be developed as an alternative source of energy on dependable lines.

Bioenergy includes those processes where biological forms of matter such as plants, vegetables, bacteria, enzyme, etc. provide the basis for energy or its conversion from one form to another. The widest use of bioenergy is the traditional way where wood plants and agricultural matter are directly burnt to provide heat.

Vegetable biomass is a new name for plant organic material wherein solar energy is trapped and stored through the process of photosynthesis in which carbon dioxide and water are transformed and form energy rich compounds. Biomass includes both terrestrial as well as aquatic matter and can be conveniently grouped into new plant growth, plant residues and wastes. The *new plant growth* includes wood, short-rotation trees, herbaceous plants, conventional crops, algae (fresh water and marine), aquatic plants. The *residues* cover not only crop materials, such as straws, husks, bagasse, corn cobs, etc. but also secondary level products such as cowdung, animal dropings, forest based residues like bark, saw dust, wood shaving etc. The term wastes has been loosely used. It is matter of vegetable origin in wrong place. It is of disposable nature like garbage, night soil, sewage solids and industrial refuse.

Table 23.1 World energy consumption since 1971 to 1987 (million tonnes of oil equivalent)

Year	Coal	Petroleum	Natural Gas	Hydroelectric	Nuclear	Total
1971	1632	2413	997	318	28	5388
1973	1668	2798	1066	332	49	5913
1975	1709	2725	1709	358	87	5959
1977	1830	2986	1162	376	132	6486
1978	1863	3082	1206	403	150	6704
1979	1976	3124	1273	413	153	6940
1980	2007	3001	1297	421	169	6895
1981	2003	2902	1321	430	198	6854
1982	2047	2825	1316	451	218	6858
1983	2101	2801	1326	475	240	6943
1984	2180	2845	1410	485	282	7202
1985	2273	2809	1494	511	348	7435
1986	2318	2899	1487	517	377	7598
1987	2387	2941	1556	524	404	7811

Source: BP Statistics, World Science News, 29(5): 19, 1992

The Planning Commission (India) has estimated production of crop residue in 1979 as 203 million tones as against the food production of 200 million tones. By 2000 A.D., the availability of crop residue would arise to the tune of 336 million tones.

Table 23.2 Crops residue

Crop	Area million ha	Production million tonnes	cal. value kcal/kg	moisture % wet basis
Rice husk	4.07	1.79	3440	19.5.
Bagasse	0.32	6.12	3800	15.0
Cotton stalk	0.79	2.39	3300	12.0
Paddy straw	4.07	22.40	3000	10.6

The urban garbage @ 0.4 kg per head amounts to 19.98 million tones per year, which can be burnt as a calorific value of 1100 kcal/kg on 50% moisture.

There are several methods to use the biomass for energy production. Some of the sophisticated methods of thermochemical conversion include pyrolysis, gassification, producer gas, bariquetting, hydrolysis and liquification.

Composition of Biomass

Plants are the major source of biomass because their cell wall is constituted by 6 components: (i) cellulose, (ii) hemicellulose, (iii) lignin, (iv) water soluble sugars, amino acids and aliphatic acids, (v) ether and alcohol-soluble constituents (e.g. fats, oils, waxes, resin and many pigments), and (vi) proteins. These components build up plant biomass. Proportion of these constituents vary in different groups of plants and even in the same group. If the concentration of sugar is high, the biomass will be sugary e.g. sugurcane, and sugar beet. Similarly, high amount of starch) present in biomass yields the starchy biomass e.g., potato and tapioca.

Cellulose

Cell wall is mainly constituted of cellulose and its fundamental unit is glucose. Formation of cellulose is a complex process. From each glucose unit, one molecule of water is removed to yield an anhydrous glucose. The anhydrous glucose units are linked end to end with β-l,4-linkage to form the long chain polymer of cellulose $(C_6 H_{10} O_5)_n$. Here n represents the degree of polymerization, the number of which varies from 5000 to 10,000. Enzymatic hydrolysis of cellulose and production of glucose are dealt ahead.

Hemicellulose

It is made of sugars (xylans) which comprises of 20-25% plant biomass on dry weight basis. It also contains glucose and several other hexoses (galactose and mannose) and pentoses (xylose and arabinose). The proportion of these constituents varies plant to plant. Degree of polymerization to yield hemicellulose does not exceed beyond 50. The polymers has branched chains. It occurs as amorphous mass around the cellulose strands. Hemicellulose are insoluble in water but easily solubilised in alkali.

Lignin

It is a complex and high molecular weight polymer and is formed by de-hydrogenation of (II), and sinapyl (III) alcohols. Presence of these alcohols differs in different alcohols, such as in angiosperm lignin is formed from coniferyl alcohols, which is formed from the mixture of coniferyl and sinapyl alcohols, and grass lignin from mixtures of coniferyl, sinapyl and coumaryl alcohols. Lignin is phenolic in nature; is very stable and difficult to isolate, since it occurs between the cells and cell-walls. It is deposited during lignification of the plant tissue and gets intimately associated within the cell walls with cellulose and hemicellulose and imparts the plant an excellent strength and rigidity. As a result of photosynthesis, an enormous amount of plant biomass is accumulated in terrestrial and aquatic systems, which are then utilized into different ways as the source of energy.

Types of Biomass

Terrestrial Biomass

Terrestrial biomass is used to fulfill the need of food, feed, vegetables, fibre, furniture and cooking purpose as well. Traditionally the need of fire/fuel was fulfilled by trees, remains of time, we totally became dependent on conventional energy sources of fossil fuel and electricity. But gradually increasing world wide human population and diminishing stock of fossil fuel have challenged us to seek out the alternative sources of energy.

Aquatic Biomass

It is obvious that the first life originated in water. Therefore, water bodies support a vast community of plant and animal. Many aquatic plants become troublesome for aquatic animals and human as well such as the aquatic weeds like water hyacinth, *Salvinia, Hydrilla, Lemna, Pistia, Wolffia,* etc. In addition to higher plants, the lower plants (especially blue-green algae and green algae) have much future prospects, as far as production of biomass conversion of aquatic biomass into biogas/hydrocarbon and abatement of pollution (in sewage oxidation) are concerned.

Salvinia

Salvinia, a member of Pteridophyta, is commonly known as water fern. It grows luxuriantly in stagnant water, for example ponds, pools and lakes. *S. molesta* is the world's worst weed known so far. In India, it predominates in Kerala, Kashmir and North-East states. Biogas production from *Salvinia* has recently been suggested.

Water Hyacinth *(Eicchornia crassipes)*

It is the most noxious weed of the world and grows abundantly in tropical regions in nonsaline water in ponds, pools, lakes reserviors, rivers and even in paddy field. It is believed that water hyacinth occupies about 2,00,000 acres land in Bihar and 30,000 acres in West Bengal. It grows luxuriantly at temperature 28-30°C. It rapidly multiplies on domestic sewage. Generally the huge amount of biomass is of no use. Now-a-days, cultivation of water hyacinth on sewage for minimizing pollution has been suggested. Use of water hyacinth in biogas production.

Wastes as Renewable Source of Energy

Waste is the spoilage, loss or destruction of either matter or energy, which is unusable to man. Gradually increasing civilization through industrialization and urbanization, has led to increase in generation of wastes into environment from various sources. Waste generation is, therefore, a necessary outcome of consumption, and also because of insufficient process, general ignorance, wasteful habits and social attitudes.

Types of Wastes

Wastes are classified into (i) energy wastes and (ii) material wastes. The main source of energy in various parts of world is petroleum oil, followed by coal. In India, about 50% oil is imported, each year. Coal mines are concentrated only in a few regions. Coal is used in generation of electricity, steam engines and fire. Most potential energy of coal is wasted during electric generation in thermal power plants. Thermal loss in India is about 20-30% because of lack of suitable technologies.

Classification Based on Chemical Nature

1. *Inorganic wastes:* Those generated by metallurgical and chemical industries, coal mines, etc.
2. *Organic wastes:* Agricultural products, dairy and milk products, slaughter houses, sewage, forestry, etc.
3. *Mixed wastes:* Those discharged from industries dealing with textiles, dyes, cake and gas, plastic, wool, leather, petroleum, etc.

The inorganic wastes may be recovered by chemical/mechanical treatment, whereas organic and mixed wastes require biological as well chemical treatments.

Classification Based on Physical State

The wastes occur in three states, the solid, liquid and gaseous ones.

1. The solid wastes can be burnt, thermally decomposed, anaerobically digested to get methane and other combustible gases or biologically converted to a variety of products.
2. Liquid wastes are most troublesome, because of the presence of non-retractable chemicals, and their further return such as NO_3, NO_2, NH_3, CO_2, SO_2, etc.
3. When concentration of these gases increases in the atmosphere they cause gaseous pollution, which has its bad impact on plant and animal lives.

The organic wastes and residues are the major sources of renewable energy.

Composition of Wastes

Waste is a general term which embraces all types of wastes irrespective of constituents and phases. Therefore, composition of waste differs with differing nature, phases and sources. It may be inorganic, organic or mixed types. Organic wastes play a major role in being is renewed and becoming a source of energy. Composition of organic materials is given under composition of biomass.

Sources of Wastes

Industries

Following industries generate various types of wastes/by products which contain sufficient amount of energy.

1. *Chemical Industries:* The chemical wastes are maleic anhydride and phthalit anhydride.
2. *Cotton Mills:* Cotton mills produce the cotton seeds and fibers as wastes.

3. *Dairy:* Dairy industry is one of the important industries which requires special attention, so far as treatment and disposal of waste are concerned. Dairy wastes contain milk whey, butter milk, unused skim milk, plant washings and traces of detergents. The waste is a dilute solution or suspension containing lactose, protein, fat and minerals. Therefore, dairy wastes serve a food substrate for production of single cell protein, lactic acid, vitamins, ethylalcohol and alcoholic beverages.
4. *Food Industry:* Waste materials of food industries are the collagen meat packaging waste and lactoserum (a by-product of cheese making food industry).
5. *Oil Refineries:* They produces wastes as gas, oil, paraffins (n-alkanes.) olefins (n-alkenes) and other hydrocarbons.
6. *Paper Mill:* The wastes are bisulphite liquor and long cellulosic pulp.
7. *Sugar Mills and Distilleries:* Sugar cane *(Sacchrum officinarum):* Sugar cane belongs to family Gramineae. It is a tall, perennial grass, the stems of which are the source of cane syrup.
8. *Sugar beet (Beta vulgaris):* Sugar beet is a member of Chenopodiaceae. It is biennial herb with fleshy leaves and swollen roots. For sugar production good crops of sugar beet can be had in Rajasthan, Punjab, Haryana and U.P.

Table 23.3 Some renewable sources of biomaterials

Plants	Residues/wastes	Products
Cashewnut	Shell, testa	Gum,tannin
Coconut	Coir, shells and pith	coirboard, coir fiber, xylose
Cotton	Fibre	fuel
Maize	Cobs	Single cell protein (SCP), Fuel
Paddy	Husk, bran, straw	Bran oil, vitamin, SCP,
Sugar beet	Pulp	Ethanol fuel
Sugar cane	Bagasse, molasses	SCP, fuel/fire, alcohol
Sunflower	Husks	SCP, Fuel (alcohol)
Tapioca (cassava)	Tubers	Ethanol
Tea		Tea waste, Caffeine
Trees	Seeds and leaves	Many products
Wheat	Straw	SCP, fuel

Importance of using bagasse as an alternative source of energy for sugar manufacturers has been realised in recent years. The Engineering Staff and College, Hyderabad organised a workshop in October, 1990 on *Strategies for the development of co-generation of electric power in sugar factory.*

The Indian Renewable Energy Development Agency (IREDA) sanctioned around 97 projects in this field. Utter Pradesh Government launched a programme for producing alternative energy from sugar cane waste in sugar mills. The first project of its kind in the country is being installed with the Sheh Road Sugar Mill in Bijnor. It will produce 6 MW electricity from the sugarcane waste. In this way fuel consumption would be reduced by 40 to 58%.

Agriculture

In agriculture, a huge amount of residues/wastes are produced which, however, are thrown into field because of non-availability of technologies for utilization at village level.

Paddy *(Oryza sativa):* Paddy plant produces paddy and straw; the products of paddy are rice, husk and bran. Husk is the outer most hard coat of paddy. Bran is the thin papery layer present between husk and rice. Out of total (about 80 million tonnes) production of paddy, about 16-18 million tonnes of husk (i,e. 20-25% of paddy) are produced per annum. Due to lack of utilization technology bran is thrown as waste, and probably in some part it is used as cattle feed. Recently, bran has become a source of rice bran oil, whether edible or non-edible. Paddy husk is used as fuel in rice mills and villages as well. Recent analysis of paddy husk of different species has shown the presence of high caloric value i.e. 3200-3500 KCal/Kg, which can replace about 10 million tonnes of coal per annum It has been found that 1 tonne of husk can replace 450 litres of furnace oil.

India has given special attention to develop technologies for utilization of paddy husk. Recently, the Central Fuel Research Institute (C.F.R.I.), Dhanbad had developed a process to produce oxalic acid from cellulosic matter and silica from, mineral matter. India had started a thermal power plant from paddy straw, which is first of its kind in the world which can generate 62 million unit of energy peranum. A rice straw-fired Thermal Power Plant was set up at a village Jalkheri (Punjab). Paddy Processing Research Centre (Tiruvarur) has also developed methods for parboiling and milling operation in rice mill to extract about 92% potential energy of husk.

Tapioca *(Manihot cscnlenta):*

Tapioca is a member of the family Euphorbiaceae. It is cultivated in Kerala, Tamil Nadu, and Karnataka. Kerala alone contributes 80% tapioca in the country. It grows in alkaline soil in marginal and infertile lands. It is a small shrub producing tubers inside the soil. Tubers arc rich in starch, therefore, cassava has been exploited commercially for starch, sago and Hour unit recently, as a source of alcohol. Tapioca can yield alcohol about 180 litres/tonne in contrast with sugarcanc which produces about 70 litres/tonne. Central Tuber Crop Research Institute (CTCRI), Tiivandrum has carried out work on alcohol production from tapioca. Starch is extracted from the tubers which is then fermented into alcohol.

Forestry:

Forests contribute a considerable amount of biomass which could be variously used as a source of energy. Lignocellulosic contents also varies from 60-75% of dry weight. Soft wood has higher lignin to cellulose ratio than hard wood. Residues/wastes generated from forestry are wood chips, saw dust, dried tree branches, tree twigs, tree-bark, leaf litter, etc.

Fire Wood as a Traditional and Future Source of Energy:

More than one-third of the world's population depends on wood for cooking and heating. Around 86% of all the wood consumed annually in the developing countries is used for fuel, and of this total atleast half is used for cooking. The situation is growing so desperate that wood is poached from forest reserves. In the face of global concern over the dwindling supply of fuel wood, the rate of forest decimation to provide basic human necessities in developing countries is alarming. No less than one and a half billion people in developing countries derived at least 90% of their energy requirements from wood and charcoal. Indeed, it has been estimated that at least half the timber cut in the world still serves its original role for mankind: as fuel for cooking and heating.

The firewood fuel is seriously threatened particularly in the developing countries. By the turn of the century, at least 300 million people will be without woodfuel for their minimum cooking and heating needs and will be forced to burn dried animal dung and agricultural crop residues, thereby further decreasing food crop yield. Because of the severity of the firewood crisis we need those firewood crops which are aggressive and quick-growing. We can also look forward for those firewood crop plants which may be slow-growing and possess low propagation property, and can be improved with the help of biotcchnological devices. While planning the biotechnological project on firewood crop development the first priority should be given to local land races. Following are the selected firewood plant species which have been identified as economical species.

Fuelwood species for Arid and Semi arid Regions

Acacia brachystaschya, A. cambagei, A. cyclops, A. nilotica, A. saligna, A. senegal. A.seyal, A. tortilis, Adhatoda vasica. Albizia lebbek, Anogeissus latifolia, Azadiruchta indica, Cajanus cajan, Cassia siamea, Colophospermum mopane, Emblica officinalis. Eucalyptus camaldulensis, E. citriodora, E. gomphocphala, E. microtheca, E. occidentalis, Haloxylon aphyllum, H. persicum, Parkinsonia aculeata, Pinus helepensis. Pithecellobium dulce, Prosopis alba, P. chilensis, P. cineraria, P. juliflora, P. pallida, P. tamarugo, Tamarix aphylla, Zizyphus mauritiana, Z. spina-christi.

Fuelwood Species for Humid Tropics

Acacia aurriculiformis, Calliandra calothyrsus, Casuarina equisetifolia, Derris indica, Gliricidia sepium, Gmelina arborea, Guazuma ulmifolia, Leucaena leucocephala, Mangroves spp., *Mimosa scabrella, Muntingia calabura, Sesbania bispinosa, S. grandiflora, Syzyjium curmini, Terminalia catappa, Terma* spp.

Fuelwood Species for Tropical Highlands

Acacia mearnsii, Ailanthus altissima, Alnus acuminata, A. nepalensis, A. rubra, Eucalyptus globulus, E. grandis, Grevillea robusta, Inga vera

Firewood crops that deserve increased recognition and research through genetic engineering and tissue culture methods to solve energy crisis of future.

Municipal Sources: High amount of municipal wastes are generated from cities, as a result of anthropogenic activity which are thrown near cities in open lands or in rivers. These wastes again causes a serious hygienic problem as it contain high amount of organic matter and pathogenic microorganisms. Municipal and domestic wastes include sewage and sludge, garbage, horse dung, cattle dung and wastes from animal slaughter houses.

1. *Sewage and Sludge:* Sewage is a product of water which are thrown away after its use. Treatment of sewage results in generation of another waste, which is called as sludge. Sludge is a solid matter in the settling tanks of sewers, and other treatment operations in a sewage treatment plant. There are 142 class I cities in India which produced around 9,000 million litres of sewage per day. However, there is no sewage treatment facilities in about 70 class I cities such as Srinagar, Ranchi, Dhanbad, Bhopal, Jabalpur. Although methods for single cell protein production from sewage oxidation pond and irrigation of agriculture crops by treated water have been developed, yet this facility is available only at limited places. Moreover, if 900 million litres of sewage is converted into sullage gas per day from the major cities, 20% of their energy demand could be met.
2. *Urban (City) Garbage:* Garbage potential of Indian cities is quite high. It is estimated that production of city garbage in India is about 41,000 tonnes per day, the annual production is about 15 million tones. Nonetheless, city garbage produced in Bombay, Madras and Calcutta is comparable to that of developed countries. In India garbage is thrown near the city. In some cities like Madras, Calcutta, Delhi, Baroda, Jodhpur, etc. Municipal solid waste composting plants are in operation. Central Mechanical Engineering and Research Institute, Durgapur has established a first pilot plant to produce electricity by using city garbage. The plant has a capacity to use about 500 kg garbage/ha, as a result of which about 5KW h electricity is generated from biogas, produced by anaerobically, combution of garbage.

Methods for Energy Production

Pyrolysis of Wood

It was first performed by Robert-Mudoch in 1792. The pyrolysis of wood results in production of gas, liquid and char which can be used as fuels. The gas can be synthesized in form of methanol or liquid hydrocarbons.

Pyrolysis is defined as the destructive distillation or decomposition of organic matter, for example, solid residues, wastes(saw dust, wood chips, wood pieces) in an oxygen-deficient atmosphere or in absence of oxygen at high temperature (200-500°C or rarely 900° C). Products of pyrolysis are gases, organic liquids and chars, depending on the pyrolysis process and temperature of reaction. The condensable liquids separate into aqueous (pyroligneous acid), oil and tar fraction (if the substrate is wood). The composition of gas is carbon monoxide (28-33%), methane (3.5-18%), higher hydrocarbons (1-3%) and hydrogen (1-3%). During pyrolysis, hydrogen content of gas increases with increase in temperature.

After pyrolysis, the amount of different products varies with the nature of wood, type of equipment and systems employed. For example, low temperature favours liquids and char; low heating rates favour gas and char. In contrast high heating rate, favours liquids and the long gas-residence time favours gas. Thus, the liquid is obtained before the solid is completely burnt to yield gases. The liquid is very useful for high energy fuel.

Wilson *et al*., (1978) have described a mobile system for the pyrolysis. The unit can move from one place to another for processing of wastes/residues. Energy transported as coal and oil would be about 2.8 times greater than transporting the wet wood waste.

Pyrolysis has been employed to produce charcoal for the last few decades. Charcoal is a smokeless and low sulphur fuel used mostly for cooking purpose. Besides wood, other residues used in pyrolysis are cotton, bagasse, ground nut shell, etc.

Gassification of Wood

When biomass reacts with steam and oxygen it produces BTU gas which consists mainly of carbon monoxide and hydrogen. This gas can be burnt as a fuel or can be used as an intermediate for the synthesis of fuels or energy, intensive chemicals, including methanol, substitute gasoline, hydrogen and ammonia.

It is a process of thermal degradation of carbonaceous material under controlled amount of air or pure oxygen, and high temperature upto around 1,000° C. As a result of gassification, high amount of gases is produced. Gassificaation of biomass is done in a gasifier designed in various ways. Success for gassification process is based on its desinging. Therefore, the design of a gasifier is an important factor in controlling gas quality which is used in a controlled manner for irrigation, pumping and electricity generation.

Following are the advantages of gassification of wood over coal:

1. much low oxygen requirements
2. practically no steam requirements
3. low cost for changing H_2/CO_2 ratios which are high in wood gas, and
4. no or very little desulfurization cost (Goldstein,1980).

When gassification of farm wastes (manure) takes place, the phenomenon is known as hydro-gassification, because gassification of organic wastes occurs in the presence of hydrogen at 500-600°C.

Three to five kilograms of biomass per hour is needed for generating power for an hour. In a year it has to run for at least 15,000 hours for which 6 tonnes of biomass is required. Therefore, it is very essential to make sure that the biomass is continuously available. Recently, United Nations Department Programme (UNDP) has recognized the India's five H.P. gasifiers as the best ones.

Bariquetting

The process consist's of drying cellulosic wastes using solar energy and then partially carboniging the material to give virtually ash-free carbon.

Liquefaction

Liquefaction involves the production of oils for energy from wood or agriculture and carbon residues by reacting them with carbon monoxide and water/steam at high pressure (4,000 Ib/in^2) and temperature (350-400°C) in the presence of catalysts. By this method about 40-50% oil can be obtained from wood. This oil serves a good source of fuel.

The Biological Process (Bioconversion)

Bioconversion involves the conversion of organic materials into energy such as fertilizer, food and chemicals through biological agency. The term biological agents means the microorganisms i.e. bacteria, actinomycetes, fungi and algae. In broad sense bioconversion involves two following steps:

1. Photosynthetic production of biomass
2. Its subsequent conversion into more useful energy forms (gaseous, liquid or solid fuel; heat and electricity).

Enzymatic Digestion

This process involves the conversion of cellulosic and lignocellulosic materials into alcohols, acids and animal feeds by using microbial enzyme e.g. cellulase, hemicellulase, amylase, pectinase, etc.

Degradation of Cellulose

It is clear that cellulose is a polymer of β-1,4 linked anhydrous glucose units, comprising 40-60% of cell wall materials of plants. Microorganisms, which produce cellulase and other enzymes in high amount, like *Cellulomonas, Trichoderma reesei, T. viride* and others are used for the production of cellulases in high amount.

There are 3 enzyme components of cellulase:

1. β-l,4-endoglucanase
2. β-l,4-exoglucanose
3. β-l,4-glucosidase

β-l,4-endoglucanase randomly attacks along cellulose chain; 6-1,4-exoglucanase splits from non-reducing end of cellulose and β-l,4-glucosidase i.e. cellobiase cleaves 2 molecules of glucose from cellobiose. Following is the sequence of cellulose degradation.

Cellulose chain —β-1,4-endoglucanase—> Smaller polymers and soluble oligomers

β-1,4-endogluconase, β-1,4-exogluconase ——> Cellobiose and Glucose —β-1,4-glucosidase (cellobiase)—> glucose (2molecules)

Degradation of Hemicellulose

Sugars constituting hemicellulose. An analogous system of enzymes is involved in the degradation of hemicellulose. This enzyme-system consists of 3 enzymes:

1. Exoxylanase
2. Endoxylanase
3. β-xylosidase (which split xylose and other short chain xylobioses)

Anaerobic Digestion

Anaerobic digestion is a partial conversion of organic substrates by microorganisms into gases in the absence of air. The gases produced are collectively known as biogas. Anaerobic digestion is accomplished in 3 stages:

1. Solubilization of complex substrates by enzymes into simple forms i.e. fatty acid, sugars, amino acids).
2. Fermentation of hydrolysed organic substrates into simplest forms e.g. organic acids.
3. Methanogenic production of methane from simple substrates by methanogenic bacteria under anaerobic conditions.

Anaerobic digestion is carried out in a digester, which is a brick-lined or concrete-lined chamber covered completely to prevent the entry of air.

Aerobic Digestion

Aerobic digestion involves the conversion of organic substrates by micro-organisms into utillizable forms in the presence of air, for example composting (biological decomposition of organic wastes/residues under controlled conditions to result in release of C, N, P, K, etc.) and oxidation systems (of sewage in oxidation ponds by bacteria and algae) to produce gases, single cell protein, fertilizers, etc.

Hydrolysis of Wood Ethanol

It is the process in which biological matter is converted into liquid fuel via hydrolysis of woody material from sugars, followed by fermentation to alcohol. It can be done in the form of hydrolysis by:

1. dilute acid, high temperature 180-200° C
2. concentrated acid, and
3. enzymatic process.

The enzymatic process is mainly concerned with microorganism which involve enzymatic or bacterial break down by microorganisms at relatively low temperature. This technique is useful for production of methane to biogas from a wide variety of plant, animal, human and industrial wastes during the process of anaerobic digestion. The process is known as fermentation. Microorganisms have, therefore, an essential role in the production of biological energy. They could also be used directly in certain energy conversion reactions as in case of *Botryococcus braunii,* to mass culture, with a view to produce hydrocarbons.

Ethanol Production for Energy

Brazilian government is the pioneer in the field of fuel alcohol production from yeast and sugarcane plants whose work is being followed as ambitious programmes in America, Canada, Australia, New Zealand, Indonesia, Kenya, and Europe. In Brazil, the Brazilian National Alcohol Programme (also known as Gas alcohol programme) was started in the first decade of the programme, the sugar cane production was raised to 30 per cent using conventional plant breeding techniques. Then the fermenter was also designed which reduced the time of fermentation from 24 hours to 6 hours. At present 2500,000,000 gallons of alcohol is produced every year, All Brazilian cars now run either on pure alcohol (95%) or on an alcohol-petrol mixture.

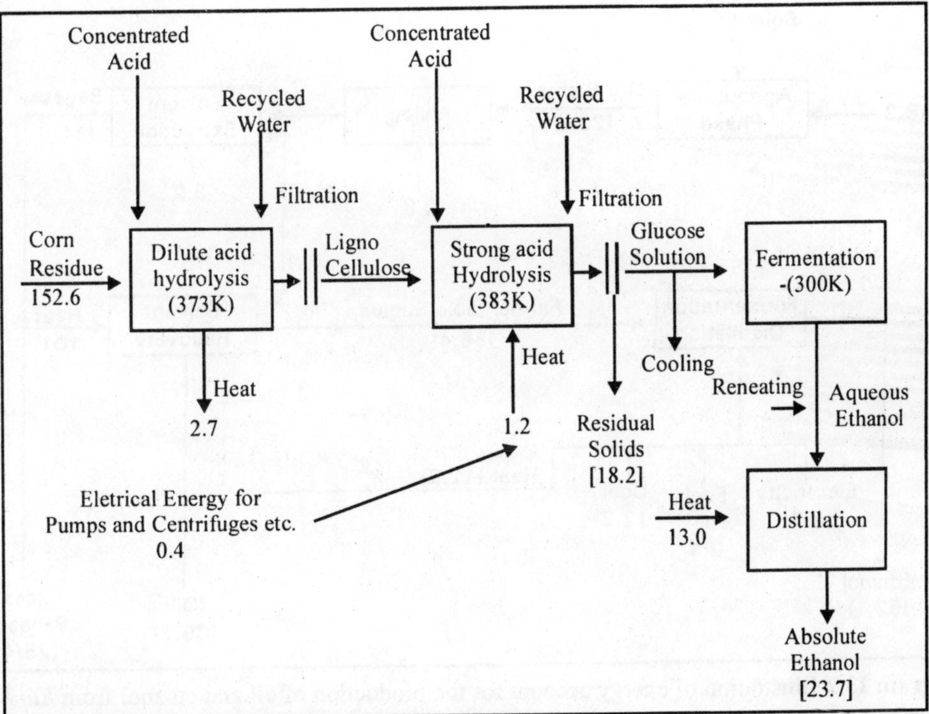

Diagram 23.1 Energy account for the production of 1 dm^3 of ethanol from the acid hydrolysis of corn residues. The unbracketed numbers are energy inputs, while the bracketed numbers are energy contents, both in MJ.

Due to high rise of oil prices, led the Brazilian government to assign a first priority to sugar cane production programme and, from 1980 to 1985, the production reached to its level of satisfaction. In United States, ethanol production has now reached to 719,000 tones (not counting alcohol beverages). The mixture of 6 to 9 parts of petrol for one part of ethanol *(gasohol)* is now commercially used in United States at about 900 service stations. It has been estimated that replacing all the petrol consumed in USA by gasohol would require a production of at least 50.6 billion liters of ethanol per year. Surplus cereals could be used for such production. The utilization of cereals in ethanol production is common in Middle West, for example, Archers Daniels Midland, located at Des Monines, Iowa (USA) has annual production capacity of 0.8 million tones alcohol. For 1995, about 50 million tones of such production is estimated.

In France, in the Bouches-du-Rhone department much research work is going on for production of alcohol from sugarbeet, Jerusalem artichoke or sugarcane. However, in 1980 the price of ethanol was 3 to 5 times higher than petrol. The Jerusalem artichoke appeared to be superior to other energy crops, due to the genetic improvement of the variety and produces large quantities of haulms which could meet the heating requirements of distillation. In France, active research is in action at Institut Francais du Petrol where extraction of inulin is carried out, as well as on the hydrolysis of this carbohydrate, and on acetonebutanol fermentation.

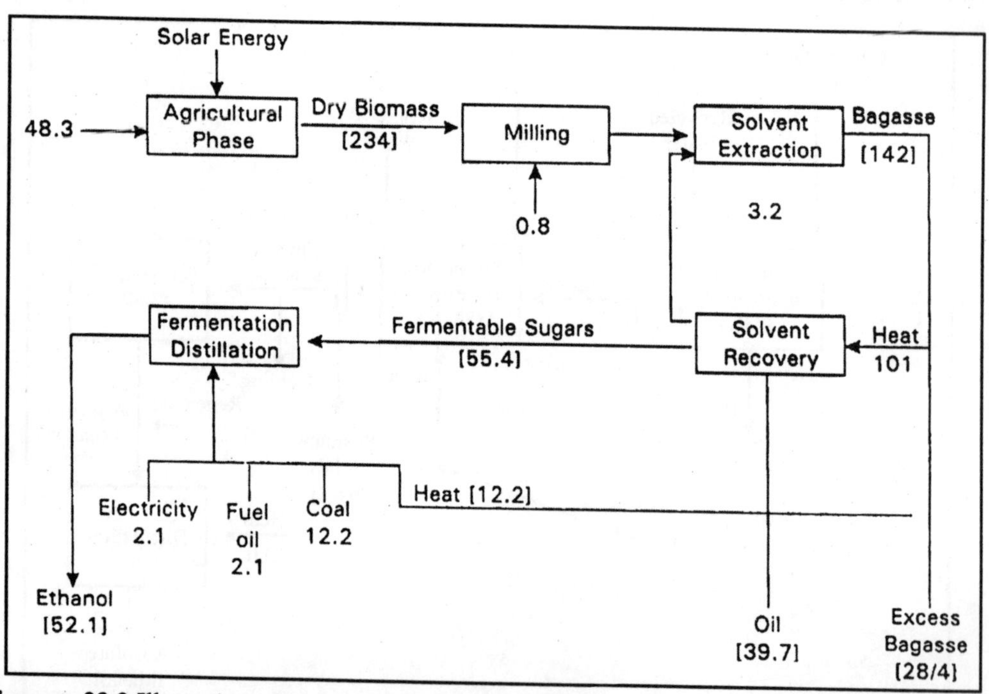

Diagram 23.2 Illustration of energy account for the production of oil and ethanol from *Euphorbia lathyris*.

In 1980, India produced 6 million hectoliters of ethanol from the fermentation of molasses, of which 80,000 hectoliters were used by the chemical industry.

Before starting industrial production of above cited crops in India, the farmers will have to be trained for this new energy source and government have to liberalise its policy of alcohol production out of these crops.

In Japan, production of fuel alcohol from sugarcane is of prime importance. Long term objective has been started to meet one-third of the petrol needs for the country. Government of Japan has envisaged co-operation with South-East Asia to build plants for production of fuel alcohol products. According to one report, an annual production of 100 million kiloliter of fuel alcohols will be produced by such co-operation.

Table 23.4 Alcohol production from some important crops.

Crop	Alcohol yield lit/ha.	Crop yield lit/tone	Alcohol production tone/ha.
Sweet Sorghum	60	1.32	80
Sugarcane	70	56.04	3923
Sugarbeet	110	30.21	3323
Sweet potatoes	125	8.36	1045
Cassava	180	8.70	1575
Molasses	245	-	-
Corn (Maize)	30	3.27	1175

In the Philippines, ten plants have been established for large scale production of alcohol. In Australia, alcohol is being produced from beet sugar or canesugar.

Conversion of Methane into the Synthetic Gas

There are two main routes through which methane can be converted into gases containing carbon monoxide and hydrogen, partial combustion and steam reforming. Partial combustion of methane, using a limited supply of oxygen, can be effected with or without the presence of a solid catalyst and the main part of the process is represented by the equation

$$CH_4 + \tfrac{1}{4} O_2 \rightarrow CO + 2H_2$$

The typical composition of the gas resulting from this process, expressed in the percentage by volume of each gas in the dried product, is 61% (hydrogen), 34.5% (carbon monoxide), 3% (carbon dioxide) and 0.5% unreacted methane. The energy content of the product gas is approximately 65% of that of the original methane.

Steam reformation of methane takes place at elevated temperatures in the presence of a solid catalyst, typically based on nickel. The central process is represented by the equation:

$$CH_4 + H_2O \rightarrow CO + 3H_2 \qquad \text{reaction I}$$

Which has an endothermicity of 206.1 kj mol[6]. Also involved is the water-gas shift reaction represented by equation:

$$CO + H_2O \rightarrow CO_2 + H_2 \qquad \text{reaction II}$$

The final composition of the product gas is governed by the chemical equilibria corresponding to equations (I) and (II). Under the typical operating conditions of a pressure of 1.2 MPa and temperature of 1088 K, the gas composition by volume is 14% (carbon monoxide), 14% (carbon dioxide), 71% (hydrogen) and 1.6 % (methane).

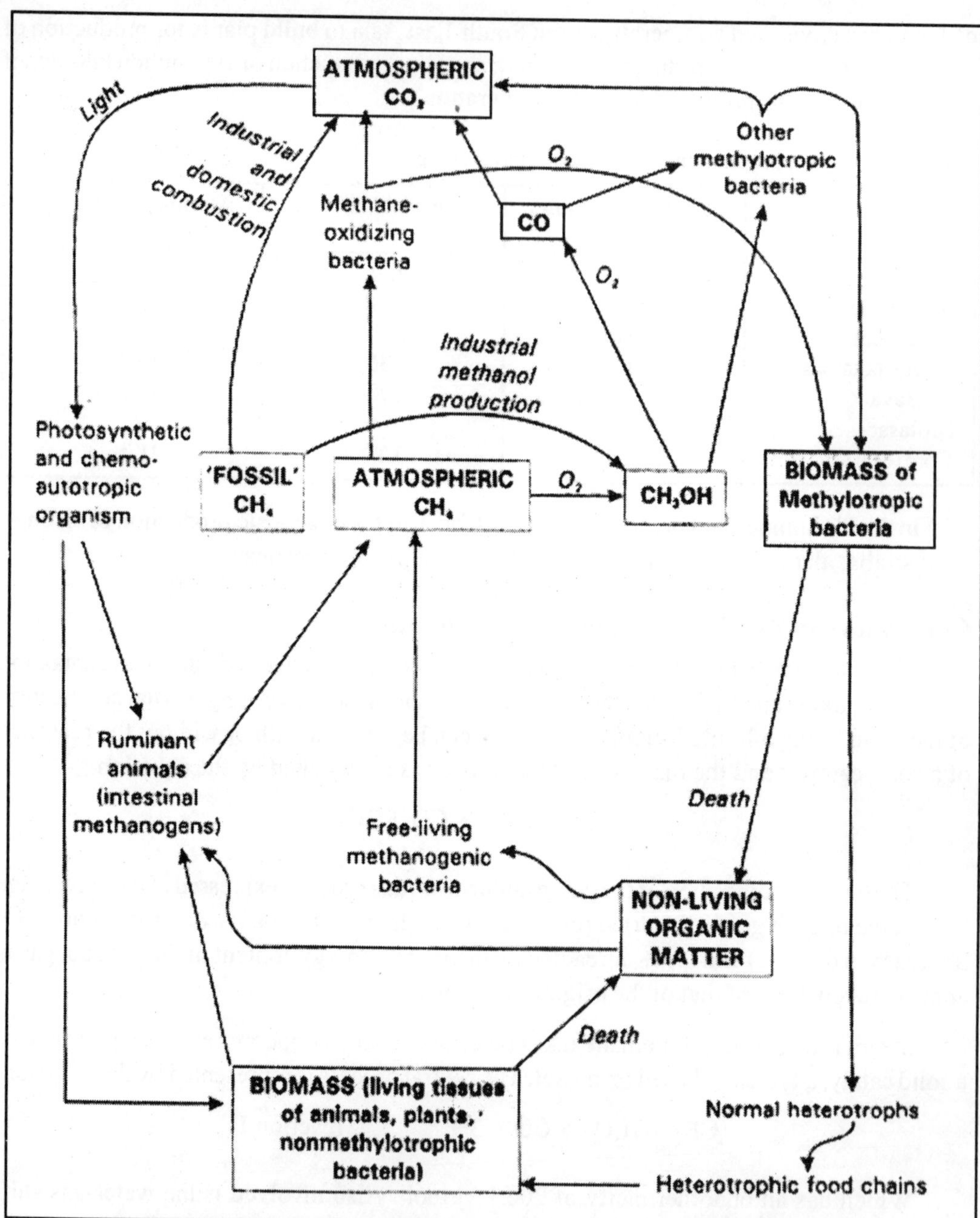

Diagram 23.3 Diagrammatic presentation of methane cycle in nature.

Factors Affecting Methane Formation

1. Addition of Algae: Ramamoorthy and Sulochana (1989) have found an enhancement in biogas production from cow dung on addition of the green algae, *Zygogonium* sp. The amount of biogas produced from the algae was twice (344 ml/g dry algae) of that obtained from cow dung (179/g dry cow dung) alone. Also, the duration of gas evolution

increased with increasing the proportion of slurry. Addition of algae holds promise to get biogas in sufficient amount even in winter season also.

2. Carbon-nitrogen (C:N) Ratio: Maximum digestion occurs when C:N ratio is 30:1. Amendment of nitrogen or carbon substrates should be done exogenously according to chemical nature of substrates used in fermentation.

3. Creation of Anaerobic Conditions: It is obvious that methane production takes place in strictly anaerobic condition, therefore, the digesters should be totally air tight. In Indian conditions, digesters are burried in soil.

4. Nitrogen Concentration:Excess amount of nitrogen inhibits growth of bacteria, and thereby lowers methane production. Therefore, use of such materials should be discouraged.

5. pH: For the production of sufficient amount of methane, optimum pH of digester should be maintained between 6-8 as the medium lowers methane formation.

6. Seeding: In the beginning seeding of slurry with small amount of sludge of another digester activates methane evolution. Sludge contains apetogenic and methanogenic bacteria.

7. Slurry: Proper solubilization of organic materials (the ratio between solid and water) should be 1:1 when it is house hold things.

8. Temperature: Fluctuation in temperature reduces methane formation, because of inhibition in growth of methanogens. In case of mesophilic digestion, temperature should be between 30°C, and 40°C, but in case of thermophilic ones, it should be between 50°C and 60°C.

The Experience of India and China

Basic and applied research is carried out, notably at the Biochemical Engineering Research Centre of the Indian Institute of Technology. It has been estimated that if 300 million tonnes of dry matter contained in the cow dung produced annually (1979), then it is possible to generate energy equivalent to 33 million tonnes of oil. India's 6th Five-year plan (1980-84) provided for the setting up of a million digesters designed to meet the energy needs of families, which corresponded to a budget of 55 million. With the developments in India, biogas can replace diesel to the extent of 80 per cent in a typical dual fuel engine without any significant modifications to the engine.

The potential of municipal sewage in India and its future prospects are described elsewhere . New techniques have been developed to make a available sullage gas for cooking purpose or industrial activities. The Okhla Sewage Disposal Works (New Delhi) receives 25 m^3 sewage per second and generates 18,000 m^3 of gas. In energy terms, it is equivalent to about 700 families of the surrounding villages. It is hoped, if whole sewage generated from the city is digested anaerobically to yield biogas, it could meet about 20% fuel demand of the city.

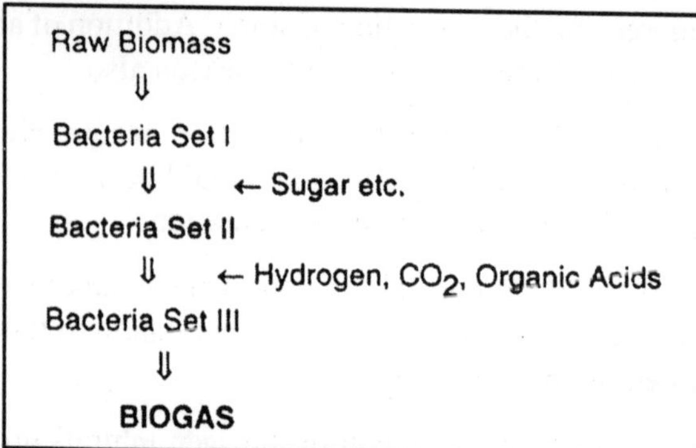

Diagram 23.4 Schematic representation of the stages of methanogenesis.

Table 23.5 Biogas plants installed up to March 31, 1992

Type	Number of plants installed
Family size biogas plants	15,75.000
Community biogas plants	401
Institutional biogas plants	359
Night-soil biogas plants	59

The Gobar Gas Scheme, a development and popularization agency, provided producers of biogas with technical assistance and is responsible for the allocation of funds for construction of digesters, provide 3-5 head of cattle, with the view to meet energy needs of villages.

According to the New China News Agency, 7.15 million biogas installations have been made at the end of 1978, arid current reports reveal that upto 1985 China have installed about 70 million of biogas plants. Government of India is also supporting in establishing regional and local offices which are responsible to establish biogas plants in every village. We are hopeful to meet out the demand of energy in near future through gobar gas.

Energy and Fuel Using Microorganisms

Hydrogen as Energy Source

Molecular hydrogen is an attractive *energy vector.* It is naturally occurring - it has to be manufactured, i.e., it is synthetic fuel. Hydrogen is the cleanest fuel, producing only water when burnt correctly, with no carbon monooxide, carbon dioxide, hydrocarbons, or solid residues formed on combustion. Hydrogen is much used in chemicals industry. Hydrogen is readily transported and stored, is easily ignited and burns smoothly and evenly in properly designed open-flame burners, both on small domestic appliances and on large industrial units. It is also well suited for catalytic oxidation. Hydrogen is now made mainly from natural gas, but it costs approximately the same to make hydrogen from coal as to make methane from coal. Hydrogen may act as a bridge between the fossil fuel age and a nuclear or solar era. Almost all types of energy source can be used in a hydrogen-energy system.

Production of Hydrogen

Most of the hydrogen produced industrial derives from fossil fuels, although some is produced by electrolysis of water.

From Fossil Fuels: The brief outline is as follows:

1. Removal of methane and other non-hydrogen constituents from refinery tail gases or coke oven gas at low temperatures
2. Reforming of natural gas (or other hydrocarbons)

$$CH_4 + H_2O \rightarrow CO + 3H_2$$

Followed by water-gas shift

$$CO + H_2O \rightarrow CO_2 + H_2$$

This is followed by CO_2 removal using physical or chemical absorption techniques

3. Direct production of synthesis gas by reaction of coal with oxygen and steam

$$3C + O_2 + H_2O \rightarrow CO + H_2$$

followed by water-gas shift and CO_2 removal

4. Partial oxidation of hydrocarbons

$$CH_4 + ½ O_2 \rightarrow CO + H_2$$

followed by water-gas shift and CO_2 removal

Of these methods, that of (2) is the most widely used.

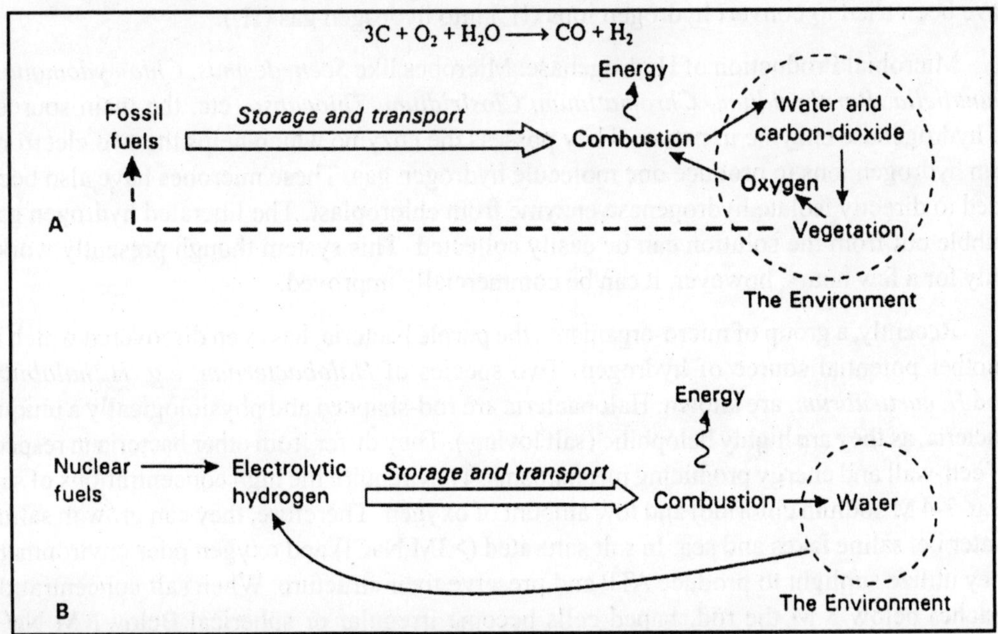

Diagram 23.5 The environmental effects of the fossil fuel and hydrogen fuel cycles, (a) The present energy cycle, (b) The hydrogen energy cycle.

Desulphurized natural gas is steam reformed in the presence of a nickel catalyst. After cooling to about 375° C, the product gases undergo the water-gas shift reaction, usually with an iron/chromium catalyst.

$$CH_4 + H_2O \rightarrow CO + 3H_2O$$
$$H_2O \downarrow$$
$$H_2 + CO_2$$
$$\text{Overall} : CH_4 + H_2O \rightarrow 4H_2O + CO_2$$

A second shift reaction may be carried out at about 200° C over a copper/zinc catalyst. The CO_2 is removed by physical or chemical adsorption. Overall, the thermal efficiency of the process approaches 70%.

It is clear that it is water which represents the inexhaustible supply of hydrogen in the future; even in the steam reforming of methane, half of the product is derived from added steam.

The electrolysis of water can be achieved by using microorganisms. They split water during the process of photosynthesis (photolysis of water). The photosynthetic machinery of green plants are being used for this purpose. The chlorophyll helps in trapping solar energy in the form of photon which is converted into ATP. In this process water is subjected to photolysis, which leads to splitting of water molecule into oxygen, electrons and hydrogen ions (H^+). The hydrogen ions donot get a chance to form hydrogen gas, but is used to form energy rich compounds like glucose. However, if these hydrogen ions can be converted into hydrogen gas (H_2) the later can be collected and used as fuel. Enzyme hydrogenase and nitrogenase have been tried to convert hydrogen ions (H^+) into hydrogen gas (H_2).

Microbial Production of Hydrogenase: Microbes like *Scenedesmus, Chlamydomonas, Dunaliella, Porphyridium, Chromatinum, Clostridium, Thiocapsa,* etc. the main sources of hydrogenase enzyme in nature. They possess the enzyme which helps the two electrons join hydrogen ions to produce one molecule hydrogen gas. These microbes have also been used to directly isolate hydrogenase enzyme from chloroplast. The liberated hydrogen gas bubble out from the solution can be easily collected. This system though presently works only for a few hours, however, it can be commercially improved.

Recently, a group of micro-organism., the purple bacteria, has been discovered which is another potential source of hydrogen. Two species of *Halobacterium, e.g. H. halobium* and *H. curtirubrum,* are known. Halobacteria are rod-shapeed and physiologically a unique bacteria, as they are highly halophilic (salt loving). They differ from other bacteria in respect of cell-wall and energy producing mechanisms. They require the high concentrations of salt (e.g. 3-4 M sodium chloride) and low amount of oxygen. Therefore, they can grow in saline water i.e. saline lakes and sea. In salt saturated (>3M NaCl) and oxygen poor environment they utilize sunlight to produce ATP and preserve their structure. When salt concentration reaches below 3 M the rod.shaped cells become irregular or spherical Below1 M NaCl helps in growth and intergrity of the cells. No salt, like sodium chloride has been found to support growth of the cells.

Microbial Production of Nitrogenase: In cyanobacteria *(Anabaena, Nostoc, Rivitlaria, etc.),* nitrogenase enzyme is the main hydrogen producing enzyme. Hydrogenase enzyme is sensitive to oxygen which is produced during photolysis of water where as nitrogenase enzyme is relatively less sensitive to oxygen. In heterocyst (nitrogen fixing organ in cyanobacteria) the concentration of oxygen is low or even nil. Therefore, the production of hydrogen becomes more efficient. The enzyme nitrogenase reduces nitrogen in to ammonia with associated release of hydrogen gas. The enzyme nitrogenase reduces nitrogen in to ammonia with associated release of hydrogen gas. The production of hydrogen gas depends on light, which provides electron donors and also activates nitrogenase and ATP formation in heterocyst.

From the biological point of view, the production of fuels and chemicals from biomass is still in an early stage of development and could profit from the screening of a large number of species of plants, algae, fungi, and bacteria for their potential Any such screening effort must be global in scope, since the organisms of potential interest are very unlikely to be encountered only in the natural ecosystems of developed countries.

Petroplants

Pertroplants or petroleum plants accumulate the photosynthetic products (hydrocarbons) of high molecular weight (10,000). They were reported by Dr. M. Calvin (1979) of the University of California. He suggested that these plants may be used as substitute for conventional petroleum sources. Calvin and coworkers screened most of the plants of Family Euphorbiaceae, especially *Euphorbia* (containing 2,000 species) which reduce CO_2 beyond the carbohydrates. About 400 plant species, belonging to different families are known which grow in different part of the country. The other plant families which have been evaluated are Asclepiadaceae, Apocyanaceae, Leguminosae, Moraceae, Dipterocarpaceae, Compositate, etc. It is hoped that petroplants can yield petroleum more than 40-45 barrel/acre. Petroplants have laticiferous canals in their stem and secrete a milky latex. The latex can be either continuously tapped like *Hevea* latex and stored or extracted from the biomass by using the organic solvents.

The product rich in hydrocrackable hydrocarbon and named as biocrudc which yields about 70.0% energy, and out of which 22% as kerosene and 44.6% as gasoline.

Hevea Rubber

Rubber plant, *(Hevea brasiliensis)* commonly known as *Hevea* rubber is the principal source of rubber which is restricted in distribution in South-East Asia. This plant meets one third of total world demand of rubber. The synthetic rubber elastomers from petroleum have not replaced the demand of natural rubber, due to its low cost. Rubber is trapped from stem of trees by making incision and collecting the latex from it. The latex is further processed to get rubber.

Plant group/families	Common names	Botanical names
Algae (Chlorophyta)		*Botryococcus* sp., *Chlorlla pyrenoidosa*
Asclepiadaceae	Aak	*Calotropis procera*
Compositae	Guayule Russian dandelion	*Parthenium arentatum. Taraxacum koksaghyz*
Dipterocarpaceae	Gurjun	*Dipterocarpus turbinatus*
Euphorbiaceae	Hevea rubber, Rubber plant, Sehund	*Hevea brasiliensis, Euphorbia abyssinica, E.resinifera. E. lathyris, E. tirucalli*
Leguminoseae	Samprani	*Copaifera langsdorfii, C. mutijuga, Hardwickia pinnata*
Myrisiicaceae		*Dialynthera otoba*
Pittosporaceae		*Pinosporum resiniferum*

Euphorbia

In Italy, Euphorbia Gasoline Refinery has been set up to tap vegetative gasoline. *Euphorbia lathyria* is an annual herb and *E. tirucalli* is a perennial one. *E. lathyris* can produce 20 tonne dry matter/ha/yr. Chemical analysis of this plant in organic solvents revealed that heptan extract and either soluble fraction constituted about 8% terperoid extract.

Guayule and Russian Dandelion

Guayule *(Partheniuum argentatum)* and *Taraxacum koksaghyz* are the members of family Compositae. These are sources of rubber. Guayule is a shrub and indigenous to North Central Mexico and South-West U.S.A. Guayule generally grows in arid, semi-arid and desert areas. The U.S. Government encouraged the cultivation of this plant after World War II to reform the economy of the country. It can tolerate temperature ranging from 32-38°C, and can grow in Indian conditions. Like Hevea guayule contains cis-polyisoprene and identical physical properties.

Aak (Calotropis procera)

Aak belongs to family Asclepiadaceae. It is a shrub of 1-2.5 meters in hight. It occurs in hot and dry regions of India on waste dry places, river beds, roadsides and forest clearings. It secretes latex which causes irritation to skin. Latex contains high amount of extractable hydrocarbons. The ratio of C,H,0 in the hexane extract has been found as 78.03%, 11.22% and 10.71% respectively. The ratio of C and H is similar to crude oil, fuel oil and gasoline. Hydrocarbon yield and energy value of C. procera are comparable to those of *E. lathyris*. Therefore, this plant can be used as a substitute of petroleum. Researches on it are being done at the Central Arid Zone Research Institute, Jodhpur.

Algal Hydrocarbons

Dead algal scum of *Boiryococcus braunii,* an unicellular alga of Chlorococcales of green algae, contains about 70% hydrocarbons. Percentage of hydrocarbon may vary. This alga is composed of proteins, carbohydrates, and lipids, the percentage of which varies. However, it has proved to be a source of hydrocarbons. As a result of metabolic activity, the hydrocarbons are synthesized during growth phase of the alga.

The algal hydrocarbons closely resemble the crude oil, and therefore, can be used as a good source of direct production of hydrocarbons. *B. braunii* grows in fresh or brakish water as well as in tropical and temperate zones. When in full growth, it becomes apparent in water as the small dots. The alga appears in two forms, as far as pigmentation and structure of synthesized hydrocarbons are concerned. The first form is of green colour and contains linear hydrocarbons with an odd number of carbon atom (25-31) low in double bonds. The second form of alga is red in colour which contains hydrocarbons with 34-38 carbon atoms and several double bonds, the *botryococcenes*. This alga accumulates hydrocarbon as globules on outer walls and cytoplasm of the cells. On cell, wall, a major portion of hydrocarbon (95%) is located, whereas a small amount (0.7%) of it is present within the cells. Hydrocarbons are recovered from the cells by centrifugation.

In addition, *Chlorella pyrenoidosa,a* fresh water alga, is known to be converted into hydrocarbons as golden liquid. Hydrogenation is done in a steel reactor at high temperature (> 400°C) and pressure (12,000 p.s.i., pound per square inch) in the persence of a catalyst (cobalt molybdate). The alga is suspended in a mineral oil in the reactor. Hydrogenation is carried out for about one hour. Consequently, 50% of algal biomass is converted into oil with a little amount (12-14%) of a byproduct, ammonium carbonate. Oil is a clear golden liquid which is separated from the reactor, blended with light gas oil in refineries and processed before its use.

□□□

CHAPTER-24

Biosensors, Biochips, Biofilms and Biosurfactents

Biosensors

The sense organs are natural biosensors such as primarily the chemical sensors of smell and taste. Biosensors technically in their various forms share a reliance of biological materials as sensing elements. Technical biosensors have been under intense development since the middle of 1960 with the prospects of commercial potential offered by biotechnology. These are the combinations of biologically active material displaying characteristic specifically with chemical or electronic sensor to convert the response into electrical signals.

Biosensors are, in fact, biocatalysts which may be purified enzyme, antibody or as whole microbial cell or as an organelle. A biosensor used as an immobolised biological molecule (usually an enzyme or an antibody) or a whole microbial cell to detect or *sense* a particular substance. The biosensor does this by reacting specifically with the substance to be detected (hence the use of enzymes or antibodies) to give a product which is used to generate an electrical signal by means of a device called transducer.

The response of biosensor is measured in terms of substrate used or product form. They are of different types like, carbon electrode, glucose electrode, ion sensitive electrode, photocell, oxygen electrode, adenosine electrode, etc.

For example a glucose electrode is constructed by immobilising a layer of glucose oxidase in polyacrylamide gel around a platinum oxygen electrode. When a solution of glucose is brought into contact with electrode, glucose and oxygen diffuses into an enzyme layer and are converted into gluconolactone and hydrogen peroxide lowering the oxygen concentration around the electrode. O_2 concentration read by electrode is proportional to glucose concentration in the sample. The effective control of the rate of reaction is ensured by high enzyme loading and limited diffusion of substrate.

Biosensor technology is progressing very fast on the front of techniques, their applications specially in the fields of analytic medicine, industry and environment, it is sometimes also useful in monitoring the presence of specific chemicals both accurately and rapidly.

1. *Analysis of Organic Compounds:* Analysis of organic compounds in fermentation and other samples can be done with the help of biosensor. Compounds such as glucose, acetic acid, lactic acid, formic acid, alcohol, methane, glutemic acid, cephalosporin antibiotic, nystatine antibiotic, nicotinic acid, vitamin B_1 etc., can be analysed.

2. *Medical Sciences:* Hepatitis antigens found in blood during infection by the virus can be detected. Abnormal amounts of urea in blood or urine in kidney diseases can be measured. Hormone gonadotropin produced during pregnancy can be tested and measured. High concentrations of creatinine produced after heart attack can be analysed.
3. *Industrial Uses:* They are used to know about thc nature of industrial products of acids, alcohols, phenols, pollutants, etc. The industrial workers can know the presence and concentrations of hazardous chemicals in the environment surrounding them. Biosensors can detect various chemical warfare agents, nerve gas, etc.
4. *Environmental Analysis (Protection):* They are used in analysis of BOD requirements, ammonia, nitrite, sulfite measurements, etc.

Biosensor technology is changing so rapidly that potentially commercializable devices quickly become obsolete as new technology emerges. Very few biosensor devices are likely to be universally applicable and thus the most appropriate sensor technology has to be linked to the market need. There has been some developments in biosensor research in the country among the institutions such as NPA, New Delhi; CECRI, Karaikud; IACS, Calcutta; TIFR, Bombay; IIT, New Delhi; CSIO, Chandigarh; CEER1, Pilani.

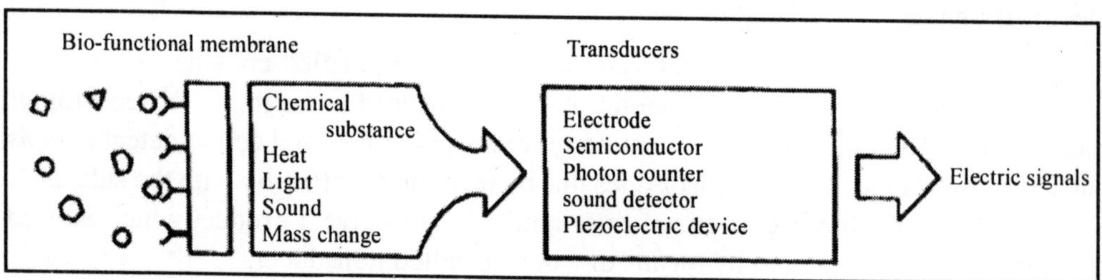

Diagram 24.1 Biosensor Principles

Conventional Biosensor

Biosensors are composed of a biofunctional material and a transducer and have been developed and applied to analytical fields, clinical analysis, food industry and environmental measurements. Biosensors have their roots in military research, as means of detecting nerve gases and other chemical warfare toxins. Their applications have branched out to include simple to use *alternate site* diagnostic devices for home, doctor's office, or drug use screening; medical and surgical monitors (small enough to fit inside a blood vessel) and environmental quality monitors.

Immobilized enzymes, microorganisms and antibodies are used as molecular recognition materials. Electrochemical devices have often been used for transducers. Various enzymes have been used as molecular recognition elements. An enzyme electrode is composed of an enzyme immobilized membrane and an electrode. The principle of an enzyme electrode is based on the detection of electroactive species produced or consumed by the enzyme reaction. For example, a conventional glucose sensor is composed of glucose oxidasc (GOD) and an electrode. GOD oxidizes glucose with the consumption of oxygen and produces gluconolaction and hydrogen peroxide. Measuring the consumption of oxygen with an oxygen electrode or production of hydrogen peroxide with hydrogen peroxide electrode, the concentration of glucose can be determined. This type of glucose sensor is in commercial use for the diagnosis of diabetes. There are many kinds of biosensors using the same principle and devices, and which are developed and used in the fields of clinical analysis and measurement of foodstuffs.

Microbial Biosensor

Microorganisms have also been utilized as molecular recognition elements. A microbial sensor consists of a microorganism immobilized membrane and an electrode. Various kinds of microbial sensors have been developed and applied to the measurement of biological compounds. The principle of a microbial sensor is based on either the change of respiration or the amount of produced metabolites as the result of assimilation of substrates by microorganisms. Furthermore, the use of auxotrophic mutants can selectively determine many kinds of substances. For example, the vitamin B12 sensor was constructed by using immobilized *Escherichia coli* 215. The *E. coli* 215 strain requires vitamin B12 for its growth. The linear relationship was obtained in the range between 5 x 109 and 25 x 109 g/ml. Within 25 days, the decrease in the response was approximately only 8 per cent.

Recently, microbial sensors using thermophilic bacteria have been developed. The use of thermophilic bacteria can possibly reduce contamination of other microorganisms by the use of high temperatures to obtain long term stability. For example, BOD and carbon dioxide sensors are constructed by using thermophilic bacteria isolated from a hot spring. Good linear correlation was observed between the BOD sensor response and BOD value in the range 1 to 10 mg/1 BOD (JIS) at 50°C. The Sensor signal was stable and reproducible for more than 40 days. For the carbon dioxide sensor, a linear relationship was obtained in $NaHCO_3$ concentration between 1 and 8 m M at 50°C and the response time was 5 to 10 min. The linear relationship was also observed in the CO_2 concentration range 3 to 8 per cent.

Microbiosensors have many advantages, as mantioned below

(i) Implantation in the human body and are suitable to *in vivo* measurement.

(ii) Can be integrated on one chip and are useful for measuring various substrates in a small amount of sample solution simultaneously.

(iii) Since semiconductor fabrication technology is applied to microbiosensors, it is possible to develop disposable transducers for biosensors through mass production.

Development of Microbioscnsor: Microbiosensors are based on ionsensitive field effect transistor (ISFET) and were first reported by Bergveld (1970). Matsuo et al. (1974) improved the ISFET using silicon nitride as the gate insulator to construct micro pH sensitive devices. They show rapid response, low power consumption, low noise and no need of a high impedance amplifier. The general circuit diagram for measuring the gate output voltage is shown in below.

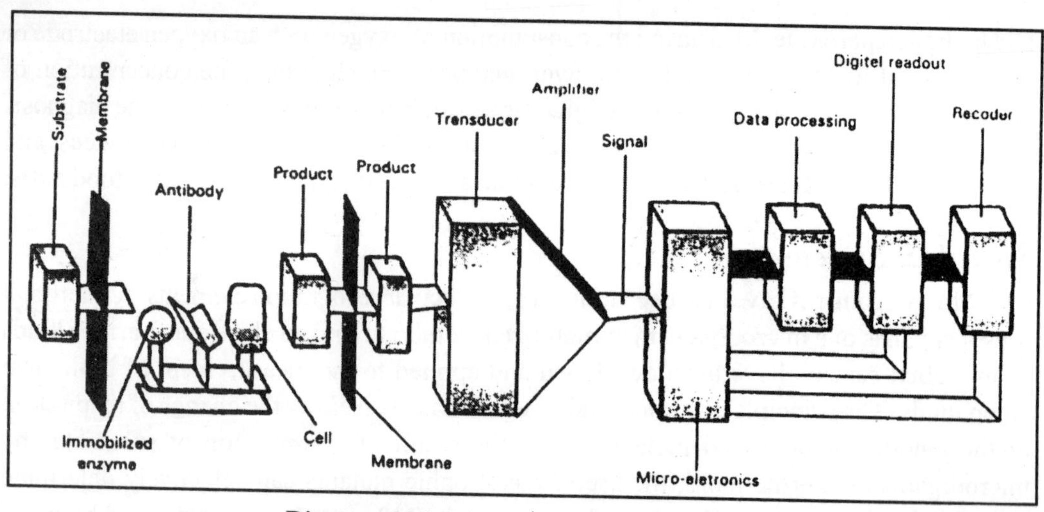

Diagram 24.2 Schematic outline of biosensor

In this circuit, the voltage between the source and drain is controlled constantly and the current between source and drain is also held constantly. The Ag/AgCl electrode is immersed in the same solution. The surface potential on the silicon nitride of the ISFET is affected by pH of the solution, with concomitant change in the gate voltage, which is proportional to the change in surface potential. Therefore, the surface potential change in the ISFET, caused by the change of pH, can be measured as the change in the gate output voltage.

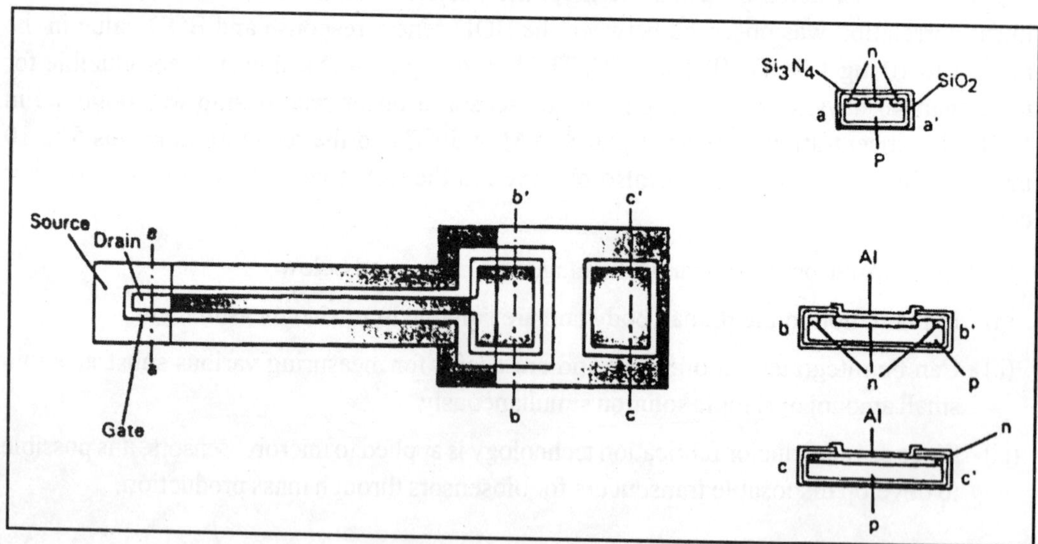

Diagram 24.3 Structure of ISFET

Fabrication: ISFETs are fabricated by using semiconductor technology. Hence, it is easy to miniaturize and integrate ISFETs on one chip. ISFET is used as a potentiometric transducer, therefore, enzymes which cause pH changes in its reaction, such as urease and oxidase, can be used.

Urea Sensor

A urea sensor consisted of a urease immobilized membrane and a pH electrode. Urease catalyzed reaction cause pH changes, so that ISFET can be used as a transducer. A micro urease sensor is fabricated as follows:

An ISFHT was laid inside a vacuum chamber and 3 amino propyltriethoxysilane (3Aptes) is vaporized at 80°C and 0.5 Torr for 30 min, followed by glutaraldehyde (GA) treatment under the same conditions. The chemically modified ISFET was covered with cellulose acetate membrane containing 1,8 diamino 4 amino methyloctane and GA. The ISFET was immersed in urease solution. The urea sensor gives the linear relationship between the initial rate of the output gate voltage and the logarithm value of urea concentration in the range 16.7 to 167 mM and can be used for 20 days with slight degradation of the enzyme activity.

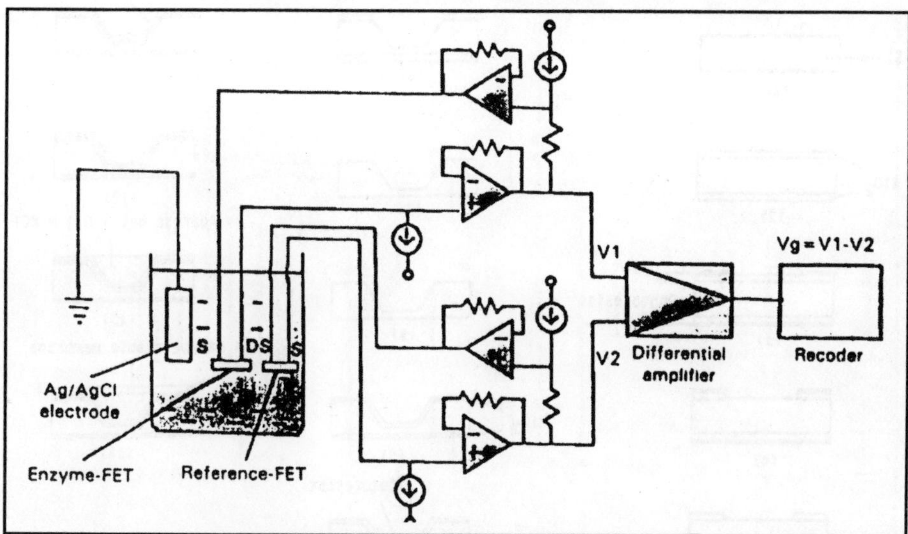

Diagram 24.4 Circuit diagram of the measuring system.

Alcohol Sensor

The study of an alcohol-sensitive microbiosensor using an ISFET and the enzyme system existing in the cell membrane is reported. The cell membrane of acetic acid producing bacteria has a complex enzyme system oxidizing ethanol to acetic acid via acetaldehyde. This system consists of membrane bound alcohol dehydrogenase (ADH), aldehyde dehydrogenase (ALDH) and an electron transfer system. This complex enzyme system, therefore, can be used for application with an ISFET.

Biosensors Using Amorphous Silicon ISFET

The ISFET device can only be manufactured by using a silicon wafer for the substrate. In recent years, devices made from amorphous silicon have received widespread attention because of their great applications potential. Various substrates such as glass and plastics can be used for preparing amorphous silicon and transistors can be fabricated with a number of different structures such as a needle of a syringe.

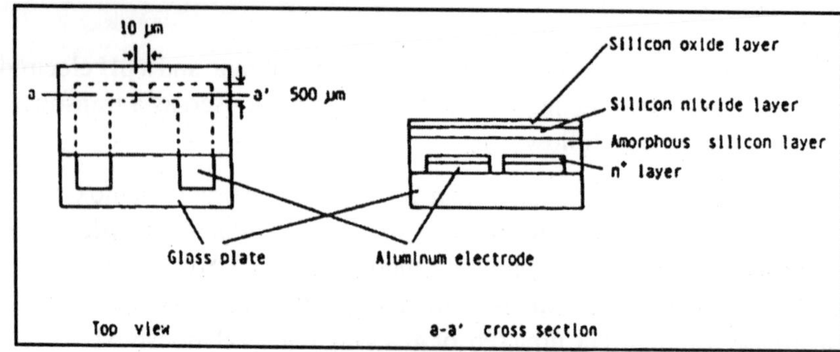

Diagram 24.5 Structure of amorphous silicon ISFET.

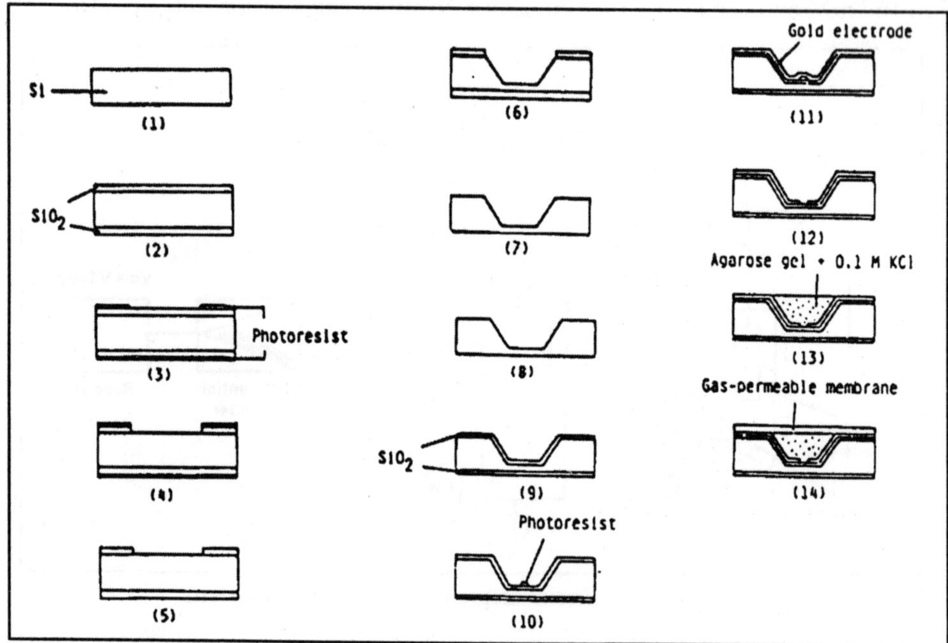

Diagram 24.6 Fabrication process of micro-oxygen electrode.

Hypoxanthine and Inosine Sensor

To maintain quality, evaluation of freshness is important in the fish industry. When a fish dies, adenosine 5' triphosphate (ATP) decomposition in the fish meat occurs and adenosine 5' diphosphate (ADP) and adenosine 5' monophpsphate (AMP) and related compounds are generated

$$\text{ATP} \rightarrow \text{ADP} \rightarrow \text{AMP} \rightarrow \text{IMP} \rightarrow \text{HxR} \rightarrow \text{Hx} \rightarrow \text{X} \rightarrow \text{U}$$

where IMP,HxR, Hx, X, and U stands for inosine 5' monophosphate, inosine, hypoxanthine, xanthine, and uric acid, respectively. Consequently, Hx accumulation with an increase in storage time can be used as an indicator of fish meat freshness. Therefore, simple and rapid methods for the determination of Hx and HxR are required in the seafood industry.

Hypoxanthine Sensor : Hypoxanthine is measured on the basis of the reaction catalyzed by Xanthine oxidase (XO). The pH change caused by uric acid is detected by using aSi ISFET.

Inosine Sensor: Inosine sensor is fabricated similarly to the hypoxanthine sensor by using nucleoside phosphorylase and XO co-immobilized on aSi ISFET simultaneously. After 90 seconds from injection of inosine solution, the gate voltage gradually increases and reaches a steady state in approximately 7 min. Xanthine formed by the decomposition of inosine catalyzed by nucleoside phosphorylase is subsequently oxidized to uric acid by XO. The linear relationship was obtained in the range 0.02 to 0.1 mM by plotting the initial rate of the gate voltage change with respect to the logarithm of inosine concentration. The oxidation of hypoxanthine to uric acid by xanthine oxidase is initiated immediately after injection. The response to inosine, however, has a time lag of 90 sec after injection, This phenomenon is attributed to the three step reaction. On the basis of this time lay. this sensor can determine inosine and hypoxanthine simultaneously.

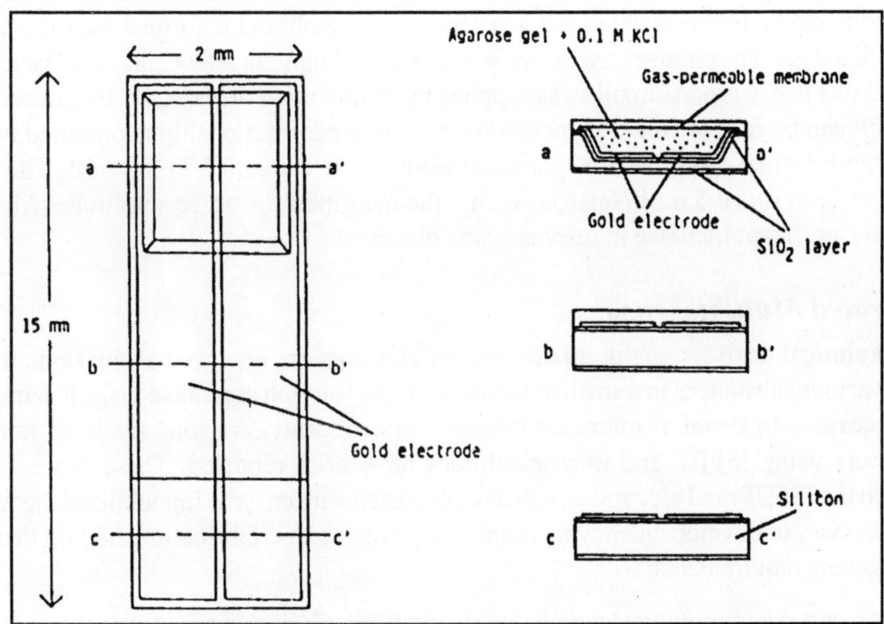

Diagram 24.7 Structure of micro-oxygen electrode

Micro-Oxygen Electrode

Clark-type oxygen electrodes have been applied to various biosensors, immobilizing either enzymes or microorganisms, which catalyze oxidation of biochemical organic compounds. At present, several oxygen electrodes, based on conventional semiconductor technology, have been fabricated by several groups, but they are not yet in mass production.The reason for this is that oxygen electrodes contain electrolyte solution, which results in difficult adhesion of the gas permeable membrane to the substrate, even if epoxy rexin is used. Thus, mass production of such devices is difficult. Recently, miniaturized and integrated biosensors have been required in clinical analysis. (Diag. 24.7).

Glucose and Carbon dioxide Sensors Using Micro-Oxygen Electrode

The glucose sensor is fabricated by immobilizing GOD on the sensitive part of the oxygen electrode by crosslinking with bovine serum albumin (BSA) and GA. The enzyme immobilized membrane is formed by dropping the sensitive part into the mixture containing 2 mg of GOD. 20 μl of 10% BSA solution and 10 μl of 25% GA solution. The glucose sensor responds as soon as the glucose solution is injected into the buffer solution and reaches a steady state in 5 to 10 min. The sensor responded almost linearly for glucose concentrations between 0.2 and 2 mM, which is comparable to conventional glucose sensors.

A microbial CO_2 sensor using this oxygen electrode is constructed by Suzuki et al. 1988). Autotrophic bacterium named as SI7, which can grow with only carbonate as the carbon source is used, (available at Fermentation Research Institute, Japan). Bacterial whole cells are immobilized on a micro-oxygen electrode. The sensitive area of the oxygen electrode is immersed in 0.2% sodium alginate solution containing S17 whole cells, then removed and immediately immersed in 5% $CaCl_2$ solution to form bacteria immobilized calcium alginate gel, The negative photo-resist as the gas permeable membrane is formed over the bacteria immobilized gel. The photo-resist in only exposed to UV light for a few minutes. The response time is 2 to 3 min. Carbon dioxide was supplied by acidification of $NaHCO_3$, the concentration of which can be related to CO_2 concentration. The linear relationship is obtained between the current decrease and $NaHCO_3$ concentration in the range 0.5 to 3.5 mM. The lowest detection limit was 0.5 mM $NaHCO_3$ within the margin of the noise amplitude. Above 3.5 mM, no significant increase in response was observed.

Integrated Multibiosensor

In clinical analysis, about 20 constituent elements are analyzed at the same time. To detect various substances in a small amount of sample solution simultaneously, it is important and necessary to develop micro multibiosensors. Recently, various kinds of integrated biosensors using ISFET and microelectrodes have been reported. These biosensors are based on the ISFETs and electrodes, which were coated with enzyme immobilized membranes. It is necessary to develop an enzyme immobilized membrane fabrication method that meets the following requirements:

1. An enzyme immobilized membrane should be precisely deposited onto a gate region or small working electrode.
2. A deposited membrane should not peel off the sensitive surface area in practical use.
3. Different enzyme membranes can be prepared without mixing.
4. Fabrication processes are applicable to a wafer and compatible with the 1C process.

Novel Biosensor Based on New Transducers

Conventional biosensors have consisted of electrochemical devices. On the other handother transducers such as optical fibers, image sensors, piezoelectric devices, and SAW devices are currently used for biosensors.

Image Sensor

Most clinical analyses are based on the determination of soluble marker substances in body such as blood and urine. The direct analysis in cell or tissue level is greatly important in clinical diagnosis. In the cancer detection, highly sensitive and rapid detection methods for abnormal cells are required. Cell diagnosis is carried out mainly by the visual inspection of trained experts or the use of a flow cytometer.

Recently, much attention has been focused on image analyzing systems composed of an image sensor and a microcomputer system. Image sensors are classified into XY address methods and the charge transfer method. In the XY address method, optical signals of each address are read by switching on the corresponding circuit. The charge transfer method was first demonstrated by using a bucket, bridge device and presently by the more advanced charge coupled device (CCD). Most image sensors in practical use are now of the CCD type.

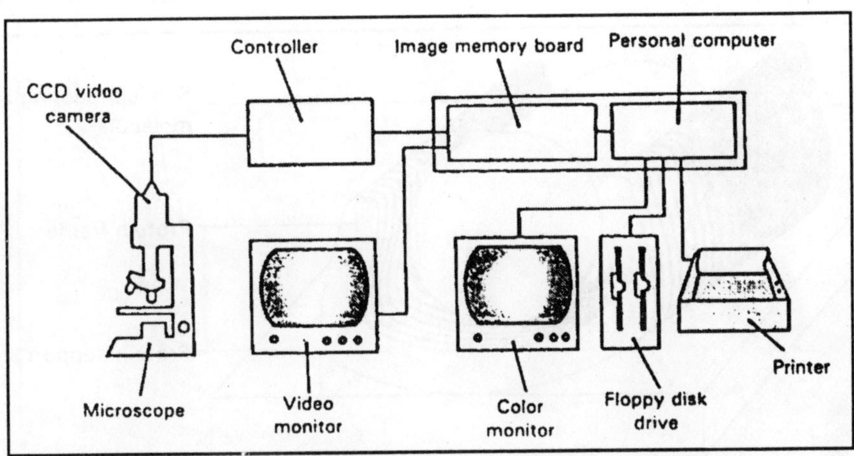

Diagram 24.8 Schematic diagram of the imaging sensor system.

The CCD is an integrated semiconductor chip composed of photodiode arrays and charge transfer circuits. Electric charges accumulated at each photodiode are transferred systematically to the output by controlling the electric potential in the chip. The output pulse height correlates with the brightness at the corresponding photodiode. Thus, a visual image focused on the CCD can be converted to a succession of analog pulses, Since the photodiodes are arranged approximately by 10μm separation, the same degree of image resolution can be obtained. There are many advantages to the solid state CCD image sensor as compared with a conventional vidicon. For example, they are compact in size, have high sensitivity, are distortionless, have no afterimage, have low power-consumption, and they have a long operational life.

Biochips

They are made from different biological materials. Biochips can control the computer by replacing silicon chips. Biomolecular computers thus made, promise to be ten to thousand times smaller than the best super computers with much faster switching times and extremely low power dissipation.

They are made up of semi-conducing molecules inserted into the protein framework and fix the whole on to a protein support. The circuit is about one molecule wide. Proteins are assembled into a predetermined three-dimensional structure. The proteins molecules take the shape similar to electrical circuits. In near future biomolecular computers will become operational.

The applications of minuscule computers based on biochips are varied. They can be used in implanting of several sorts in human body, like regulation of heart beats, responses to nerve impulses by artificial limbs (bearing such computer device), overcoming of blindness and deafness, etc. Biochips may be infected by microbes since they are made up of proteins. Biochips are also damaged if the information of protein digesting enzyme is wiped out. These are some expected dangers and problems in biochips technology.

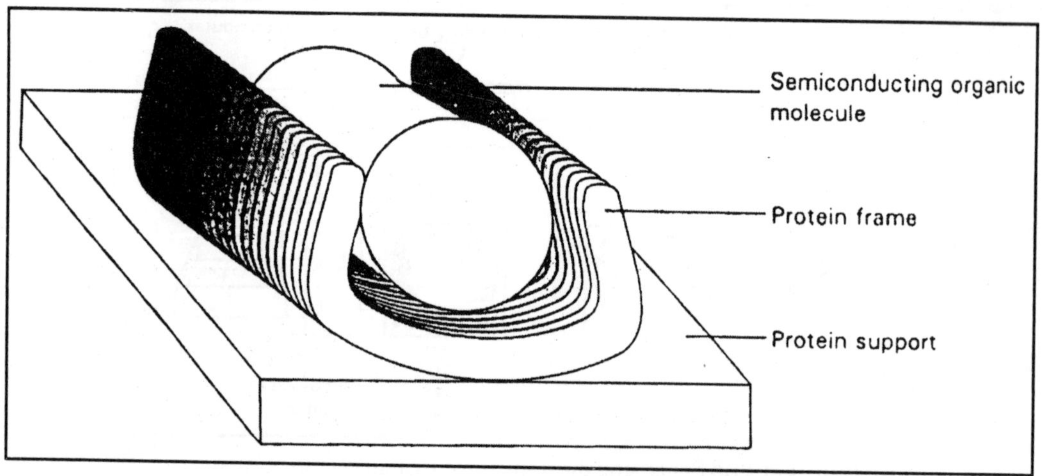

Diagram 24.9 Basic principles behind biochip research

Biofilms

A biofilm is an accumulation of microbial cells and inorganic components held together in a polymeric matrix and firmly attached to a substratum. Accumulation of bio films is encountered in many natural and modulated environments. It may be fundamental to process performance i.e. for fixed-film biological waste water treatment, sudden deterioration of water quality and deterioration of substrata.

Biosurfactents

There has been a great deal of interest in biosurfactents, i.e., surface active compounds produced by microorganisms. Biological surfactants possess a number of potential advantages over their chemically manufactured counterparts, including lower toxicity, biodegradability, a vide variety of possible structures, and ease of synthesis from inexpensive, renewable feed stocks. Consequently, biosurfactents may have applications in numerous areas, particularly for enhance oil recovery and in foods, beverages, cosmetics, pharmaceutical preparations, ,etc.

Biosurfactents, however, suffer from two serious drawbacks. First, the structure of the biosurfactents produced by a given microorganism is genetically determined, thus usually not permitting any variation. Second, the very nature of biosurfactents makes their recovery from the fermentation broth and subsequent purification difficult and costly.

❑❑❑

CHAPTER-25

Biotecnology and Environmental Protection

Over the course often thousand years human have successfully learned to ex ploit ecological system at sustainable level. Humans out of ignorance, short sightedness, greed or desperation have polluted air and water, undermined the productivity of the land through accelerated soil erosion, creeping deserts, increased flooding and decline soil fertility. They, thus, destroy the basis of their own livelihood and they violated the limits of natural systems. The global concern for environmental protection found expression in UN conference on the 'Human Environment', held at Stockholm in June 1992.

Pollution - A Global Problem

Environmental pollution is a global problem. The decline in environmental quality as a consequence of pollution is evidenced by loss of vegetation cover and biological diversity, excessive concentration of harmful chemicals in the ambient atmosphere and in food grains, growing risks of environmental accidents and threats to life support system. Those substances, which cause pollution, are known as pollutants.

Pollutant may be defined as any chemical (radionuclide, organo-phosphorus compound or trace gases) or geo-chemical substance (dust, sediment, grit, etc), biotic component or product (pollens or products of microbial activity), or physical agent (heat, sound, etc.) that released intentionally or inadvertently by man into the environment in such concentration that may have adverse, harmful or unpleasant and inconvenient effects. It is also defined as any solid, liquid or gaseous substance present in such concentration as may be or tend to be injurious to environment [Indian Environment (protection) Act, 1986]. The term pollutant is usually applied to non-living, man-made substance or other nuisances or excess in a specific location. The important pollutants are: Gaseous pollutants [Oxides of nitrogen, sulphur compounds (SO_2, H_2S), carbon compounds (CO_2, CO), ozone, halogens (chlorine, bromine, iodine, etc.) and chloroflurocarbons, fluoride compounds, metals (Mercury, lead, iron, zinc, nickel, tin, cadmium, etc.), agricultural pollutants (pesticides, herbicides, fungicides and fertilizers), complex organic pollutants (benzene, benzpyrens, acetic acid, ether, etc.), biotic component (pollens), deposited matter (soot, smoke, tar, dust, grit, etc.), solid wastes, radioactive wastes, noise, heat, etc. Soot is a black substance, essentially carbon from the incomplete combustion of wood, coal, natural gas, etc., as deposited on the inside of chimneys and other surfaces.

Detergents, pesticides and biocides, chlorofluorocarbons, plastics and plasticizers, solvents, fuel, paints, dyes, medicines, fuel and food additives, etc., are some examples of the multiplicity of chemical products made and disseminated for the benefit of man. All these have the inherent capacity to disturb the life support system.

Biotechnology and Pollution Control: Some Aspects

Unfortunately, biotechnology, being still in the process of development, does not possess a sharp and easily defined form. Only in last 10 years, microbiologists and genetic engineers have made progress, and we are hopeful to solve pollution problems in future. Mineral ore deposits are also becoming more scarce and expensive to recover from earth's crusts. Microorganism can be used to enhance the recovery of metals from low-grade ores and effluents containing undesirable quantities of heavy metals or other toxins. Enzyme technology is an area of considerable current interest and development. Enzymes are biological catalysts and have been used for many years as isolated agents particularly in food industry, e.g., rennin, papain, and invertase. Most of the enzymes are lost in product or the effluent after process; therefore, re-usability is essential to reduce the cost of enzyme production and to improve the economics of enzyme catalysed reactions in industry. This can be possible through biotechnological applications. The conversion of cellulose to sugar is a challenge for chemists for many years. The knowledge of biotechnology has made it possible to elucidate the requirements for biological enzymatic hydrolysis of such materials as straw and wood to a carbohydrate for use as a substrate in fermentation. The important uses are in the field of: enzyme technology, fermention technology, sewage disposal, food industry, waste, heavy metal industry, pesticides and herbicides industry, uranium mining, oil recovery, biodegradation of plastics, etc.

Biotechnology deals with living processes as it is inherently less likely to give rise to major environmental problems. In fact, biotechnology may be helpful to solve existing environmental problems; the ability of microorganisms to bind heavy metals and to detoxify highly toxic substances will help us to solve the problem. The other pollutants like oil spills; pesticides, herbicides, chemical effluents and gases are some of the hazards for the society. Biotechnology can be used as a tool to solve this problem in two ways:

1. The root cause can be attacked by the introduction of biotechnological production methods, which are intrinsically less polluting.
2. Microbes can be deployed as voracious scavengers, removing all kinds of pollutants.

Water Pollution

Water serves as the second natural medium for the growth of microorganisms and stands next to soil. The growth of microorganisms in water mainly depends on the amount of available mineral nutrients and the dissolved oxygen present in it. It has been observed that as the amount of organic matter increases in water, the number of microorganisms also increases but upto certain limit. The number of bacteria and other microbes will always be higher in river passing by thickly populated cities then of the villages, because persons living in cities, are continuously disposing sewage water and other waste products in rivers which contain a very high amount of mineral nutrients- a medium for their growth. Moreover, the pH, temperature range and inorganic phosphate content as well as the situation of the lake and river also support the growth and cause a dense population of microorganisms. These organisms (bacteria, blue green algae, etc.) form heavy blooms under these conditions. The possible factors responsible for limiting the growth and density of microorganisms are the available amounts of zinc, copper, and the poor quantity of nitrate, nitrogen, etc. It has been noticed that the excess of calcium is harmful for the luxuriant growth of microorganisms, especially to algae in general.

Microbiology of Lake

Stored water, especially Lake Water is often classified into three major types based on their suitability for the support of living matter.

1. *Eutrophic lake:* The lake, which is well nourished, receives water from a stream draining areas, and rich in plant-vegetation are referred to as eutrophic lake. This kind of lake contains a high amount of organic matter and minerals, which support the luxuriant growth of micro and macroscopic aquatic organisms.
2. *Oligotrophic lake:* Such lakes are poorly nourished and are less productive. The organic and mineral contents of such lakes are very small and are not capable of supporting the growth in aquatic life.
3. *Dystrophic lake:* These lakes contain high organic matter of special type, such as the remains of incomplete digestion of matter in the area draining into the lake. Dystrophic lakes are often dark in colour, and acidic in nature. Due to the acidic nature a very few species of bacteria and other forms of aquatic life are able to grow.

System of Saprophytism in Stored Water

It is rather difficult to make a complete list of biological population of stored water because each kind of stored water differs from the other in its biological associations. Therefore, one can find a mixed microbic population in stored water. The different groups of microflora of stored water, especially of water reservoirs mainly depend on the rate of population of water and source of population that is dead or living organic and inorganic matter. The physiological properties of water reservoir may be included under the term saprophytism. This property depends on the amount of total organic matter present in water and the mineralization of organic matter. Mineralization helps in self-purification of water reservoirs. On the basis of this property, water reservoirs are divided broadly into three following zones:

Pleosaprophytic Zone: This is the heavily polluted zone of water-reservoir and contains sufficient amount of organic residues of plants and animals. Chemically, the organic residue is rich in proteins, soluble carbohydrates, cellulose, pectin substances, and glycogen, fats and other compounds. Microorganisms, that are present in this zone, are of anaerobic type and grow on previously mentioned organic matters after their proteolytic action. Certain species of microorganisms, which are resistant to hydrogen sulphide, carbon dioxide and methane, are also present in this zone. These microorganisms perform activities like putrefactive deaminization of amino acids, decomposition of cellulose and pectic substances. The amount of available dissolved oxygen is insignificant or nil in this zone.

Mesosaprophytic Zone: This zone is relatively less polluted due to the high rate of mineralization and oxidation of organic matter. The process such as the oxidation of ammonia into nitric acid, hydrogen sulphide into sulphuric acid and oxidation of organic substances are more common in this zone. The bacterial population is very high; sometimes it reaches upto 100,000 per ml.

Oligosaphrophytic Zone: This zone contains relatively pure and clear water and is free from organic substances. It contains, therefore, a very small number of bacteria and other microorganisms. Iron bacteria constitute major part of the microbic population of this zone and oxidise ferrous oxide into ferric oxide.

Sewage

Sewage is defined as *the water supply of the used water of community.* It contains dilute water-borne wastes from residence, business houses, and industries. The chemical composition of sewage varies from day-to-day or even from hour. It also varies considerably between different cities, because they produce the wastes of different characters. Sewage water contains inorganic wastes, which create a problem of disposal, but apart from inorganic waste, undesirable organic matters, which are offensive and dangerous, are also present. Inorganic compounds of sewage water support the growth of harmful bacteria and other microorganisms, which sometimes lead to the epidemics among the human beings.

Biological Oxygen Demand

The rate of removal (that is consumption) of oxygen by microorganisms in aerobic degradation of the dissolved or even paniculate organic matter in water is called biological oxygen demand (BOD) and it is used as an index of organic pollution in water. More the oxidizable organic matter present in water, more the amount of oxygen required to degrade it biologically, hence more the BOD. It is evaluated by measuring oxygen concentration in sample before and after incubation in the dark at 20°C for 5 days.

The BOD value is useful in stream pollution control managements and in evaluating the self-purification capacity of a water body. Along with the BOD, the quantity of oxygen in a body of water i.e. dissolved oxygen (DO) is indicated by the kind of biotic life, which lives there. When dissolved oxygen is reduced below 4 to 5 ppm of water, fishes are scare. Further reduction of the amount of oxygen results in an increase in anaerobic bacteria. However, BOD values should not be used equivalent to the organic load regardless of the presence of non-degradable organic matter, presence of toxins and local changes in populations of microorganisms. BOD test should be restricted to only suitable wastes in management of treatment plants.

BOD Sensor

Biosensors are the combinations of biologically active material displaying characteristic specificity with chemical or electronic sensor to convert the response into electrical signal.

They are, in fact, biocatalysts and are used as purified enzyme, antibody or as whole microbial cell or as an organelle. The response of biosensor is measured in terms of substrate used or product form. They are of different types like, carbon electrode, glucose electrode, ion sensitive electrode, photocell, oxygen electrode, thermister, adenosine electrode, etc.

Table 25.1 BOD strengths of effluents

Effluent	**BOD**
Domestic sewage	350
Sulphite liquor from paper mills	20,000-45,000
Beer	
(a) spent grains press	15,000
(b) hop press liquor	7,430
(c) mash filter cloth wash	4,930
(d) yeast wash water	7,400
(e) spoilt beer	Upto 100,000
(f) Bottle washing	550
Maltings	
(a) suspended solids	1,240
(b) wastes	20-204
(c) grain washings	1,500
Industrial alcohol stillage (molasses)	10,000-25,000
Distillery stillage	10,000-25,000
Yeast production	3,000-14,000
Antibiotic wastes	5,000-30,000
Penicillin	
(a) wet mycelium from filter	40,000-70,000
(b) filtrate	2,150-10,000
(c) wash water	210-13,800
Streptomycin spent liquor	2,450-5,900
Aureomycin spent liquor	4,000-7,000
Solvents	Up to 2,000,000

BOD is a test for the measurement of organic pollution. Usually, BOD test requires a five-day incubation period. The results of the test are dependent on the skill of the operator. Biosensor is used for rapid test. *Trichosporon cutaneum* (yeast) is used as biosensor. The organism is sandwitched between an oxygen permeable Teflon membrane and a porous membrane. Then the membrane is directly fixed on the surface of platinum cathode of an oxygen probe. A continuous flow system using biosensor has been developed for automatic estimation of 5-day BOD.

When the sample solution containing glucose and glutemic acid is injected into the system, organic compounds permeated through the porous membrane. Then the immobilized microorganisms assimilate them. Consumption of oxygen by immobilized microorganisms starts and then they cause a reduction in the dissolved oxygen around the membranes. This results into the reduction in electrode current with time until a steady state is reached within 18 min. The steady state current depends upon BOD of the sample solution.

There is a relationship between the current decrease and 5-day BOD of standard solution below 60 mg/l. The current is reproducible with ± 6 percent of the relative error when BOD 40 mg/l of the standard solution is employed.

Immobilized Microbial Cells and Waste Water Treatment

Pollutants either of natural origin (lignified wood) or of artificial origin (organophosphates, chlorobenzenes) are not as recalcitrant as what they are thought to be. Microbes have evolved in metabolic activities either to use these chemicals as sources of nutrient or to detoxify their surrounding. These chemicals are degraded by their catabolic activity. There are certain pollutants, which are tough, new strains of microbes can be developed to degrade them by mutations and genetic transfers.

Whole Cell Enzymes and Biodegradation

The enzymes are responsible for degradation of pollutants in nature. Use of complete cells as enzyme source is more effective than purified enzymes because the enzymes within the cells are better protected from denaturation by the intact cell wall and membrane. Degradation activity is usually composed of a series of reactions; therefore, it is difficult to develop different systems for different enzymes working together. Isolation and purification of the component enzymes is rather expensive and there are chances of denaturation. Sometimes, induced enzyme production occurs in cells, which is required for degradation activity. The rate of degradation and microbes resistance to toxic pollutants remain better when the mixed culture cell population is used for the degradation rather than individual, independent enzyme system.

Further, it is also observed that use of immobilized cells instead of free suspended cells has additional advantages because over all capability of cells becomes more enhanced. The cells do not contaminate, the outgoing reaction mixture, and are more evenly dispersed by immobilization (this minimize the diffusional restrictions). The immobilized cells are used with more ease to exploit the kinetic features of continuously stirred and packed bed type of reactors, minimization of product inhibition.

Immobilized Cells and Enzymes for Waste Water Treatment

Immobilized cells and enzymes systems are mainly used in BOD/COD reduction, in specific pollutant detoxification, as biosensor and to get the maximum useful products from the waste.

Nitrate Reduction

Immobilized cells are used in removing the particular substances from wastewater. At the East Hide Sewage Works, waste water is forced through a sand bed on which species such as *Hypomicrobium* are grown in presence with added methanol to cause nitrate reduction.

Nitrate pollution of ground water is a health hazard. The old methods for removing nitrate are ineffective and impracticable. Use of living microorganisms for denitrify water is slow, incomplete, difficult to set up and maintain. Mobite GmbH, (Germany) and the University of Michigan have developed a new technique by a series of oxido-reductase enzyme system. The redox-process, driven by electric current results in complete conversion of nitrate into gaseous nitrogen. Such electro-bioreactors can also be used for the removal of other water contaminants like pesticides after identification of appropriate enzymes and cofactors.

In an another plant at Rye Meads, processing 81000 m^3 of water per day, uses a similar technique but without methanol addition in which nitrate reductions between 50 to 90 per cent have been obtained. *Micrococcus denitrificans* cells encapsulated in liquid membranes are also used for the reduction of nitrate to nitrite. The immobilized cells are also resistant to 10^4M $HgCl_2$ while free cells were sensitive.

Oxidation of Ammonia in Waste Water

Immobilized cells of *Nitrosomonas europaen* in alginate have been used to oxidise ammonia in wastewater to NO_2 and NO_3 for reduction of BOD and prevention of algal growth. Soluble nitrates have been provided with an exogenous carbon source and reduce nitrates and nitrites completely to gaseous products.

Biological Oxygen Demand (BOD)

Immobilised cells are also used as biosensors to continuous evaluatation of BOD of waste water. These analytical procedure usually work by measuring the loss of enzyme activity caused by exposure to the pollutant. The degree of inhibition of the enzyme being proportional to the concentration of the pollutant present in the sample under test.

Wheat Strach and Paper mill Industry

Immobilised α-amylase enzyme is used for treatment of wastewaters from the wheat strach industry. This enzyme is also used for clarification of colloidal starch-clay suspensions of white waters from the paper mill industry.

Pulp Mill

The effluents of pulp mill are brown and contain partially chlorinated lignin derivatives known as kraft lignins. These are normally treated in aerated lagoons or activated sludge systems and BOD/COD is reduced. This treatment unables to reduce the colour. However, it may be decolourised by using white-rot fungus *Coriolus versicolor* entrapped in calcium alginate beads and supplemented with various carbon and energy sources. Lignolytic fungus-*Phanerochaete chrysosporium* is also used for this purpose. About 80% of the colour is removed after incubation of immobilised cells in the medium for 3 days in presence of sucrose. This method is useful because it eliminates carcinogenic chlorinated lignins.

Degradation of Phenols

Phenols degradation in waste waters of hospitals, laboratories and coal processing to coke can be achieved either by using the fungus *Aureobasidium pullulans* adsorbed by fibrous asbestos or by using cells to *Pseudomonas* spp. either adsorbed to anthacite coal or entrapped on alginate gel. This treatment is advantageous because it avoids damage to the conventional biological treatment systems.

Methane Formation

CH_4 can be continuously formed for 90 days from the waste waters by using a population of methanogenic cells entrapped in agar, collagen or polyacrylamide membranes. Methanogenic bacteria have been used to degrade waste waters from the alcohol fermentation factory to CH_4.

Glucose Production

Glucose can be produced from waste cellulose. This is also one of the means of disposing of waste material and also producing valuable products. However, no practical process using immobilisation techniques has yet to be emerged in this direction.

Lactose Hydrolysis

Immobilised enzymes are also used to hydrolyse lactose from waste water of cheese-making industry where BOD is high.

Detoxification from Cyanide

Cyanide is present in aqueous wastes and can be detoxify by using immobilised mycelia of *Stemphylium loti* which contain cyanide hydralase. Cyanide is converted into formamide.

Sewage Disposal and Treatment

The methods used to treat sewage before it is discharged into rivers, lakes or the ocean are not so expensive. The procedures are considered to be effective for the purpose of rendering the waste moderately safe in so far as its bacterial content is concerned. The virus load and toxic chemical content of sewage discharges have not been as thoioughly evaluated.

Sewage Disposal

Natural wastes include domestic sewage and animal slurries. Sewage may be contaminated with industrial chemicals and animals slurries with agricultural chemicals, but their polluting effect otherwise arises from organic matter, whose biological destruction consumes oxygen, and ammonia. Sewage effluent is normally discharged to a water course. Its polluting load is destroyed naturally and, if the flow of the receiving waters is sufficient in relation to the discharge, no significant effects on the river can be detected. The problems of chronic pollution of water course are less technical than economic, but research is needed on the biological effects of chronic low levels of pollution. The morden sewage plant is based on the activity of *Bacillus, Vorticella* and other microbes. They break down organic matter to carbon dioxide and water in 'activated ludge digester' or trickling filter. In activated sludge digester air is forced through the system to encourage aerobes. In a trickling filter, liquid is run over a gravel bed.

The sewage processing plants are depend on the action of microbes to purify waste water by consuming a wide variety of solid materials. However, how sewage is broken down is still not clear. This is because of the composition of sewage which varies considerably and there are many different species of microbe at work in sewage ponds.

The waste disposal is undoubtedly the largest single application of microorganisms by man. The success of this application depends upon the enormous metabolic diversity of microbes.

There are various methods which are used to treat sewage water. The suitability of the method is mainly dependent on the area to be served and by the density of population. The sewage water may be treated by physical, chemical or biological processes to remove organic matter.

Physical Treatment

Using mechanical devices, various suspended and floating solids from sewage can be removed physically. Different types of screening devices are used. After screening, the sewage is passed through grit chambers to allow the solid matter to settle. Then it is passed through sedimentation tanks. The polluted water is kept in these tanks for few hours. The supernatant liquid or effluent is drawn off. Chlorine is passed through it and finally discharged into a body of water, or into porous soil.

Biological Treatment

The method is widely used for sewage treatment The water from toilet, bath-tubes, kitchen sinks etc. (sanitary sewage) is processed in septic tank which can handle at the most a few hundred gallons per day to municipal plants which handle of gallons.

Septic Tanks

This type of tank is commonly used in rural areas where public sewers are not available. It is constructed below the ground level near the house or within the house. It is about 3' 5' x 5' in area and about 750 gallon of sewage water may be filled in it. The domestic waste entering the tank is retained long enough to permit sufficient decomposition and sedimentation. Due to activity of anaerobic bacteria, most of the organic matter is hydrolyzed and get fermented. Sugars, alcohols, organic acids, amino acids, amines, glycerol, fatty acids and other chemicals including gases like hbS, CH_4, CO_2, and H_2. The final products are unstable. The effluent from the tank is distributed under the soil surface through perforated pipes. This procedure, however, does not eliminate pathogenic microorganisms from the sewage. Therefore, it is essential that drinking water supply should be at the safe distance from the septic tank and disposal field. It is essential to remove undigested material which resist microbial actions are termed sludge. This should be taken out periodically through pumps or by any other means. The sludge may be mixed with soil (away from houses) which can serve as a good source of humus.

Trickling filters

In this process sewage water, from city sewares is first carried to soil compartments where it undergoes physiochemical changes with the soil. It than gradually trickles down the soil to a depth of nearly 2 meters. Such filteration procedure leads to the separation of suspending particles and organic matter from it. These are then decomposed by purifying bacteria. The filtered water, then, drained out and reaches the natural pure-water-reservoirs like well. This is the most simple and cheap system of water purification where the sewage water is purified by both mechanical and biological property of soil.

Based on the above system, trickling filters have been designed. It consists of a large bed of stones 8 to 12 feet deep. The liquid sewage is sprayed over the surface of the bed either by a series of sprinklers or by a rotating sparger. The spraying of sewage water is done intermittently so that as a layer of water goes down the filter, which draws a layer of air above it down through the filter. This helps to develop aerobic condition in the bed and the filtering medium becomes coated with a living film of aerobic microorganisms. They help in hydrolyzing and oxidizing the organic matter of sewage. This process is relatively slow. Finally, clear effluent is drained off at the bottom through the bed.

Oil Recovery

Oil is the fuel around which the world has built its major infra structures and on which highly mobile lifestyles are based. It is almost indispensable in today's societies and yet the impact of various predictions over the last few years that oil supplies will ultimately run out has only received serious consideration since the 1973 oil crisis.

The organic deposits of petroleum oil and natural gas were probably formed from tiny microorganisms rather than from the debris of large plants, as was the case with coal. When these microscopic creatures settled on ocean floor and were later covered with mineral sediments, tiny droplets of body oil were squeezed out of them. This oil was trapped into large deposites during formation and was altered chemically by heat and pressure. Most of the oil in such reservoir is found not as vast pools of liquid, but-as a coating on grains of rock. The oil sticks tenaciously of these grains and must be dislodged before it can be brought to the surface. Ordinary water is too thin a liquid to budge most of this oil; it simply flows past the oil-coated grains. In tertiary oil recovery (also known as enhanced oil recovery) materials are mixed with water to make viscous, and one such material is called xanthan gum.

Xanthan gum is a polysaccharide produced by the bacterium *Xanthomonas campestris*. The gum is an inert compound, which thickens water and improves its ability to dive out oil trapped underground. It is, when mixed with drilling muds, it also serves as a lubricant for the giant drill as they penetrate the rock.

Under field conditions, bacteria like *Bacillus* and *Clostridium* have also been tested for-oil recovery in plant of *Xanthomonas* The preliminary result are encouraging. Sugars and other nutrients are fed to these bacteria while they are deep below ground. Here they grow, produce chemicals that help wash the oil free, and yield carbon dioxide and other gases which also assist in forcing the oil to the surface.

It is interesting to note that these bacteria can tolerate high pressure and temperature, lack of water and oxygen and large quantities of salt and sulphur. The qualities of microbes will enhance the commercial use of these organisms in oil industry.

Oil Pollution

Various species of *Pseudomonas* have the property to consume available hydrocarbons from oil. Each species has the capacity to consume specific type of hydrocarbon, therefore, it is essential to adopt the following techniques:

1. A mixture of different strains of *Pseudomonas* can be used. This method has been successfully used to clear up oil-contaminated water in derelict ships and in cleaning up water supplies.
2. Transformation of oil consuming genes in one strain of *Pseudomonas.* Such newly developed strains have been successfully tested under field conditions. Most of the biotechnologists agree that it is unlikely that bacteria could ever cope with the large oil spills produced by tanker. But it needs careful watching.

The transformation of oil consuming genes has been done by Ananda Chakrabarty of General Electric. He created a superbug which would be able to mop up all types of hydrocarbon in oil spills. He introduced it in plasmids of several different strains of *Pseudomonas* into a single cell. The newly developed strain has not been used commercially. We are still hopeful to receive new stable strains of *Pseudomonas* to over come the problem of oil pollution in water and takners.

Air Pollution

It this known that the major part of air pollution of artificial origin is derived from the combustion products of industry, traffic, households concentrated in urban areas. The degree of air pollution is generally proportional to the industrial development of the country. Out of many harmful gases, sulphur dioxide is an air pollutant produced by burning of coal. In air it combines with water to form sulphuric acid (acid rain). Its low pH causes release soluble aluminium salts into the environment and causes toxic effects. Research are on the way to develop the new strains of sulphur bacteria to clean up sulphur dioxide before it is discharged from chimneys.

Pesticides and Herbicides Pollution

Pesticides are biologically active chemicals used for killing plants and animals. The first modern chemical pesticides were arsenic compounds used on potato, cotton and apples. To day, the delebrate or accidental discharge of noxious organic chemicals (pesticides and harbicides) into the enviqanment is very common. Biotechnologists are trying to develop the strains of bacteria which can survive on these toxic chemicals! Ananda Chakrabarty and his co-workers announced that they have developed a microbe that attacked 2,4,5-T, a very persistant herbicide and the main ingredient of the agent Orange, which was used to destryo vast areas of jungle in Vietnam. Although, the suitability of the biotechnologically developed microbes has yet to be proved.

Degradation of Herbicides, Insecticides and Pesticides

Pseudomonas putida contains dehalogenases enzyme which degrade mono -and dichloroacetates and propionates from herbicides and pesticides. Dalpon degradation activity is stable for 20,000 hrs. in chemostal. Mixed microbial population when grown on parathion was found to hydrolyse a number of organophosphate insecticides. The population contain several enzymes which could be immobilised with operational half-life of 280 days.

Heavy Metals

Few elements are directly toxic to organisms. They are tend to be heavy and have a high atomic mass (in the form of small particles less than 5 micron or fumes). Among them are mercury, lead, cadmium, nickel, gold, platinum, silver, bismith, arsenic, selenium, vanadium, cromium and thalium. These elements have their effects on organisms and are available as the pollutants in environment. The efforts have been made to use bacteria to eliminate the toxic effects of these heavy metals.

The strains *of Pseudomonas aeruginosa* is known to accumulate great quantities of uranium (as much as half of the total weight of each cell excluding its water content). *Thiobacillus* bacteria also leach metals out of rock ores and can accumulate silver. Silver is also available in waste waters of film factories and other industrial sites. These bacteria can reduce the losses of this expensive resource.

Toxic Metal Accumulation

Zoogloea ramigera accumulates copper from factory waste. Microbes also have potential to remove toxic metals in wastes including uranium.

Microbes in Leaching of Metals

It is only the last two decades that people came to know the important role of microbes in hydrometallurgical industries for the extraction of metals from their ores. The hydrometallurgical method of leaching of metals from low grade ores by passing through stacks of ores has been known from ancient times. The sulphide metals are oxidised to sulphate and sulphuric acid. The rate of this process can be increased by the activities of iron oxidising bacteria. Biological leaching is brought about by the bacteria like *Thiobacillus thiooxidants. Ferrobacillus terrooxidants, F. sulfooxidants,* etc. These bacteria have been reported to be helpful in the leaching of copper, nickel, molybdenum, zinc, arsenic, antimony and germinium from their sulphates.

Recovery of Copper and Cobalt by Biopolymer Gels

Many biopolymers derived from microbes and plants are known to bind metals strongly. The use of biopolymers as absorbents for hazardous metals or strategically important metals has been a topic of intense research in recent years. Due to an excellent selectivity for certain metals by biopolymers and the low cost of producing biopolymers under mild condition, this method can offer an alternative to conventional methods of metal recovery like precipitation, electrowinning and ion exchange.

Decay of Organic Matter

Most of the microorganisms (bacteria, fungi) live saprophytically on dead organic matter of soil. They decompose complex organic matter into simple one and make it available for plants. In fact. they bring about decay by their various digestive and respiratory processes. Thus they form an important constituent of soil called humus. They also prevent leaching of many inorganic substances. Green manure added to the soil does not improve the physical and chemical structure of soil but various microorganisms like bacteria and fungi synthesize long chain polysacchride polymer containing uranic acid units which act as a cementing material in soil. These cementing substances are responsible for the production of soil aggregates or soil crumbs. The gum producing fungi are also important in synthesizing cementing material.

Plastic Industry

Plastic industry has a temptic market for biotechnological establishments. The acetone, glyccrol and butanol are produced along with ethanol- a major product of alcohol factories. Ethanol serves as a vital starting material for the manufacture of detergents, dyes, adhesives and resins for synthetic fibres.

Reports form the University of Warwick, England in 1981 opened new direction in the field of biotechnology by using microbes in plastic industry. Microbes can able to add oxygen to alkenes. When supplied with propylene or ethylene gases, the bacterium *Methylococcus capsulatus* can insert an oxygen atom into the molecules to produce propylene oxide or ethylene oxide. This bacterium can live happily at about 45°C (113° F) and at this temperature the alkane oxides are in gaseous form. It is easier to collect a gas than that of liquid from fermentation vessel.

Biotechnological techniques have made possible the biodegradation of plastics. The disposal of waste plastic is a serious problem because it degrades very slowely. The storage compound of most of the bacterium is polydeoxybutyrate. This can be processed using bacterium like *Alcaligenes eutophus* which degrade plastic more quickly, so making disposal easier.

Blue green algae like *Nostoc, Anabaena, Scytonema* are often employed in the reclamnation of *alkaline usar lands.*

□□□

CHAPTER-26

Neoplasia

Introduction

Neoplasm means localized cell multiplication or tumor; the cells have undergone genetic transformation, so that control of genetic material on cell functionning has lost. These cells differ in structure and function from the origin cell type. This act of cell division is known as neoplasia (= literally known as new growth). A neoplasm, as defined by Willis, is an *abnormal mass of tissue, the growth of which exceeds and is uncoordinated with that of the normal tissues and persists in the same excessive manner after the cessation of the stimuli, which evoked the change.*

In common medical usage a neoplasm is often referred to as a tumor and the study of tumours is called Oncology (i.e. oncos means, tumor; and logos means to study). The tumours may be benign or malignant. Malignant tumours are collectively referred to as cancers. Different types of tumours and differences between benign and malignant ones have been depicted in Tables 26.1 and 26.2 respectively.

Table 26.1 Nomenclature of tumors

Tissue of Origin	Benign	Malignant
1. Composed of One parenchymal Cell Type		
A. Tumors of mesenchymal origin		Sarcomas
I. Connective tissue and derivatives	Fibroma	Fibrosarcoma
	Lipoma	Liposarcoma
	Chondroma	Chondrosarcoma
	Osteoma	Osteogenic sarcoma
2. Endothelial and related tissues		
Blood vessels	Hemangioma	Angiosarcoma
Lymph vessels	Lymphangioma	Lymphangiosarcoma
Synovium		Synovioma (Synoviosarcoma)
Mesothelium		Mesothelioma
Brain coverings	Menigioma	Invasive meningioma
3. Blood cells and related cells		
Hematopoietic cells		Leukemias
Lymphoid tissue		Malignant lymphomas
4. Muscle		
Smooth	Leiomyoma	Leiomyosarcoma
Striated	Rhabdomyoma	Rhabdomyosarcoma
B. Tumors of epithelial origin		Carcinomas

Tissue of Origin	Benign	Malignant
i. Stratified squamous	Squamous cell papilloma	Squamous cell or epidermoid carcinoma
ii. Basal cells of skin or adnexa		Basal cell carcinoma
iii. Epithelial lining Glands or ducts	Adenoma Papilloma	Adenocarcinoma papillary carcinoma
iv. Respiratory passages		Bronchogenic carcinoma Bronchial "adenoma"
v. Neuroectoderm	Nevus	Melanoma (melanocarcinoma)
vi. Renal epithelium	Renal tubular adenoma	Renal cell carcinoma
vii. Liver cells	Liver cell adenoma	hepatocellular carcinoma
viii. Urinary tract epithelium transitional	Transitional cell papilloma	Transitional cell carcinoma
ix. Placental epithelium	hydatidiform mole	Choriocarcinoma
x. Testicular epithelium (germ cells)		Seminoma Embryonal carcinoma
II. More Than One Neoplastic Cell Type-Mixed Tumors-Usually Derived from One Germ Layer		
1. Salivary glands tumor of salivary gland origin	Pleomorphic adenoma (mixed origin)	Malignant
2. Renal anlage		Wilms' tumor
III. More Than One Neoplastic Cell Type Derived From More Than One Germ Layer Teratogenous		
1. Totipotential cells in gonads or in embryonic rests	Mature teratoma, dermoid cyst	Immature teratoma, teratocarcinoma

Cancer

The term cancer is derived from Latin word for crab. Cancer can be defined as a disease involving heritable defects in cellular control mechanism resulting in the formation of malignant and usually invasive tumours. Cancer is one of the major problems of modern medicine, it is also a fascinating biological problem. In biological terms it is the manifestation of changes in one of the more general properties of the cells, their ability to adjust their growth rate to the architectural requirements of the organism. A cancer arises from a single cell that undergoes permanent hereditary changes and consequently, multiples, giving rise to billions of similarly altered cells.

There are two main changes in a cancerous cells:

1. One change is defined as being of a regulatory nature. The multiplication of the cells of an animal is carefully regulated; multiplication takes place only when it is required.

The cancer cell, on the other hand escapes the regulatory mechanisms of the body and is continuously in a multiplication cycle.

2. The other change concerns the relations of cancer cells with neighbouring cells in the body. Normal cells are confined to certain tissues, according to rules on which the body's overall architecture depends. The cancer cell is not confined to its original tissue but invades other tissues where it proliferates; i.e., metastasis.

Table 26.2 Comparison of benign and malignant tumors

Characteristics	Benign	Malignant
Differentiation/anaplasia	Well-differentiated; structure may be typical of tissue origin	Some lack of differentiation with anaplasia; structure is often a typical
Rate of growth	usually progressive and slow; may come to a standstill or regress; mitotic figures are rare and normal	Erratic and may be slow to rapid; mitotic figures may be numerous and abnormal
Local invasion	Usually cohesive and expansile, well demarcated masses that do not invade or infiltrate the surrounding normal tissues	Locally invasive, infiltrating the surrounding normal tissues; sometimes may seem cohesive and expansile
Metastasis	Absent	Frequently present; the larger and less differentiated the primary, the more likely are metastases
History	Big	Short

The two phases of carcinogenesis as suggested by Berenblum (1941, 1949) are:

Initiation of the Tumour: This is associated with alkylation of DNA and a transient inhibition of DNA synthesis. At this level no morphological markers are seen, whereas biochemical markers characteristic of altered DNA are demonstrable. Thus, the initiators induce the changes in the nature of cells at molecular level (biochemical level) and these may remain latent for the life span of the animal without showing any morphological characteristic to distinguish them from the neighbouring non-neoplastic cells in the tissue.

Promotion of the Tumour: This is associated with the modification in the gene activity, causing a reversal of cell differentiation and after initiation phase, a promoting factor is required to allow expression of their neoplastic phenotype and growth to form a clinical or a frank neoplasm.

Raick (1974) proposed that the ability of the promoter to modulate gene activity, perhaps depressing genes and inducing reversal of the cell differentiation, might be the alteration that is essential for the activation of the gene loci that code for the neoplastic phenotype. The cell proliferation and hyperplasia are the most conspicuous changes induced by the promoters.

The cancer cells have some characteristic features, which distinguish them from the neighbouring normal cells. But these features may be-different in different types of tumor cells.

Morphology of the Cancer Cells

The Intermitotic Nucleus

Shape and volume

The nucleus of the cancer cell is characterized by the irregularity of its contents and the presence of deep fissures. Altmann (1852) described this condition as the Kernspalten, which sometimes results in a markedly lobulated appearance. Many neoplastic nuclei appear to be sprouting. deformed by bulby protrusions, constricted at their base or even completely separated from the central nuclear mass. The presence of deep indentations of the nucleus increases the surface, also possibly increasing the nucleo-cytoplasmic interchanges.

The cancer cells are also characterized by the increase in nuclear and nucleolar volume. Nuclear hypertrophy, chromatin lumps or chromocenters, increased DNA have been stressed as features of malignancy but in actual fact none are typical of cancer. Nuclear hypertrophy in malignancy may be the result of multiplication of chromosomes, the number of which is in multiple of the original set. The cancer cells are generally polyploids.

Contents

There are reports to confirm that the nucleus of the cancer cell contains more chromocenters than that of the normal cell (chromocenters = chromatin lumps + heterochromatic region). This is the classic conception of the cancer cell nucleus having numerous heterochromatin masses, either dense and voluminous or minute, giving the nucleus a dusty appearance.

In the tumour cells DNA content is higher than in homologus normal tissue. During the development of tumor, the successive mitosis evidently imply an increase in the total quantity of DNA. Increase in DNA content correlates with the gradual transformation of the normal cells into the cancer cells. Variation in DNA content within the same tumor is correlative with changes in nuclear volume and chromosome count.

RNA synthesis in the nucleus also increases during the tumor formation. Nuclear proteins play a predominant role in the determination of nuclear size. Studies on squamous cell carcinoma have demonstrated that increase in protein shows geometrical progression (two or four times).

Nucleolus

Cancer cells contain a very hypertrophic nucleolus, among other signs, a characteristic aid in the diagnosis of cancer. Page et al. (1938) drew attention to the very high frequency of nucleolar inclusions (argentophile vacuoles and granules)- nucleolini in malignant tumours. The number of nucleoli in cancer cells is frequently increased as compared with normal homologus tissues. Perhaps this is related to the polyploidy frequently found in cancers. The shape may also be irregular.

In view of the primary importance of the nucleolus in protein synthesis, derangement of its normal activity might be followed by serious chemical changes and consequently by disturbances in the biological behaviour of the cancer cells.

Nuclear membrane: The nuclear membrane of cancer cells is often described as thickened and irregularly folded.

Mitotic Nucleus

Hansemann (1890) drew attention to the abnormal behaviour of chromosomes in cancer cells. Genetic instability is a remarkable feature of cancer cells, even in human tumours and the chromosomal change a constant occurrence as a tumour strain evolves. The range of chromosome variation is said to be narrower in tumours derived from single cell clones. As a rule chromosomal behaviour stamps the cancer cell a aneuploid with more chromosomes than is usual. A tumour of diploid predominance may show a shift to polyploidy.

The form of chromosomes is extremely variable in cancer cells. These may be very short squat and contracted (contraction chromosomes). Other types of chromosome are elongated, filamentous and thin. The range between extreme chromosomal sizes is wider in cancer cells than in normal cells. Dicentric and annular chromosome are also observed.

The anomalies of mitosis to be found in cancer cells are of extraordinary variety:

1. Anomalous behaviour of chromosomes: Sometimes the chromosomes of the equatorial plate spread away from the centre, thus forming a pattern known as hollow metaphase. At the time of anaphase certain chromosomes fail to follow the movement of others towards the poles but remain behind. Very often, the chromosomes, owing to their stickiness, do not completely separate, and at the time of anaphase, the two adherent chromosomes form chromosomal bridges.
2. Anomalous behaviour of the spindle fibres: This anomaly may manifest itself in the absence of spindle fibres. In this condition the arrangement of chromosomes at the equatorial plate is absolutely irregular. The chromosomes divide but do not move toward the opposite poles. The nucleus is reconstituted after a certain time and then contains a double quantity of chromosomes. The lack of spindle may be partial and thus involve very unequal division of chromosomes among the daughter cells. Over production of spindles particularly by prolongation of the metaphase may permit additional multiplication of centromeres. Multiploar mitosis thus occur, which are very common in neoplastic cells.
3. Uncompleted mitosis: The normal mitosis may be interrupted at any stage. The abnormalities may be described as follows-

   i. Lack of Cytodieresis: This is frequently a sequel of cytoplasmic disturbance and leads to the formation of multinucleated giant cells, a characteristic finding in cancer tissues.

   ii. Lack of Karyodieresis: This involves a division of chromosomes within the intact nuclear membrane. Its various stages are known as an endoprophase, an endometaphase and an endotelophase. No equatorial plate is formed. The chromosomes are scattered throughout the nuclear area. The achromatic figure is absent. It produces a nucleus with a double number of chromosomes.

iii. *Lack of chromosomal separation:* Endoreduplication involves duplication without separation of the chromosomes. It results in nuclear hypertrophy. It is not recognised at the time of its occurrence and its results do not become apparent until the next complete division.

iv. *Duration and frequency of mitosis.* Levis (1951) observed and confirmed the findings of Lambert (1913) that mitosis in rat sarcoma cells are of much longer duration than those in normal corresponding fiboblasts. A high level of mitotic activity is a failure of the neoplastic state.

The Cytoplasm

The cytoplasm becomes shrinked. Cytoplasmic-nuclear ratio decreases.

1. *Cell membrane:* The contact between individual cells in normal cultures is continuous and discrete and even slight overlapping is rare. Tumour cells rarely show well defined cell membrane, edge to edge contacts nearly always overlap. Tumour cells show more pseudopodial processes than normal cells. These cells may show considerable, poor or no phagocytic and pinocytic activity.
2. *Endoplasmic reticulum:* ER is customarily deficient in malignant tissues as opposed to normal cells and less differentiated tumour cells show the greatest loss of membranous ergastoplasm. In most anaplastic cells neither the membranes nor the granules may be present. The electron microscope has disclosed groups of 3-20 straight or curved membranes within certain tumour cells. These fenestrated structures have no special distribution within the cytoplasm and their nature is unknown.
3. *Golgi complex:* No constant specific features of these structures have been discovered in tumour cells. It seems to be present in all neoplasms. Very hypertrophied in certain benign or malignant tumours, they are only occasionally visible in highly dedifferentiated tumours. It may be inert in anaplastic tumour cells.
4. *Centriole:* The centriole may be hypertrophic in certain rumours. In some it may lose its polarity (displacement).
5. *Mitochondria:* Mitochondria are thought to decrease in number with cellular dedifferentiation. Cancer cells often contain smaller mitochondria than normal cells. In some, these may be larger because they are swollen. The mitochondria become rounded and more or less swollen progressively lose the majority of their internal crests, become clear, and are finally transformed into apparently empty sacs. The cristae also become irregular.

Metabolism of Cancer Cells

The essential feature of the change of normal cell into the cancer cell is that in the neoplasm, the capacity for growth has supplanted that of function. In the cancer cell there is a tendency to converge towards a common enzymatic pattern of activity, whereas the cells of normal organs possess specific chemical characteristics. Certain chemical functions are lost in the transformation of normal to neoplastic cell. These changes can be considered under the following headings-

1. *Sources of energy:* Neoplastic cell has a full complement of oxidative enzymes but have a reduced capacity for oxidation. Anaerobic energy yielding mechanisms are predominant. No new pathways of metabolism in tumours are distinguishable and no new enzymes or co-enzymes are present.
2. *The energy-substrate systems:* A normal cell is more concerned with function, involving energy releasing catabolism than with growth involving anabolic energy. Tumour cells are concerned primarily with growth, not with function. The specific functional enzyme substrate systems therefore become superfluous in comparison with those required for the anabolic processes of growth and cell division. Normal cells possess diversified metabolic activities, whereas, in tumour cells there is much greater biochemical uniformity.
3. *Competitive struggle:* A large neoplasm acquires nitrogenous building blocks from the body stores to satisfy the continual demand for protein synthesis, but the supply of these blocks is not unlimited. Cancer cells appear to exercise priority over the demands of normal tissues for amino acids thus constituting a nitrogen trap. This trap accounts for the wasting and cachexia that are often so characteristic, a feature of die in later stages of malignancy.

Immortalization

Neoplastic cell cultures can grow indefinitely, whereas, normal cell cultures die after some generations, e.g. human cell cultures die after 50 generations.

Loss of Contact Inhibition

Cancer cells apparently lack proper recognition and communication. Contact inhibition is a process when the cells do not move and grow in a culture, this is because of the contact of plasma membranes of different cells and formation of gap junctions. Cancerous cells lose the property of contact inhibition. Such cells divide even after forming a monolayer. Loss of contact inhibition enables the cells to dissociate from neighbouring cells and to infiltrate other organs. Such cells pass over or under one another, they grow on top of one another, and they frequently form gap junctions.

Invasions

The transformed cells have the ability to invade other tissues. Normal cells lack this property. The transformed cells first invade extracellular matrix and then enter blood circulation. These pass through the wall of vessel and form metastasis tumour. The invasion and metastasis are biologic hallmarks of malignancy.

Loss of Anchorage Dependence

Most normal cells must be attached to a rigid substratum (i.e. they must be anchored) in order to grow. Transformed cells can grow even when they are not attached to the substratum, as for example when they are suspended in a semisolid medium containing agar or methyl cellulose. The cells that have lost anchorage dependence generally form tumours with high efficiency when they are injected into animals that can not immunologically reject the cells.

Increased Sugar Transport

Tumour cell consumes much more glucose than normal cells, because they have to grow and multiply. There is a great increase in the rate of sugar transport across the surface cell membrane after transformation. This increases the intake of glucose by the transformed cells. More transporters (Permeases) are available on cell surfaces, perhaps due to the increased glycolytic activity of transformed cells, which give them a higher requirement for sugar transport.

Disorganisation of the Cytoskeleton

Normal cells have a cytoskeleton of microtubules and microfilaments. These have a regular arrangement and bring about co-ordinated cell movement. In transformed cells the fibres are few in number, thinner and disorganised. These undergo depolymerization and disaggregation.

The loss of cytoskeletal elements has been considered a possible cause of the increased mobility of cell surface proteins.

Protease Secretion

Cancerous cells secrete a protease called plasminogen activator, which cleaves a peptide bond of plasminogen (a serum protein), converting it to the protease plasmin. It results in loss of actin microfilaments, growth stimulation etc. Increased plasmin may help the cells penetrate the basal lamina, thus, facilitating invasiveness of transformed cells.

Release of Transforming Growth Factors

Transforming growth factors (TGFs) are proteins secreted by transformed cells that can stimulate growth of normal cells.

Loss of Capacity for Growth Arrest

Normal cells suppress their own growth when the concentrations of any of the many critical nutrients or factors falls below a threshold value. When the concentration of isoleucine, phosphate, epidermal growth factors (EGF) or another substance that regulates growth falls below the necessary level, normal cells go into quiescence. Transformed cells are deficient in their ability to respond by growth arrest to lowered nutrient or factor concentrations.

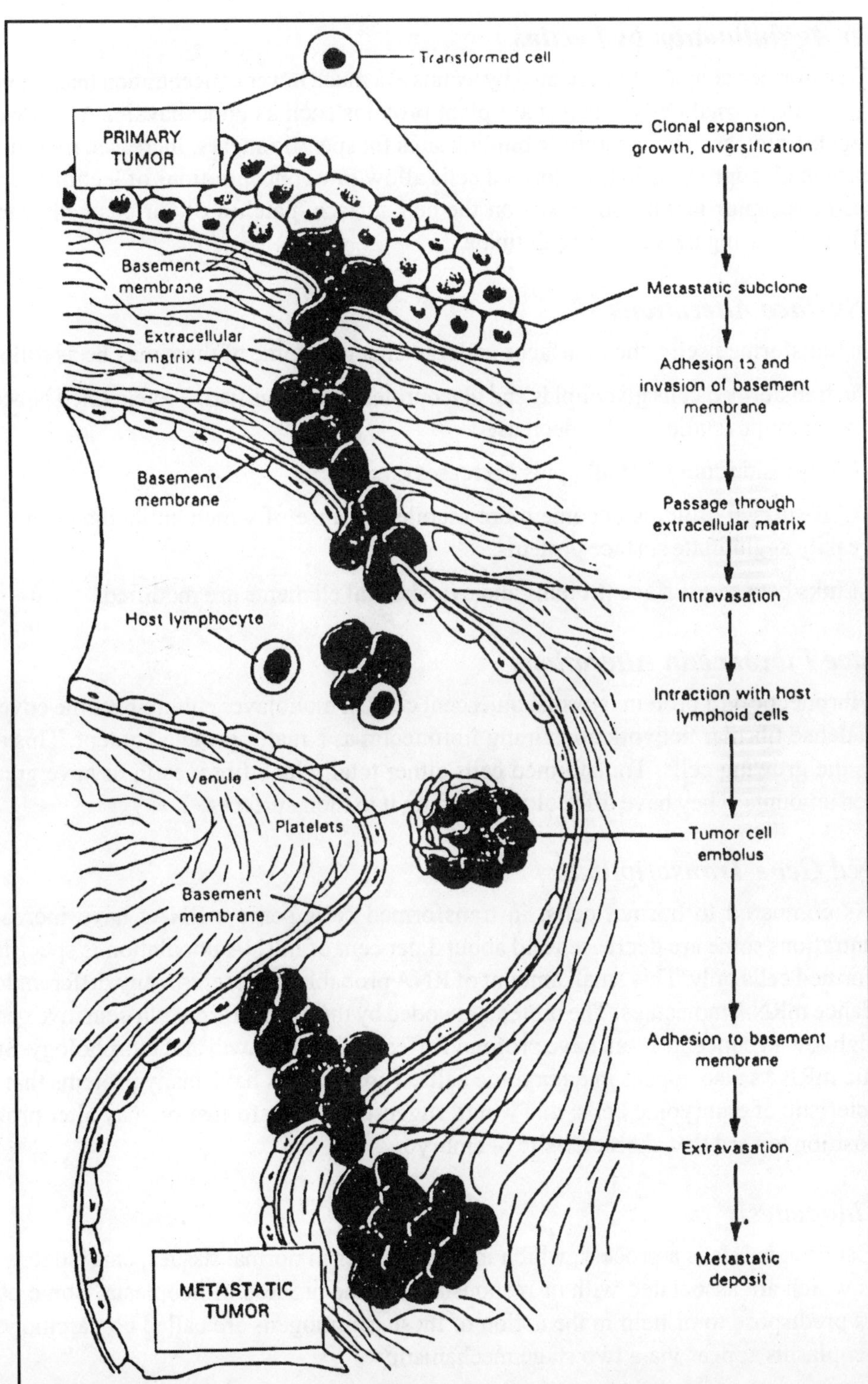

Diagram 26.1 The metastatic cascade. A schematic illustration of the sequential steps involved in the hematogenous spread of *a* tumor.

Easier Agglutination by Lectins

Transformed cells are agglutinated by lectins at a much lower concentration than required to agglutinate normal cells. Lectins are plant proteins such as concanavalin A and wheat germ agglutinin; these have multiple binding sites for specific sugars. Increased mobility of cell surface glycoproteins in transformed cells allow low concentrations of lectius to make patches of receptor-lactin complexes on the cell surface, patches of various cells can be cross-linked causing the cells to agglutinate.

Cell Surface Alterations

In transformed cells, their surfaces undergo changes, some of them may be as follows:

1. In transformed cells glycolipids and glycoproteins are modified. Protein lined N-acetyl neuraminic (sialic) acid is decreased.
2. Ganglioside content of all lipids decreases.
3. Cell surface proteins become more mobile, because of which antibodies can more easily agglutinate surface proteins.
4. Links between surface proteins and cytoskeletal elements are modified.

Surface Fibronectin Alterations

Fibronectic is a protein. Normal quiescent cells in monolayer culture become covered with a dense fibrillar network containing fibronectin as a major protein content. This also covers the growing cells. Transformed cells either totally lack fibronectin or have greatly reduced amounts. They have difficulty in binding it to their surfaces.

Altered Gene Transcriptions

As compared to normal cells, in transformed cells some mRNAs have increased concentrations some are decreased and about 3 per cent of mRNA population is specific to transformed cells only. This small amount of RNA probably represents many different low-abundance mRNA molecules. The proteins encoded by these transformation-sensitive genes, although low in concentration, have profound effects on cell growth and morphology. Such specific mRNAs also appear in embryonic cells. Tumour cells have many proteins that are characteristic of embryonic cells; this would suggest that transformation may alter protein composition toward that characteristic of embryo.

Carcinogenesis

Carcinogenesis is a process, which induces cancer in normal tissues, carcinogens are agents which are associated with or responsible for the process of neoplasia. Some other factors predispose to or help in the action of these carcinogens are called co-carcinogens. The neoplasms appear via a two stage mechanism:

1. Initiation.
2. Promotion.

During the phase of initiation single cells or group of cells undergo a biochemical change without showing any histologically recognisable morphological alterations. Upto certain limits such changes are reversible. These cells henceforth capable of being stimulated by even non-specific and non-carcinogenic agents (called Promoters) to neoplastic proliferation, which now becomes irreversible. Promotion refers to that part or period of the process, which covers events from the time the irreversible biochemical subcellular changes have occurred upto the time when a tumour becomes clinically visible.

Molecular Basis of Cancer (Carcinogenesis)

Some of the fundamental principles of molecular basis of cancer may be as follows-

1. Nonlethal genetic damage (mutations) lies at the heart of Carcinogenesis. Such genetic damage may be acquired by the action of environmental agents such as chemicals, radiations, viruses or it may be inherited in the germ line.
2. Two classes of normal regulatory genes-the growth-promoting protooncogenes and the growth-inhibiting cancer suppressor genes (antioncogenes)-are the principal targets of genetic damage. Mutant alleles of protooncogenes are dominant. In contrast, both normal alleles of the tumour suppressor genes must be damaged for transformation to occur, so this family of genes is sometimes referred to as recessive oncogenes.
3. Carcinogenesis is a multistep process at both the phenotypic and generic level. A malignant neoplasm has several phenotypic attributes, such as excessive growth, local invasiveness and the ability to form distant metastases. These characteristics are acquired in a stepwise fashion a- phenomenon called tumour progression.

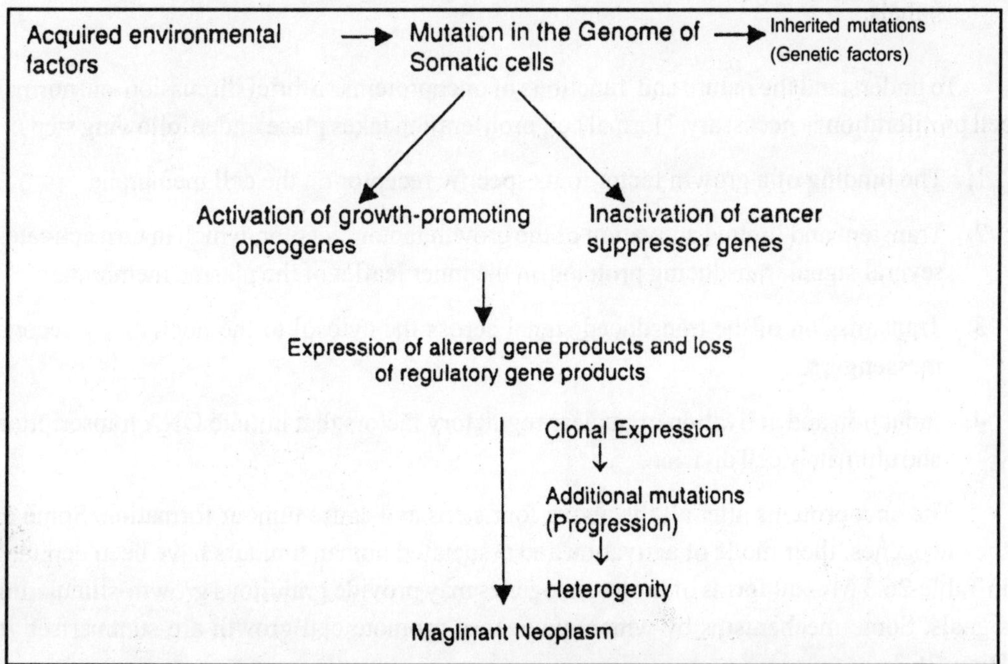

Diagram 26.2 Flow chart of a simplified scheme of cancer pathogenesis.

Oncogenes and Cancer

In 1989, J. Michael Bishop and Harold Varmus received the Noble prize for their pioneering contributions in the field of proto-oncogenes and their potential to transform into cancer causing oncogenes. Protooncogenes are cellular genes. Molecular hybridization revealed that the viral oncogene (V-onc) sequences were almost identical to sequences found in the normal cellular DNA. Proto-oncogenes are named after their viral homologues because these sequences were first discovered in viruses.

Each V-onc is designated by a three-letter word that relates the oncogenes to the virus from which it was isolated. Thus, the V-one contained in feline sarcoma virus is referred to as V-fes, whereas the oncogene in simian sarcoma virus is called V-sis. The corresponding proto-oncogenes are referred to as-fes and-sis by dropping the prefix.

The proto-oncogenes may become oncogenic by retroviral transduction (V-oncs) or by influences that alter their behaviour in situ, thereby converting them into cellular oncogenes (C-oncs).

Oncogene Products

The proteins encoded by oncogenes are called oncoproteins. The oncoproteins resemble the normal products of protooncogenes with the following exceptions.

1. Oncoproteins are devoid of important regulatory elements.

   Their production in the transformed cells is not dependent on growth factors or external signals.

To understand the nature and functions of oncoproteins, a brief discussion on normal cell proliferation is necessary. Normal cell proliferation takes place under following steps:

1. The binding of a growth factor to its specific receptor on the cell membrane.
2. Transient and limited activation of the growth factor receptor, which in turn activates several signal transducing proteins on the inner leaflet of the plasma membrane.
3. Transmission of the transduced signal across the cytosol to the nucleus via second messengers.
4. Induction and activation of nuclear regulatory factors that initiate DNA transcription and ultimately cell division.

The oncoproteins alter all the above four steps and cause tumour formation. Some of the oncogenes, their mode of activation and associated human tumours have been depicted in Table-26.3 Mutant forms of proto-oncogenes may provide gratuitous growth-stimulating signals. Some mechanisms by which oncogenes promote cell growth are summarised in Diag. 26.3.

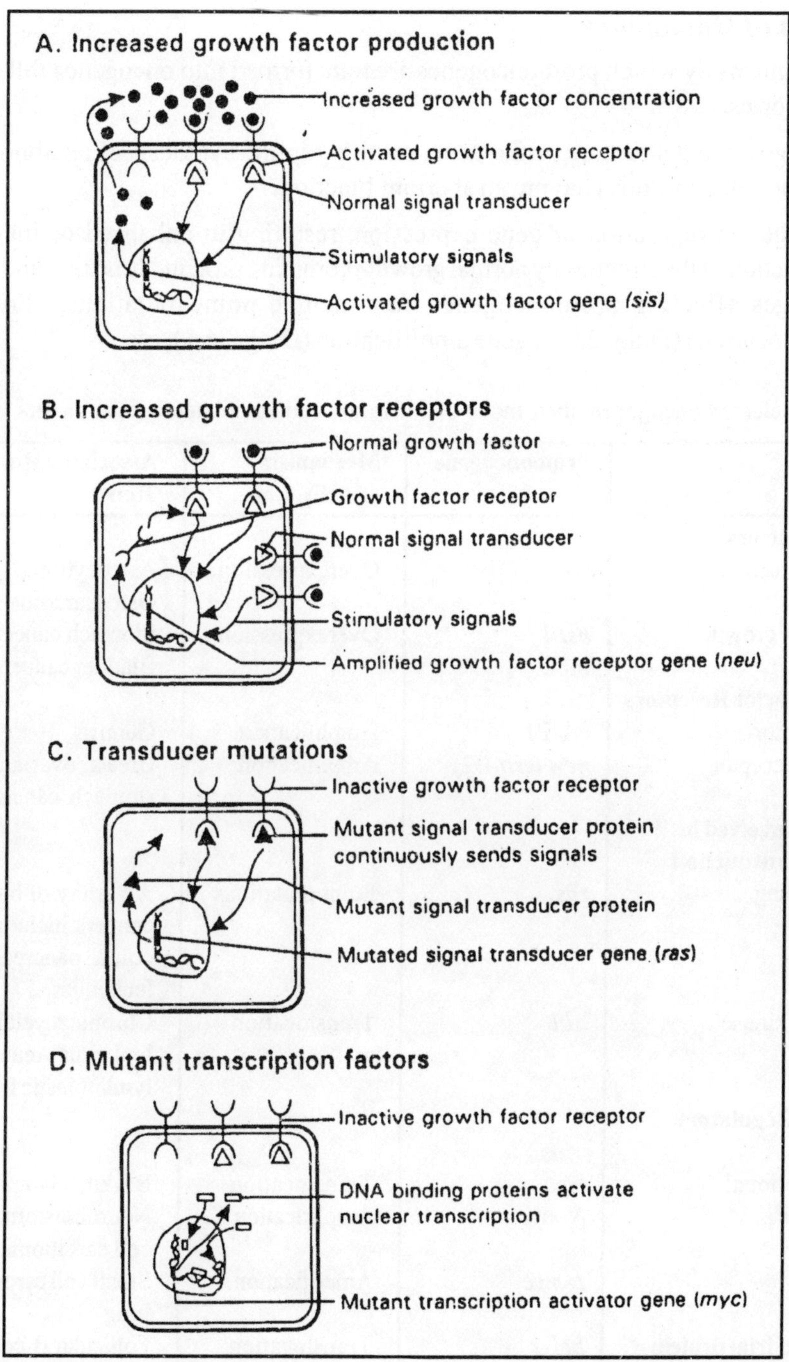

Diagram 26.3 Mechanisms by which an oncogene may promote cell growth. (A) It may code for a growth factor that stimulates the tumor cell by autocrine mechanisms. (B) It may encode a growth factor receptor and be amplified, thus increasing the number of receptors on tumor cells. (C) It may encode for defective signal transducers that transmit growth-promoting signals without an external trigger. (D) It may encode a transcription factor that binds to DNA and stimulates cell growth.

Activation of Oncogenes

Mechanisms by which protooncogenes are transformed into oncogenes fall under rwo broad categories.

1. Changes in the structure of the gene, resulting in the synthesis of an abnormal gene product (oncoprotein) having an aberrant function.
2. Changes in regulation of gene expression, resulting in enhanced or inappropriate production of the structurally normal growth-promoting protein. Structural and regulatory changes affecting protooncogenes may include point mutations, chromosomal translocations (Diag. 26.4), gene amplification (Diag. 26.5), etc.

Table 26.3 Selected oncogenes, their mode of activation and associated human tumors

Category	Protooncogene	Mechanism of Activation	Associated Human Tumor
Growth Factors			
PDGF-β chain	*sis*	Overexpression	Astrocytoma Osteosarcoma
Fibroblast growth factors	*hst-l* *int-2*	Overexpression	Stomach cancer Bladder cancer
Growth Factor Receptors			
EGF receptor	erb-Bl	Amplification	Gliomas
EGF-like receptor	*new (erb*-B2)	Amplification	Breast, ovarian, and stomach cancers
Proteins Involved in Signal Transduction			
GTP-binding	*ras*	Point mutations	A variety of human cancers including lung, colon, pancreas, many leukemias
Tyrosine kinase	*abl*	Translocation	Chronic myeloid leukemia Acute lymhoblastic leukemia
Nuclear Regulatory Proteins			
Transcriptional activators	*myc*	Translocation	Burkitt's lymphoma
	N-myc	Amplification	Neuroblastoma small cell carcinoma of lung
	L-myc	Amplification	Small cell carcinoma of lung
Mitochondrial protein	*bcl-2*	Translocation	Follicular B-cell lymphoma

Cancer Suppressor Genes

Many genes have been identified which, when present in a mutant form, cause cells to become tumorigenic. Where their function is known, the normal counterparts of such protooncogenes have often been found to play a role in the control of cellular proliferation. They may be changed into a functional oncogene by mutations that alter their biochemical properties or that result in a change in their level of expression. Thus a point mutation that alters the *RAS gene* into an oncogene changes the rate at which it hydrolyses GTP, the nucleoside triphosphate that regulates its activity. The mutation that alters the *MYC* transcription factor into an oncogene in some forms of leukemia is a chromosome rearrangement that probably results in it aberrant expression.

In order to render primary tissue culture cells oncogenic it is often necessary to introduce two different oncogenes that act co-operatively to induce cellular transformation. The *RAS* and *MYC* oncogenes form such a co-operating pair. The fact that multi-hits appear to be necessary for oncogenic transformation is believed to explain the fact that some forms of naturally occurring cancers, such as bowel cancer, are progressive diseases. It also produces a satisfying explanation for the fact that the frequency of occurrence of cancer shows a strikingly non-linear dependence upon age.

The *RAS* and *MYC* oncogenes were isolated because of their ability to transform cells in culture. Evidence that they actually play a role in naturally occurring cancers derives from the presence of a potentially oncogenic form of each of these genes in human tumors. Because the point mutation in the *RAS* oncogene normally occurs within the same codon, it is possible to use oligonucleotide probes, in Southern transfer or in PCR, to detect a change at this position. The re-arrangement around the *MYC* gene can readily be mapped using Southern transfer.

RAS and *MYC* are dominat oncogenes; when the oncogenically activated form on the gene is introduced into an immortalized cell line it becomes tumorigenic, despite the fact that a normal copy of the gene is also present. There are also recessive oncogenes or tumor suppressor genes, such as the retinoblastoma *(RB)* gene, that appear to act as suppressors of cell proliferation. Only if both copies of the gene are inactivated by mutation do cells, become tumorigenic. Biochemical studies have shown that in normal cells the *RB* protein is complexed with several proteins that are potential activators of cellular proliferation. It is assumed that these proteins are rendered inactive when bound to *RB* and that deletion of the *RB* gene, or mutations that inhibit its binding to other proteins, are oncogenic because of these properties. The *P53* gene encodes another cellular protein that acts as a tumor suppressor, apparently by a similar mechanism. Remarkably, point mutations in the P53 gene have been found in almost half of the human cancers characterized so far, including cancer of the colon, breast, lung and liver.

These results have major implications for the diagnosis of cancer. It should be possible to design diagnostic probes that will allow identification of the lesions present in particular cancers, as they first present and during their progression. These can be used to give a much more coherent and meaningful categorization of tumor type, based upon the oncogenes that are active within the cell rather than upon the cell's histological appearance. A body of information can thus be built up describing the likely prognosis for the various categories, and this could be of great value in planning therapy.

Knowledge of oncogene re-arrangements is already being used in the treatment of hematopoietic cell cancers such as leukemias and lymphomas. These are treated by whole body irradiation, or with massive doses of selectively cytotoxic drugs.

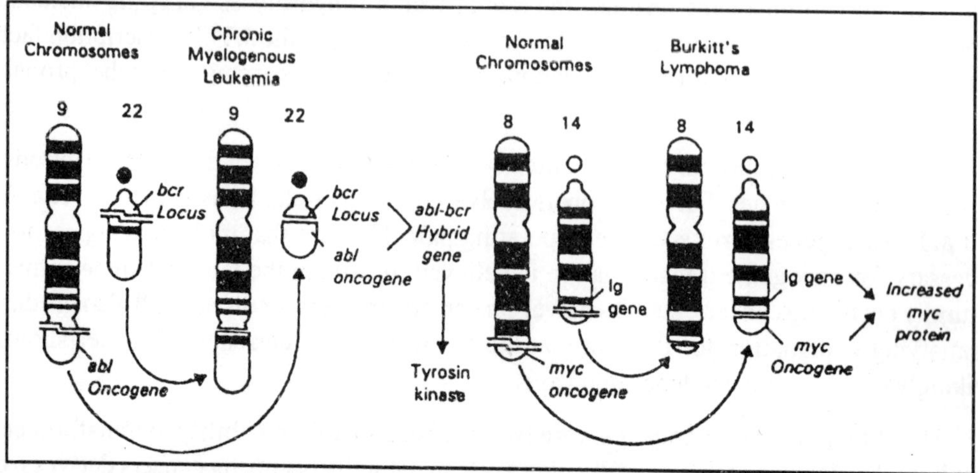

Diagram 26.4 The chromosomal translations and associated oncogenes in Burkitt's lymphoma and chronic myelogenous leukemia.

The aim is to destroy all the hematopoietic stem cells in the bone marrow population, so that it can be repopulated with healthy cells. It is, done by grafting. The marrow froms a compatible donor and keeping the patient under immunosuppressive therapy, in order to avoid a graft versus host reaction. This type of surgery is extremely difficult and the success rate is low. A graft versus host reaction can occur at any time, sometimes years after the transplantation.

It is, therefore, preferable to use the patient's own marrow. The explanted marrow is purged of cancer cells by chemotherapy or irradiation. Provided a genetic alteration in the cancer cells. Know, it is possible to design a PCR probe that will distinguish cancer cells from the non-malignant cells. Such a genetic difference will often be provided by the rearrangement of DNA that led to the cells becoming oncogenic. When it is clear that no cancer cells survived the purging, the marrow can be safely re-implanted.

A similar approach can be used to analyze blood for residual cancer cells after the less extreme drug treatments used to treat some forms of leukemias. It is then possible to assess the efficacy of the therapy and predict the likely extent of remission.

By genetically typing the cells when the disease is first diagnosed, it is also possible to discriminate between a blastric crisis (a recurrence of the disease due to sudden expansion of the previously existing cancer cell population) and a new (primary) manifestation of the same disease.

We have chosen to discuss cancer in some detail because there have been such great strides in our understanding. However, the power of molecular biology is such that most of the important human diseases are being studied using its methods. While it is beyond the scope of this chapter to discuss them individually another excellent example of the revolution brought about by molecular techniques is in the understanding and treatment of infectious and parasitic diseases.

Exogenous Carcinogens

(a) Chemical carcinogens

Many chemical carcinogens have been identified, some of them are listed in Table-26.4. The following pertinent observations have emerged from the study of chemical carcinogens.

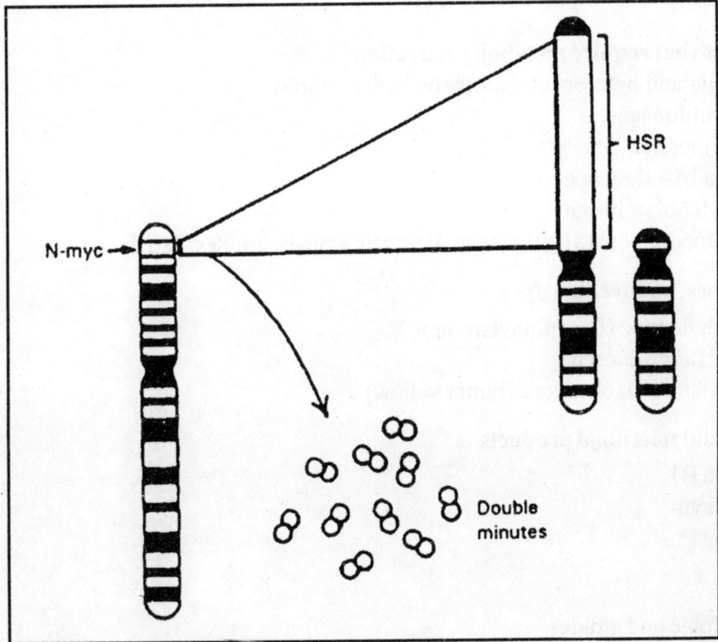

Diagram 26.5 Amplification of the N-myc gene in human neuroblastomas.

1. Chemical carcinogens are of extremely diverse structure and include both natural and synthetic products.
2. Some are direct reacting and require no chemical transformation to induce carcinogenicity, but others are indirect reacting and become active only after metabolic conversion. Such agents are referred to as procarcinogens and their active end products are called ultimate carcinogens.
3. All chemical carcinogens are highly reactive electrophiles (have electron-deficient atoms) that react with the electron-rich atoms in RNA, cellular proteins, and mainly DNA.
4. The carcinogenicity of some chemicals is augmented by agents that by themselves have little if any cancerous activity. Such augmenting agents have traditionally been called promoters However, many carcinogens have no requirement for promoting agents.

5. Several chemical carcinogens may act in concert or with other types of carcinogenic influences (e.g., viruses or radiations) to induce neoplasia.

Table 26.4 Major chemical carcinogens

Direct-acting carcinogens
Alkylating agents
anticancer drugs (Cyclophosphamide, chlorambucil, nitrosoureas, and others)
Acylating agents
1-Acetyl-imidazole
Dimethylcarbamyl chloride
Procarcinogens that require metabolic activation
Polycyclic and heterocyclic aromatic hydrocarbons
Benz(a)anthracene
Benzo(a)pyrene
Dibenz(a,h)anthraccne
3-Methylcholanthrene
7, 12-Dimethylbenz(a)anthracene Aromatic amines, amides, azo dyes
Aromatic amines, amides, azo dyes
2-Naphthylamine (B-naphthylamine)
2-Acctylaminofluorene
Dimethylaminoazobenzene (butter yellow)
Natural plant and microbial products
Aflatoxin B1
Griseofluvin
Betel nuts
Others
Nitrosamine and amides
Vinyl chloride, nickel, chromium
Insecticides, fungicides
Polychlorinated biphenyls (PCBs)
Arsenic
Asbestos

Hydrocarbons

These are mostly coal-tar derivatives. Attention to tar in relation to skin cancer was drawn by Percival Pott in 1775. Chimney-sweeps cancer was also prevalent in those days. In 1932, the active agent in coal tar was found out to be Benzpyrene. Since then many other chemical carcinogens have been discovered and even synthesized. Some of the important ones are dibenzanthracene (DBA), Methyl cholanthrene (MC). The most powerful carcinogen belonging to this group is 9:10, dimethyl, 1:2 benzanthracene (DMBA).

Chemical carcinogen acts both as initiators and promoters although the relative activity differs in different carcinogens. DBA is a potent initiator but a weak promoter, benzpyrene is moderately potent both as an initiator and promotor.

Aromatic amines

Naphthylamines during their excretion have a carcinogenic action on the mucosa of urinary bladder of dog and man. Thus, aniline dye workers have a serious occupational hazard in this respect. The organ specificity suggests that the carcinogenic action is due to metabolites of the amines excreted in urine rather than to the amines themselves. Benzidine, a hardening agent used in rubber industry, is also a carcinogenic amine. Its use in the laboratory, therefore, for testing occult blood has been abandoned and replaced by guaic acid.

Azo-dyes

Agents such as scarlet red, butter yellow have been shown to produce carcinoma of liver in rats. Butter yellow acts when there is deficiency of riboflavin whereby liver can not detoxify the dye.

Other Chemicals of carcinogenic properties are urethane (a potent initiator) and alkylating agents like nitrogen mustard.

Physical Carcinogens

Radiations

Ionizing radiations and ultra-violet rays in direct sun light are mutagenic and thereby carcinogenic. Thus incidence of malignancy is higher amongst radiologists due to exposure to X-rays over long periods and farmers in whom the skin remains exposed to direct sunlight over long periods of life (farmer's cancer). Such carcinomas as are mentioned here and carcinomas associated with ones occupation are also known as occupational or social carcinomas. Radioactive material inhaled or ingested accidentally may produce carcinoma e.g. miners and persons handling radioactive chromium or uranium. They have a higher incidence of lung cancer. The workers engaged in preparing watch dials with luminous paint containing radio-active material developed osteogenic sarcomas of bones due to their practice of pointing the paint brush by licking with tongue and lips, whereby, the material entered the body and got selectively accumulated in bones. Exposure to atomic radiations increases the incidence of malignancy in the population affected.

Heat

Heat as a physical carcinogen plays its role indirectly by inducing necrosis and regeneration. Examples are met with following:

1. In hypertrophied scars (Keloids) following burns especially if subjected to chronic irritation. Carcinomas supervening on scars are also called Marjolin's ulcer and though in the majority, they are squamous cell carcinomas but sometimes they may also be basal cell carcinomas if sweat glands or hair follicles are present in the scars.

2. Kangri cancer met within Kashmir valley where to ward off excessive cold, people keep specially designed baskets (called kangri) containing live charcoals under the

clothes directly in touch with the skin. Such tumours are, therefore, seen over the skin of chest and abdominal wall.

3. Chhota cancer seen in some natives of South India, who have the practice of smoking with the lighted end of the cigar inside the mouth.

Viral Oncogenesis

Both RNA and DNA viruses can be transforming agents, such viruses are often called tumour viruses. Tumour viruses cause transformation as a consequence of their ability to integrate their genetic information into the host cells DNA; most often they cause the chronic production of one or more proteins called transforming proteins, which are responsible for maintaining the transformed state of the infected cells. These proteins are synthesized under the direction of transforming genes in an integrated viral genome. For DNA viruses, the transforming genes are integral parts of the viral genome. For retroviruses (RNA viruses), the transforming genes are normal or slightly modified cellular genes that are either appropriated from or hyperactivated in the host cell.

RNA oncogenic viruses: Animal retroviruses transform cells by following two mechanisms:

1. Some called, transforming viruses, contain a transforming viral oncogene, such as src; abl, or myb.
2. Others, called slow transforming viruses (e.g., mouse mammary tumour virus) do not contain a V-onc (viral oncogene) but the proviral DNA always found to be inserted near a protooncogene. Under the influence of strong retroviral promoter, the adjacent normal or mutated protooncogene is over expressed.

DNA oncogenic viruses: The well studied DNA viruses causing human cancers are- Papilloma virus, Epstein-Barr virus (EBV), and hepatitis B virus (HBV). General comments relating to transformation by DNA viruses are:

1. Transforming DNA viruses form stable associations with the host cell genome. The integrated virus is unable to complete its replicative cycle because the viral genes essential for completion of replication, are interrupted during integration of viral DNA.

2. Those viral genes that are transcribed early (early genes) in the viral life cycle are important for transformation. They are expressed in transformed cells.

Hormones as Carcinogens

Hormones are known to exert a trophic influence on their target organs by controlling their proliferative and secretory activities. In endometrium and breast, phases of proliferation alternate with those of secretory activity due to a balanced secretion of estrogens and progesterone, in turn controlled by pituitary gonadotrophins. Such hyperplasia involution cycles also occur in thyroid gland. Testosterone has a trophic action on prostate. The neoplastic action of excessive hormonal stimulation is due, in some cases to a promoting action on initiated cells; but in the majority, it is due to long continued hyperplasia leading to changes. Examples include nodular swellings in breast in mammary dysplasia; adenomas and nodular enlargements of thyroid, prostate, etc. such tumours (bening or malignant) solely induced by excess hormonal stimulation are also called hormone dependent tumours.

The role played by raised hormonal levels in tumorigenesis has been beautifully illustrated by the technique of parabiosis by which two or more animals of the same genetic constitution are sutured side by side so that they possess a common blood supply. Thus, when a normal female is joined in parabiotic union with either a castrated male or a female, the gonadotrophic hormone excreted in excess by the castrate is transferred to the intact partner and stimulates its ovaries. Although more estrogens are produced, they are rapidly broken down by the liver and consequently do not pass over to the castrate, whose pituitary gland remains uninhibited. Granulosa cell tumours have been reported to develop in this manner.

Endogenous Carcinogens

The structural carbon-ring nucleus of sex hormones, sterols, bile acids and cholesterol is similar to that of carcinogenic hydrocarbons. The endogenous carcinogens may originate through some error in synthesis or metabolism of sex hormones or bile acids; such errors may be genetic based appearing at various age periods or governed by mutations. Methyl cholanthrene (a known carcinogen) has synthesized from deoxycholic acid, a normal constituent of bile.

Co-carcinogens

Diet

Direct (experimental) and indirect (observations on different human populations) evidences indicate that dietary factors especially vitamins play a significant role in the genesis of some carcinomas. Thus, cancer of liver in rats fed on rice and butter-yellow can be prevented by adding to the diet yeast or vitamin B-complex, in particular riboflavin. Choline deficiency in rats over a prolonged period may result in liver cancer.

Age

No age is exempt from neoplasm. However, they are more common in individuals beyond 40-45 years of age, which may be called as cancer bearing age. It is likely that long continued proliferation of cells over a period of years induces spontaneous mutations and higher the age, the greater the proportion of such mutant cells.

Heredity

There is no gene, specific for neoplasms. However, it is well known that cancer not only runs in family generations but often has a predilection for the same organs or tissues in members of various generations in the family group, e.g. cancer of breast, stomach, intestine, cervix, neuroblastomas, etc. What actually is inherited appears to be an increased potentiality of certain tissues or organs to give off mutant cells more regularly and in progressive earlier years. In some cases, heredity plays its part in producing some anatomical congenital error of development, which in later years gets superimposed by malignancy, e.g. intestinal multiple polyposfs.

Trauma

The role of trauma is only indirect, by producing local pain. It may draw attention to an already existing neoplasm at the site of trauma which was unnoticed by the patient prior to trauma; or trauma by inducing haemorrhage in an already existing tumour causes it to enlarge suddenly, thus, making it more apparent and even painful.

Chronic Irritation

This factor operates by inducing chronic inflammation with repeating episodes of epithelial necrosis, regeneration, and so on. Long continued regenerative activity will in due course of time give rise to mutant cells. Examples of appearance of malignant tumours at the site of chronic irritation are met with gall stones and cancer gall bladders; renal stones and carcinoma of renal pelvis. Squamous cell or basal cell carcinomas supervening on Keloids; margins of chronic ulcers and sinuses not responding to routine therapeutic measures and in case of Schistosoma haematobiuin and carcinoma urinary bladder due to chronic irritation caused by the spinous processes of the eggs of the parasite. In gall bladder, renal pelvis, etc., metaplastic changes may precede onset of a carcinoma, which will now be of the squamous cell type.

Pre Cancerous Conditions

There are certain diseases or malformations, which are superimposed by malignant changes with great frequency so that detection of such diseases or malformations increases the chances of malignant tumors appearing in the patient. These include leukoplakia, syphilitic glossitis, long standing mammary dysplasia, xeroderma pigmentosum, intestinal multiple polyposis, senile keratosis, kraurosis vulavae, etc.

❑❑❑

CHAPTER-27

Biotechnology and Anti-microbial Drugs

Introduction

The control of microorganisms is critical for the prevention and treatment of disease. The chemical and physical agents are used to treat inanimate object in order to destroy microorganisms or inhibit their growth. Microorganisms grow on and within other organisms, and microbial colonization, lead to disease, disability, and death. Thus the control or destruction of microorganisms residing within the bodies of humans and other animals is of great importance.

When disinfecting or sterilizing an inanimate object, one naturally must use procedures that do not damage the object itself. The same is true for the treatment of living hosts. The most successful drugs interfere with processes that differ between the pathogen and host, and seriously damage the target microorganism while harming its host as little as possible.

Chemotherapeutic Agents

Modern medicine is dependent on chemotherapeutic agents, chemical agents that are used to treat disease. *Chemotherapeutic agents destroy pathogenic microorganisms or inhibit their growth at concentrations low enough to avoid undesirable damage to the host.* Most of these agents are antibiotics [Greek *anti,* against, and *bios,* life], microbial products or their derivatives that can kill susceptible microorganisms or inhibit their growth. Drugs such as the sulfonamides are sometimes called antibiotics although they are synthetic chemotherapeutic agents, not microbially synthesized.

Antibiotics are chemical substances excreted by some microorganisms which inhibit the growth and development of other microbes. Some of these drugs (chemicals) that were obtained naturally were put to chemical modifications (by removing some chemical groups and adding other) in attempts to enhance beneficial effects while minimizing the toxic effects. The resultant modified product is termed as semisynthetic antibiotics. Most antibiotics currently used are semisynthetic. The chemist has synthesized many drugs that have got the antibacterial property and less toxicity. These drugs are called synthetic antibiotics drugs. Naturally occurring antibiotics, their semisynthetic derivatives and synthetic antibiotics have got the same target i.e., anti-microbial action. Hence all these drugs were put together to be called anti-microbial agents.

The anti-microbial agents may broadly be classified according to the types of organisms against which they act.

1. Antibacterial drugs
2. Antiviral drugs
3. Antifungal agents
4. Antiprotozoal agents
5. Antihelminthic drugs

Chemotherapeutic agents, like disinfectants, can be either cidal or static. Static agents reversibly inhibit growth; if the agent is removed, the microorganisms will recover and grow again. The antibacterial drugs are often described as:

Bacteriostatic

A substance that causes a cessation of growth of the microorganism which is reversed when the chemical is removed. Such agents acting on bacteria are called bacteriostatic. The term bacteriostatic describes a drug the temporarily inhibits the growth of a microorganism. The effect is reversible. When the drug is removed, the organisms will resume growth and infection or the diseases may recure. Typical bacteriostatic drugs are the tetracyclines and sulfonamides.

Bactericidal

The substance which kills microorganisms is termed bactericide. It is described as drug that attaches to its receptor and causes the death of the microorganism. Typical bactericidal drugs are the betalactams (penicillins, cephalosporins) and the aminoglycosides. Many of the bacteriostatic drug in higher doses act as bactericidal agents. When the growth of bacteria is stopped by bacteriostatic drugs, the body defence mechanisms will kill the bacteria.

General Characteristics of Anti-microbial Drugs

An Ehrlich os clearly saw, to be successful a chemotherapeutic agent must have selective toxicity: it must kill or inhibit the microbial pathogen while damaging the host as little as possible. The degree of selective toxicity may be expressed in terms of (1) the therapeutic dose, the drug level required for clinical treatment of a particular infection, and (2) the toxic dose, the drug level at which the agent becomes too toxic for the host. The therapeutic index is the ratio of the toxic dose to the therapeutic dose. The larger the therapeutic index, the better the chemotherapeutic agent (all other things being equal).

A drug that disrupts a microbial function not found in eucaryotic animal cells often has a greater selective toxicity and a higher therapeutic index. For example, penicillin inhibits bacterial cell wall peptidoglycan synthesis but has little effect on host cells because they lack cell walls; therefore penicillin's therapeutic index is high. A drug may have a low therapeutic index because it inhibits the same process in host cells or damages the host in other ways. These undesirable effects on the host, called side effects, are of many kinds and may involve almost any organ system. Because side effects can be severe, chemotherpeutic agents should be administered with great care,

Drugs vary considerably in their range of effectiveness. Many are narrow-spectrum drugs- that is, they are effective only against a limited variety of pathogens. Others are broad-spectrum drugs and attack many different kinds of pathogens. Drugs may also be classified based on the general microbial group they act against: antibacterial, antifungal, antiprotozoan, and antiviral. Some agents can be used against more than one group; for example, sulfonamides are active against bacterial and some protozoa.

Table 27.1 Properties of Some Common Antibacterial Drugs

Drug	Primary Effect	Spectrum	Side Effects*
Ampicillin	Cidal	Broad (gram +, *some* •)	Allergic responses (diarhea, anemia)
Bacitracin	Cidal	Narrow (gram +)	Renal injury if injected
Carbenicillin	Cidal	Broad (gram+, many-)	Allergic responses (nausea, anemia)
Cephalosporins	Cidal	Broad (gram+, some -)	(Allergic responses, thrombophelbitis, renal injury)
Chloramphenicol	Static	Broad (gram +, -, rickettsia and chlamydia)	Depressed bone marrow function, allergic reactions
Cipprofloxacin	Cidal	Broad (gram +, -)	Gastrointestinal upset, allergic responses
Cindamycin	Static	Narrow (gram +, anaerobes)	Diarrhea
Dapsone	Static	Narrow (mycobacteria)	(Anemia, allergic responses)
Erythromycin	Static	Narrow (gram +, mycoplasma)	(Gastrointestinal upset, injury)
Gentamicin	Cidal	Narrow (gram -)	(Allergic responses, nausea loss of hearing, renal damage)
Isoniazid	Static or cidal	Narrow (mycobacteria)	(Allergic reactions, gastrointestinal upset, hepatic injury)
Methicillin	Cidal	Narrow (gram +)	Allergic responses (renal toxicity, anemia)
Penicillin	Cidal	Narrow (gram +)	Allergic responses (nausea anemia)
Polymyxin B	Cidal	Narrow (gram -)	(Renal damage, neurotoxic reactions)
Rifampin	Static	Broad (gram +, mycobacteria)	(Hepatic injury, nausea, allergic responses)

Drug	Primary Effect	Spectrum	Side Effects*
Streptomycin	Cidal	Broad (gram +, -, mycobacteria)	(Allergic responses, nausea loss of hearing, renal damage)
Sulfonamides	Static	Broad (gram +, -,)	Allergic responses (renal and hepatic injury, anemia)
Tetracylines	Static	Broad (gram +, -; rickettsia and chlamydia)	Gastrointestinal upset, teetch discoloration (renal and hepatic injury)
Trimethoprim	Cidal	Broad (gram +, -,)	(Allergic responses, rash, nausea, leukopenia)
Vanocomycin	Cidal	Narrow (gram +)	Hypotension, neutropenia, kidney dmage, allergic reactions

* Occasional side effects are in parentheses. Other side effect not listed may also arise.

Chemotherapeutic agents can be synthesized by microorganisms or manufactured by chemical procedures independent of microbial activity. A number of the most commonly employed antibiotics are natural- that is, totally synthesized by one of a few bacteria or fungi. In contrast, several important chemotherapeutic agents are completely synthetic. The synthetic antibacterial drugs in Table 27.1 are the sulfonamides, trimethoprim, chloramphenicol, ciprofloxacin, isoniazid, and dapsone. Many antiviral and antiprotozoan drugs are synthetic. Semisynthetic antibiotics are natural antibiotics that have been chemically modified by the addition of extra chemical groups to make them less susceptible to inactivation by pathogens. Ampicilline, carbenicillin, and methicillin are good examples.

Determination the Level of Anti-microbial Activity

Minimum Inhibition Concentration (MIC)

It is that concentration of a drug, which under the experimental conditions is the lowest. This concentration will inhibit growth (as measured by observed turbidity). If the MIC of the drug is found doubled, this means that the concentration of the organism has doubled. It is done in the laboratory by the double dilution method, starting with the lowest drug concentration and then the doubling.

Minimum Bactericidal Concentration (MBC)

This is the minimum concentration of the drug that kills the organisms completely. It is done in the laboratory after the MIC take a loop from all the tubes that show no turbidity in the MIC test and plate it. The first drug concentration in the tube showing no growth is the MBC

How to Make a Concentration

To get any concentration of a compound from any higher concentration given as a stock, measure out the required concentration value in ml and then make up the solution to the volume in ml of the stock solution concentration.

Determination of anti-microbial effectiveness against specific pathogens is essential to proper therapy. Testing can show which agents are most effective against a pathogen and give an estimate of the proper therapeutic dose. It may be determined by two methods:

1. Dilution Susceptibility Tests, 2. Disk Diffusion Tests.

Currently the disk diffusion test more often used is the Kirby-Bauer method, which was developed in the early 1960s at the University of, Washington Medical School by William Kirby, A.W. Bauer, and their colleagues.

Mechanisms of Action of Anti-microbial Agents

It is necessary to know something about the mechanisms of drug action because such knowledge helps to explain the nature and degree of selective toxicity of individual drugs and sometimes aids in the designing of new chemotherapeutic agents.

Table 27.2 Mechanisms of Antibacterial Drug Action

Drug	Mechanism of Action
Cell Wall Synthesis Inhibition	
Penicillin	Inhibit transpeptidation enzymes involved in the cross-linking of the polysaccharide chains of the bacterial cell wall peptidoglycan. Active cell wall lytic enzymes.
Ampicillin, Carbenicillin Methicillin, Cephalosporins	
Vancomycin	Binds directly to the D-Ala-D terminus and inhibits transpeptidation.
Bacitracin	Inhibits cell wall synthesis by interfering with action of the lipid carrier that transports wall precursors across the plasma mcmbrune.
Protein Synthesis Inhibition	
Streptomycin	Binds with the 30S sub-unit of the bacterial ribosome to inhibit protein synthesis and causes misreading of mRNA.
Gentamicin	
Chloramphenical	Binds to the 50S ribosomal subunit and blocks peptide bond formation through inhibition of peptidyl transferase.
Tetracylines	Binds to the 30S ribosomal subunit and interfere with aminoacyl-tRNA binding.
Erythromycin and clindamycin	Binds to the 30S ribosomal subunit and interfere with aminoacyl-tRNA binding.
Fusidic acid	Binds to EF-G and blocks translation.
Nucleic Acid Synthesis Inhibition	
Ciprofloxacin and other quinolones	Inhibit bacterial DNA gyrase and thus interfere with DNA replication, transcription, and other activities involving DNA.

Drug	Mechanism of Action
Rifampin	Blocks RNA synthesis by binding and inhibiting the DNA-dependent RNA polymerase.
Cell Membrane Disruption	
Polymyxin B	Binds to the plasma membrane and disrupts its structure and permeability properties.
Metabolic Antagonism	
Sulfonamides	Inhibit folic acid synthesis by competition with p-aminobenzoic acid.
Trimethoprim	Blocks tetrahydrofolate synthesis through inhibition of the enzyme dihydrofolate reductase.
Dapsone	Interfers with folic acid synthesis.
Isoniazid	May disrupt pyridoxal or NAD metabolism and functioning. Inhibits the synthesis of the mycolic acid *cord factor.*

Anti-microbial drugs can damage pathogens in several ways. The most selective antibiotics are those that interfere with the synthesis of bacterial cell walls (e.g., penicillins, cephalosporins, vancomycin, and bacteriacin). These drugs have a high therapeutic index because bacterial cell walls possess a unique structure, not found in eucaryotic cells.

Streptomycin, gentamicin, spectiomycin, clindamycin, chloramphenicol, tetracyclines, erythromycin, and many other antibiotics inhibit protein synthesis by binding with the procaryotic ribosome. Because these drugs discriminate between procaryotic and eucaryobic ribosomes, their therapeutic index is fairly high, but not as favorable as that of cell wall synthesis inhibitors. Some drugs bind to the 30S (small) subunit, while others attach to the 50 S (large) ribosomal subunit. Several different steps in the protein synthesis mechanism can be affected: aminoacyl-tRNA binding, peptide bond formation, mRNA reading, and translocation. For example, fusidic acids to EF-G and blocks translocation, whereas mucopirocin inhibits isoleucyl-tRNA synthetase.

The antibacterial drugs that inhibit nucleic acid synthesis or damage cell membranes often are not as selectively toxic as other antibiotics. This is because procaryotes and ecuaryotes do not differ as greatly with respect to nucleic acid synthetic mechanisms or cell membrane structure. Good examples of drugs that effect nucleic acid synthesis or membrane structure are quinolones and polymyxins. Quinolones inhibit the DNA gyrase and thus interfere with DNA replication, repair, and transcription. Polymyxins act as detergents of surfactants and disrupt the bacterial plasma membrane.

Antimetabolitie Drugs

They block the functioning of metabolic pathways by competitively inhibiting the use of metabolites by key enzymes. Sulfonamides and several other drugs inhibit folic acid metabolism. Sulfonamides and several other drugs inhibit folic acid metabolism. Sulfonamides (e.g., sulfanilamide, sulfamethoxazole, and sulfacetamide) have a high therapeutic index because human cannot synthesize folic acid and must obtain it in their diet. Most bacterial pathogens synthesize their own folic acid and are therefore susceptible to inhibitors of folate metabolism. Antimetabolite drugs also can inhibit other pathways. For example, isoniazid, it interferes with either pyridoxal or NAD metabolism.

Factors Influencing the Effectiveness of Anti-microbial Drugs

It is crucial to recognize that drug therapy is not a simple matter. Drugs may be administered in several different ways, and they do not always spread rapidly throughout the body or immediately kill all invading pathogens. A complex array of factors influence the effectiveness of drugs.

First, the drug must actually be able to reach the site of infection. The mode of administration plays an important role.

1. A drug such as penicillin G is not suitable for oral administration because it is relatively unstable in stomach acid.
2. Some antibiotics, for example, gentamicin and other aminoglyocosides- are not well absorbed from the intestinal tract and must be injected intramuscularly or given intravenously.
3. Other antibiotics (neomycin, bactracin) are applied topically to skin lesions. Normal routes of administration often are called parenteral routes. Even when an agent is administered properly, it may be excluded from the site of infection. For example, blood clots or necrotic tissue can protect bacteria from a drug, either because body fluids containing the agent may not easily reach the pathogens or because the agent is absorbed by materials surrounding it.

Second, the pathogen must be susceptible to the drug. Bacteria in abscesses may be dormant and therefore resistant to chemotherapy, because penicillins and many other agents affect pathogens only if they are actively growing and dividing. A pathogen, even through growing, may simple not be susceptible to a particular agent. For example, penicillins and cephalosporins, which inhibit cell wall synthesis, do not harm mycoplasmas, which lack cell walls.

Third, the chemotherapeutic agent must exceed the pathogen's MIC value if it is going to be effective.

Finally, chemotherapy has been rendered a less effective and much more complex matter the spread of drug resistance plasmids.

OTHER ANTIBACTERIAL DRUGS

Sulfonamides or Sulfa Drugs

A good way to inhibit or kill pathogens is by use of compounds that are structural analogues, molecules structurally similar to metabolic intermediates. These analogues compete with metabolites in metabolic processes because of their similarity, but are just different enough so that they cannot function normally in cellular metabolism. The first antimetabolites to be used successfully as chemotherapeutic agents were the sulfonamides, discovered by

G. Domagk, Sulfonamides or sulfa drugs are structurally related to sulfanilamide, an analogue of paminobenzoic acid. The latter substance is used in the synthesis of the cofactor folic acid.

Table 27.3 Chemotherapy of Some Representative Bacterial Pathogens

Pathogen	Representative Diseases	Drugs of Choice
Gram-Positive Bacteria		
Corynebacterium diphtheriae	Diphtheria	Erythromycin, penicillin G
Staphyolcoccus, pencillinase-positive	Boils, pneumonia, wound infections	A cephalosporin, cloxacillin, dicloxacillin.
Staphylococcus, pencillinase-negative	Boils, pneumonia, wound infections	Pencillin G or V, a cephalosporin, vancomycin
Streptococcus, hemolytic	Strep throat, skin infections, sepsis, rheumatic fever Pneumonia	Penicillin[b] G or V, erythromycm, or a cephalosporin
Streptococcus pneumoniae	Pcnumonia	Penicillin Penicillin[b] G or V, erythromycin, or a cephalosporin
Gram-Negative Bacteria		
Bordetella pertussis	Whoping cough or pertussis	Erythromycin, ampicillin
Escherichia coli	Urinary tract infections	Sulfonamide, cephalosporin, ampicillin
Haemophilus influenzae type b	Menigitis, pneumonia	Cefotaxine, ceftriaxone, ampicillin
Klebsiella pneumonias	Pneumonia, urinary tract infections	Newer cephalosporins[c], gentamicin
Logionella pneumophila	Pneumonia	Erythromycin
Neisseria gonorrhoeae	Gonorrhea	Amoxicillin, ceftriaxone, spectinomycin
Pseudomonas aeruginosa	Urinary tract and burn infections, pneumonia	Carbenicillin or ticarcillin
Salmonella typhi	Typhoi fever, septicemia, gastroenteritis	Ceftriaxone, chloramophenicol, ampicillin
Shigella dysenteriae	Dysentery	Trimethoprim-sulfamethoxazole, quinolones, ciprofloxacin
Vibrio cholerae	Cholera	Tetracycline, trimethoprim-sulfamethoxazole
Acid-Fast Bacteria		
Mycobacterium-avium complex	Pneumonia	Rifampin plus ethambutol
Mycobacterium tuberculosis	Tuberculosis	Isoniazid plus rifampin ± pyrazinamide
Other Bacteria		
Chlamydia trachomatic	Nongonococcol urethritis, trachoma	Doxycycline, azithromycin, erythromycin
Mycoplasma pneumonias	Pneumonia	Tetracycline, erythromycin
Rickettsia spp.	Rocky Mountain spotted fever, typhus fever	Tetracycline, chloramphenicol
Treponema pallidum subsp. *pallidum*	Syphilis	Penicillin G tetracycline, ceftriaxone

When sulfanilamide or another sulfonamide enters a bacterial cell, it competes with p-aminobenzoic acid for the active site of an enzyme involved in folic acid synthesis, and the folate concentration decreases. The decline in folic acid is detrimental to bacterium because folic acid is essential to the synthesis of purines and pyrimidines, the bases used in the construction of DNA, RNA, and other important cell constituents. The resulting inhibition of purine and pyrimidine synthesis leads to cessation of bacterial growth or death of the pathogen. Sulfonamides are selectively toxic for many pathogens because these bacteria manufacture their own folate and cannot effectively take up the cofactor. In contrast, humans cannot synthesize folate and must obtain it in the diet; therefore sulfonamides will not affect the most.

The effectiveness of sulfonamides is limited due to the increasing sulfonamide resistance of many bacteria. Furthermore, as many as 5% of the patients receiving sulfa drugs experience adverse side effects, chiefly in the form of allergic responses (fever, hives, and rashes).

Quinolones

A second group of synthetic anti-microbial agents are increasingly used to treat a wide variety of infections. The quinolones are synthetic drugs that contain the 4-quinolone ring. The first quinolone, nalidixic acid, was synthesized in 1962. More recently a family of fluoroquinolones has been produced. Three of these- ciprofloxacin, norfloxacin, and ofloxacin-are currently used in the United States, and more fluoroquinolones are being synthesized and tested. Quinolones are effective when administered orally. They sometimes cause adverse side effects, particularly gastrointestinal upset.

Quinolones act by inhibiting the bacterial DNA gyrase or topoisomerase II, probably by binding to the DNA gyrase complex. This enzyme introduces negative twists in DNA and helps separate its strand. DNA gyrase inhibition disrupts DNA replication and repair, transcription, bacterial chromosome separation during division, and other cell processes involving DNA. It is not surprising that quinolones are bacterial.

Antibiotics

Classification of Antibiotics

Antibiotics can be classified in following groups:

1. Penicillin, 2. Streptomycin and Dehydrostreptomycin, 3. Chloramphenicol, 4. Tetracycline, 5. Microlides, 6. Antifungal 7. Miscellaneous

Requirements of Antibiotics

1. Antibiotics should not be toxic.
2. They should not precipitate serum proteins.
3. They should not cause haemolysis.
4. They should not affect the vague sites adversely.
5. They should not be pyrogenic.

6. They should not cause histamine like responses.
7. They should preferably, be soluble in water
8. They should reasonably stable.
9. They should be effective against pathogens.
10. They should be well tolerated in the doses required.
11. They should have few undesirable side-effects.

Not all antibiotics fulfil these requirements; but the most likely one to cover most these requirements is penicillin. Some antibiotics are toxic clinically but useful in research work, e.g. antibiotics that are lethal to nucleic acids. Some antibiotics are useful topically but toxic internally, e.g. bacitracin.

Mode of Action of Antibiotics

Antibiotics interfere with growth and metabolism of microorganisms, which supress the growth and other metabolic activities of cells. Waksman described various modes of action of antibiotics substitutes are as follows:

1. Acting as a substituent for essential nutrient.
2. Interfering substance during vitamin utilization.
3. Modifying the metabolism of bacterial cell.
4. Competing for enzyme required by bacteria.
5. Interfering with respiratory mechanism, specially hydrogenase system.
6. Favouring certain lytic mechanism of cell.

How Antibiotics Act?

There are five sites of action of antibiotic agents.

The cell wall: The Peptidoglycan component of the cell wall in bacteria is very important for its integrity (not present in man). Penicillin and Cephalosporius interfere with the formation of peptidoglycan layer so that the bacterial cell absorbs water and brusts.

The cytoplasmic membrane: Inside the cell wall is the site of most of the cell's biochemical activity. Colistin act at this level, it disorientates the molecule so that the membrane becomes permeable and vital metabolites escapes.

Inhibition of protein synthesis: Tetracyclin, Chloramphenicol, Erythromycin and aminoglycosides interfere with the build up of peptide chain on the ribosomes.

Nucleic acid metabolism: Rifampicin actionomycin and Nalidixic acid interfere with RNA and DNA metabolism.

Intermediary metabolism: Sulfonamide, Trimethoprim, PAS and Isoniazed interfere with bacterial metabolism.

Use of Anti-microbial Drugs

1. Choice of drug can be made on clinical diagnosis; e.g. segmental pneumonia in a young person usually caused by streptococcus pneumonia and can be treated by Penicillin, leprosy by leprae, bacilli treated by Dapsone and Rifampicin.
2. Choice of drug on bacteriological identification and sensitivity test is done when: (i) infecting organisms not identified on clinical examination eg. meningitis or U.T.I. (ii) when though the infecting organisms can be identified by the clinical diagnosis but their sensitivity is not known e.g., carbuncle, gonorrhoea etc.

Action on the cell wall

The mode of action of antibiotic on bacteria on staining property, depends on their chemical composition of cell wall. Gram-negative bacteria have a high lipid content and on hydrolysis they produce a complete range of amino acids, found in protein whereas a Gram-positive bacteria give lipids in small amount and 4 to 5 types of amino acids. Those antibiotics which act on the cell wall are penicillin, N-cephalosporine, novobiocin vancomycin etc. The selective action of streptomycin on the tubercle bacillus is due to die fact that the permeability of cell membranes in the bacilli and in the tissue cells of animals and man differs due to the dissimilar chemical structure of the cytoplasm of these organisms.

On protoplast membrane

The membrane of the living cell acts as osmotic barrier keeping the concentration of metabolites and nutrients in the cell and site for biosynthesis of other cell component is the site of action of following antibiotics: streptomycin, polymyxin, novobiocine, etc. Penicillin affects the formation of L-forms in the shape of pleomorphic protoplasmic structures.

On Nucleic acid and enzyme biosynthesis

Antibiotics like, streptomycin, actinomycin-D, chloramphenicol, tetracyclines, etc., act directly on DNA dependent polymerase synthesis. Chloramphenicol is a specific inhibitor of the biosynthesis of bacterial protein. It comes into action with the peptidyl transferase areas of SOS ribosome. Competing with the aminoacyl end of the aminoacyl tRNA, chloramphenicol blocks the formation of the peptide bond. Other antibiotics like tetracyclines, lincomycin, erythromycin, kanamycin, neomycin, spectinomycin sparsomycin, fucidine and others belong to the group of antibiotics which inhibits protein biosynthesis in bacteria at the ribosome level. The antibiotic rifampin supresses protein biosynthesis by inhibiting the activity of RNA polymerasse.

Streptomycin inhibits the incorporation of some amino acids in protein synthesis and attacks the bacterial enzyme with the participation of which the introduction of pyruvic acid into the tricarboxylic acid cycle by its union with oxalacetic acid takes place. Streptomycin also inhibits the activity of biotin-containing enzymes catalysing the union of carbon dioxide with carbonic acids. This disturbs reading of the genetic code and synthesize leucine instead of alanine.

The antifungal antibiotics impair the intactness of the cytoplasmic membrane in fungi. Antineoplastic antibiotics supress the synthesis of nucleic acid in bacterial and animal cells and bind with DNA which serves as the matrix for RNA synthesis. Bruneomycin leads to sharp inhibition of the synthesis of DNA or to its destruction.

Several theories and hypothesis have been given to understand the mechanism and action of antibiotics but still we lack the exact and precise mechanism which can solve the question completely.

Absorption, Metabolism and Extraction of Antibiotics

Absorption of antibiotics depends on the route of administration and the diffusion capacity of the drug from the site. Drugs administered parentally are more quickly adsorbed than orally or locally routed. The onset of drug action depends upon the route of administration and its adsorption. Delay in administration and its adsorption depends on the fact that the drug must be convened to an active component or to the products, which are secondly responsible for the action. The drug (antibiotic) is metabolised by one of the following ways:

1. Kidneys are the most important channel for the elimination of drugs.
2. They are also excreted through faeces, sweat and through the respiration, specially the volatile drugs.

Route of Administration

Local application: They are applied mainly to skin, mucous membranes of alimentary canal, respiratory, genitourinary tract and to the cornea. Even the tissue and organs sealed deeply can also be treated with the local injection of antibiotics.

Oral: They can be taken orally.

Parentral: Antibiotics are injected intra-muscularly, interavenously, intrathecally (in spinal cord) and in the bonemarrow etc. This route of administration is referred to as parentral. Interavenously route is less painful, less irritant and fast results, larger quantities of drugs can be given (bolus dose) while intera-muscularly route is cheaper and less complicated.

Inhalation: Antibiotics are also used in the form of aerosole that is they are dispersed into the small molecules and are sprayed in respiratory passages.

Combinations of Antibiotics

When do you want to use the combination of two or more anti-microbial agents? The indications for combination of anti-microbial agents are:

1. to obtain potentiation eg. Contrimoxazole, Penicillin and Gentamicin.
2. to delay development of drug resistance; eg. in treatment of tuberculosis.
3. to broaden the spectrum of activity; when mixed infection is there.
4. to reduce severity or incidence of adverse reactions. In combination, lower therapeutic dose of each drug reduces adverse reactions.

Bactericidal drugs act on rapidly dividing organisms. Thus a bacteristatic drug, by reducing multiplication, may protect the organisms from bactericidal drugs.

When a combination of antibiotic is required, it is better to use two bacteristatics or two bactericidal drugs.

Doses of Antibiotics

A dose of an antibiotic or a drug depends on the termination of drug action, which depends on the elimination from the body according to the proportion of the plasma concentration. The value of concentration approaches zero as the item increases. For example an antibiotic attains maximum concentration after 2 hours and disappears from that place after 8 hours; therefore, the drug should be repeated after 6 hours or 8 hours.

Demerits of Antibiotic Therapy

1. Antibiotics develop drug-resistant strain in body.
2. They develop narrow spectrum property of antibacterial activity in body.
3. Body failure to respond in certain infection caused by Gram-negative bacili and other microorganisms.
4. They develop nutritional deficiency in body.

Side Effects of Antibiotics

Large doses of penicillin and streptomycin have a neurotoxic action. Tetracyclines affect the liver, chloromycetin has a toxic effect on the haemotopoietic organs, and chlortetracycline and oxytetracycline upon intravenous injection may lead to collapse with a lethal outcome. After injection of penicillin and streptomycin a rash, contact dermatitis, angioneurotic odema, anaphylactic reactions or allergic asthma may occur. On local application of antibiotics, allergic reaction may arise. *Staphylococcal colitis* proceeds very severally, and is characterized by profuse diarrhoea, dehydration of the body, toxic phenomena, shock and collapse. Penicillin treatment may cause a rapid drop in blood pressure, cyanosis, superficial breathing, loss of consciousness, and convulsions are observed, and in some cases death may occur.

Sometimes antibiotics may cause skin allergy reaction. This is caused by action of streptomycin. Quite often allergic manifestations are observed in the mucous membranes such as hyperaemia and oedema of the pharynx and tongue.

Antibacterial agents may induce genetic disorder in microorganism's cells and cause chromosomal aberration; some of them possess a teratogenic effect leading to the development of foetal monstrosities if they are taken in the first days of pregnancy.

General Principles for Use of Antibiotics

1. Make correct diagnosis, find the site of infection and organisms responsible then use antibiotic agent after finding the sensitivity.

2. Remove barrier eg. lack of free drainage of abscesses (so drain the abscess before starting antibiotic)
3. Decide whether antibiotic agent is really necessary.
4. Select the best drug i.e. activity of the drug should match with the infecting organisms. The drug should be capable of reaching the site of infection.
5. Antibiotics should be given in optimum dose with proper frequency and through most appropriate route.
6. Therapy should be continued till apparent cure is achieved and three days more to avoid relapse.
7. In Typhoid, tuberculosis and endocarditis the drugs should be continued for a longer time after apparent clinical cure.
8. Microbiological examination to be done; to prove the complete cure after withdrawal of antibiotic agent.
9. Prophylactic chemotherapy for surgical and dental procedure should start at the time of surgery and continue for 48 hours to reduce the risks of development of resistance.

Production of Antiobiotics

Antibiotics are obtained by special methods employed in the medical industry for the production of antibiotics strains of fungi, actinomycetes and bacteria are used, which are seeded in a nutrient substrate. Antibiotics are secreted by some outside of the cells and into the surrounding environment, others are largely retained within the cells and must be separated by extraction. After a definite growth period the antibiotic is extracted, purified and concentrated, checked for action. Common antibiotics are Penicillin, Streptomycin, Gramicidin, Aureomycin, Chloramphenicol and Terramycin.

Mode of Administration

Antibiotic drugs can be administered by mouth and also by intramuscular and intravenous injections. They are used not only in the treatment of infections but also in the prevention of certain infections. The number of antibiotics is being constantly increased. Some are useful clinically others are not satisfactory for clinical application but are more useful for other purposes. Antibiotics are used as growth stimulants in poultry and live stock feeds. Powerful antibiotics such as penicillin has proved to be of such tremendous importance for the destruction of organisms especially those capable of producing disease.

Side Effects of Antibiotics

Of the many hundreds of antibiotics discovered, only few are of wide application in medicine. They are Penicillin, Streptomycin, Chloramphenicol and Tetracycline. Prolonged use may weaken the body's natural defences against invading germs and may have undesirable side effects. Certain antibiotics, used without medical supervision, cause the patient to develop rashes, swellings and other symptoms of severe allergic reactions. Excessive doses damage the kidneys, in the case of streptomycin. They sometimes cause complete and permanent deafness. In every case, the appropriateness, choice, and dosage of antibiotics should be decided by the physician. Large doses of penicillin and streptomycin have a neurotoxic action, tetracyclines affect the liver, chloromycetin has a toxic effect on the haematopoietic (blood cell forming) organs, and chlorotetracycline and oxytetracycline upon intravenous injection may lead to collapse with lethal outcome. Allergic reactions arise during local application of antibiotics. A severe complication is anaphylactic shock from the use of penicillin in which a rapid drop in blood pressure, cyanosis superficial breathing, loss of consciousness and convulsions are observed. Contact dermatitis is caused by the action of streptomycin using over long periods.

Antibiotics should be used only on physician's prescription, not in selfmedication. The popularity of antibiotics is due to their ability to destroy many kinds of pathogens and to their relatively nontoxic properties when given systematically.

Rational Use of Antibiotics

There are many factors responsible for the development of resistance of microorganisms to antimicrobial agents. Inappropriate selection of an anti-microbial agent in the absence of bacteriological culture sensitivity results plays an important role in the development of resistance. Under such circumstances it becomes very essential for the physician to have a working knowledge of the common pathogenic microorganisms which would be involved in a particular disease and the specific antibiotic/antibiotics which could be used in order to obtain desired response.

Antibiotics and Plant Diseases

For example. Bacterial canker of cherry caused by *Pseudomonas morsprunorum* has made it impossible to grow cherry on a large scale in the U.K. Bordeaux mixture may control the disease, but the mixture kills the soft tissues of plants and has a deleterious effect on the spraying machinery. A small beginning was made to use antibiotics on an experimental scale. Presently antibiotics, can check as many as forty diseases.

Of these, Fire Blight of Apple in U.S.A..caused by *Erueinia amylovora* and Stone-fruit Blast in New Ireland, of which the causal agent is *Pseudomonas syringal,* are some of the diseases which have been controlled by antibiotic like streptomycin. It is sprayed for 2 to 3 times at different stages of fruit development.

Antibiotics and Animal Diseases

Cows usually suffer from a disease called mastitis, which is caused by streptcocci and staphylococci. The characteristic symptoms are the inflammation of the udder and decrease in the yield of milk. This disease is cured by giving local injections of penicillin. Antibiotics are used as antiprotozoal agents in many animal diseases. However, it may be mentioned that this aspect of antibiotic therapy is still needs research.

Antibiotics and Animal Feed

In Western countries, pigs are reared in large numbers. It is generally desired that the young ones should be fed on some food, other than the cow's milk (preferably cow's milk) so that the cow could bear another litter in the shortest possible time. This is done by maintaining the young ones on cow's milk to which is added some antibiotic like terramycin, aureomycin, streptomycin, or penicillin.

In America, antibiotics are also added to poultry feed which result in higher rate of growth. Antibiotics suppress the growth or completely elimnate the undesirable microbial agents in the gut of the poultry, which would otherwise inhibit the optimal growth of birds. Antibiotics also help in the vitamin economy of birds, though the actual mechanism involved is not yet clearly understood. Perhaps it is due to the altered microflora of the gut. Laying of eggs increases in hens after administration of antibiotics. It is not certain whether this is due to the suppression of undesirable microbial flora or the vitamin sparing effect of antibiotics.

Antibiotics and Growth of Animals

Antibiotics aid in the growth of other animals too. For example ruminants like lambs, calves etc. are weaned at a very early age of a few weeks. Such administration of antibiotics stimulate their growth. At the time of weaning the ruminant has only one functional compartment for its stomach but the administration of antibiotics is found to hasten the typical ruminant stage with the fully developed reticulorumen.

Antibiotics and Food Preservation

Antibiotics preserve foods too. Meat is among the most expensive and perhaps the most relished item of food. Even if it is slightly contaminated with bacteria during slaughtering.and distribution, putrefaction and decay set in very rapidly, causing a great deal of economic loss. Therefore, in order to prevent such spoilage antibiotics such as streptomycin and penicillin are used. They are injected into the blood vessels of the animal either before or after slaughter. They can also be sprayed on the surface of the raw meat.

Greater care is required in the preservation of fish in comparison to meat. From the time of catching till it is sold at the counter, it takes considerably long time.

Even when kept in ice, they cannot normally be preserved for more than seven days. This causes heavy loss to the fish industry. By the time it reaches die counter another food bit of it will have decayed. Since a number of bacteria thrive even at temperatures as low as 0° to 5° C fish is still open to spoilage in spite of the relatively expensive preservation in ice. Hence people turned their attention to antibiotics. Aureomycin and terramycin have been successfully used for this purpose. This fish when still on board the ship is dipped in the antibiotic solution. It is then packed in ice containing the antibiotic and stored away in chilled (-1 to 1° C) sea water. USA. has built fishing ships which are specially fitted with antibiotic containing tanks and refrigerated stores.

Since the yeasty, old or stale, smell of antibiotics disappears during cooking the objection to food preservation using antibiotics disappears. Though primarily of chemotherapeutic uses, the antibiotics have already given sufficient proof of their value to humanity in other fields as well.

Secretion of Antibiotic from Cyanobacteria

Some blue-green algae, such as *Nostoc,* secrete antibiotics called bacteriocins, that kill related strains of the alga. A bacteriocin is a proteinaceous antibiotic that is active against procaryotic strains closely related to the organism that produces the antibiotic. Others blue-green algae secrete antibiotics that are active against a wide range of blue-green and eucaryotic algae. *Scytonema hofmanni* produces such an antibiotic. This antibiotic, called cyanobacterin, is a chlorine-containing gamma-lactone. All of these antibiotics probably play an active role in the survival of the producing organism by inhibiting growth of competing organisms.

Some blue-green algae respond to low concentrations of iron in the environment by excreting compounds called solubilizing chelators or siderophores, which solubilize iron in the environment. These siderophores are secondary hydroxamates and renere insoluble iron into an available form that can be taken up into the cell.

Secretion of Antibiotic from Cyanobacteria

[illegible]

Glossary

Abaxial: The side or on the distant from the axis; as on the lower surface of a leaf.

Aberrant: A growth that deviates from the normal or usual type; as exceptional growths (aberrations) which occur in some tissue cultures.

Abort: To stop or fail to reach full development (maturity) or to develop imperfectly; as some[1] hybrids are prone to embryo abortion or as some attempts to culture a plant may fail (abortive attempt).

Acclimate or acclimatize: To adapt or adjust physiologically to a new environment or climate. The process is acclimation or acclimatization and is regulated by natural processes or human cultural practices. Usage of acclimatize may be limited to adaptive responses involving changes in a selected environmental feature, such as light intensity.

Adenine ($C_5H_5N_5$ mw 135.14) : A white crystalline purine base present in DNA, RNA and nucleotides like ADP and ATP. A B group vitamin [84] generally available as $C_5H_5N_5$. 3 H_2O mw 189.13. It is added to some tissue media, as adenine sulphate, to promote shoot formation and for its weak cytokinin effect. It may reinforce the effects of other cytokinins. It is present in plant tissues combined with niacinamide, phophoric acids and D-ribose.

Adsorb: The adhesion of a liquid, gas or dissolved material to a solid surface, resulting in concentration of the adsorbate [the adsorbed material | to the adsorbent [the adsorbing agent, such as activated charcoal]. The process is adsorption.

Adventitious: 1. Produced in an abnormal or unusual position, or at an unusual time of development or away from the natural habitat; as when plant organs (buds, shoots, roots, others develop on callus or nonzygotic embryos (embryoids) develop without an ovary or fertilization on callus (adventive embryony). 2.A mariner of growth relying on adventive processes.

Aerate: To supply with or mix with air or gas. The process is aeration.

Aerobic or aerobiotic: Refers to an organism that lives in or a process occurring in the presence of molecular oxygen.

Agar or agar-agar: A gelatinous polysaccharide obtained from the red alga *Gelidium corneum* and from several other red algae. It is a solidifying agent that mixed with nutrient media (0.6-1 %), forms a gel for growing tissue cultured plants and for other purposes. It ranges in quality from relatively inert to very impure (complex, undefined). Its firmness as a gelling agent is affected by medium pH [it is softer when the medium is more dilute). Agar gels melt at about 100°C but solidify at about 44°C.

Age: 1. The period in the life cycle of an organism; the process of growing older, mature. 2. The state of being old or senescent. 3. Culture age is a function of the number of subcultures and the time after subculture.

Agenesis: The absence of development.

Agglutinate: 1. To cause to unite, adhere; as if glued. 2. To gather into a clump or mass; as protoplasts and bacteria, in the presence of specific antibodies, tend to stick to one another. The process is agglutination.

Agglutinin: An antibody capable of clumping bacteria or other cells.

Agrobacterium tumefaciens: The bacterium causing crown gall disease of plants, inducing tumors to form. Tumors may be cultured in vitro and are useful in the study of this disease. The Ti plasmid of A. *tumefaciens is* known to cause the disease and a small portion of it is used as a vector in the genetic modification of higher plant cells. Novel DNA sequences are spliced into the plasmid DNA segment, the DNA is circularized and introduced into the cultured plant cells.

Aliquot or aliquot part: An evenly divided unit, portion or sample (fraction) of the whole.

Alkaline: A basic solution with a pH above 7.0.

Amino acid: One of many organic acids containing a basic amino group [NH_2], and an acidic carboxyl group [COOH]. Of several hundred naturally occurring amino acids only 20 are commonly found in protein molecules, linked by peptide bonds. Some are commonly added to plant tissue culture media, especially glycine and glutarmine.

Anther culture: Refers to culture of single pollen grains or of the anther, containing the male gametophytes or microspores, with the objecive of producing monoploid plants.

Anthesis: The flowering period or efflorescence. This is the time of full bloom which lasts till fruit set.

Antiauxin: A chemical that interferes with the auxin response. They may or may not involve prevention of auxin transport or movement in plants. Some said to promote morphogenesis in vitro; as 2,3,5-triiodobenzoate (TIBA mw 499.81), or 2,4,5-trichlorophenoxyacetate (2,4,5-T mw 255.49), which stimulate the growth of some cultures.

Antibiotic: One of many natural organic substances [or their synthetic analogues] secreted by plants or microorganisms that are toxic to other species, retard or prevent their growth and presumably function as defense mechanisms; as bacitracin, gentamycin, mycostatin, nystatin, penicillin, phosphomycin, rifampicin, streptomycin and terramycin etc. The phenomenon is antibiosis. Antibiotics are sometimes included in plant tissue culture media, with varying results, ranging from dramatic culture stimulation to induction of chromosomal instability.

Antioxidant: A substance (such as ascorbic acid, . citric acid or others] which is sometimes added to the sterilizing solution or to isolation medium:. to inhibit or prevent oxidative browning of the culture medium, due to bleeding of phenolic exudates. The latter may lead to tissue necrosis and death. Ascorbic acid [100 mg/1] and citric acid [150 mg/1] are most commonly used in sterilizing solutions.

Apex: The most extreme point of growth of a plant; as the apical shoot and root tips are located at the apexes [apices], and contain (apical meristem.

Artificial seed: Encapsulated or coated somatic embryos (embryoids) that are planted and treated like seed.

Ascorbic acid or vitamin C($C_6H_8O_6$ mw 176. 12): A water soluble vitamin present naturally in some plants and also synthetically produced. Aside from its use as a vitamin, it is used as an antioxidant in plant tissue culture; included in disinfection solutions and sometimes in media.

Aseptic: Free of pathogens, contaminants, algae, bacteria, fungi, viruses, etc; absence of all microorganisms, asepsis is a fundamental requirment for plant tissue culture (aseptic culture).

Asparagine [Asn, $C_4H_8N_2O_3$ mw 132.12]: An amino acid occassionally included in plant tissue culture media, as a source of reduced nitrogen.

Aspirate: To draw something in or out, up or through using suction or a vacuum; as aspiration [vacuum] may be used in the disinfection process to draw disinfectant into the surface layers of plant tissue.

Assay: 1. To test or evaluate. 2. The substance to be analyzed or the process of examining or testing it [chemically or by other means].

Autoclave: 1. An closed chamber in which to heat substances under pressure; above their boiling points, to sterilize utensils, liquids, glassware, etc., using steam. The routine method uses steam pressure of 103.4 X 10 Pa at 121° C for 15 minuts, or longer for large volumes to reach temperature. 2. A pressure cooker. These employed in medium and instrument sterilization for plant tissue culture work. Over-sterilization degrades culture media constituents and caramelizes sugars, so is to be avoided. Under-sterilization results in culture media or equipment contamination. Heat labile components of tissue culture media cannot be autoclaved. These are commonly filter-sterilized. 3. To carry out the process of autoclaving.

Auxin: One of a large class of plant hormones [phytohormones] allegedly produced in the growing tips of stems and roots.

Auxin-cytokinin ratio: The relative proportion of auxin to cytokinin prsent in plant tissue culture medium. Varying the relative amounts of these two hormone groups in tissue culture formlas affects the proportional growth of shoots and roots in virto. As the ratio is increased [increased auxin or decreased cytokinin contact], roots are more likely to be produced, and as it is decreased root growth declines and shoot initiation and growth are promoted. This relationship was first recognized by C.O. Miller and F. Skoog in the 1950's.

Auxotroph: A microorganism or plant cell line possessing a typical nutritional requirements for some item[s] [growth factors] in addition to those required by the control or wild type which can synthesize them; as auxotrophic cells requiring certain amino acids.

Availability: A reflection of the form and location of nutritional elements and their suitability for plant absorption. In plant tissue culture media this is related to the abundance of each nutritional element, the osmotic concentration and pH of the medium, the stability and solubility of the item in question, the presence of adsorbing agents in the media and other factors.

Axenic: A pure culture of one species. This implies that cultures are free of microorganisms [aseptic or germ free].

Axillary bud proliferation: Propagation in culture by protocol and media which promotes axillary [lateral shoot] growth. This is a technique for mass production [micropropagation] of plantlets in culture, achieved primarily through hormonal inhibition of apical dominance and stimulation of lateral branching.

Axis: The main plant stem.

B vitamin: One of a complex of vatimins made in plants and though to be essential for healthy growth. Some are routinely added to plant tissue culture media to promote growth. The latter include thiamine [B_1], niacin [B_3], adenine [B_4] and pyridoxine [B_6].

B5: See Gamborg, O.L., R.A. Miller, and K. Ojima [1968].

BA or BAP: See N-[phenylmethyl]-lH-purin-6-amine or 6-benzylaminopurine.

Bactericide: A substance or agent that kills bacteria [usually rapidly]; is bactericidal.

Bacteriostat: A chemical or other agent that does not kill but prevents bacterial growth and multiplication; is bacteriostatic.

Bacticinerator: An electrical appliance, capable of generating elevated temperatures, into which metal instruments are inserted and held while sterilization occurs. Bacticinerators are generally used to facilitate aseptic operations within laminar air flow cabinets, and may replace the bunsen burner, especially in areas without access to natural gas.

Ball, E. [1946]: The first to obtain plant from cultured shoot apices.

Base: 1. A water soluble chemical compound that may dissociate releasing hydroxyl ions, but not hydrogen ions; reacts with acid to form a salt; turns litmus paper blue; has a pH of more than seven; and is used to dissolve auxins and gibbeiellins and to adjust the pH of plant tissue culture media. 2. A building block of nucleic acid molecules; purine or pyrimidine base.

Batch culture: A cell suspension grown in liquid medium of a set volume. Inocula of successive suhcultures are of similar size and cultures contain about the same cell mass at the end of each passage. Cultures commonly exhibit five distinct phases per passage; a lag phase follow inoculation, then an exponential growth phase, a linear growth phase, next a deceleration phase and finally a stationary phase.

Bergmann, L. [1960]: Was first to obtain callus by plating cells from suspension cultures onto solid medium.

Bioassay: A biological assay or assessent procedure, performed on living cells or on a living organism; sometimes used to detect minute amounts of substances which influence or are essential to growth.

Biosynthesis: Biological synthesis, the building or forming of biochemical compounds in a living organism.

Biotechnology: The industrial use of biological processes; as yeast ferrmentation for alcohol production or plant cell culture for extraction of secondary products. The scope of biotechnology has been dramatically increased through genetic engineering.

Biotin ($C_{10}H_{16}O_3N_2S$ mw 244.31: Part of the vitamin B complex. It is also called vitamin H. It is present in all living cells, bound to polypeptides or proteins, and is of import in fat, protein and carbohydrate metabolism. It is a common addition to plant tissue culture media.

Bleach: A fluid, powder or other whitening [bleaching] or cleaning agent. Bleach contains calcium hypochlorite or sodium hypochlorite and is a common disinfectant used for cleaning working surfaces, tools and plant materials in plant tissue culture.

Bridge: A filter paper or other substrate used as a wick and support structure for a plant tissuer culture when liquid media are used.

Buchner funnel: A porcelain funnel with a perforated flat circular base employed in vacuum filtration. After E. Buchner [1860-1917].

Bud: 1. An embryonic or undeveloped, enemerged stem, leaf or flower, often enclosed by reduced or specialized leaves called bud scales. The beginning of incipient growth or development. 2, To graft onto another type of tree, plant. 3. A vegetative outgrowth from a yeast.

Bud-sport or bud mutation: A somatic mutation in a bud giving rise to a branch, fruit or flower that is a typical of the plant on which they occur. These characters [if beneficial] may be retained through vegetatrive propagation from the affected area.

Buffer: A substance in solution, or system that resists pH change, [despite addition of some small amount of acid or base] and withstands shock; as buffered media can resist pH drift. The degree to which pH change is resisted is a measure of the solution's buffering capacity.

Bung: A stopper or closure.

Burgeon: To flourish, bud, sprout or grow.

Cabinet [growth cabinet]: A cupboard suitable for incubating a small number of culture vessels or tubes under controlled environments. The degree of control over light, temperature or relative humidity is a function of quality of the cabinet but is generally less than diat implied by die use of the term incubator or laboratory incubator.calcium alginate: A hydrophyllic colloid obtained from seaweed. A derivative of alginic acid $[C_6H_8O_6]_n$. It reversibly immobilizes plant protoplasts in which condition they can tolerate a reduced osmotic potential.

Calcium hypochlorite [Ca[OCl]$_2$ 4H$_2$O mw 215]: An algicide, bactericide and fungicide capable of disinfecting and bleaching. It is used in dilute solution [5-10% w/v] as a plant tissue disinfectant often when agitating or using a vacuum. Tissue damage occur if contact is prolonged. Thorough washing with sterile water generally follows treatment.

Calcium pantothenate or pantothenic acid {Ca[C$_9$H$_{16}$NO$_5$]$_2$ mw 476.54}: The calcium salt of vitamin B_5. An occasional vitamin additive to plant tissue culture media. Solutions of this vitamin are not stable to autoclaving.

Callus, pl. *calli or* calluses: 1. Wound tissue, tissue formed on or below a wounded surface. 2. Disorganized tumor-like masses of plant cells that form in culture. These proliferate in an irregular tissue mass, and vary widely in texture, appearance and rate of growth; a function of the tissue type [species and explant] and the composition of the medium, the process of callus formation is callogenesis.

Callus culture: The cultivation of callus, usually on soldified medium and initiated by inoculation of small explants or sections from established organ or other cultures (inocula). Callus may be maintained indefinitely by regular subdivision and subculture. It may be used as the basis for organogenetic [shoot, root] cultures, cell cultures or proliferation of embryoids.

Cannula: A small tube for insertion into tissue; as a cork borer.

Caplin, S.M. and F.C. Steward: These men demonstrted [194's, 1950's] the promotive effect of coconut milk used with synthetic auxins for previously difficult to grow tissues and species.

Carbon source: A source of the non-metallic element carbon [C]; as organic substances like sugars taken up and metabolized by plant tissue cultures.

Casamino acid: An amino acid obtained through the digestion [acid hydrolysis] of the milk protein casein. Protein digest yields casamino acids and other substances. These are undefined constituents of some plant tissue culture media.

Casein: The principal protein of milk, a phosphoprotein.

Casein hydrolysate [edamin]: A milk protein digest product composed of amino acids [casamino acids] and other substances. This complex [undefined] product is sometimes used as an addition [0.02-2.10%] in nutrient solutions used in plant tissue culture as a non-specific source of organic nitrogen.

Caulogenesis: Shoot formation; as de novo shoot development from callus.

Cell count: The number of cells per unit suspension volume or callus weight. Tissue is treated with chromic acid [5-8%] or pectinase [0.25%) for up to 15 minutes followed by mechanical dispersion, then cell numbers are estimated with a hemocytometer [haemocytometerl.

Cell culture: The culture of single or groups of cells on solid or dispersed in liquid nutrient media [cell suspension cultures] usually following a set of defined growing conditions [protocols].

Cell hybridization: Formation of synkaryons, viable cell hybrids produced through cell fusion, they are identifiable by their increased chromosome number compared to the parent cells and the possession of characters found in one or another of the parental cells.

Cell line: Developmental history or descent, through cell division from a single original cell; as callus may constitute a group of cell, all descendants from a single cell plating. Numerous cell lines may be present in a culture. Any deviation in culture technique may favor one cell line over another.

Cell number: The absolute number or approximation of the number of cells per unit area of a culture or medium volume.

Cell selection: Selection within a group of genetically different cells; involves competition between cells often under some stress. Selection criteria may involve cell viability;! a unique phenotype or biochemical activity; or another basis for choice. Select cells or cell lines are generally relocated to fresh medium for continued selection and often are exposed to an increased level of the stress agent. The final ojective is usually to regenerate plants from those select cells, in hope that the plants will exhibit the traits selected for at the cellullar level.

Cell suspension: Cells and small aggregates of cells suspended in a liquid medium; as in cell suspension cultures. Explants, or callus derived from them are transferred to liquid medium and the cultures are then agitated on a mechanical shaker. The ensuing single cells and small cell clusters are used for a number of purposes in plant tissue culture; as in single cell cloning.

Cellulose or 4-glucanohydroIasc- An enzyme complex that hydrolyzes cellulose to the smaller fragments, cellobiose, by degrading (1-4) linkages. These in turn are hydrolyzed by cellobiase to glucose. Cellulase is used alone or, with other enzymes to digest plant cell walls to produce protoplasts.

Cellulysin: A brand name for cellulase isolated from *Trichoderma viride*; an enzyme used to degrade cellulose in the cell wall digestion process used in protoplast formation.

Centrifuge: An apparatus used for separating particles from suspension using centrifugal force. Balanced tubes containing the suspension are rapidly rotated causing sedimentation of particles to the bottom of the tubes (pelleting) based on their weight and the density of the suspending medium.

Chelate: A chemical compound with which metal atoms may be combined in order to prevent them from precipitating out of solution and so being unavailable to plants; as ethylenediamine tetra-acetic acid, disodium salt [Na_2EDTA] is complexed to iron [and other metal ions] in nutrient solutions used for plant tissue culture. The process is chelation.

Chemically defined medium: A nutrient medium for plant tissue culture in which all of the chemical components are fully known and stated. Quality [high purity] chemicals are utilized, and no undefined constituents are included.

Chemostat: An open, continuous culture system wherby the inflow of fresh medium including a growth limiting compound is monitored to maintain constant culture growth. Balancing the fresh medium inflow is a regulated outflow of cells and spent medium.

Chemotherapeutant: A chemical used to pretreat diseased source plants prior to excision or incorporated into media to support some theraperutic objective; as malachite green or virazole [ribavirin] are used to eliminate, virus in meristem tip culture. The process is chemotherapy.

Chimera : A plant or tissue composed of more than one kind of genetic tissue. In a periclinal chimera one genotype occurs in a superficial layer, covering a genetically different core, in a mericlinal chimera one genotype occurs in a localized pan of the plant and the rest is occupied by another genotype. In a sectorial chimera two genotypes share distinct sectors of the plant. Chimeras can occur naturally; he established artificially by grafting; or developed through somatic mutation by chemicals such as colchicine or by other means inplant tissue cultures.

4-chlorophenoxyacetic acid or para-chlorophenoxyacetic acid [pCPA, $C_8H_7O_3$Cl mw.186.59]: A synthetic hormone of die auxin type which is sometimes used in plant tissue culture media.

Choline chloride [C_5H_{14}ClNO mw 139.63]: An alkaloid, [B complex vitamin], occasionally added to plant tissue culture media. Choline, found in many plant organs, is die basic constituent of lecithin.

Chromic acid or chromium trioxide [CrO_3 mw 100.01]: This compound exists only in solution and is made by mixing potassium dichromate and concentrated sulphuric acid. A dilute solution may be used to prepare tissues for hemocytometer [haemocytometer], that helps to determine cell numbers in cell suspension cultures. Concentrated solutions were once used for cleaning new or dirty glassware but have now been supplanted by detergents.

Clonal multiplicaiion or clonal propagation: Vegetative [asexual] propagation from a single cell or plant. See clone

Clone: 1. A basic category of cultivar. The original cultivar or variety. Not to be confused with source clone, indicating different origins within the clone. Members have a common origiin are the extension of a single cell [via mitosis or plant and have been produced by vegetative means only. Genetic uniformity is accepted. A clone may have one or more sources. 2. While the expectation is that members of a clone are both phenotypically and genotypically identical, this assumption is not always true and members may not be genetically or phenotypically equivalent [varients]. The objective of much tissue culture propagation is to propagate very large numbers of selected plants with the same genotype. The process is cloning. Prefix mearly reflects the explant used to initiate the clone [meristem tip source clone]. In a calli, cloning involves the callus stage. 3. Specific DNA sequences are said to be cloned when isolated and propagated.

Closed continuous culture: A cell suspension culture with a continuous influx of fresh medium, maintained at constant volume by the efflux of spent medium. All cells are retained within the unit. See continuous culture.

Cocking, E.C. [I960]: Obtained the first higher plant protoplasts using root cells subjected to fungal cellulase and demonstrated new cell wall regeneration on protoplasts from tomato fruit locule tissue.

Coconut milk or ceconut water: Liquid endosperm from the center of the coconut seed. A complex, undefined addendum of variable quality and effects in some nutrient solutions [2-15%, v/v] for plant tissue culture. It has growth promoting effects and cell division factors. It is replaceable in some cases by cytokinins and/or sugar. It was first used by van Overbeek et al. [1941] to stimulate Datura embryo cultures.

Co-culture: The joint culture of two or more types of cells; as a plant cell and a microorganism or two types of plant cells; as is done in various dual culture systems or *the* nurse culture technique.

Colony: 1. A group of interdependent cells or organisms. 2. An aggregate of cells developed from a single cell; as in single cell platings, a clone composed of one cell line.

Complex substance: A complicated and undefined substance; as in some addenda to nutrient solutions used in plant tissue culture. Examples include protein hydrolysate, yeast or malt extracts, endosperm from corn or coconuts, juices like oranges or tomato and others.

Conditioning: 1. A phenotypic alteration resulting from the action of external agents during critical developmental stages. 2. An undefined interaction between media and tissues encouraging the growth of single cells or small aggregates. Conditioned medium may be used to induce growth of cells plated at low densities [below the minimum inoculum size] or cells unable to grow for unknown reasons. Conditioning may be accomplished by immersing cells or callus contained within a porous material [such as dialysis tubing) into fresh medium for a time dependent on cell density and a volume related te the amount of fresh medium.

Contaminant: An undesirable bacterial, fungal or algal microorganism accidently introduced to a culture or culture medim. It may overgrow the plant cells or inhibit their growth through release of toxic metabolities. Rigorous exclusion of potential contaminants must be practiced in plant tissue culture.

Contaminate: To accidentally introduce a substance or organism [contaminant] into a medium or culture. The process is contamination.

Continuous culture: A cell suspension culture with a continuous influx of fresh medium, maintained at constant volume by the efflux of spent medium [closed continuous] or with die efflux of cells and spent medium [open continuous].

Control: 1. The untreated plant *or* unchanged [standard] protocol or treatment for comparison with the experimental treatment. 2. To direct or regulate; as in induction of organogenesis in cultures through hormone regulation.

Controlled environment: A chamber, room or situation in which the environmental parameters of light, temperature, and relative humidity are controlled. The partial pressure of gases may also be controlled.

Corn milk: Corn endosperm tissue. An undefined complex addendum to some plant tissue culture media.

Cotrans formation: The transfer of a well characterized plant trait, expressed in vitro, which is linked to another plant trait or traits not expressed in culture.

Criticl concentration: The chemical concentration above or below which reaction or developmental process will not proceed.

Crown gall: A disease of plants in which tumors form. The causal agent is the bacterium, *Agrobacterium tumefaciens*. Gall tissue has been grown in vitro. A tumor inducing portion of the bacterial genome [the Ti plasmid] my be used experimentally as a genetic tool to incorporate [vector] genetic information into plant cells.

Cryoprotectant: An agent able to prevent freezing and thawing damage to cells as they are frozen or defrosted. These substances have high water solubility and low toxicity. They are classified either as permeating [glycerol and dimethyl sulfoxide] or non-permeating [sugars, dextran, ethylene glycol, polyvinyl pyrolidone and hydroxyethyl starch] agents.

Culture: 1. A general term for the cultivation of microorganisms, animals, plants or their cells in vivo or in vitro especially in order to improve the breed. 2. In vitro aseptic cultivation of microorganisms or of cells, tissues, or organs of plants, animals in or on prepared nutrient media under controlled, aseptic conditions.

Culture alteration: A persistent change in a cultured cell or tissue anatomy or physiology, including a change in one or more nutritional requirements or a change in its proliferative capacity.

Culture medium: A prepared nutrient solution (substrate) that may be chemically defined for growing plant tissues or other organisms in vitro.

Culture room: A controlled-environment room [light, temperature, relative humidity, often air conditioned, etc.] for incubation of plant cultures.

Culture tube: A test tube used to contain medium and an explant or culture derived from it, usually incubated in a controlled environment.

Culture vessel: A container used to hold medium and an explant or culture derived from it, usually incubated in a controlled environment.

Cybrid: A cell or plant cytoplasmic hybrid [heteroplast] with the nucleus of one and cytoplasmic organelles of another, or of both cells or plants.

Cysteine [$C_3H_7NO_2S$ mw 121.13]: An amino acid in proteins and precursor of coenzyme A and cysteine.

Cysteine hydrochloride [$C_3H_7NO_2S.HCl$ mw 157.63]: An occasional, amino acid [cysteine salt] additive in plant tissue culture media. It is included primarily for its antioxidant properties.

Cytochimera or chromosomal chimera: Cells within a tissue possessing different chromosomal numbers, as when a callus consists of diploid and polyploid regions of cells.

Cytodifferentiation: A functional and/or morphological differentiation occurring in cells during ontogenesis, affecting their phenotype.

Cytokinin: One of a large class of plant hormones [phytohormones] or synthetic growth substances. Cytokinins and many of their analogs have been chemically synthesized. They promote cell division and enlargement in cultures in the presence of auxins and have other effects such as control of organ [bud, root] differentiation; influence auxin transport; inhibit senescence, abscission of leaves and breaking of dormancy and apical dominance in buds. They resemble kinetin [K] [the type member] in physiological activity. They are N^6-substituted aminopurine compounds. In tissue culture, these hormones are employed to stimulate cell division and induce axillary bud proliferation. Cytokinin withdrawal stimulates rooting of plantlets. The most common cytokinins used in plant tissue culture are K [N-[2-furanylmethyl-lH-purin-6-amine], N-[phenylmethyl]-lH-purin-6-amine or 6-benzylaminopurine [BA or BAP] and N-[3-methyl-2-butenyl]-lH-purin-6-amine or 2-isopentyladenine (2-iP). Stock solutions of these are prepared by dissolving in HC1 acid [ca. 1 M] then making up to volume with water. These stock solutions are usually refrigerated.

Cytolysis: Cell dissolution or disintegration.

Cytoplasmic hybrid or cybrid: A cell or plant hybrid [heteroplast] with the nucleus of one and the cytoplasmic organells of another or of both cells or plants.

Cytoplasmic variant: A maternally inherited change in a cellular trait.

Cytotoxic: A chemical or other agent toxic to cells.

De novo: [Latin: *from the beginning, a new*]. Arising, sometimes spontaneously, from unknown or very simple precursors.

Decereration phase: The declining growth rate phase following the linear phase and preceeding the stationary phase in most batch suspention cultures.

Decontaminate: To free from contamination or surface sterilize.

Dedifferentiation: The resumption of meristematic activity by more or less mature cells through a reversal of the process of cell or tissue differentiation. Cell division leading to the formation of small, microvacuolate, isodiametric cells with prominent nuclei, that are capable of organized development; as when adventitious buds or roots develop on mature tissue.

Deficiency: 1. An insufficient supply or unusable form, of one or more nutritional or environmental requirements, for plant development, growth or physiological function. The absence of adequate conditions for plant growth or performance, resulting in disease. 2.The deletion of a gene or a series of genes. This may result in a mutant phenotype or may be lethal.

Dchumidifier: An apparatus that removes moisture from the air.

Deionized water: Water that has been passed through an ion exchange device to remove soluble minerals and some organic salts. The process is deionization.

Deletion: 1. Omission, removal or cancellation of some factor. 2. The lost portion of a chromosome or a nucleotide sequence in nucleic acid.

Derepression: The mechanism by which repression is alleviated; as when a repressing metabolite is removed, resulting in the increased level of a protein or enzyme.

Detergent: A cleaning agent; one of numerous synthetic cleaning preparations chemically different from soap. Detergents are commonly used to prepare work areas for aseptic plant tissue culture manipulations and to remove dirt and microorganisms from plant material prior to explanation.

Determination or topophysis: 1. The process of commitment to a specific pathway of development [cell, tissue, organism, growth form]. This commitment may occur well in advance of the appearance of the phenotype.The perpetuation of the phenotype provides an example of the stability of this process. 2. A phenomena in which explanted meristems and ensuing growth derived from them in culture, taken from different areas of a plant representing different phases, perpetuate the different phase phenotypes of that plant; as meristems from thorny [juvenile] citrus give rise to thorny individuals unlike meristems explanted from adult tissue that perpetuate the thornless condition in propagules derived from them.

Dewar flask: A double-walled glass flask used to keep liquids at other than ambient temperatures; as a thermos does.

Dicamba or 3,6-dichloro-2-methoxybenzoic acid [$C_8H_6O_3Cl_2$ mw 221.04]: A synthetic growth regulator of the auxin type with herbicidal properties. It is used to promote in vitro callus growth and as a herbicide. It is dissolved in base [Ca 1M KOH or NaOH] but is not very soluble. Alcohol is often used to dissolve 2,4-D. However, many tissues are sensitive to alcohol at low concentration

Dimethyl sulfoxide [DMSO, C_2H_6OS mw 78.13]: A highly hygroscopic liquid with little odor & colour. It is an organic cosolvent sometimes used in small quantities to dissolve neutral organic substances in plant tissue culture media preparation. Readily penetrates skin. An extremely powerful solvent. It can usually be replaced by a less toxic solvent. It also has uses as a cryoprotectant.

Direct embryogenesis: Embryoid formation directly on the surface of zygotic or somatic embryos or on seeding plant tissues in culture, without an intervening callus phase.

Direct organogenesis: Organ formation directly on the surface of relatively large intact explants, without an intervening callus phase.

Dispense: To give, deal or porton out; as nutrient medium is portioned into glassware for plant tissue culture.

Dissect: To cut, expose, separate or divide the parts of a plant or animal into sections.

Distilled water: Water that has been converted to steam and the vapour recondensed. This process removes dissolved materials, particulates and microorganisms. The process may be repeated [double distilled] for added purity, often in a glass apparatus [glass distilled] to minimize metal recontamination. Water of this purity is commonly used to make nutrient media for plant tissue culture.

Donor plant [mother plant]: The source that plant used for propagation, whether an explant, graft or cutting. Source plants used for micropropagation are usually pathogen-tested (disease-free).

Dry ice: Frozen [solid] carbon dioxide [CO_2]. It is commonly used as a refrigerant.

Dual culture: A culture system that includes plant tissue and one organism [such as a nematode species] or microorrganism [such as a fungus). Dual cultures are used to study host-parasite interactions or in the production of axenic cultures for a variety of purposes. Usually the microorganism selected is an obligate parasite.

Embryo culture : 1. Denotes a culture in which the explant was an embryo. Embryo cultures have been used to obtain viable offspring from seeds with a tendency for embryo abortion or when viable seed are limited in number. 2. Cultures in which embryos are induced to form [embryogenesis] whether in suspension or on a variety of explants or cultures on solidified media. More correctly, these nonzygotic or somatic embryos are termed embryoids.

Enation: Outgrowth on a plant surface.

Environmental chamber: A controlled environment cabinet [incubator] in which temperature, light quality, intensity and duration; and preferably also the relative humidity and airflow are controlled.

Enzyme: Any one of a number of spoecilized proteins produced in living cells, that speed the rate of specific chemical reactions, even at very low concentrations [organic catalyst], but is not used up in the reaction.

Equimolar: The same amount of solute per litre of solutiion, the same molar concentration.

Erlenmeyer flask: A wide-necked variety of conical, flat-bottomed flask commonly used for medium preparation. Smaller versions are used as culture containers. Named after E. Erlenmeyer [1825-1909].

Established culture : 1. Achievement of an aseptic viable explant, synonymous with Stage I culture [culture establishment] or transplant, synonymous with Stage IV culture [establishment in soil]. 2. A suspension culture subjected to several passages, adjusted so that cell number per unit time is constant from subculture to subculture.

Ethylenediaminetetraacetic acid, disodium salt [$Na_2EDTA, C_{10}H_{14}N_2O_8Na_2.2H_2O$ mw 372.25]: A chelating agent which reversibly binds positive ions such as iron, magnesium, etc. It is commonly added to plant tissue culture media to keep iron and other salts available by releasing them slowly into the medium as required.

Ethylmethanesulfonate [EMS, $C_3H_8SO_3$ mw 124.14]: A very potent chemical mutagen frequently used in mutagenic studies. It acts by adding ethyl groups to guanine, subsequently causing base paring errors as it binds to adenine.

Ex vitro: [Latin; from *glass*]. Organisms removed from culture and transplanted, generally to soil or potting mixture.

Excise: To cut, sever or otherwise remove or extract an organ or a segment of tissue from a plant or plant part; as the surgical removal of shoot tips. The process is excision.

Explant: The excised plant portion used to initiate a tissue culture. The process of dissection and removal to culture of these small organs or tissue sections is explantation. Explant choice, the timing of excision and pretreatment are important determinants of culture success.

Explant donor: The source plant or mother plant from which the explant used to initiate a culture is taken.

Exponential phase: A phase in culture in which cells undergo their maximum rate of cell division. It follows the lag phase and preceeds the linear growth phase in most batch propagated suspesion cultures.

Exude: To discharge slowly, leak liquid material [exudate] sometimes through pores or cuts but often by diffusion into the medium. This process of exudations associated with a lethal browning of explants in some woody plant species. These in vitro exudates have not been chemically characterized but are often referred to as tenins or oxidized polyphenols.

Factor: 1. A condition, influence or constituent to be taken into account. 2. A genetic character determinant [gene].

Ferric ethylenediamine tetraacetate, sodium salt [NaFe. EDTA $C_{10}H_{12}FeN_2NaO_8$ mw 367.07]: A microelement iron and sodium salt added to some plant tissue culture media. It may be substituted for $FeSO_4$ + Na_2.EDTA however, if substituted on an equimolar basis, while the same amount of Fe will still be available, half the Na will be lacking and may have to be added in the form of another salt

Field test: An evaluative test whereby the field performance of experimental plants is assessed in comparison to controls.

Filter sterilize: To sterilize by passing a solution through a porous material capable of separating out suspended microbes or their spores; as heat-labile components of nutrient media are sterilized.

Fixative: A compond that stabilizes, sets or fixes other compounds or structures securely so that their structural integrity is retained; as the process of fixation utilizes chemical agents to permanently prepare cells or tissues for microscopy.

Flame sterilize: To sterilize tools or instruments usually by heating them in a flame until they glow. This procedure is used in conjunction with ethanol [70%] emersion or ethanol (95%) dip. The most common means of sterilizing forceps and scalpels used in sterile tissue culture manipulations.

Floccule: An aggregation [coalescence] of microorganisms or colloidal panicles floating in or on a liquid. Flocculation is seen in some contaminated liquid media appearing as a cloud.

Fluid drilling: A mechanical procedure for planting seed; pre-germinated seeds are suspended in a gel and sowed through a fluid drill seeder. This technology is potentially adaptive for sowing artificial seed [somatic embryos or embryoids].

Fluorescence: the absorption of light of a specific wavelength followed by emission of light of a longer wavelength.

Folic acid [$C_{19}H_{19}N_7O_6$ mw 441.40]: A member of the vitamin B group [B_6]. It is also known as vitamin M. It is present in green leaves and has some coenzyme activities. It is occasionaly added to plant tissue culture media.

Freeze-dry or lyophilize: To dry in a frozen state under vacuum; as tissues are freeze-dried to obtain a dry weight or to preserve them for analysis.

Fresh weight or wet weight: The weight of a plant or plant part including the water content.

Furfuryladenine: N-[2-furanylmethyl]-lH-purin-6-amine or 6-furfurylaminopurine or furfuryladenine or kinetin [K, sometimes kN, $C_{10}H_9N_5O$ mw 215.22]: A growth hormone of the cytokinintype isolated from animal and plant DNA, and capable of promoting plant cell division. It is often used in plant tissue culture media. It is soluble in dilute HC1 [ca. 1M). The first cytokinin discovered was extracted from herring sperm DNA by Miller et al., 1955.

Fusogen: A fusion-inducing agent used for protoplast agglutination in somatic hybridization studies; as is polyethylene glycol.

GA_3: gibberellic acid.

Gametoclone: A plant regenerated from a tissue culture originating from gametic tissue.

Gelatin: A glutinous, proteinaceous material produced by boiling, which partially hydrolyzes the collagen,of animal connective tissues. It is sometimes used to gel or solidify nutrient solutions for plant tissue culture. Other gelling agents, specifically agars, are usually preferred.

Gelrite: The brand name fo a synthetic [Pseudomonas-derived) refined polysaccharide used as a gelling agent and agar substitute.

Generation time: The time between successive generations of cells or organisms within a population.

Generative: Refers to a somatic cell or tissue.

Genetic engineering or recombinant DNA technology: Technology involving man-made changes in the genetic constitution of cells [apart from selective breeding]. This technology usually employs a vector (such as the Ti plasmid of *Agrobacterium tumefaciens*) for transferring useful genetic information from a donor organism into a cell or oganism that does not possess it. Genetic engineering has many potential uses.

Genetic selection: Selection of genes, cells [cell selection], clones, etc., by man within populations or between populations or species. The usual purpose is to alter a specific phenotypic character. Such selection usually results in differential success rates of the various genotypes, reflecting many variables including selection pressure and genetic variability in populations.

Genetic transformation: The transfer of extracellular DNA [genetic information] among and between species; as, for example, with the use of bacterial or viral vectors.

Genetic variation: Differences in individuals derived from the same genotype in distinction to differences caused by the environment. Genetic variance denotes the proportion of phenotypic variance caused by differences in the genetic make-up of an individual.

Genome: The basic haploid chromosome set of an individual; the sum of its genes.

Genotoxic: Carcinogenic; toxic to the chromosomes.

Geotaxis: Plant orientation with respect to gravity.

Germplasm or germ plasm: 1. The reproductive body tissues distinct from somatic [nonreproductive tissues]. The genetic material, basis of heredity of an organism, passed on through previous generations. 2. An individual representing a type of species or culture that may held in a repository for agronomic, historic [or other) reasons.

Gibberellic acid [gibberellin A_3 or GA_3, $C_{19}H_{22}O_6$ mw 346.37]: One of the gibberellins, a group of growth hormones promoting cell division and elongation. The first of the group to be isolated and the most widely used gibberellin in plant tissue culture. Isolated from the fungal pathogen *gibberella fujikuroi.* It dissolves in base [ca. 1M KOH or NaOH].

Glutamic acid or glutamate [glu, $C_5H_9NO_4$ mw 147.13]: An amino acid involved in nitrogen metabolism. It is added to some plant tissue culture media as a source of reduced nitrogen.

Gnotobiotic: The growth of an organism alone [sterile] or in the presence only of known organisms.

Gro-lux: The brand name for a type of wide spectrum fluorescent lamp.

Growth inhibitor: Any substance [its own or that of another organism] that inhibits the growth of an organism. The inhibitory effect can range from mild inhibition [growth retardation] to severe inhibition or death [toxic reaction]. Two hormones that may act as inhibitors are ethylene and abscisic acid (ABA). The concentration of the substance and the length of exposure to it [dose] determines its effects, as do other factors such as the relative susceptibility of the organisms exposed to it.

Guha, S. and S.C. Maheshwari [1966]: Were among the first to obtain haploid [n] plantlets from anther culture.

Haberlandt, G.: An inspirational pioneer botanist who was the first to attempt to culture plant cells at the turn of the 20th century. He predicted the existence of hormones that would stimulate cell division [auxins] and suggested the possibility of exploiting the totipotentiality of plant cells.

Hanging droplet technique: See microdroplet array technique.

Hemicellulase: An enzyme from *Aspergillus niger,* available as a commercial preparation, that degrades hemicellulose to galactose.

HEPA filter: An acronym for high efficiency particulate air filter. A filter capable of screening out particles larger than 0.3 m. They are used in laminar air flow cabinets [hoods] for sterile transfer work.

Heterograft or xerograft: An interspecific graft. The likelihood of success of this graft is proportional to the degree of relatedness between the donor and recipient.

Heteroplasmon or heteroplast: A cell with a mixture of two types of cytoplasm [cytoplasmic hybrid]; cell with foreign organells.

Hexitol: A six carbon sugar alcohol, such as [meso-, myo-, or i-] inositol.

Hood: Laminar air flow cabinet.

Hydrate: To add or incorporate water. The process is hydration.

Hypertrophy: An abnormal increase in cell size causing irregular swelling oir growth. Hypertrophy is usually a disease or other stress-induced response.

Hypochlorite: A salt of hypochlorous acid [sodium hypochlorite, potassium hypochlorite or calcium hypochlorite]. All are oxidizing agents used for disinfecting and for bleaching.

IAA: lH-indole-3-acetic acid.

IBA: H-indole3-butanoic acid.

In situ: In the natural, original place or position; as in the location of the explant on the mother plant prior to excision.

In vitro: [Latin = *in glass*]. Experimentation on organisms or portions thereof in glassware or culture; growing under artificial conditions as in tissue culture. The antonym of in vivo.

In vivo: [Latin = *in life*] Experimentation on organisms under natural conditions within intact living organisms, the antonym of in vitro.

Incubate: To maintain under conditions favorable for development, often in an incubator. The process is incubation.

Incubator: An apparatus providing controlled environmental conditions, [light, photoperiod, temperature, humidity, etc.] suitable for [incubating] plants or plant cultures. This term is sometimes used as as synonym for cabinet [growth cabinet] but usually implies a greater degree of environmental control.

Indirect embryogenesis: Embryoid formation on cellus tissues derived from zygotic or somatic emtryos. seedling plant or other tissues in culture. The antonym is direct embryogenesis.

Indirect organogenesis: Organ formation on callus tissues derived from explants. the antonym is direct organogenesis.

IH-indole-3-acetic acid [IAA, $C_{10}H_9NO_2$ mw 175.18]: A naturally occurring plant growth hormone and the principal plant auxin. It is commonly used in plant tissue culture media. It dissolves in base [KOH or NaOH ca. 1 M]. It is unstable to light so is usually stored in the dark.

IH-indole-3-butanoic acid [IBA, $C_{12}H_{13}NO_2$ mw 203.23]: A naturally occurring plant growth hormone of the auxin type. It is synthetically produced and commonly used in plant tissue culture media, and in horticulture to promote rooting of cuttings. It dissolves in base [ca. 1 H KOH or NaOH].

Induction media: 1. Media that can induce organs or other structures to form. 2. Media that will cause variation or mutation in the tissues exposed to it.

Infestation: Occupation by threatening numbers of insects, mites or potential disease agents; as mite infestations in incubators. An infesting organism can cause contamination of tissue cultures. This should not be confused with infected, as infection is not necessarily implied.

Infra red gas analyzer [IRGA]: An instrument for measuring the proportion of a particular gas in a mixture; as changes in CO_2 concentration can be monitored to evaluate a plant's photosynthetic or respiratory activity.

Inoculate: 1. To deliberately introduce something into; as inoculum is placed into [or onto] medium to initiate a culture. This process is inoculation, but should not be confused with contaminate. 2. Vaccinate.

Innoculum: 1. Material introduced [inoculated] onto or into a host or medium. 2. Potential inculum implies that it is infective and may by chance result in natural inoculation of a host.

Innoculum size: A critical minimum volume [minimum inoculumsize] is necessary to initiate some culture growth due to the diffusive loss of cell materials into the medium. Inoculum size depends on medium volume and culture vessel size, and affects the subsequent culture growth, cycle. Use of conditioned medium, that is medium that has supported culture growth, can decrease the minimum inculum size.

Inositol; myo, meso, or i-inositol or hexahydroxycyclohexane [$C_6H_{12}O_6$ mw 180.16]: A widely distributed sterioisomeric sugar alcohol in plants and essential in animals. Inositol is included by many in the B complex vitamin group. It is involved in the synthesis of phospholipids, cell wall pectins and membrane systems in cell cytoplasm. It is commonly added to the nutrient medium used for plant tissue culture [ca. 100 mg/l] for its growth promoting effects.

Insertion element: Generic term for DNA, insertion sequences found in bacteria capable of genome insertion. Postulated to be responsible for site-specific phage and plasmid integration.

Ionizing radiation: High energy protoplasm-injring radiation that can break covalent chemical bonds or remove electrons from atoms, attaching them to other atoms producing charged ion pairs. This radiation includes, U.V., x-rays, γ rays and β-particles.

Irradiate: I. To illuminate. 2. To emit or expose to waves of light, heat or nuclear emissions. 3. To treat by exposure to radiation. The process is irradiation.

Irradiance: The total radiation on an exposed surface.

Isograft or syngraft: A graft or transplant among isogenic [genetically identical] individuals; as on the same organism.

Isolation medium: A medium suitable for explain survival and development. It may be synonymous with Stage I medium, or contain antioxidant[s], reduced hormone concentration, bacterial indicators, and preceed Stage I culture.

Isopropanol [C_3H_8 mw 60.09]: An alcohol that is sometimes used for disinfecting purposes as a less costly alternative to ethanol.

Jiffy pots: A brand name for peat pots tliat are sometimes used for ex vitro transplantation.

Juvenile: The immature, usually non-reproductive (reproductively incompetent), sometimes phenotypically distinct phase of plant growth. Juvenility should not be confused with young in age although young plants may also be in a juvenile phase. Juvenile phase tissue has repeatedly been found to perform better in culture than adult phase tissue. It is more likely to grow and to form callus, embryoids, shoots etc.

Kinetin [K, sometimes Kn]: [2-furanylmethyI]-lH-purin-6-amine.

Kinin: The original class name for substances promoting cell-division to which the prefix *cyto* has been added [cytokinins] to distinguish them from kinins in animals systems.

Lag phase: 1. Used to describe the first of five growth phases of most batch propagated cell suspension cultures in which inoculated cells in fresh medium adapt to new environment and prepare to divide.

Laminar air flow cabinet or hood: A structure in which a uniform flow of filtered air prevents settling of particles in the work area. The air is filtered through prefilter [furnace-filter quality], then a high efficiency particulate air [HEPA] filter that strains out particles greater than 0.3 m. These hoods are commonly employed for aseptic plant tissue culture manipulations and are designed to accommodate one to the several persons, depending on the model.

Layering: 1. Covering stems, runners or stolons with soil causing adventitious roots to form at the nodes enabling propagation by root cuttings. This procedure is used commercially to propagate many plants, such as the brambls. 2. In vitro layering involves the horizontal placement on agar of cultured shoots [with or without leaves] or nodal segments to promote axillary bud proliferation.

Lecithin: A naturally occurring, choline-containing phospholipid present in animal and plant tissue. Chemically lecithins are similar to fats, but they also contain phosphorus and nitrogen.

Limiting requirement or factor: An environmental variable whose absolute level at a given time limits the growth or other activity of an organism.

Linear phase: The constant increase in cell number following the exponential growth phase and proceeding the deceleration phase in most batch suspension cultures.

Liquid culture: The culture of plant cells on liquid medium, in suspension or on supports. Cultures are either held steady [stationary culture] or are shaken [agitated culture or shake culture].

Liquid medium: Medium not solidified with a gelling agent. Liquid media are used for suspension cultures and for a wide range of research purposes. They are also useful for Stage I cultures in some micropropagation protocols. Usually a support structure or wick is used to hold the tissue above the nutrient medium.

Lyophilize or freeze dry: To freeze rapidly then dehydrate under high vacuum, the process is lyopholization.

Lysis: Cell rupture or destruction, as through enzymatic action.

Macerate: To disintegrate or separate tissues through cutting, soaking, enzymatic or other action, resulting in cell dissociation.

Macerase: A brand name for pectinase. Pectinase is useful in the isolation of intact cells and protoplasts of higher plants.

Magnetic stirrer: An apparatus used for stirring solutions. A magnetic stir bar is placed at the bottom of the container and is twirled in the solution that requires mixing by an electrically driven, rotating magnet below the platform of the apparatus.

Malachite green [$C_{22}H_{25}N_2Cl$ mw 364.66]: A chemical used as a seed disinfectant in agriculture. It is included in some tissue culture media as an antiviral agent but has received mixed reviews.

Malt extract: An extract from dried and ground, germinated grain seed [usually barely, sometimes rice or corn]. It is an undefined constituent of some plant tissue culture media.

Mannitol [$C_6H_{14}O_6$ mw 182.17]: A sugar alcohol widely distributed in plants and often employed as a nutrient and osmoticum in plant tissue culture work; as in suspension medium for plant protoplasts.

Mannose [$C_6H_{12}O_6$ 180.16]: A hexose component of many polysaccharides and mannitol. It is occasionally employed as a carbohydrate source in plant tissue culture media.

Maternal inheritance: Inheritance controlled by extrachromosoml [cytoplasmic] hereditary determinants.

MDA: Microdroplet array technique.

Medium: The substrate for plant growth; as nutrient solution, soil, sand, etc. This is a general term for the liquid or solidified formulation upon which plant cells, tissues or organs develop in plant tissue culture.

Medium formulation: One of many available, usually derived formulas that are used in plant tissue culture. The formulas commonly contain the macroelements and microelements [high and low salt formulas are available], some vitamins [B vitamins, inositol], hormones [auxin, cytokinin and sometimes gibberellin], a carbohydrate source [usually sucrose or glucose] and often other substances such as amino acids [the most common addition is glycine] or complex growth factors. Media may be liquid or solidified with agar, the pH is adjusted [ca. 5-6] and the solution is sterilized [usually by filtration or autoclaving]. Some formulations are very specific in the kind of explant or plant species that can be maintained, some are very general.

Mercuric chloride or mercury bichloride [$HgCl_2$ mw 271.52]: This compound was once commonly used as an antiseptic and fixing agent [0.05-0.10%] for plant material. However, it is highly toxic and is now little used for this purpose.

Mericlinal: A meristem tip source clone. This term can be synonymous with calliclone or may not involve callus.

Meristem: A localized region of continuing mitotic cell division [meristematic cells], of protoplasmic synthesis and tissue initiation. From these undifferentiated tissue new cells arise that differentiate into specialized tissues. Meristems are located at the apical [shoot and root tips], axillary, marginal or lateral [cambia] and other growing points. In addition, in vitro meristems may occur within more or less differentiated callus tissue and are termed meristemoids. The meristematic dome without any leaf primordial tissue is sometimes used as an explant in virus elimination work but usually a meristem tip [meristematic dome plus one pair of leaf primordial] is the explant.

Meristem culture: The culture of meristems; meristematic dome tissue without adjacent leaf primordia or stem tissue. In practice this usually means meristem tip culture. This may more generally imply the culture of meristemoidal regions of plants or of meristematic growth [associated with or sharing the characteristics of the meristem] in culture.

Meristem tip: A common explant comprised of the meristem [meristematic dome] and usually one pair of leaf primordia. Also refers to apical [apical meristem tip] or lateral [lateral or axillary meristem tip] origin. The term meristem tip is often confused with the term shoot tip, which is much larger and usually has more immature leaves and some stem tissue.

Meristem tip culture: Cultures derived from meristem tip explants and are excisd for virus elimination or axillary shoot proliferation purposes, less commonly for callus production.

Meristemming: The utilization of or process of utilizing meristem tips for explants. This term is often used to refer to micropropagation via axillary shoot proliferation.

Meristemoid: A cluster of small, isodiametric meristematic cells with a meristem, or cultured tissue, with the potential for developmental [totipotential] growth. These cells have a dense microvacuolated cytoplasm and a high nucleo-cytoplasmic ratio. They may give rise to plant organs [shoots, roots] or entire plantlets in culture.

Methionine [Met, $C_5H_{11}NO_2S$ mw 149.21]: An amino acid precursor in ethylene synthesis and occasionally added to plant tissue culture media.

Methyl methanosulfonate [MMS, $C_2H_6SO_3$ mw 110.13]: A frequently used, very potent chamical mutagen which acts by adding methyl groups to guanine and subsequently causes base pairing errors as it binds to adenine.

N-[3-methyl-2-butenyl]-1H-purin-6-amine or isopentenyladenosine [2ip or IPA, $C_{10}H_{13}N_5$ mw 203.20]: A synthetic cytokinin similar in structure to zeatin and commonly used in plant tissue culture media. It dissolves in acid [HC1 ca. 1M].

Microclimate or microhabitat: The climate in the immediate vicinity or surrounding an organism or its parts.

Microcutting: A tiny cutting, as an explant such as a meristem tip removed for culture with the use of a dissecting microscope.

Microdroplet array or multiple drop array [MDA] or hanging droplet technique: Introduced by Potrykus, Harms and Lorz [1979], this technique is used to evaluate large numbers of media modifications, employing small quantities of medium into which are placed small numbers of cells. Droplets of liquid culture [medium and suspended cells or protoplasts] are arranged on the lid of a petridish, inverted over the bottom half of the dish containing a solution with a lower osmotic pressure, and the dish is sealed. The cells or protoplasts form a monolayer at the droplet meniscus and can easily be examined.

Micrograft: An in vitro graft. This procedure involves placing a meristem tip or shoot tip explant onto a decapitated rootstock, that has been grown aseptically from seed or micropropaged. The purpose may vary, from virus elimination, using meristem tips and virus-free rootstocks, to viral assays or to erxpedite grafting that is usually done in the greenhouse. The explanted scion is placed directly on the cambium of the severed stock or inserted into a T or inverted T-shaped epidermal incision on the stock. This technique has been used to eliminate viruses, viroids and mycoplasmas from Citrus species.

Micromole [M]: One millionth [0.000001] of mole [10^{-6} M]. The unit of concentration used for microelements, vitamins, hormones and some other organic addenda used in plant tissue culture media.

Micropropagation: Refers to propagation in culture by axillary or adventitious means. It is a general term for vegetative [asexual] in vitro propagation. It sometimes refers specifically to axillary bud proliferation.

Microwave oven: An oven in which heating results from microwave radiation of the food or contents. Such ovens are commonly emloyed to melt agar in making up plant nutrient media and may be useful to sterilize liquids and some plastic or glass items.

Minimum effective cell density: The inoculum density below which the culture fails to give reproducible cell growth. The minimum density is a function of the tissue [species, explant, cell line] and the culture phase of the inoculum suspension. Minimum density decreases inversely to the aggregate size and division rate of the stock culture. It is important to know this ratio when plating suspension cultures for selection of single cell lines so that colonies will not overgrow one another prior to obtaining readily manipulable size.

Minimum inoculum size: The smallest inoculum that can be successfully used for subculture.

Mitotic index [MI]: The ratio of nuclei undergoing mitosis [including prophase] to total nuclei. MI - [Number of nuclei in mitosis/Total number of nuclei examined in the sample] X 100. A MI of 0.3 means that 30% of cells in the population are observed in mitosis.

Mitotic nondisjunction: Occurs when sister chromatids fail to migrate to opposite poles of the cell during mitotic anaphase. The result is daughter cells with hyperploid and hypoploid chromosome counts.

Mixoploid: Refers to cells with variable [euploid, aneuploid] chromosome numbers; as mosaic or chimaeral components that differ in chromosome number, or the result of a variety of mitotic irregularities.

Mold: Any fungus growth; as on plants or in cultures.

Monoculture: A one-crop agriculture system. An intensified monoculture could refer to a clonal [one genotype] agricultural system.

Mononucleotide: A nucleic acid basic-building-block. Composed of a pentose, a phosphoric acid plus a purine or pyrimidine base.

MS: An abbreviation for the Murashige, T. and F. Skoog [1962] medium formulation. MS is a defined medium formulated for tobacco callus culture and now used more than any other for the culture of a very wide range of plants species. It is characterizd by higher levels of nitrogen [both nitrate and ammonium], potassium and calcium than most other plant culture media.

Mutability: The propensity of an individual's genes or genotype to undergo heritable mutation.

Mutagen: A chemical or physical treatment or agent capable of producing genetic mutation. Causes insertion, deletion or alteration of a base or part of the nucleic acid chain [DNA]. Artificial mutations can be induced in cells using chemicals or ionizing radiation. The process is mutagenesis. Electromagnetic and partical radiation, such as U.V., X-rays and β particles all increase mutation frequency. So do chemical mutagens such as ethylmethanesulphonate, acridine, nitrous oxide and many others.

NAA: 1 -naphthaleneacetic acid.

NAA_m or -naphthaleneacetic acid [NAA, $C_{12}H_{10}O_2$ mw 186.20]: A synthetic hormone analog of the auxin type. It is frequently used in plant tissue culture media and in horticulture to promote rooting of cuttings. It dissolves in base [ca. 1M KOH or NaOH].

Naphthaleneacetic acid: 1-naphthaleneacetic acid.

2-naphthaleneacetamide acetic acid or 2-naphthoxyacetic acid or b-naphthoxyacetic acid [NOA or b NOA, $C_{12}H_{10}O_3$ mw 202.20]: A synthetic hormone analog of the auxin type. It is sometimes used in plant tissue culture media and in horticulture to promote rooting of cuttings. It dissolves in base [ca. 1M KOH or NaOH].

Natural complex: A complicated, non-synthetic, often undefined addendum in plant tissue culture media, such as orange or tomato juice or coconut milk.

Neoplasm: Localized cell multiplication or tumor; a collection of cells which have undergone genetic transformation. These cells differ in structure and function from the original cell type.

Neoteny: The occurrence [retention] of juvenile characters in the adult state or vice versa.

Newcommer's fluid: A fixative used in chromosome analysis composed of 6 parts isopropyl: 3 parts- propionic acid: 1 part petroleum ether: 1 part acetone: 1 part dioxane [with or without ferric acetate].

Niacin or riicotinic acid [$C_6H_5NO_2$ mw 123.11]: Vitamin B_3 forming part of a respiratory coenzyme. It is a common micronutrient addition to plant tissue culture media. It is sometimes also known as vitamin PP.

Niacinamide or nicotinamide [$C_6H_6N_2O$ mw 122.12]: An amide of niacin occasionally added to plant tissue culture media.

Nipple flask: A flat bottomed, round sided flask with many side projections or nipples. Suspension cultures are agitated and aerated in these flasks as medium flows into and out of the projections while the apparatus rotates. The flasks and rotating apparatus were designed by Steward et al. ca. 1952.

Nitsch, J.P. and C. Nitsch [1969]: Developed a now widely used plant tissue culture medium formulation NN, for use in anther culture.

NN: An abbreviation for the Nitsch, J.P. and C. Nitsch [1969]medium formulation.

NOA: [2-naphthalenyloxy] acetic acid.

Nobecourt, J.R. [1937]: He was among the first to describe the prolonged culture of carrot callus.

Non-competent: Cells or tissues incapable of undergoing morphogenesis. The antonym is competent.

Nucellar embryony: The process by which individual cells of the nucellus give rise to somatic embryos [embryoids]. Plants on which this occurs naturally are considered good candidates for tissue culture of somatic embryoids.

Nucellus: An enzyme that can hydrolyze internucleotide linkages of a nucleic acid.

Nucleo-cytoplasmic ratio: The ratio of cell nuclear to cytoplasmic volume. This ratio is elevated in meristematic cells and low in differentiated cells.

Nurse: A culture technique or the callus upon which a filter paper is placed separating single cells from the callus in the paper raft technique of Muir [1953]. The callus [nurse tissue] releases growth factors and nutrients that induce growth in a single cells supported by the filter paper (those being nursed] and sharing the communal environment.

Nutrient: A nutritive substance or ingredient, as are the major and minor mineral elements [macroelements and microelements] necessary for plant growth and development as well as the organic addenda, sugars, vitamins, amino acids and others employed in plant tissue culture media.

Nutrient film technique [NFT]: Hydroponic plant growth whereby plant roots are suspended in shallow, slowly circulating nutrient solutions delivered as a continuous film of liquid rather than in bursts of liquid.

Nutrient gradient: A diffusion gradient of nutrients and gases is set up in tissues where only a portion of the tissue is in contact with the medium. This is common in callus culture and causes differential growth rates and senescent regions of callus. Similarly diffusion gradients of nutrients are set up in the medium adjacent to the tissue. Gradients are less likely to form in liquid media.

Ontogeny: The developmental [ontogenctic] life history of an organism, its development from fertilized egg [zygote] to seed-forming adult [reproductive phase], with emphasis on embryonic development.

Open continuous culture: A cell suspension culture with a continuous influx of fresh medium, maintained at constant volume by the efflux of cells and spent medium.

Organ culture: The growth in aseptic culture of plant organs such as roots or shoots, beginning with organ primordi or segments and maintaining the characteristics of the organ.

Organic cosolvent: Compound used to dissolve some neutral organic substance, as in media preparation. These include alcohol [usually ethanol], acetone and dimethylsulfoxide [DMSO].

Organized growth: The in vitro development of organized explants such as meristem tips or shoot tips, floral buds or organ primordia, or their de novo formation from unorganized tissues.

Organogenesis: The initiation [de novo] and growth of organs [roots and shoots, usually] from cells or tissues; as in organ culture. Organs may form on the surface of explants [direct organogenesis] or upon an intervening callus phase [indirect organogenesis[.

Organoid: An anomalous organ-like structure formed in culture; as on leaves, roots or callus.

Ortet: The original mother plant or donor plant from which vegetatively propagated plants are derived.

Osmium tetroxide or osmic acid [OsO_4 mw 254.20]: A fixing agent used to prepare tissues for electron microscopy.

Osmolarity: The total molar concentration of the solutes affecting the osmotic potential of a solution or nutrient medium.

Osmometer: A device for measuring the osmotic pressure [water potential] of solutions.

Osmoticum: An agent such as glucose or sucrose employed to maintain the nutrient medium osmotic potential equivalent to that of the cultured cells [isotonic]. This prevents cell damage in vitro.

Ovulary culture: A culture in which the explant is an ovary containing the ovule[s]. The ovules may be fertilized in culture [in vitro fertilization]. This technique is used primarily when the ovary is essential to proper embryo development.

Ovule culture: A culture derived from an explanted ovule; which may be fertilized in culture [in vitro fertilization]. This technique is used to study development of zygotes and young embryos and is sometimes used to rescue embryos susceptible to abortion when embryo culture is not possible.

Oxygenase: An enzyme enabling an organism or system to utilize atmospheric oxygen.

Packed cell volume [PCV]: A quantitative method of estimating cell growth. It is based on the total cell volume in an aliquot of suspension culture. The aliquot is centrifuged for 5 minutes at 200 g after which the packed cell volume is expressed as a percentage of the aliquot volume.

Panicle culture: Aseptic culture of grain panicle segments, usually in an effort to induce microspore development.

Pantothenic acid [C9$H_{17}NO_5$ mw 219.23]: Also known as vitamin B_5 it is of wide occurrence in most plant and animal tissue, is essential for cell growth and is an important coenzyme in fat metabolism. It is added to some plant tissue culture media as the calcium salt.

Papain: A water soluble proteolytic enzyme [protease] extracted from papaya fruit and used especially as a meat tenderizer.

Paper raft technique: A technique developed by W.H. Muir (1953) to promote development of single cells taken from suspension cultures in which cells are placed onto filter paper squares set on actively growing callus [nurse tissue]. Growth factors and nutrients from the callus tissue diffuse through the filter paper, promoting cell growth and development.

Para-amino-benzoic acid [Paba, $C_7H_7NO_2$ mw 137.12]: An occasional addition [vitamin Bx] to plant tissue culture media.

Parahormone: A substance with hormone-like properties that is not a secretory product; such as ethylene or carbon dioxide.

Parasexual hybridization: Refers to genetic recombination by means other than normal fertilization of germ cells [parasexual] leading to hybrid cells or individuals; as in hybrid cells or plants derived from somatic cell fusion.

Particle radiation: Refers to - particles [positively charged] and β-particles [negatively charged], electrons, protons and neutrons. These particles are used to produce mutant cells or organisms in plant tissue culture.

Parts per million[ppm]: This term has largely been replaced by the equivalent mg/1 of solution or mu 1/1 for liquids or gasses.

Passage number: The number of subculture intervals. Culture age is a function of passage number and the culture dilution ratio.

Passage time: The interval between subcultures or the culture period.

PCV: packed cell volume.

Pectinase: An enzyme that degrades pectin, the adhesive material that cement cells together. It is used alone or with other enzymes to digest the polygalacturonic acid of plant cell walls (to sugar and galacturonic acid) in protoplast production.

Peptide: A compound consisting of two or more amino acids covalently linked. Peptides with three or more amino acids are polypeptides.

Periclinal chimera: Refers to a condition in which genotypically or cytoplasmically different tissues are arranged in concentric layers.

pH: A measurement of the degree of acidity or alkalinity of a solution. The negative log of the hydrogen ion concentration, in moles per litre (M/1). On the pH scale from 0-14, there is a 10 fold difference for each unit of change of pH. The lower the value, the more alkaline [more hydroxyl ions it contains]. Most culture media are adjusted in the range pH 4-6 using 1M NaOH or HC1.

pH drift: A shift in pH that occurs as cultures grow and is related to the buffering capacity of the medium.

Phenocopy: A non-hereditary phenotypic change that is environmentally induced, during a limited developmental phase of an organism, that mimics the effect of a known genetic mutation.

Phenolic oxidation: The process by which many plant species which contain phenolics blacken through oxidatin when they are wounded; as in vitro, soon after explanation or subculture. This may lead to growth inhibition or, in severe cases, to tissue necrosis and death. Antioxidants are incorporated into the sterilizing solution or isolation medium to prevent or reduce oxidative browning.

Phenols or phenolics: A class of organic compounds with hydroxyl group[s] attached to the benzene ring forming esters, ethers and salts. Phenolic substances may bleed from newly explanted tissues, oxidizing to form colored compounds visible in nutrient media or on the filter wicks supporting the tissue.

Phloretic acid: A phenolic acid formed by the oxidation of a hydroxyl [OH] of 1,3,5-trihydroxybenzene. An occasional constituent of some plant tissue culture media.

Photoheterotroph: A photosynthetic organism that requires an organic hydrogen source. The condition is photoheterotrophy.

Photometer: An instrument for measuring the luminous intensity of light sources.

Photosynthetically active radiation [PAR]: The photon flux expressed as moles or microeinsteins per meter square per second [$molm^{-2}$ or Em^2S^{-1}] or as watts per meter squared [Wm^{-2}] over the wavelength range of 400-700 nm; the part of the light spectrum which is primarily absorbed by plants and used in pohotosynthesis.

Phytokinin: An absolute term for cytokinin.

Phytotron: A controlled environment building or chamber for studying plant growth under defined conditions.

Plant tissue culture: A general term encompassing the in vitro culture of plant cells, tissues, organs and whole plantlets.

Plantlet: A shoot, sometimes exclusively a rooted shoot, growing in culture or derived from culture.

Plasmid: A small circular molecule of double stranded DNA occurring naturally in bacteria and yeast. They replicate independently as the host cell proliferates. They often carry vital genes such as may confer resistance to environmental, biological or chemical factors. While some plasmids have large and difficult to isolate DNA, others are small in size and can easily be separated from the host cell, genetically modified via gene insertion, and proliferated in a bacterial [usually] cell culture. By means of an appropriate vector [for plants this could be *Agrobacterium rumefaciens*], the modified plasmid genome (or part of it) can be introduced into individual plant cells or protoplasts in vitro. The cells containing the new genetic information can be used to regenerate plants possessing the desired genetic information. New plant cultivars possessing novel characteristics such as resistance to pathogens, herbicides and environmental stress; morphological or physiological characteristics are theoretically possible by these means.

Plasmotype: A cell type displaying features which are expressions of cytoplasmic, rather than nuclear inheritance.

Plasticity: The range of environmentally-inducible phenotypic expressions.

Plating efficiency: An estimate of the percentage of viable cell colonies developing on an agar plate relative to the total number of cells spread onto the plate. Plating efficiency is a function of the tissue [species, explain and cell line); medium composition; plating density and the phase of the stock culture.

Pleiotropy: The condition in which several characteristics are effected by a single gene.

Pollen culture: The culture of pollen grains, which germinate in vitro. Such cultures may eventually form monoploid callus, from which shoots or embryoids develop into monoploid plants.

Polyethylene glycol [PEG] or carbowax: A fusion-inducing agent [fucogen] for agglutinating protoplasts which is used in somatic hybridization studies. This compound is also sometimes used in media as a non-metabolite osmoticum and is available in various molecular weights, ranging from ca. 200 to 6,000.

Polygalacturonic acid: A long chain sugar acid polymer composed of galacturonic acid and hexose subunits.

Polygenes: Systems of genes associated with quantitative character variation in which each gene individually effects the phenotype in a minor way.

Polypropylene: A strong, flexible, transparent thermoplastic formed by the polymerization of propylene. It is used in many labware products.

Polyvinylpyrrolidone [PVP, $[C_6H_9NO]n$ mw [111.145]n]: An occasional constituent of plant tissue culture isolation media. It is of variable molecular weight and has antioxidant properties so is used to prevent oxidative browning of explanted tissues. It is less frequently used as an osmoticum in culture media.

Population density: Cell number per unit medium area or medium volume.

Potassium hypochlorite [KCIO mw 90.6]: A water soluble bleaching or disinfecting agent used in plant tissue culture practices.

Potato extract: A common [undefined] organic addendum to plant tissue culture media in monocot and another culture systems.

Preadaptation: In possession of advanageous characters enabling an organism to be well suited [adapted] to environmental conditions previously unknown to it.

Precipitin test or microprecipitin test: A serological assay in which visible particulates [precipitates] form when soluble antigen and antibody react. This precipitin reaction detects and identifies antigens.

Prefilter: A coarse filter [furnace filter] such as those used in a laminar air flow cabinet to screen out large particles before air is forced through a much finer filter [HEPA filter].

Premix: To mix before use; as nutrient mixtures are commercially available as dry powders of preweighed ingredients, which are put into solution when required.

Primary culture: 1. May be synonymous with Stage I culture. 2. A recent culture not yet subcultured for the first time.

Probe DNA: A radioactively lebeled [usually DNA molecule used to detect complementary-sequence nucleic acid molecules by molecular hybridization. To localize the probe DNA sequence and reveal the complementary hybridization sequence autoradiography is often used.

Promeristem or protomeristem: The embryonic meristem-containing organ initials or foundation cells.

Propagule: The form or portion of an organism used for reproduction or propagation; as new shoots or callus derived from explants are subdivided into propagules and reculturd for further multiplication.

Propylene oxide [C_3H_6 mw 58.08]: A coluorless, flammable, volatile liquid used as a chemical sterilant for plant material and soil, or as a solvent for resins during the preparation of plant material for embedding (light or electron microscopy)

Pretcase: A group of water soluble protein-degrading enzymes [proteolytic] that function by breaking peptide linkages. In this group are pepsin (principal protease in the gastric juice of vertebrates), trypsin [protease produced in the pancreas] and papain (thiol protease derived from papaya fruit).

Protein digest: The enzymatic hydrolysis of proteins to yield their building block components, amino acids and short chain peptides [chains consisting of two to several amino acids normally without enzymatic function].

Protoclone: Distinct phenotypic regenerants from a plant protoplast. A clone initiated from a protoplast or protoplast-fusion product.

Protocol: A sequence of activities, techniques or procedures linked to achieve a specific goal; as a schedule for protoplast preparation.

Protoplast culture: The isolation and culture of plant protoplasts by mechanical means or by enzymatic digestion of plant tissues or organs, or cultures derived from these. Protoplasts are utilized for selection or hybridization at the cellular level and for a variety of other purposes.

Protoplast fusion: The coalescence of the plasmalemma and cytoplasm of two or more protoplasts in contact with one another. Initial adhesion Ls a random process but coalescence may be promoted in various ways (induced fusion). When adhesion occurs between adjacent protoplasts during enzymatic wall degradation or between freshly isolated protoplasts in the absence of a fusion agent. it is termed spontaneous fusion.

Pure culture: Axenic culture.

Pure line: All cell or individual members are homozygous for one or more characters or genes and will give rise to more cells or oganisms with the character (s) under consideration.

Pyrex: A brand name for borosilicate glass. Pyrex is a high temperature (autoclavable), impact, abrasion and chemical resistant, stable glass of which much laboratory glassware is made.

Quiescent: Quiet, at rest, not necessarily dormant and having the potential for resumed activity; can apply to cells unlikely to divide, the non-meristematic cells.

Radiation: Rays of heat, light or particles in wave form.

Ramet: The vegeiatively propagated offspring of an ortet. An individual of a clone.

Reaction norm: The range of phenotypic responses of a particular genotype in response to the environmental influences.

Reciprocating shaker: A plateform shaker with a back and forth action used for agitating culture flasks at variable speeds.

Recombinant DNA: Genetic material with novel gene sequences produced by cross-overs, chromosome reassortment, by other natural means or through genetic engineering.

Recombinant DNA technology: Genetic engineering.

Reconstructed cell or recon: A viable cell hybrid, cybrid or a transformed cell resulting from genetic engineering.

Reculture or subculture: The aseptic transfer of a culture or a portion (division when necessary) of that culture (inoculum) to fresh nutrient medium.

Redifferentiation: Cell or tissue reversal in differentiation from one type to another type of cell or tissue.

Redox: Oxidizing and reducing reactions involving transfer of an electron from a donor to an acceptor molecule; the donor becomes oxidized and the acceptor is reduced. Arranged in order of intensity these reactions are known as the redox series.

Regenerate: 1. To form or create again; reform. *2,* To give or gain new life or to renew [heal] by a new growth of tissues; as regenerants [shoots, plantlets] form from explants in plant tissue cultures. The process is regeneration.

Reinert, J. [1958]: Among the first to observe adventive embryogenesis in cell cultures of carrot.

Rejuvenation: 1. Synonymous with dedifferentiation. 2. Treatment that leads to culture invogoration [such as subculture] or revival [dormancy breaking].

Repelcote: The brand name of a material [dimethyl dichloro-silane] used to coat glassware [soda glass]; for tissue cultur containers and for other purposes.

Repression: Altered gene expression resulting in the failure of a specific protein synthesis.

Resistance transfer factor: A plasmid, present in certain types of bacteria such as E. coli, that can impart resistance to antibiotics in animals exposd to them.

Resistivity: The degree of resistance [in ohm-cm or ohm-m] or interference to the flow of an electrical current or the movement of particles from place to place. A measure of water purity; the purer the water, the greater its resistivity. The reciprocal of conductivity.

Rhizogenesis: Root formation and growth; as in root development de novo from callus.

Rhodamine isothiocyanate or tetramethylrhodamine isothiocyanate [$C_{25}N_3SO_3H_{21}$ mw 536.10]: A dye used to stain plant protoplasts in the identification of fusion products.

Riboflavin or lactoflavin [$C_{17}H_{20}N_4O_6$ mw 376.36]: A water soluble, vitamin [B_2] essential to cellular respiration. It is important in carbohydrate metabolism and implicated in photooxidation of auxins and perception of phototrophic stimuli. Riboflavin is an occasional addition to plant tissue culture media. It is sometimes called vitamin G.

Rogue: 1. To critically evaluate and eliminate unwanted plants from a population; as undesirable phenotypic variants, weeds and diseased plants are destroyed in plant propagation practices. 2. A variant plant in a population [short].

Root culture: Isolated root tips of apical or lateral origin may produce in vitro root systems with indeterminate growth habits. These were among the first kinds of plant tissue cultures [White, 1934] and remain important research tools in the study of developmental phenomena; mycorrhizal; symbiotic and plant-parasitic relationships.

Rotary shaker: A platform shaker with a circular motion used for agitating culture flasks at variable speeds.

Rotating culture or rotator: A wheel-like device for slowly [ca.1 rpm] rotating and gently agitating cultures, usually in a vertical plane.

Schiff's reagent: A mixture of pararosaniline. HCl [an analine dye] and sodium bisulfite [$NaHSO_3$], used in staining chromosomes and nuclear material. See Feulgen's test.

Scion or cion: The portion of a bud or shoot used for grafting onto another plant [the stock].

Screen house: A metal or plastic mesh enclosure similar in function to a greenhouse but with more exposure to the elements.

Scutellum: That part of the cotyledon in Gramineae [grasses] that absorbs food from the endosperm at germination. An explant source for plant tissue culture.

Selection culture: Utilizes difference [s] in environmental conditions or more usually in culture medium composition, such that preferred variant cells or cell lines [presumptive or putative mutants] are favored over other variants or the wild-type.

Selection pressure: The measur of the effectiveness of natural or experimental selection in altering the genetic composition of a population.

Selection unit: Single cells or small clusters, units of optimum size for isolating and regenerating variants or mutants; the minimum number of cells effective in the screeing process.

Selective advantage: Implies the possession of increased fitness within an individual or a population.

Selective agent: An environmental or chemical agent that impose a lethal or sublethal stress on growing plants, or portion there of in culture, enabling selection of resistant or tolerant individuals.

Semicontinuous culture: The maintenance of cells in a culture vessel in an actively dividing state by periodically draining the medium and adding fresh medium.

Sequestrene 330 Fe: The brand name for an iron chelate used in some tissue culture media.

Serial float culture: Sunderland's [1977-1979] technique of floating anthers on liquid medium and subculturing them to new medium at several day intervals an anther dehiscence, pollen release and development occur, increasing anther productivity.

Shake culture: An agitated suspension culture. Usually a flask [commnly an Erlenmeyer flask] containing the culture is attached to a horizontal or platform shaker, or agitated with a magnetic stirrer, to provide adequate aeration for cells in the liquid medium.

Shaker or platform shaker: A platform fitted with clips for grasping flasks [usually Erlenmeyer flasks], or with surfaces suitable for attaching flasks, with set or variable speed control. It is desirable to adjust the shaking speed for gentle, even agitation of suspension cultures.

Shoot tip graft or micrograft: The grafting of a very small shoot tip or meristem tip onto a prepared seedling or micropropagated rootstock in culture. Meristem tip grafting is used for in vitro virus elimination with Citrus and for other plants as an alternative to grafting in the greenhouse.

Skoog, F. and C. Tsui [1948]: Established that shoot and root formation were chemically [hormonally] regulated in tobacco callus cultures.

Sodium hypochlorite [NaOH mw 74.44]: A frequently used plant tissue sterilant at 0.5-2% w/v. The usual source is dilute [10-21 % v/v] commercial laundry bleach, sometimes with surfactant is common. Tissue damage may occur if contact with tissue;is prolonged. Thorough washing with sterile water generally follows treatment.

Solid or solidified medium: A medium solidified with agar, a synthtic starch polymer or some other gelling agent. Solid media are widely used in plant tissue culture. For suspension cultures and for many research purposes liquid media are preferred.

Somaclone: A plant regenerated from a tissue culture originating from somatic tissue.

Somatic cell embryogenesis: The production of embryos from somatic cells of explants [direct embryogenisis] or by induction on callus formed by explants [indirect embrygenesis]. These two processes may not be materially different in results.

Somatic cell variant or embryoid: An organized embryonic structure morphologically similar to a zygotic embryo but initiated from somatic [non-zygotic] cells. These develop into plantlets in vitro through developmental processes that are similar to those of zygotic embryos.

Somatic hybrid: A cell or plant product of somatic cell fusion; as the result of cell or protoplast fusion and implying genomic integration. The process is somatic hybridization.

Somatic mutation: Mutation occurring in vegetative cells or tissues.

Somatic organogenesis: The production of shoots, roots or other organs on somatic tissues of explants [direct organogenesis] or by induction on callus formed by explant [indirect organogenesis].

Sorbitol [$C_6H_{14}O_6$ mw 182.17]: A sugar alcohol which is the main translocatable carbohydrate in some plants. Occasionally it is added to plant tissue culture media.

Source clone: A source that originates from a-single plant or explant within a clone; as from a specific virus tested [SVT] or specific pathogen tested [SPT] individual explant or plant.

Source plant: A mother plant or donor plant from which an explant used to initiate a culture is taken.

Spent medium: Medium discarded when a culture is subcultured. The implication is that the medium has been depleted of nutrients, dehydrated or accumulated toxic metabolic products.

S phase: The cell cycle phase during which DNA synthesis occurs.

Spontaneous fusion: Uninduced protoplast fusion which may occur between freshly isolated protoplasts or following adhesion of adjacent cells during enzymatic cell wall degradation.

Spontaneous variation: The variation in plant populations derived from tissue cultures not exposed to mutagens but occurring as a result of the culture conditions.

Stages of culture [I-IV]: Stage I: Aseptic explantation or establishment of the explant in culture. Stage II: Multiplication of the propagules. Stage III: Rooting of the propagules and preparation for transplant to soil. Soil IV. Establishment of Stage II or III propagules ex vitro-in soil or potting mix.

Stationary culture: A non-agitated culture, the antonym is shake culture.

Sterilization: The process of making things sterile through 1. Rendering plants non-reproductive. 2. Killing or excluding microorganisms or their spores withheat, filters, chemicals or other sterilants. Dry heat sterilization is useful for metal instruments and glassware. Foil-wrapped items are subjected to 150°C for a minimum of 3 hours in a hot air oven. Steam sterilization or autoclaving is useful for nutrient media, distilled water, paper products and glassware. Solutions contained in glass flasks, plugged with cotton and capped with foil are subjected to 1.05 kg/cm^2 [121°C] for 10 to 20 minutes [depending on the total volume and how it is distributed]. Foil wrapped capped glassware are autoclaved in the same way. Filter sterilization through a "Millipore-type" membrane is useful for thermolabile solutions such as those containing vitamins and urea. Chemical sterilization [most commonly sodium hypochlorite] is useful for plant materials in preparation for excision [surface sterilization] and for working surfaces. Some medium constituents that are thermolabile or insoluble in water may be sterilized through dissolving in organic solvents, such as chloroform or alcohol. They are then dispensed to sterile filter paper for solvent evaporation and the filter paper with the residue is added to or below the sterile medium. Metal instruments are flame sterilized by immersion in 70 % ethahol until required, then flaming.

Stock: 1. The root and a portion of the stem [rootstock] of a plant to which is grafted a part of the same or another plant [scion]. 2. A group of closely related plants.

Stock plant: The source plant from which cuttings or explants are made. These are usually maintained carefully in an optimum state for [sometimes prolonged] explant use. Preferably they are certified, pathogen-free plants.

Stock solution: A solution, usually concentrated [10 to 100 times the final medium-concentration), of select medium constituents that regrouped for compatibility to avoid precipitation and prepared before hand to save time during medium preparation. Usually they are frozen or stored in the refrigerator and portions are utilized as media are prepared.

Subculture or passage: A culture derived from another culture or the aseptic division and transfer of a culture or a portion of that culture [inoculum] to fresh nutrient medium. Subculturing is usually done at set time intervals, the length of which is called the subculture interval or passage time.

Subline: A cell line regenerated from a unique cell line of a hybrid callus colony.

Supraoptimum: An amount [level] greater than required; as an inhibitory concentration of an exogeneous growth factor.

Surface sterilization: The removal of plant surface microflora prior to aseptic excision of explants. Surface sterilization is accomplished by immersion of tissue in one of many sterilants, such as calcium hypochlorite or sodium hypochlorite, hydrogen peroxide, mercuric chloride, silver nitrate or bromine water for an empirically determined period of time. Sometimes a preceeding ethanol dip or spray is useful, as is the addition of a surfactant to the sterilant. Agitation or vacuum may improve the efficiency of sterilization. Thorough washing in sterile distilled water usually follows, to remove the sterilant, although in some cases this is not necessary [hydrogen peroxide].

Surfactant: A surface active agent or wetting agent; as is Tween 20 or Tween 80, Teepol, Lissapol F, Alconox, etc. These agents act by lowering the surface tension so are common addenda to solutions used to surface sterilize plant materials prior to aseptic excision of explants.

Suspension culture: Cells and group of cells [aggregates] dispersed in an aerated,usually agitated, liquid culture medium. These are obtained by adding friable callus to the medium. The plant species, explant type and treatment; the composition of the nutrient medium; and many other features determine the size and nature of the cell aggregates, which may appear different during each culture growth phase. These cultures are used to study cell division, differentiation and metabolism, in secondary product synthesis or form the basis for single cell lines, callus cultures, somatic cell embryogenesis and many other culture purposes.

Synchronized cells: Synchronized mitosis is a group of cells in culture by natural or artificial means.

Synchronous culture: A plant cell or microbial culture treated in such a way as to have all [or most] cells or individuals in the same stage of development or mitosis. This can be achieved in various ways including via temperature variation and nutrient limitation.

Tent: The enclosure or process of including plants, such as exvitro plantlets, under plastic or glass to maintain elevated relative humidity.

Teratogenic: Refers to an agent which induces or increases the incidence of gross structural abnormalities [teratomata] in an individual or a population; as do x-rays.

Test tube fertilization: Pollination, followed by fertilization, occurring in vitro.

Tetrazolium or triphenyl [2,4,5] tetrazolium chloride [TCC, $C_{19}H_{15}NCl$ mw 334.81]: Used in a quantitative assay of cell viability. It becomes reduced to a red, water-insoluble form, formazan, in the presence of oxidative metabolism. It is then dissolved in alcohol and measured spectrophtometrically.

Thermolabile or heat labile: Destroyed or altered during heating; as some viruses are killed during heat therapy [thermotherapy] of infected plants, or as some medium components which destroyed by autoclaving and must be filter sterilized.

Thiamine or aneurine [$C_{12}H_{17}ON_4SCl$ mw 300.82]: Vitamin B_1, a coenzyme involved in carbohydrate metabolism. A common constituent of nutrient media used for plant tissue culture. It is usually added as thiamine hydrochloride.

Thiamine hydrochloride or aneurine hydrochloride [$C_{12}H_{17}ON_4SCl.HCl$ mw 337.28]: A thiamine [vitamin B_1] salt commonly included in plant tissue culture media.

2-thiouracil: An anti-viral agent sometimes included in initiation medium for virus elimination from infected explants. Of uncertain value, its mode of action is nuclear, but may involve inhibition of viral RNA synthesis.

Thiourea [CH_4N_2S mw 76.12]: An occasional additive in plant tissue culture media as a source of reduced nitrogen.

Tissue explant: An excised plant portion of tissue used to initiate a culture.

Tocopherol [$C_{29}H_{50}O_2$ mw 430.69]: Vitamin E, an occasional additive in plant tissue culture media.

Tocopheryl acetate [$C_{31}H_{52}O_3$ mw 472.73]: A form of vitamin E occasionally used in plant tissue culture media.

Totipotency: The potential [totipotential] or inherent capacity of a plant cell or tissue to develop into [recreate] an entire plant if suitably stimulated. Totipotency implies that all the information necessary for growth and reproduction of the organism is contained in the cell. Although theoretically all plant cells are totipotent the meristematic cells are best able to express it.

Toxic: Poisonous; as are some chemicals [toxicants] or any substances present to excess and detrimental to normal plant function or growth.

Trace element: Microelement

Tracer: A substance that can be followed within a reaction or an organism; as radioactive isotopes and certain dyes.

Transcription or genetic transcription: DNA dependent RNA synthesis. The process by which cellullar RNA molcules are synthesized as determined by homologous DNA sections where their sequence recurs in complementary form [the template or transcription strand].

Transfer: 1. Culture initiation; the first placement of an explant in culture. 2. Subculture; relocation of cultures to fresh nutrient medium. 3. chamber; laminar air flow cabinet or hood [transfer hood] or room [transfer room] in which this is accomplished. 4. the act or process of transferring.

Transfer room: A small room, sometimes sterilized internally by a bactericidal lamp [U.V. irradiation] and provided with clean [filtered] air; employd for sterile transfer work.

Transplant: 1. To relocate or remove to a new growing place. 2. The cultured tissue or explant, relocated or transferred to a new site [in vitro]. 3. Stae IV; the transfer of plantlets or shoots ex vitro from aseptic culture to soil. The process is transplantation.

Transplant shock: Refers to the stress involved when Stage II or Stage III cultures are transplanted to soil [Stage IV]; many or all of the regenerated shoots or plantlets may die if suitable care is not taken to acclimatize them gradually to the soil environment, particularly prevention of water stress.

2,3,5-triiodobenzoate [TIBA mw 499.81]: An inhibitor of auxin movement or transport [antiauxin], sometimes included in plant tissue culture media for its growth promoting effects.

True to type: Applied to a plant or propagation source this term denotes correct cultivar identification and lack of variation in productivity or performance. Verification is determined visually by an expert or through biochemical, serological or other means.

Tryptamine or 3-[2-aminoethyl] indole [$C_{10}H_{13}N_2$ mw 161.23]: An occasional addition to plant tissue culture media as a source of reduced nitrogen.

Tryptophan [Trp, $C_{11}H_{12}O_2N_2$ mw 204.23]: An amino acid precursor of indole compounds such as IAA. Occasionally added to plant tissue culture media.

Trypan blue [$C_{34}H_{24}N_6Na_4O_{14}S_4$ mw 960.83]: A dye used in a quantitative method to assay cell viability. The method is based on the ability of live protoplasts to exclude the dye while dead protoplasts cannot.

Tumor inducing principle [TCP]: The plasmid carried by Agrobacterium tumefaciens, the crown gall organism. Through incorporation into the host genome the host tissue is transformed into tumor tissue.

Turbidostat: An open continuous culture system wherein the inflow of fresh medium is controlled by the turbidity of the culture, a function of the amount of cell growth. Balancing the fresh medium inflow is a regulated outflow of cells and spent medium, restoring the original turbidity level.

Tween 20 [polyoxyethylene sorbitan monolaurate mw 1227.54]: The brand name for a wetting agent or surfactant which breaks the surface tension of tissues. Commonly added to disinfecting solutions to make them more effective.

Tyrosine [Tr, $C_9H_{11}NO_3$ mw 181.19]: An amino acid, sometimes used in plant tissue culture media as a source of nitrogen.

Ubiquitous: Occurs everywhere, as do bacteria in the environment.

Ultrasonic cleaner: A device to include high frequency vibration of materials, removing adhering substances from surfaces by mechanical action. This device is useful for cleaning glassware and for disinfecting plant material.

Ultraviolet light [U.V.]: Radiation with wavelength [100-400 nm) at the voilet end of the visible spectrum. Generated by mercury vapor lamps, it is sometimes used in tissue culture for its mutagenic properties or to reduce ambient contaminants in work areas due to its bactericidal properties. Exposure is harmful to the eyes and skin and is to be avoided by workers in plant tissue culture facilities.

Unorganized growth: In vitro formation of tissues with few differentiated cell types and lacking recognizable structure; as with many calli.

Valine [Val, $C_5H_{11}NO_2$ mw 117.1: An amino acid found in seeds and proteins. Occasionally added to plant tissue culture media.

van Overbeek, J., M.E. Conklin and A.F Blakeslee [1941]: First to demonstrate the growth-promoting effect of coconut milk [for excised Datura embryos].

Vermiculite: Any of many minerals [commonly altered micas] whose granules can expand greatly, becoming highly absorbent. Used alone or more commonly as a component of potting mix.

Vernalin: An hypothetical hormone-like substance found in plant meristematic regions, produced by vernalization. This substance is apparently graft transmissible, but has not yet been identified. Different cold requiring species may form different substances during vernalization.

Victorin: A host specific toxin produced by *Helminthosporium victoriae* utilizable for screening oat cells for resistance to the pathogen.

Virus elimination: Thermotherapy, chemotherapy and meristem or meristem tip culture, used alone or in combination have been employed for the elimination of systemic viruses from plants. Plants must repeatedly test negatively for the virus in question for assurance that it has been eliminated.

Virus-tested or virus-free: A plant that appears healthy and repeatedly tests negatively for the presence of one or more identifiable viruses. Such a plant may then be used as a stock or donor plant [explant source] for propagation purposes, and may be certified as virus tested [certified virus tested]. The term virus-free is incorrect in most cases, as such a plant may contain one or more viruses which have not been assayed.

Vitamin B complex: Includes a large group of water soluble vitamins that function as coenzymes; thiamine [B_1], riboflavin or vitamin G [B_2], niacin or nicotinic acid [B_3], pathothenic acid [B_5], pyridoxine [B_{12}], biotin or vitamin H, folic acid, or vitamin M [BC], inositol, choline and others.

Vitamin B_X: para-amino-benzoic acid. Vitamin C: ascorbic acid.

Vitamin D: A fat soluble vitamin composed of a group of related steroids. Present in all plants.

Vitamin E: See tocopherol.

Vitainin K: A fat-soluble quinone involved in photosynthetic electron transfer in plants and blood clotting in animals.

V/V: May indicate simple proportion; as 3;1 [v/v]. May indicate percent volume in volume; as the number of cm^3 [mls] of constituent in 100 cm^3 [mis] solution.

Water of hydration: The amount of water chemically bound to a substance. This amount may be variable and must be taken into account when solutions of salts are prepared; as in medium preparation.

Water potential: Governs the conduction of plant water from regions of high to low water potential [towards increasing solute concentration] and is a measure of the free energy status of water in a system. The total water potential of system is the sum of the osmotic component [dependent on solute concentration], the matric component [dependent on water-building substances and surface forces] and the pressure component [usually turgor pressure].

White, P.R.: Produced the first successful plant tissue cultures [organ cultures] in the, early 1930's. He isolated and grew tomato root tip in a liquid nutrient solution composed or inorganic salts, sucrose and yeast extract [1934]. He showed that yeast could be replaced with the B vitamins, thiamine, pyridoxine and niacin [1937]. He wrote A Handbook of Plant Tissue Culture [1943], The Cultivation of Animal and Plant Cells [1954]. In 1963 he published a low salt nutrient culture medium widely used in plant tissue culture.

Wick: A filter paper, chromatography paper or a strip of some other material used to support plant tissue above a liquid medium and to transport the water and nutrients to the tissue.

Wild type: 1. The genotype or phenotype of an organism predominating in the control [standard or wild] population, in its natural element. 2, A specific gene predominant in this population.

w/m^2 or watt per square metre: A common unit of light measurement.

Woody plant medium [WPM]: A modified MS [1962] medium developed for woody plant species by G. Lloyd and B.McCown [19801. It has less nitrogen [both ammonium and nitrate], sodium, potassium and chloride than does MS [1962] medium. It has been increasingly used for the commercial propagation of ornamental trees and shrubs.

Wound tissue: See callus.

W/v: Weight in voluume; as the number of grams of constituent in 100 cm^3 [mls] solution.

Xerograft: See heterograft.

X-ray or Roentgen ray [r]: Electromagnetic radiation of short wave length produced by high speed electrons impacting on a metal object under vacuum.

Xylose [$C_5H_{10}O_5$ mw 150.1]: An aldopentose sugar [wood sugar] present in the woody tissues of many plants as polymeric xylan. It is occasionally used as a carbohydrate source in plant tissue culture media.

Yeast extract: A complex, undefined, B vitamin-containing addendum to some plant tissue culture media.

Yield test: Productivity assessment.

Zeatin or 2-methyl-4-[lH-purin-6-ylamino]-2-buten l-ol [Z, $C_{10}H_{13}N_5O$ mw 215.21]: A natural cytokinin first isolated from corn. It is sometimes included in plant tissue culture media. It dissolves in acid [HC1 ca. 1 M].

Zeatin riboside [mw 351.4]: A naturally occurring cytokinin sometimes used in plant tissue culture media.

Zephiran: The brand name for a disinfecting agent for plant material containing benzalkonium chloride.

Ziehl's stain: Carbolfuchsin.

APENDIX - 1

Reagents and Stains

2iP Stock Solution

N^6-[2-isopentyl] adenosine (2iP)	100.0 mg
NaOH. 1 M solution	5.0 ml

Dissolve the 2iP in 5 ml of HCl solution. Adjust the final volume of the solution to 100 ml using distilled water. Each ml contains 1 mg 2iP.

α-Amylase Solution

a-Amylase (barley malt)	0-2
Phosphate buffer (0.2 M pH 5.9)	10ml

Dissolve the enzyme completely in the buffer solution. If barley malt α-amylase is unavailable, fungal or bacterial enzyme can be substituted.

α-Napthol Solution

α-Napthol	10.0 mg
NaOH (1 M solution)	100.0 ml

Dissolve the α-napthol in 100 ml 1 M NaOH solution. Store at room temperature. The solution should be made shortly before use.

Agar Medium

Agar	15.0 g

Add the agar to 1 liter of distilled water. Dissolve the agar by placing the mixture in a steam bath. Autoclave for 25 min at 121° C.

Alginate Solution

Na alginate	4.0 g

Disperse the alginate salt in 100 ml of warm distilled water. Dissolve the alginate by placing the mixture in a steam bath.

Ammonium Acetate Solution

Ammonium acetate	38.54 g

Dissolve the acetate salt in 100 ml of distilled water.

BAP Stock Solution

Benzylaminopurine (BAP)	100.0 mg
HC1 1 M solution	5.0 ml

Dissolve the kinetin in 5 ml of HCl solution. Adjust the final volume of the solution to 100 ml using distilled water. Each ml contains 1 mg of BAP.

Benedict's Reagent

$CuSO_4 5H_2O$	1.73g
Na_2CO_3 anhydrous	10.00 g
Na citrate	17.30 g

Dissolve the Na citrate and Na_2CO_3 in 60 ml of distilled water. Dissolve the $CuSO_4$ separately in 20 ml of distilled water. While stirring the citrate-carbonate solution, add the $CuSO_4$ solution. Adjust the final volume of the solution to 100 ml by adding distilled water.

Bromphenol Blue/Glycerol Solution

Bromphenol blue dye	250.0 mg
Glycerol	50.0 ml

Dissolve the bromphenol blue in 50 ml of water, then add the glycerol.

$CaCl_2$ Solution

$CaCl_2 2H_2O$	14.7 g

Dissolve the $CaCl_2$ in 1 liter of distilled water.

Cauliflower Homogenization Solution

Na lauryl sulfate	100.0 g
Na citrate	2.94 g
NaCl	8.18g

Dissolve the Na citrate and NaCl in 1 liter of distilled water. Add the Na lauryl sulfate and mix until all the detergent is emulsified.

Cell Wall Lysis Solution

Cellulase	2.0 g
Mannitol	10.27 g

Dissolve the mannitol in 100 ml of distilled water, then dissolve the cellulase. Filter-sterilized solution through a 0.45 μm filter. Store the solution in a sterile bottle at 4°C.

Chloroplast Lysis Solution

Tris-HCI buffer, 0.1 M pH 8	100.0 ml
EDTA	1.5 mg
NaCl	0.58 g
Sodium dodecyl sulphate	1.0 g
β-Mercaptoethanol	78.0 mg

Dissolve all the ingredients except the mercaptoethanol in 100 ml of tris-HCl buffer. Working under a chemical fume hood, add the mercaptoethanol.

Dibasic Potassium Phosphate Solution

K_2HPO_4	4.56 g
H_3PO_4	(as needed)

Dissolve the K_2HPO_4 in 75 ml of distilled water. Adjust the solution pH to 8.0 by adding H_3PO_4 as neded. Adjust the final volume of the solution to 100 ml.

Enzyme Extraction Buffer

Potassium phosphate buffer 0.2 M pH 7.5	99.0 ml
TritonX-100	1.0 ml
2-Mercaptoethanol	780.0 mg
Na_2EDTA	370.0 mg

Combine all the ingredients. Store in a sterrile bottle at 4°C.

Enzyme Visualization Solution

Tris-HCl buffer 0.05 M pH 7.5	100.0 ml
Hypoxanthine	700.0 mg
Nicotinamide adenine dinucleotide (NAD)	30.0 mg
Nitrotetrazolium blue	20.0 mg
Phenazine methosulphate (PMS)	4.0 mg

Heat the hypoxanthine in the tris buffer until it dissolves. Cool the solution to room temperature and add the other reagents.

Ethanol 70% Solution

Ethanol 95 % solution	735.0 ml

Mix the ethanol with 265 ml of distilled water to make 1 liter of solution. Store at room temperature in a sealed bottle or jug.

Evan's Blue Stain Solution

Evan's blue stain 250.0 mg Mannitol	12.7 g

Dissolve the mannitol in 100 ml of distilled water, then dissolve the Evan's blue stain. Store-refrigerated at 4°C.

Formic Acid Electrophoresis Buffer

Formic acid	30.0 ml
Glacial acetic acid	60.0 ml
Distilled water	910.0 ml

Wearing protective clothing, gloves, and eyewear, slowly add the two acids to the distilled water. The final pH of the buffer should be approximately 1.9.

Glutaraldehyde Solution

Glutaraldehyde is normally sold as a 25% stock solution by most chemical suppliers. Use the stock solution as it is.

H_2O_2 Buffered Solution

H_2O_2 3 % solution	100.0 ml
Phosphate buffer (0.2 M pH 7.0)	800.0 ml

Mix the two solutions together. Adjust the final pH of the solution to 7.0 by adding 1 M HCl or 1 M NaOH as required. Adjust the final volume of the solution to 1 liter.

H2O2 Sterilization Solution

H_2CO_2 3% solution	100.0 ml
Triton X-100 detergent	0.5 ml

Mix the peroxide solution and detergent thoroughly. Dispense the mixture in 150-ml dilution bottles.

Hypochlorite Solution

NaOCl 5.25 % v/v solution	50.0 ml
Ethanol 90% v/v solution	50.0 ml

Mix the two solutions together. Store in an amber or light-proof bottle at room temperature.

IAA Stock Solution

Indoleacetic acid (IAA)	100.0 mg
NaOH 1 M solution	5.0 ml

Dissolve the IAA in 2 ml of NaOH solution. Adjust the final volume of the solution to 100 ml using distilled water. Each ml contains 1 mg of IAA.

IBA Stock Solution

Indolebutyric acid (IBA)	100.0 mg
NaOH 1 M solution	5.0 ml

Dissolve the IBA in 2 ml of NaOH solution. Adjust the final volume of the solution to 100 ml using distilled water. Each ml contains 1 mg of IBA.

Iodine Solution

KI	5.0 g
I	0.5 g

Dissolve the Ki and I in 100 ml of distilled water. Store in an amber or foil-wrapped bottle to prevent deterioration due to light. Prepare a working stock solution to test for the presence of starch by diluting 1 ml of iodine solution in 99 ml of distilled water.

Kinetin Solution

Kinetin	50.0 mg
HC1 1 M solution	5.0 ml

Dissolve the kinetin in 5 ml of HC1 solution. Adjust the final volume of the solution to 100 ml using distilled water. Each ml contains 0.5 mg of kinetin.

Loading Buffer

Glycerol	50.0 ml
EDTA	2.9 g
Xylene cyanol stain	0.1 g
Bromphenol blue stain	0.15 g

Dissolve the EDTA, xylene cyanol, and bromphenol blue in 50 ml of distilled water. Completely mix with the glycerol.

Lugol's Solution

KI	6.0 g
I	4.0 g

Dissolve tin. Ki and I in 100 ml of distilled water. Store in an amber or foil-wraped bottle to prevent deterioration of the solution due to light.

Lysine Solution

Lysine	0.05 g

Dissolve the lysine in 500 ml of distilled water.

Methylene Blue Solution

Methylene blue stain	0.02 g

Dissolve the methylene blue in 100 ml of distilled water.

Nopaline Solution

Nopaline	1.0 g

Dissolve the nopaline in 2 ml of distilled water. Store refrigerated at 4°C.

Phosphate Buffer (0.2 M pH 5.9)

$NaH_2PO_42H_2O$	24.8 g
Na_2HPO_4 anhydrous	2.8 g

Dissolve the two Na salts in 900 ml of distilled water. Adjust the final pH to 7.0 by adding 1 M HCl or 1 M NaOH. Adjust the final volume of the solution to 1 liter.

Phosphate Buffer (0.2 M pH 7.0)

$NaH_2PO_42H_2O$	10.76 g
Na_2HPO_4 anhydrous	17.32 g

Dissolve the two Na salts in 900 ml of distilled water. Adjust the final pH to 7.0 by adding 1 M HC1 or 1 M NaOH. Adjust the final volume of the solution to 1 liter.

Potassium Acetate Solution

K acetate	49.07 g

Dissolve the acetate salt in 100 ml of distilled water.

Protoplast Fusion Solution

Polyethylene glycol (PEG) 1500-2000 (MW)	50.0 g
Glucose	1.8g
$CaCl_2 2H_2O$	150.0 mg
K_2HPO_4	12.0 mg

Dissolve the glucose, $CaCl_2$ and K_2HPO_4 in 100 ml of distilled water. Slowly add the PEG to this solution. Autoclave the mixture at 121° C for 20 min to dissolve the PEG.

Sodium Acetate Solution

Na acetate	40.83 g

Dissolve the acetate salt in 100 ml of distilled water.

Sodium Alginate Solution

Na alginate	4.0 g

Add the alginate slowly to 100 ml of distilled water. Steam or autoclave the mixture to fully dissolve the alginate. The solution is very viscous and is difficult to manipulate with pipettes.

Sodium Azide Solution

NaN_3	20.0 mg

Disolve the azide in 100 ml of distilled water. Combine 1 ml of this stock solution with 99 ml of distilled water to make a working stock for mutagenesis work. Autoclave the solution and store at room temperature.

Sodium Chloride Solution: NaCl (81.8 g), dissolve the NaCl in 1 liter of distilled water.

Sodium Citrate Solution: Na citrate (8.18 g), dissolve the Na citrate in 1 liter of distilled water.

Sodium Hypochlorite Sterilization Solution

Household bleach (5.25% NaOCl solution)	100.0 ml
Triton X-100 detergent	0.5 ml

Dilute the household bleach with 900 ml of distilled water to make 1 liter of sterilized solution. Add the detergent and mix thoroughly.

Soluble Starch Solution

Soluble starch	0.5 g
Phosphate buffer (0.2 M pH 5.9)	10 ml

Dispcrsc thc starch in the buffer solution as thoroughly as possible. Steam the solution for at least 1 hour, then filter the solution to remove any undissolved starch.

Spinach Homogenization Buffer

Tris-HCl buffer, 0.05 M pH 8.0	500.0 ml
EDTA	0.44 g
Bovine serum albumin	0.5 g
PVP-40	5.0 g
β-Mercaptoethanol	0.35 g
Mannitol	31.9 g

Dissolve all the ingredients except the mercaptoethanol in 500 ml of tris-HCl buffer. Working under a chemical fume hood, add the mercaptoethanol.

Sterile Distilled Water: Fill 150 ml capacity dilution bottles with 100 ml of distilled water. Autoclave at 121°C for 15 min.

Sucrose Gradient Solution 30%

Tris-HCl buffer 0.05 M pH 8.0	100 ml
Sucrose	30.0g
EDTA	0.73 g
Bovine serum albumin	0.1 g

Dissolve all the ingredients in 100 ml of tris HCI buffer.

Sucrose Gradient Solution 52%

Tris-HCl buffer 0.05 M pH 8.0	100 ml
Sucrose	52.0 g
EDTA	0.73 g
Bovine serum albumin	0.1 g

Dissolve all the ingredients in 100 ml of tris HCI buffer.

TBE Buffer (l0 x stock)

Tris	10.8 g
Boric acid	5.5 g
EDTA	0.93 g

Dissolve the ingredients in 100 ml of distilled water.

TE 10 Buffer

Tris-HCl buffer 0.05 M pH 8.0	20.0 ml
EDTA	0.03 g

Dilute the tris-HCl buffer in 80 ml of distilled water. Dissolve the EDTA in the diluted tris-HCl buffer

TE 50 Buffer

Tris-HCl buffer 0.05 M pH 8.0	100.0 ml
EDTA	0.29 g

Dissolve the EDTA in the tris-HCl buffer.

Tetracycline Solution: Tetracycline (1.0 g), dissolve the antibiotic in 10 ml of distilled water. Sterilize the solution using a 0.45-μm pore. Acrodisc ™(Gelman) filter and syringe. Collect the sterile solution in a sterile screw-cap test tube and store at 4° C; it will last up to 1 month if refrigerated. Just prior to use, dilute 1 ml of this solution in 9 ml of sterile distilled water. Dispense it in 20 x 150 mm culture tubes.

Tris-Glycine Buffer (5x Stock)

Tris	7.5 g
glycine	36.0 g

Dissolve the tris and glycine in 500 ml of distilled water. Adjust the solution pH to 8.3 using 1 M HCI or 1 M NaOH. Store the concentrated stock at 4°C. Mix 1 part of the concentrated stock solution with 4 parts of distilled water just prior to use.

Tris-HCl Buffer 0.05 M pH 7.5: Tris (6.6 g), dissolve the tris in 500 ml of distilled water. Adjust the pH to 7.5 by adding 1 M HC1. Add distilled water to adjust the final volume to 1 liter.

Tris-HCl Buffer 0.05 M pH 8.0: Tris (6.6 g), dissolve the tris in 500 ml of distilled water. Adjust the pH to 8.0 by adding 1 M HC1. Add distilled water to adjust the final volume to 1 liter.

Tris-HCl Buffer 0.1 M pH 8.0: Tris (13.2 g), dissolve the tris in 500 ml of distilled water. Adjust the pH to 8.0 by adding 1 M HC1. Add distilled water to adjust the final volume to I liter.

Wash Buffer

Tris-HCl buffer 0.05 M pH 8.0	100.0 ml
Sucrose	10.3 g
EDTA	0.58 g
Bovine serum albumin	0.1 g

Dissolve all the ingredients in the tris-HCl buffer.

Alexander Stain (Alexander, 1980)

1. Test the viability of cells/embryos by using Alexander stain (Alexander, 1980)
2. Cut the section of embryos (5-10 μM) for microscopic examination of bipolar embryogenesis.

1. The stain mixture is prepared by adding following chemicals-

i. Ethanol 95%	20ml
ii. malachite green	20 mg (2 ml of 1 % solution in 95 % alcohol)

iii. Distilled water 50ml
iv. Glycerol 40ml
v. Acid fuchsin 100 mg (10 ml of 1 % aqueous solution)
vi. Phenol 5g
vii. Lactic acid 0.5-1.0 ml

2. Prepare by adding above items as mentioned above, shaking after the addition of each item
3. Store in coloured bottle for 8-10 days before use
4. Add 1 g of phenol to the acid fuchsin stock solution to prevent contamination
5. Prepare destaining reagent by adding glycerol (35 ml) and lactic acid (15 ml) in 50 ml distilled water to remove excess of stain taken by the specimen.

□□□

APENDIX - 2

Culture Media for Protoplast Culture

Protoplast Culture Media for PC I Group

F.5 Medium (Frearson et al., 1973)

Constituents	Concentrations (mg/1)
$CaCl_2.2H_2O$	850.0
$CuSO_4.5H_2O$	0.025
$CoSO_4.7H20$	0.15
$FeSO_4.7H_2O$	13.9
H_3BO_3	3.1
KH_2PO_4	353.6
KI	0.498
KNO_3	525.0
$MgSO_4.7H_2O$	739.0
$MnSO_4.4H_2O$	11.15
Na_2EDTA	18.75
Na_2MoO_4-$2H_2O$	0.125
NH_4NO_3	412.25
$ZnSO_4.7H_2O$	4.3
BAP	1.0
Biotin	0.05
Folic acid	0.5
Glycine	1.0
Mannitol	130,000.0
Meso-inositol	100.0
NAA	2.0
Nicotinic acid	5.0
Pyridoxine HCl	0.5
Sucrose	10,000.0
Thiamine HCl	1.0
pH	5.8

Nagata and Takeba Medium (1971)

Constituents	Concentrations (mg/l)
CaCl2.2H_2O	220.0
$CuSO_4.5H_2O$	0.025
$CoSO_4.7H_2O$	0.030
$FeSO_4.7H_2O$	27.8
H_3BO_3	6.2
KH_2PO_4	680.0
KI	0.83
KNO_3	950.0
$MgSO_4.7H_2O$	1233.0
$MnS0_4$.4H20	22.3
Na_2EDTA	37.3
$Na_2MoO_4.2H_2O$	0.25
NH_4NO_3	825.0
$ZnSO_4.4H_2O$	8.6
BAP	1.0
Mannitol	130,000.0
Meso-inositol	100.0
NAA	3.0
Sucrose	10,000.0
Thiamine HC1	1.0
pH	5.8

V 47 (Binding, 1974)

Constituents	Concentrations (mg/1)
$CaCl_2$ $2H_2O$	735.0
$CuSO_4.5H_2O$	0.015
$CoCl_2$-$6H_2O$	0.015
H_3BO_3	2.0
KH_2PO_4	68.0
KI	0.25
KNO_3	1480.0
$MgSO_4.7H_2O$	984.0
$MnSO_4.H_2O$	5.0
Na_2EDTA	37.0
$Na_2MoO_4.2H_2O$	0.1
NH_4NO_3	1444.0
$ZnSO_4.7H_2O$	1.5
BAP	0.4

Constituents	Concentrations (mg/1)
Biotin	0.04
Folic acid	0.4
Glycine	1.4
Glucose	99,000.0
Mannitol	8,19,000.0
Meso-inositol	10.0
NAA	1.52
Nicotinic acid	2.0
Pyridoxine HC1	0.7
Sucrose	1,70,000.0
Thiamine HC1	4.0
pH	5.8

Murashige and Skoog Medium

For protoplast culture, use MS culture medium as described for tissue culture supplemented with 13% mannitol (w/v).

Modified B5 Medium (Kartha et al., 1974)

Constituents	Concentrations
+ Sorbitol	4.55% w/v
Mannitol	4.55% w/v
d-glucose	0.25% w/v
D-ribose	0.125% w/v
N-Z anime	0.015% w/v
$CaCl_2.2H_2O$	0.0875%
2,4-D	2.3 x 10^{-6}M
BAP	10-6M
NAA	10-6M
Calculated molarity	0.52

Protoplast Culture Media PC II Group

Media calculated under PC I Group with minor changes. The changes are - the concentration of mannitol/sorbitol present in media individually or in combination reduced by half. Concentration of agar in the medium is 0.8%.

Proroplast Culture Media PC III Group

Media described PC I Group with following minor changes:

1. Mannitol/sorbitol/D-glucose/D-ribose individually or in combination are omitted in the media.

2. The sucrose concentration in all media is adjusted to 2% w/v.
3. PC III Group media may be prepared with 1.6% or 0.8% agar depending on the mode of protoplast culture.

Composition of the digestion mixture, wash medium and regeneration medium used for isolation and culturing of protoplasts.

Digestion Mixture

Salts and Vitamins of Murashige & Skoog (MS)

Cellulysin	1.0%
Macerase	0.5%
Rhozyme	1.0%
$CaCl_2$	4.5 mM (0.5g/l)
Mannitol	0.3 M (54.7 g/1)
Sorbitol	0.3 M (54.7 g/1)
MES	0.3 mM (640 mg/1)
pH	5.7

Filter Sterilize Washing Medium (MS)

$CaCl_2$	4.5mM
Sorbitol	0.3 M
Mamiitol	0.3 M
pH	5.7

Autoclave Regeneration Medium (MS)

Sucrose	20 g/1
2,4-D	0.5 mg/1
Kinetin	1.0 mg/1
Coconut water	2.0% (v/v)
Inositol	100 mg/1
Casein hydrolysate	200 mg/1
Sodium pyruvate	5.0 mg/1
Citric acid	10.0 mg/1
Malic acid	10.0 mg/1
Fumaric acid	10.0 mg/1

Vitamins (pH 5.7)

Nicotinamide	1.0
Pyridoxine HCl	1.0
Thiamine HCl	1.0
D-calcium pantothenate	0.5
Folic acid	0.2

p-aminobenzoic acid	0.01
Choline chloride	0.5
Riboflavin	0.1
Ascorbic acid	1.0
Vitamin B12	0.01
Sugar and Sugar Alcohols (g/1)	
Fructose	0.125
Fibose	0.125
Xylose	0.125
Mannose	0.125
Cellobiose	0.125
Sorbitol	54.7
Mannitol	54.7

Laboratory Help

WEIGHING

Many different kinds of balances are used in chemistry laboratories, including (a) triple-beam balances, which require manual adjustment of weights on each of three balance beams; (b) top-loading automatic balances; and (c) analytical balances, which are accurate to closer than 1 mg. Strictly speaking, such balances measure mass rather than weight; however, the word weight is commonly used for mass when this will not cause error or confusion. Your instructor will explain or demonstrate the operation of the balances used in your laboratory.

Weighing Solids

Most solids can be weighed in glass containers (such as vials or beakers) or on special glossy weighing papers. Filter paper and other absorbent papers should not be used for accurate weighings, since a few particles will always remain in the fibers of the paper. If you are weighing an indeterminate amount of a solid (such as the product of a reaction), its container or the weighing paper should first be weighed separately and the mass recorded (unless the balance has a taring system). Then the solid is added, the total mass is recorded, and the mass of the container or weighing paper is subtracted. If the balance has a taring system, the mass readout scale is set to zero (using the taring control) while the container or weighing paper is on the balance pan. Then the solid is added, and its mass is read directly from the scale.

If you are measuring out a specified quantity of a solid, the expected total mass of the solid plus its container should be calculated and the balance should be set to approximately the value. Then the solid is added in small portions, using a clean scoop or spatula, until the desired reading is attained. If the balance has a taring system, its readout scale is zeroed with the container on the pan and is then set to the desired mass of the solid.

Most products obtained from a preparation are transferred to vials or other small containers, which should be weighed empty and then reweighed after the product has been added. As a general practice, the container should be weighed with its cap and label on and this tare weight recorded. Then the mass of the contents at any given time can be obtained by subtracting the tare weight from the total mass of container and contents.

Note: Allow a moistened label time to dry before weighing a labelled container.

Weighing Liquids

The mass of an indeterminate quantity of a liquid is measured as described above for a solid. A tared container should be used, and it should be kept stoppered during the weighing to avoid loss by evaporation. To weigh out a specified quantity of a liquid from a reagent bottle, you should first measure out the approximate quantity of the liquid by volume and then weigh that quantity accurately in a closed container. For example, if 9.0 g of chloroform (d = 1.5 g/ml) is required for a synthesis, a little more than 6 ml (9.0 g/1.5 g/ml) of the liquid is measured into a graduated cylinder and transferred to a tared vial. If the measured weight is not close enough to that required, liquid can be added or removed with a clean dropper. If the liquid is provided in dropper bottles, it can be transferred directly from the bottle to the container on the balance pan *(Care: Do not get any thing on the balance)* without pre-measuring.

Note: Never withdraw liquid directly from a reagent bottle with your dropper or pipet, as you may contaminate the liquid.

Calculate the volume from : V = mass/density

Clean up any spills in the vicinity of the balance immediately.

MEASURING VOLUME

A given volume of a liquid can be measured using either a graduated cylinder, a pipet, or a syringe, depending upon the quantity and accuracy required. Burets and volumetric flasks are also used to measure liquid volumes accurately. Their use is discussed in most general and analytical chemistry laboratory manuals.

Graduated Cylinders

Graduated cylinders are not highly accurate, but they are adequate for measuring specified quantities of solvents and wash liquids as well as liquid reactants that are present in excess. The liquid volume should always be read from the bottom of the liquid meniscus.

Pipets

Graduated or volumetric pipets can be used to accurately measure relatively small quantities of a liquid. Suction is required to draw the liquid into a pipet; however, using mouth suction is unwise because of the danger of drawing toxic or corrosive liquids into the mouth. An ordinary ear syringe works quite well as a pipetting bulb. Another convenient pipetting-bulb assembly.

1. The top end of the pipet is inserted into the pinch-cock valve.
2. The pinchcock is opened by pinching it at the glass bead and the bulb squeezed to eject the air.
3. The pipet tip is placed in the liquid and the pinch cock is squeezed open to fill the pipet to just above the calibration mark.

4. The bulb is removed from the narrow end of the dropper and the pinchcock is carefully opened until the liquid falls to the calibration mark.

5. The liquid is then delivered into another container by opening or detaching the pinchcock valve.

Most volumetric pipets are calibrated to deliver (TD) a given volume, meaning that the measured liquid is allowed to drain out by gravity, leaving a small amount of liquid in the bottom of the pipet. This liquid is not removed, since it is accounted for in the calibration. Graduated pipets are generally filled to the top (zero) calibration mark and then drained into a separate container until the calibration mark for the desired volume is reached. The remaining liquid is either discarded or returned to its original container. The maximum indicated capacity of some graduated pipets is delivered by draining to a given calibration mark and of others by draining completely. It is important not to confuse the two, since draining the first type completely will deliver a greater volume than the indicated capacity of the pipet.

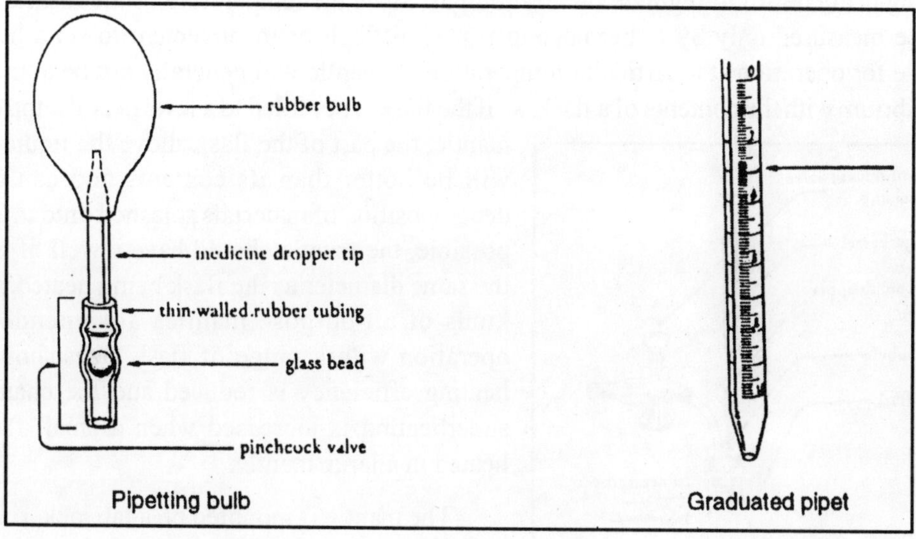

Syringes

Syringes are most often used for the precise measurement and delivery of very small volumes of liquid, as in gas-chromatographic analysis < OP-32 >. A syringe is filled by placing the needle in the liquid and slowly pulling out the plunger until the barrel contains a little more than the required volume of liquid. Then the syringe is held with the needle pointed up and the plunger is pushed in to eject the excess pointed up and the plunger is pushed in to eject the excess sample. Excess liquid is wiped off the needle with a tissue.

Syringes should be cleaned immediately after use by rinsing them several times with a volatile solvent, then removing the plunger and letting the barrel dry. Microsyringes can be dried rapidly by aspiration. The needle is inserted carefully through the dropper bulb and the aspirator is turned on for a minute or so. The pinchclamp is then opened to release the vacum, the aspirator is turned off, and the syringe is removed.

HEATING MANTLES

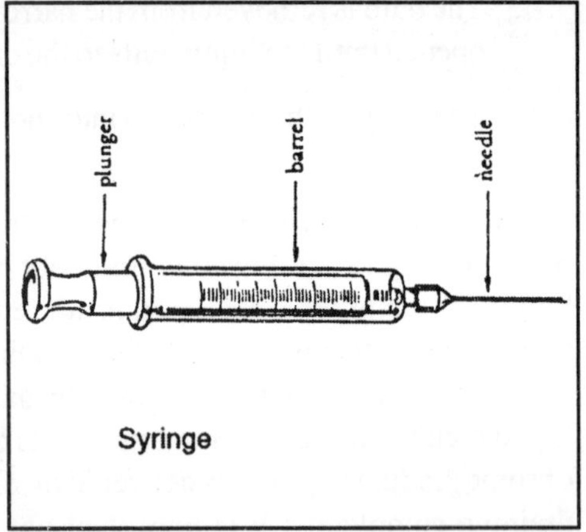

Syringe

A heating mantle is perhaps the most satisfactory laboratory device yet developed for heating over a wide temperature range. Unlike a Bunsen burner, a heating mantle can be controlled precisely. Unfortunately, most heating mantles are costly, take a long time to warm up, and accommodate only a narrow range of flask sizes. It is also difficult to monitor the operating temperature of a heating mantle. Despite these disadvantages, heating mantles are highly recommended for most heating operations.

A heating mantle must be used in conjunction with a variable transformer or a time-cycling heat control to regulate the heat output. Since the temperature of the mantle itself can be measured only by a thermocouple, it is difficult or inconvenient to-set a heating mantle for operation at a particular temperature. A mantle will generally not be at thermal equilibrium with the contents of a flask, so if the flask is not filled to a level near the top of the mantle, the part of the flask above the liquid level will be hotter than its contents and can cause decomposition of materials splashed onto it. When possible, the mantle should have a well of nearly the same diameter as the flask being heated. Some kinds of all-purpose mantles are intended for operation with a range of flask sizes; however, heating efficiency is reduced and the chance of superheating is increased when a small flask is heated in a large mantle.

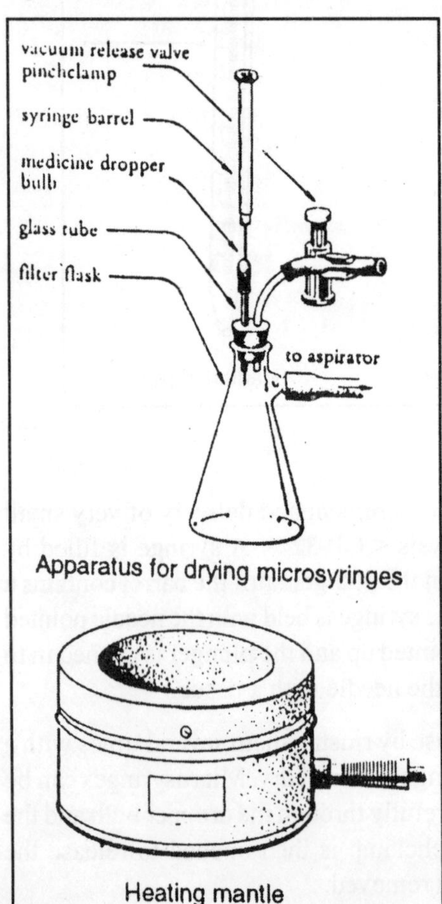

Apparatus for drying microsyringes

Heating mantle

The mantle is mounted on a lab jack, ring, set of wood blocks, or some other support so diat it can be lowered and removed quickly if the rate of heating becomes too rapid. The flask is clamped in place so that it is in direct contact with the heating well, and the heating control dial is adjusted until the desired rate of heating is attained. Because a heating mantle responds slowly to changes in the control setting, it is easy to overshoot the desired temperature by turning the control too high at the start. If this occurs, the mantle should be lowered so that it is no longer in contact with the flask. The voltage input should then be reduced and the mantle allowed to cool down. Further adjustment may be required to maintain heating at the desired rate.

MIXING

Reaction mixtures are frequently stirred, shaken, or otherwise agitated to promote efficient heat transfer, improve contact between the components of a heterogeneous mixture, or mix in a reactant that is being added during the course of a reaction. If the reaction is being carried out in an Erlenmeyer flask, this can be accomplished by manual shaking and swirling, or by using a stirring rod. If the apparatus is not too unwidely and *the* reaction time is comparatively short, ground glass assemblies can sometimes be manually shaken for adequate mixing. This is most easily done by clamping the assembly securely to the ringstand and carefully sliding the base of the ringstand back and forth. But when more efficient and convenient mixing is required, particularly over a long period of time, it is necessary to use some kind of magnetic or mechanical stirring device.

Magnetic Stirring

A magnetic stirrer consists of an enclosed unit containing a motor attached to a magnet, underneath a platform. As the magnet inside the unit rotates, it can in turn rotate a teflon- or glass-covered stirring bar inside a container placed on (or above) the platform. The rate of stirring is controlled by a dial on the stirring unit. Since no moving parts extend outside this container, a magnetically-stirred reaction assembly can be completely enclosed if necessary.

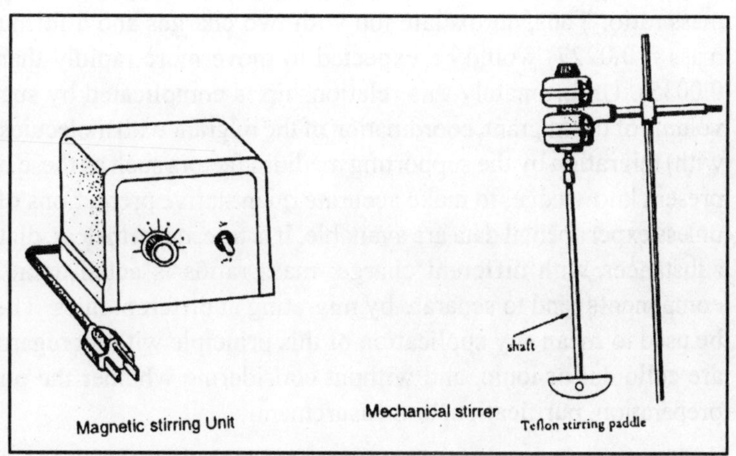

Magnetic stirrers can be used with heating mantles or heating baths that are not constructed of ferrous metal. They work particularly well with oil baths, since they can be used to stir the oil and a reaction mixture simultaneously. The reaction flask must be positioned close enough to the bottom of the oil bath to allow sufficient transfer of magnetic torque from motor to stirring bar. When a copper or aluminium steam bath is used for heating, the flask should be clamped inside the rings, close to the bottom of the steam bath. Some hot plates have an integral magnetic stirrer, and these units can be used (with a heating bath) to simultaneously heat and stir a reaction mixture.

When magnetic stirring is used during a reaction, the heat source is set directly on the stirring unit and a stir bar is placed in the reaction flask in place of boiling chips. The stirring motor should be started and cooling water for the reflux condenser turned on (if applicable) before heating is begun.

Mechanical Stirring

Mechanical stirring utilizes a stirring motor connected to a paddle or agitator by means of a shaft extending through the neck of the reaction vessel. A glass sleeve or bearing is used to align the shaft, which is ordinarily made of glass to reduce the likelihood of contamination. Mechanical, stirrers can exert more torque than magnetic stirrers, and are preferred when viscous liquids or large quantities of suspended solids must be stirred. A variety of stirring paddles made of teflon, glass, and chemically resistant wire are available.

ELECTROPHORESIS

Any charged particle suspended between the poles *of* an electrical field tends to travel toward the pole that bears the charge opposite to its own. The rate at which it travels is conditioned by a number of factors, including the characteristics of the particle, the properties of the electrical field, and environmental factors, such as temperature and the nature of the suspending medium. The mobility of a particle is approximately proportional to its charge: mass ratio. Thus, an oxalate ion with two charges and a formula weight of 88.1 (charge/ mass = 0.0227) would be expected to move more rapidly than a stearate ion (1/283.5 = 0.0035). Unfortunately this relationship is complicated by such factors as the molecular volume of the migrant, coordination of the migrant with molecules of solvent, and interference with) migration by the supporting medium factors such as these make it impossible, with our present knowledge, to make accurate quantitative predictions of electrophoretic mobilities unless experimental data are available. It is true, nevertheless, diat when a solution containing substances with different charge: mass ratios is acted upon by an electrical field, the components tend to separate by migrating at different rates. The word electrophoresis will be used to mean any application of this principle without regard to whether the substances are colloidal or ionic, and without considering whether the purpose of the application is preparation, purification, or measurement.

Electrophoretic Mobilities

The rate at which a particle moves under a controlled set of circumstances is reproducible, making it possible to calculate how far it will travel during an electrophoresis, once the necessary data have been accumulated. Let it be emphasized that mobilities can be established solely by experimentation, and that they are reproducible only when all conditions are controlled. Only voltage gradient and time of migration can be treated as variables if mobility calculations are to be valid. Variations in pH, temperature, ionic strength, medium, and the like, have not successfully been taken into account in mobility calculations. Knowing the mobilities of the components of a mixture

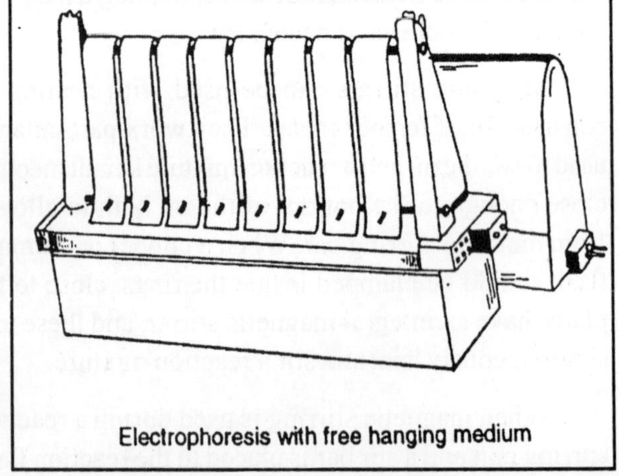
Electrophoresis with free hanging medium

enables one to predict the positions of the components after arbitrary time intervals or in response to varying field strengths. This is useful for locating and identifying components after separations have been obtained and for calculating the time necessary to effect complete separations.

Conventionally, mobility is defined as the distance a particle will travel in a unit of time per unit of strength of an electrical field. Distance of travel is customarily stated in centimetres and time in second, field strength is expressed as the voltage gradient, in volts per centimetre, along the electrophoretic. From this it follows that the dimensions of mobility, u, are cm/sec divided by volts/cm.

which simplifies to : $u = \text{cm}^2/\text{volts x sec}$

If this formula for mobility seems strange remember that it does not describe velocity but is instead a factor intended for use in calculating velocity under defined conditions in response to any given voltage gradient.

Instrumentation

The area upon which electrophoretic separations occur, called the bed, can *be* composed of any of a number of materials including gels, films, and powders. It is moistened with an electrolyte solution (usually a buffer). The ends of the bed are immersed in more of the electrolyte contained in two chambers designed to hold electrodes that are connected to a dc power supply. Provision is made for adjusting the electrolyte in the electrode chambers to equal levels so that siphoning action does not occure through the bed. The entire apparatus, excluding the power supply, is enclosed in an an airtight chamber to prevent excessive evaporation of buffer. When a spot of sample mixture is applied to the bed and the power is turned on, those components of the mixture whose particles are charged will migrate toward the electrode having the opposite polarity.

APENDIX - 4

Culture Media and Preparation

The success in cell, tissue and.organ culture technology is related to the selection or development of the culture medium. As no single medium will support the growth of all tissue cultures therefore modifications in the nutritional component including growth regulators are often necessary for different types of growth responses in a single explant material.

Various media compositions which are frequently used for tissue culture. A literature search is useful for selecting the appropriate culture medium as a starting points in developing a medium for specific purpose such as callus induction, axillary bud proliferation, organogenesis, somatic embryogenesis, anther culture etc. A nutrient medium generally contains inorganic salts, vitamins, growth regulators, a carbon source and gelling agent. Other components added for specific purposes include organic nitrogen compounds, hexitols, amino acids, antibiotics and plant extracts. The Murashige-Skoog medium (MS) (1962), Revised Murashige-Skoog medium (Raj Bhansali and Arya, 1978), White's medium (1963), Linsmaier and Skoog (LS) (1965), B5 (Gamborg et al, 1968), Nitsch and Nitsch (1969), Woody plant medium (Llyod and McCown, (1981), Somatic embryogenesis medium (Raj Bhansali, 1988, 1990) and derivatives of these media have wide applications for different plant species and for different culture objectives. The decision on using type of media for the metabolic needs of the cultured cells and tissues, is a major factor of success in plant regeneration process.

Media Components

Inorganic Salts

A relatively small number of mineral salts are used as component of media for plant tissue culture. The inorganic salt formulations can vary in various reported media, however MS formulation is most widely used with or without modifications. The distinguishing feature of MS inorganic salts is their high content of nitrate, potassium and ammonium in comparison to other salt formulations. The stocks are prepared at 100 X (times) the final medium concentration. Each stock is added at the rate of 10 ml per 1000 ml of medium prepared.

The Na-FeEDTA stock should be protected from light stored in bottle that is amber coloured or wrapped in aluminium foil. Concentrated salt stocks enhance the accuracy and speed of media preparation. Guidelines for maintaining stock solutions.:

1. All salt stocks should be stored in the refrigerator and are stable for several months
2. Always prepare stocks with glass distilled or demineralized water
3. Label the stock solutions clearly with date

4. Reagent grade chemicals should be used to ensure maximum purity
5. The nitrate stock usually precipitate out and must be heated until crystals are completely dissolved before using
6. Any stock showing cloudy or has bacterial or fungal growth should be discarded
7. Do not combine the stock to other stocks unless they are stable and compatible.

Plant Growth Regulators

The four classes of growth regulators are commonly used in tissue culture media i.e., auxins, cytokinins, gibberellins and abscisic acid (Appendix-1). The type of growth regulators and concentration used will vary according to the cell culture purpose. A list of the mostly commonly used growth regulators and their abbreviations and molecular weight is given in Table .

An auxin (IAA, NAA, 2,4-D or IBA) is required for the induction of cell division and root initiation in cultured tissues. The auxins are mostly used in combination with cytokinins. The 2,4-D is used for callus induction where as IAA. IBA, and NAA are used for root induction. Auxin stocks are usually prepared by dissolving in ethanol, I N NaOH or 1 N KOH until crystals dissolved (not more than 0.3 ml/lO mg of auxin), rapidly adding 90 ml of distilled water and increasing the volume to 100 ml in a volumetric flasks. Though the auxins are thermostable, however IAA is destroyed by low pH, light, oxygen and peroxidases. The NAA and 2,4-D are most stable form of auxin.

The cytokinins (KN, BAP, 2iP and Zeatin) are adenine derivatives, promote cell division, shoot proliferation, organogenesis and somatic embryogenesis. They have essential role in differentiation and micropropagation of most plant species.

The cytokinin stocks are prepared in a few drops of 1N HC1 and water to dissolve crystals. Gentle heating is usually required for complete dissolving of crystals. Bring the stock up to the desired volume by adding double distilled water in a volumetric flasks. Cytokinin stocks can be stored for several months in the refrigerator. Cytokinins are thermostable during autoclaving in media.

The gibberellins are infrequently used in plant tissue cultures as it can inhibit callus growth but for meristem culture after shoot primordia formation are used in plant regeneration and elongation.

The stock solution of GA_3 can be prepared by dissolving in water and adjusting the pH 5.7. They are not thermostable therefore should be filter sterilized.

The abscisic acid is useful in embryo culture and somatic embryogenesis. Abscisic acid, is heat stable but light sensitive. Stock solutioas can be prepared in double distilled water and stored in coloured bottle in refrigerator. Dilution of stock solutions may be as per the requirerrlent.

Vitamins: Vitamins have catalytic functions in enzyme reactions. The most commonly used vitamins in tissue culture media are nicotinic acid (B_3), thiamine (B_1) and pyridoxine

(B_6). They are added in medium before autoclaving. The stocks are usually prepared in water at 100 X or 1000 X (10 ml per 1000 ml medium or 1 ml per 1000 ml medium) and stored in a freezer.

Carbon Source

The carbohydrates in form of sucrose or glucose (2-5% W/V), as a carbon source are essentially required in tissue culture as cells or tissues are generally not photosynthetically active. Lower levels of a carbohydrate may be used in protoplast culture but higher levels are required for embryo or anther culture.

Sugars undergo caramelization prolonged on autoclaving (too long period) and will react with amino acid compounds. Sugars are degraded and form melanoidin, which are brown, high molecular weight compounds that can inhibit cell growth.

Hexitols: Among hexitols, myo-inositol has been found very important ingredient in tissue cultures, it is considered as growth promoter in tissue cultures. This has an action like carbon source as well as vitamin like. Mannitol or sorbitol are good osmotica for protoplast isolation. It is water soluble and stock can be made up at the strength of 100 X (10 ml aliquots are used for 1000 ml medium).

Gelling Agent

In tissue cultures, washed or purified agar of TC grade or Difco-bacto agar grade is used. The agar must be kept in motion while dissolving otherwise it will burn on the bottom of the flask. The agar must be completely dissolved before it is dispensed into the culture vessels.

The agar can also be melted in a autoclave or in a foil capped Erlenmeyer flask for 15 min at 121°C and dispensed aseptically into sterile containers by using laminar air-low bench before solidification of agar.

Amino Acids and Amides

The amino acids and amides are very important in tissue cultures specially for the morphogenesis. All L forms of amino acids are commonly used, as L-tyrosine can contribute to shoot initiation, L-arginine can facilitate rooting, and L-serine can be used in haploid embryos induction in microspore cultures. L- cysteine is used for controlling phenol leaching from explant tissues. Amides such as L-glutamine and L-asparagine can induce somatic embryogenesis.

Antibiotics

The various fungicides and bactericides are used in case plant explants on cultures excessively contaminated. These chemicals are toxic not only to contaminants but also to cultures or explant materials so restricted use should be made for additions into the culture medium. The antibiotics are soluble in water should be made fresh and be added to the medium after autoclaving by filter sterilization.

Natural Complexes

The natural complexes such as coconut (endosperm) milk (CM), yeast extract (YE), malt extract (ME), tomato juice, potato extract, casein hydrolysate (use enzyme digest) and fish emulsion are used in tissue cultures for various purposes. Addition of these complexes in the medium make the medium undefined, since variation in growth promoting or inhibiting compounds in these complexes.

Antioxidants: The antioxidants such as citric acid, ascorbic acid, pyrogallol, phloroglucinol and L-cysteine are used in tissue culture to reduce excessive browning of the explants. Adsorbents like PVP and activated charcoal are also used for checking excessive browning.

ADDITIONAL REQUIREMENTS

Quality of water, chemicals and natural complexes: Demineralized or double distilled water of high purity are used in making stocks and medium. Glass distilled water is most desirable and stored in clean containers.

Callus-induction medium

Murashige and Skoog medium	1.01
2,4-D	1.0 mg
Agar	8.0 g

Prepare 1 litre of standard MS medium. Add the 2,4-D and adjust the medium pH to 5.5 using 0.1 M NaOH. Dissolve the agar in the medium in a steam bath. Dispense the medium into culture tubes or vessels and autoclave for 20 min at 121°C.

Chlorate Selection Medium

$MC1O_3$	600.0 mg
$Ca(NO_3)_2 4H_2O$	118.0 mg
$MgSO_4 7H_2O$	19.5 mg
K_2HPO_4	19.7 mg
P-IV trace metal solution	3.0 ml

Dissolve the salts in 900 ml of distilled water. Add the P-IV trace metal solution and adjust the medium pH to 7.0. Add additional distilled water to make 1 litre of medium.

Chlorate Selection Medium Overlay

Chlorate selection medium	500.0 ml
Gel-rite ™ gelling agent	4.0 g

Slowly add the Gel-rite a little at a time to the chlorate selection medium while stirring the mixture with a magnetic stirrer. Set the mixture in a steam bath to dissolve the Gel-rite. Dispense 4 ml of the overlay medium into each culture tube; 4 ml should spread out as a very thin layer over the surface of the media in plates.

Embryo Culture Medium

Murashige and Skoog medium	1.01
Agar	8.0 g

Prepare 1 litre of standard MS medium. Adjust the medium pH to 5.5 using 0.1 M NaOH. Dissolve the agar in the medium in a steam bath. Autoclave for 25 min at 121°C. Allow the medium to cool to 50°C in a temperature controlled water bath. Pour the medium into sterile 100 mm petri plates.

L/10 Green Algae Medium

$Ca(NO_3)_2 4H_2O$	0.118/g
$MgSO_4\ 7H_2O$	0.0195 g
K_2HPO_4	0.0197 g
P-IV trace metal stock	3.0 ml

Dissolve all of the salts in 1 litre of distilled water. Adjust the medium to pH 7.0 by adding 1 M HC1 or 1 M NaOH.

Micropropagation Medium

Murashige and Skoog medium	1.01
Indolebutyric acid (IBA)	1.0 mg
Benzylaminopurine (BAP)	3.0mg

Prepare 1 litre of standard MS medium. Add the IBA and BAP and adjust the medium pH to 5.5 using 0.1 M NaOH. Dissolve the agar in the medium in a steam bath. Dispense the medium into culture tubes or vessels and autoclave for 20 min at 121°C.

Murashige and Skoog (MS) Medium

Macro salts

NH_4NO_3	1.65g
KNO_3	1.90g
$CaCl_2\ 2H_2O$	0.44 g
$MgSO_4\ 7H_2O$	0.37 g
KH_2PO_4	0.17g

Micro salts

$FeSO_4\ 7H_2O$	27.80 mg
$Na_2EDTA\ 2H_2O$	33.60 mg
KI	0.83 mg
H_3BO_4	6.20 mg
$MnSO_4 4H_2O$	22.30 mg
$ZnSO_4\ 7H_2O$	8.60 mg
$Na_2MoO_4\ 2H_2O$	0.25 mg

$CuSO_4$ $5H_2O$	0.025 mg
$CoCl_2$ $6H_2O$	0.025 mg
Organic supplements	
Myoinositol	100.00 mg
Nicotinic acid	0.05 mg
Pyrodoxine HCl	0.05 mg
Thiamine HCl	0.05 mg
Glycine	0.20mg
Sucrose	20.00g

Dissolve the salts and organics in 800 ml of distilled water. Adjust the medium pH to 5.7 by adding 1 M NaOH. Add additional distilled water to adjust the final volume to 1 litre.

MS/C Medium

MS salts and organic supplements	
IAA solution	8.0 ml
Kinetin solution	2.5ml
Agar	8.0 g

Dissolve the salts and organics in 800 ml of distilled water. Add the IAA and kinetin solutions. Adjust the medium pH to 5.7 by adding 1 M NaOH or 1M HC1. Add additional distilled water to adjust the final volume to 1 litre. Add the agar and heat the medium on a hot plate or in a steam bath until the agar melts. Stir the medium occasionally until all the agar is dissolved and the solution is clear. Do not let the medium boil. Dispense 8 ml aliquots in 20 X 150 mm culture tubes (approximately 120 tubes of medium).

***Myriophyllum aquaticum* Shoot-Induction Medium**

Murashige and Skoog medium	1.0 litre
[2-isopentenyl] adenine (2iP)	2.0 mg
Agar	8.0 g

Prepare 1 litre of standard MS medium. Add the 2iP and adjust the medium pH to 5.7 using 0.1 M NaOH. Dissolve the agar in the medium in a steam bath. Dispense the medium into culture tubes or vessels and autoclave for 20 min at 121 °C.

***M. aquaticum* Stock Plant Medium**

Murashige and Skoog medium	1.01
Agar	8.0 g

Prepare 1 litre of standard MS medium. Adjust the medium pH to 5.7 using 0.1 M NaOH. Dissolve the agar in the medium in a steam bath. Dispense the medium into culture tubes or vessels and autoclave for 20 min at 121 °C.

P-IV Trace Metal Solution

Na_2EDTA	0.750 g
$FeCl_3$ $6H_2O$	97.0 mg

$MnCl_2 4H_2O$	41.0 mg
$ZnCl_2$	5.0 mg
$CoCl_2$ $6H_2O$	2.0 mg
Na_2MoO_4	4.0 mg

First dissolve the Na_2EDTA in 500 ml of distilled water, then dissolve the remaining metal salts. For greater accuracy, it may be easier to prepare 10 X concentration stock solutions of the Zn, Co, and Mo salts and add 1/10 of these stocks to the P-IV stock solution.

***Pandorina* Ammonium Medium**

NH_4CI	27.0 mg
$CaCl_2$ $2H_2O$	100.0 mg
$MgSO_4$ $7H_2O$	19.5 mg
K_2HPO_4	19.7 mg
P-IV trace metal solution	3.0 ml

Dissolve the salts in 900 nil of distilled water. Add the P-IV trace metal solution and adjust the medium pH to 7.0. Add additional distilled water to make 1 litre of medium.

***Pandorina* Nitrate Medium**

$NaNO_3$	35.0 mg
$CaCl_2$ $2H_2O$	100.0 mg
$MgS0_4 7H_2O$	19.5 mg
K_2HPO_4	19.7 mg
P-IV trace metal solution	3.0 ml

Dissolve the salts in 900 nil of distilled water. Add the P-IV trace metal solution and adjust the medium pH to 7.0. Add additional distilled water to make 1 litre of medium.

***Pandorina* Nitrite Medium**

$NaNO_2$	35.0
$CaCl_2$ $2H_2O$	100.0 mg
$MgSO_4$ $7H_2O$	19.5 mg
K_2HPO_4	19.7 mg
p-iv trace metal solution	3.0 ml

Dissolve the salts in 900 ml of distilled water. Add the P-IV trace metal solution and adjust the medium pH to 7.0. Add additional distilled water to make 1 litre of medium.

***Pandorina* Hypoxanthine Medium**

Hypoxanthine	68.0 mg
$CaCl_2$ $2H_2O$	100.0 mg
$MgS0_4$ $7H_2O$	19.5 mg

K_2HPO_4	19.7 mg
P-IV trace metal solution	3.0 ml

Dissolve the salts in 900 ml of distilled water. Add the P-IV trace metal solution and adjust the media pH to 7.0. Add additional distilled water to make 1 litre of medium.

***Pandorina* Uric Acid Medium**

Uric acid	84.0 mg
$CaCl_2 2H_2O$	100.0 mg
$MgSO_4 7H_2O$	19.5 nig
K_2HPO_4	19.7 mg
P-IV trace metal solution	3.0 ml

Dissolve the salts in 900 ml of distilled water. Add the P-IV trace metal solution and adjust the medium pH to 7.0. Add additional distilled water to make 1 litre of medium.

Potato Dextrose Agar

White potatoes, sliced	250 g
Dextrose	20 g
Agar	15 g

Boil die potatoes in 500 ml of distilled water for 15 min until soft. Filter this mixture through cotton to remove most of the paniculate matter. Dissolve the dextrose in 200 ml of the potato infusion. Add 800 ml of distilled water. Adjust die final pH to 3.5-4.0. Dissolve the agar in a steam bath or on a hot plate. Autoclave at 121°C for 25 min.

Trypticase-Soy Broth Medium

Trypticase	17.0g
Phytone	3.0g
NaCl	5.0 g
K_2P0_4	2.5 g
Glucose	2.5 g

Dissolve the ingredients in 1 litre of distilled water. Adjust the medium pH to 7.3 by adding 1 M NaOH. Dispense 10 ml aliquots in 20 X 150 mm culture tubes, (approximately 100 tubes of medium).

Yeast Extract Broth (YEB)

Yeast extract	1.0 g
Beef extract	5.0 g
Peptone	5.0 g
Sucrose	5.0 g
$MgSO_4$ $7H_20$	0.5 g

Dissolve all the ingredients in 1 litre of distilled water. Adjust the medium pH to 7.0 by adding 1 M NaOH. Dispense and autoclave for 25 min at 121°C.

Yeast Extract Indicator Medium (YI)

Yeast extract	1.0 g
Lactose	10.0 g
Agar	20.0 g

Dissolve the yeast extract and lactose in 1 litre of distilled water. Adjust the medium pH to 7.0 by adding 1 M NaOH. Add the agar and dissolve it by heating the mixture in a steam bath or on a hot plate. Autoclave for 25 min at 121°C.

Related Procedures

Ultraviolet Light

UV light may be divided into three wave length groupings near UV (315-400 nm), mid range UV (280-315 nm) and far UV (200-280 nm). Maximal sensitivity in humans is at about 280 nm. Exposure to direct or indirect mid-range UV can cause acute eye irritation after a latent period of 2-24 hrs. Because retina is not sensitive to UV eye damage may result without the subject being aware of the exposure. Skin is also sensitive to UV which may cause for skin cancer. Hence protect your eyes and skin from the effects of UV irradiation by wearing goggles with side shields by clothing, and by limiting exposure.

Preparation of Phenol

All crystalline phenol must be redistilled at 160° C to remove contaminants that cause or cross linking of DNA or RNA. Soon after distillation add 0.1 % hydroxyquinoline. The melted phenol is extracted several times with an equal volume of 1.0 M Tris pH 8.0 followed by 0.1- M Tris pH 8.0 and 0.2% β-mercaptoethanol, until pH of the aqueous phase is 7.6. Phenol is stored in aliquots at 4° C under equilibration buffer for periods upto 1 month.

Phenol is widely used as a disinfectant and germicide. It is a dangerously toxic materrial that can produce poisoning when ingested, inhaled or absorbed through the skin. The toxic effect include headache, dizzines, nausea, weakness, difficulty in breathing, unconciousness and death. Phenol is corrosive to skin, initially producing a softened area followed by severe burns.

1. If phenol is spilled on the skin, flush immediately with large amounts of water. Do not use ethanol.
2. If eyes are contaminated, wash them with running water for about 15 min, call for medical help.

Working with 32p Labelled Compounds

β-particles with an energy of 1.71 MeV (6.1 meter range in air) is emitted by 32p Hence 32p labelled compounds must be handled carefully with much caution using shields. When 15 particles hit targets, electromagnetic radiation known as Bremsstrahlung is produced, the yield of which is directly proportional to the density of material used for shielding. Therefore, a low density material may be added to absorb the Bremsstrahlung emitted. Always wear gloves (two pairs if necessary), protect eyes and use dosimeters. Always cover the work area with absorbant papers and use survey meter to check spillage. Eating, smoking or drinking while handling radioactive compounds should be banned. Use special tape to label containers and tubes in which radioactive materials are kept. The maximum permissible burden of 32p is 30 uCi but the maximum permissible burden for bone is only 6 uCi.

Silanization of Plastic and Glassware

Place the plastic and glasswares to be silanized in a desiccator. Add about 1 ml of dichlorodimethyl silane in a small container in the desiccator. Pull vacuum on the system until dichlorodimethyl silane boils. Close the system and allow to sit for about 2 hrs. Open the desiccator in a fume hood and allow to air out several hours. Rinse the plastic and glassware with water and autoclave.

10 X Restriction Endonuclease Buffers (Refrigerate)

Low	100 mM Tris-HCl (pH 7.5) 100 mM $MgCl_2$ 10 mM Dithiothreitol (DTT)
Medium	500 mM NaCl 100 mM Tris-Cl (7.5) 10 mM DTT
High	1M NaCl 500 mM Tris-Cl pH 7.5 100 mM $MgCl_2$ 10 mM DTT

Preparation of Dialysis Tubing

Cut dialysis tubing into convenient length. Boil them for 10 min in a large volume of 2% sodium bicarbonate and 1 mM EDTA. Cool and rinse thoroughly in distilled water and again boil for 10 min in distilled water. Cool and store in refrigerator submerged in water. Just before use rinse with water. Wear gloves while handling the dialysis tubing.

Lengths and Molecular Weights of Common Nucleic Acids

Nucleic Acid	Number of Nucleotides	Molecular weight
LAMBDA DNA	48,502 (Circular, dsDNA)	3.0×10^7
pBR322 DNA	4,363 (ds DNA)	2.8×10^6
28SrRNA	4,800	1.6×10^6
23S rRNA	3,700	1.2×10^6
18S rRNA	1,900	6.1×10^5
16S rRNA	1,700	5.5×10^5
5S rRNA	120	3.6×10^4
tRNA (*E. coli*)	75	2.5×10^4

Standards

1 kb of ds DNA (sodium salt) — 6.6×10^5 Daltons

1 kb of ss DNA (sodium salt) — 3.3×10^5 Daltons

1 kb of ss RNA (sodium salt) — 3.4×10^5 Daltons

The average MW of a deoxynucleotide base — 324.5 Daltons

Common conversions of Nucleic acids and Proteins

I. Spectrophotometric Conversions

IA_{260} unit of ds DNA — 50 µg / ml

IA_{260} unit of ss DNA — 33 µg / ml

IA_{260} umt of ss DNA — 40 µg / ml

II. Protein Molar Conversions

100 pmoles of 100,000 MW protein — 10 *µg*

100 pmoles of 50,000 MW protein — 5 µg

100 pmoles of 10,000 MW protein — 1 µg

III. Protein/DNA Conversions

1 kb of DNA = 333 amino acids of coding capacity — 3.7×10^4 MW

10,000 MW protein — 270 bp DNA

30,000 MW protein — 810 bp DNA

50,000 MW protein — 1.35 kb DNA

100,000 MW protein — 2.7 kb DNA

Half life of Important Radioisotopes Used

Radionucleotide	Half life
Tritium	12.43 years
Carbon-14	5.730 years
Sulphur-35	87.4 years
Phosphorous-32	14.3 years

APENDIX - 6

Problems and Possible Solutions in Plant Tissue Culture Work

Symptoms	Possible Causes	Possible Solution
Culture contamination	Source heavily infested/ infected with micro organisms Poor sterilization	Improve sterilization method (70 % Ethanol, drop of detergent, antimicrobial agents) clean plant material/select unexposed tissue.
Explant dies	Strong disinfectants	Use weaker disinfectant / change sterilizing agents
	Media too strong	Use ½ or ¼ strength
	Wrong stage of growth	Obtain explants at different stage of growth
Culture blackens and dies	Contaminated by micro-organisms	Discard with care. Review sterile technique and sanitation
	Bleeding	Subculture immediately. Transfer more frequently Try different agar concentrations
	Agar problem Water problem Wrong formula media constituents	Check water purity Try different formula
Explant live but no growth	Dormant	Chill for a month
	Media too harsh	Use explant at different stage of growth
	Wrong formula	Lower salts and hormones try different formula
Culture live but no growth	Too cold	Increase temperature
	Wrong medium	Try different medium
Shoots too long (leggy), and poor multiplication leaves are yellow, waterly	Too little cytokinin	Increase cytokinin Change Cytokinin/auxin ratio
Leafy shoots too short	Hormones too strong	Decrease or omit hormones
No multiplication	Too little cytokinin	Increase cytokinin
	Needs chilling	Cold store 4-8 weeks
	Too cold	Run cytokinin / auxin grid
	Requires dormacy period	Increase temperature Cold treat 3-8 weeks
Fat stems, small and pale leaves	Too much cytokinin	Decrease cytokinin/increase auxin

Symptoms	Possible Causes	Possible Solution
Unwanted callus	Wrong hormones	Decrease or omit hormones Run cytokinin/ auxin grid
Leaves chlorotic	Contaminant Too hot Wrong formula	Index for contaminants Decrease temperature Try different medium
Leaves succulent (watery), virtrification, abnormal stem, embryos	Osmotic potential upset	Decrease temperature
	Too high cytokinin Wrong agar Culture too old	Increase agar strength Decrease hormones Try different agar Transfer more often
Premature rooting	Wrong hormone balance	Transfer more often Run cytokinin/auxin grid Increase cytokinin/decrease auxin
Red stems / embryos / cells	Stress	Change light and / or temperature Lower sucrose content Increase nitrate (NO_3) Transfer more frequently
Non-friability of callus	Too much sugar Not enough NO3 Culture too old Media composition Plant source	Using number of media having range of hormones. Select most rapidly growing callus. Change to liquid culture and after friable to agar medium.

Source : Bhansali, R.R. (1995)

APENDIX - 7

Use and Storage of Coconut Water

Coconut water has been shown to stimulate shoot proliferation in many species of plants. Coconut water is prepared from selected coconuts by filter sterilized and frozen prior to use. This precipitation should not effect the growth of the plant tissue. Coconut water can be divided into smaller aliquots, corresponding to standard medium batch size, and refrozen until needed. Coconut water should be used at a concentration of 5-20% (v/v).

Sterilization of Media

Tissue culture media are generally sterilized by autoclaving at 121°C and 1.05 kg/ cm^2 (15-20 psi). The time required for sterilization depends upon the volume of medium in the vessel. Dispense medium in small aliquots whenever possible as many media components are broken down on prolonged exposure to heat. Medium exposed to temperatures in excess of 121°C may not properly get or may result in poor cell growth.

Minimum Autoclaving Time

Volume of Medium per Vessel (ml)	Minimum Autoclaving (min)	Volume of Medium per Vessel (ml)	Minimum Autoclaving (min)
25	20	500	35
50	25	1000	40
100	28	2000	48
250	31	4000	63

Minimum autoclaving time includes the time required for the liquid volume to reach the sterilizing temperature (121° C) and 16 min at 121° C. Times may vary due to differences in autoclaves. Validation with system is recommended. Several medium components are considered thermolabile and should not be autoclaved. Stock solutions of the heat labile components are prepared and filter sterilized through a 0.22 µm filter into a sterile container. The filtered solution is aseptically added to the culture medium which has been autoclaved and allowed to cool to approximately 35-45°C. The medium is then dispensed under sterile conditions.

❑❑❑

How to Make your own Gene Bank

Gene banking is the practical and effective method to combat plant extinctions. It is a kind of freezer that preserves seed and pollen. This technology does not require extensive knowledge or specialized training, and expensive equipment. Gene banks are generally easy to construct and maintain, although a few problems may arise during or after construction of the bank.

Many people mistakenly assume that governmental agencies or educational institutions are the most appropriate groups to operate gene banks. Actually, these groups can be unreliable over the long term because their funding priorities are not always permanent.

Who are the Gene Bank Operators?

Reliable gene bank operators are often plant enthusiasts, who are not only devoted to furthering their preferred group of plants, but are also interested in conservation. Species collected eagerly by plant enthusiasts may receive minimal attention at arboreta and botanic gardens. Furthermore, changes of arboretum directors may signal a change in emphasis on which plants are to be saved. Amateur groups, therefore, can have a powerful effect on the effort to preserve some species of plants. Some groups, such as those that breed and cultivate ferns, cacti and alpine plants, already operate some form of a seed or spore bank. For such groups, creation of an actual cryogenic gene bank would simply extend the preservation techniques they already operate.

What is Gene Banking?

The theory behind gene banks is really quite simple. Cryogenic preservation is no more than suspended animation at sub-freezing temperatures. The process requires little more than a household chest freezer, storage containers made of glass, metal or plastic, a simple desiccant or drying agent to dehydrate the seed and the seeds themselves.

Seeds are tiny organisms that stay alive so long as there is a food reserve inside the seed to fuel vital chemical reactions that keep the seed alive. These chemical reaction rates are dependent on the surrounding temperature- the higher the temperature, the faster the reactions will take place and the less time the food reserves will last. The converse also is true. For every 5° C drop in temperature, the life of the seed will double. For example, onion seeds with ten per cent moisture are viable for 16 weeks at 35° C but will live for 78 years at 0°C. Dropping the temperature down to -15°C would increase longevity to 624 years, provided the ice crystals caused by the ten per cent moisture level do not harm the seed when frozen. Clearly, if seeds could be stored at sub-freezing temperatures, they might be viable for centuries.

How cold is cold enough?

Two factors dictate the storage temperature: the desired cost of the freezer and the desired longevity of the seed. The freezer cost increases as the temperature inside the apparatus decreases. Calculating desired longevity is not so simple. Cryogenics theory suggests that extremely low temperatures allow seeds to last for many centuries, but such tremendous life-spans may not be necessary. If human beings somehow manage to persist through the next century or two, we suspect they will have come to terms with their planet's ecological problems. One or two hundred years of storage, therefore, probably is sufficient.

Besides the storage temperature, the other major concern is water. Too much water in a seed will form ice crystals as the temperature dips below the freezing level. Seeds with a high water content are plagued by large ice crystals that rupture individual cell membranes and destroy the seed. Many seeds have a naturally reduced water content and can be frozen soon after they have ripened without danger of rupture. Some seeds, however, require extra care to reduce their water content. While this usually is a fairly easy task, species with extremely fleshy or oily seeds simply cannot be frozen and must be preserved by other \ methods.

How much water should be reduced?

The life of a seed will double with each one per cent decrease in water content. A minimum of four per cent water content appears to be required by seeds to stay alive. International standards for long-term seed storage suggest than an average seed should contain five per cent water and be stored at -18°C. When both temperature and water content are reduced, the two factors multiply- when temperature drops 5°C and water decreases by one per cent, the seed lives four times as long.

Collecting, Cleaning and Drying

The first step in collecting for the gene bank is to make certain that the seeds are alive. A variety of biochemical tests and stains can be used, but the most dependable method is also the easiest. Simply plant a known number of seed and see how many germinate. Only on rare occasions will all of the seeds begin to grow. Plant scientists usually strive for about 95 per cent germination. Some plants have naturally low germination rates. For example, we have found that *Aloe albida* germinates at a rate between 25 and 30 per cent. This species is endangered, perhaps because of this highly inefficient rate of germination.

Because space in a cryogenic gene bank is usually at a premium, do not waste space by storing dead seed, chaff or pieces of seed capsules and stalks. Cleaned seed takes up less space in the bank and is easier to inspect and observe. At our gene bank in Irvine, we usually harvest pods as soon as they start to ripen or split, put the seeds in a container, and expose them to air for a few weeks. When first extracted from pods, the water content in seeds is quite high. However, in areas with low humidity the seeds can mature and dry naturally in the air. This occurs because seeds match their moisture content with the water vapour in the air. They either lose or absorb water, depending on whether the water vapour in the air is higher or lower than the water in the seed.

Seed Drying

Dry air results in dry seed. Air-dried seeds harden and can be handled safely. The seed held in the gene bank at the Royal Botanic Gardens, Kew, is dried artificially in a conditioning room with a temperature of 16° C and a relative humidity of 14 per cent. Seeds can be dried simply in humid areas by storing them for several days in a closed container with a desiccant or drying agent. The most familiar desiccant is Silica Gel, a product available in hardware and photographic shops. Another product which readily absorbs moisture from the air is Drierite™, a calcium salt. Drierite™ comes with a colour indicator: when it turns blue, it can still absorb moisture; when it turns pink, it is hydrated and can absorb no more. One caution, is always remembered that only a fresh desiccant can absorb moisture. It is important to test the desiccant before using it to dry seed. A handy test for freshness is to dip a strip of filter or blotting paper in a saturated solution of cobalt chloride. When the paper dries, it will turn blue if the air is dry, and pink if the air is moist. When you discover that the desiccant is stale, rejuvenate it by baking it on a Teflon-coated baking sheet for two or three hours at 60° C or higher. This will dry it back to its original state.

Now that the desiccant is ready to use, find any container that can be sealed. About 2.5 cm (1 inch) of desiccant should be placed on the bottom of the container and covered with either cheese-cloth or wire gauze, permitting air to reach the drying agent. Open it at least once a day to stir the seed.

During the first two days the seeds will lose most of their water content. A little more is lost during the next five or six days. Within a week, most seeds can be safely frozen. This is the most practical method of drying seed for a home gene bank.

How Much Dry is Too Dry?

Overdried seed can result if proper care is not taken. Air-drying at 35 °C for several days is also recommend. This process has been used routinely to dry many different kinds of seed. During the drying period, the seeds are subjected to considerable temperature stress and an acceleration of the metabolic processes. Drying the seed at the gentler room temperature is desirable.

Freeze-drying is a process that rapidly freezes tissue and evaporates most, if not all, of the water. Seeds subjected to freeze-drying probably would die due to considerable ice crystal damage and the strong vacuum used to dry the material. There are reports, however, that pollen has been stored successfully following freeze-drying treatment. For seed, the safe two-step process of first drying and then freezing is recommended.

Storage Containers

There is just one cardinal rule about storage containers: *they must be absolutely airtight so that already dried seeds will not take up water from the air.* The air within a freezer contains water molecules which attach to dried seed if the container is not airtight. Doubters can place a piece of cobalt indicator paper into a freezer and see the results. The ice build-up in a freezer is further proof.

Containers for Seed Storage

Many kinds of containers can be used to store seed, but the best are those with seals caused by the fusion of the material in the container. Three varieties- metal, glass, and plastic - appear to be almost foolproof.

Metal cans are probably the sturdiest containers, but without a machine to weld the lids into place, these are both troublesome and expensive. Large samples of big seeds (such as beans and peas) store very well in metal.

Glass containers, the second option, are more fragile but this method is relatively cheap. Inexpensive disposable glass test tubes are easy to get from scientific or medical supply houses. They are sealed at the end of the tube with an inexpensive propane torch. They are available at most of the hardware stores.

With a little practice the tube can be sealed within a few seconds without significantly heating up the seed sample.

Plastic container: The third type of container is an envelope made of foil-laminated plastic that can be heat-sealed. These envelopes have an inner lining of polyethylene and an outer lining of aluminium or some other metal. While the polyethylene is slightly pemeable to gasses and could allow some water molecules to cross into the envelope, the metal lining is impermeable arid ensures that the container will be airtight. The plastic will melt and seal when heat is applied to the open end. You can add a vacuum device to suck most of the air out of the package. These envelopes take up less space and are easiest to store, but they are not easy to find.

Labelling and Access

Imagine a freezer full of packages, tubes or cans of seed from which you must retrieve a specific packet. This is one of the most difficult problems associated with small gene banks. Bankers need to figure out the most efficient way of storing large quantities of small packets so that one can be retrieved without defrosting the entire collection. A system of numbered shelves within the freeze seems to be the most practical way around the problem. Drawers are impractical and should be avoided because ice build-up could freeze the drawers shut. It is important to label all samples clearly because the label must resist sub-freezing temperatures for long periods of time. You must use ink that will not fade or become brittle, and the paper must not self-destruct. A dual system of labels is probably the best bet. One label should be placed on the outside of the sample and another should be placed inside. As an extra precaution, a record of the sample's shelf position should be recorded in a book or mini-computer. The name of the sample - should be imprinted on to a metal foil label. This is one of the best possible long-term labels. The internal label is the most important one as it will always accompany the seeds. It is also a good idea to number the samples. This number could be correlated with the year of the sample and stored with the permanent record.

Record-keeping is obviously a vital part of operating a gene bank. Records should be precise and document the quality and viability of the seed. The original source of the seed and the date of processing are further useful bits of information to retain. The more complete the record, the more valuable the sample. Notes on other parameters would be useful, but do not get overwhelmed by records. Seeds should be the prime focus of the bank, not paperwork. In the final analysis, the last seed of a species is more important, even without records, than a very fine file about an extinct species.

How Large a Seed Sample?

There is no simple formula to determine how large a seed sample should be. The size should depend on the ultimate use of the sample. If the sample is needed to conserve the entire gene pool, that is. the full range of variation within the population, then the sample should contain nearly 10,000 individuals or seeds. If only a representative sample is needed to retrieve a few plants for illustrative specimens, breeding experiments, or even genetic examination, small samples of 20 to 100 seeds might be sufficient. Because the seeds are stored in sealed containers, samples should be easy to extract. Removing a small sample directly from the freezer is better than taking out a large sample, defrosting it to remove the needed amount, then resealing and refreezing the remainder. Seeds can handle some defrostings and refreezings, but we do not know how many times they can suffer this treatment. It is more prudent to save defrostings for powercuts.

Ideally, the sample of a particular species should have several components. First, it should contain a basic foundation sample that should not be touched except in dire circumstances. Second, it should hold another large sample that can be used to search for a particular genetic variant. Finally, smaller samples should be reserved to fulfil requests for seeds from scientists, institutions or their individuals who might wish to examine representative plants or products.

Because gene banks are susceptible to the variety of accidents and catastrophes that can affect any institution or building, gene bankers should distribute duplicate samples to other gene banks. Fire, earthquakes, tornadoes or floods could devastate a collection of hundreds of species. Having back-up copies in other collections, and allocating a portion of your own freezer for the maintenance of duplicates from others, is a sensible approach.

The Freezer

A few points should be made regarding the type of freezer to use. Upright freezers are a poor idea because cold air rushes out when the door is opened. Chest freezers are the best choice for small gene banks. These range in price depending on the desired temperature, the highest-priced freezers having the lowest temperatures. Freezers with temperatures below-

18°C are unnecessary since the longevity of the seed is already longer than the working life of the freezer. Actually, even a regular household chest freezer with a temperature of -15°C is adequate for a gene bank. Technology and science are still unable to prevent ice build-up; the best way we know to handle this problem is to chip away the ice at regular intervals.

An important addition to the freezer is a battery-powered alarm connected to the freezer's thermostat that indicates when the temperature rises above acceptable levels. Such an alarm is invaluable in signalling powercuts or other technical problems. Another good precaution is a small, portable, gas-driven generator- kept in good working order- that could be used during long-term power failures.

Pollen and Spore Banks

Seeds are the most likely candidates for gene banks, but this cryogenic technique may be useful in preserving pollen from flowering or cone-bearing plants, and spores from non-flowering plants such as ferns and mosses as well. People sometimes ask why we bother with pollen, now that the cryogenic technology has been worked out for seeds. One reason is that pollen grains are so tiny that thousands and even millions of grains can be stored in a very small vial. Obtaining great diversity is quite simple. Botanists are able to use pollen grains from certain species to grow whole plants. This is still a difficult, tedious procedure with routinely perform these transformations. Pollen grains can also be used when needed to help create hybrids. When seed is used to develop hybrids, the seeds must be retrieved from the bank and then planted. It takes several years of maturation for the plant to be useful in hybridization. Because pollen can be used as is, it is much more economical and efficient. A pollen bank can be an extremely powerful tool in plant breeding since it frees breeders from the tyranny of time. The process can be streamlined and quickened by crossing pollen from plants that normally flower at the end of a season onto flowers that appear at the beginning of the season.

In Irvine we have found another use for pollen banks. Several specimens we grow at the arboretum are self-sterile, they set seed only if pollinated with pollen from another individual. We had managed to raise *Cytanthus obliquus,* a relatively rare African amaryllid, to maturity. This process took ten years and produced five plants, all of which became virus-infected. We decided to replace the diseased plants and also increase our stock from seed. Seedlings are usually virus-free. Here we ran into trouble. Each of the five plants flowered but never two at the same time. Even though each plant produced abundant pollen, none of the plants would set seed from its own pollen. We finally hit upon the perfect solution: storing pollen from this species in the bank. We now have no trouble pollinating the plants and we can generate as many seedlings as we require. The original five plants were replaced by several hundred offspring.

Fern spores, which are very similar in size to pollen, can he treated just like pollen. Even the spores of tropical terns like staghorns, *Plytycerium* species, can he germinated after drying and freezing.

The biggest problem with pollen is its susceptibility to water damage. Rain, heavy dew or water from any other source that comes into contact with pollen is liable to kill it. Pollen should be collected only from fresh flowers that have not been exposed too much to the elements.

We store pollen in gelatin capsules, the type used to hold medicine. These capsules can be purchased from the local pharmacist. The stamen, which contains pollen, is removed from the flower, inserted into the capsule and shaken several times to deposit the dust-like pollen on the capsule wall. The stamen is then withdrawn. The species name and date may be written directly on the capsule with a waterproof laundry marker. We dry the pollen by exposing the capsules to the air inside a frost-free refrigerator for 24 hours. The moderately dry air circulating in these refrigerators will cause sufficient water loss to permit the capsules to be frozen safely. Water is able to move out through the walls of a gelatin capsule. We store the capsule inside individual plastic containers that have a little Drierite[R] in them to prevent moisture from moving through the gelatin and then place these containers in the bank. Relevant data are written on the wall of the plastic container. Pollen samples withdrawn from the freezer have limited viability and should be used within two days of withdrawal. Little is presently known about the longevity of pollen in cryogenic storage. We have tested pollen that had been frozen for eight years and found it to be viable, but no one knows yet whether pollen will he as hardy as seed seems to be.

The data presented here and elsewhere on the longevity of seeds are extrapolation figures obtained by keeping large samples at different temperatures, withdrawing seeds and germinating them in successive years and noting the decline in viability with each test. After a few years the rate of the decline and a forecast of the percentage of the seeds that will stay alive at any particular time can be determined. These studies have brought forth a few adverse factors. A main problem is that with time, mutations and chromosomal abberations accumulate. These are probably caused by interactions between the background cosmic radiation, which is everywhere, and the genetic material in the cells of the seed. Since the background radiation appears to be more or less constant, the probability of a mutation occurring depends on the amount of time the seed is exposed to the radiation. As the seed's longevity is increased, so is the chance that a mutation will appear.

Other long-term effects can also occur. The proteins in the seed may degenerate. Also, despite the low temperatures, some chemical reactions still occur. Finally, molecular events that we know nothing about may take place. The effects of these changes do not appear on our extrapolation curves. Nevertheless, we are totally convinced that cryogenic banking is presently the most effective and efficient way to preserve species.

As we mentioned earlier, not all seeds can be stored cryogenically. Some plants have fleshy seeds with high water content. The seeds of some *Crinum* species for example, resemble small potatoes and simply cannot be dried. *Nerine,* another related genus, has seeds that begin to germinate before they are released from the mother plant. Dehydration would kill the seed. Seeds with very hard seed coats may not be dryable either and seeds with very high oil content can stubbornly resist drying. These are isolated problems; most species can be processed for cryogenic storage. Further experimentation should show us how to deal with some of the fleshy and oily seed. Viability of seed in storage may also be affected by the amount of oxygen or other gasses available, but these effects and other relatively negligible factors should not concern the new gene Banker.

Despite what seems to be a complicated series of steps, anyone with determination and a little effort can create a gene bank. Extensive know-how or technology called from futuristic science fiction is not necessary. Gene banking is an elegantly simple solution for a vexing and important problem.

Other Banks

Since cryogenic gene banks are relatively easy and inexpensive to operate, why are there so few? Many people do not realize that cryogenic banks involve such simple techniques. For them, cryogenics conjures up images of technicians in white coats wheeling gleaming vats of liquid nitrogen. As the simplicity of these banks becomes widely known, more and more facilities will be developed. Already, some of the institutions and groups that currently use gene banks without sub-freezing temperatures have been persuaded to adopt the more efficient, cryogenic techniques.

The value of gene banking was first discovered in the late 1960s and early 1970s, when plant scientists began looking into the possibility of saving the genetic variability in agricultural crops. The concept was recently expanded to employ sub-freezing temperatures and include non-agricultural plants. In this chapter we will take a look at the history of gene banking and at some of the different gene banks around the world.

Natural Gene Banks

Gene banking may be a relatively new concept for mankind, but natural gene banks are a regular part of nature. These natural ones lend support to the theory that seeds preserved today in man made banks will be viable hundreds, if not thousands, of years into the future. Soil is a kind of gene bank to which seeds are added year after year. When wild plants produce seed each season, not all of the seeds will grow the following year. Much of the seed drops to the ground and then waits for appropriate conditions to permit germination. Most seed gets covered by either dirt or dust and is eventually buried. The number of seeds that accumulate over the years can be prodigious, as seen in a study made in Denmark. One block of topsoil- 1 in square and 20 cm deep- yielded roughly 135,000 seeds. Of these, about 50,000 were living and could germinate under the correct conditions. Compared to other studies, this is an extreme example, but it is not unusual to recover thousands of viable seeds from a square metre of land. If the plants in an area change, then the species in the natural bank also will change. Deep in the soil there may be species no longer common. At the University of California, Irvine Arboretum, once decided to add a few small terraces and so removed soil from one area to a depth of 60 cm. Surprisingly, a field of stinging nettles sprang up after the next rains. Until this time, no nettles had ben found on the grounds, nor in the surrounding area. The seed must have been buried decades before and was waiting patiently for the right conditions. These natural gene banks are often seen when a forested area is cleared. All sorts of plants suddenly germinate from seed that has been long buried. The plants may be totally unrelated to the normal flora of the forest.

The length of time that seeds remain viable in the natural seed bank depends on a variety of conditions. To begin with, the colder the climate, the better the chances of survival. An interesting study involved seed dug up from beneath Danish churches built centuries ago. The soil beneath the church (1,700 years old) yielded viable seeds from two weeds, *Spergula,* commonly called corn spurry, and *Chenopodium,* or goose foot. Another 600-year-old, church had soil laced with 13 different viable species. *Lotus* sees dug up from a peat bog and found to be 1,040 years old have grown, as have the *Lotus* seeds rescued from a 237 year old herbarium specimen. The record is held by lupin seeds dug up from frozen soil in the Arctic. The seeds were radiocarbon-tested to be about 10,000 years old and they germinated when given the correct conditions. While, some seeds seem to be able to last almost indefinitely, we should remember that there are many species with seed that does not stay viable for more than a few months or years. Nevertheless, the ability of many species to remain viable for centuries validates the theory of artificial gene banking.

Besides the need for cold temperatures, seeds need to be relatively dry to survive in a gene bank- whether it is natural or artificial. Many ripe seeds tend to have seed coats impermeable to water. Before the germination process can begin, the seed coat must break down so that the seed can absorb moisture. Dry seeds are safe from germination and may remain in this state of suspended dormancy for many years.

□□□